1st International Symposium on Engineering and Technology (ISETech 2019)

IOP Conference Series: Materials Science and Engineering
Volume 767

Kangar, Malaysia
23 December 2019

ISBN: 978-1-7138-4705-2
ISSN: 1757-8981

Printed from e-media with permission by:

Curran Associates, Inc.
57 Morehouse Lane
Red Hook, NY 12571

Some format issues inherent in the e-media version may also appear in this print version.

Institute of Physics
Dirac House, Temple Back
Bristol BS1 6BE UK

Phone: 44 1 17 929 7481
Fax: 44 1 17 920 0979

techtracking@iop.org

Additional copies of this publication are available from:

Curran Associates, Inc.
57 Morehouse Lane
Red Hook, NY 12571 USA
Phone: 845-758-0400
Fax: 845-758-2633
Email: curran@proceedings.com
Web: www.proceedings.com

TABLE OF CONTENTS

Preface

Proceeding of the First

International Symposium on Engineering and Technology 2019

Putra Regency Hotel at Kangar,Perlis, Malaysia

23rd December 2019

The International Symposium on Engineering and Technology is a bi-annual symposium, which first edition has been organized in 2019. The symposium is organized by Centre of Excellence Advanced Communication Engineering (ACE) and the School of Computer and Communication Engineering, Universiti Malaysia Perlis (UniMAP). The theme for this year's ISETech is "Driving Engineering and Technology towards Humanity". The theme is chosen in view of the current revolution and rapid developments in the Industrial Revolution Four (IR 4.0).

This symposium focuses on research works which highlight the contribution to humanity in the broad areas of Engineering and Technology, specifically Electrical and Electronic (EE), Civil Engineering (CE), Chemical & Natural Resources (CNR), Mechanical & Manufacturing (MM), Science & Mathematics (SM) and Information & Communication Technology (ICT).

The first edition of ISETech has been successfully organized with the support of academician, industry representative and postgraduates with the presentation of 71 papers. These papers have been reviewed by academia and industrial experts from multiple countries across the world which includes Australia, India, USA, UK, Thailand, Singapore and many others. We appreciate their support and will be looking forward to organize the next edition of the conference in 2021.

ISETech 2019 PATRON & ADVISORS

Patron

Prof. Ir. Ts. Dr. R. Badlishah bin Ahmad

Conference Advisor

Prof. Dr. Uda bin Hashim
Prof. Ir. Ts. Dr. Syed Alwee Aljunid bin Syed Junid

ISETech 2019 ORGANIZING COMMITTEE

General Chair
Assoc. Prof. Dr. Azremi Abdullah Al-Hadi
Ir. Rosdisham bin Endut

Conference Chair
Assoc. Prof. Ir. Dr. Mohd Rashidi bin Che Beson

Conference Secretary
Ts. Dr. Saidatul Norlyana Azemi
Dr. Amiza Amir

Treasurer
Ts. Dr. Siti Zuraidah binti Ibrahim
Dr. Wee Fwen Hoon

Technical Program Committee
Assoc. Prof. Ts. Dr. Soh Ping Jack
Dr. Zahereel Ishwar Abdul Khalib
Dr. Muhammad Solihin bin Zulkifli
Dr. Mohd Hafizuddin bin Mat
Dr. Dedet Candra Riawan
Widhia Oktoeberza KZ

Local Arrangement
Assoc. Prof. Dr. Ruzelita Ngadiran

Publicity
Dr. Lee Yeng Seng
Dr. Liyana Binti Zahid
Dr. Mohd Aminudin bin Jamlos
Dr. Mimi Diana binti Ghazali

Advisory Committee

Prof. Dr. Jugdutt (Jack) Singh, La Trobe University, Melbourne

Prof. Dr. Lim Chee Peng, Deakin University, Australia

Prof. Dr. N.S Murthy, NIT Warangal, India

Prof. Dr. Borhanuddin Mohd. Ali, UPM, Malaysia

Prof. Dr. Ivan Andonovic, University of Strathclyde, United Kingdom

Prof. Dr. Wayne Rowe, RMIT University, Australia

Prof. Dr. Prayoot Akkaraekthalin, King Mongkut's University of Technology North Bangkok (KMUTNB), Thailand

Prof. Dr. Xiaodong Yang, Xidian University, China

Prof. Dr. Mohammad Rakib Uddin, Universiti Teknologi Brunei (UTB), Brunei Darussalam

Assoc. Prof. Ts. Dr. Muhamad Shahbani Abu Bakar, Universiti Utara Malaysia, Malaysia

Assoc. Prof. Ir. Dr. Hanung Adi Nugroho, Universitas Gadjah Mada, Indonesia

Peer review statement

All papers published in this volume of *IOP Conference Series: Materials Science and Engineering* have been peer reviewed through processes administered by the proceedings Editors. Reviews were conducted by expert referees to the professional and scientific standards expected of a proceedings journal published by IOP Publishing.

IOP Conf. Series: Materials Science and Engineering **767** (2020) 012001 doi:10.1088/1757-899X/767/1/012001

Application of FSS for Microstrip Antenna for Gain Enhancement

F.N.M.Zaid, S.N.Azemi, and C.B.Mrashidi

Advanced Computer Engineering Center (ACE),
School of Computer and Communication Engineering,
Universiti Malaysia Perlis (UniMAP),
02600 Arau, Perlis, Malaysia

farinanurnajwa@gmail.com, snorlyana@unimap.edu.my.
rashidibeson@unimap.edu.my

Abstract. In this present paper, a dual-band microstrip patch antenna (MPA) using frequency selective surface (FSS) is proposed. The dual-band MPA is designed using the rectangular shaped patch with an elliptical slot at the center of the patch and full ground plane. The FSS in this paper act as a reflection plane which is loaded onto the dual-band MPA in order to improve the gain of the antenna at higher band. The FSS is obtained by the periodic array of 3 x 5-unit cell in the x-y direction, which based on the amalgamation of a square and circular loop elements. After merging the FSS, the gain and return loss at 5 GHz increased to 0.915 dB and -25.08 dB respectively. The dual-band MPA is simulated using the FR-4 substrate with a thickness of 1.6 mm and a relative permittivity of 4.4. The overall size of the MPA and FSS is relatively compact and easy to fabricate. The proposed antenna can be used in WiMAX and WLAN applications.

1. Introduction

In a wireless communication system, the antenna is one of the most important components as both of the transmission and reception of radio frequency signals are using it. There are varieties types of antennas but microstrip patch antenna is the favored type of printed antenna among people. The substrate that mainly used for this type of antenna is FR-4 because it is universal and cheap as well as for education purpose. Microstrip patch antenna plays a very significant role in the maintenance of wireless communication systems which are very simple in construction that using a conventional microstrip fabrication technique. Latest applications nowadays such as mobile communications, WLAN, Global Positioning System (GPS), WiMAX, Bluetooth, and radars are using microstrip antenna.

A basic patch antenna only capable to operate in one frequency. Nevertheless, nowadays demands require an antenna to operate in multiple or different frequency band in one single antenna. Due to limited space of the device, at least the antenna is demanded to operate in dual frequency band such as WiMAX and WLAN. Dual band operation is activated by simultaneously stimulating the first and second resonant modes. The designed antenna can generate sufficient impedance bandwidth to cover the Worldwide Interoperability for Microwave Access (WiMAX) system at 3.5 GHz as well as the Wireless Local Area Networking (WLAN) system at 5.0 GHz by selecting slot dimensions appropriately. The structure of dual-band antenna is not a new idea especially dual-band microstrip patch antenna. Many manufacturers have merged multiple elements for years to establish antennas that operate in two different bands.

An alteration of the basic patch antenna makes it capable to operate in dual band, such as creating a slot on the design. By creating a slot on the design, a new current path is created hence producing another frequency band. However, by producing another current path, the antenna will suffer the frequency gain. Where now the gain already been divided into two pathways. During the 1960s, the technology of frequency selective surface (FSS) were only owned by military to be used in applications such as radar absorber materials, stealth aid and filter design [1]. FSS is a planar periodic structure organized on a dielectric substrate with homogenous two-dimensional array of elements. Determine by the characteristic of the array elements, any plane waves that approaching the FSS will either be transmitted, reflects entirely or partly. The phenomenon when the wave is transmitted is passband while the other one is called stopband. These happened when the frequency and resonant frequency of EM wave and FSS elements were matched. As stated by the circuit theory, spatial filters or referred to as FSS structures are equivalent to the microwave filters. FSS is the repetitive surface that is designed to improve antenna's performance and parameters, for instance, gain, bandwidth, radiation pattern, reflection coefficient, transmission coefficient, and VSWR by merging the FSS with an antenna.

In this present work, a proposed dual-band microstrip patch antenna (MPA) with elliptical slots at the center is introduced in the rectangular patch and the full ground plane at resonant frequency 3.5 GHz and 5 GHz. Then, the designed FSS is combined with this MPA to improve the gain of the antenna. The simulations and measured results of the return loss and gain of the designed antenna over the frequency band of interest are presented.

2. Microstrip Patch Antenna (MPA) design and development

The designed antenna is printed on both sides, one side is the patch and another side is ground plane. The feeding technique that is used in this project is a microstrip line feeding technique. This technique is when a conducting strip is connected to the edge of the patch. The feed also etched on the substrate. One of the advantages of microstrip line feed is that it is easier to fabricate compared to another method. Microstrip line feed can be consider as extension of patch as it is just conducting strip connecting to the patch.

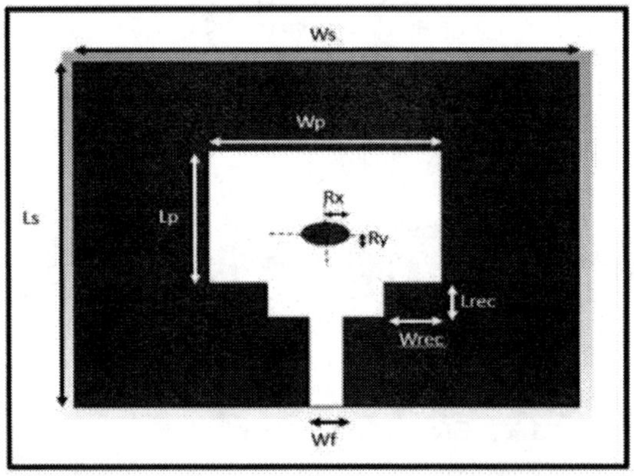

Figure 1. Schematic Proposed of dual-band MPA

Table 1. Design Parameters of Proposed dual-band Antenna.

Parameters	Description	Dimension (mm)
Ws	Width of Substrate	86
Ls	Length of Substrate	62
Wp	Width of Patch	40
Lp	Length of Patch	30
Wrec	Width of Rectangular Slots	10
Lrec	Length of Rectangular Slots	8
Wf	Width of Feed	6

The performance of the microstrip antenna eventually relies on upon its dimension. As the antenna is simulated with the selected substrate which is FR-4 with a 1.6 mm thickness and relative permittivity of 4.4, the value of width and length of the patch can be found using the following formula that had been provided.

Patch width (*W*)

$$W = \frac{c}{2f_r} \cdot \sqrt{\frac{2}{\varepsilon_r + 1}}$$

Patch length (*L*)

$$L = \frac{c}{2f_r\sqrt{\varepsilon_{eff}}} - (2\Delta l)$$

Effective dielectric constant (*εeff*)

$$\varepsilon_{eff} = \frac{\varepsilon_r + 1}{2} + \frac{\varepsilon_r - 1}{2}\left(1 + \frac{12h}{W}\right)^{-\frac{1}{2}}$$

Length of extension (Δl)

$$\Delta l = 0.412h \left[\frac{(\varepsilon_{eff} + 0.3)\left(\frac{W}{h} + 0.264\right)}{(\varepsilon_{eff} - 0.268)\left(\frac{W}{h} - 0.8\right)}\right]$$

[Where, f_r is resonant frequency, εr is relative dielectric constant, h is substrate height. Using above expressions width of patch (W) = 26 mm and length of patch (L) = 20 mm are calculated.]

Figure 1 shows proposed dual-band MPA having rectangular shaped patch. The proposed antenna is altered in order to achieve the resonant frequencies which are 3.5 GHz and 5 GHz. It is fabricated on FR-4 substrate of dimension 86 mm × 62 mm, with the relative permittivity of 4.4. Here the proposed antenna is designed using two rectangular slots in each bottom edge of the patch and an elliptical slot which is located in the center of the rectangular patch. The geometrical dimensions of proposed dual-band microstrip patch antenna design used in the simulation are given in Table 1. In figure 2 a dual-band MPA is designed using CST studio Suite 2017 software tool, where the return loss without FSS is observed are −23.447 dB and −22.142 dB at a resonant frequency of 3.5 GHz and 5 GHz respectively which is shown in the figure 3. The gain of the dual-band antenna without FSS at 5 GHz is shown in figure 4. The designed dual-band MPA then going through the fabrication process. The fabricated antenna then soldered with a 50Ω SMA connector and measured with network analyzer as shown in figure 5.

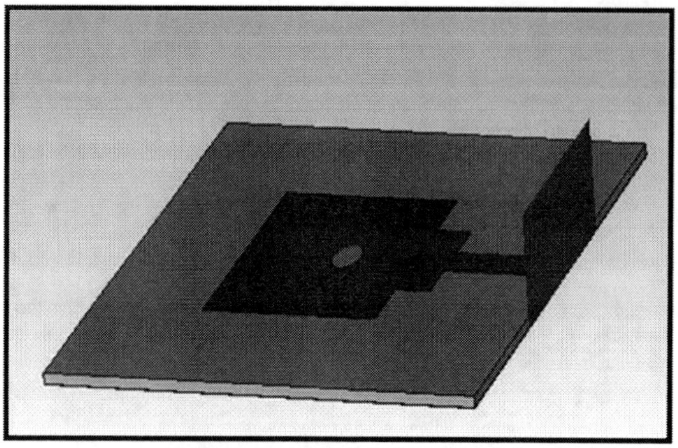

Figure 2. 3D view of dual-band MPA

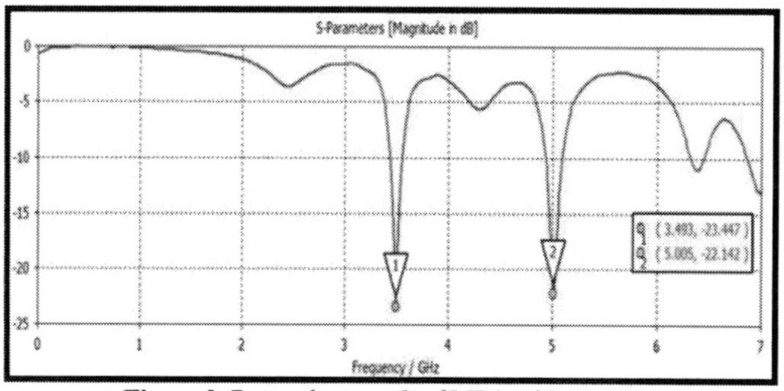

Figure 3. Return loss result of MPA without FSS.

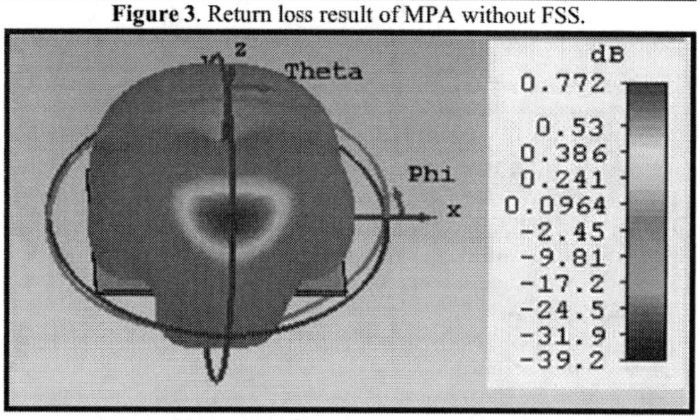

Figure 4. Gain at 5GHz without FSS

IOP Conf. Series: Materials Science and Engineering **767** (2020) 012001 doi:10.1088/1757-899X/767/1/012001

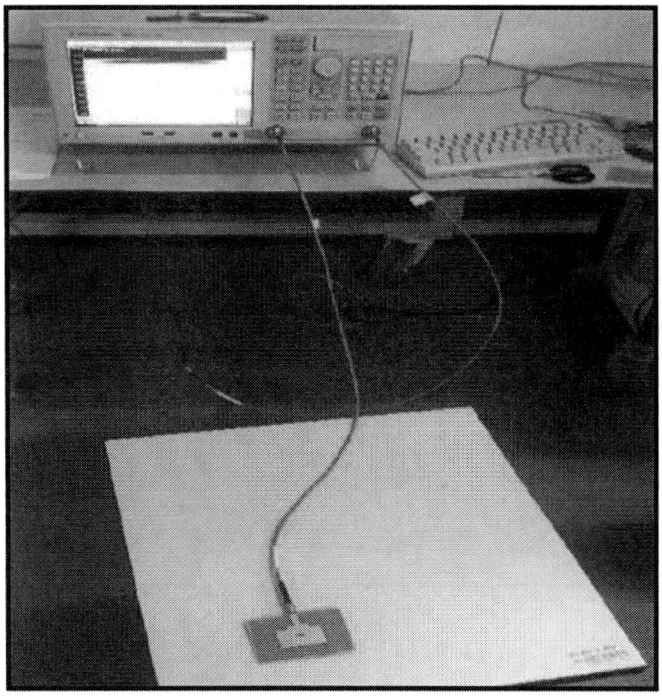

Figure 5. Measurement of dual-band MPA without FSS

3. Frequency Selective Surface (FSS) design and development

The TE / TM excitation of frequency selective surface (FSS) generates resonance attitudes for specific frequencies. Throughout this project, the FSS is designed based on the combination of rectangular patch and ring layout. This FSS is designed with the aim to enhance the gain and return loss of the dual-band MPA at 5 GHz only. So, this FSS is band-pass type. This is because the gain at 3.5 GHz is already high. The width and length of the full FSS design is same as the dimension of the MPA which is 86 mm x 62 mm. The geometrical dimensions and parameters of FSS unit cell is shown in figure 6 and Table 2. The FSS is obtained by the periodic array of 3 x 5-unit cell in the x-y direction. The return loss, S11 of the FSS is shown in figure 7. The designed FSS then going through the fabrication process. The fabricated FSS is shown in figure 8.

Figure 9 shows the designed dual-band MPA merged together with FSS. The air gap between the dual-band MPA and FSS is 42mm. Lately the use of FSS as an antenna gain enhancer is being extensively researched. After implanting the new FSS with the rectangular patch antenna, it is found that the gain has been improved from 0.772 dB to 0.915 dB at 5 GHz as the designed FSS is only focused on the higher frequency. This is because the gain at lower frequency band is already high which 1.59 dB. In figure 10, the return loss of the dual-band MPA with FSS is observed, which the results are −17.042 dB and −25.08 dB at a resonant frequency of 3.5 GHz and 5 GHz respectively. The gain of the dual-band antenna with FSS at 5 GHz is shown in figure 11. The fabricated dual-band MPA and FSS are merged together and measured using Network Analyzer as shown in figure 12.

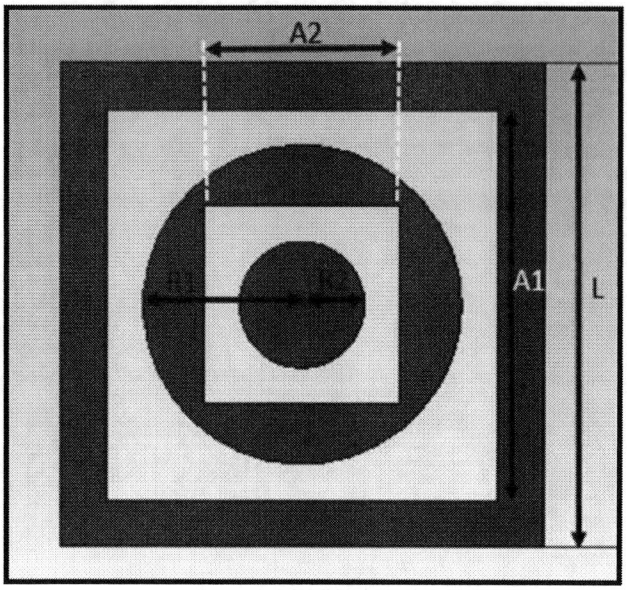

Figure 6. FSS Unit Cell

Table 1. Design Parameters of Proposed FSS

Parameter	Dimension (mm)
Side Length, L	16.67
Outer Square Loop Length, A1	13.50
Outer Ring Radius, R1	5.5
Inner Square Loop Length, A2	6.83
Inner Ring Radius, R2	2.165

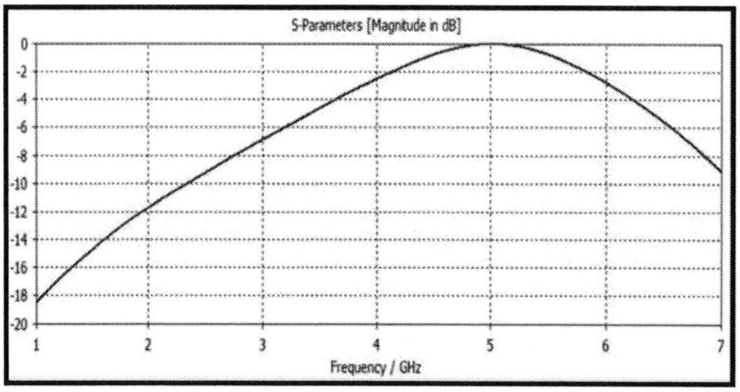

Figure 7. Return Loss (S11) of FSS

Figure 8. Fabricated FSS

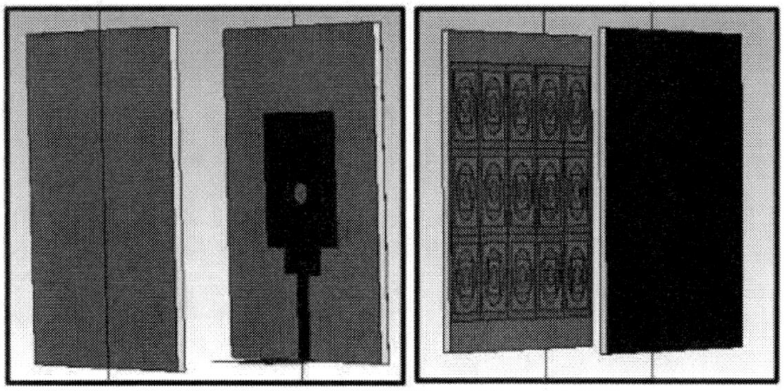

Figure 9. Designed Dual-Band MPA with FSS

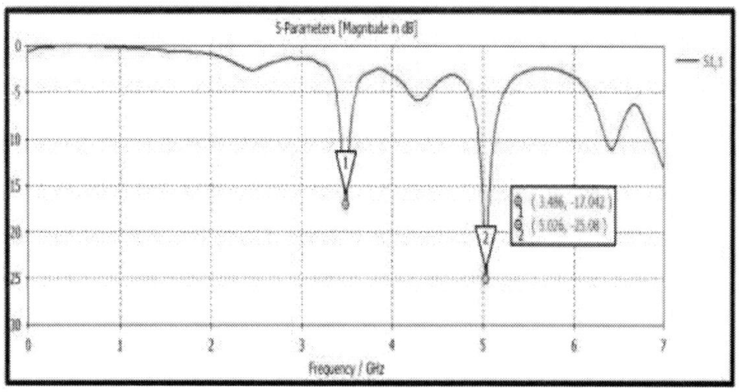

Figure 10. Return Loss of Dual-Band Antenna with FSS

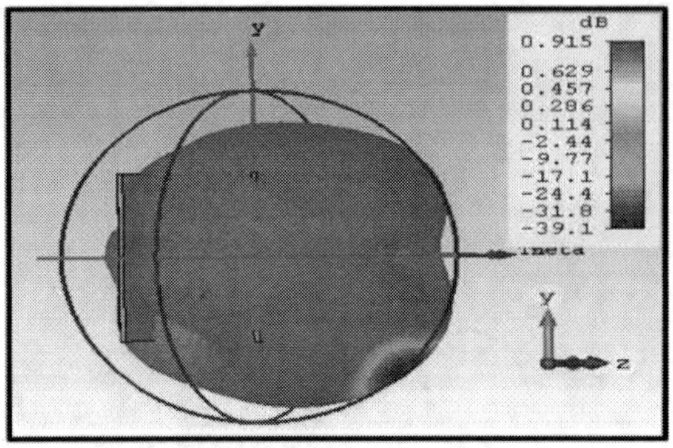

Figure 11. Gain of Dual-Band Antenna with FSS

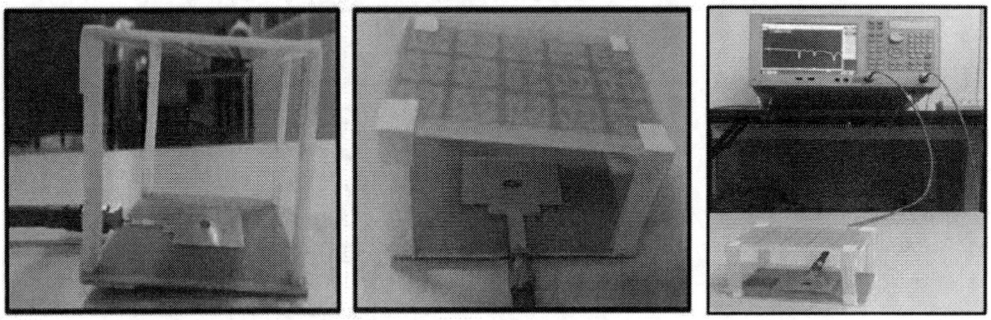

(a) (b) (c)

Figure 12. (a) Combination of MPA and FSS from side view, (b) FSS antenna and MPA from front view, (c)Measurement of dual-band MPA with FSS

4. Simulation and measurement result

Simulation result and measured result of dual-band MPA with and without FSS are shown in figure 13 and figure 14 respectively. Meanwhile Table 3 shows the comparison value of the results mentioned above.

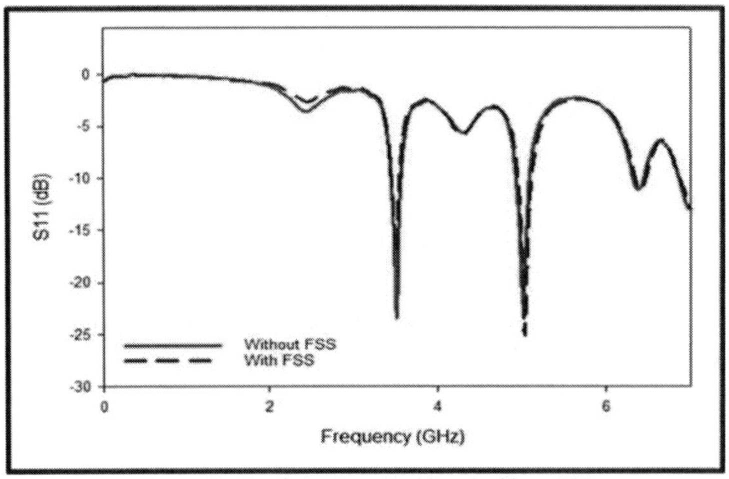

Figure 13. Simulation Result with and without FSS

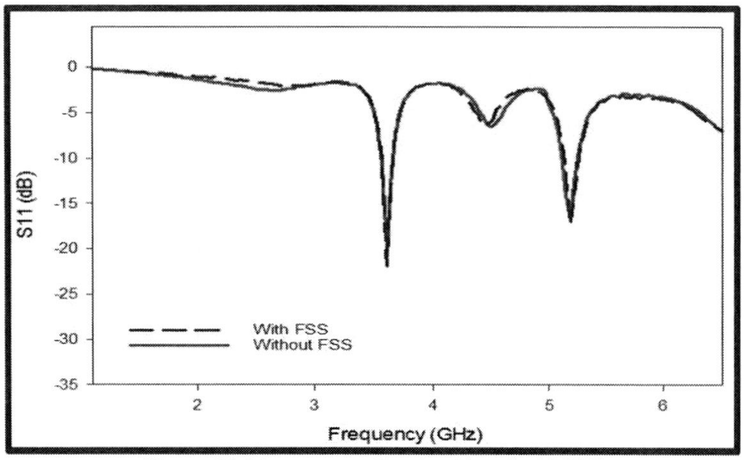

Figure 14. Measurement Result with and with FSS

Table 2. Comparison of Measured and Simulation Result

	Simulation		Measurement	
Frequency (GHz)	3.5	5	3.623888	5.189327
S11 Result without FSS (dB)	-23.447	-22.142	-20.552	-16.418
S11 Result with FSS (dB)	-17.042	-25.08	-14.409	-17.001

5. Conclusion

In this paper, a new FSS is designed to improve the gain of the dual-band microstrip patch antenna. The centre frequency of this FSS is at 5.0 GHz. The proposed FSS structure acted as the reflection plane which improved the performance which are gain and return loss of the dual-band at WLAN (5 GHz) frequency successfully. The 5 GHz return loss after loading FSS is better which is from -22.142 dB to -25.08 dB. Although there are a slightly different value due to some error in fabrication, the measured return loss after loading FSS also increasing which is from -16.418 dB to -17.001 dB. The gain after merging the MPA with FSS also higher which is from 0.772 dB to 0.915 dB. So, the proposed FSS is proved to improve the gain of dual-band MPA at 5 GHz frequency.

References

[1] A. Mellita, C. Ds, and S. S. Karthikeyan, "Gain Enhancement of a Microstrip Patch Antenna Using a Novel Frequency Selective Surface," pp. 4–7, 2017.

[2] K. Pengthaisong, P. Krachodnok, and R. Wongsan, "Design of a dual-band antenna using a patch and frequency selective surface for WLAN and WiMAX," 2013 10th Int. Conf. Electr. Eng. Comput. Telecommun. Inf. Technol. ECTI-CON 2013, no. c, pp. 1–4, 2013.

[3] I. Ali and R. Y. Chang, "Design of dual-band microstrip patch antenna with defected ground plane for modern wireless applications," *2015 IEEE 82nd Veh. Technol. Conf. VTC Fall 2015 - Proc.*, 2016.

[4] A. Mounsef, Y. El Gholb, N. El, and A. El, "Design and simulation of a dual band microstrip patch antenna for an emergency medical service system," pp. 0–3, 2017.

Personal Shopper – Mobile Phone Applications

F N Zamani, S N Azemi, Amiza Amir, Naimah Yaakob

Advanced Computer Engineering Center (ACE),
School of Computer and Communication Engineering,
Universiti Malaysia Perlis (UniMAP),
02600 Arau, Perlis, Malaysia

snorlyana@unimap.edu.my

Abstract. This project is focused on the development of personal shopper mobile phone application. The purpose of this system is to help part-timer and full-timer personal shoppers gather in one platform and also for the customer to hire a personal shopper to buys their items. This project will implement a rating system, give reward to the customer, accept customer requests, view customer requested details and contact the customer through Whatsapp application. This application is very flexible for the customer and personal shopper where the customer can become a personal shopper and vice versa. This entire project is developed according to software engineering methodology with the waterfall model. The tool used to create this project is Android Studio with Java, PHP, XAMPP Server, and MySQL database.

1. Introduction

Nowadays, two-thirds of the world's 7.6 billion inhabitants now have a mobile phone [1] which is approximate to 5.16 billion users. Based on 2015 statistics, there are currently 50 billion Android apps were downloaded [2]. Both statistics show that mobile applications are globally accepted worldwide. The mobile application is mostly developed using the Android operating system. This is because the Android operating system is compatible with almost all mobile devices compared to the iPhone operating system.

In Malaysia, there is an increasing number of people who are becoming personal shoppers, either on a full-time or part-time basis supported by readily social media platforms and higher demand of local consumers [3-4]. The concern is how to gather these people in one platform and help customers to run errands). This project proposed a mobile application using the Android operating system to gather all of these people who want to become personal shoppers and customer who wants to find a personal shopper to run errands. The system will use an Android Studio as a software platform.

2. Personal shopper mobile phone application design and development

A methodology is a system of procedures may be derived to understand the different situation or solving different problems within the scope of the discipline. The methodology is a set of practices to achieve research goals. In this project, the waterfall model has been chosen. This is due to the waterfall model requires planning for the long-term which requires complete clarity in requirements, quality control activities like testing are performed towards the end of the project, this model is more process oriented and this model is used only when the requirements are very well known, clear and fixed. In the waterfall model, there are 5 stages which are Requirement, Design, Implementation, Verification, and

Maintenance. All these stages are carried out in system development. The requirement phase involved the understanding of the system design and its function. This phase can be interpreted as the literature review phase. All the literature review is based on the feature needed in order to design a personal shopper application and the technique needed to construct the system. After this phase complete, the process will run to the next process [5-6].

The design phase happens after the first phase is completed. During this phase, after research on the first phase, the software architecture is designed based on the requirement and feature. The database design also being done during this stage. Interfaces for a personal shopper mobile application are designed during this stage. The implementation phase happens after the design phase completed. This is the phase where visions and plans become a reality which means the action is needed in order to complete this project. The coding is done in this phase to prepare for the next phase. The verification phase involved system testing and system deployment. The coding is integrated and tested to ensure if it works as expected. Testing and correction are done repeatedly until it meets the system requirement. Once it is tested, the system is deployed in the respective environment. A sanity checks on the environment is performed after the system is deployed to ensure the application does not break. The maintenance phase occurs after the verification phase. This phase involves making the modification to the system and improves system performance. This phase occurs due to errors discovered after the system has publicly published.

The application framework is presented in this section. The proposed can be divided into four parts, which is Android Application, APIs, Web Server, and Database Server [7]. Figure 1 shows the basic working system of the frameworks. For this personal shopper application, API level 21 is being used. Reason for using API level 21 is because they are a certain function of this application cannot be used for API level 20 and below. Based on the data below, cumulative distribution for API level 21 is 85%, which means this API level 21 cannot cover 100% android phone user. It cannot be helped since API level 20 and below cannot support a certain part of this personal shopper application. Based on the data, API level 21 is being used for Android version 5.0 in 2014 ("Android - History," n.d.), that means they are a possibility that, less android user is using API level 20 and below since its already 2018 and the new API already been released which is API level 28. For this application, XAMPP server is used to run a web server.

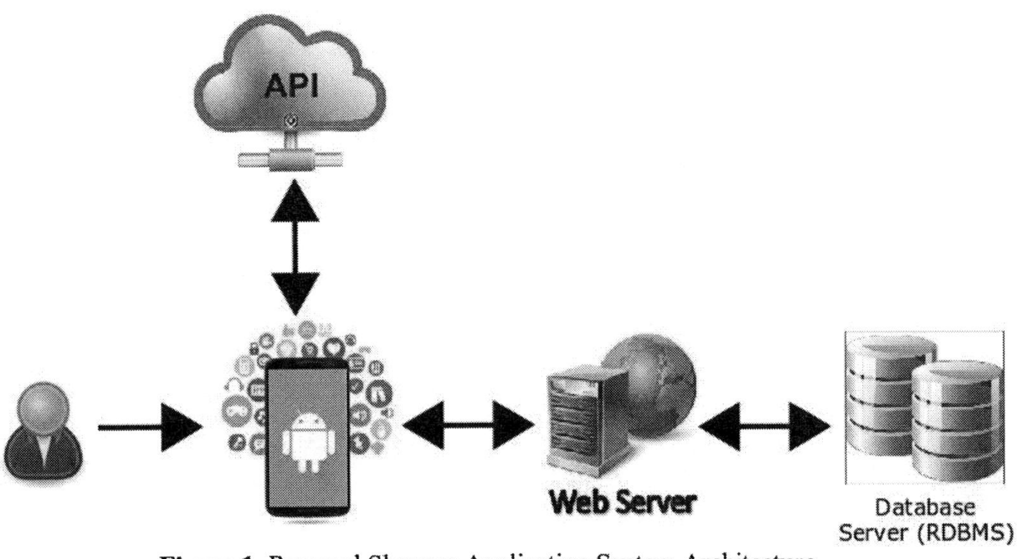

Figure 1. Personal Shopper Application System Architecture

The web server will collect the requested PHP page from its document root which is "htdocs" folder in XAMPP. MySQL is chosen as the database to store the data as it offers a graphical user interface (GUI) for users to create, design and browses database schemas. It is easier to use as it only requires basic knowledge of SQL to operate. Requirements analysis is the process of determining user expectations for a product. It can be divided into two, which is functional requirement and non-functional requirement.

A use case diagram is a graphic illustration of the cooperation among the components of a system. A use case is a methodology used in system analysis to distinguish, clear up and sort out the system requirements as shown in figure 2. In this project, there is only one role available in the use case diagrams which is the user. User can act as a customer and as a personal shopper. User must register before logging into this application. After the user has logged in successfully, the user can access the main application. Customer can place an order for groceries, shopping, food and part-timer. Customer can view their orders after the personal shopper accepts their orders. Customer can delete the jobs placed, rate the personal shopper in-charges of those jobs and press the finished button after the job is done. A personal shopper can accept jobs that the customer has placed. A personal shopper can view their accepted jobs and use the Whatsapp application that links from this application to contact the customer. Users can view their profile and edit their profile picture, phone number and name. Based on Entity Relationship Diagram, this project shows that one user can request many orders for groceries, shopping, food and part-time. It also showed that one user can accept many orders for groceries, shopping, food and part-time.

An Entity Relationship (ER) Diagram is a kind of flowchart that outlines how' entities,' such as individuals, objects, or concepts, are interrelated within a system [8]. ER diagrams are often used to design or troubleshoot relational databases in the fields of software engineering, education, business data systems and research. Also known as ERDs or ER models, they use a characterized set of images such as diamonds, ovals, rectangles and connecting lines to illustrate the interconnections of entities, relationships and their attributes. Based on ERD shown in figure 3, this project shows that one user can request many orders for groceries, shopping, food and part-time. It also showed that one user can accept many orders for groceries, shopping, food and part-time.

3. Personal shopper mobile phone application result

The project outcome is based on the accomplishment of building an application that can help customer to find personal shopper. Figure 4 (a) is the first interface that the user will see after the first launch of the "UShopper" application. Existing users can insert their email and password to open the account in this panel. If your email and password match the email and password stored in the database, the user will be sent to another panel once the login is successful, while a new user will need to click the register button to create a new account. When the register button is clicked on the login panel, the user will be directed to a new panel which is a register panel where the user is required to insert their details as shown in Figure 4(b). The details to be filled in are the name, telephone number, e-mail, password and country. If the user successfully signs into the application, the homepage panel will be displayed as shown in figure 4(c). If the user clicked on the order menu in the application menu list, this panel will be shown to the user. There are several menus in the menu of orders, including food, grocery, shopping and part-time menus.

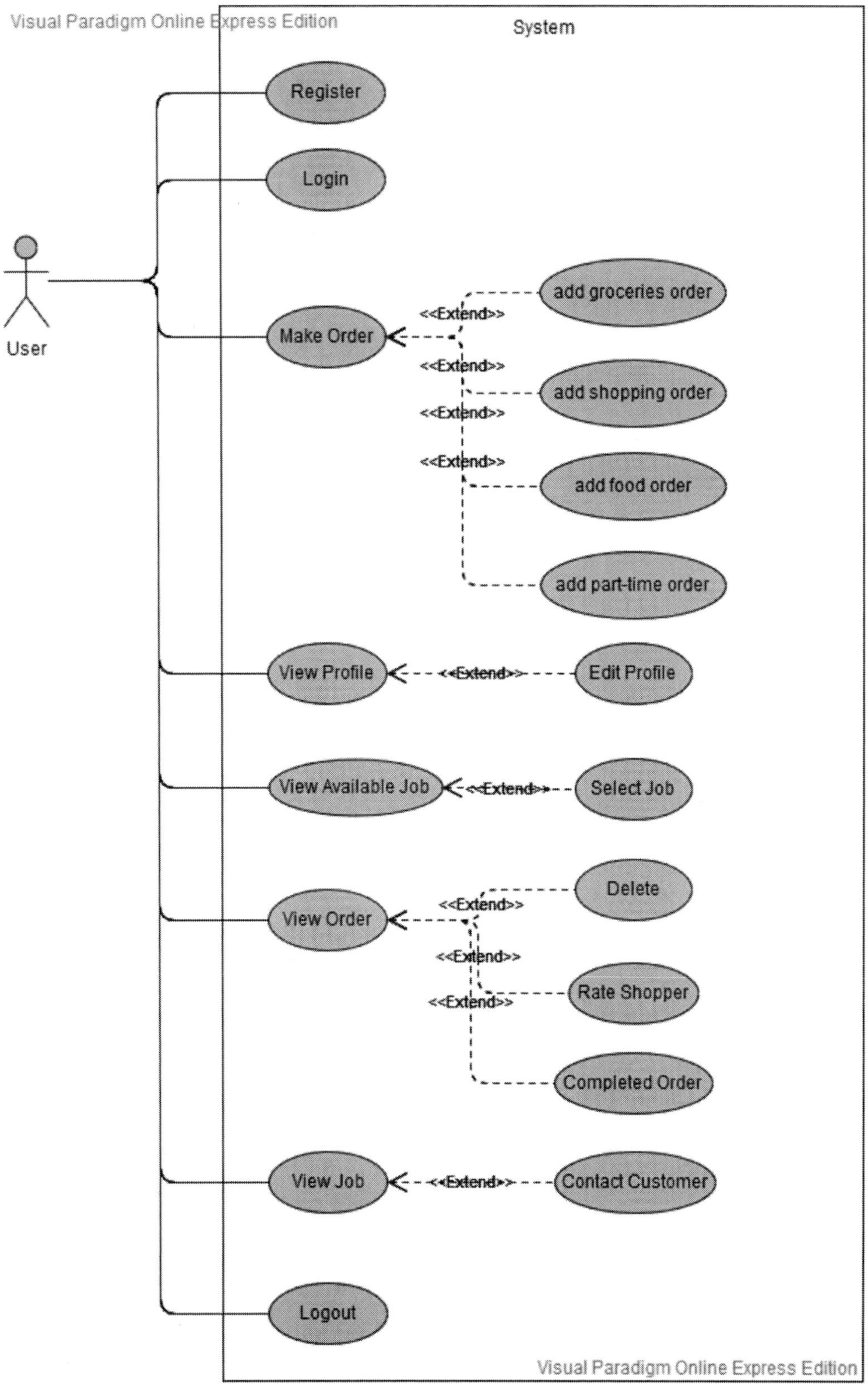

Figure 2. System Use Case Diagram

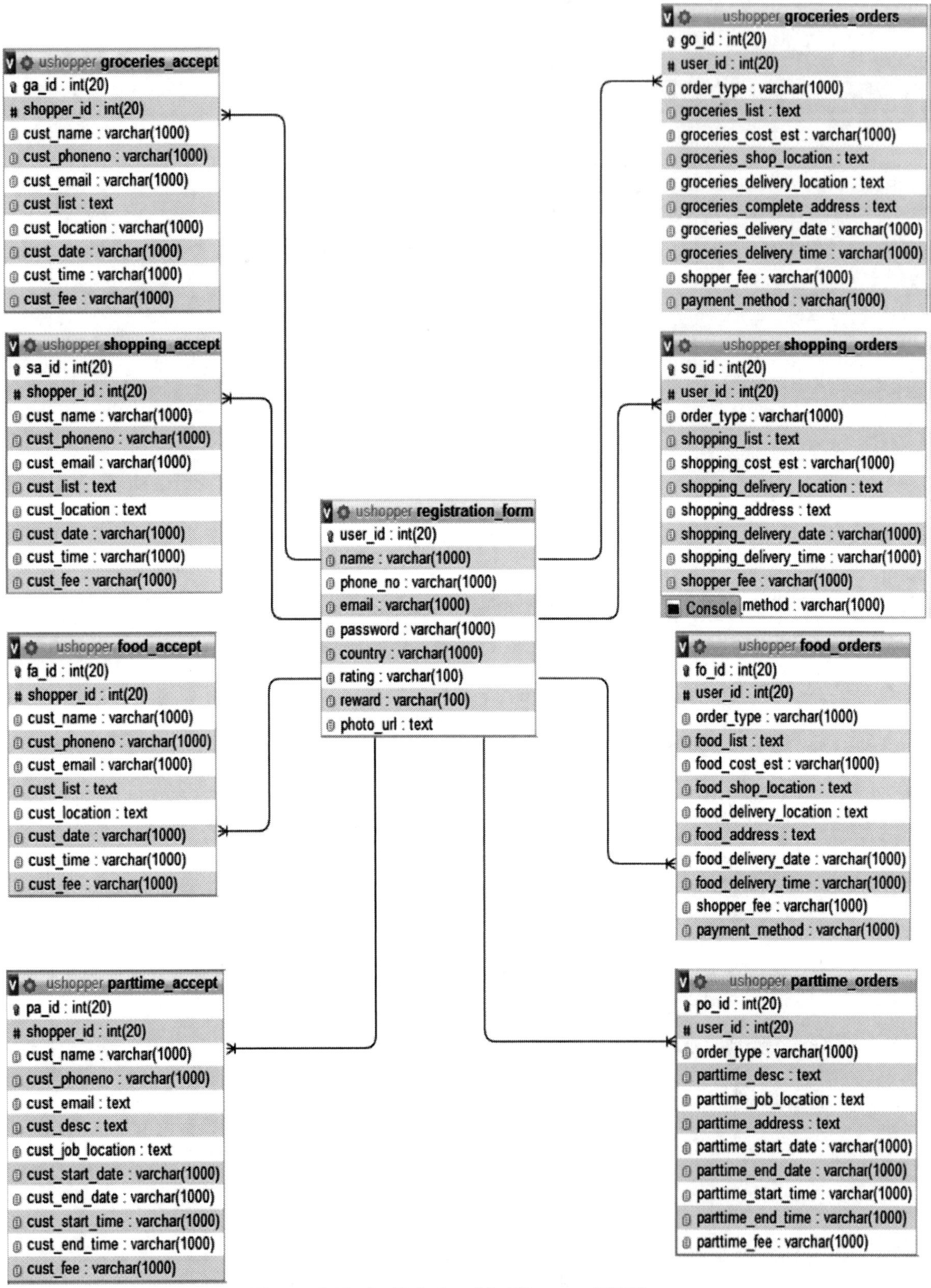

Figure 3. Entity Relationship Diagram (ERD)

IOP Conf. Series: Materials Science and Engineering **767** (2020) 012002 doi:10.1088/1757-899X/767/1/012002

(a)

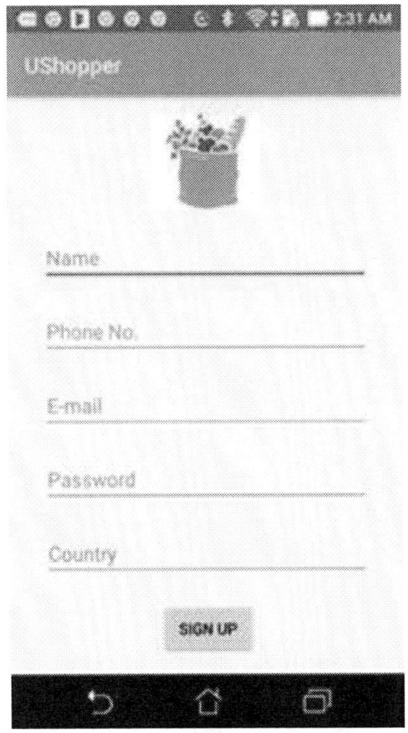

(b)

(c)

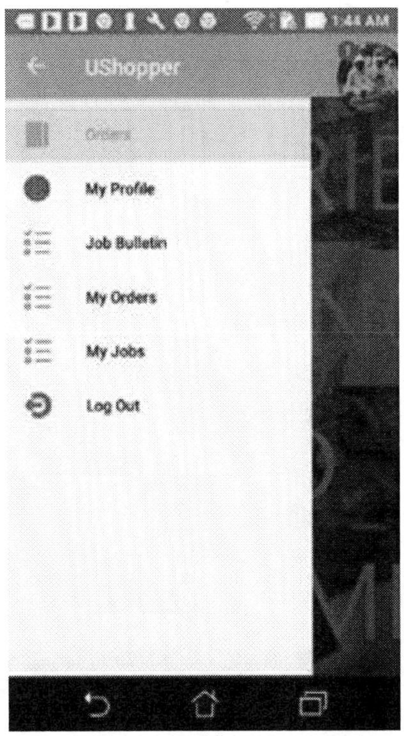

(d)

IOP Conf. Series: Materials Science and Engineering **767** (2020) 012002 doi:10.1088/1757-899X/767/1/012002

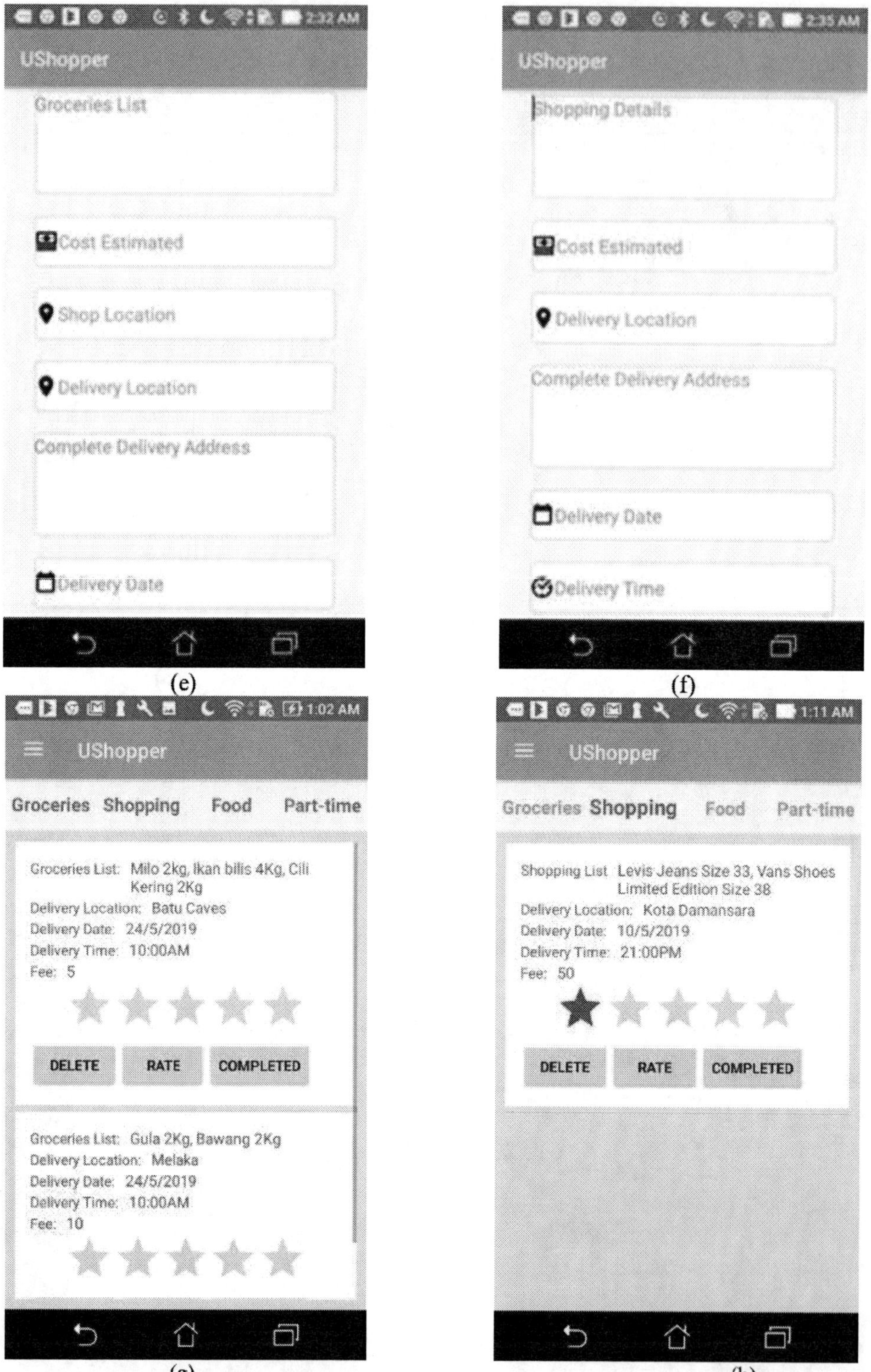

Figure 4. (a) Login Interface, (b) Register Interface, (c) Homepage Interface, (d) Application Menu Interface, (e) Order Menu Interface, (f) Shopping Order interface, (g) My Order Menu Interface, (h) Lists of Current Requested Groceries Order Interface

4. Conclusion

The aim of this project is to build a mobile application for customer convenient as well as for a personal shopper using Android Studio. Fortunately, this aim is achieved but there is a biggest issue in this system which is security. In this modern technology era, security issue has always been the biggest issue in a mobile phone application. Thus, we should look in depth to study the security policy if a system is involved with privacy information, such as user phone number, name and address. The system developed should have a security team, make sure to consult the security team whenever there is a change to plan on the app or a major revision, so they know what to do if something unexpected happens. However, nothing is perfect. This system has a lot of improvements that needs to be made as it is only equipped with basic features for customer to request order, rate the personal shopper, delete current order and completed the current order and personal shopper to accept the order and contact the customer through Whatsapp application. Thus, future research of this system should be focused on customer and personal shopper interaction in this application and customer can choose groceries based on hypermarket catalogues and food item based on popular orders. Besides that, current system does not implement online payment system for customer to paid personal shopper using credit card, online banking or PayPal. In future, the capability of this system over internet should be upgraded with machine learning where the system can detect the frequent items order by customer. Also, it is better if this system is able to hire security team in the future to ensure that user privacy information can be protected. In conclusion, there are some modifications and improvements can be added to a better result from the overall development of the project which are this system can be develop using machine learning to increase level of practically and accuracy, hire security team after system releasing to the market, so security team can protect user privacy information, create a messaging system to enable customers and personal shopper to communicate directly in the application rather than using Whatsapp, link application to the hypermarket catalogue so that users can select items based on the hypermarket catalogue, create a popular food suggestion system in the application, so users can choose food order based on popular order and implement online payment system, allowing users to pay personal shopper using credit card, online banking, and PayPal instead of just using cash on delivery payments.

5. References

[1] Lara Srivastava 2005 Mobile phones and the evolution of social behaviour *Behaviour & information technology* p. 111-129 2005.

[2] Green, Eileen, and Carrie Singleton 2009 Mobile connections: an exploration of the place of mobile phones in friendship relations *The Sociological Review* p. 125-144 2009.

[3] Harn, Adeline Chua Phaik, Ali Khatibi, and Hishamuddin bin Ismail 2006 E-Commerce: A study on online shopping in Malaysia *Journal of Social Sciences 13* p. 231-242 2003.

[4] Khatibi, Ali, Ahasanul Haque, and Khaizurah Karim 2006 E-Commerce: A study on internet shopping in Malaysia *Journal of Applied Sciences* p. 696-705 2006.

[5] Gong, Lei, and Cong Zhou 2008 Development and Research of Mobile Termination Application Based on Android *Computer and modernization 8* p. 85-88 2008.

[6] Abrahamsson, Pekka, Antti Hanhineva, Hanna Hulkko, Tuomas Ihme, Juho Jäälinoja, Mikko Korkala, Juha Koskela, Pekka Kyllönen, and Outi Salo. 2004 Mobile-D: an agile approach for mobile application development *Companion to the 19th annual ACM SIGPLAN conference on Object-oriented programming systems, languages, and applications* p. 174-175 2004.

[7] Holla, Suhas, and Mahima M. Katti 2012 Android based mobile application development and its security *International Journal of Computer Trends and Technology 3* p. 486-490 2012.

[8] Genero, Marcela, Geert Poels, and Mario Piattini 2008 Defining and validating metrics for assessing the understandability of entity–relationship diagrams *Data & Knowledge Engineering 64* p.534-557 2008.

IOP Conf. Series: Materials Science and Engineering **767** (2020) 012003 doi:10.1088/1757-899X/767/1/012003

Defected Ground Plane of Wearable Circular Patch Antenna

A ALI[1], S N Azemi[1], Jusoh M[2] and Sabapathy T[2]

[1]Advanced Computer Engineering Center (ACE),
School of Computer and Communication Engineering,
Universiti Malaysia Perlis (UniMAP),
02600 Arau, Perlis, Malaysia
[2]Bioelectromagnetic (BioEM) Research Group,
School of Computer and Communication Engineering,
Universiti Malaysia Perlis (UniMAP),
02600 Arau, Perlis, Malaysia

snorlyana@unimap.edu.my

Abstract. In this paper, Circular patch antenna design with defected ground plane is designed. The antenna is designed with circular patch and defected ground plane to allow it to operate in lower frequency as well as improved the antenna gain. The antenna has a feeding port and two defected ground slots to affect the frequency of the antenna. The proposed antenna is designed and simulated using Computer Simulation Technology (CST) Studio Suite 2016. The designed antenna has an operating frequency of 2.42GHz. A prototype has been fabricated and tested using network analyser and the measured results are in a good agreement with simulations.

1. Introduction

In recent years, there has been growing interest in wearable wireless technologies that enable solution for developing so-called smart clothes. In the future, the clothing will facilitate communication among object and people [1]. However, the major challenge of designing wearable antenna is to make the technology invisible to the users. Another key aspect is the robustness of the antenna performance to the mechanical solicitation, operating scenario, and operation such as ironing and washing. Furthermore, the wearable antenna operates in proximity or direct contact with the human body hence strongly effect the antenna performance [2].

An important feature of wearable devices is that they must allow wireless communications from or to the body via conformal and wearable antennas. Since these antennas must be located on body, their design strategy may be quite different compared to conventional antennas. Hence, the characteristics of the antenna is fabrics, which are wearable, durable and flexible, made it suitable to be integrated into clothing. Wearable antennas are generally used for body-centric communications within the sphere of personal area network (PAN) and body area network (BAN). There are number of specialized occupation segments that apply body centric communication systems, such as fire fighters, military and paramedics [3].

Since a flat conformal structure with a larger ground plane is required, patch antennas are usually chosen as the basis of wearable antenna designs [4-6]. Here, we proposed defected ground plane wearable antenna that are compact, flexible, light weight, low-cost with the polarization perpendicular to the body. The antenna consists of two slots in the ground plane that improve the antenna gain as well and antenna efficiency. It is also reduced the total size of the antenna compared with the non-slotted ground plane.

2. Antenna design development and specification

Several simulations were run in order to optimize each of the antenna parameters. The final dimensions of the antenna were obtained after a complete parametric study of all the conductive parts. The antenna includes a single layer of patch, one feeding line port and the ground plane which has been defected in two cutting slots by using the subtract method in CST. Firstly, for perfect match of load impedance of microstrip feedline to the radiation patch, a circular microstrip patch antenna was designed with inset line feed. Then the ground plane is defected by inserting two slots. The result of frequency before cutting the ground plane was 2.9 GHz. However, by introducing the defected ground plane, the frequency is shifted to the lower frequency which is at 2.45 GHz that has been targeted. In order to improves the antenna impedance matching, the width of the feed line is line is reduced and longer. While making the slot along the side of the feeding line shorter and narrower. These changes lead to the desirable value of S_{11} of our proposed antenna design. Shown in figure 1 is the schematic circular patch antenna with a radius of 24mm. Whilst shown in figure 2 is the fabricated antenna using Conductor (shieldit) with a thickness of 0.17mm, conductivity of 1.18×10^5 and substrate(felt) with a thickness of 3mm, permittivity 1.44 and loss tangent of 0.044.

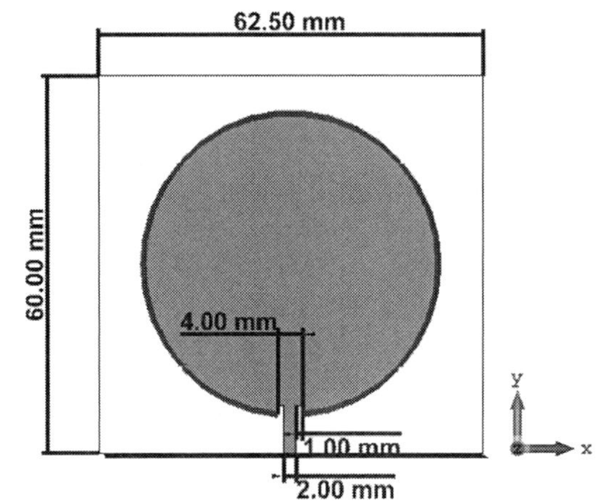

(a)

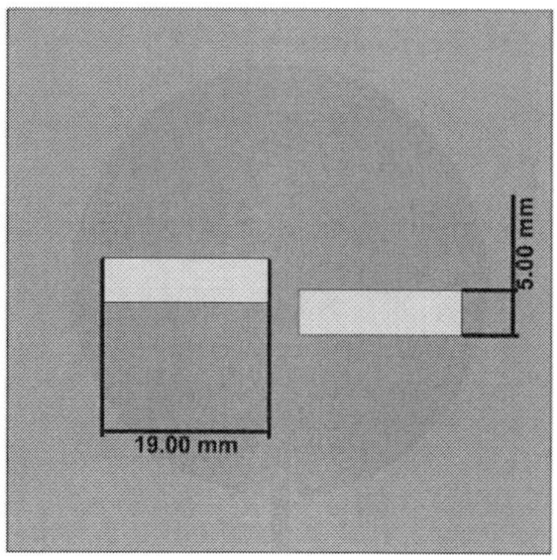

(b)

Figure 1. Simulated design of wearable circular patch antenna. (a) Front View, (b) ground plane view

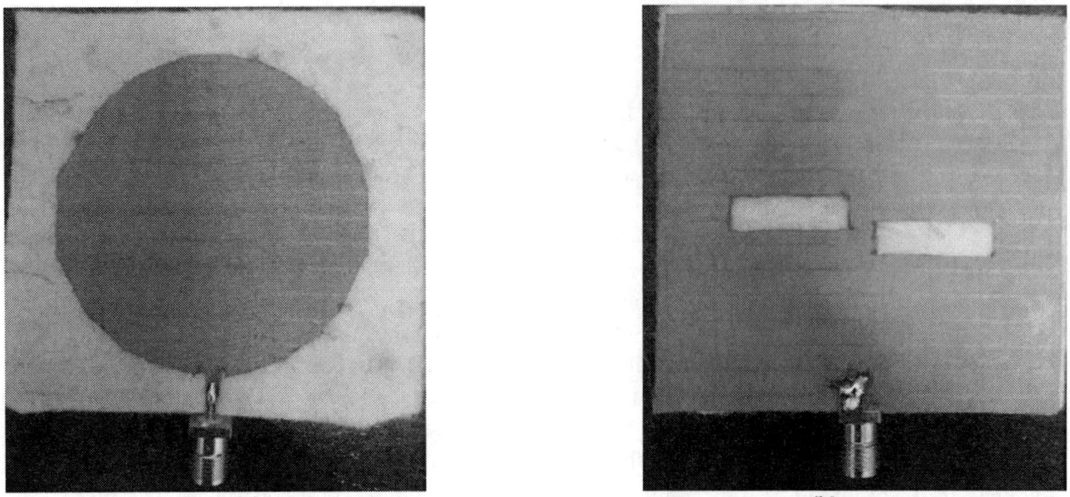

(a) (b)

Figure 2. Fabrication of circular patch antenna using conductor shieldit material with felt substrate. (a) Front view, (b) Antenna rear view.

3. Result and Discussion

A circular shaped Microstrip patch antenna with single band characteristics is investigated in this paper by using relevant standards of various principles. A software tool named CST simulator is used to construct this proposed design. Defected ground plane with two slots are used to improve the impedance matching of the scheduled antenna. This report provided an insight of circular patch antenna with and without the defected ground. The antenna developed using the CST 2016 software to simulate the results. A prototype has been fabricated using the conductor shieldit material with felt substrate. The input return loss is represented by S11 parameters results describing the electrical behavior between the ports, by measuring the input signal and the output signal of the source we can have an accurate way for getting the return loss value. Shown in figure 3 is the effect of introducing the defected ground plane on the antenna. The effect is clearly noticeable on the antenna performance where the frequency shifted from 2.9GHz to 2.45GHz.

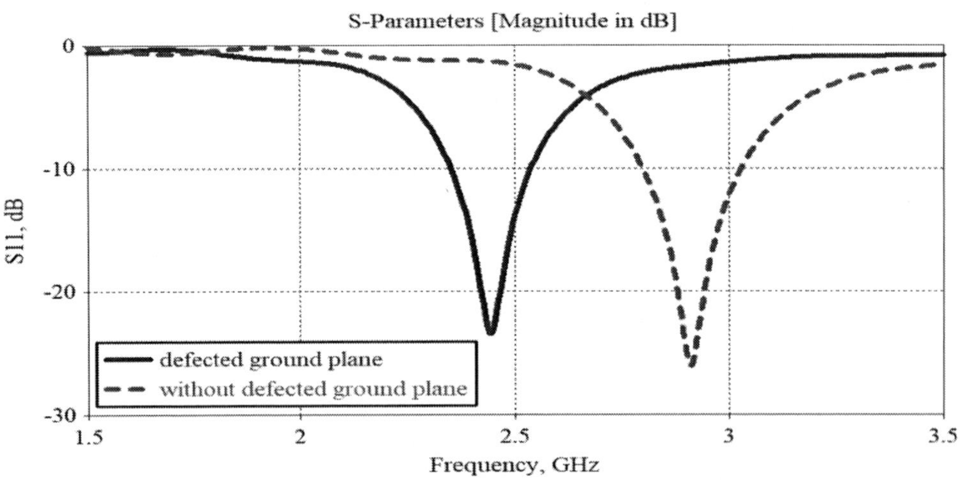

Figure 3: S11 result of without defected ground plane and with defected ground plane.

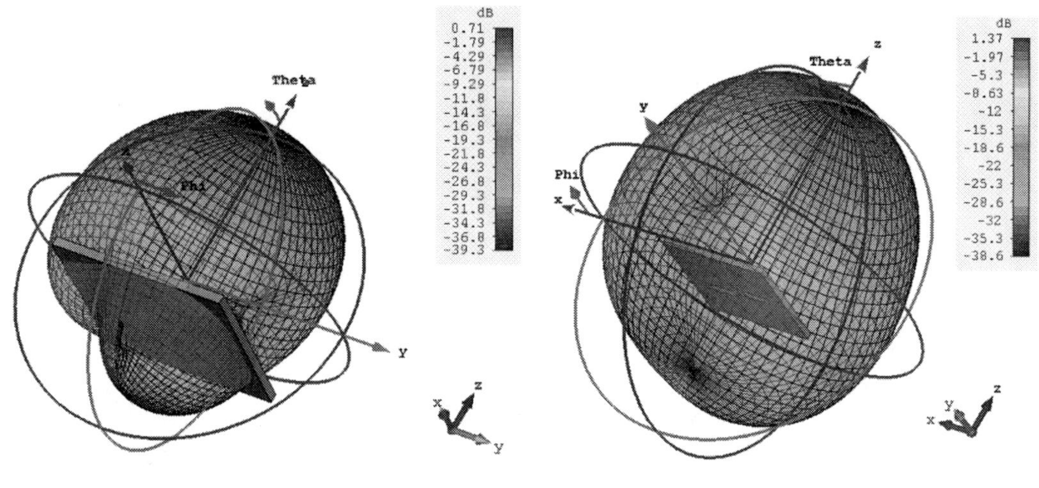

(a) (b)

Figure 4: Farfield result of (a) Circular antenna without defected ground plane, (b) circular antenna with defected ground plane.

Shown in figure 4 is the 3D view of the farfield and the impact on the gain that the effected ground did to the antenna. The antenna gain increased form 0.71dB to 1.43dB which is almost double from the original gain. It is considered significant improvement. It is believed by introducing slot at the back of the ground plane has affected the main lobe radiation of the antenna hence become more directive consequently increase the antenna gain. Figure 5 and 6 is the result of the antenna simulation versus measurement and VSWR respectively. It is clearly showing that the measurement results are identically the same as simulation.

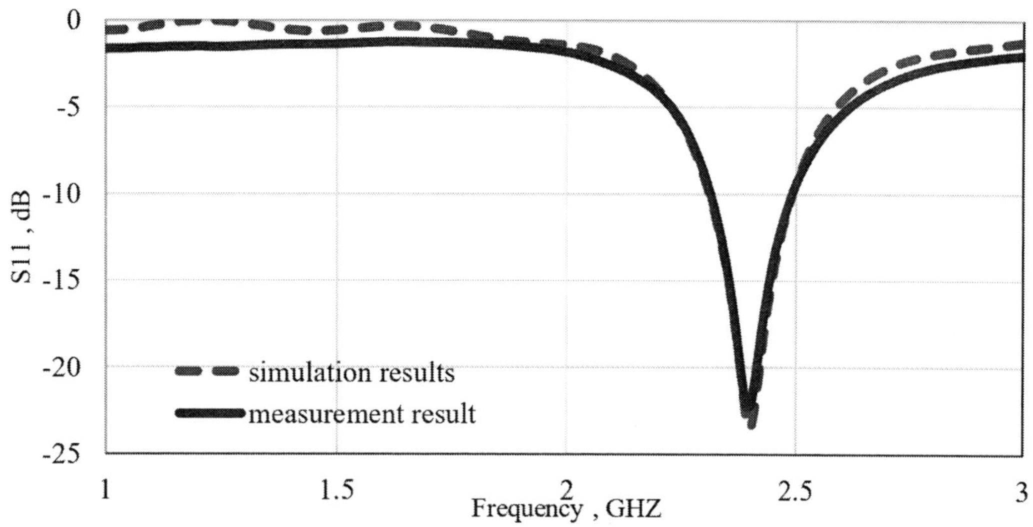

Figure 5: S11 Measurement versus simulation result

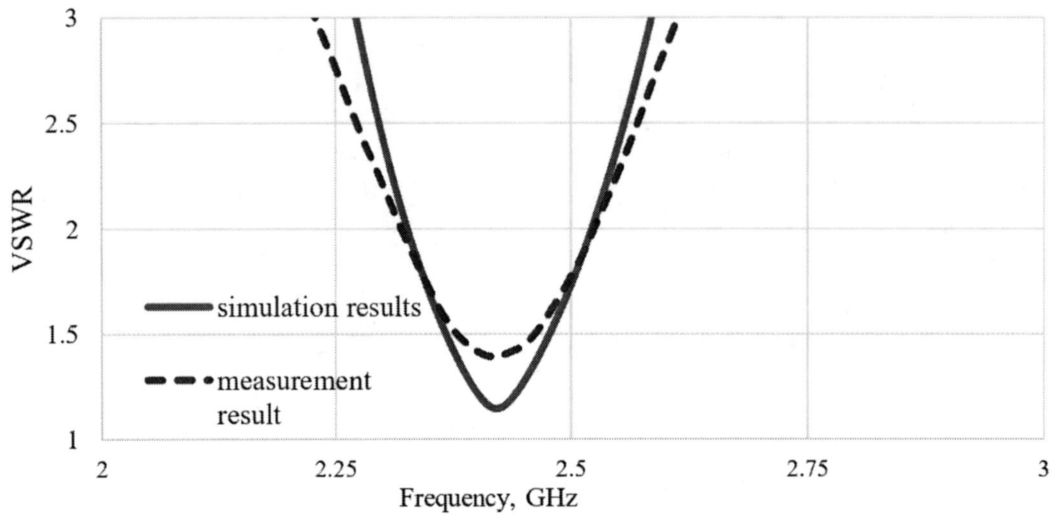

Figure 6: Voltage Stranding Wave Ratio (VSWR) Measurement versus simulation result

4. Conclusion

In this paper a circular patch wearable antenna has been designed and simulated using CST software for W. A defected ground plane has been introduced to the original design to acquire the targeted frequency which is at 2.45 GHz. This antenna can be used in the WLAN application and it has bandwidth of 67MHz. Two slots have been cut from the ground plane of the antenna to affect the resonant frequency and some changes has been done to the feeding line to improve the S_{11} parameters. The integration of two slots on the ground palne resulted in stable radiation patterns and a low backward radiation. A prototype has been fabricated and tested using network analyzer and the simulated and measured reflection coefficients are shown in figure 5 and 6, with a satisfactory agreement. The antenna has a low profile and suitable as wearable antenna for wearable application.

5. References

[1] Benito Sanz-Izquierdo, John C. Batchelor, M. I. Sobhy, "Button antenna on textiles for wireless local area network on body applications", Microwaves Antennas & Propagation IET 4, vol. 11, pp. 1980-1987, 2010 \
[2] N. H. M. Rais et al., "A review of wearable antenna", Antennas & Propagation Conference 2009. LAPC 2009., 2009.
[3] Rofan Aziz, Karsid, Dedi Suwandi, Basari, "Design of Leaf Textile Antenna for Body Area Network in Biomedical Applications", Applied Science and Technology (iCAST) 2018 International Conference on, pp. 318-321, 2018
[4] S. Yan, P. J. Soh, and G. A. E. Vandenbosch, "Made to be worn," Electron. Lett., vol. 50, no. 6, pp. 420, Mar. 2014.
[5] B. Ivsic, D. Bonefacic, and J. Bartolic, "Considerations on embroidered textile antennas for wearable applications," IEEE Antennas Wireless Propag. Lett., vol. 12, pp. 1709–1711, Jan. 2014.
[6] P. J. Soh, G. A. E. Vandenbosch, S. L. Ooi;, and N. H. M. Rais, "Design of a broadband all-textile slotted PIFA," IEEE Trans. Antennas Propag.,vol. 60, no. 1, pp. 379–384, Jan. 2012.

IOP Conf. Series: Materials Science and Engineering **767** (2020) 012004 doi:10.1088/1757-899X/767/1/012004

Simulation of differential relay for transformer protection

N S Azizan[1], C L Wooi[1], B Ismail[1], S N M Arshad[1], M Isa[1], W A Mustafa[2] and MNK Rohani[1]

[1]Centre of Excellence for Renewable Energy (CERE), School of Electrical System Engineering, Pauh Putra Main Campus, Universiti Malaysia Perlis, 02600 Arau, Perlis, Malaysia.
[2]School of Engineering Technology, Kampus Sg. Chuchuh, Universiti Malaysia Perlis, 02100 Padang Besar, Perlis, Malaysia.

nursyuhadazizan@gmail.com

Abstract. In electric power system, transformers are an important element and their protection are critical. As the transformer might be damaged during a fault, faster and more selective protection was required. For this purpose, the design of differential protection was utilized. In this study, the differential relay simulation has been successfully done using MATLAB/Simulink. The essential approach is to protect the power transformer against internal faults and avoid interference in the power system. The evaluation has been accomplished at different fault conditions. Several types of faults including double line-to-ground, single line -to-ground and line-to-line faults are studied. Simulation studies were performed and the relays performance with different system parameters and conditions were investigated. The obtained results illustrated for the differential relay represents an appropriate action.

1. Introduction

The appearance of protective relays in a power system is to detect system anomaly and to assassinate proper ways to isolate swiftly only the faulty element from the healthy system. The flashover of lightning and other faults in the transmission line leads to shortage of electricity [1]. The possibility of fault occurrence on the overhead lines is much more due to their greater lengths and liability to atmospheric conditions [2]. There are several methods for the protection of transmission and distribution lines [3-7]. Transmission and distribution networks have moved from being a local entity to becoming a continent spanning machine, which consisted of power plants, wind turbine farms, country spanning transmission line and etc. The joints connecting all these elements were the transformers. The transformers scale the voltages to connect the lower level distribution networks to the highway of electricity: the transmission network [8].

However, the transformer might be damaged during a fault. Therefore, a faster and more selective protection was required for the continuity operation of the power system. For this purpose, differential protection was utilized. By defining its zones using current transformers, differential relay could provide 100% selectivity, and tripping time could be controlled under one cycle. This was done by comparing the currents flowing into the transformer [9]. Any internal error will show up as a differential current and caused a trip. In the ideal system, this could be done easily. Under the internal and external faults, any transformer protection should be able to operate. The main function of differential relay was to send trip signal to circuit breaker when there is a certain difference current value between primary side and secondary side of the transformer [1]. Current Transformers (CTs) were used to convert the primary current to secondary current and the differential relay can be used to check if there is any differential current.

Under the normal operating condition, the power output to the transformer should be equal to the power input to the transformer. If internal faults happened in the power system, there will be a differential current generated. In order to design a proper protection, the faults need to be located within the protection zone [10]. Then, proper action should be done to isolate the faulty zone from the rest of the power system. Current differential relays were one of the protection devices used to protect the electrical equipment against the internal faults. This is due to the reason that current differential relays have high sensitivity and simplicity. This paper aimed to study the modelling and simulation of differential relay for transformer protection operation using MATLAB/Simulink.

2. Methodology

The power system model was simulated and performed using Simulink/MATLAB. In this model, there are 11/33 kV power system with current relay protection scheme between the transmission line and transformer. Figure 1 showed that the power transformer was connected to a power system consisting of a three-phase power generator. In order to check the switching effect, it feeds the load over this transformer between two circuit breakers, CB1 and CB2.

The differential relay was used to detect the internal fault and generated a trip signal to the associated circuit breakers to segregate the faulted section from the rest of the power system [10]. This protection minimized the harm to transformer and power system. Equation 1 explained the operation of differential relay where the differential relay will operate once the differential current measured from CT1 and CT2 exceed certain value. The differential current that exceed certain value was called the pickup current. In order to make system protected, the breaker will not send a tripping signal if the measured differential current does not exceed the relay's pickup current. The relay will send a tripping signal if there were internal faults, the differential current formed in relay was higher than the pickup current. After the tripping signal received, the breaker will disconnect and isolate the fault from power system.

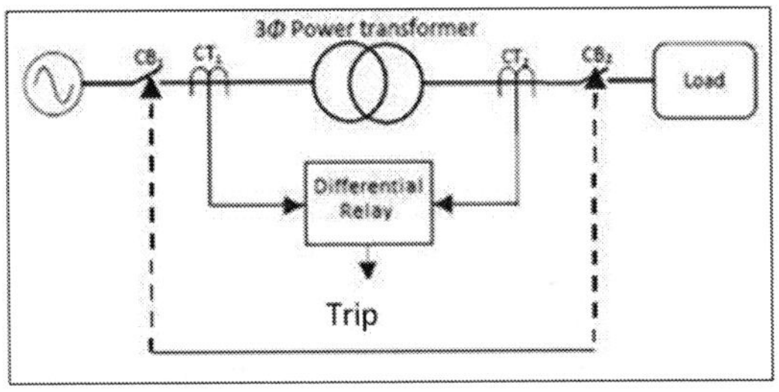

Figure 1. Basic configuration of differential relay.

$$\Delta I = |I_1 - I_2| \quad (1)$$

where
ΔI = The differential current
I_1 = The secondary current measured by CT1
I_2 = The secondary current measured by CT2

Figure 2 showed in detail on how the simulation of differential relay has been carried out. Basically, when there was no power loss in ideal transformer, there was no operating current. But in practical, there was an eddy current loss and core loss in transformer. There was also imbalance in phase shifting, tap-changing, CTs ratio and transformer ratio because of the operated current will not be zero. This differential relay was designed to respond to the circuit breaker when there is difference in current from both CTs of transformer in the term of its inconsequential relation to the current flowing from the protecting zone as shown in figure 1.

First, it read the current in transformer of primary and secondary then it will identify the differences between the current measured from the primary side and secondary side of transformers. Then, the fault was created to test the system. The relay will sense any changes in current flow through the transformer. Finally, the information is collected from the results of tripping in the system.

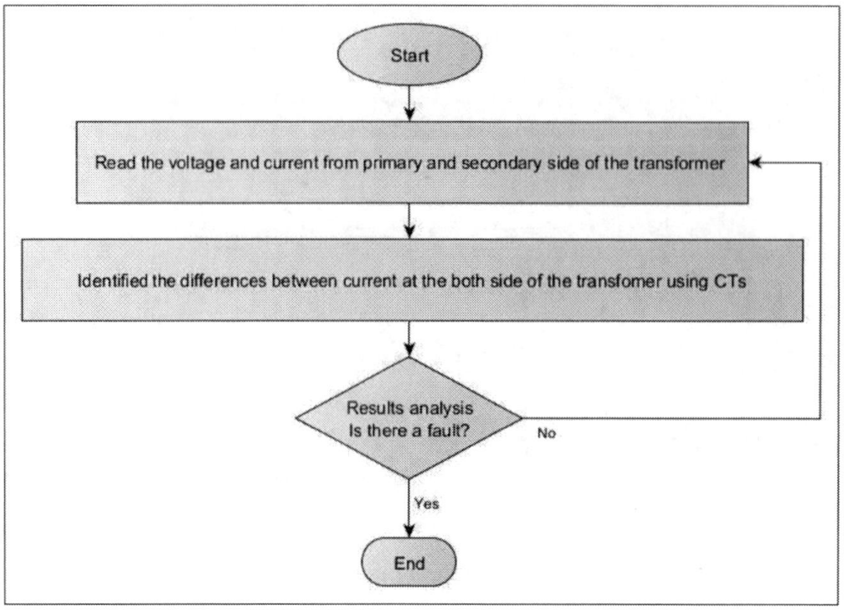

Figure 2. Simulation process flowchart of differential relay.

2.1. Circuit Operation

A power system model has been developed using MATLAB/Simulink software. Figure 3 showed the differential relay in power system model. The general overview of the power distribution system consisted of three-phase source connected with three-phase breaker. Then, transmission line was connected to the breaker. In between them, there were three-phase V-I measurement which is used to measure the value of load current. Lastly, a three-phase series RLC load was connected at the end of transmission line which resembled the load for the system.

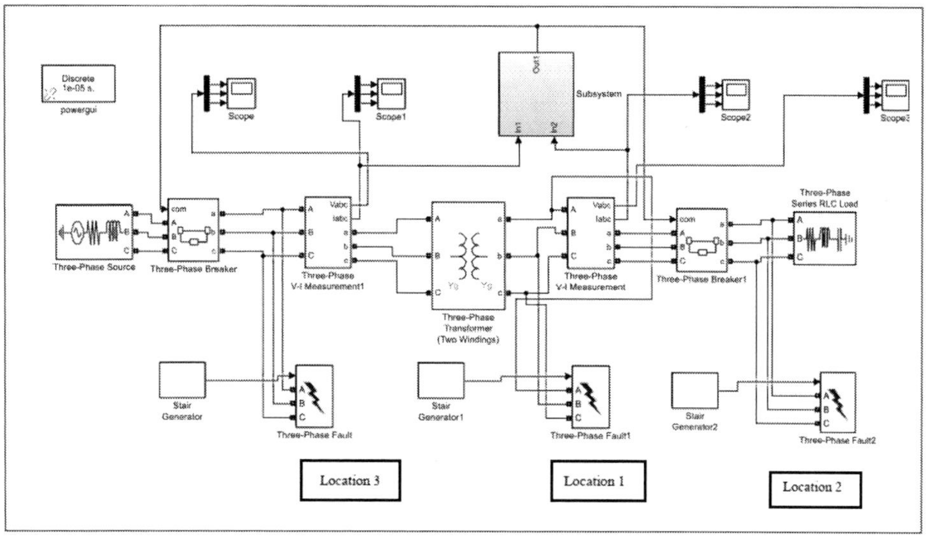

Figure 3. Differential Relay Model in Simulink.

Figure 4 showed the design of differential relay in Simulink which consisted of two input portals In1 and In2. The In1 and In2 represented the output currents of the CT1 and CT2 respectively. In order to analyse the system, these two input signals would be divided into three parallel paths. All the signals entered a block named amplitude comparator. The first signals entered a block named amplitude comparator. The second signals made the signals to be impressed in the harmonic test and send the result to a block named harmonic comparator.

The amplitudes of both input current In1 and In2 will be compared by amplitude comparator. Then, the amplitude comparator sent an indicating signal to the comparator block and showed the difference between the two signals. At the same time, the harmonic comparator block will receive two signals from two blocks named subsystem. Then, it will send the result to the comparator. The subsystem block will evaluate the input signal to investigate if harmonics exist in this signal.

Figure 4 also illustrated the contents of the subsystem block. There were two input entering the subsystem. There were gain three at input two as the secondary winding of transformer is 33kV, in order to balance the current on the first input of 11kV. The purpose of gain three is to increase the secondary voltage side to become less in transformer.

The control system of relay is based on the digital operating system. The relational operator is used to compare the input current with the pickup current as desired. The output of the relational operator was connected to the SR flip flop.

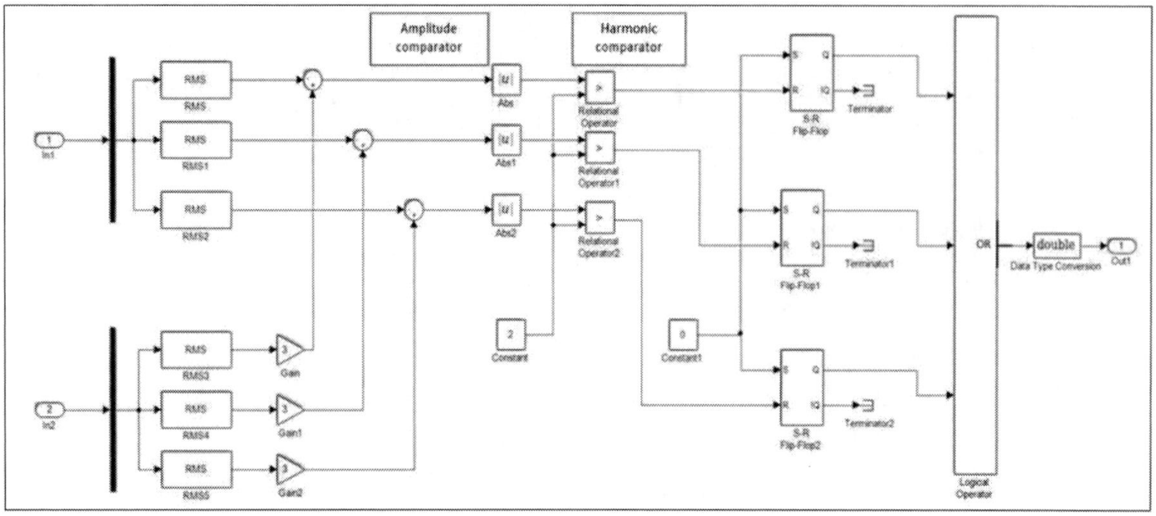

Figure 4. Differential Relay in Subsystem.

Table1. Truth Table of SR Flip Flop.

S	R	Q	State
0	0	Previous State	No Change
0	1	0	Reset
1	0	1	Set
1	1	X	Forbidden

The basics of SR flip flop was shown in Table 1. Based on the truth table, to make the circuit breaker tripped, all the input "R" of the flip-flop should be connected to '0'. All the 'S' inputs should be connected to the output of relational operators. When a fault happened, the relational operator will become 1 and with input 'R' equal to 0, the flip-flop output (Q) will become 1 (SET). Then, trip signal will be sent.

3. Results and Discussions

The normal conventional relay was simulated in MATLAB using the Simpower system of SIMULINK. The external and internal fault cases were simulated. When external fault appeared and the circuit breakers were closed to connect the transformer, the currents were generated. In designing the differential relay, there must be a relay before circuit breaker that assigns the tripping signal to it after correlate input signals. The currents were measured by current transformers and then introduced to the relay. The faults were created at three different condition; at three phase fault 1, three phase fault 2 and three phase fault 3.

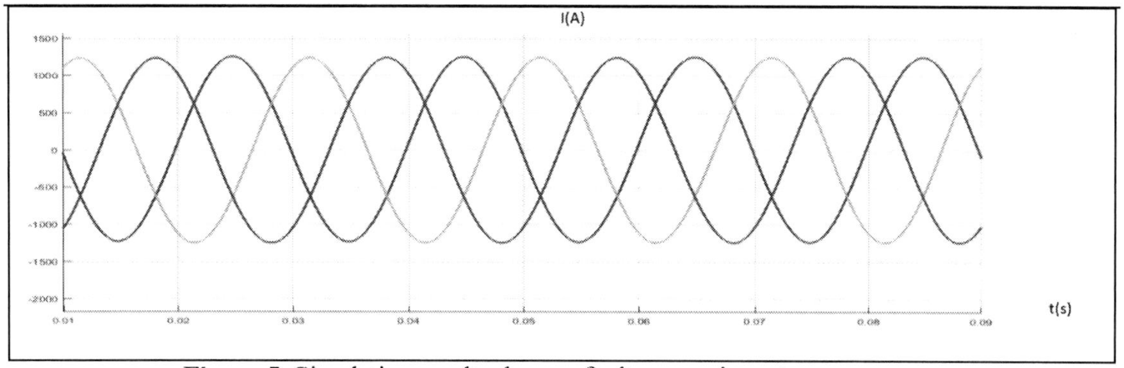

Figure 5. Simulation result when no fault occurs in system.

Figure 5 showed the result for no fault condition in the system. Under this condition, relay did not assign any tripping signal to the breaker hence the signal was in normal condition. Figures 6 and 7 showed the results for three-phase fault created at t=0.1s when fault occurred at location 1 which relay sent a signal to the circuit breaker. Faults was created when three-phase faults at location load 2 trip the circuit breaker at t=0.1s which showed that there were internal faults. Consequently, the transformer was isolated from the grid. Figure 8 showed that there were no fault and no circuit breaker functioned for three-phase I(A) fault at location 3 as it is located as an external three-phase faults.

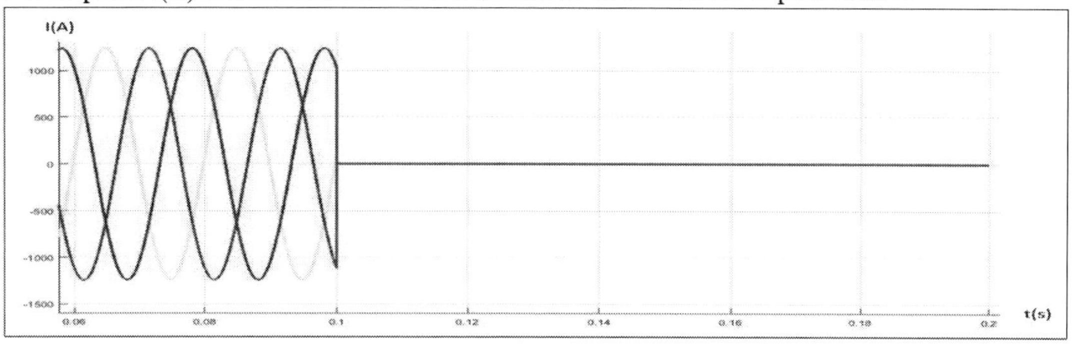

Figure 6. Simulation result when three-phase fault is occurred at location 1 (L-L-L).

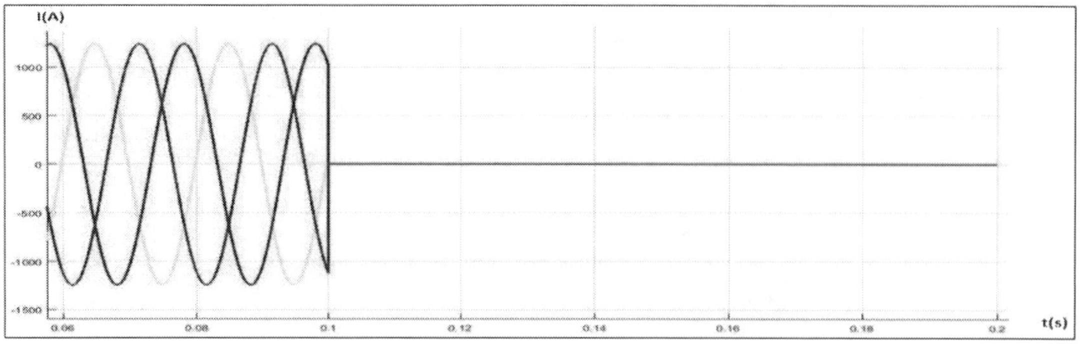

Figure 7. Simulation result when three-phase fault is occurred at location 2 (L-L-L).

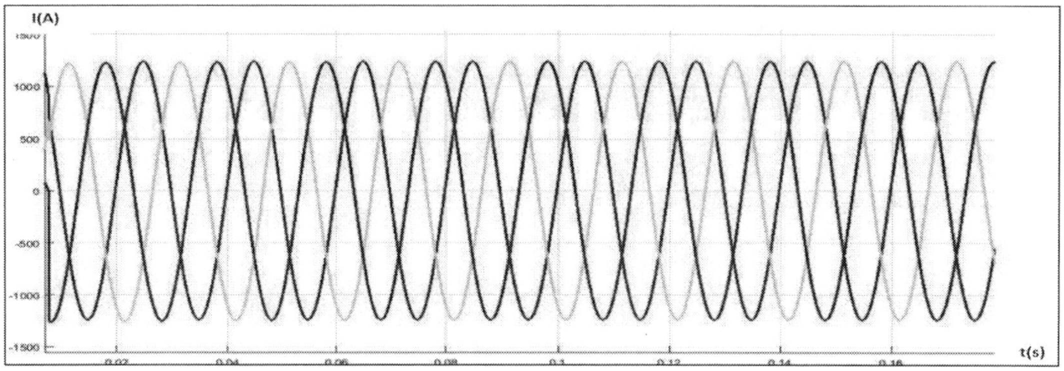

Figure 8. Simulation result when three-phase fault is occurred at location (L-L-L).

Figures 9 and 10 showed the results for LLG fault created at t=0.1 s in the system where the relay sent signals to the circuit breaker. Faults were created at location 1 and 2, trip the circuit breaker at t=0.1 s shows it was an internal LLG fault. Figures 10 and 11 showed the results for the internal fault LLG. The internal fault LLG was located inside the protected zone and therefore the corresponding relay sent a tripping signal to the circuit breaker.

It was observed that in case of L-L-G fault, only the faulty phase was disconnected by the circuit breaker. Thus, transformer was segregated from the grid. Double line to ground included 15 % of power system faults. Figure 11 showed that there were no fault and no circuit breaker function at three phase fault 3 because it was located as outside protection zone and it was an external L-L-G faults.

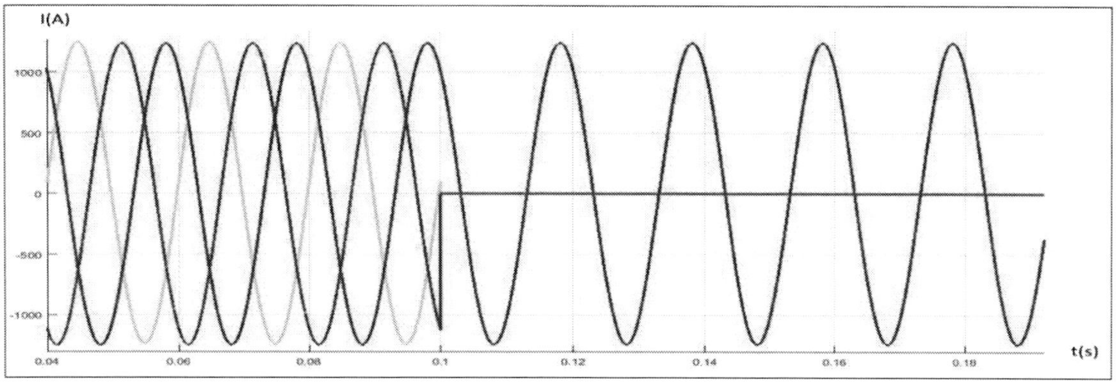

Figure 9. Simulation result when LLG fault is occurred on phase A and B at location1.

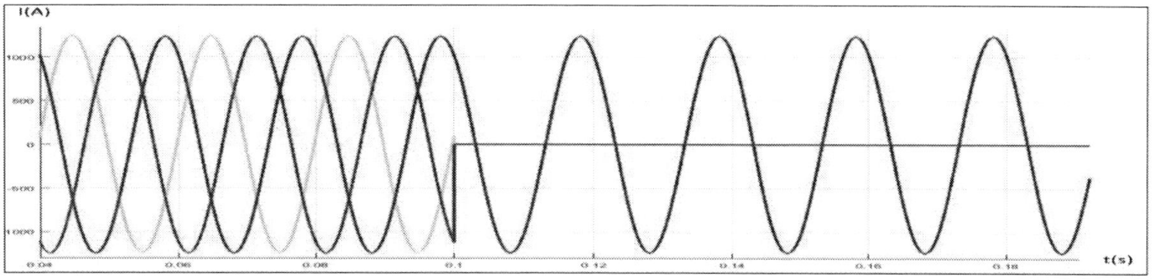

Figure 10. Simulation result when LLG fault is occurred on phase A and B at location2.

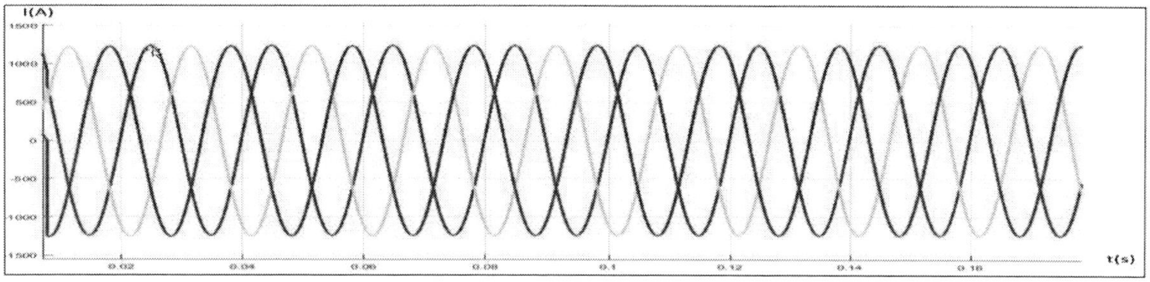

Figure 11. Simulation result when LLG fault is occurred on phase A and B at location 3.

Referring to Figures 12 and 13, there were small phase difference between the two currents which caused by different source angles at t= 0.1 s. For an internal fault, with the polarity of the current transformers, the current flow into the protected zone from both line terminals. For L-L at phase A and phase B, the relay responses the fault and selectivity open one of the faulted phases so that other phases able to supply power. Line to line included 15 % of power system faults.

For an external fault, the current flow into the line and out of the other. Figure 14 showed the result for external L-L fault on system hence no tripping signal from the differential relay to the circuit breaker. If the fault occurs outside the bounded area, it is not desirable to disconnect the transformer.

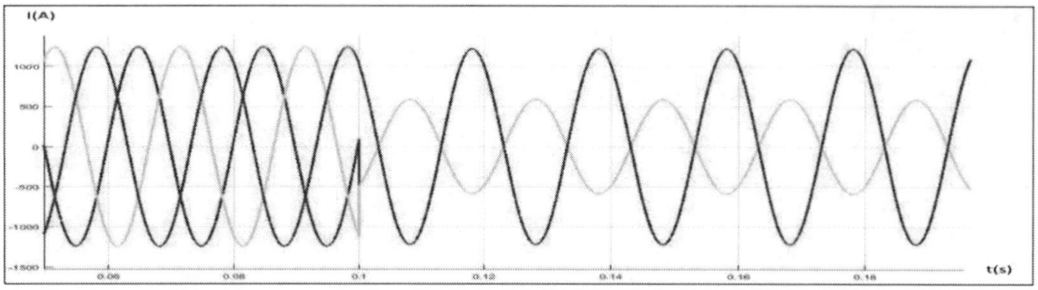

Figure 12. Simulation result when fault is occurred on phase A and B at three phase fault 1 (L-L).

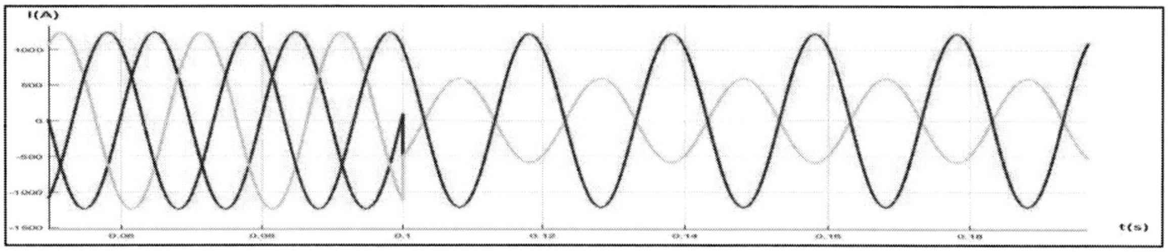

Figure 13. Simulation result when fault is occurred on phase A and B at three phase fault 2 (L-L).

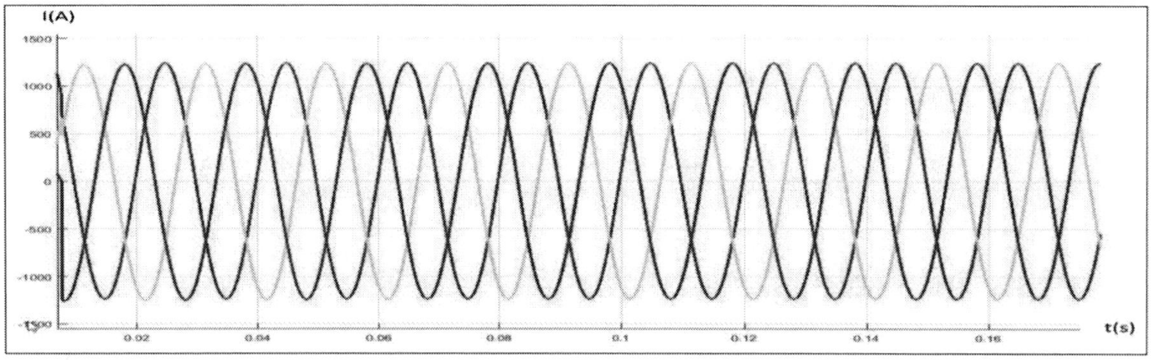

Figure 14. Simulation result when fault is occurred on phase A and B at three phase fault 3 (L-L).

Figures 15 and 16 showed the results for internal L-G fault, phase A to ground, at three phase fault 1 and 2. There was an internal fault inside the protected zone relay. The relay sent a tripping signal to

IOP Conf. Series: Materials Science and Engineering 767 (2020) 012004 doi:10.1088/1757-899X/767/1/012004

the circuit breaker at created fault t= 0.1s as there are differences between the currents. Simulation result for L-G was shown in Figures 16 and 17. The proposed relay allowed a trip signal to the circuit breaker as fault happen. It was noticed that in case of L-G fault, only the faulty phase was detached by the circuit breaker over relay. The transformer was disconnected from grid as it opened all ungrounded conductors of the faulted circuit.

Figure 17 showed the result for external L- G fault, phase A to ground, at three phase fault 3. On the system, there are no tripping signal from the differential relay to the circuit breaker. The external faults showed that the fault outside the protection zones of the differential relay. In external fault condition, the current over transformer was changed but a decisive differential relay must not give a tripping signal to the circuit breaker. The summary of fault and relay tripped is simulated as stated in Table 2.

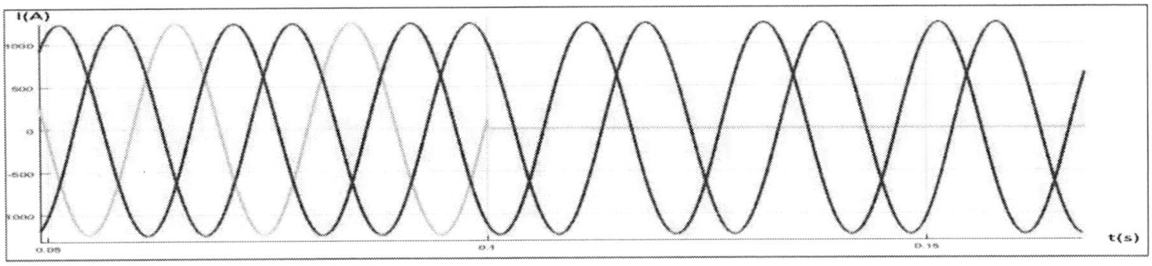

Figure 15. Simulation result when fault is occurred on phase A and ground at three phase fault 1 (L-G).

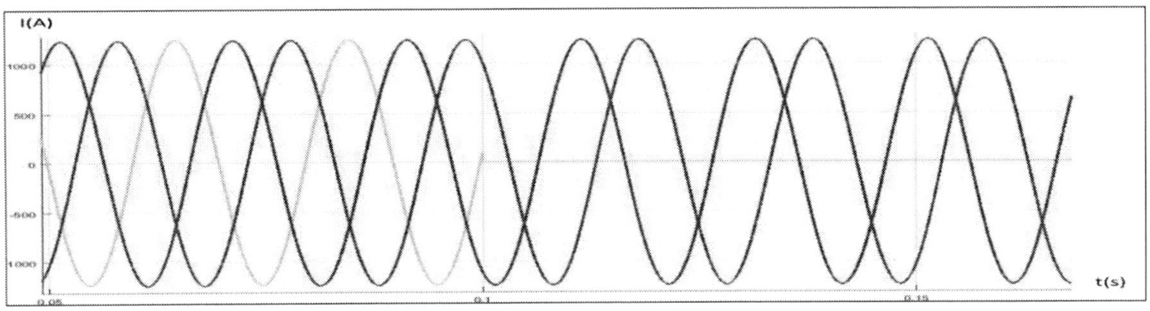

Figure 16. Simulation result when fault is occurred on phase A and ground at three phase fault 2 (L-G).

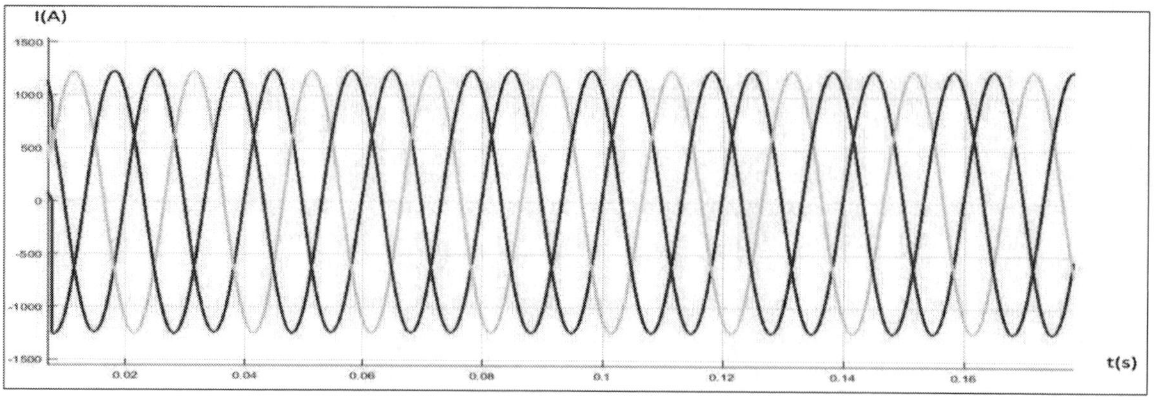

Figure 17. Simulation result when fault is occurred on phase A and ground at three phase fault 3 (L-G).

Table 2. Type of fault and clearing scheme.

Type of Fault	Fault Location	Relay Tripped
Three Phase	1	Tripped
	2	Tripped
	3	Not Tripped
Double Line to Ground	1	Tripped
	2	Tripped
	3	Not Tripped
Line to Line	1	Not In Phase
	2	Not In Phase
	3	Not Tripped
Single Line to Ground	1	Tripped
	2	Tripped
	3	Not Tripped

Circuit breaker worn for the protection purpose in power system cannot work by its own determination. The circuit breaker sends the tripping signal after correlate input signal with the logic. The result is compared as in Table 2. The tests are conducted in two situations, external faults and internal faults by phases. For faults inside the protection zone, the relay has reacted accurately, at the presences of the located fault. In addition, for external fault in different conditions normally the relay has not operated when fault located at three phase fault 3.

4. Conclusions

As conclusion, the differential relay simulation had been successfully done and the results were amazing as all the types of fault tested, triggered the tripping system. The basic approach is to assure the power transformer across internal faults at created fault t=0.ls and prevent interruption due to other operating conditions. In three phase faults, as fault happen, the transformer is consequently isolated from the grid. In double line to ground and single line to ground, it is observed that only the faulty phase is detached by the circuit breaker over the relay. In line to line fault, there are small differences between two current which caused by different sources angle. The evaluation has been accomplished at different fault conditions. Simulation studies are performed and the relays performance with different system parameters and conditions is investigated. The obtained result illustrate that the proposed differential relay represents an appropriate action.

Reference

[1] R. Haas *IEEE Power Engineering Review* **12** 48-49.
[2] *IEEE Guide for Protective Relay Applications to Power Transformers.*
[3] C.-H. Einvall and J. Linders A three-phase differential relay for transformer protection *IEEE Transactions on Power Apparatus and Systems* **94** 1971-1980.
[4] J. Thorp and A. Phadke A microprocessor based three-phase transformer differential relay *IEEE Transactions on Power apparatus and Systems* **2** 426-432.
[5] C. Mathews An improved transformer differential relay *Electrical Engineering,* **73** 648-648.
[6] A. L. Orille-Fernandez, N. K. I. Ghonaim, and J. A. Valencia A FIRANN as a differential relay for three phase power transformer protection *IEEE Transactions on Power Delivery* **16** 215-218.
[7] I. Hall *et al.* New line current differential relay using GPS synchronization *Power Tech Conference Proceedings 2003* Bologna 2003 **3** 8.
[8] IEEE Standard Electrical Power System Device Function Numbers, Acronyms, and Contact Designations - Redline *IEEE Std C37.2-2008 (Revision of IEEE Std C37.2-1996)* 1-62.
[9] W. Jiekang Microprocessor-based fundamental frequency tracking system for digital protection relaying in *Industrial Technology IEEE International Conference 2004* **3** 1276-1280.
[10] N. S. Jadhav and A. Thorat Design of a differential relay for 1000-kV transmission line using MATLAB in *2013 International Conference on Energy Efficient Technologies for Sustainability* 2013 1164-1168.

Acknowledgement

The authors wish to thank the Centre of Excellence for Renewable Energy (CERE), School of Electrical System Engineering, Universiti Malaysia Perlis and Ministry of Education (MOE) for the financial aid.

IOP Conf. Series: Materials Science and Engineering **767** (2020) 012005 doi:10.1088/1757-899X/767/1/012005

Development of U-Shape Slot Wearable Antenna for In-Body Communications

N A Kamaruddin, S N Azemi, S Z Ibrahim and C B Mrashidi

Advanced Computer Engineering Center (ACE),
School of Computer and Communication Engineering,
Universiti Malaysia Perlis (UniMAP),
02600 Arau, Perlis, Malaysia

snorlyana@unimap.edu.my

Abstract. The ability to have a communication with devices implanted inside a human body will cause a great improvement in current wireless medical applications technology. Wireless Capsule Endoscopy (WCE) is a medical device that could send images from inside of human's intestines to the sensor outside the body. However, this device has few disadvantages like its location cannot be detected once it entered the body and it also cannot be control from outside the body. Considering these factors, an antenna that has the ability to penetrate into human body tissues for in-body communication is proposed. UWB system has considered as the solution for future in-body communication devices since current standard does not allow high data rate wireless connections between implanted nodes. Low part of UWB frequency band which is 3.1 GHz to 5.0 GHz is used in this research in order to reduce the attenuation through the body tissues as the frequency increase. The design of this antenna has taken in consideration of the propagation medium which is the human body tissues. Simulation for the designed antenna was done in CST Software. The size of this antenna is designed to be compact and wearable on human body. The substrate used for this antenna is cotton to ensure comfort once it is placed on the human body. The results that are considered in this research are the S11, directivity and gain of the antenna. Both simulation results and measured results are compared to evaluate the ability of this antenna.

1. Introduction

The ability to establish a link between outside the human body and inside the human will increase the opportunity to detect diseases or abnormalities suffered by patients. People suffering from chronic diseases have been increasing by year and the development of new technology is needed in order to treat these diseases efficiently and low in cost. Wireless Capsule Endoscopy (WCE) is a medical device used to detect diseases inside the gastrointestinal (GI) tract without causing any painful experience to the patient. However, this device still has few limitations which cause it to be less effective. The location of this capsule cannot be identified once it is swollen by the patient. This is said to be the main disadvantage of WCE technology [1][2].

WCE is very reliable and the most effective way to diagnose GI tract and it is the only method that able to capture color images of the small intestines. Despite from its abilities to do such things, WCE cannot be controlled from the outside. The medical instructor cannot control WCE to go back to certain position for re-diagnosis and the location of the WCE also cannot be detected. Doctors can only diagnose the patient after few hours after the capsule was swollen. Due to these drawbacks, few researches on antenna performance to detect the location of the WCE and to have communication inside the body have been made. On-body and in-body antenna means that one antenna is placed inside the human body, while the other will be placed on the human body [3]. Challenges in designing this type of antenna are both implanted antenna and on-body antenna cannot be placed too far from each other to ensure a proper link between them. Thus the size of the antenna needs to be small enough to be suitable to be placed on human abdomen. Besides, the antenna matching is important and should be ensured within the whole frequency band of interest. Also, the propagation of radio waves through the human body tissues is found as a challenge for this type of antenna [4].

Ultra Wide Band (UWB) has few main advantages like large bandwidth, low profile of its devices, low power consumption and so on which make it a potential candidate to be used for in-body communication [5]. The low part of UWB frequency is the best choice to overcome the problem [6] in which the attenuation through the human body tissues increases dramatically as the frequency increases as proclaimed by [7]. Therefore, this shows that not whole UWB frequency is needed for this project. Achieving large bandwidths with low-profile structure inside UWB is the advantage of slotted patch antenna [8]. Therefore this type of antenna is predicted to be the best candidate for this research. Apart from that, patch antenna can be made into compact size that is small enough to be put on human abdomen in order to make it comfortable for on-body purpose.

This project will be focused on the antenna used on-body that has the ability to penetrate through human body tissues. The frequency band used is UWB and results will be taken from the S11 reading. The directivity and the gain of the antenna will also be observed to ensure the efficiency of the designed antenna and the suitability to be used on the human body.

2. Antenna Development

The antenna was designed by using CST Software so that the results can be simulated directly. The simulation of CST Software is done in two conditions, free space condition and on human body condition. The purpose of having the antenna tested in two conditions is because the lack of computer's capability to simulate the antenna if too many layers are present. Type of antenna use is patch antenna and its frequency band is low part of UWB which is from 3.1 GHz until 5.0 GHz. The design specification of the antenna is presented in Table 1. The design of the antenna is shown in Figure 1. The parameters for the antenna are tabulated in Table 2. Substrate that is used for this antenna is textile cotton. The conductive material that is used for the patch is Shieldit Super with thickness of 0.17 mm and electric conductivity of $1.18x10^5$ and the substrate is Pure cotton with thickness of 1.6 mm and dielectric constant of 1.6.

Table 1. Design Specification

Parameter	Details
Antenna type	Patch antenna
Technique	Slotted
Frequency (free space condition)	Range of 3.1 – 5 GHz
Frequency	4.2 GHz
Bandwidth	600 MHz
Radiation pattern	Directional

Table 2. Dimensions for the antenna

Parameter	Dimension (mm)	Parameter	Dimension (mm)
G_w	80	P_2	36
G_l	80	P_3	46
S_w	70	P_4	24
S_l	15	P_5	24
P_1	9	P_6	6

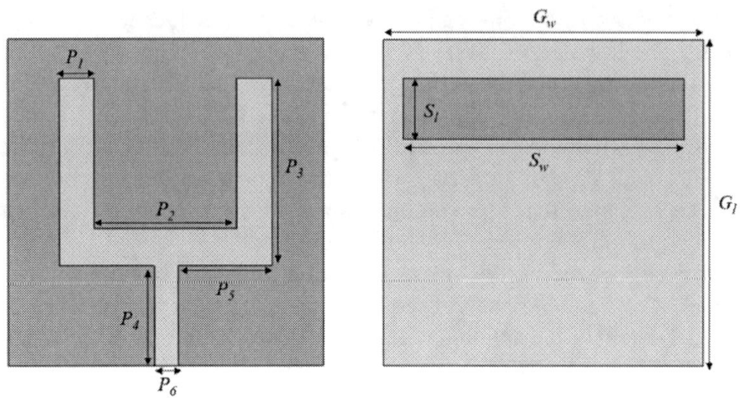

(a) Front view (b) Back view

Figure 1. Design of the antenna

By using CST Software, the antenna is designed in order to simulate the S11 results. In this part of process, as long as reflection coefficient of -10 dB can be obtained in any 3.1 to 5.0 GHz frequency, the result is acceptable. Figure 2 shows the interface of CST Software with the designed antenna on its menu. Once the desired frequencies are obtained, human tissues model is added onto the design. This is where the important part is since the objective of this project is to having an antenna that can work through human tissues. Since the objective of this antenna is to enable the connection inside the human body, human tissues model is simulated inside the CST Software according to the human tissues layers proposed by [9] and by using the dielectric constant for every layers by [7]. The simulated human tissue layers are as shown in Figure 3 and the parameters of every layers are tabulated in Table 3.

Table 3. Electrical properties for the tissues model

Layer	Dielectric constant	Conductivity (S/m)	Height (mm)
Skin	33.13	4.420	1.5
Sub-cutaneous fat	4.194	0.330	17
Muscle	47.21	7.313	8.0
Visceral fat	4.194	0.330	15
Small intestines	47.83	9.368	0.5

IOP Conf. Series: Materials Science and Engineering **767** (2020) 012005 doi:10.1088/1757-899X/767/1/012005

Figure 2. Designed antenna on free space condition

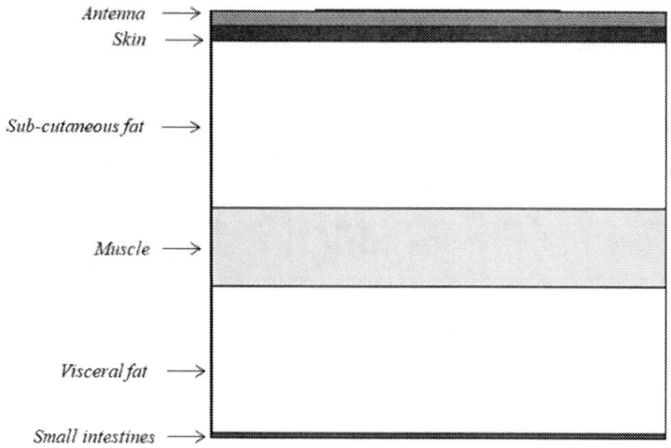

Figure 3. Simulated human body tissue layers

The designed antenna was fabricated once the results from the simulation were as predicted. The designed antenna was tested out to see if it can properly work in real environment or not. The designed antenna was placed on a human body and a network analyzer was used as the device to measure the performance of the fabricated antenna. The reflection coefficient, S11 was measured.

The illustration of the measurement set up is shown in Figure 4. The illustration shows is the set up for measurement on human body condition. The antenna has been designed under two conditions, thus measurement on free space condition is also conducted.

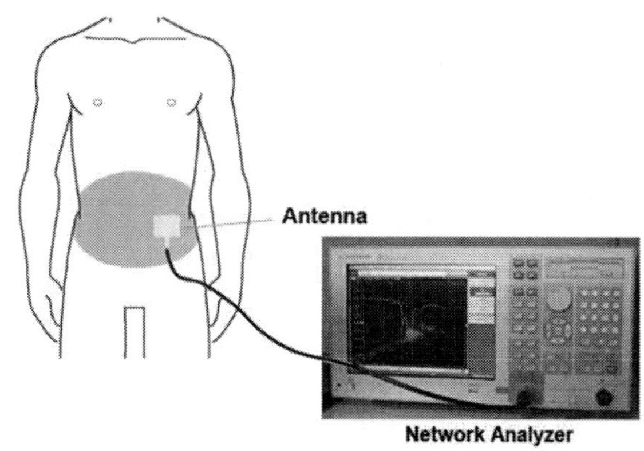

Figure 4. Illustration on the measurement set up on human body

Since the antenna is designed under two conditions, which are free space condition and on human body tissues condition, the measurement for testing is also done in both condition. This is to able comparison between simulation and measurement in both conditions can be done. Measurement is only done for measuring the reflection coefficient, *S11*.

2.1. Free e Space Condition Measurement Setup

The antenna is placed on a piece of Polystyrene foam to avoid a direct contact with the table that could affect the measurement of the antenna. Polystyrene foam has zero permittivity so it is the best item to be used during measurement of the reflection coefficient. Figure 5 shows the setup of the antenna for the measurement on free space.

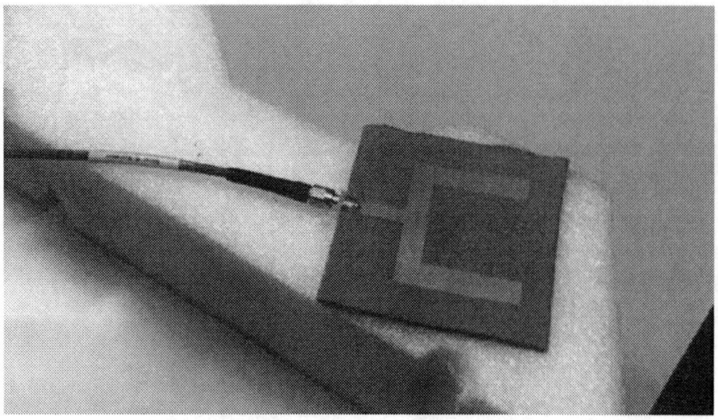

Figure 5. Measurement set up for free space condition

2.2. On Body Condition Measurement Setup

Three persons have volunteered to be used as human sample for this project. All the humans are adult males. The antenna is placed on the abdomen of the human. Figure 6 shows the set up for the measurement on a human body. The results for the measurement on the human body are the one that will determine whether this antenna is suitable to be used for in-body communication or not. The results for these measurements and the comparison between simulated and measured results are presented in details in next section.

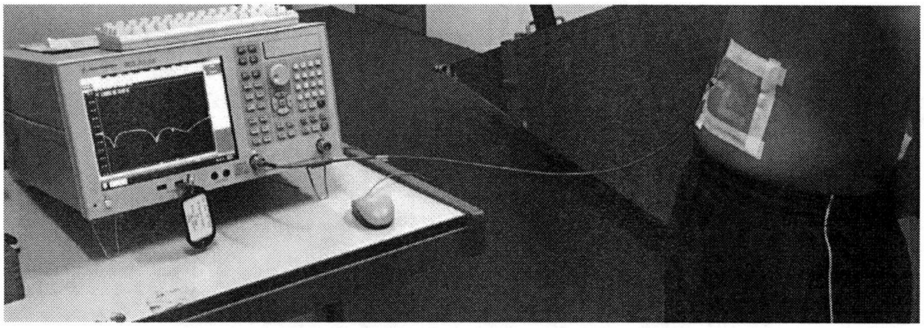

Figure 6. Set up for the measurement on human body

3. Results and Discussion

Results from the experiment are presented and discussed in this section. For simulation results, reflection coefficient and radiation pattern are presented while for measurement results, reflection coefficient for the fabricated antenna is presented.

The designed antenna has been tested in two different conditions, which are free space and on human body condition. Figure 7 shows the results for both conditions during simulation using CST software. Frequency shift occurred after the antenna is placed on the human body. This is due to the loss in the tissues layers. However, this antenna is aimed to work at 4.2 GHz with reflection coefficient of lower than -10dB when it is placed on a human body. The frequency band obtained was from 3.9 – 4.7 GHz and the impedance bandwidth (BW) is 800 MHz (44%). Impedance bandwidth is determined by -10 dB reflection coefficient. Simulation results for S11 on human body condition show a positive feedback.

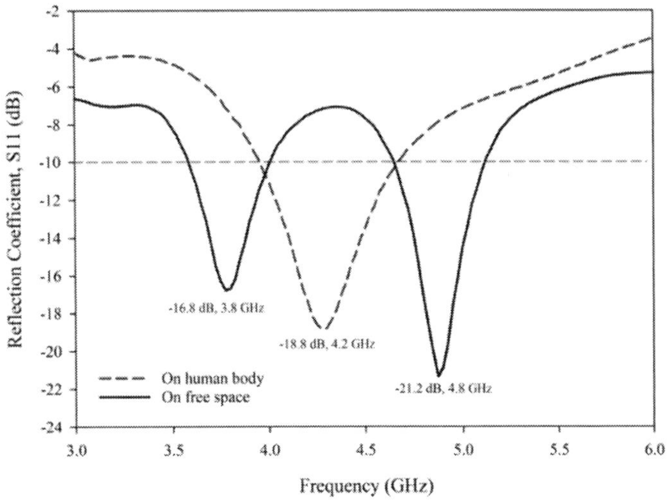

Figure 7. *S11* result for simulation on human body and on free space condition

Figure 8 shows the *S11* result for simulation and measured of free space condition while Figure 9 (a) until (d), shows the *S11* for on human body condition. For on human body condition, the antenna has been tested on three different human samples. Figure 10 shows the combination of simulation and measurement for both conditions. In Figure 8, Reflection coefficient results for simulation and measurement for free space is presented. The measurement result shows better performance of the antenna. Apart from that, the operational frequencies between simulation and measurement remained the same which are at 3.8 GHz and 4.9 GHz.

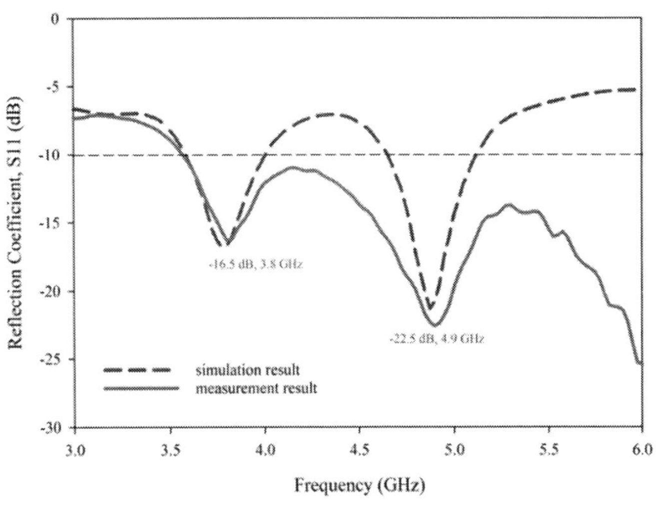

Figure 8. *S11* result for free space condition

For on human body condition, Figure 9 (a) to (b) show the results when the measurement is compared with simulation for every human sample. Figure 9(d) is the comparison of all samples with the simulation results. Based from Figure 9(d), it can be seen that there are not much different between simulation and measurement for all three samples. When tested, the antenna managed to work on 4.2 GHz for all three samples. This is a great achievement since every human has different thickness for their body tissues and it may disturb the performance of the antenna. Table 4 summarized the performance of the antenna for on body condition.

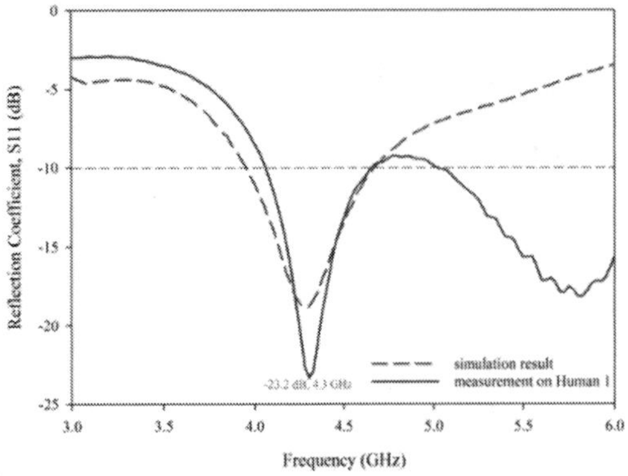

(a) Simulation and measurement result for human sample 1

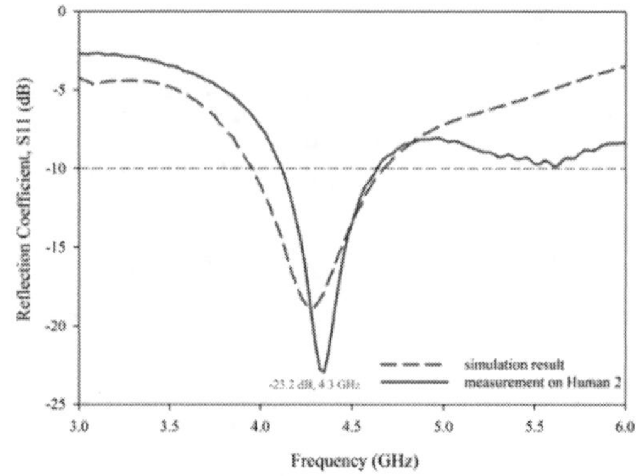

(b) Simulation and measurement result for human sample 2

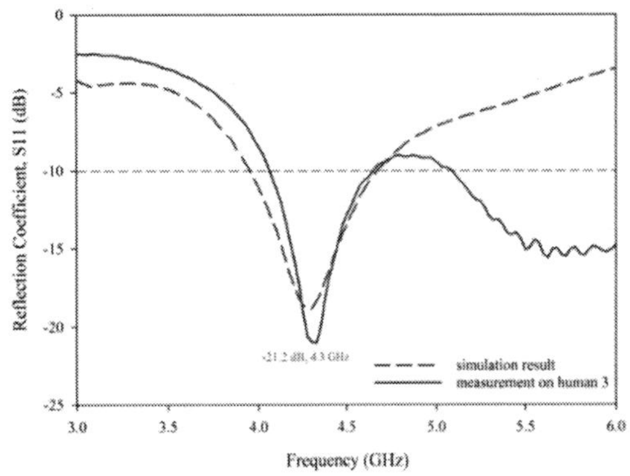

(c) Simulation and measurement result for human sample 3

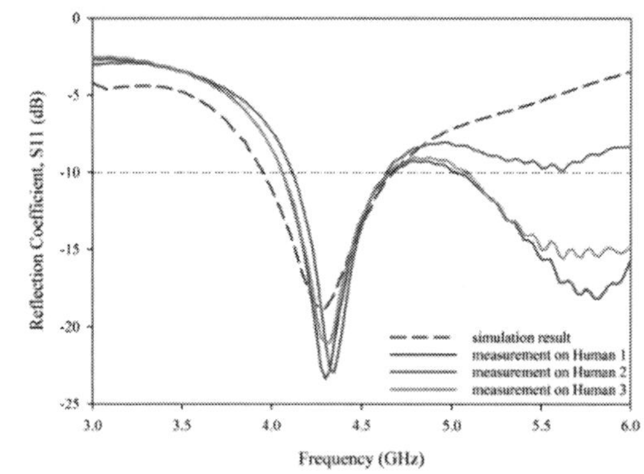

(d) Simulation and measurement for all three human samples

Figure 9. *S11* result for on human body condition.

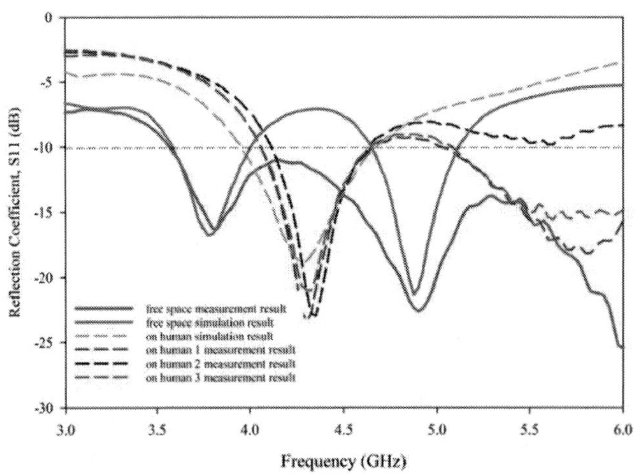

Figure 10. Combination of *S11* results for both conditions

Table 4. Bandwidth for Measured *S11* Results for On Body Condition

Condition	Operating Frequency	Frequency Band (MHz)	Bandwidth (MHz)
Human Sample 1	4.2 GHz	4000 - 4600	600
Human Sample 2	4.2 GHz	4100 - 4660	560
Human Sample 3	4.2 GHz	4050 - 4650	600

One of the main antenna properties that were analyzed when considering the ability of the designed antenna in free space was the radiation pattern. Radiation pattern of an antenna is represented by the energy radiated. Radiation pattern results that are presented in this section are simulated results from free space condition. Since in free space condition, two frequency bands were obtained. Those frequencies are at 3.8 GHz and 4.8 GHz. Figure 11 shows the results at 3.8 GHz while Figure 12 shows the results at 4.8 GHz.

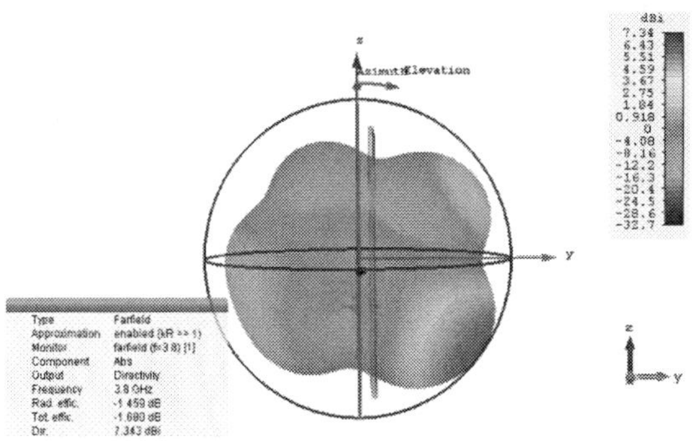

(a) Simulated 3D radiation pattern at 3.8 GHz for directivity

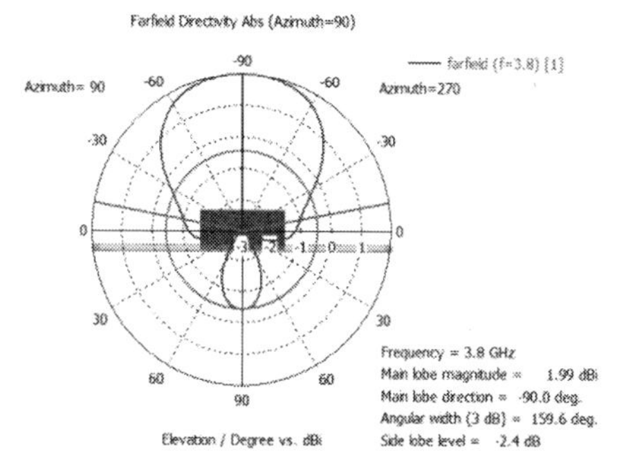

(b) Simulated polar radiation pattern at 3.8 GHz for directivity

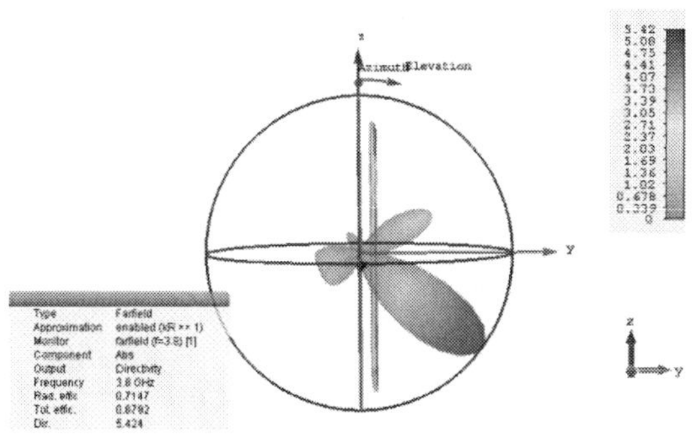

(c) Simulated 3D radiation pattern at 3.8 GHz for total efficiency

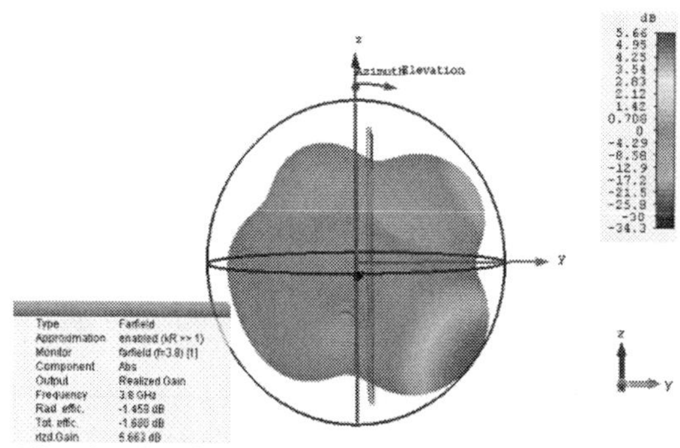

(d) Simulated 3D radiation pattern at 3.8 GHz for realized gain

Figure 11. Simulated radiation pattern at 3.8 GHz

Figure 11 shows the simulated radiation pattern at 3.8 GHz for free space condition. The directivity of the antenna at that frequency can be seen in Figure 11(a) and Figure 11(b). Figure 11(a) is the 3D version of the radiation pattern while (b) is the polar version of the radiation pattern. From the figures, it shows that the antenna has a directional directivity of 7.343 dBi. For Figure 11(c), the 3D radiation pattern is plotted in linear scaling. From the figure, it shows that the total efficiency of the antenna is

0.6792 which is 67.92 % efficient. Figure 11(d) shows the gain of the antenna. From the figure, it shows that the antenna has a gain of 5.663 dB. Figure 12(a) until Figure 12(d) are the radiation pattern results for frequency at 4.8 GHz in free space condition.

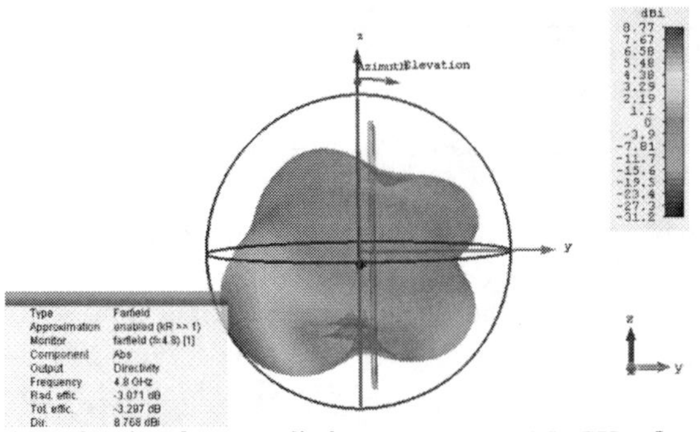

(a) Simulated 3D radiation pattern at 4.8 GHz for directivity

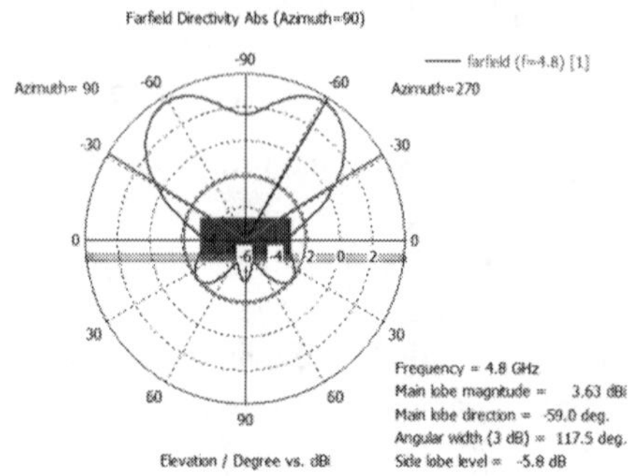

(b) Simulated polar radiation pattern at 4.8 GHz for directivity

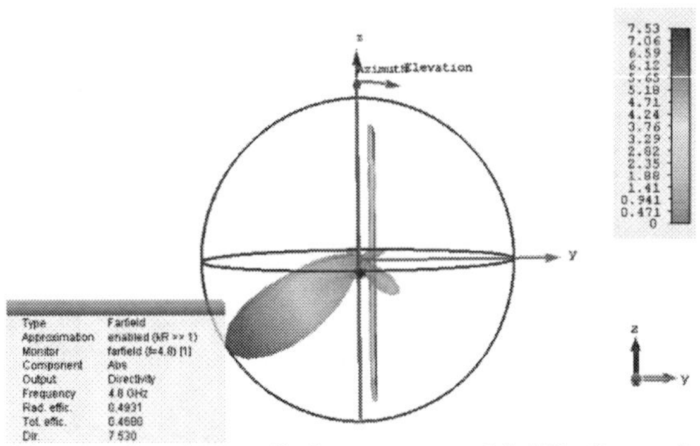

(c) Simulated 3D radiation pattern at 4.8 GHz for total efficiency

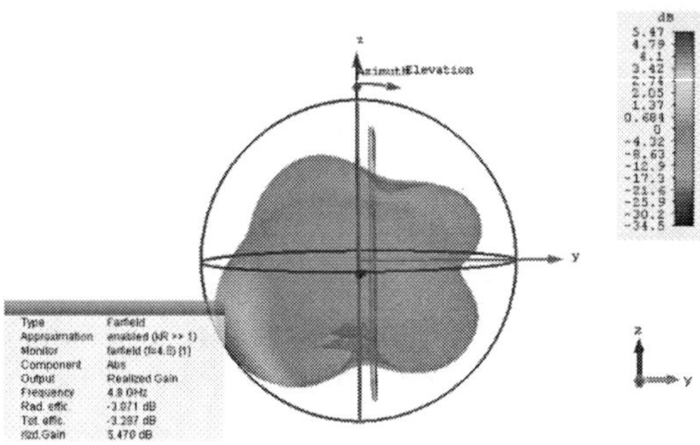

(d) Simulated 3D radiation pattern at 4.8 GHz for realized gain

Figure 12. Simulated radiation pattern at 4.8 GHz

Figure 12 shows the simulated radiation pattern at 4.8 GHz for free space condition. The directivity of the antenna at that frequency can be seen in Figure 12(a) and Figure 12(b). Figure 12(a) is the 3D version of the radiation pattern while (b) is the polar version of the radiation pattern. From the figures, it shows that the antenna has a directional directivity of 8.768 dBi. This value is a bit higher compared to at 3.8 GHz. For Figure 12(c), the 3D radiation pattern is plotted in linear scaling. From the figure, it shows that the total efficiency of the antenna is 0.4678 which is 46.78 % efficient. The total efficiency drops around 20% compared to at the 3.8 GHz. Figure 12(d) shows the gain of the antenna. From the figure, it shows that the antenna has a gain of 5.470 dB.

4. Conclusion

The designed antenna is aimed to able communications inside the human body. In order to reduce attenuation through the tissues, lower band of UWB which is 4.2 GHz is used as the operational frequency. Based on the measurement *S11* results on the human body, frequency of 4.2 GHz is successfully obtained. The size of the antenna is also small enough to be placed on human abdomen and it was made from cotton substrate. Cotton substrate is very soft and light in weight so it is suitable to be used on body. The objective of this research is successfully achieved because this antenna has been fabricated and the coefficient results, *S11* shows positive results as desired. Even after it has been tested on three different human samples, the results are acceptable and barely any different from the simulation.

5. References

[1] G. Pan and L. Wang, "Swallowable wireless capsule endoscopy: Progress and technical challenges," Gastroenterol. Res. Pract., vol. 2012, 2012.

[2] H. Mateen, R. Basar, A. U. Ahmed, and M. Y. Ahmad, "Localization of Wireless Capsule Endoscope: A Systematic Review," IEEE Sens. J., vol. 17, no. 5, pp. 1197–1206, 2017.

[3] E. Miralles, C. Andreu, M. Cabedo-Fabres, M. Ferrando-Bataller, and J. F. Monserrat, "UWB on-body slotted patch antennas for in-body communications," 2017 11th Eur. Conf. Antennas Propagation, EUCAP 2017, pp. 167–171, 2017.

[4] C. Garcia-Pardo et al., "Ultrawideband Technology for Medical In-Body Sensor Networks: An Overview of the Human Body as a Propagation Medium, Phantoms, and Approaches for Propagation Analysis," IEEE Antennas Propag. Mag., vol. 60, no. 3, pp. 19–33, 2018.

[5] R. Chávez-Santiago and I. Balasingham, "Ultrawideband signals in medicine," IEEE Signal Process. Mag., vol. 31, no. 6, pp. 130–136, 2014.

[6] R. Chavez-Santiago, "Propagation models for IEEE 802.15. 6 standardization of implant communication in body area networks," IEEE Commun. Mag., no. August, pp. 80–87, 2013.

[7] C. Gabriel, "Compilation of the Dielectric Properties of Body Tissues at RF and Microwave Frequencies." Brooks Air Force, N.AL/OE- TR- 1996-0037, San Antonio,TX, 1996.

[8] M. Klemm, S. Member, I. Z. Kov, G. Tröster, and S. Member, "Novel Small-Size Directional Antenna for UWB WBAN / WPAN Applications," IEEE Trans. Antennas Propag, vol. 53, no. 12, pp. 3884–3896, 2005.

[9] P. Ara, M. Heimlich, and E. Dutkiewicz, "Antenna performance for localization of capsule endoscope," Int. Symp. Med. Inf. Commun. Technol. ISMICT, pp. 1–5, 2014.

Acknowledgments
The authors wish to acknowledge sincere thanks to Collaborative Research in Engineering, Science and Technology (CREST) for awarding their Undergraduate Research Project Grant to fund the research.

IOP Conf. Series: Materials Science and Engineering **767** (2020) 012006 doi:10.1088/1757-899X/767/1/012006

Characteristic of melted slag from wrought iron industry as green concrete material

Zainul Faizien Haza, M. Afif Shulhan, Karolus Boromeus Kadis, and Arbianto

Civil Engineering Department, Faculty of Engineering, Universitas Sarjanawiyata Tamansiswa, Jalan Miliran 16, Yogyakarta, Indonesia

zainul@ustjogja.ac.id

Abstract. Utilization waste material on concrete ingredients or its manufacturing that did not harm the environment is the need for the environment sustainability. It is the background of research in concrete design for more environmental friendly as often called green concrete. This work aims to investigate the usability of the melted slag, obtained from local wrought iron industry, as aggregates replacement according to waste material utilization. In particular, it focused on studying the feasibility of replacing the aggregates using melted slag by investigating the basic physical characteristic. Slag is known have a high percentage of silica and ferric oxides, which gives pozzolanic activities that suitable to use as cement blended product. Slag waste used in this work is obtained from the wrought iron industry from Pakis district, Magelang Regency, Central Java, Indonesia. The slag chunk extracted from the furnace then crushed to smaller particle sizing as fine aggregates and coarse aggregates. The aggregates characteristic is observed in relation to gradation test, specific gravity, bulk density, clay lumps percentage, and Los Angeles abrasion value. All of observation parameters are taken consecutively based on SNI and ASTM specification to evaluate its compatibility as concrete aggregates. The current work remarks that the fine and coarse aggregates generated from wrought iron slag do not reveal complete fulfilment in the aggregate specifications of any standard. Even though, the slag fine aggregates could be used in standard or lightweight concrete with further treatment such as crushing to smaller particle size, washing to minimize the clay lumps, and mixing with natural aggregates.`

1. Introduction

Concrete is the most used building material worldwide with global production at about 12 billion per year. It is estimated that the average consumption at about 1.8 tonnes per year per every living human being, which much higher compared to steel (roughly 0.19 tonnes/year per person) [1]. There are many reasons why concrete was remarkably reliable, such as high strength, high durability, and environmentally low impact compared to steel [2]. However, due to its massif production volume (12 billion/year), the total carbon emissions of concrete become significant than steel or other materials [1].

The essential ingredients of concrete are aggregates, portland cement, and water, are considered having an environmental impact. The extraction of aggregates having land use implication, while the manufacturing of cement is facing a carbon (CO2) emission issue [3,4]. The portland cement was responsible for 74 % to 81 % of the total CO2 emissions of concrete. Globally, cement industries contribute about 5 % of the global (CO2) carbon emissions [5]. Emissions number of cement processing are coming from energy consumption of the kiln process and the chemical reaction (CaCO3→ CaO + CO2) to decompose limestone. The chemical reaction itself contributes about 0.5 tonnes of CO2 per tonne CaO produced [5]. This reason encourages the need for concrete that is more environmentally friendly as often called green concrete.

Green concrete expressed as concrete that utilizes waste material on its ingredients, or its manufacturing did not harm the environment [6]. Green concrete can be easily identified using such parameters: the amount of cement substitution materials, manufacturing process, and life cycle analysis [7]. As reported in previous studies, there are two significant ways to produce green concrete. First, expand the use of residual or waste product, then the utilization of new types of cement which have lower CO2 emissions [8].

Since the 1990s, the use of waste material was encouraged to reduced cement and natural aggregates [9]. The use of Fly ash waste confirmed an excellent result to cut down the amount of portland cement used in concrete [7]. Furthermore, the by-product of foundries (foundry sand and cupola slag), glass, wood ash, sludge successfully introduced as substitution material for aggregates replacement [9,10]. The use of new cement materials such as alkali-activated cement, magnesia cement, polymer cement, and sulfoaluminate cement was found later [7].

However, proper treatment of slag waste still a big task for Indonesia, due to its classification as toxic and hazardous waste (B3). The study case is a local wrought iron industry in Pakis district, Magelang Regency, Indonesia, which produces about 3 tonnes slag waste/week from the heating process of wrought iron. Currently, slag waste was just thrown away as a landfill, which creates groundwater pollution. Thus, there is a need to utilize slag waste as a useful product to solve this environmental problem altogether giving added value to a particular product.

Based on the presented reason, this work aims to investigate the usability of the melted slag, obtained from local wrought iron industry, as aggregates replacement. In particular, it focussed on studying the feasibility of replacing the aggregates using melted slag by investigating the basic physical characteristic.

2. Application of slag

Slag defined as a by-product of steel manufacturing that originated from the molten steel that separates from dirt in the furnace. Steel slags categorized as a non-metallic residue that remains in when the molten steel has extracted from the furnace [11]. The slag consists of calcium silicates, calcium aluminoferrites, and oxides of some minerals (calcium, iron, magnesium, and manganese) [12]. However, along with the technological development, the term "slag" has been expanded. Slag is not only associated with steel manufacturing waste but also represents slag that originated from molten waste material (indicated in Fig.1) [13]. Thus, new kind of slag has developed, such: de-sulphurization slag from metal processing, and synthetic slag extracted from municipal solid waste incineration [14].

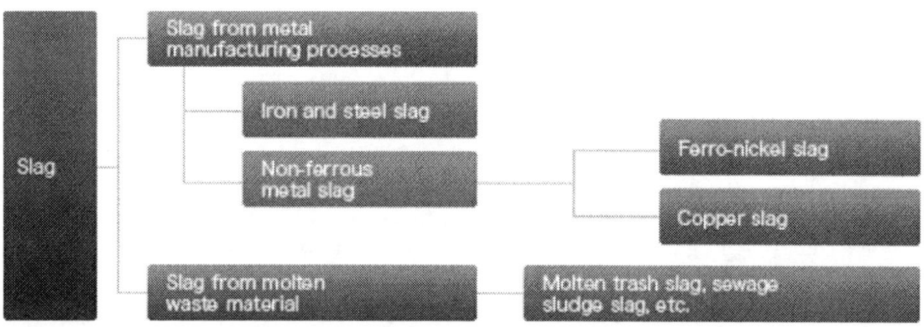

Figure 1. Classification of slag [13]

The chemical composition depends on the types of furnace used in steel manufacturing, such: blast furnace (BF), Electric-arc Furnace (EAF) and Basic oxygen furnace (BOF) [15, 13]. Both of BOF and

BF slag have a high percentage of silica and ferric oxides, which gives pozzolanic activities that suitable to use as cement blended product [16]. In contrast, EAF slag possess low pozzolanic activity and having expansive nature of the aggregates, which is causes crack to use on the cement-based product [11, 16].

The use of BF slag as aggregates replacement on concrete gives improvement on freezing, thawing, sulfuric acid, and salt attack resistance with the strength that conforming Japan standard [17,18]. Furthermore, along with the addition of ash filler, the utilization of BF slag could produce high strength concrete [19]. Previous research also presented that EAF slag has successfully used on concrete pavements [20], hot mix asphalt, and bituminous treated sub-base [11]. Even though EAF slag has an expansion problem, it proved that the use of EAF aggregates replacement could produce higher compressive strength of concrete than natural aggregates [16, 21].

3. Experimental Methods

Slag waste that being assessed obtained from the wrought iron industry from Pakis district, Magelang Regency, Central Java Indonesia (Fig. 2). The slag chunk extracted from the furnace then crushed to smaller particle sizing as fine aggregates and coarse aggregates. The aggregates characteristic assessed for such parameter: gradation test, specific gravity, bulk density, clay lumps percentage, and Los Angeles abrasion value. All the parameter investigations are taken consecutively based on SNI and ASTM norm [22-25]. The evaluated parameter then compared to ASTM [22,23] and SNI [24] specification of concrete aggregates to know whether it was suitable to use as concrete aggregates or not.

Figure 2. Wrought iron slag waste

4. Result and Discussion

The use of grained material on concrete has prescribed on ASTM [22-23] for normal and lightweight concrete. Locally, there is also Indonesian Standard or SNI [24] that regulate the specification of aggregate for regular concrete for application in Indonesia. All three standards focus on establishing measures of acceptance regarding grading, physical properties, abrasion resistance, and deleterious material content. The value of those parameters of aggregates should comply with the limit listed on the norm. However, due to aggregates heterogeneity, there is a possibility that it will fit several parameters and fail to meet some criteria in a particular standard. Hence, the codes ensure that the use of the fails aggregates was not prohibited. This exception applies if the concrete strength made from fails aggregates has not a significant difference compared to concrete from adequate aggregates.

4.1. Aggregates grading

The grading of aggregates being examined to ASTM [22-23] to check the conformity to standard concrete specification (Table 1 and 2). Figure 3(a) reveals that the slag fine aggregates gradation curve was not fit with the ASTM C33 [22] range for normal concrete. The evidence addressed by the percentage of the granules which passed in 1.18 and 2.36 mm sieve below the lower limit of ASTM C33 (Table 1). Thus, according to Kosmatka et. al. [26], this gradation curve could be classified as gap-graded aggregates, as several particle sizes are missed (1.18 and 2.36 mm particles) from the range. The use of gap-graded aggregates could lead to concrete that is likely to segregate or honeycomb [26]. Consequently, it needs higher cement volume to produce the concrete compared to well-graded aggregates. Contradiction with ASTM [22], toward Indonesian standard [25], the slag fine aggregates fortunately categorized as coarse sand class and still can be applied for 15-20 MPa concrete constituent.

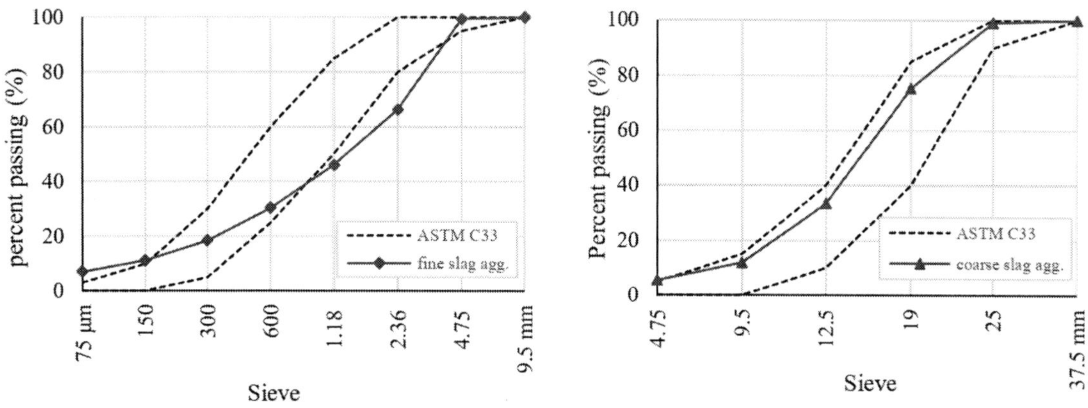

(a) (b)

Figure 3. Gradation curve of (a) slag fine aggregates and (b) slag coarse aggregates compared to ASTM C33 specification

Table 1. Slag fine aggregates gradation

Sieve	Percent passing (%)	
	ASTM C33 requirement	Fine slag agg.
9.5 mm	100	100
4.75 mm	95 - 100	99.4
2.36 mm	80 - 100	66.4
1.18 mm	50 - 85	46.0
600 µm	25 - 60	30.3
300 µm	5 - 30	18.4
150 µm	0 - 10	11.2
75 µm	0 - 3	7.0

Table 2. Slag coarse aggregates gradation

Sieve	Percent passing (%)	
	ASTM C33 requirement	Coarse slag agg.
37.5 mm	100	100
25 mm	90 - 100	99.1
19 mm	40 - 85	75.4
12.5 mm	10 - 40	33.4
9.5 mm	0 - 15	11.1
4.75 mm	0 - 5	5.5

4.2. Physical parameters

As shown in Table 3 and Table 4, the physical properties for the fine and slag coarse aggregates do not conform to ASTM [22,23] and Indonesian standard (SNI) [24] specification. However, not all the obtained result was evaluated cause the norm does not give the exact limitation of some aggregates parameter. Only four parameters that could be compared to ASTM and SNI, such as fineness modulus, Los Angeles abrasion, clay lumps, and bulk density.

The slag fine aggregates exhibit that the number of fineness modulus is about 4.21, which categorized as coarse sand. This finding in line with the gradation analysis delivered earlier. Nevertheless, the fineness modulus does not comply with any limit given by the codes (Table 3). This result means that the slag gradation still too coarse to be classified as fine aggregates. Therefore, to be used as fine aggregates, the slag aggregates must be further treated. The treatment that possible to carry on is combining the slag aggregates with natural fine aggregates or crushing the slag to smaller grain size.

Table 3. Physical characteristic of slag fine aggregates

Parameters	Slag fine aggregates	Code specification limit		
		ASTM C33, concrete agg.	ASTM C330, lightweight conc.	SNI 04 1989, concrete agg
Fineness modulus	4.213	2.3 - 3.1	-	1.5 - 3.8
Clay lumps	8.51 %	3%	2%	5%
Bulk density	1.158 gr/cm^3	> 1.120 gr/cm^3	< 1.120 gr/cm^3	-
Water absorption	13.63 %	-	-	-
Bulk specific gravity	1.76	-	-	-
SSD specific gravity	2	-	-	-
Apparent specific gravity	2.3	-	-	-
Moisture content	6.276%	-	-	-

Table 3 also stated that slag fine aggregates retain a high amount of clay (about 8.51 %) which the standard gives a 2 - 5 % limit. This fact points out that the quality of the bond between aggregates and cement paste will be affected. Furthermore, the bulk density and water absorption (Table 3) gives an illustration that the fine aggregates is very porous and tends to comply with ASTM C33 (standard concrete).

Table 4. Physical characteristic of slag coarse aggregates

Parameters	slag coarse aggregates	Code specification limit		
		ASTM C33, concrete agg.	ASTM C330, lightweight conc.	SNI 04 1989, concrete agg
Fineness modulus	3.747	-	-	6 - 7.10
Los Angeles abrasion	35.0 %	50	-	-
Clay lumps	2.8 %	-	2%	1%
Bulk density	1.017 gr/cm^3	> 1.120 gr/cm^3	< 0.880 gr/cm^3	-
Water absorption	5.102 %	-	-	-
Bulk specific gravity	1.89	-	-	-
SSD specific gravity	1.986	-	-	-
Apparent specific gravity	2.555	-	-	-
Moisture content	1.122 %	-	-	-

In contrast with the data presented in Table 3, the physical characteristic in Table 4 shows that the slag coarse aggregates have extremely smooth particles (FM around 3.74) compared to Indonesian standard (FM between 6-7.10). Nevertheless, concerning that the ASTM standard does not give any restriction to Fineness modulus, the slag coarse aggregates still comply with the ASTM rule. Furthermore, the slag coarse aggregates exhibit a remarkable result concerning abrasion value with the Los Angeles resistance is about 35 % (ASTM limit 50%).

5. Concluding Remark

This paper presents the properties of slag waste from the wrought iron industry that were experimentally examined based on the specification standard of concrete aggregates. According to the result of conducted experiments, the followings concluded:

1. The fine and coarse aggregates obtained from wrought iron slag does not reveal complete fulfilment in the aggregate specifications of any standard.
2. The slag fine aggregates could be used in standard or lightweight concrete with further treatment such as crushing to smaller particle size, washing to minimize the clay lumps, and mixing with natural aggregates.
3. The slag coarse aggregates pose a satisfactory result to be used as concrete aggregates. However, the manufacturers should give proof that the strength of concrete made from these aggregates has a similar strength to concrete with natural aggregates.

Acknowledgments

Authors wishing to acknowledge LP3M Universitas Sarjanawiyata Tamansiswa for the funding support from Penelitian Terapan Unggulan (PTU) Scheme.

References

[1] Domone P and Illston J 2010 *Construction Materials : Their Nature and Behaviour* 4th edition (New York: Spon Press)
[2] Struble L and Godfrey J 2006 *How Sustainable is Concrete?* International Workshop on Sustainable Development and Concrete Technology (Illinois: University of Illinois at Urbana-Champaign) p 201.
[3] Humpreys K and Mahasenan M 2002 *Toward a Sustainable Cement Industry* (Geneva: World Business Council for Sustainable Development. Substudy 8, Climate Change)
[4] CIF 2003 *Cement Industry Environment Report* (Australia:Cement Industry Federation)
[5] Flower DJM and Sanjayan JG 2007 *Green House Gas Emissions Due to Concrete Manufacture* International Journal of Life Cycle Analysis vol 12. p 282
[6] Obla KH 2009 *What is Green Concrete?* The Indian Concrete Journal. p26
[7] Suhendro B 2014 *Toward Green Concrete for Better Sustainable Environment* Procedia Engineering 95 (Amsterdam: Elsevier) p 305
[8] Glavind M and Jepsen MT 2002 *Evaluation of Green Concrete Types* XVIII Symposium on Nordic Concrete Research (Helsinger)
[9] Naik TR 2002 *Greener Concrete Using Recycled Material* Concrete International. p 49
[10] Kumar R and Naik TR 2010 *Sustainable Concrete with Industrial and Post-Consumer By-Product* Int. Conf. On Sustainable Constuction Materials and Technology (Ancona : Univesita Politecnica delle Marche)
[11] WSDOT 2015 *WSDOT Stategies Regarding Use of Steel Slag Aggregate in Pavement* Report (Washington: Washington State Departement of Transportation)
[12] National Slag Association – *Steel Slag Base and Sub Base Aggregates Product Information* (Coatesville: National Slag Association)
[13] Nippon Slag Association 2003 *Types of iron and steel slag* (Japan: Nippon Slag Association)
[14] Galan MR, Farinas BA, Moreno FMB, Leiva C, Navarette B, and Vilches LE 2019 *Synthetic Slag Production Method Based on a Solid Waste Mix Vitrification for the Manufacturing of Slag-Cement* Journal of Materials vol 12 p 208

[15] Dippenaar R 2004 *Industrial uses of slag-The use and re-use of iron and steelmaking slags* VII int. Conf. On Molten Slags Fluxes and Salts (The South African Institute of Mining and Mettalurgy) p 57
[16] Alizadeh R, Chini M, Ghods P, Hoseini M, Montazer S, and Shekarchi M 2003 *Utiliztion of Electric Arch Furnace Slag as Aggregates in Concrete-Environmental Issue* Int. Conf. On Recent Andv. In Concrete Technology (Romania) p 451
[17] Matsumoto T, Tobo H, and Watanabe K 2018 *Iron And Steel Slag Product And New Effective Utilization Technologies* JFE Technical Report no 23 p 62
[18] Miyamoto T, Akahane K, Torii K, and Hayashiguchi S 2015 *Production and Use of Blast Furnace Slag Aggregates for Concrete* Nippon Steel & Sumitomo Metal Technical Report no 109 p 102
[19] Kaarolina R and Putra ALA 2018 *The Effect of Steel Slag as a Coarse Aggregate and Sinabung Volcanic Ash as Filler on High Strength Concrete* IOP Conf. Ser.: Materials Science and Engineering vol 309
[20] Verian KP, Panchmatia P, and Olek J 2017 *Investigation Of Use Of Slag Aggregates And Slag Cements In Concrete Pavements To Reduce The Maintenance Cost* Joint Transportation Reseacrh Program Publication (West Lafayette: Purdue University)
[21] Faleschini F, Brunelli K, Zanini MA, Dabala M, and Pellegrino C 2015 *Electric Arc Furnace Slag as Coarse Recycled Aggregate for Concrete Production* Journal of Sustainable Metallurgy vol 2 p 44
[22] ASTM C33/C33M 2013 *Standard Specification for Concrete Aggregates* (Pensylvania: ASTM International)
[23] ASTM C330/C330M 2009 *Standard Specification for Lightweight Aggregates for Structural Concrete* (Pensylvania: ASTM International)
[24] SNI S-04-1989-F 1989 *Building Material Specification* (Jakarta: Departemen Pekerjaan Umum)
[25] SNI T-15-1990-03 1990 *Procedures for The Mix Design of Normal Concrete* (Jakarta: Departemen Pekerjaan Umum)
[26] Kosmatka SH, Kerkhoff B, and Panarese WC 2002 Design and Control of Concrete Mixtures 14th ed. (Illinois: Portland Cement Association)

A Review of Oil Palm Fruit Ripeness Monitoring Using Microwave Techniques in Malaysia

K Y You[1] , F H Wee[2] , Y S Lee[3] ,Abbas Z[4], K Y Lee[5], E M Cheng[6], C S Khe[7] and M F Jamlos[8]

[1]School of Electrical Engineering, Faculty of Engineering, Universiti Teknologi Malaysia.
[2]Bioelectromagnetic Research Group (BioEM), School of Computer and Communication Engineering, Universiti Malaysia Perlis.
[3]Bioelectromagnetic Research Group (BioEM), Department of Electronic, Faculty of Engineering Technology, Universiti Malaysia Perlis.
[4]Department of Physics, Faculty of Science, Universiti Putra Malaysia.
[5]Department of Electrical and Electronic Engineering, Lee Kong Chian Faculty of Engineering and Science, Universiti Tunku Abdul Rahman.
[6]School of Mechatronic Engineering, Universiti Malaysia Perlis.
[7]Department of Fundamental & Applied Sciences, Universiti Teknologi Petronas Malaysia.
[8]Faculty of Mechanical Engineering & Automotive, Universiti Malaysia Pahang.

kyyou@fke.utm.my

Abstract. Palm oil products are one of the most important sources of economic and export products in Malaysia. Furthermore, now, the evolution of industry 4.0 has begun to take place globally. The development of sensors has played a role in this evolution. In this paper, various microwave sensors used in oil palm monitoring are reviewed. The advantages and disadvantages of each microwave sensor are discussed in detail. However, conventional and optical monitoring methods are also briefly described.

1. Introduction

This article focuses on the research and measurements of microwave sensors for oil palm fruit ripeness processing applications. Normally, sensory devices that operate in the microwave frequency range (300 MHz to 300 GHz) can be called as microwave sensors. The microwave sensor techniques have been widely used in the food industry and agriculture processing. Due to different configuration of sensors, the operating frequencies and the measurement techniques involved with sensors are different as well. Measurement methods can be grouped into either resonant technique or non-resonant technique according to the structure and sub-principle of the measurements. Free-space and reflection / transmission measurement methods are classified into the category of non-resonant techniques. To be specific, the free space method appeared as a far-field measurement and a horn antenna is mainly used as the sensor or radiator; whereas circular, rectangular or coaxial waveguides are applied in reflection / transmission measurement methods and the sample is contacted directly.

During the time that choosing for appropriate method, there are a few factors which need to be considered albeit different types of measurement methods are available to utilize, for instance cost, accuracy, operating frequency and shape of sample. Most of those measurement methods are applied in order to analyze the moisture content, *m.c.* in the foods under test and agriculture products. It is because of the polarization of water molecules which comprised in the agri-foods products which are sensitive and showed an evident response when the water molecules are exposed to microwaves. The microwave device can be applied as a measuring method to detect the moisture content, *m.c.* in agri-foods products that contain water due to the inclination of water to absorb microwave energy. Therefore, when there are changes in the amount of moisture among agricultural products, a change in the transmission / reflection coefficient or resonant frequency (from the change in dielectric permittivity) will be measured by the sensor. Here, authors attempt to provide comprehensive guidelines in microwave sensor design, which is particularly for oil palm fruits measurement.

2. Oil Palm in Malaysia

The first commercial planting of the oil palm (*Elaeis guineensis*) (from African) in Malaysia was planted at Rantau Panjang, Kuala Selangor in 1911-1912 by Fauconnier and were ready for harvesting in 1917. Nonetheless, the planted region remained small until 1965, the private sector and statutory bodies, for instance the Federal Land Development Authority (FELDA) had implemented large-scale plantings; this significant increase had also led FELDA becomes the world's second-largest oil palm plantation company. Recently, Malaysia has becomes the world's second-largest producer of palm oil. The primary palm oil importers of Malaysian are Pakistan, India, China, the United States and the European Union.

The Malaysia production of crude palm oil for 2018/19 was reached 20.5 million metric tons. A rich source of vitamin E and α and β-carotene can be provided by palm oil, which serves a vital role in cancer inhibition, suppression of cholesterol production and blood coagulation; thus, palm oil has gradually becomes popular. Nutrition, likewise the environmental preservation, will figure prominently in consumer's option of edible oils consumption. Apart from its value of nutrition, it is increasingly popular as raw materials for the bio-fuel for automobiles and industry of oleochemical.

3. Oil Palm Fruit Bunch

The fruit bunch, as shown in Figure 1 is void and might reach 35 cm in breadth and 50 cm in length. The bunch comprises the spikelets, spines, stalks and fruit. The composition of the fruits bunch is shown in Figure 2. The oil palm fruit is a drupe, 2 to 5 cm long and weighs about 4-20 g. It consists of a kernel which is encapsulated by the pericarp, which comprises endocarp (shell), mesocarp (pulp) and exocarp (skin). The mesocarp contains 16 % of fibers [1], which run in a longitudinal manner through the oil-bearing tissue from the base towards the fruit tip.

A palm or fruit may be explained as being tenera, pisifera, or dura variety according to the thickness of shell. The pisifera is shell-less, thus the pisifera is not vital in terms of commercial value. The variety which carries thick shell is described as dura; meanwhile tenera has a high mesocarp content and thin shell. The kind of fruit which is mostly preferred for commercial purpose is the tenera variety, due to its high mesocarp to oil potential.

Figure 1. A tenera bunch halved to show (1) the stalk plus (2) spikelets and fruits, comprising (3) the mesocarp, (4) the shell, and (5) the kernel.

4. The Ripeness of Oil Palm Fruits

The mesocarp of the oil palm fruits produces the palm oil. Usually, the first fruit bunch is produced by a palm tree in approximately three years. According to Ariffin *et al.* (1990) [2], during the early stage of fruit establishment, the amount of moisture content, *m.c* appeared to be higher. When the fruit ripens, the excessive water in the mesocarp is diminished, whereby occurs simultaneously with oil accumulation which approaches around 14 to 15 weeks after anthesis. There is about 80 % of water in the fresh mesocarp; at around 20 to 24 weeks after anthesis, the amount will be reduced swiftly to approximately 30-40% in the ripe fruit. The reduction in moisture content, *m.c* from 15 to 20 weeks after occurs in relation to the increased accumulation of oil in the mesocarp as shown in Figure 2 (a) [3]. The fruit will fall to the ground when the fruit is loose as the oil content approaches maximum level. This optimum time for oil palm harvesting. In addition, Hartley, (1977) [1] proposed that the relationship between oil and moisture content, *m.c* in a mesocarp may be expressed as:

$$Oil = 87.38 - 1.08\ Water \tag{1}$$

where *'Oil'* and *'Water'* are the percentage of oil per mesocarp and percentage of water per mesocarp, respectively. There is a possibility given by the close relationship between oil contents and moisture, whereby the percentage of moisture content, *m.c* can be used as a parameter in order to determine the mesocarp ripeness. The relationship between the oil content and moisture content, *m.c* in mesocarp of oil palm fruit is shown in Figure 2 (b).

5. A Measure of Ripeness of Oil Palm Fruits

The visual indications for instance flotation method, number or percentage of detached fruits per bunch and changes in colour of the fruits have helped to develop different methods to determine the ripeness of oil palm fruits. Fruit ripeness is usually characterized by the variation in colour from reddish-orange or black to orange as shown in Figure 3. Unfortunately, the colour change that takes place on ripening is not a consistent feature and differs from individual palm types and geographical regions.

The percentage or number of detached fruits per bunch appeared to be a usual practice that helps to gauge the ripeness of bunch in the oil palm plantation. This is due to the fruit which is about to be separated from the bunch will lead to the maximum oil production. However, ripening probability among a fruit bunch is not uniform. Different grading plans of bunch ripeness have been proposed in order to resolve this issue [1] for instance 2 detached fruits per kilogram of bunch weight, 20 to 40 detached fruits per bunch or 20-25% detached fruits to total fruits.

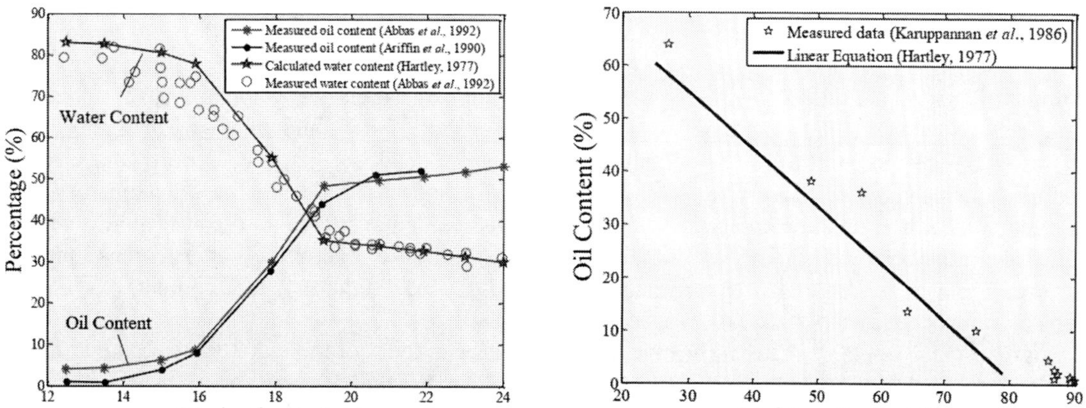

(a) (b)

Figure 2. (a) Difference in moisture content, *m.c* and oil content with weeks after Anthesis. (b) The relationship between the oil content and moisture content, *m.c* in oil palm mesocarp.

Figure 3. Various ripeness stages of oil palm fruits

This technique does not correlate with oil quantity and quality, thus it is inaccurate. In addition, it is unable to determine differentiation between fruits that are easily detached by hand and detached fruits that are on the ground. Furthermore, due to the wide range of bunch size within a population of oil palms, it will cause quite large variations in percent-detached bunch / fruits.

The particular gravity of the mesocarp of the ripened fruit which is much lower than water is the fundamental concept behind the flotation method [2]. Usually, a particular gravity of 0.89 will be obtained by the segment of mesocarp around the fruit tip. The method is quick and simple to apply, however the entire oil content in the fruit is unable to determine. The amount of moisture content, *m.c* of the mesocarp can be a foundation of a feasible parameter, which is useful to gauge the degree of fruit ripeness.

Recently, many advanced indirect methods are proposed, namely using its natural physics parameters such as refractive index, *n* (Imaging and optical techniques) and relative complex permittivity (Microwave techniques) of the oil palm fruits to correlate with oil and water contents, *m.c* in the oil palm fruits. To be specific, the optical sensor system comprises primarily of a receiver for receiving light (such as optical spectrum analyzer) and an emitter (such as Mid-infrared source) for emitting light. When the emitted light is reflected or interrupted by the sensing oil palm sample, the amount of light that reaches at the receiver will be changed [3-4]. Additionally, the receiver detects these changes and correlates to desired measured oil or moisture content, *m.c* of the oil palm fruits. Also, some researchers use imaging methods to distinguish fruit maturity depending on the ripe fruit being reddish-orange, and the unripe fruit is black [5-7] as shown in Figure 2. Besides, the bulk nuclear magnetic resonance (NMR) and magnetic resonance imaging (MIR) methods also are applied for monitoring ripeness of oil palm fruit in accord with the change of spin-spin relaxation times respect to the ripeness stages [8]. Recently, wire coil resonators [9] are used to gauge the ripeness of the oil palm fruit based on the shifting of resonating frequency.

Similar to the optical method, oil and moisture content, *m.c* using microwave technique can also be referred to the shifting resonance frequency/wavelength or the change in the reflection/transmission coefficient. Different microwave sensors were utilised to gauge the ripeness of oil palm fruits, for example microstrip sensor [10], coplanar sensor [11], rectangular dielectric waveguide [12], monopole [13], rectangular waveguide [14], open-ended coaxial probe [15-18], and microstrip ring resonator [19] as illustrated in Figure 4. The moisture content, *m.c.* in the oil palm mesocarp under test is monitored using most of those ripeness measurement methods. It is due to the oil palm mesocarp contains polarization of water molecules that is sensitive and demonstrated a crucial response when it is exposed to microwaves. The inclination of water to absorb microwave energy, enabling the microwave device to be utilized as a measuring method to observe the moisture content, *m.c* in agri-foods products comprising water.

Microstrip and coplanar sensors (in Figure 4 (a), (b)) the amount of moisture content, *m.c* in the oil palm fruits has been successfully estimated according to the attenuation measurement. The characteristics and microstrip and coplanar line discontinuities when the line was filled by oil palm mesocarp. The microstrip and coplanar line sensor measure the ripeness of the oil palm mesocarp, in accord with the correlation between the measured attenuation and the *m.c* inside the mesocarp fruit. For rectangular dielectric waveguide measurements, the oil palm mesocarp is placed in the sample holder between the two dielectric waveguides, as illustrated in Figure 4 (c). Nevertheless, laborious sample preparation is required by the sensors since the fresh mesocarp of the oil palm fruit has to be detached from the nut and crumbled into a form of semi-solid sample. It is undeniably time-consuming in order to prepare the fruit sample.

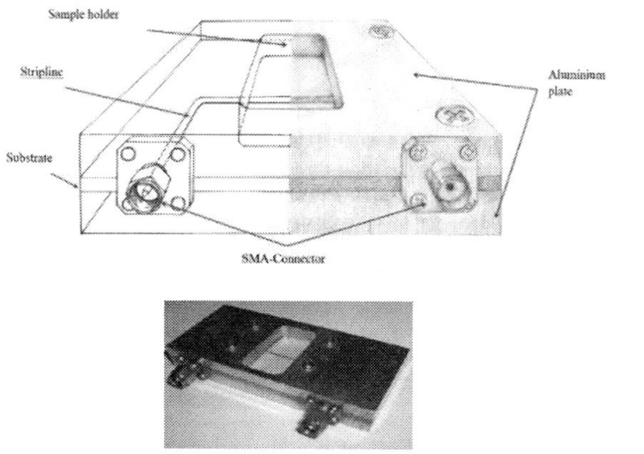

(a)

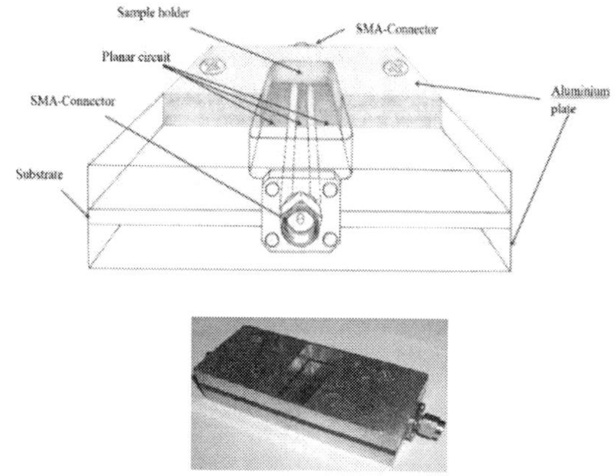

(b)

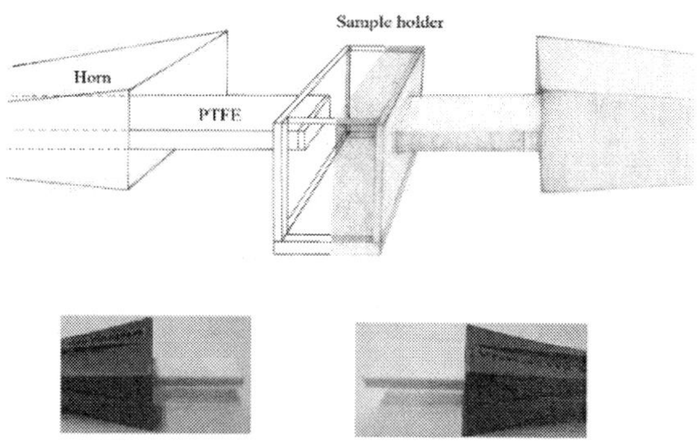

(c)

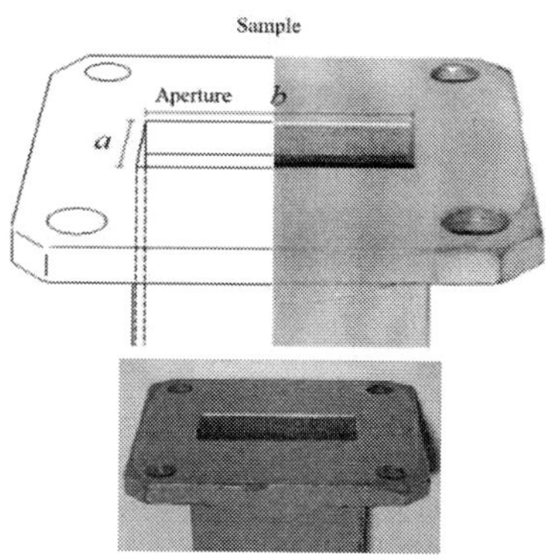

(d)

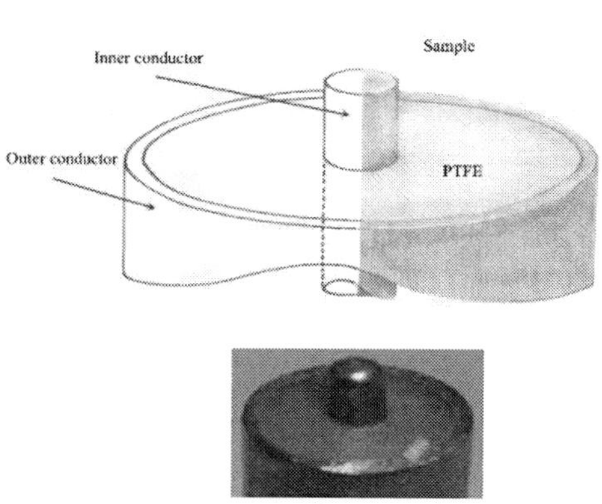

(e)

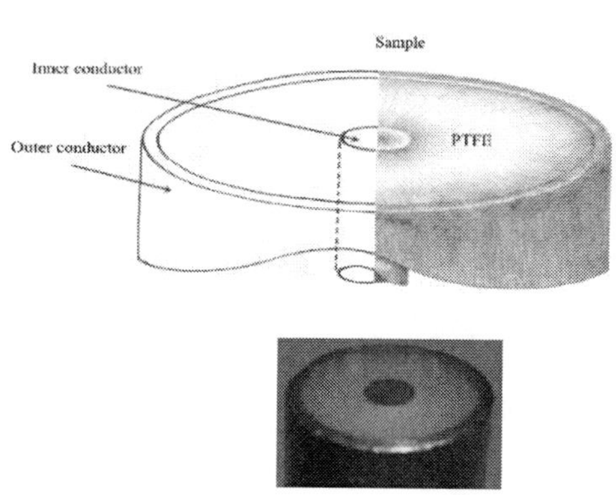

(f)

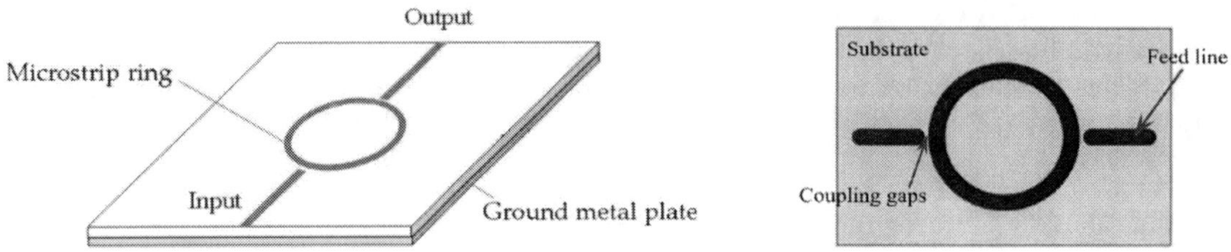

(g)

Figure 4. (a) Microstrip waveguide (MWG) [11]. (b) Multilayer conductor-backed coplanar waveguide (CBCPW) [12]. (c) Rectangular dielectric waveguide (RDWG) [12]. (d) Rectangular waveguide (RWG) [14]. (e) Monopole [13]. (f) Open-ended coaxial waveguide (OECWG) [16-18]. (g) Microstrip ring resonator [19].

Table 1. Comparison of previous works features for oil palm fruits measurements using microwave sensors

Type of sensor	**Comparison features of a microwave sensor for oil palm fruits measurements**							
	Operating frequency	Set-up speed	Nondestructive test	Sensitivity	Cost of sensor	Precision	One-port measurements	Single fruit measurements
MWG [10]	10.7 GHz	Slow	No	Good	Low	Good	No	No
CBCPW [11]	2.2 GHz	Slow	No	Good	Low	Good	No	No
Commercial HP 85070B OECWG [15]	0.13 – 20 GHz	Slow	No	Good	Expensive	Good	Yes	No
RDWG [12]	8 – 12 GHz	Slow	No	Good	Low	Good	No	No
Monopole [13]	0.5 GHz	Very fast	Yes	Very good	Very low	Poor	Yes	Yes
RWG [14]	8 – 12 GHz	Fast	No	Good	Low	Good	Yes	No
OECWG[16-18]	1 – 5 GHz	Very fast	Yes	Good	Very low	Good	Yes	Yes
MRR [19]	2.2 – 3 GHz	Fast	Yes	Good	Low	Fair	No	Yes

In general, the structure of a monopole (in Figure 4 (e)) is similar to that of the coaxial waveguide (in Figure 4 (f)), except for the protruding center conductor. The monopole measurement is in accord with the concept whereby a reflected signal through the coaxial line, in which the extended center conductor is buried into the oil palm mesocarp. This type of sensor is much more sensitive than many other types of slot sensors. Nevertheless, the measurement precision is less due to its scattering and radiating tendency. Figure 4 (f) type measurements, solid sample and parallel interfaces which carefully machined are required, likewise the whole cross-section of the coaxial line must be perfectly filled. Recently, the microstrip ring sensor (in Figure 4 (g)) for oil palm fruit measurement is proposed by [19]. Although, the microstrip ring resonator is sensitive in monitoring the maturity of the palm fruit, but the precision of measurement is influenced by the size of the fruit is measured. The comparison of previous work features for oil palm fruit measurements using microwave sensors was tabulated in Table 1.

6. Relationship between Dielectric Properties and Moisture Content

The relationship between a microwave with agri-foods materials can be expressed by the relative complex permittivity, ε_r

$$\varepsilon_r = \varepsilon_r' - j\varepsilon_r'' \tag{2}$$

in which the real part, ε_r' is the dielectric constant, whereas the imaginary part, ε_r'' is the dielectric loss factor. Furthermore, the electric field distribution is influenced by the dielectric constant, ε_r' and thus the phase of waves travelling through the material. On the other hand, the attenuation or energy absorption of the material is influenced by the loss factor, ε_r''. Since water absorbs most of the electromagnetic energy at microwave frequencies, hence both ε_r' and ε_r'' are closely related with moisture content [20] and also, the relative permittivity of the material is predominantly influenced by the amount of moisture in the total volume of the material. It is because of the relative permittivity of water (ε_r = 80 at dc frequencies) which usually being much greater than the other elements in the oil palm fruits (fiber: 2.2, oil: 2.5). Therefore, a change in resonant frequency (from the change in dielectric permittivity) or transmission / reflection coefficient will be measured by the sensor when there are changes of the amount of moisture in the oil palm mesocarp; in other words, it can be directly related to a change in *m.c* of the mesocarp that was acquired from oven drying technique. The relative complex permittivity, ε_r of oil palm mesocarp is well described by using the following model as [18]:

$$\varepsilon_r^* = \varepsilon_\infty + \frac{\varepsilon_s - \varepsilon_1}{\left(1+\left(j\omega a_1 \tau_1\right)^{a_2}\right)^{a_3}} + \frac{\varepsilon_1 - \varepsilon_\infty}{1+\left(j\omega \tau_2\right)^{a_4}} - \frac{j\sigma}{2\pi\varepsilon_o f} \tag{3}$$

where a_1, a_2, a_3, and a_4 are the empirical constant, which determined by optimizing measured data. The τ_1 and τ_2 are the relaxation time for low and high-frequency processes, respectively. The ε_1 is the height of the intermediate plateau from the baseline as well as the intersection between the line and ε_r'-axis yields the value of ε_s. Genereally, the optical permittivity, ε_∞ is independent of temperature, T and frequency, f [21]. The conductivity term, σ is determined by optimized with the measured loss factor, ε_r'' at low frequency (≤ 5 GHz). Moreover, the approximate values of conductivity, σ are tabulated in Table 2.

However, equation (3) is limited for room temperature, T (25 ± 1 °C). The estimated values of ε_s, ε_1, τ_1 and τ_2 correlate with the different percentages of water content, *m.c* are displayed in Table 2. The value of the relative complex permittivity of the oil palm mesocarp is mainly contributed by the water content, *m.c*, since the permittivity of water is more significant than other substance in the mixtures (such as oil and fiber) and it is sensitive to the microwave frequencies, particularly at 2.45 GHz and 10 GHz [21]. The displacement of water by the fiber contents and oil has caused the decrease in relative complex permittivity of the oil palm mesocarp.

The oil palm mesocarp has higher relaxation time, τ below the frequency of 10 GHz for various water content, *m.c*, while the relaxation of oil palm mesocarp is approximated to 8.377 ps relaxation of free water for frequencies that are above 10 GHz. The high value of relaxation time, τ of oil palm mesocarp at low frequencies is caused by the binding of water molecules in the mixtures which has slowed down the orientation polarization of polar molecules, such as the water adsorption by fibers and the capillary water between the fibers. However, the bound molecules are broken down above 10 GHz. As a matter of fact, the overall effect on water structure in oil palm mesocarp is extremely complex.

The correlation between both dielectric constant, ε_r' and loss factor, ε_r'' of oil palm mesocarp and frequency (2.6 GHz to 5 GHz) for the different percentage of water content, *m.c* is demonstrated in Figure 6 and Figure 7. The dielectric constant, ε_r' were measured using commercial Agilent 85070E dielectric probe. From Figure 6, dielectric properties are quite linear at moisture content below 40 %, suggesting that at low moisture, there is higher oil-water emulsion concentration. When the sensor is pressed firmly to the sample of fruits, water begins to accumulate at the surface between fruit and sensor, hence higher uncertainty measurements occurred when the moisture is above 70 %. Based on Figure 7, the relationship between percentage of moisture content in mesocarp and dielectric constant, ε_r' are listed in Table 3. In all measured fruits, during low frequency, the loss factor, ε_r'' decreased, and a broad minimum between 1 and 3 GHz is reached, subsequently increased again as shown in Figure 7. Since dielectric properties in agriculture is considered primarily depended on ionic conductivity of fluids comprised in their cellular structure and water activity, thus ionic conductivity, σ at lower frequencies is the factor which influences this behaviour.

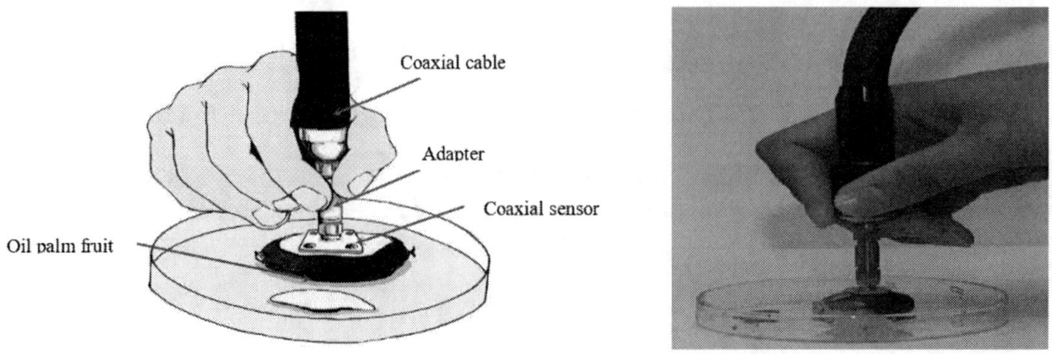

Figure 5. Part of the open-ended coaxial waveguide (OECWG) experimental set-up.

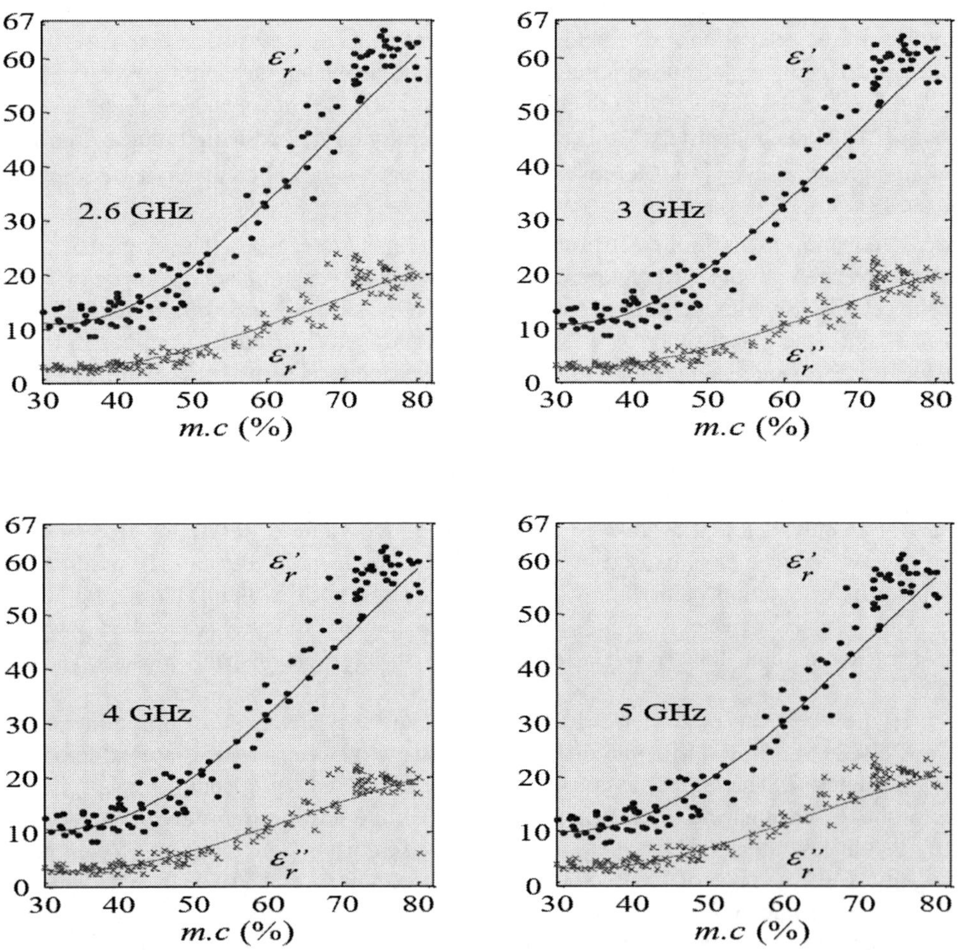

Figure 6. Variation in dielectric constant, ε_r' and loss factor, ε_r'' with $m.c$ of 25 °C [17].

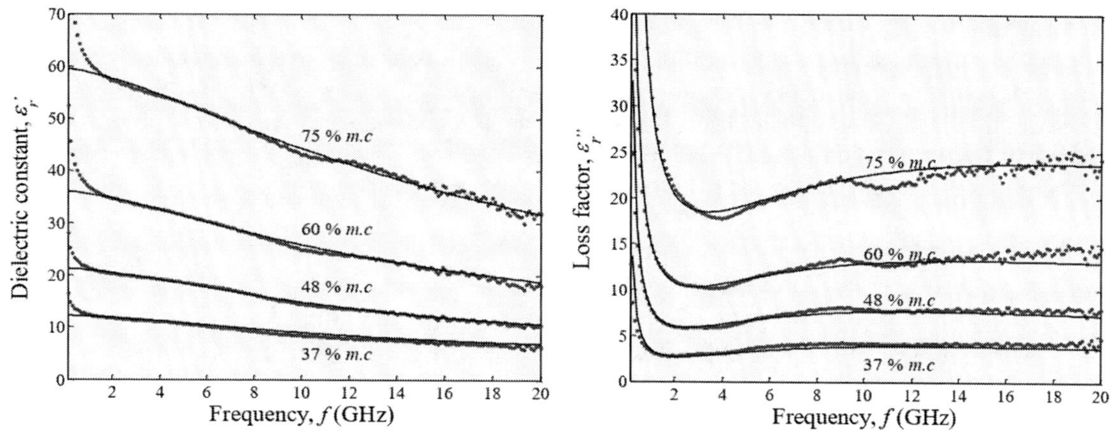

Figure 7. Variation in dielectric constant, ε_r' and loss factor, ε_r'' with frequency at room temperature (25 ± 1)$^{\circ}$C for different water content, *m.c* in oil palm fruits [18].

Table 2. The Debye and Cole-Cole parameters of oil palm mesocarp as a function of percentage water content, *m.c* (%) at (25 ± 1)$^{\circ}$C [18].

Cole-Cole parameters of oil palm mesocarp (30 % ≤ *m.c* ≤ 80 % ; pure water) ; (0.13 GHz ≤ *f* ≤ 20 GHz) ; (25 ± 1) °C	
$\varepsilon_s = -3.9899\times10^{-4}(m.c)^3 + 8.0046\times10^{-2}(m.c)^2 - 3.8988(m.c) + 67.258$	±0.9792
$\varepsilon_1 = -1.8394\times10^{-5}(m.c)^4 + 3.8076\times10^{-3}(m.c)^3 - 0.27371(m.c)^2 + 8.7394(m.c) - 98.783$	±0.5093
$\varepsilon_\infty = v_{water}\varepsilon_\infty^{water} + (0.84 - v_{water})\varepsilon_\infty^{oil} + (0.16)\varepsilon_\infty^{fiber}$, where $\varepsilon_\infty^{water} = 4.9$, $\varepsilon_\infty^{oil} \approx 2.5$, $\varepsilon_\infty^{fiber} \approx 2.2$	
$v_{water} = \dfrac{0.01m.c(0.16\rho_{fiber} - 0.16\rho_{oil} + \rho_{oil})}{\rho_{water} - 0.01\rho_{water} + 0.01\rho_{oil}}$, $\rho_{water} = 0.9959\ \text{gml}^{-1}$, $\rho_{oil} = 0.9073\ \text{gml}^{-1}$, $\rho_{fiber} = 0.92\ \text{gml}^{-1}$	
$\tau_1 = -1.1394\times10^{-17}(m.c)^3 + 1.3309\times10^{-15}(m.c)^2 - 1.0209\times10^{-13}(m.c) + 1.6660\times10^{-11}$	±1.119ps
$\tau_2 = -5.5307\times10^{-17}(m.c)^3 + 8.7071\times10^{-15}(m.c)^2 - 4.6620\times10^{-13}(m.c) + 1.6004\times10^{-11}$	±0.478ps
$\sigma = 6.2861\times10^{-4}(m.c)^2 - 3.6521\times10^{-2}(m.c) + 6.3931\times10^{-1}$	$\pm0.02482\ \Omega^{-1}\text{m}^{-1}$
$a_1 = -2.2890\times10^{-4}(m.c)^2 + 3.6771\times10^{-2}(m.c) + 1.4734\times10^{-1}$	±0.0008234
$a_2 = -3.5634\times10^{-6}(m.c)^2 - 4.5790\times10^{-4}(m.c) + 1.0124$	±0.00002904
$a_3 = 4.0888\times10^{-5}(m.c)^2 - 8.7443\times10^{-3}(m.c) + 1.1543$	±0.0008381
$a_4 = 3.5374\times10^{-5}(m.c)^2 - 2.5052\times10^{-3}(m.c) + 8.7535\times10^{-1}$	±0.004263

Table 3. Relationship between % water content, *m.c*, dielectric constant, ε_r' and conductivity, σ of oil palm mesocarp for 3 GHz at $(25 \pm 1)^{\circ}$C [18].

Water Content, *m.c* (%)	Dielectric Constant, ε_r'	Conductivity, σ ($\Omega^{-1}\text{m}^{-1}$)
30 – 40	10 – 20	0.11 - 0.17
40 – 50	20 – 25	0.17 - 0.38
50 – 60	25 – 35	0.38 - 0.7
60 – 70	35 – 55	0.7 - 1.1
70 – 80	55 – 65	1.1 - 1.7

7. Conclusion

In this paper, various microwave sensors particularly for oil palm moisture measurement have been reviewed. In addition, the permittivity formulation for oil palm fruits has also been revisited in which the permittivity values of oil palm fruits is very important to optimize the pretreatment process of mill crude palm oil using microwave and solvent extraction methods. Hopefully the microwave sensor for oil palm fruit measurements will be commercialized in the near future.

Reference

[1] Hartley C W S 1977 *The Oil Palm* (London: Longman Group Limited).
[2] Ariffin A, Mat S R, Banjari M and Wan O W E 1990 *Palm Oil Research Institute of Malaysia Bulletin.* **21** 30-34.
[3] Saeed O M B, Shariff A R M, Mahmud A R B, Shafri H Z M, Alfatni M S, Aouache M and Ehsani R 2013 *Advanced Science Letters.* **19** (12) 3468-3472.
[4] Saeed O M B, Sankaran S, Shariff A R M, Shafri H Z M, Ehsani R, Alfatni M S and Hazir M H M 2012 *Computers and Electronics in Agriculture.* **82** 55-60.
[5] Abdullah M Z, Guan L C and Mohd Azemi B M N 2001 *Food Bioprod. Proc.* **79** 223-231.
[6] Ishak W I W and Razali M H 2010 Hue optical properties to model oil palm fresh fruit bunches maturity index *World Multi-Conference on Systemics, Cybernetics and Pattern Recognition.*
[7] Makky M 2016 *Agriculture and Agricultural Science Procedia.* **2016** 230-240.
[8] Shaarani S M, Cardenas-Blanco A, Gao Amin M H, Soon N G and Hall L 2010 *International Journal of Agriculture and Biology.* **12** 1560-8530.
[9] Aliteh N A, Misron N, Aris I, Sidek R M, Tashiro K and Wakiwaka H 2018 *Sensors.* **18** 2496.
[10] Khalid K and Abbas Z 1992 *J. Microwave Power and Electromag. Energy.* **27** 3-10.
[11] Khalid K and Teoh L H 1998 *Meas. Sci. Technol.* **9** 1191-1195.
[12] Abbas Z, Mokhtar R, Khalid K, Hashim M and Aziz S A 2007 *Eur. Phys. J. Appl. Phys*. 40 207-210.
[13] Lee K Y, Abbas Z, Jusoh M A, You K Y and Cheng E M 2011 *Sensors.* **11** 4073-4085.
[14] Ali A 2006 A study on relationship between reflection coefficient and moisture content in oil palm fruits using rectangular waveguide (Malaysia: MSc Thesis Universiti Putra Malaysia).
[15] Zakaria Z 1998 Dielectric properties of oil palm mesocarp at various moisture content (Malaysia: MSc Thesis Universiti Putra Malaysia).
[16] Abbas Z, K Y You, Shaari A H, Zakaria A and Khalid K 2005 *Japan. J. Appl. Phys*. 44 5272-5274.
[17] Abbas Z, K Y You, Shaari A H, Khalid K, Hassan J and Saion E 2005 *IEEE Sensors Journal.* **5** (6) 1281 - 1287.
[18] You K Y, Abbas Z and Khalid Kaida 2010 *Measurement Science Review.* **10** (1) 7-14.
[19] Ahmad A F, Abbas Z, Abrass H A, You K Y 2019 *BioResources.* **14** 1819-1837.
[20] You K Y, You L L, Yue C S, Mun H K and Lee C Y 2017 Physical and chemical characterization of rice using microwave and laboratory methods: Rice ed K Y You (London: InTech) 81-99.
[21] Nyfors E and Vainikainen P 1989 *Industrial Microwave Sensors* (Norwood, MA: Artech House, Inc).
[22] Karuppannan I, Omar O and Ariffin A 1986 Profiles of biochemical changes during the development of oil palm mesocarp (Elaesis guineensis). *Proc. Mal. Biochem. Soc. Conference.* **11** 129-134.

IOP Conf. Series: Materials Science and Engineering **767** (2020) 012008 doi:10.1088/1757-899X/767/1/012008

A study of electronic payment system

Marwah Naeem[1], Methaq Hameed[2], and Mustafa Sabah Taha[3*]

[1]School of Computing, UUM, College of Arts and Science

[2]Basrah Oil Training Institute, Ministry of Oil, Iraq

timimymustafa@gmail.com

Abstract. The essential objective of this study examines the rising consciousness pertaining to varying notions that are associated with regards to electronic payment systems, related to the benefits, challenges and security concerns. By utilizing software as a service platform, the vendors of the payment processing system work on this template which will channel the payment traffic from the single payment towards the multiple payment schemes for the clienteles. Consumers frequently disclose data that is confidential that belongs to them for example, names, particulars about their cards and numerous other information during their online transactions to perform a payment. A method that assists in the transference of money electronically defines the online payment system. This payment manner generally entails the array of a network of computers, the internet and various other repositories for digitized value methods. By acquiring whatsoever means of payment via the internet entails that the user of the system has received payment made online, and has a mutual knowledge of certain private data with the vendor or the entity that provides the service. This study initiates a comprehensive holistic review encompassing the whole components of payments made online or electronically, that focuses on an investigation over the various researches done pertaining to payment methods that are executed electronically. The current most up-to-date researches were examined to garner an insightful comprehension of payment methods that are electronically employed.

1. Introduction

The swapping of products and services which is bilateral was initiated prior to recorded chronicles; however as the interchanges of products among individuals became more complicated, value of products and services are equated in a conceptual manner, which was initiated through the barter system, and went across a continuum from authorised money, to payment orders, to the various cards inclusive of debit or credit cards, and the prevalent method of payment done electronically or through an electronic payment gateway [1]. The conventional technique of payment encounters particulars problems such as falsified signatures, counterfeit money, and cheques that could not be honoured. Nevertheless, an e-payment system that is meticulously organized may resolve these security concerns and offer an added benefit such as flexibility in utilization [2, 3].

This mode of payment has been well received and recommended as the result of their capabilities of expediting interchanges of currency, security assurance, and swiftness of accessibility to means of securing wealth [4–6]. The reduced volume of transactions in terms of financial values had inversely increased the cost of supporting the system by the scale of the economy, thus payment methods by means of cash had become costlier. Additionally, the operational value of running a cash payment mode of mechanisms via the internet is much cheaper in comparison to the most minimum cash transaction when done manually [7]. The general public has become accustomed to online mode.

On the other hand, certain problems had cropped up during the on-going transactions that involve cash that was channelled through the internet of which the conventional techniques of payment are not able to resolve. Due to this, specialists in the field of finance have initiated an investigation of variant techniques of electronic payment that focus on the concerns pertaining to payment methods in the digital and electronic forms [3]. Every individual proceeding goes through procedures via the payment passageways that acts as the point of entry into various financial entities. Precise information with regards to the payment that was undertaken between distinctive entities and establishments dealing with financial matters is verified via the respective gateways of payment [8]. This current study explicates the heightened consciousness pertaining to electronic payment systems (EPS). The arrangement of this study is the following: the numerous denotations and facets pertaining to EPS are covered in Section 2, meanwhile, the EPS benefits and prevalence are discussed in Section 3 and 4. In addition, the challenges pertaining to payment mechanisms are covered in Section 5, and imminent EPS security concerns are discussed in Section 6. Ensuring that, Section 7 covers the review of related works pertaining to EPS, and finally, Section 8 forms the conclusion of this study.

2. Electronic payment system

Prevalently, the traditional payment methods that utilize cash are progressively being demoted [9] and this role has been taken over by the payment method that utilizes digital means, as the development of goods transactions increases amongst the various trades via the internet. Several methods of electronic payments for transactions to pay for services via the internet are seen in [10, 11]. Among the forms of Inter-Organizational Information System (IOS) is an Electronic Payment System that is devoted to enabling proceedings that are linked with money among the clients and varying establishments to be carried out. A necessity could arise for complicated communications exchanges amongst technologies, the environ, and associates in ensuring the EPS efficiency.

The particular EPS/IOS aspects enable it to be distinctive from the conventional information methods that are formed on the basis of internal mechanisms in terms of technology and organization, [12–14]. It emphasizes the requirement for a collaboration between the various methodologies towards enabling an effective technique [15].

Throughout the years, the world has witnessed a worldwide rise in cashless proceedings yearly via mobile transactions and electronic payment systems. However, the year 2012 has seen a 0.9% drop in the annual growth rate from 8.9% in the previous year 2011 it has dropped to 7.7% in the year 2012 [15]. Nevertheless, the year 2014 saw a rise in payment transacted electronically worldwide to a value of 8.9% worldwide, achieving 387.3 billion transactions, which signifies the utmost marked an important rise from the inaugural publication of the World Payments Report. The increment was essentially caused by the accelerated progress in the freshly emergent markets that are related to finance. A predicted projection of 10.1% progress globally was expected in the year 2015 and it was anticipated then that worldwide paperless exchange volume will increase to 426,300,000,000 [17].

Electronic payment systems (EPS) are considered as useful and user-friendly and is also secure by unique users and assorted users, who perceived the means as an entry point towards progress in technology in terms of the economy worldwide [18]. EPSs have evolved into significant promoters of electronic commerce which are relied on by businesses done electronically. They have diminished the number of fraudulent attempts and activities within the payment system worldwide [11, 19].

2.1. History of Electronic Payment Systems (EPSs).

Electronic mode of payment originates back to the year 1918 on the occasion of Federal Reserve Bank of the United States of America initiated channeling of money through the telegraphic method. Nonetheless, prior to 1972, the adoption of these mechanisms was not widespread in the USA pending the consolidation of the Automated Clearing House (ACH) in that year. Henceforth, the widespread use of the system in the US escalated which enabled its commercial banks and central treasury to present it as an optional means of payment in addition to the traditional payments using cheques [11].

Credit card emergence can be traced back to the year 1914 with the issuance of a card to clients by shopping centers, oil companies, commercial accommodation places and the Western Union as a mode of paying for products and services rendered. Four decades after that, the world witnessed the extensive utilization of credit cards which has turned into a recognized model of an alternative form of payment.

Prior to the 1990s, credit cards were an alternative form of payment based on paper. However, after that, they evolved into systems that are executed electronically. The utilization of credit card has escalated tremendously in terms of its pace in the industry and has been the precursor of the debit card system. Each of the two is prevalently utilized as a mode to pay for services and products [10, 20].

2.2. *Definitions of EPS*

The intricate Electronic Payment Systems are defined by the multitudinous modes of payment that is executed electronically. It is multi-functional attributes have been inaccurately described in literature at times. Pertaining to capacities. Electronic payment can function as electronic banking, electronic cash, banking via internet, online banking and many others. Taking all of these into consideration, there are current researches in providing an extensive definer of electronic payment [5]. The description of Electronic Payment Systems can be found in [9] and is explained as responsible undertakings in the financial scope which enable the client and the trader to converge and utilize equal platform through methods electronically. Moreover, [15] electronic mode of payment is perceived to be interconnecting relationships among the general public and organizations driven by establishments that offers financial dealings.

As stated by [21], a payment that is done electronically entails whatsoever shape of a system that enables the interchange of money via the internet. Likewise, [22] expounded that a system that enables payment to be made electronically related to services offered on the web.

A supplementary description regarding e-payment is that it involves whatsoever kind of payments or interchanges transacted through electrical means [23]. Additionally, according to researcher [24] perceived e-payment as a bilateral system that allows the interchange of money via online transaction between two individuals. Furthermore, [25] presented e-payment as the interchanges involving money executed electronically. An additional description pertaining e-payment entails whatsoever kind of payment involving an interchange of data done electronically for example credit and debit card data as an alternative mode of payment such as cash money or cheque [26].

According to [27]. electronic payment is request made for fiscal interchange by a client to another individual as payment. As per [28], e-payment entails the acts of paying payments via transmissions made electronically, through a clearinghouse done automatically, or through card mechanisms that are commercially implemented. It was further described by [29] as whatsoever kind of fiscal interchange through whatsoever means electronically. In addition, according to [30] defines e-payment as credit or debit bank accounts receiving transactions executed via signals transmitted electronically.

Furthermore, according to [31], it is whatsoever kind of payment that does not involve physical money excluding paper cheques. Further to this, [32] defines e-payment as whatsoever interchanges done electronically that is perceived as a kind of payment for products and services executed through e-payment means, which provides consumers with long distant accessibility with their accounts through systems did electronically. Typically, EPS is described as whatsoever kind transference between consumers and providers through online channels assisted by a financial instrument done digitally[33].

2.3. *Types of E-payment systems*

There are numerous kinds of EPS created in the financial mechanisms worldwide, for example, cheques in electronic form, fund transfers are done electronically, electronic cash, debit and credit cards [11,34]. There are two kinds of online payments; i) ones that utilise Internet Banking Payment Gateway (IBPG), and ii) the ones that utilise external payment mode of the platform. Firstly, the payment mode is straight and precise due to the fact that the consumer is informed of payment made online through an e-business channel connected to a banking platform. The second form of online payment entails the transmission of money from the consumer's account to the provider's account via a payment system that is based outside. In the first instance, the internet is connected to the IBPG which is linked to a banking process system; its design is explicitly for administering and authorising the payments. The bank is connected to the buyer with the seller through this process. The online payment that utilises the IBPG type of online payment technique which is dependent upon gateways to transmit payments[35].

Four EPSs collectives that were categorised by [36] are; electronic cash, a tiny amount of payments, online credit card payment, and cheques in electronic forms. It was emphasised in greater detail by the study that individual system possesses particular advantages and disadvantages. The researchers expounded that each of the collectives can be evaluated from the facets of technology economy, social and legalities. In actuality, there are only two particular types of payment systems [3,37]:

2.3.1. Electronic-based transaction. There are four methods of payment systems that utilise the internet as a basis:

a) *Cyber cash.*
This online type of service involves the credit card data of the customer to undergo through a process, be claimed for costs incurred, and put into the dealer's account by electronic techniques. Cyber cash operates as a gateway for payment to be channelled. It is dependent on digital signatures to guarantee the procedures for paying in terms of security concerns [37]. Even though e-money covers a more comprehensive area that entails the whole fund transactions that utilise the internet, cyber cash, on the other hand, focuses on the entire money interchange systems undertaken via the Internet. There is a marked difference between e-money and cyber cash although it might be hard to identify because money in the form of cash can be acquired from e-money and will merge into it [38].

b) *Secure electronic transaction (SET).*
This type of e-money transmitting system is an ensured secure payment system that could be undertaken online for money exchanges transacted via the internet. Developed by Master Card and VISA, it is a technology platform that is open for the utilisation of businesses [39]. SET guarantees the secure transmission of money that is based on a card via the Internet. It utilises certificates that are in digital form to validate the authorization status of a provider or a cardholder [37].

c) *First virtual holdings.*
It is among the earliest form of platforms that facilitate payment transacted via the internet, which is dependent on the external validation techniques to channel payments made online. This mode of payment system is independent of encryption, utilised solely for information selling via the Internet, as opposed to the interchange of products and services. It only compiles data pertaining to payments made by the user by the use of a telephone system that is automated, and it is independent of cryptographic methods and digital signatures. Essentially, it is specifically dependent on the meticulous tracking of sales and purchases transacted to decrease fraudulent occurrences [40]. Being an open system, its core feature entails certain information is restricted to be online as they are sensitive data pertaining to the customer's financial data. Thus it utilises the first virtual PIN issued by the initial virtual entity for with the objective of exchanges transactions minus the utilisation of credit card numbers. As the respective PIN numbers act as identification, it enables their transmission via the Internet with no charges incurred on the client's account as a receipt of a validation email from the customer is needed to authorize whatsoever payment form [37].

Since the year 2010, the world has witnessed an enormous development in the utilisation of cards as a method of payment. This is attested by the drop in transactions that utilise cheques in the last thirteen years. Amongst the systems that are based on the utilisation of cards, debit cards form the greatest number of users of approximately 45.7% of the cashless money transactions worldwide, demonstrating debit cards as the most aggressive accounting to 12.8% of the payment choice in the year 2014. The implies the quality of security and ease-of-use of the system that uses cards, which is superior over other modes of payment choice [17]. Payments can be transacted electronically through platforms that use mobile appliances; numerous smartphones that are Android-based offers payment to be made by using online platform.

Electronic Payment Systems can also be transacted through the use of Desktop computers. Other mobile phones can be used by customers, which runs on numerous platforms to implement financial proceedings. Customers are able to convey PIN numbers or use WAP to execute payment transactions electronically via the mobile internet. In the execution of an electronic payment made by a customer, proceedings using the debit or credit card, authentication can be done by a vendor through the attachment of a gadget linked to the mobile phones. In the United States of America, a composite of Power Swipe, that is linked to a Nextel telephone, of 3.1 ounces in weight, which consists of a reader that can read magnetic stripe, is inserted into a connector that is used for reloading the battery of a mobile phone, and for printing purposes it utilises an infrared port [41].

d) *Net bill*
The net bill payment system independent of internet usage to enable online proceedings that are secured. It is a mode of payment at the micro-level, with servers that retain customers' and sellers' accounts, which enable customers to pay for the goods via the system. Seeking for information, for example, a database query, text from a page, or a software program can be executed in whatsoever external architecture, through the interchanges of bits with the clients. Clients are charged for the number of things they use, and clients are also offered unlimited access if required [42]. Financial tools software will examine the goods receipts. Thus, it enables the transmission of information between the financial tools software and the server belonging to the dealers [37].

2.3.2. Internet-Based Payment System. There are four essential techniques in payment systems that utilise the internet

a) *Debit card*
It is the dominantly used electronic payment form of the platform; it utilizes banking through the internet and the Automatic Teller Machine (ATM) card [43]. Users of the debit card are enabled to make immediate payment for products bought via the bank. Money is secured in banks and will be withdrawn only when the users of the Debit Card makes any purchases through its use. There are two kinds of Debit Cards; online and offline [37].

b) *Smart card*
A card that is made of plastic and embedded with a microchip, with information such as the amount of money, information of an individual and can be pre-loaded, with the ability to execute immediate payments is called a smart card or known as a chip card [44]. The user is provided with a validated PIN by the service provider. In order to assure the protection of data within these cards, the encoded information is stored within them and possess great processing speed. Examples of this kind of card are the VISA and Mondex cash cards [45].

c) *Credit card*
The credit card is circulated to customers by financial institutions to execute payments via online means [44]. It is the prevalently utilized e-payment system. However, it is not suitably used for making payments with petty value [43].

d) *E-cash*
It was initiated and presented as an option to credit cards for making payments to buy anything via the internet [43]. It is a form of EPS that enables users to execute transactions through a gadget via online means, with the funds in a repository. E-cash is money in a digital format, utilizing a pre-installed software on the customer's computer that enables transactions to be

made [44]. It is a much-favored choice among customers as it enables transactions that are small in amount and it possesses a minor cost [43, 46].

3. Security of Electronic Payment system

Security in terms of data and information is of the utmost importance in the entire information systems [46]. Figure 1 illustrates aspects of security of data assurance that encompass the methods, practices, and technology involved.

- Unintentional modifications or transformations (integrity).
- Unauthorised accessability (confidentiality).
- Demand-based ease of accessibility for authorised customers (availability).

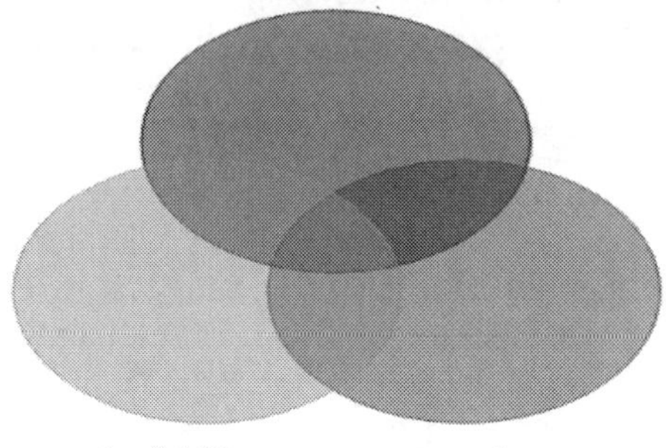

Figure 1. Depiction of Information Security (CIA)

The stated aspects related to security are required to be functional in an electronic payment system, whereby an EPS that is not secure will build a feeling of distrust in customers, which will defeat the purpose of assuring their confidence to receive the system.

According to [47], there exist various security concerns within the environment electronic banking and electronic payment systems as they are dependent on the essential Information Communications and Technology platforms that are susceptible to various weaknesses in economics and businesses

3.1. Security Demands in EPSs.

Specific necessities that must be adhered to by all electronic payment system are:

a) *Integrity and Authorization.*
System integrity is defined as its preciseness, credibility, and wholeness that are grounded on the quality of businesses and aspirations. This means that a customer will only be charged upon the client's affirmed consent. Moreover, without the client's approval of any payment, it is obligatory that sellers are denied acceptance of any payment [48].

b) *Confidentiality*
It is established as the protection of confidential data from unapproved disclosure. Several establishments have made confidentiality an integral element of their proceedings. Within this scenario, confidentiality implicates the covert nature of the information pertaining to whatsoever type of proceedings. Generally, the concerns of customers are towards the protection of their proceedings [49]. If a situation arises for the necessity for obscurity and secrecy, only a specific group of people will receive particular information.

c) *Availability and Reliability.*
It ensures that the information frameworks and data are there when required; it needs a constant and efficient framework transmission time. The entire system with all its elements should be in a state of constant readiness for the transmission or receiving of payments as and when required [49].

3.2. Enhancing EPS Security.
According to [55,50-53], the strategy that is generally utilized for Electronic Payment System security assurance is based on covert electronic systems for example encryption, watermarking, digital signatures and steganography as described in Figure 2, When applied, the technologies have enabled stealing and falsification. Henceforth, a concise description of the stated security techniques that makes electronic payments more secure is as follows:

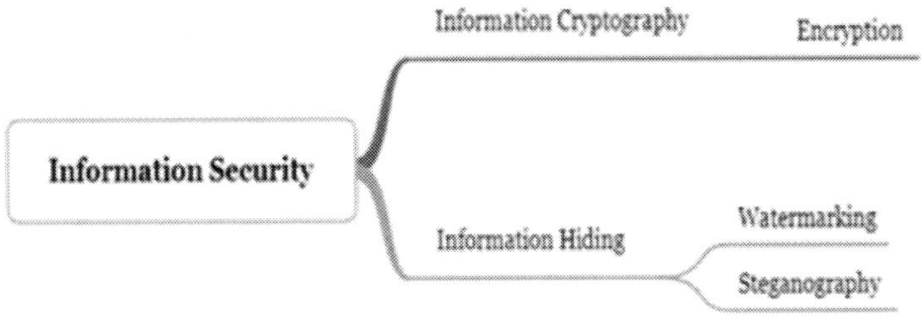

Figure 2. Categorization of Information Security

3.2.1. Information security - The utilization of the internet has transformed electronic payment, however, open mode of transmission has presented certain problems such as ensuring secured data. Figure 2 illustrates various methods of area security systems in resolving the challenges. A method of scrambling data that makes it uninteresting to an authorized intrusion is known as cryptography or data encryption. However, at times, encrypting electronic payment data can be complicated. Hence, an undetected method of transmission that hides the presence of covert data is essential.

Due to this, a technique that hides data is needed. In concealing data, there are two parts; steganography and watermarking [54] The two methods assures the concealment of covert data and are associated with each other but have differing aims. The principal objectives of concealing the presence of covert information and safeguarding the covert data are known as steganography. In contrast, watermarking focuses on safeguarding the integrity of the covert information by concealing or not concealing the presence of the covert data from intruders. The main objective of watermarking is to safeguard the intellectual property of the content.
Keep your text and graphic files separate until after the text has been formatted and styled. Do not use hard tabs, and limit use of hard returns to only one return at the end of a paragraph. Do not add any kind of pagination anywhere in the paper. Do not number text heads-the template will do that for you.

4. Related Works
In [55] has recommended the BLAH algorithm to identify credit card falsifications. It is a hybridization of BLAST and SSAHA structures. The structures are competent algorithms for the sequencing array towards the identification of fraudulent credit cards. Additionally, the BLAHFDS system identifies dubious falsification proceedings through the utilization of a profile analyzer and deviation analyzer.
The fraudulent identification analyzers utilize BLAH as an instrument to align sequence. The said model was recommended so that synthetic proceedings can be generated in order analysis can be executed on the BLAHFDS achievements. It demonstrated a satisfactory achievement of a fraudulent identification system exhibiting great precision and swift speed of processing. However, it is unable to identify duplicated credit card or replicated proceedings. Fraudulent activities in other industries for example in the telecommunications sector can be thwarted through the utilization of BLASTSSAHA hybridization method.

According to [56] has forwarded a Fraud Detection System (FDS) that merges the outcome attained from a client's current and previous conduct. In the FDS, there are 4 constituents which are Rule-based filter, Transaction history database, Dempster–Shafer adder, and Bayesian learner. The degree of doubtfulness in individuals performed through the utilization of Rule-based Filter looking at its differences from the spending norm. The Dempster–Shafer adder merges the entire doubtful proceedings acquired through the Rule-based filter to determine a primary belief. In order to resolve certain issues a typical tool that uses statistics was utilized, which is the Hidden Markov Model (HMM) .

Ashphak et al emulated the operational order engaged in the credit card transaction procedure through the utilization of HMM. In [57], the correlations between the transition probability and the hidden state and examinations were deliberated. It was recommended by Sasirekha et al for Intrusion Detection Systems (IDS) which combines abnormality, abuse, and models that assist in the making of decisions to be utilized in order to obtain a reduced rate of incorrect positive alerts and greater identification precision. By using the abnormality identification module, an incorporated IDS that was based on the HMM approach, identification of assaults in the credit card system is enabled. Additionally, a card user's attitude is taken into account as attributes and dubious proceedings are identified from the spending profile of the user.

Suspicious unauthorized proceedings are transmitted to an abuse identification system. Through different research done in [58], expounded that faith in the online electronic commerce environ is crucial because it ascertains the client's readiness to utilize a system that uses online proceedings. In order to manage or avert fraudulent hazards, and for making transactions which use online mode of transaction, digital certificates, and digitized signature are typically utilised [58].

Additionally, from another research in [59], the researcher explicated about electronic commerce for products and services that utilizes online proceedings. However, it was noted that the very policies that were put in place to safeguard and guarantee security and dependability offered by service providers are actually the hindrances to shopping being executed online.

Consequently, a client's feedback with regards to online shopping are issues pertaining to the sharing of confidential and private data, unsought communications from the shopping system done online, and the monitoring of their shopping behavior [60].In addition to concerns regarding the security of the system, customers are worried about unlawful bridging gadgets which has their own protection through technological means, are able to obtain customer's private, monetary or confidential data that is linked to the transactions.

They are also anxious data shared with sellers who deal with orders and purchases through the internet, in addition to being swindled by sellers who do not deliver the ordered products which have been settled in full, are amongst the probable dangers and risks perceived pertaining to purchases done through the internet. Enhanced security design for purchases done via the internet enables users; misconduct to be reduced to a minimal level with the execution of proceedings implemented through the internet [61].

User's banking information data and information about the card must not be disclosed whilst and ensuing the transactions made via the internet, as it will make it susceptible to unlawful usage and abuse. If on any occasion the data is disclosed, it can be abused by an intruder for vested interests. Thus the system that deals with payments executed through the internet could be enhanced through the introduction of strong rules and regulations, integrating legislative, strict conditions for the protection of data, inclusive of the issuance of attestation from reliable third parties [61].

Moreover, in different research in [62], the researcher improved on the protective layers to the shopping system executed via the internet in order to stimulate customers ' online purchases or electronic commerce involvements, in addition, to generate consciousness amongst the Libyan financial sectors. Customers will be undaunted and free from worries when they utilize a channel via the internet in the event that their monetary and private data are appropriately safeguarded and shielded [62]. Additionally, it is essential that online portals incorporate aspects that promote a reliable connection between consumers and online portal sites so as to boost product purchases and draw consumers. It is imperative for the portals to guarantee that each proceeding must be grounded on accords that imperative to be satisfied [63]. The safeguarding of consumers' information and security will result in consciousness and confidence in the Libyan financial sectors.

Eben explained the influence of data protection for e-businesses that focuses on the probable damages and hazards to security that might incur as a result of these weaknesses. There are six scopes in e-business security that entails trust, not corrupted, authorized usage, monitoring, availableness, and non-rejecting. The research advised an extensive and holistic security policy that will guarantee the protection of a business should be designed [64]. Cheng, Hamid, and Cheng [65] recommended five discerned security hazards as tangible, performance, time, psychological, and monetary loss.

Tangible hazards encompass loss of money or card, meanwhile, performance hazards constitute hazards of added costs incurred upon usage. Psychological hazards entail the hazards which will influence the perception of the customer's image during the transactions of such payments; time loss hazards are dangers that might arise from a procedure that takes a long time to complete in comparison to another method of payment. Monetary hazards entail endeavors that will evoke a monetary loss that cannot be reimbursed when executed. Srinivasan [66] is convinced that the favorable outcome of e-businesses is dependent on certain aspects. According to him, it is important for electronic businesses to endeavor in creating confidence traversing a time span. He stated that certain aspects that promote in acquiring consumers' confidence constitute goods or services provisions, Website attractiveness, branding, trustworthy stamps, and service value. Believe and confidence could be ascertained in various ways, such as the content of data, goods, proceedings, technology, and establishment.

The researcher examined 'believe and confidence' from transactional perspectives and emphasized the actions that electronic businesses are obligated to implement in promoting consumer confidence in this paper. It is difficult to evaluate aspects that promote a sense of belief and confidence due to the fact that they evolve through the span of time. Individuals often depend on prior events in order for them to be confident in any business; they are equally dependent on third party feedback and suggestions. It is finalized that electronic businesses are accessible almost everywhere at any time, however particular aspects may hinder the development and preserving trust.

The main challenge of this research is that there are gaps to fill such as on the methods that an establishment may enhance online security whilst proceedings are ongoing. Pereay Monsuwé, et al. recommended a structure in enhancing the authors' comprehension pertaining to customers' perception and behavior with regards to online proceedings in the United States of America and the European countries. The structure mobilizes the Technology Acceptance Model (TAM) constructions which form the foundation, extending its exterior aspects, with its application on an online shopping environment.

It is noted that the consumers' perceptions and behavior toward transactions executed via the internet are influenced but not confined to the benefits and effortlessness in the utilisation of the platform, but are also affected by exterior aspects for example customer behavior, goods features, aspects that are dependent on situations, previous purchasing events, and confidence in proceedings that are executed via the internet [67].

Abdul et al., evaluated the influence of protection and believe, on the customer's readiness to engage in proceedings executed via the internet amongst the urbanites of Morocco. This research evaluates consumer's readiness to make purchases via the internet by taking into account aspects in demography, and aspects of believing confidence and security. Through findings from the logistics regression analysis executed, it is discovered that consumers' readiness to engage in purchasing via the internet, which is dependent on aspects of trust, security, age, consciousness, and fraud. It was also discovered that a majority of the subject cohorts will make purchases via the internet willingly. The outcome of this research is most significant for suppliers that execute businesses through the internet and as well as for the government towards an enhanced comprehension pertaining to online purchases. The result of this research promotes individuals who are responsible for making decisions to create improved online purchasing conveniences with the competitive edge in technology for the consumers who make purchases via the internet and the suppliers [68]. The electronic banking effect on the economy in Bangladesh was researched by Baten and Kamil; in addition, they researched the advantages and range of electronic payments in Bangladesh [69]. In other respects, Salehi and Alipour embarked on research in electronic banking in an emergent economy with the objective of probing to offer proof empirically from Iran [70]. Through the utilization of Statistical Package for Social Sciences (SPSS), James examined the reception towards electronic banking in Nigeria. The result from the research demonstrates that the receptiveness towards electronic banking is essentially dependent upon the age, earnings, level of education, conceived advantages, conceived hazards, and conceived user-friendliness [71].

5. Conclusion

From the evolving current inclinations, it very clear that electronic payment systems in the world trading and commercial sectors have become crucial. Their magnitude spans from proceedings worth one dollar up to millions of dollars' worth. This research provides an extensive comprehension of EPS and the protection of payment factors. It is recommended that an electronic payment system that is protected for the purpose of executing purchases via the internet through the utilisation of information security systems such as steganography or cryptography, or by hybridization of the methods; in safeguarding a consumer's information and keep it to the minimal the abuse or fraudulent acts on the retailer's or seller's end. In deliberating on prior researches done before this, there exist the potential to acquire priceless comprehension on the advantages and disadvantages of prevalent existing electronic payment systems, and the accessible protection techniques involving EPSs. In addition, the current research of this paper examines the electronic payment systems from various points of view pertaining to security with the objective of providing enhanced consumer knowledge and contentment.

6. References

[1] Ko, Samuel CK, and Ross D. Murch 2001 Compact integrated diversity antenna for wireless communications *IEEE Transactions on Antennas and Propagation* p. 954-960.

[2] Jibin, Huang, and He Yigang 2019 A Compact Dual Band Antenna for TD-LTE/WiMAX Applications *Proceedings of the 2019 International Symposium on Signal Processing Systems. ACM,* 2019.

[3] K. Pengthaisong, P. Krachodnok, and R. Wongsan 2013 Design of a dual-band antenna using a patch and frequency selective surface for WLAN and WiMAX *2013 10th Int. Conf. Electr. Eng. Comput. Telecommun. Inf. Technol. ECTI-CON* 2013 p. 1–4

[4] I Ali and R Y Chang Fall 2015 Design of dual-band microstrip patch antenna with defected ground plane for modern wireless applications *2015 IEEE 82nd Veh. Technol. Conf.* VTC

[5] A. Mounsef, Y. El Gholb, N. El, and A. El, 2017 Design and simulation of a dual band microstrip patch antenna for an emergency medical service system *2017 International Conf. on Wireless Networks and Mobile Communications (WINCOM)* p. 0–3, 2017.

[6] A. Mellita, C Ds, and S S Karthikeyan Gain Enhancement of a Microstrip Patch Antenna Using a Novel Frequency Selective Surface *2017 Twenty-third National Conference on Communications (NCC)* p. 4–7, 2017

[7] Shinozaki, Yuka, and Hiroyuki Arai. 2018 Microwave Absorber by FSS Using Paper Instead of Dielectric Substrate *2018 IEEE International Workshop on Electromagnetics: Applications and Student Innovation Competition (iWEM)* p. 1-2, 2018.

[8] Wu, Te-Kao 2018 Improved Broadband Bandpass FSS Filters For 5G Applications *IEEE International Symposium on Antennas and Propagation & USNC/URSI National Radio Science Meeting*, p. 2033-2034, 2018.

[9] Luo, Shengyuan, Yingsong Li, Tao Jiang, and Beiming Li 2019 FSS and Meta-material Based Low Mutual Coupling MIMO Antenna Array *IEEE International Symposium on Antennas and Propagation and USNC-URSI Radio Science Meeting* p. 725-726 2019.

[10] Sharma, Somia, and Mahesh P. Abegaonkar 2019 Gain and Aperture Efficiency Improvement using Uniplanar EBG Superstrate *IEEE Indian Conference on Antennas and Propogation (InCAP)* p. 1-3 2018.

[11] Machado, Gabriel Gonçalves, Robert Cahill, Vincent Fusco, and Gareth Conway 2019 Resistively Loaded FSS Clad Thermal Blankets for Enhanced RF Space Communications *Interrnational Conference on Electromagnetics in Advanced Applications (ICEAA)* p. 0048-0052 2019.

Implementation of Sine Cosine Algorithm (SCA) for Combinatorial Testing

J M Altmemi[1,2], R R Othman[1], R Ahmad[1] and A S Ali[1]

[1]School of Computer and Communication, Universiti Malaysia Perlis, Kampus Pauh Putra, Arau, Perlis, Malayisa
[2]Iraq University College, Basra, Iraq

jalalaltmemi@gmail.com

Abstract. Before being released to the market, software should be screened to ensure that the quality assurance measurement goals have been attained. To achieve this, one of the types of testing sorts is combinatorial testing (CT) aimed at discovering the faults that occur by interacting with the software. A minimization strategy for test cases is indeed important for optimizing test cases and reducing time. As NP hard (where NP is a non-deterministic polynomial) is the problem of generating the minimum test suite of combinatorial interaction testing (CIT). this paper discusses the implementation, and validation of an efficient strategy for t-way testing. The main contribution of the sine cosine algorithm SCA is to show that the strategy was sufficiently competitive as compared to other strategies in terms of the generated test suite size. Unlike most paper. The main contribution of SCA is to show the generation of test data for a high coverage strength ($t < 12$).

1. Introduction

Computer use enters every corner of social life and is the major driver of economic and social advancement. Software has become a main problem influencing social and economic development as the centre of information technology. Several software testing techniques have been suggested and examined to enhance the quality of software. Different test techniques are aimed at varying kinds of faults and can be efficient for different test scenarios. For example, mutation testing realizes the concept of using artificial flaws to encourage testing. Many empirical studies demonstrate that an efficient interaction space, e.g. targeting detection of faults, only constitutes a tiny percentage of the general interaction space in real software systems[1],[2]. In addition, there is a comparatively tiny amount of variables engaged in these efficient interactions, of which 4 to 6 are generally the upper boundaries.[3] Mutation is typically used as a manner of assessing the adequacy of test suites, guiding test case generation and supporting experimentation. But mutation has reached maturity and is gaining popularity gradually in both academia and industry[4].

Also, metamorphic testing (MT) is a technique conceived to alleviate the oracle problem. It is based on the idea that often it is simpler to reason about relations between outputs of a program, than it is to fully understand or formalise its input output behavior. But the overall effectiveness of MT is still lacking or poor [5]. Furthermore, some previous MT experiments have yielded contradictory results.

In this article we are focusing on a promising test which is combinatorial testing (CT), also known as combinatorial interaction testing (CIT)[6, 7]. CT has been shown to be efficient in detecting system interaction flaws. CT is helpful in identifying certain flaws because CT can be regarded as offering a kind of software testing shortcut. However, it has its own pitfalls, like not testing all possible combinations of parameters. If the parameters and their values are not correctly chosen, CT's ability to detect defects will be reduced[6], the aforementioned test also called In this paper we implement the sine cosine algorithm for CT to enhance the performance of the testing method.

Recent attempts to improve CT have concentrated on the implementation of meta-heuristic algorithms as the foundation of t-way approaches (where t indicates interaction strength) by formulating interaction testing as an optimization issue (e.g. genetic algorithms (GA)[8], particle swarm optimization (PSO)[9], harmony search algorithm (HS)[10], ant colony algorithm (ACO)[11], simulated annealing (SA)[12] , flower pollination algorithm[13], and cuckoo search (CS)[14]). The implementation of these meta-heuristic strategies seems to be efficient in achieving alternates of excellent quality. As recorded in t-way testing benchmarking studies[15, 16], the sine cosine algorithm outperforms all the aforementioned algorithms[17].

The rest of this paper consists of

- Introduction
- Combinatorial testing
- Sine cosine algorithm
- Result and discussion
- Conclusion

2. Combinatorial testing

The purpose of functional testing and specification-based testing is to define test cases that can exercise an application's functional requirements by systematically choosing representative combinations of parameter values. In reality, a software system generally relies on many parameters, like input parameters, state variables, and environments for the configuration. This makes it difficult to test the system taking into consideration all possible parameter value combinations. in the literature various techniques have been suggested over the last 35 years to tackle this problem, such as equivalence partitioning, boundary value analysis, as well as combinatorial testing [18]. To be used in software testing, they select representative and appropriate combinations of parameter value combinations. In this paper we use CT to solve the software testing issue. CT usually involves data intensive systems, system settings, and GUI-based testing for example [18-23]. CT is based on the assumption that a subset of parameter combinations is considered instead of testing all possible combinations of parameters, which satisfies some predefined combinations strategies. That can generate combinations of parameters that allow them to cover all parameter pairs (triples, etc.) called t-way combinations. Therefore, in t-way testing for example (2-way, 3-way) it is necessary to identify test suites so that all their value combinations are covered by at least one test case given any t input parameters. The names of these test suites called cover arrays [18, 24].

2-way testing (also called pairwise testing) is the most prevalent form of t-way testing. In 2-way testing the test suite must contain test cases so that each combination of their values is covered by at least one test case given each of the two input parameters. For example, let us think that we have a system with four input parameters (A, B, C and D), each with an integer varying from 0 to 50. In order to test such a system using all input value combinations, it is necessary to call system call 504 times. Alternatively, we simply assume all combinations of pairs of input values by implementing pair testing, for example, [A, B], [A, C], [A, D], [B, C], [B, D], [C, D] etc. Through this method, we are reducing the number of systems calls to about 502, which is two orders of magnitude less than the exhaustive generation of all combinations. To clarify the concept, let's consider a system with k parameters, each with v values. We need v^k test cases to test such a system for all input combinations, one for each possible combination, whereas the number of test cases increases logarithmically in k [25, 26] with pairwise testing. The fundamental rationale for CT is that not each parameter leads to every fault, and that interactions among a several parameters often cause a fault [27]. Empirical experience [28] has shown that CT can become very efficient in decreasing the size of test cases without wasting too much coverage of application code for example [29-31]. In fact, fault revealing suite capabilities (2-way and 3-way testing can capture 50-75% of faults [31-34]). CT issues have been addressed by several researchers through the development of different strategies based on many different approaches or categories, including pure computational approaches (e.g. PICT[35, 36], density[37], TVG[38], and ITCH[39]) and pure search approaches (e.g. SA[40], ACA[41], and HSS[42]). This article aims to review the works involved in implementing the SCA.

3. Sine cosine algorithm

The SCA is a meta-heuristic algorithm based on the population [5]. Accordingly, as its name suggests, the SCA exploits the functions of sine and cosine to update the positions of its population. Each position is a vector handle. The positions are updated to be specific depending on:

$$X_i^{(t+1)} = x_i^t + r_1 * \text{Sin}\,(r_2) * |\text{r}3P_i^t - X_i^t|, r_4 < 0.5 \quad (1)$$

$$X_i^{(t+1)} = x_i^t + r_1 * \text{Cos}\,(r_2) * |\text{r}3P_i^t - X_i^t|, r_4 \geq 0.5 \quad (2)$$

In the above equations, x i^{th} refers to the current solution's position in the ith dimension and the ith iteration; r_1, r_2, r_3 and r_4 selected as random numbers between 0 and 1; Pi is the best destination point position in the ith dimension, and | | refers to the absolute value. The remaining parameters r_1, r_2, r_3 and r_4 require more elaboration due to their significance for the exploration and exploitation of the SCA. The r_1 parameter refers to the search circle radius. Through the iteration phase, r_1 can also be adaptively and dynamically varied using:

$$r_1 = a\frac{1-t}{T} \quad (3)$$

where
t = current iteration,
T = maximum iteration number,
a= constant.

Because of the cyclic nature of sine and cosine, the r_2 parameter determines the updating distance whether the movement is inward (the direction of exploitation when sine and cosine are negative) or outward (exploration direction when sine and cosine are positive), as shown in Figure 1 and Figure 2 .

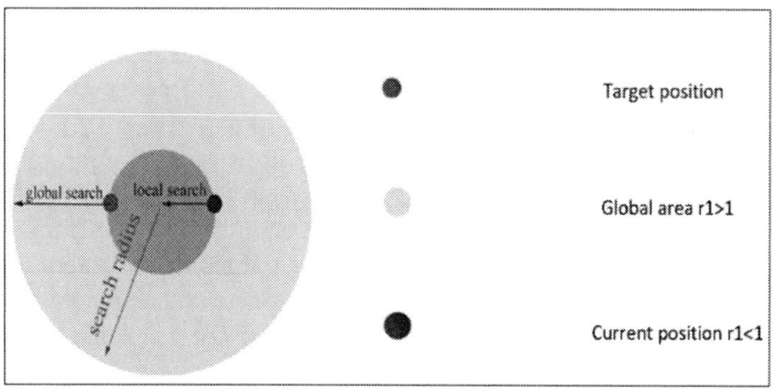

Figure 1. Effects of Sine and Cosine on search radius.

r_3 makes a suitable trade-off between emphasizing or deemphasizing the impact of desalination in the distance definition by allocating random weights to the best solution so far acquired. Finally
r_4 chose equation 1 or 2 depends on a random value between 0 and 1 if r4 less then 5.0 take equation 1 else take equation 2.

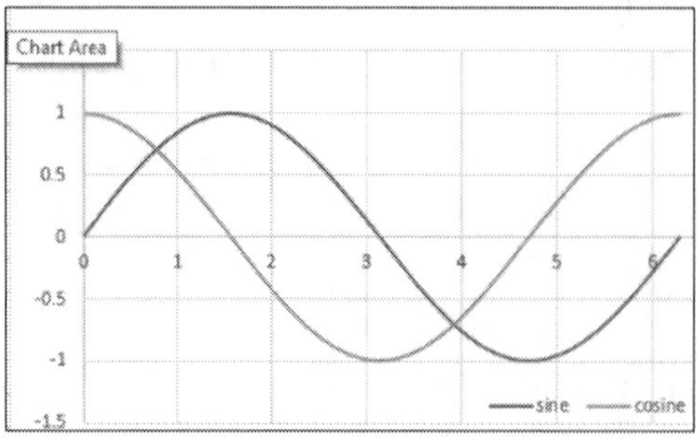

Figure 2. Sine and cosine with range of [-1,1].

To clarify, Algorithm 1 provides the general pseudo-code for the SCA.

3.1 Algorithm 1: SCA Pseudo Code

Initialize a set of search agents (solutions)
Initialize the parameter T (maximum number of iterations) and coefficient A
Select the fittest solution Xbest from the set of solutions
Initialize the iteration count t=0 While t< T If r_4 <0.5 go to
Update each solution vector with the help of equation (1)
Else Update each solution vector with the help of equation (2).
Update A value help with equation (3)
Compute the fitness of each updated solution vector
Update the coefficient A T=t+1

4. Results and Discussion

In the area of optimization using meta heuristics and evolutionary algorithms, several test cases should be used to determine the algorithm's efficiency. Our studies concentrate on two associated objectives: first characterizing the SCA's efficiency by repeating the experiment for choosing the best test ten times, and second benchmarking the SCA and other meta-heuristic approaches. The results show that the SCA introduced performs better than other meta-heuristic algorithms. We have divided our experiments with four groups to achieve these objectives. In table 1, we take a system configuration t=2 with 2,3,4,5,6 V-valued where parameters are varied from 3 to 12. In table 2 , we take a system configuration with t=3 and Value 3,4,5,6 and 7 where parameters are varied from 4 to 12. In the table 3, system configuration t=4 and V varies from 3 to 6 and parameters varies from 5 to 12 .all these experiments shows implement SCA with 200 iteration and we select the best one from the 10 experiments. As mentioned in table 4 , we made comparisons with some existing strategies as preliminary evaluation of SCA in terms of the generated test size. The strategies selected for comparison are Jenny [43], PICT [44], tConfig [45], IPOG-D, IPOG [46, 47], DPSO [48, 49], PSTG [50], and CS [15, 51]. We compare our strategy with others by taking parameters (varying from 3 till 12 with 3 values) as a system configuration and we changed the interaction strength (t) from 2 to 11. Table 4 demonstrates the results obtained for the above experiment. The darkened cells show the better results obtained for a specific test configuration. Cells marked NS (not supported) indicate that the tool could not generate the test case for a specific configuration. And cells marked Day means it took a long time. Concerning the size of the generated test suite, Jenny, TConfig, PICT and PSTG did not produce good results with an optimum test suite. Nonetheless T=7 IPOG, IPOG-D, PSTG and CS were not supported. We note that DPSO and SCA produced optimal results. Where t>9 DPSO took a long time. Finally, we can note that our SCA strategy compares well against other strategies). As mentioned, that SCA has produced competitive test suites for most strategies and tools.

Table 1. Generated Test Suite Size With T=2.

Configuration T=2 Iteration 200	1st	2nd	3rd	4th	5th	6th	7th	8th	9th	10th	Best	Mean
CA(N; t, 3^3)	9	10	10	9	10	10	9	10	10	9	9	9.7
CA(N; t, 3^4)	12	10	10	10	10	9	12	10	9	10	9	10.2
CA(N; t, 3^5)	11	13	13	14	13	14	12	12	13	14	11	12.8
CA(N; t, 3^6)	15	15	14	15	15	15	14	15	14	14	13	14.7
CA(N; t, 3^7)	16	16	16	16	15	15	14	15	15	16	14	15.4
CA(N; t, 3^8)	17	16	16	16	16	19	15	16	16	15	15	16.3
CA(N; t, 3^9)	16	17	17	16	17	17	17	16	17	17	16	16.7
CA(N; t, 3^10)	18	17	17	18	17	18	17	19	18	18	17	17.7
CA(N; t, 3^11)	17	18	18	19	18	19	18	18	19	18	17	18.2
CA(N; t, 3^12)	18	17	18	18	19	18	19	19	19	19	17	18.3
CA(N; t, 4^10)	30	30	30	30	30	31	30	30	30	30	30	30.1
CA(N; t, 5^10)	48	48	48	46	47	47	46	47	48	47	46	47.2
CA(N; t, 6^10)	67	66	66	66	67	66	66	66	66	67	66	66.2
CA(N; t, 2^7)	8	8	7	7	7	7	7	7	8	7	7	7.3
CA(N; t, 3^7)	15	15	15	15	15	15	15	15	15	15	15	15.0
CA(N; t, 4^7)	25	26	24	25	28	26	26	25	26	26	24	25.7
CA(N; t, 5^7)	39	41	38	37	40	39	39	38	39	39	37	38.9
CA(N; t, 6^7)	54	56	54	54	54	56	55	56	57	54	54	55.1
CA(N; t, 7^7)	74	74	75	73	74	74	73	75	73	76	73	73.9

Table 2. Generated Test Suite Size With T=3.

Configuration T=3 Iteration 200	1st	2nd	3rd	4th	5th	6th	7th	8th	9th	10th	Best	Mean
CA(N; t, 3^4)	32	34	31	31	31	29	33	32	30	27	27	31.4
CA(N; t, 3^5)	41	40	41	40	39	41	41	41	43	40	39	40.8
CA(N; t, 3^6)	44	43	47	45	46	45	44	45	47	45	43	45.1
CA(N; t, 3^7)	50	51	50	50	51	52	49	50	51	52	49	50.4
CA(N; t, 3^8)	53	55	56	55	54	54	53	55	54	54	53	54.3
CA(N; t, 3^9)	59	59	58	60	58	57	59	60	59	58	57	58.8
CA(N; t, 3^10)	62	63	62	62	64	63	62	62	62	64	62	62.4
CA(N; t, 3^11)	65	65	66	64	67	66	66	66	67	66	64	65.8
CA(N; t, 3^12)	71	68	70	69	67	68	69	68	70	71	67	68.9
CA(N; t, 2^7)	15	15	14	15	15	16	12	14	14	15	12	14.5
CA(N; t, 4^7)	118	119	120	120	118	117	118	118	121	118	117	118.5
CA(N; t, 5^7)	231	228	229	232	228	226	227	231	229	231	226	229.0
CA(N; t, 6^7)	396	398	392	390	394	393	398	395	388	395	388	394.5
CA(N; t, 7^7)	622	617	621	615	626	618	618	622	623	622	615	619.9

Table 3. Generated Test Suite Size With T=4.

Configuration T=4 Iteration 200	1st	2nd	3rd	4th	5th	6th	7th	8th	9th	10th	Best	Mean
CA(N; t, 3^5)	103	100	98	98	101	100	106	97	103	96	96	100.7

IOP Conf. Series: Materials Science and Engineering **767** (2020) 012009 doi:10.1088/1757-899X/767/1/012009

CA(N; t, 3^6)	138	134	133	135	130	132	134	136	131	137	130	134.1
CA(N; t, 3^7)	155	155	158	159	154	158	158	156	157	155	152	156.7
CA(N; t, 3^8)	176	176	176	174	175	174	177	176	177	176	174	175.7
CA(N; t, 3^9)	193	195	195	196	197	196	191	194	193	195	191	194.4
CA(N; t, 3^10)	214	214	210	211	211	209	213	216	215	214	209	212.6
CA(N; t, 3^11)	230	229	230	227	231	229	232	231	230	231	227	229.9
CA(N; t, 3^12)	250	247	249	250	246	253	249	251	251	248	246	249.6
CA(N; t,4^10)	730	735	726	728	729	733	726	734	734	730	726	730.6
CA(N; t,5^10)	1750	1741	1749	1746	1752	1736	1738	1744	1742	1746	1736	1744.2
CA(N; t,6^10)	3495	3590	3488	3489	3491	3497	3496	3495	3499	3494	3488	3504.4

Table 4. Size performance for CA(N; 4, 3P) where k is varied from 3 to 12.

T	P	V	Jenny	tConfig	PICT	IPOG-D	IPOG	DPSO	PSTG	CS	SCA
2	3	3	9	10	10	15	9	9	9	9	9
2	4		13	10	13	15	9	9	9	9	9
2	5		14	14	13	15	15	11	12	11	12
	6		15	15	14	15	15	14	13	13	13
	7		16	15	16	15	15	14	15	14	15
	8		17	17	16	15	15	15	15	15	15
	9		18	17	17	15	15	15	17	16	16
	10		19	17	18	21	15	16	17	17	17
	11		17	20	18	21	17	17	17	18	17
	12		19	20	19	21	21	16	18	18	17
3	4	3	34	32	34	27	32	27	27	28	27
	5		40	40	43	45	41	41	39	38	38
	6		51	48	48	45	46	33	45	43	43
	7		51	55	51	50	55	48	50	48	48
	8		58	58	59	50	56	52	54	53	52
	9		62	64	63	71	63	56	58	58	56
	10		65	68	65	71	66	59	62	62	62
	11		65	72	70	76	70	63	64	66	64
	12		68	77	72	76	73	65	67	70	67
4	5	3	109	97	100	162	97	81	96	94	96
	6		140	141	142	162	141	131	133	132	131
	7		169	166	168	226	167	150	155	154	150
	8		187	190	189	226	192	171	175	173	171
	9		206	213	211	260	210	187	195	195	191
	10		221	235	231	278	233	206	210	211	206
	11		236	258	249	332	251	221	222	229	227
	12		252	272	269	332	272	237	244	253	246
6	7	3	1089	921	1015	1201	921	729	977	973	957
	8		1466	1515	1455	1763	1493	1409	1402	1401	1398
	9		1840	1931	1818	2526	1889	1682	1684	1689	1698
	10		1260	Day	2165	2834	2262	1972	1980	2027	1980
	11		2459	Day	2496	3886	2607	2250	2255	2298	2282
	12		2757	Day	2815	4087	3649	2512	2528	2638	2537
7	8		3402	day	3143	Ns	Ns	2250	Ns	Ns	3020
	9		4746	day	4618	Ns	Ns	4451	Ns	Ns	4414
	10		6030	day	5884	ns	Ns	5443	Ns	ns	5443
	11		7194	Day	7116	N	Ns	6471	Ns	Ns	6467
	12		8296	Day	8314	Ns	Ns	Day	Ns	Ns	7615
8	9		10572	day	9763	Ns	Ns	6667	Ns	Ns	9302
	10		14999	Day	14599	Ns	Ns	13933	Ns	Ns	13956
	11		19369	Day	18859	Ns	Ns	Day	Ns	Ns	17800

	12	23620	Day	23112	Ns	Ns	Day	Ns	Ns	21670
9	10	32474	Day	30181	Ns	Ns	21466	Ns	Ns	28620
	11	47009	Day	45521	Ns	Ns	Day	Ns	Ns	43834
	12	61764	Day	59966	Ns	Ns	Day	Ns	Ns	56480
10	11	99560	Day	92435	Ns	Ns	Day	Ns	ns	87789
	12	147004	Day	141990	Ns	Ns	Day	Ns	Ns	136091
11	12	305797	Day	278993	Ns	Ns	Day	Ns	Ns	267700

5. Conclusion and Future Directions

In this paper, we have mentioned an experimental study on meta-heuristic strategy and acceptance mechanism for the generation of combinatorial (t-way tests). The SCA evaluation focuses on the strategy's effectiveness in generating a better size of test suites against current strategies that support uniform.

As shown in table 8 their different sets of experiments were implemented. The main aim here was to show that the strategy was sufficiently competitive compared to other strategies in terms of the generated test suite size. Finally, the results show that in supporting uniform interaction, SCA can compete with other current strategies.

References

[1] Kuhn, D.R. and M.J. Reilly. An investigation of the applicability of design of experiments to software testing. in 27th Annual NASA Goddard/IEEE Software Engineering Workshop.

[2] Kuhn, D.R., D.R. Wallace, and A.M. Gallo, Software fault interactions and implications for software testing. IEEE transactions on software engineering, 2004. 30(6): p. 418-421.

[3] Hariri, F., et al. Comparing Mutation Testing at the Levels of Source Code and Compiler Intermediate Representation 12th *IEEE Conference on Software Testing, Validation and Verification (ICST) 2019.*

[4] Papadakis, M., et al., Mutation testing advances: an analysis and survey *Advances in Computers. 2019* Elsevier. p. 275-378.

[5] Chen, T.Y., S.C. Cheung, and S.M. Yiu, Metamorphic testing: a new approach for generating next test cases. 1998, Technical Report HKUST-CS98-01, Department of Computer Science, Hong Kong.

[6] Ma, L., et al., Combinatorial testing for deep learning systems. arXiv preprint arXiv:1806.07723, 2018.

[7] Ghandehari, L.S., et al., A combinatorial testing-based approach to fault localization. *IEEE Tran. Software Engineering*.

[8] Lee, C.K.H., A review of applications of genetic algorithms in operations management *Eng. Applications of Artificial Intelligence* **76** 1-12.

[9] Jain, N., U. Nangia, and J. Jain, A review of particle swarm optimization. *J. Inst. Eng. (India)*: Series B 2018. 99(4) 407-411.

[10] Zhang, T. and Z.W. Geem, Review of harmony search with respect to algorithm structure. Swarm and Evolutionary Computation 2019 **48** 31-43.

[11] Dorigo, M. and T. Stützle, Ant colony optimization: overview and recent advances *Handbook of metaheuristics*. 2019, Springer. p. 311-351.

[12] Streif, M. and M. Leib, Comparison of QAOA with Quantum and Simulated Annealing. arXiv preprint arXiv:1901.01903, 2019.

[13] Abdel-Basset, M. and L.A. Shawky, Flower pollination algorithm: a comprehensive review. Artificial Intelligence Review, 2018: p. 1-25.

[14] Joshi, A., et al., Cuckoo Search Optimization-A Review. Materials Today: Proceedings, 2017. 4(8): p. 7262-7269.

[15] Ahmed, B.S., T.S. Abdulsamad, and M.Y. Potrus, Achievement of minimized combinatorial test suite for configuration-aware software functional testing using the cuckoo search algorithm. Information and Software Technology, 2015. 66: p. 13-29.

[16] Zamli, K.Z., et al., An experimental study of hyper-heuristic selection and acceptance mechanism for combinatorial t-way test suite generation. Information Sciences, 2017. 399: p. 121-153.

[17] Mirjalili, S., SCA: a sine cosine algorithm for solving optimization problems. Knowledge-Based Systems, 2016. 96: p. 120-133.

[18] Ammann, P. and J. Offutt, Introduction to software testing. 2016: Cambridge University Press.

[19] Calvagna, A. and A. Gargantini. IPO-s: incremental generation of combinatorial interaction test data based on symmetries of covering arrays *International Conference on Software Testing, Verification, and Validation Workshops* 2009.

[20] Yu, L., et al. Acts: A combinatorial test generation tool *IEEE Sixth International Conference on Software Testing, Verification and Validation* 2013.

[21] Ghandehari, L.S., et al. BEN: A combinatorial testing-based fault localization tool *IEEE Eighth International Conference on Software Testing, Verification and Validation Workshops (ICSTW) 2015.*

[22] Yuan, X., M.B. Cohen, and A.M. Memon, GUI interaction testing: Incorporating event context *IEEE Tran. Soft. Eng.* 2010 37(4): 559-574.

[23] Yuan, X., M. Cohen, and A.M. Memon. Covering array sampling of input event sequences for automated GUI testing *Proc. twenty-second IEEE/ACM international conference on Automated software engineering* 2007.

[24] Souto, S. and M. d'Amorim *Time-space efficient regression testing for configurable systems. Journal of Systems and Software* 2018. 137: p. 733-746.

[25] Torres-Jimenez, J. and I. Izquierdo-Marquez Survey of covering arrays 1*5th International Symposium on Symbolic and Numeric Algorithms for Scientific Computing* 2013.

[26] Lawrence, J., et al., A survey of binary covering arrays *elect. j. of combinatorics* 2011 18(1): p. 84.

[27] Pradhan, D., et al., Search-based test case implantation for testing untested configurations. Information and Software Technology, 2019. 111: p. 22-36.

[28] Nasser, A.B. and K.Z. Zamli. Self-adaptive Population Size Strategy Based on Flower Pollination Algorithm for T-Way Test Suite Generation. in International Conference of Reliable Information and Communication Technology. 2018. Springer.

[29] Soh, Z.H., et al., Code Coverage of T-way Test Suite Data on Distributed Environment. Int. J. Advance Soft Compu. Appl, 2018. 10(1).

[30] Fujimoto, S., H. Kojima, and T. Tsuchiya, Controlling Occurrence Frequencies of Parameter Values in Pair-Wise Testing. International Journal of Software Engineering and Knowledge Engineering, 2018. 28(07): p. 983-1000.

[31] Koc, U. and C. Yilmaz, Approaches for computing test-case-aware covering arrays. Software Testing, Verification and Reliability, 2018. 28(7): p. e1689.

[32] Drave, I., et al. Model-Based Testing of Software-Based System Functions. in 2018 44th Euromicro Conference on Software Engineering and Advanced Applications (SEAA). 2018. IEEE.

[33] Kleine, K. and D.E. Simos, An efficient design and implementation of the in-parameter-order algorithm. Mathematics in Computer Science, 2018. 12(1): p. 51-67.

[34] Zamli, K.Z., A.A. Alsewari, and M.I. Younis, T-Way Testing Strategies: Issues, Challenges, and Practices, in Computer Systems and Software Engineering: Concepts, Methodologies, Tools, and Applications. 2018, IGI Global. p. 2011-2024.

[35] Ahmed, B.S., Test case minimization approach using fault detection and combinatorial optimization techniques for configuration-aware structural testing. Engineering Science and Technology, an International Journal, 2016. 19(2): p. 737-753.

[36] Choi, E.-H., et al. Test effectiveness evaluation of prioritized combinatorial testing: A case study. in 2016 IEEE International Conference on Software Quality, Reliability and Security (QRS). 2016. IEEE.

[37] Wang, Z., B. Xu, and C. Nie. Greedy heuristic algorithms to generate variable strength combinatorial test suite. in 2008 The Eighth International Conference on Quality Software. 2008. IEEE.

[38] Ahmed, B.S. and K.Z. Zamli, A variable strength interaction test suites generation strategy using particle swarm optimization. Journal of Systems and Software, 2011. 84(12): p. 2171-2185.

[39] Alsewari, A.A. and K.Z. Zamli. Interaction test data generation using harmony search algorithm. 2011 IEEE Symposium on Industrial Electronics and Applications.

[40] Cohen, M.B., et al. A variable strength interaction testing of components *Proceedings 27th Annual International Computer Software and Applications Conference. COMPAC 2003.*

[41] Chen, X., et al. Variable strength interaction testing with an ant colony system approach. in 2009 16th Asia-Pacific Software Engineering Conference. 2009. IEEE.

[42] Alsewari, A.R.A. and K.Z. Zamli, Design and implementation of a harmony-search-based variable-strength t-way testing strategy with constraints support. Information and Software Technology, 2012. 54(6): p. 553-568.

[43] Zamli, K.Z., et al., Design and implementation of a t-way test data generation strategy with automated execution tool support. Information Sciences, 2011. 181(9): p. 1741-1758.

[44] Czerwonka, J., Pairwise testing in the real world: Practical extensions to test-case scenarios. Microsoft Corporation, Software Testing Technical Articles.

[45] Williams, A.W., Determination of test configurations for pair-wise interaction coverage *Testing of Communicating Systems* 2000 Springer p. 59-74.

[46] Lei, Y., et al., IPOG/IPOG-D: efficient test generation for multi-way combinatorial testing *Software Testing, Verification and Reliability,* 2008. 18(3): p. 125-148.

[47] Lei, Y., et al. IPOG: A general strategy for t-way software testing *14th Annual IEEE International Conference and Workshops on the Engineering of Computer-Based Systems (ECBS'07).*

[48] Omran, M., A.P. Engelbrecht, and A. Salman, Particle swarm optimization method for image clustering *International Journal of Pattern Recognition and Artificial Intelligence 2005* 19(03): p. 297-321.

[49] Wu, H., et al., A discrete particle swarm optimization for covering array generation. IEEE Transactions on Evolutionary Computation 2014. 19(4): p. 575-591.

[50] Ahmed, B.S. and K.Z. Zamli. PSTG: a t-way strategy adopting particle swarm optimization Fourth Asia International Conference on Mathematical/Analytical Modelling and Computer Simulation 2010

[51] Li, X. and M. Yin, Modified cuckoo search algorithm with self adaptive parameter method *Information Sciences* 2015. 298: p. 80-97.

IOP Conf. Series: Materials Science and Engineering **767** (2020) 012010 doi:10.1088/1757-899X/767/1/012010

Numerical Study of Fiber Reinforced Polymer Reinforced Normal Strength Concrete (FRPNSC) under Hydrocarbon Fire

N I M Shukeri[1,], **N Rusli**[1]

[1] Institute of Engineering Mathematics, Universiti Malaysia Perlis (UniMAP), Kampus Pauh Putra, 02600 Arau, Perlis, MY.

izhamshukeri@gmail.com

Abstract. Main safety requirements in concrete structural are the fire resistance requirements. One of the structural components is Fiber Reinforced Polymer Reinforced Normal Strength Concrete (FRPNSC). FRP reinforcement has been used as the replacement for conventional steel due to anti-corrosion and lightweight characteristics. Severe degradation on chemical bond properties for FRP will be effected when the temperature is rises. It is important to understand the minimum concrete cover thickness and concrete aggregates types to achieve fire resistance requirements. Standard fire equations are commonly used for fire simulations study. However, studies with hydrocarbon (HC) fire equations which fire ignited from petrochemical are limited. Therefore, in order to study the fire resistance of FRPNSC under hydrocarbon fire, temperature at the reinforcement needs to be predicted. In this study, explicit finite difference method (EFDM) used to solve the heat transfer model. The numerical algorithm of EFDM heat transfer model was constructed and used to analyse the concrete thickness and aggregates to achieve fire resistance requirement. The temperature result obtained by the EFDM model successfully validated with test data. FRPNSC under HC for carbonate aggregates give significant effect on the fire resistance compared to standard fire. The carbonate aggregates types also shows better fire performances compared to lightweight aggregates.

1. Introduction

Concrete is the major materials used for building structures. Its advantages compared to among other building materials are on fire-resistive properties. However, concrete structures must still need to be designed for fire effects. The design should consider the structural components which must be able to withstand dead and live loads without collapse even though the rise in temperature causes a decrease in the strength and modulus of elasticity for concrete [1]. Normal Strength Concrete (NSC) was used widely in building constructions. The conventional concrete mixture, NSC types used in reinforcement is classified into three different types, which is carbonate, lightweight and siliceous aggregate. Carbonate aggregate types have been reported as intermediate range in fire resistance performance, while lower performance on fire resistance has been specified to siliceous and lightweight aggregate due to their heat thermal properties [2].

The concrete structure that reinforced by FRP called as FRP reinforced normal strength concrete (FRPNSC). FRP composites are made by incorporating two or more materials into one with the intent of suppressing undesirable properties by fabricated a polymer matrix, for example epoxy, vinylester, or polyester and reinforced with various types of carbon, glass, and/or aramid fibers [3]. The application of FRP reinforcement in concrete replace the traditional steel reinforcement become interesting in civil engineering members [4]. The advantages of FRP over steel such as strength, anti-corrosion, and durables have shown better integrity on concrete strength. FRP reinforcement also has enhanced the flexural and shear capacity of the concrete structure. In term of cost, FRP has better values compared to steel, then increase the reliability of FRP reinforcement in civil engineering applications [5]. Similar to other building structures, FRP reinforced structural members should also design materials to fulfill the fire resistance requirements and structural requirements based on building codes [6]. The major issue concern is about the weakness of FRP deal with high temperature. The glass transitions temperature (T_g) for FRP within 65°C-82°C where the FRP begin to change its properties such as bond degradation. The polymer resin will be soft and rubbery after the temperature exceeds the glass transition temperature [7]. This will give the impact to concrete reinforcement and the stability of concrete.

In the fire resistance studies, the structural components which used to strengthen the concrete such as FRP also must be taken into account. Most of the fire performances studies applied the (American Society Testing Materials) ASTM standard fire as guidelines for the fire sources. The test procedure was performed by using ASTM fire are widely used among researchers in the world. The fire resistance for FRPNSC defined as where the materials able to withstand the temperature rising before it reach 250°C within 2 hours [8]. At this temperature, the material properties of FRP in the reinforcement reported will change and give a significant effect on the whole of concrete stability. Previous experimental study was successfully conducted by [9] to clarify the fire performance of FRPNSC slab under ASTM fire and provide fire resistance guidelines for one of Japan company, NEFCOM. Beside types of concrete, another factor such as types of fire also gives a significant effect on concrete during fire. For example, a fire which ignited by fuel, petroleum or chemical will produce hydrocarbon (HC) fire [10]. The temperature of HC fire increase over time is different compared to standard fire [11], thus the fire resistance performances also different.

In fire performances studies, temperature across concrete must be determined. However, it is difficult to measure the temperature across the concrete structure with various factors such as types of fire, concrete thickness and aggregate types of concrete [12]. Previous research for theoretical and experimental had been conducted to understand the fire resistance requirements for FRPNSC materials but the factors improve over years. Its time and cost problems to conducts multiple experiments.

In order to encounter these problems, the temperature across the concrete will be represented by heat transfer equations. The numerical method based on explicit finite difference method (EFDM) had been used to solve the heat transfer equation and predict the temperature across FRPNSC [2, 8, 13]. Thermal properties for NSC have been set up in EFDM models and was successfully produced the temperature across FRPNSC under standard fire. However, in the previous models, the authors did not consider HC fire as the source of temperature. The experimental studies for FRPNSC under HC fire are also limited in previous study since hard to simulate the real HC fire. In addition, the numerical study for the new FRPHSC also limited.

Therefore, the fire-resistance study for FRPNSC and FRPHSC under HC fire will be conducted. In this present study, 1D heat transfer equations will be used to represent the heat transfer across concrete slab during the fire. Explicit Finite Difference Method (EFDM) models will be used to solve 1D heat transfer equations to predict the temperature across FRPNSC thickness. The temperature result obtained from the models will be used to determine the required concrete thickness for FRPNSC achieved the fire resistance. Finally, the summary between types of fire and concrete thickness data will be used to provide the new fire resistance guidelines for FRPNSC.

2. Governing heat transfer equations

The heat transfer across concrete structure can be represented by the heat transfer equations. The following governing Equation (1) obtained by considering that the net rate of heat flow should be equal to the rate of internal energy increase. The heat flow was given by the Fourier law related to temperature gradients presented by [2],

$$\frac{\partial}{\partial x}\left(k\frac{\partial T}{\partial x}\right)+q=\rho C_p\frac{\partial T}{\partial t} \quad (1)$$

IOP Conf. Series: Materials Science and Engineering **767** (2020) 012010 doi:10.1088/1757-899X/767/1/012010

where k is the thermal conductivity of the concrete, T is the temperature, t is time, ρ is the material's density, C_p is the heat capacity and q is a term representing the heat generation within the material. The physical significance of the terms in Equation (1) presents the mathematical description of the conservation energy principles. On the left-hand side, the terms represent heat transfer into or out of the differential of plane due to heat conduction while on the right hand sides represent the energy stored generated by heat. The Equation (1) can be used by right simplication and boundary conditions to optimize the usage of heat transfer equations in this study. The contributions of heat transfer in internal reinforcing steel assumed to be minimal and neglected due to more portion of concrete than steel.

The change in energy was stored and represented as (ΔQ_{store}) in difference of time on the right hand-side of equation. At the element surface, the energy changes must be equal to the difference between the heats into the element. It is because the radiation process represented by heat generation (Q_{gen}) and the heat out of the element (Q_{cond}) caused by the conduction in the x-direction. The difference presented as,

$$\Delta Q_{store} = Q_{gen} + Q_{cond} \tag{2}$$

In Equation (2), the convection is ignored as one of mechanism for heat transfer from surrounding to the surface of slab. The convection contributes less than 10% for heat transfer at the surface of slab in standard fire endurance tests [14]. Therefore, it can be ignored for numerical fire modelling. The Equation (1) and Equation (2) can be described as,

$$\Delta Q_{store} = \rho C_p \frac{\partial T}{\partial t} \tag{3}$$

and can be rewrite as,

$$\Delta Q_{store} = \rho\, C_p \frac{\Delta T}{\Delta t} \tag{4}$$

where ΔT is the difference change in time interval Δt, $\rho\, C_p$ is the specific heat capacity of concrete (based on the aggregate types). For a differential across the element, the heat conduction equation, for x-direction by using the relation presented in Equation (5),

$$\Delta Q^{x}_{cond} = -kA\frac{\partial T}{\partial x} \tag{5}$$

Since the heat transfer problem is one-dimensional, the conduction carries out the heat from elementary layer of material per unit time is given by,

$$\Delta Q^{x}_{cond} = -kA\frac{\Delta T}{L} \tag{6}$$

where k is the thermal conductivity, surface area A and thickness L. The heat transferred into a surface element with surface area A, due to radiation can be approximated as follows

$$Q_{gen} = A\,\sigma\varepsilon_{surf}\left[\left(T_{surr}+273\right)^4-\left(T_{surf}+273\right)^4\right] \tag{7}$$

where σ is the constant of Stephan-Boltzman, is the emissivity of the slab surface, is the surroundings temperature, and is the temperature surface being heated.

3. Numerical Method

3.1. Discretization Process for Time Derivatives

For the surface element, the time derivation in Equation (1) is discretized with finite difference scheme given in [8] for (ΔQ_{store}) presented as,

$$\Delta Q_{store} = \rho\, C_p \frac{\left[T_i^{j+1} - T_i^j\right]}{\Delta t} \tag{8}$$

where T_1^{j+1} is the temperature for the element of current time step and T_1^j is the temperature for the element of previous time step.

3.2. Discretization Process for Spatial Derivatives

From Equation (5), A is equal to 1 and the spatial derivatives which consider heat conduction along x-direction is as in [12] which given by

$$\Delta Q_{cond}^{x} = -\left[\frac{k_i^j + k_{i+1}^j}{2}\right]\left[\frac{T_i^j - T_{i+1}^j}{(\Delta x)^2}\right] \tag{9}$$

where T_{i+1}^j the temperature for elementary layer next to surface element at previous time step k_i^j, and k_{i+1}^j are the thermal conductivities of the surface and elementary layer next to surface at their previous time step respectively.

3.3. Initial Conditions, Boundary Conditions and Time Steps

Initial conditions provided in this study follow from the previous successful study [15], which is 20°C.

$$T(x,0) = 20°C \tag{10}$$

In this study, the temperature is in elevated situation, which will give different values of boundary conditions. Radiation from the fire, transfer the heat to boundary at room temperature. After several arrangements, two equations are obtained to predict the temperature at the boundary. For the boundary exposed by fire at $x = 0$, the formulation presented in Equation (11), where ε_n is the emissivity of concrete given by 0.9 and σ is Stefan Boltzmann constant given by 5.67 x 10^{-8} W·m $^{-2}$·K $^{-4}$.

$$T_i^{j+1} = T_1^j + \frac{2\sigma\varepsilon_n}{(\rho C_p)_1^j \Delta x}\left[\left(T_f^j + 273\right)^4 - \left(T_1^j + 273\right)^4\right]\Delta t$$
$$- \frac{\left(k_1^j + k_2^j\right)}{(\rho C_p)_1^j \left(\Delta x\right)^2}\left(T_1^j - T_2^j\right)\Delta t \tag{11}$$

On the other side, FRPNSC has been exposed to air. The convective process between concrete and air situation at $x = m$ are imposed by the following equations

$$T_m^{j+1} = T_m^j + \frac{\left(k_{m-1}^j + k_m^j\right)}{(\rho C_p)_m^j (\Delta x)^2}\Delta t - \frac{2\sigma\varepsilon_n\left[\left(T_m^j + 273\right)^4 - \left(T_0^j + 273\right)^4\right]\Delta t}{(\rho C_p)_m^j \Delta x}$$
$$+ \frac{\gamma\left(T_m^j - T_0^j\right)^{1.25}}{(\rho_n c_n)_m^j \Delta x} \tag{12}$$

where

$$\gamma = 3600(1.32)\left[\frac{\left(T_m^j - T_0\right)}{10.7}\right]^{0.25} \tag{13}$$

represents the coefficient of convective heat transfer between concrete to air. Both Equations (11) and (12) will be used to obtain the boundary temperatures by taking the ρC_p as the thermal capacity for concrete.

Common fire resistance study will use standard fire ASTM as the source of temperature. However, in different cases, this study consider the fire ignited by fuel, petroleum or chemical where hydrocarbon fire will be produced and increased over time [10]. In 2000, the hydrocarbon gas exploded and ignited the fire which destroyed the structure of British Petrol (BP) [16]. Therefore, the standard fire ASTM is not reliable to be applying for all situations. The application of HC fire curve has been presented as follows:

$$T_{hcf} = 20 + 1080 \times (1 - 0.325 \times e^{-0.167t} - 0.675 \times e^{-2.5t}) \tag{14}$$

where T_{hcf} is the temperature from HC fire and t time in hours. Therefore, the expression in Equations (14) will be used in this study.

4. Test problem

4.1. Validation of FRPNSC under ASTM fire

The validation process for this numerical model performed against experimental data has been done in our previous works [17]. The duration of fire follows the fire protection guidelines which is 2 hours protection [9]. 120 mm FRPNSC slab sample has been exposed to standard fire [11] for 120 minutes. Thermal properties relations for Carbonate aggregate have been applied in this model.

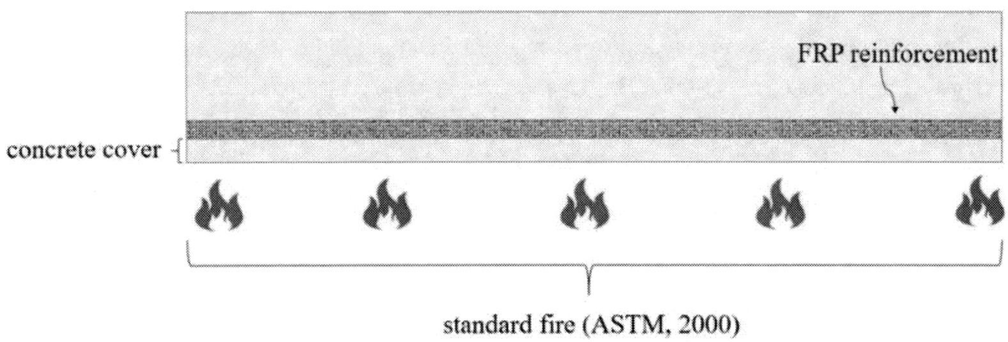

Figure 1. FRPNSC slab with carbonate aggregates under standard fire

The temperature predicted through the concrete thickness had been compared with the test data [9]. In order to avoid concrete spalling, the temperature will be simulated at minimum thickness of concrete.

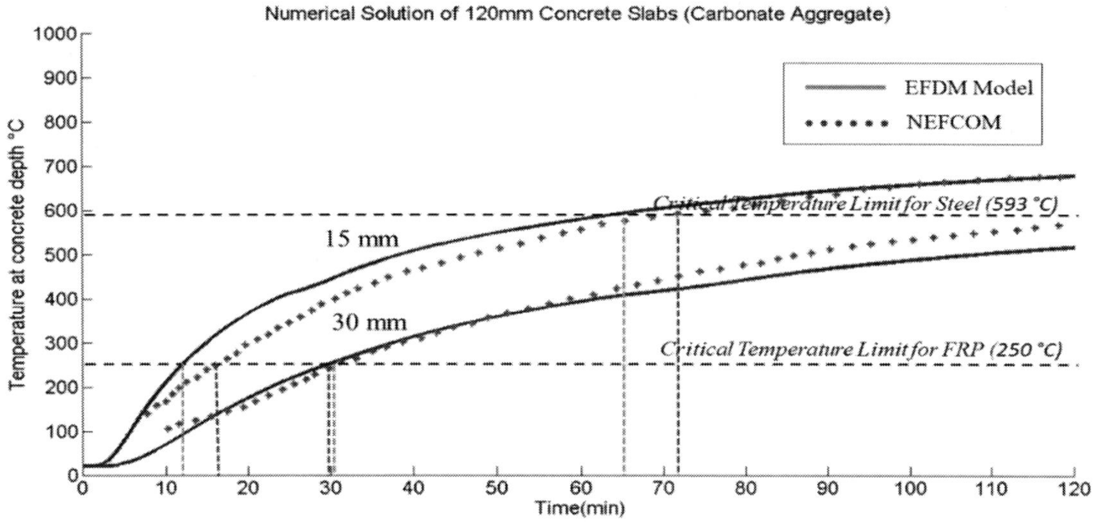

Figure 2. The validation of EFDM model against NEFCOM data (FRPNSC under Standard Fire)

From figure 2, small discrepancies between EFDM against NEFCOM data have been shown during 120 minutes of simulations. The critical temperature limit for FRP at 250 °C has been reached within 12.12 minutes and 29.732 minutes for 15 mm and 30 mm concrete cover thickness respectively. The percentage of errors at 250 °C are 24.72 % for 15 mm concrete cover thickness and 4.41 % for 30 mm concrete cover thickness. EFDM model has overestimated the temperature for 15 mm concrete cover thickness.

EFDM has predicted the material failure occurs at steel temperature limit, 593 °C in 64.97 minutes after heating for 15 mm concrete cover. While for 30 mm depth of concrete, the steel temperature limit does not even reach. It is because of the larger concrete cover thickness factors that delay the heat transfer across materials. A good agreement between numerical and test data through concrete cover thickness shown in figure 2. The validation was successfully conducted and EFDM model able to be used for further discussion. The investigations continue to predict the temperature for FRPNSC with carbonate and lightweight aggregate types under HC fire.

4.2. Simulation of FRPNSC under HC fire

The EFDM models has been used to predict the temperature of FRPNSC under HC fire. The results show that the carbonate aggregates types show poor fire performance under HC compare to the simulation under standard fire (figure 1). The poor fire performances causes by the excessive heat generation produce by HC fire compare to standard fire. 3 different depth of concrete applied; and show the critical temperature limit for FRP reach at 5.92, 19.84 and 42.72 minutes respectively.

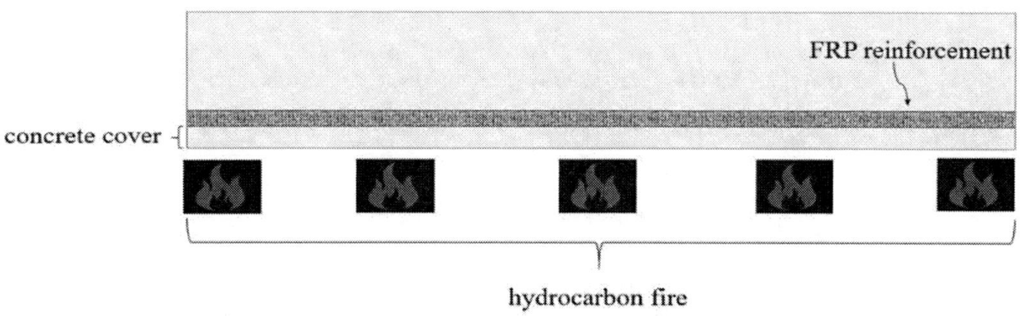

Figure 3. FRPNSC slab with carbonate aggregates under HC fire

IOP Conf. Series: Materials Science and Engineering **767** (2020) 012010 doi:10.1088/1757-899X/767/1/012010

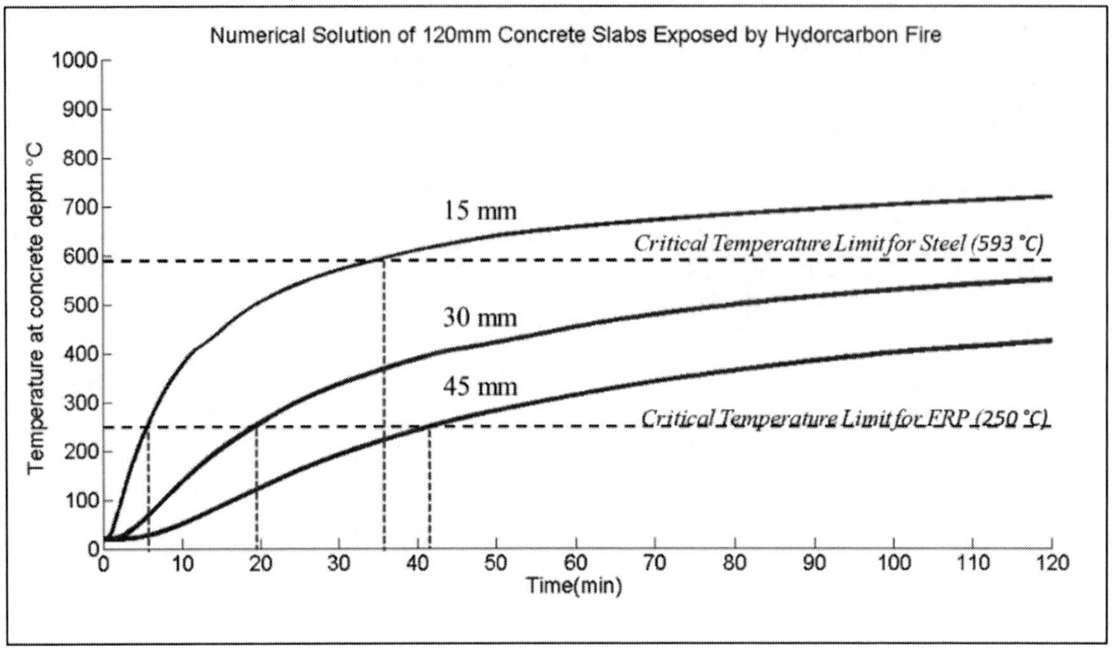

Figure 4. Temperature at concrete depth of FRPNSC under hydrocarbon fire (Carbonate aggregates)

For material failure at critical temperature of steel in 36.32 minutes for 15 mm, while for 30 mm and 45 mm concrete cover, the temperature does not reach within 120 minutes under the exposure of HC fire. Therefore, 30 mm concrete cover thickness is enough to protect the structural within 120 minutes under HC fire. Furthermore, EFDM also used to simulate the temperature for lightweight aggregates concrete slab.

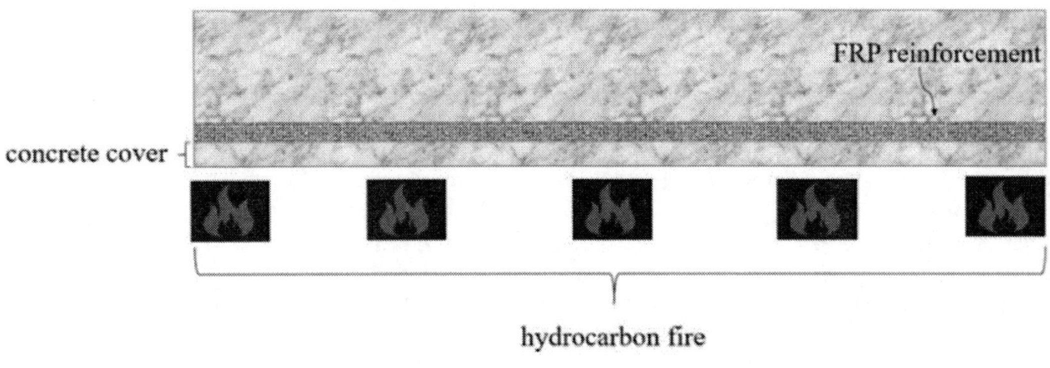

Figure 5. FRPNSC slab with carbonate aggregates under HC fire

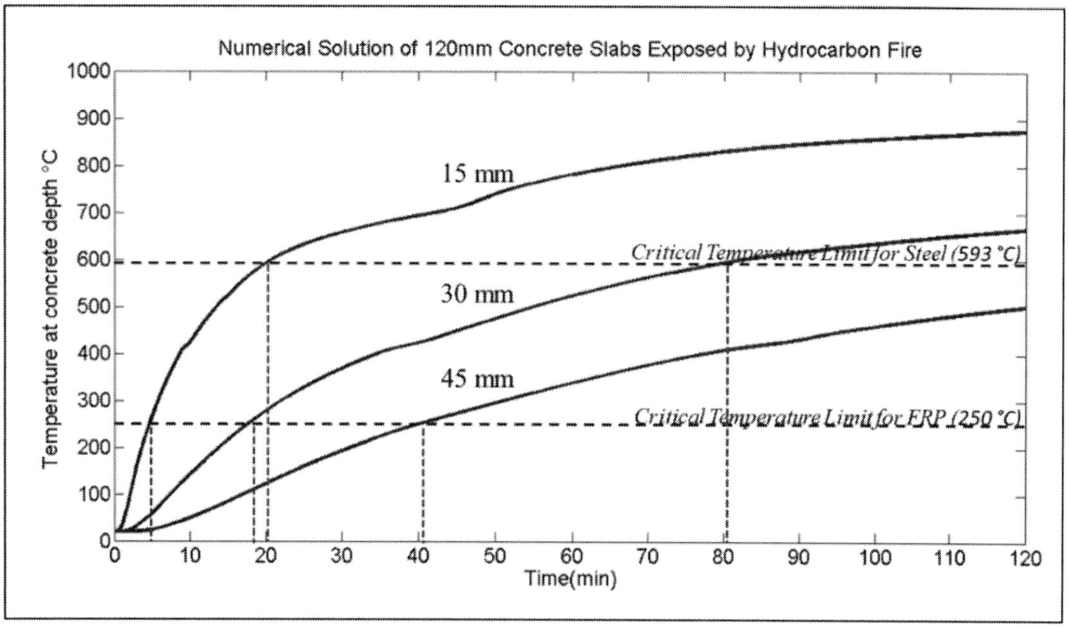

Figure 6. Temperature at concrete depth of FRPNSC under HC fire (Lightweight aggregates)

In figure 6 shows that, the FRP start to degrade at 15 mm, 30 mm and 45 mm within 4.80, 18.24 and 41.12 minutes respectively. The temperature increases rapidly across the concrete thickness.

From all concrete thickness presented, only 45 mm concrete cover thickness does not reach critical temperature of steel. Therefore, it can be determined as the minimum concrete cover thickness needed for lightweight aggregate of FRP RC exposed with HC fire.

5. Conclusion

In conclusion, the EFDM in this study has been successfully predicted the temperature across FRPNSC slab under both standard and HC fire. The temperature profile obtained from FRPNSC with carbonate aggregate show better fire performances compare to lightweight aggregate under HC fire conditions. The results also show the poor fire performances of concrete slab under HC due to excessive temperature produced compare to standard fire. Therefore, EFDM models developed can be further used in fire resistance study for FRP reinforced concrete study.

6. References

[1] Khoury G A. Effect of fire on concrete and concrete structures. *Prog Struct Eng Mater* 2000; 2: 429–447.

[2] Kodur V K R, Bisby L A. Evaluation of Fire Endurance of Concrete Slabs Reinforced with Fiber-Reinforced Polymer Bars. *J Struct Eng* 2005; 131: 34–43.

[3] Lee L S, Jain R. The role of FRP composites in a sustainable world. *Clean Technol Environ Policy* 2009; 11: 247–249.

[4] Rafi M M, Nadjai A, Ali F. Fire resistance of carbon FRP reinforced-concrete beams. *Mag Concr Res* 2007; 59: 245–255.

[5] Adelzadeh M, Hajiloo H, Green M. Numerical Study of FRP Reinforced Concrete Slabs at Elevated Temperature. *Polymers (Basel)* 2014; 6: 408–422.

[6] Firmo J P, Correia J R, Bisby L A. Fire behaviour of FRP-strengthened reinforced concrete structural elements: A state-of-the-art review. *Compos Part B Eng* 2015; 80: 198–216.

[7] Williams B, Bisby L, Green M, et al. An Investigation On Fire Performance Of FRP-strengthened R/C Beams. *Fire Saf Sci* 2005; 8: 247–258.

[8] Bisby L A, Kodur V K R. Evaluating the fire endurance of concrete slabs reinforced with FRP bars: Considerations for a holistic approach. *Compos Part B Eng* 2007; 38: 547–558.

[9] Kodur V K R, Baingo D. Fire resistance of FRP reinforced concrete slabs. *Intern Rep No 758* 1998; 37.
[10] Blagojević M D, Pešić D J. A new curve for temperature-time relationship in compartment fire. *Therm Sci* 2011; 15: 339–352.
[11] ASTM E119. Standard Test Methods for Fire Tests of Building Construction and Materials, ASTM International, West Conshohocken, PA. *ASTM Int West Conshohocken, PA*; 552. Epub ahead of print 2016. DOI: 10.1520/E0119-14.
[12] Bisby L A, Green M F, Kodur V K R. Response to fire of concrete structures that incorporate FRP. *Prog Struct Eng Mater* 2005; 7: 136–149.
[13] Williams B, Bisby L A, Kodur V K R, et al. Fire insulation schemes for FRP-strengthened concrete slabs. *Compos Part A Appl Sci Manuf* 2006; 37: 1151–1160.
[14] Lie T T. Calculation of the fire resistance of composite concrete floor and roof slabs. *Fire Technol* 1978; 14: 28–45.
[15] Firmo J P, Correia J R, Bisby L A. Fire behaviour of FRP-strengthened reinforced concrete structural elements: A state-of-the-art review. *Compos Part B Eng* 2015; 80: 198–216.
[16] Reader T W, Oconnor P. The Deepwater Horizon explosion: Non-technical skills, safety culture, and system complexity. *J Risk Res* 2014; 17: 405–424.
[17] Shukeri N I M, Yahya Z, Rusli N. Application of FDM heat transfer model to study the reinforced concrete in thermal aspects. In: *AIP Conference Proceedings*. 2018. Epub ahead of print 2018. DOI: 10.1063/1.5041548.
[18] Bisby L A, Green M F, Kodur V K R. Response to fire of concrete structures that incorporate FRP. *Progress in Structural Engineering and Materials* 2005; 7: 136–149.

IOP Conf. Series: Materials Science and Engineering **767** (2020) 012011 doi:10.1088/1757-899X/767/1/012011

Comparison of Laguerre-Gaussian and Hermite–Gaussian Modes for Optical-CDMA over Multi-Mode Fiber

Alaan Ghazi [1,2] S A Aljunid [2] Syed Zulkarnain Syed Idrus [1] Alaa Fareed [3] C B M Rashidi[2] Aras Al-dawoodi [4]

[1] School of Human Development & Techno Communication University Malaysia Perlis Malaysia.
[2] School of Computer and Communication Engineering University Malaysia Perlis Malaysia.
[3] School of Computing University Utara Malaysia Malaysia.
[4] Computer Science Department College of Computer Science and Information Technology Kirkuk University Iraq

drsyedalwee@gmail.com

Abstract. Optical fiber became essential in telecommunication networks long-distance data transmissions because of its great features such as large bandwidth and long transmission distance. However different profile mode in the optical fiber communication system has different features that provide better results through Multi-mode fiber (MMF). This paper proposes a comparison of different modes for optical code division multiple access (Optical-CDMA) based on one dimension zero cross-correlation (ZCC) code over multi-mode fiber (MMF) transmission system for short haul-local area network. Furthermore Optical-CDMA system transmitted data based on Spatial VCSEL that operate over wavelength 1550 nm. The system is carried out by using three Laguerre-Gaussian modes (LG 01 LG 02 and LG 0 3) and compare with three Hermite–Gaussian modes (HG 01 HG 02 and HG 0 3) based on 1D-ZCC codes. Over different distance comparison of different modes in Optical-CDMA has evaluated the performance based on eye diagrams bit-error rates (BER) and Q-Factor measurement. The findings show that the proposed system demonstrates good performance over more distance MMF based on Laguerre-Gaussian modes.

1. Introduction

In the last decades the demand for faster internet becomes crucial due to a large amount of internet connected device. Furthermore with the increased use of multimedia applications such as live sharing video and streaming and direct to home television services the demand for bandwidth requirement by end-users has also increased [1-3]. The scholars paid attention to find clever solutions on how to meet the need. These efforts led to the inventing fiber optic technology that enables achieving high capacity demand [4]. In an optical fiber there are two kinds of medium such as single-mode fiber (SMF) and multi-mode fiber (MMF). SMF differs than MMF with a small core diameter; however MMF allows more light paths to propagate. In MMF the different paths are known as modes in waveguides. Modes are amplitude distribution and spatial phase that propagate unaltered in the waveguide and are orthogonal to one another. The modes can then be exploited as independent channels [5]. According to Shannon limitation [6 7] the capacity increases with a signal to noise ratio [8]. Similarly over the last ten years the growth of per capita IP and Internet traffic has followed a curve of steep growth. Globally beginning with (8GB) in 2014 while per capita Internet traffic it is expected to become (22GB) by 2020

[1 2]. In this respect the monthly per capita Internet traffic was 1GB in 2008 contrasted with 10 MB in 2000 [1]. The communication system using SMF will reach capacity margins in the near future [2 3]; this will increase the needs to use the MMF due to its extremely high bandwidth. Therefore to transfer a larger amount of data it is necessary to enhance the capacity of fiber optic. Nowadays several techniques are available such as time-division multiplexing access (TDMA) orthogonal frequency division multiple access (OFDMA) wavelength division multiple access (WDMA) and optical code division multiple access (optical-CDMA) [9-13]. Optical-CDMA has been implemented to enhance the capacity of data transmission and security for fiber optic communication [14 15]. The optical-CDMA is a scheme that assigns an individual sequence or optical code to every user enabling simultaneous access of multiple users to the optical channel. Specifically the assigned optical code is used with different spectrum modulation to spread the data. Before transmitting over the optical channel this information is code division multiplexed and then receive it the multiplexed information and correlates it with the authorized signature sequence [15]. Multiple users accessing the channel are only able to extract the original information when their optical orthogonal codes match [15 16]. Different modes are needed to be multiplexed onto optical fiber to increase the total throughput for Optical-CDMA. Consequently leading to improve the capacity and bandwidth demand in future information networks and to utilize the optical fiber's available bandwidth. Laguerre-Gaussian (LG) [11] and Hermite–Gaussian (HG) [17-22] are the most common modes used in optical communication. In this paper different modes are investigated to compare the performance of the MMF in Optical-CDMA based on 1D-ZCC code. Additionally two systems with different modes LG and HG modes are compared. The rest of the paper is structured as follows. Section 2 describes Optical-CDMA simulation. Section 3 presents analysis results and discussions of LG & HG modes: Optical-CDMA. The conclusion of this work is provided in Section 4.

2. Optical-CDMA Simulation and Description

Figure 1 illustrated the comparison of LG and HG modes for Optical-CDMA based on one-dimension ZCC code over MMF transmission system. We designed and simulated the system using opti-system [23-26] and MATLAB [27] software.

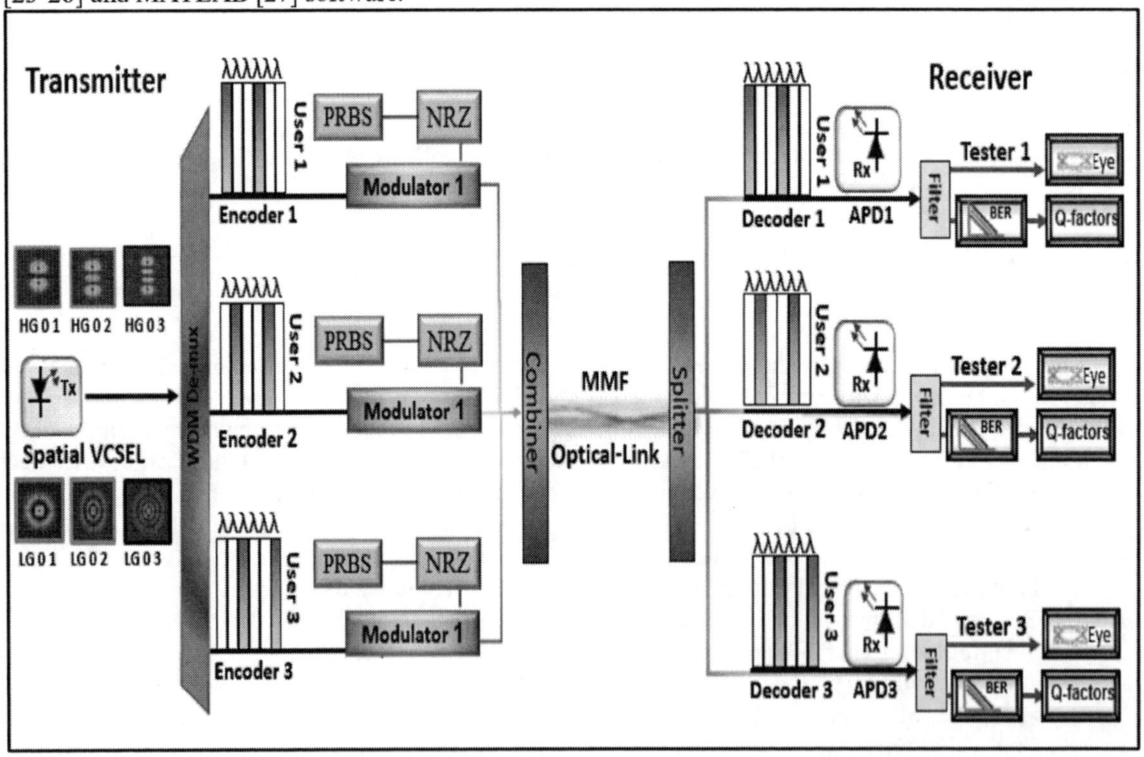

Figure 1. Simulation setup of Comparison of different modes in Optical-CDMA.

IOP Conf. Series: Materials Science and Engineering **767** (2020) 012011 doi:10.1088/1757-899X/767/1/012011

The comparison of different modes in Optical-CDMA system consists of three parts: 1) Transmitter; 2) MMF; and 3) Receiver. Specifically the transmitter parts are divided into five components as follows:

1. Spatial vertical-cavity surface-emitting laser (spatial VCSEL).
2. Encoding.
3. Pseudo random binary sequence (PRBS) generator.
4. NRZ pulse generator.
5. Modulator and power combiner.

Briefly we now describe the components of the transmitter. Firstly spatial VCSEL is used to generate two different sets of modes that operate over one wavelength (1550 nm). Figure 2 presents the generated modes LG 0 1 LG 02 and LG 03 (i.e. LG modes). Mathematically the LG mode can be defined as follows [28]:

$$\psi_{m,n}(r\varphi) = \left(\frac{2r^2}{W_0^2}\right)^{\left|\frac{n}{2}\right|} L_m^n\left(\frac{2r^2}{W_0^2}\right) exp\left(\frac{r^2}{W_0^2}\right) exp\left(j\frac{\pi r^2}{\lambda R_0^2}\right)\begin{Bmatrix} Sin(|n|\varphi)n \geq 0 \\ Cos(|n|\varphi)n \geq 0\end{Bmatrix} \quad (1)$$

where m and n represent the X (describes the azimuthal index) and Y (describes the radial index) indexes; R indicates to the radius of curvature; W_0 is spot size and L_m^n refers to the Laguerre Polynomial.

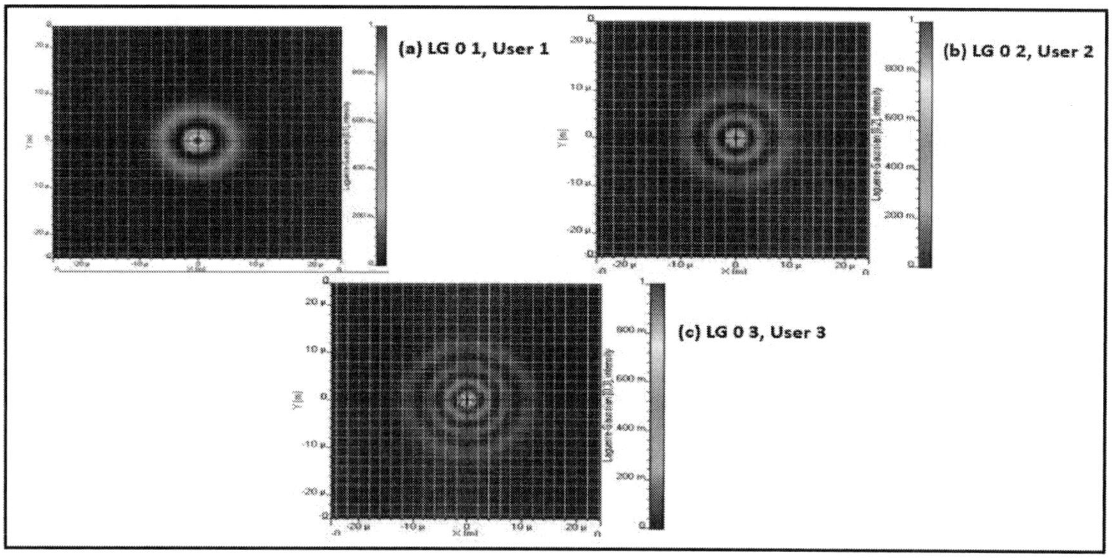

Figure 2. Spatial visualizer of three LG modes.

Additionally as shown in Figure 3 the other generated set of modes (by spatial VCSEL) are HG 0 1 HG 0 2 and HG 0 3 (i.e. HG modes). The HG mode is mathematically described as follows [28].

$$\psi_{m,n}(r\phi) = H_m\left(\frac{\sqrt{2x}}{W_0^2 x}\right)\exp\left(\frac{-x^2}{W_0^2 x}\right)\exp\left(j\frac{\pi x^2}{\lambda R_0 x}\right)H_n\left(\frac{\sqrt{2x}}{W_0 y}\right)\exp\left(\frac{-y^2}{W_0^2 y}\right)\exp\left(j\frac{\pi y^2}{\lambda R_0 y}\right) \quad (2)$$

Similar to LG mode m and n represent the X and Y index that describe the mode dependencies for the X and Y-axis; R indicates to a radius of curvature and W_0 is the spot size. However H_m and H_n are the Hermite polynomials.

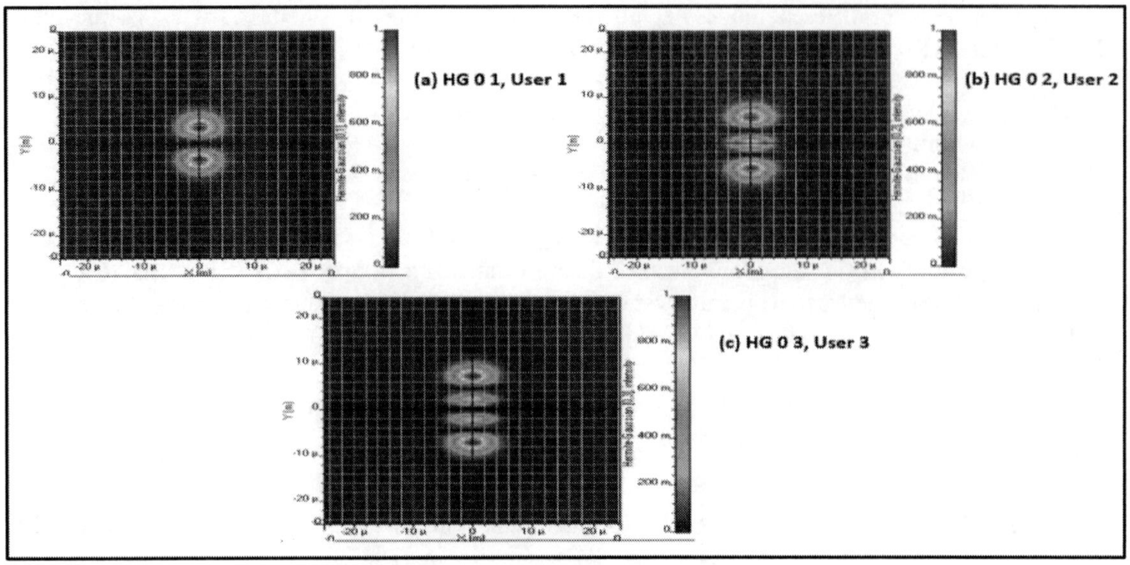

Figure 3. Spatial visualizer of three HG modes.

The second component of the transmitter is encoding which is based on one-dimension ZCC code (used for three users). PIIN is strongly related to MAI in Optical-CDMA system because of the overlapping spectra from different users [14 15]. The 1D-ZCC code can be formulated in a matrix form (K × L) where K row is a number of the users and L element is written as the code length. Additionally the ZCC codes are implemented for three users. This code can be written in a unit code matrix as follows:

$$ZCC(w=2) = \begin{bmatrix} 1 & 0 & 0 & 1 & 0 & 0 \\ 0 & 1 & 0 & 0 & 1 & 0 \\ 0 & 0 & 1 & 0 & 0 & 1 \end{bmatrix} \quad (3)$$

The input signal is generated using three PRBS at the third component of the transmitter. It is generated from the data generator at 1.866 Gbps. The fourth component is a non-return-to-zero (NRZ) signal generating sequence. Lastly Mach-Zehnder modulator and the power combiner are allocated at the transmitter as a fifth component.

As mentioned earlier the second part of our system is the fiber optic link. Here we used MMF and the maximum distance link is 8 km. Lastly the third part of the system is a receiver that consists of the following components:

1. The incoming signal from MMF is power splitter into three users and three decoders.
2. Photo-detector (PIN) is used to perform the conversion from optical to electrical domain.
3. Low pass Bessel filter (LPBF).
4. The last component is the performance analyzer.

Finally Table 1 summarizes the parameters of comparison of different modes in Optical-CDMA

Table 1. Parameters of comparison of different modes in Optical-CDMA.

Component Name	Parameters	Value
Spatial VCSEL	Power	1dBm
	Wavelength	1550 nm
	Laguerre-Gaussian	0 1 0 2 0 3 modes
	Hermite–Gaussian	0 1 0 2 0 3 modes
Multimode	Attenuation	0.25 dB-Km
	Core radius	50 nm
APD	Responsivity	1 A/W
	Dark Current	10 A

3. Optical-CDMA Results and Description

In the experiments we measured BER Q-Factor and eye diagram the comparison of different modes in Optical-CDMA based on 1D-ZCC code MMF transmission system. Now we analyze these performance parameters to determine how much improvement can get from our system.In the beginning the BER and Q-Factor of LG modes in Optical-CDMA based on 1D-ZCC code are calculated in Figures 4 and 5. Specifically we calculate both BER and Q-Factor of (USER 1: LG 0 1 USER 2: LG 0 2 and USER 3: LG 0 3) over the distances (2 Km 3 Km 4 Km 5 Km 6 Km 7 Km 8 Km and 9 Km) respectively. The Figures demonstrate that BER and Q-Factor are accepted over the distances (from 2 Km until 8 Km) whereas they are not accepted at distance of 9 Km.

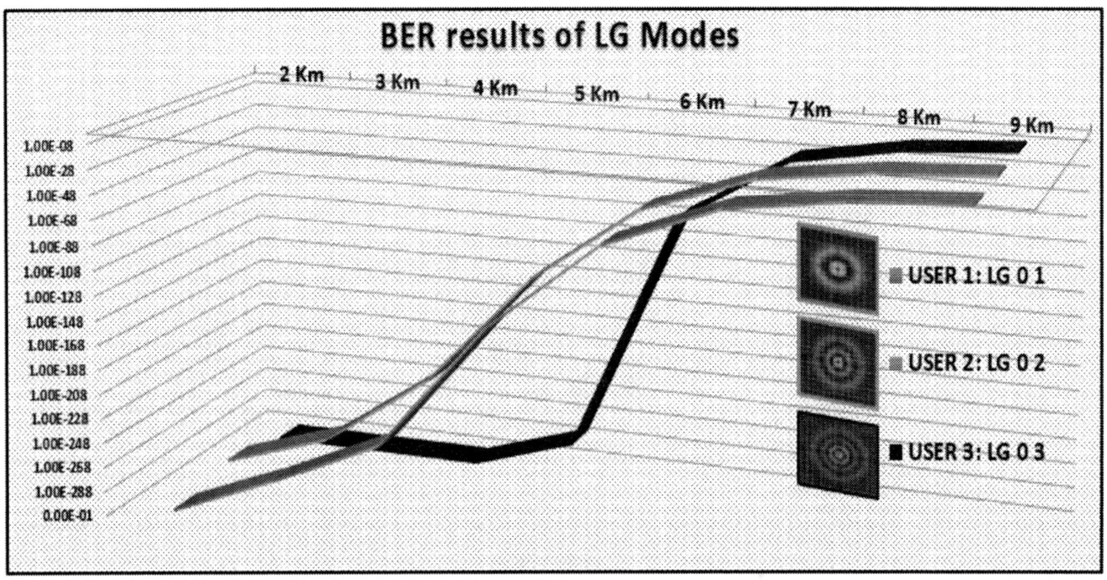

Figure 4. BER results of (USER 1: LG 0 1 USER 2: LG 0 2 and USER 3: LG 0 3) over different distances.

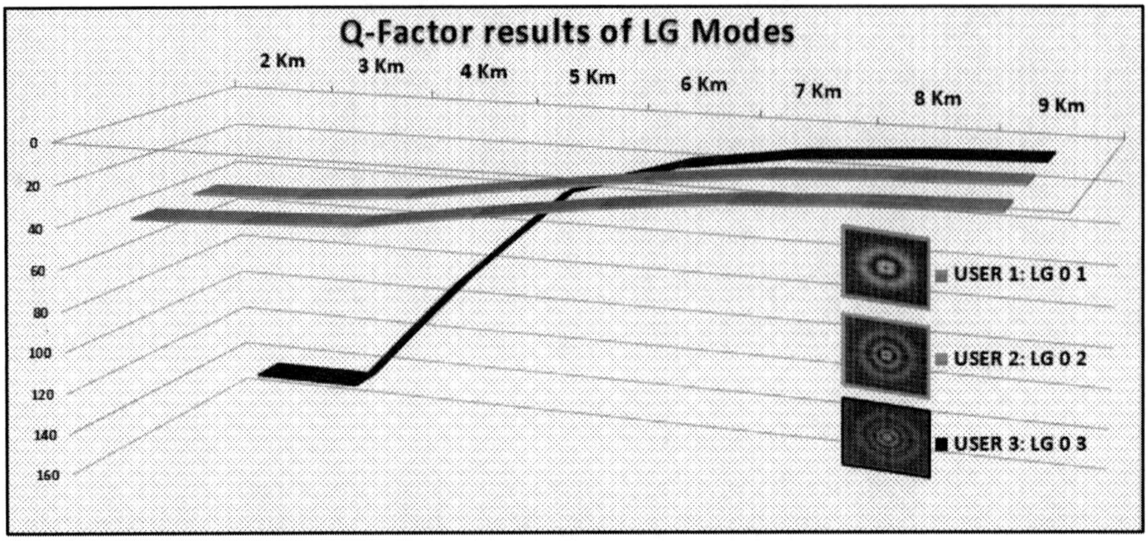

Figure 5. Q-Factor results of (USER 1: LG 0 1 USER 2: LG 0 2 and USER 3: LG 0 3) over different distances.

In the second part of the experiments as shown in Figures 6 and 7 we measured BER and Q-Factor of HG modes in Optical CDMA over the distances 2 km to 9 km respectively. Specifically we evaluated the performance for USER 1: HG 0 1 USER 2: HG 0 2 and USER 3: HG 0 3 over the distances 2 km to 9 km. Similar to LG modes BER and Q-Factor showed an accepted performance over the distances 2km to 8km. However the present unaccepted performance at a distance of 9 km.

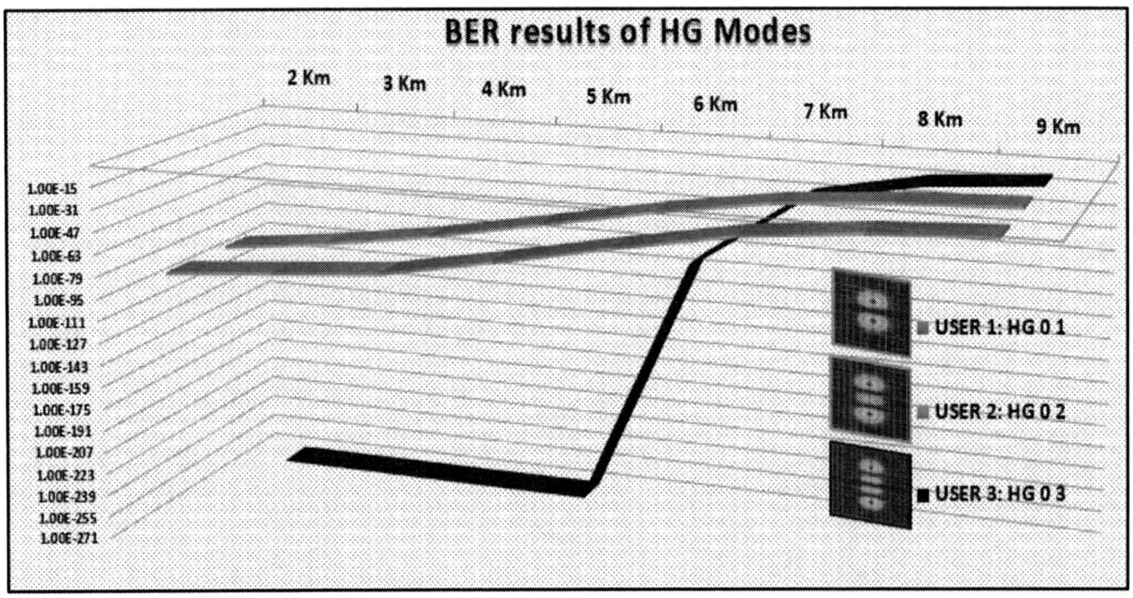

Figure 6. BER results of (USER 1: HG 0 1 USER 2: HG 0 2 and USER 3: HG 0 3) over different distances.

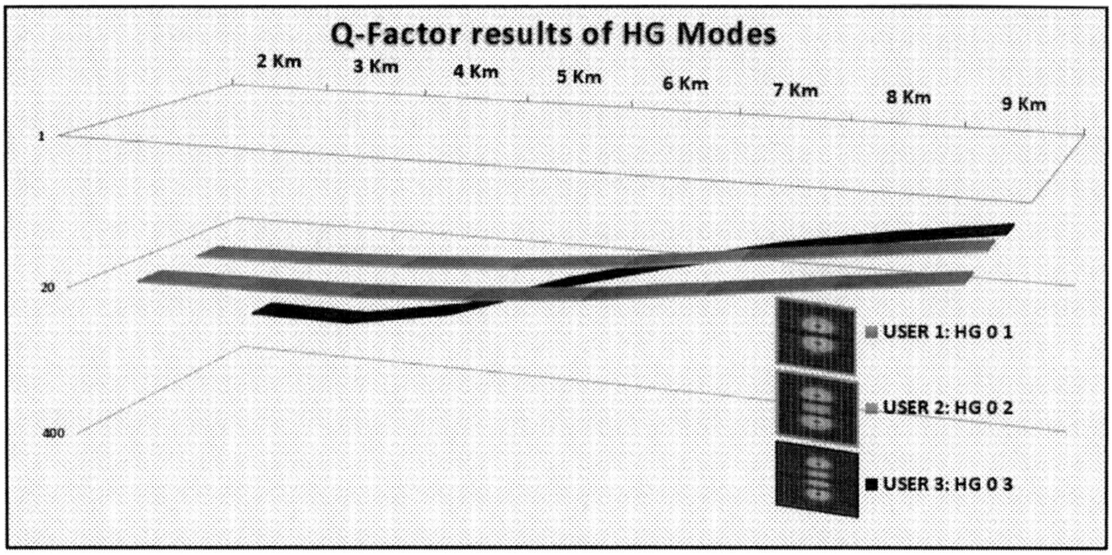

Figure 7. Q-Factor results of (USER 1: HG 0 1 USER 2: HG 0 2 and USER 3: HG 0 3) over different distances.

As shown in Figure 8 we evaluated the eye diagram of the comparison of different modes in Optical-CDMA over MMF at a distance of 8 km. We calculate the eye diagram of (USER 1: HG 0 1 USER 2: HG 0 2 and USER 3: HG 0 3) and (USER 1: LG 0 1 USER 2: LG 0 2 and USER 3: LG 0 3) respectively. It is important to mention that all channels have clean and open eyes. The outcomes demonstrate that the results are better with using of LG mode for Optical-CDMA based on 1D-ZCC code compared to HG modes.

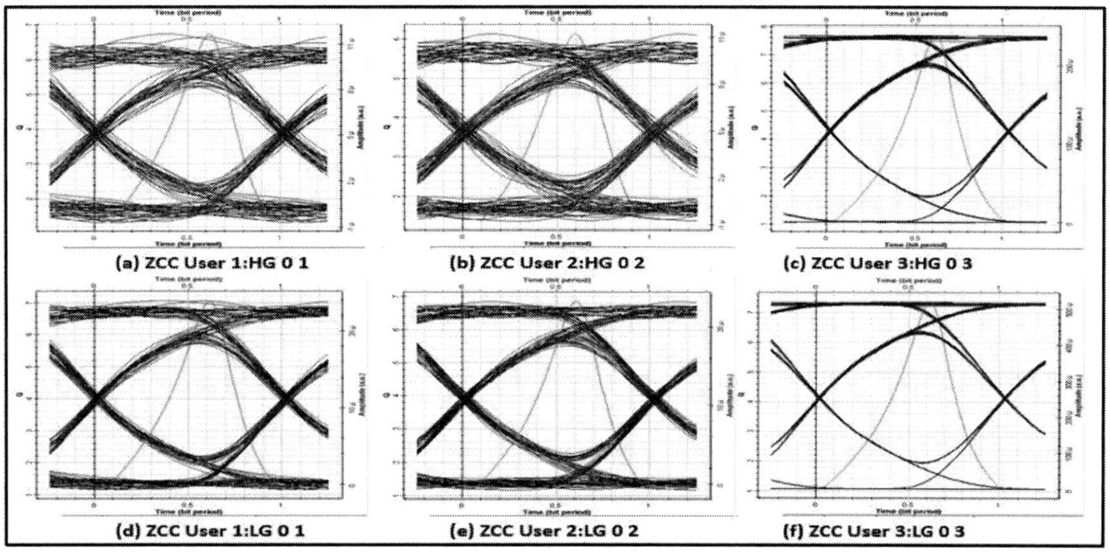

Figure 8. Eye diagram of (USER 1: HG 0 1 USER 2: HG 0 2 and USER 3: HG 0 3) and (USER 1: LG 0 1 USER 2: LG 0 2 and USER 3: LG 0 3) at a distance of 8 km.

Finally in Figures 9 and 10 we measured BER and Q-Factor of LG and HG modes in Optical-CDMA based on 1D-ZCC code over MMF. The results demonstrate that LG modes are better than HG modes in Optical-CDMA system based on 1D-ZCC code over MMF.

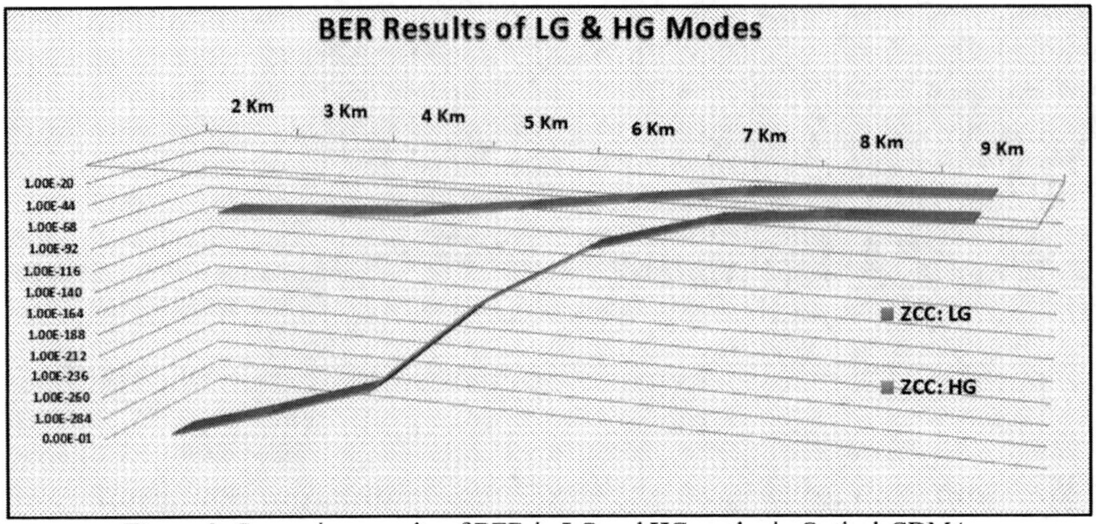

Figure 9. Comparison results of BER in LG and HG modes in Optical-CDMA.

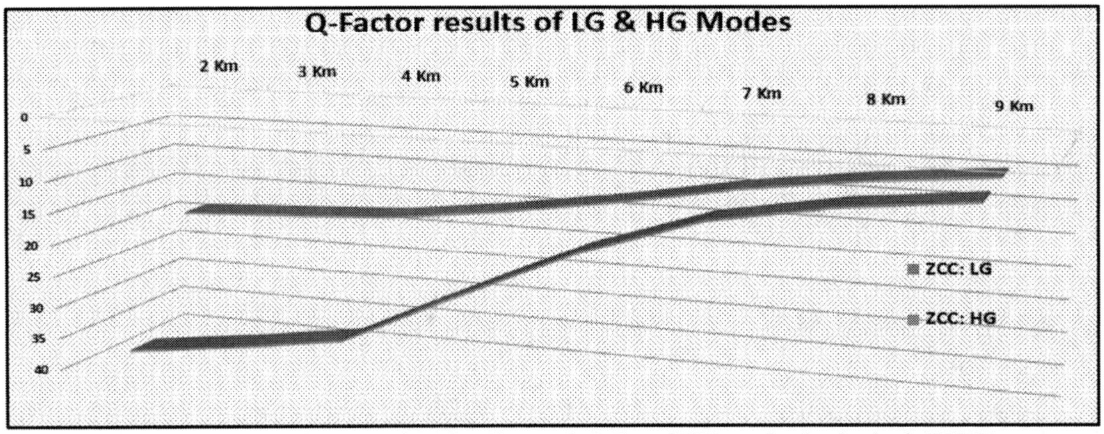

Figure 10. Comparison results of Q-Factor in LG and HG modes in Optical-CDMA.

4. Conclusion

In this paper we studied the comparison of LG and HG modes for Optical-CDMA based on one-dimension ZCC code over MMF transmission. Different modes can be used as a data carrier with different 1D-ZCC code in Optical-CDMA to improve the bandwidth. These modes should be selected carefully to get the longest possible distance with the lowest BER and Q-Factor. The outcomes demonstrated that it is possible to get a successful transmission data over 8 km MMF link with acceptable BER Q-Factor and eye diagrams. However at a distance of 9 km the results are not acceptable due to nonlinearity and distortion in MMF. This issue can be solved by using artificial intelligence (AI) to overcome the nonlinearity issue and enable the MMF to transmit the data over more distance. Therefore as future work we will work on applying AI algorithms in our system.

5. Reference

[1] Index CVN forecast and trends 2017–2022 2018 White Paper 2018 in Editor Ed ^ Eds Book forecast and trends 2017–2022 White Paper
[2] Index CVN Forecast and Trends 2017–2022 2019 White Paper Cisco San Jose CA USA
[3] Luo T 2011 Digital Equalization Of Fiber-Optic Transmission System Impairments

[4] Fareed A Amphawan A Fazea Y Sajat MS and Chit SC Channel Impulse Response Equalization based on Genetic Algorithm in Mode Division Multiplexing Journal of Telecommunication Electronic and Computer Engineering 2018 JTEC **10** (2-4) p 149-154
[5] Lyubopytov V S, Gizatulin A R, Tlyavlin A Z and Sultanov A K 2014 Optical-domain mode coupling compensation for mode division multiplexing systems in Editor (Ed)^(Eds) Book Optical-domain mode coupling compensation for mode division multiplexing systems (International Society for Optics and Photonics edn) pp 915604
[6] Essiambre R-J Kramer G Winzer PJ Foschini GJ and Goebel B Capacity limits of optical fiber networks Journal of Lightwave Technology 2010 28 (4) p 662-701
[7] Bai N Ip E Huang Y-K Mateo E Yaman F Li M-J Bickham S Ten S Liñares J and Montero C Mode-division multiplexed transmission with inline few-mode fiber amplifier Optics express 2012 20 (3) pp 2668-2680
[8] Essiambre R-J Nonlinear capacity limit to optical communications in Editor (Ed)^(Eds) Book Nonlinear capacity limit to optical communications (Optical Society of America 26–31 July 2015 edn) pp NTu2A 3
[9] Noori A Amphawan A Ghazi A and Ghazi SA Dynamic evolving neural fuzzy inference system equalization scheme in mode division multiplexer for optical fiber transmission Bulletin of Electrical Engineering and Informatics 2019 8 (1) pp 127-135
[10] Grammel G and Jones KP Wavelength provisioning for customer premise equipment (CPE) in dense wavelength-division multiplex (DWDM) networks in Editor (Ed)^(Eds) Book Wavelength provisioning for customer premise equipment (CPE) in dense wavelength-division multiplex (DWDM) networks (Google Patents 2019 edn) pp
[11] Al-Dawoodi A Fareed A Masuda T Ghazi A Fakhrudeen A Aljunid S Idrus S and Amphawan A Comparison of Different Wavelength Propagations over Few-Mode Fiber based on Space Division Multiplexing in Conjunction with Electrical Equalization International Journal of Electronics and Telecommunications 2019
[12] Nawawi N Anuar M and Junita M: Cardinality improvement of Zero Cross Correlation (ZCC) code for OCDMA visible light communication system utilizing catenated-OFDM modulation scheme Optik 2018 170 pp 220-225
[13] S A Aljunid and Syed Zulkarnain Syed Idrus Sara Alshwani AMF Mohammed Nasih Ismael Aras Al-Dawoodi Alaan Ghazi: Hermite-Gaussian Mode In Spatial Division Multiplexing Over Fso System Under Different Weather Condition Based On Linear Gaussian Filter International Journal of Mechanical Engineering and Technology (IJMET) 2019/2/2 10 (1) pp 1095-1105
[14] Anuar M Aljunid S Arief A and Saad N: LED spectrum slicing for ZCC SAC-OCDMA coding system in Editor (Ed)^(Eds): Book LED spectrum slicing for ZCC SAC-OCDMA coding system (IEEE 2010 edn) pp 128-132
[15] Krishnan P: Performance Analysis of Hybrid RF/FSO System Using BPSK-SIM and DPSK-SIM Over Gamma-Gamma Turbulence Channel With Pointing Errors for Smart City Applications IEEE Access 2018 6 pp 75025-75032
[16] Ahmad N A Junita M Aljunid S Rashidi C and Endut R: Analysis of hybrid subcarrier multiplexing of OCDMA based on single photodiode detection in Editor (Ed)^(Eds): Book Analysis of hybrid subcarrier multiplexing of OCDMA based on single photodiode detection (EDP Sciences 2017 edn) pp 01028
[17] Amphawan A Ghazi A and Al-dawoodi A: Free-space optics mode-wavelength division multiplexing system using LG modes based on decision feedback equalization in Editor (Ed)^(Eds): Book Free-space optics mode-wavelength division multiplexing system using LG modes based on decision feedback equalization (2017 edn) pp
[18] Masunda T Amphawan A and Al-dawoodi A: Effect of decision feedback equalizer taps on 3× 6-channel mode-wavelength division multiplexing system performance in multimode fiber in Editor (Ed)^(Eds): Book Effect of decision feedback equalizer taps on 3× 6-channel mode-wavelength division multiplexing system performance in multimode fiber (EDP Sciences 2017 edn) pp 01012
[19] Amphawan A Chaudhary S Samad H and Ahmad J: Mode division multiplexing of LG and HG

modes in Ro-FSO Proceeding of the Electrical Engineering Computer Science and Informatics 2015 2 (1) pp 133-137

[20] Amphawan A Fazea Y and Ibrahim H: Investigation of channel spacing for Hermite-Gaussian mode division multiplexing in multimode fiber in Editor (Ed)^(Eds): Book Investigation of channel spacing for Hermite-Gaussian mode division multiplexing in multimode fiber (IEEE 2015 edn) pp 34-39

[21] Mori T Sakamoto T Wada M Yamamoto T and Yamamoto F: Few-mode fibers supporting more than two LP modes for mode-division-multiplexed transmission with MIMO DSP Journal of Lightwave Technology 2014 32 (14) pp 2468-2479

[22] Alaan Ghazi Ahmed M Fakhrudeen Mohammed Nasih Ismael AA-D Sara Alshwani: SDM Over Hybrid FSO Link Under Different Weather Conditions and FTTH Based On Electrical Equalization International Journal of Civil Engineering and Technology 2019/2/2 10 (1) pp 1396-1406

[23] 23 Guide OU: RSOFT Design Group Inc 2010

[24] 24 Guide OU: USA: RSoft Design Group in Editor (Ed)^(Eds): Book USA: RSoft Design Group (Inc 2008 edn) pp

[25] Aras Al-Dawoodi Hm Sara Alshwani Alaan Ghazi Ahmed M Fakhrudeen S A Aljunid Syed Zulkarnain Syed Idrus Aso Ahmed Majeed Kameran Ali Ameen: Investigation Of 8 X 5 Gb/S Mode Division Multiplexing -FSO System Under Different Weather Condition Journal of Engineering Science and Technology 2019 14 (2) pp 674_681

[26] Ghazi A Aljunid S Noori A Idrus SZS Rashidi CRB and Al-Dawoodi A: Design & investigation of 10x10 gbit/s MDM over hybrid FSO link under different weather conditions and fiber to the home Bulletin of Electrical Engineering and Informatics 2019 8 (1) pp 121-126

[27] Release M: The mathworks Inc Natick Massachusetts United States 2013 488

[28] Ghatak Aa Ghatak A Thyagarajan K and Thyagarajan K An introduction to fiber optics (Cambridge university press 1998 1998)

An IoT platform for seizure alert wearable device

Emelia Akashah P A[1] and Noor Shita A[1]

[1] Department of Computer & Information Sciences,Universiti Teknologi Petronas
32610 Seri Iskandar, Perak Malaysia

emelia.akhir@utp.edu.my

Abstract. This research aims to propose the system architecture for a wearable device that is able to detect seizure attacks for children aged 15 years old and below. The objective of this research is to alert parents or relatives of the patients whenever the seizure attack is detected. In many cases, seizure attack happened in absences of the parents or relatives. Therefore, the wearable device proposed in this research should be able to monitor their children heart rate and motion remotely. Notification also will be sent to parents or relatives via mobile apps should there is any trigger of seizure detection. Data were collected via interview. The IoT wearable device architecture was then designed before being implemented.

1. Introduction

Internet-of-Things (IoT) seems to be one of appropriate solutions to many things such as logistics, traffic management, smart houses etc. This technology allows a wide range of devices to communicate with each other and gather data to be analysed [1]. According to [2], one of the interesting domains to apply IoT is medical and health care. IoT enables medical applications to provide early intervention to the users. IoT also able to provide remote monitoring that could benefits many people. For instance, a person with chronic disease can always being monitored by their parents or relatives whenever they are alone [3].

In this research, the concept of IoT is applied to develop a device that could help to detect seizure attack for children aged 15 years old and below. The device should be able to alert the parents or relatives in case the attack happens when the children are alone.

A seizure is a short scene of symptoms that caused by a burst of abnormal electrical activity in the brain. Normally, a seizure endures from a few moments to a couple of minutes. The mind contains many nerve cells (neurons). Regularly, the nerve cells are continually sending modest electrical messages down nerves to all parts of the body. Distinctive parts of the mind control diverse parts and elements of the body. In this manner, the indications that happen amid a seizure rely on upon where the irregular burst of electrical activity happens.

There are several types of seizures but they are broadly divided into two main types which are generalised and focal (used to be called partial):

- Generalised seizures: This type of seizures occurs if you have a burst of abnormal electrical activity which spreads throughout the brain. It affects consciousness and may cause a convulsion to a person.
- Focal seizures: In focal seizures, the burst of electrical activity starts in, and stays in, one part of the brain. Therefore, you tend to have localised (focal) symptoms. Various parts of the brain control different functions, so symptoms depend on which part of the brain is affected. Focal

seizures are said not affecting consciousness but may affect emotions, sensations, behaviours, muscles or combinations of these.

In this research project, the author has chosen the generalised seizures to study further.

2. Literature Review

2.1. Seizure Prediction and Detection

Early detection of seizure among children is crucial in order to prevent long-term neurological injury [4]. Patients with epilepsy regularly portray seizures as happening "out of nowhere" which emphasizes the evident sudden, unanticipated path in which seizures tend to strike [5]. This represents to a standout amongst the most disable parts of the disease [6], particularly for those not able to accomplish complete seizure control. It can prompt to a serious sentiment defencelessness that strongly affects the regular daily existence of a patient. Also, unexpected scenes of gazing, loss of muscle control or loss of awareness can represent a genuine damage chance and can even be life-debilitating on the off chance that they happen while the patient is driving, crossing a bustling road, washing, swimming or climbing stairs. In this manner, a strategy fit for estimating the event of seizures could significantly enhance the personal satisfaction for epilepsy patients [7]. For a long time, epileptic seizures were thought to strike suddenly however there is currently mounting proof that seizures create minutes to hours before clinical onset. Clinical findings authenticating the presence of a pre-seizure state incorporate a significant increment in blood flow in the epileptic transient flaps minutes before seizure onset [8] [9], increment in cerebral oxygenation hours before the ictal occasion [10], increment in blood oxygen level dependent practical MRI (fMRI) signals [11] and additionally changes in RR interims on the ECG minutes before seizure onset [12]. Since Viglione and colleagues the visionary work on predicting epileptic seizures in the 1970s, they has concentrated on extracting characteristic features that are predictive of an impending seizure from intracranial electroencephalogram (EEG) signals [13]. In any case, there is mindfulness that solitary quantitative methods are probably not going to foresee seizures in all patients. As of late, seizure expectation work has diversified and there is an expanding enthusiasm for strategies for determining seizures from other physiological or non-physiological factors [13]. Regardless of whether seizures can be heartily anticipated, a gadget that can distinguish seizures and trigger a caution has vital utility. Since seizures frequently cause loss of cognizance, most patients especially children below 15 years experiences difficulty to reporting the event of seizures.

2.2. Seizure Detection using Wearable Device Attached and Information Apps

Current advancements for procuring signals from the patient's body are very much. Numerous sensors are accessible which can recognize the heart beat and strong developments non-intrusively and precisely. Such non-invasive technique for measuring heart beat in this project is bioimpedence sensors in pebble wristband. This sensor can accurately monitor the heartbeat of the patients as this sensor measures resistance of skin tissue to the electrical currents in your skin to give you real-time heart rate readings. Bioimpedance sensors is also capture both your Resting Heart Rate and Passive Heart Rate for a holistic view of your heart.

Heart beat are to be checked consistently. Any sudden variety in heart beat which is created by the onset of epileptic seizures is recognized and confirmed or notify by push notification on the smart phone. At the point when the seizure is confirmed, notification is transmitted to the smart phones for initiating necessary protective measures for the patient. The device is invented as wireless, wearable and individual gear. Motions and movement collected using precision motion sensor and powerful algorithm to passively track and quantify your activities, distance, calories, active time, and idle time. The band is made of hypoallergenic rubber and encased in medical review. It is smooth, adaptable and totally latex-free. The band inside contains advanced technology that the band should treat with care and keep away from superfluous bending. The top and base casings are made of anodized aluminium. The device that compatible with this kind of band are IOS and Android, thus we created an information apps that is compatible with the type of phones.

The device developed in this research is used primarily to detect repeated shaking movement during seizure event. It can inform close-by family or caregiver when a seizure happens through cautions, telephone calls or text-based notifications. A parental figure can then help the individual amid and after the seizure. For instance, they can help re-position the individual, ensuring they are on their side in the event that they are not cognizant. They can likewise ensure somebody doesn't nod off on their stomach a short time later. On the off chance that breathing or different issues happen, they can call for medicinal offer assistance. They may also can give protect solutions or call for an ambulance if the seizure keeps going too long or the individual has repeated seizures.

Our research has found some limitations on the existing seizure detection that are available in the market:

- Seizure alert devices may not be functional for individuals who live alone or who don't need others to keep an eye on them. The information apps in this research are giving the information in the apps itself on how to handle the seizures when occurred in case they are collapse in public area.
- Complex partial seizures are typically not detected by devices that identify shaking developments.
- The devices as of now accessible are not affirmed by the Food and Drug Administration (FDA). The confirmation that these gadgets work to identify seizures regularly originates from studies performed in hospital epilepsy monitoring units and in a few circumstances, in home settings. A portion of the devices have not been considered in a methodical manner so we don't know how well they function in the 'real world'.
- The devices do not alert caregivers of breathing issues or changes in heart rate, which might be imperative in SUDEP. In this research project, the changes in heart rate is considered to detect the occurrence of the seizures.
- More scientific evidence is expected to confirm how well seizure alert devices function and on the off chance that they are useful in detecting SUDEP.

3. Methodology

3.1 Literature Survey

A literature review has been conducted by reading, analysing and comparing between existing projects from various sources. A clear idea of components to use in developing this project is obtained after reviewing relevant literatures. This is important as to ensure the existence of the research project as well as its feasibility within the time frame given.

3.2 Collecting Data from Children and Infants, Parents and Doctor (interview)

An interview has been conducted with the respondent who are the patient (children), parents and doctor. The purposes of this interview are to get the overview of the seizures, how the parents detect the seizures occurred and the problem, challenges and opinion from these three respondents on the implementation of information apps. This interview also conducted at the respondent's house which located at Bangi, Selangor. Before the interview session, the author did some research based on readings and design the questions to be interview. All the questions to be asked during interview is designed based on the literature review. The interview was recorded and conducted in formal.

3.3 Prototype cycles (Develop, Demonstrate and Refine)

For this phase, few activities will be conducted such as designing the system flow which contain use case and activity diagram, design the system component where the system architecture has been developed and identify tools and equipment, develop prototype, design user interface and coding mobile application. To complete this prototype phase, the author uses PhoneGap and jQuery to develop the information apps. PhoneGap and jQuery helps the author to create software application for Android operating system (OS) and iOS. The firebase framework also will be used for the development of the information apps.

3.4 Testing

The testing will be conducted after the prototype has been developed. The application will undergo two types of test which is system test and user acceptance test in this testing phase. System test or also known as functional test is conduct to make sure that all the function in retrieving the data from database MySQL (Structured Query Language) to mobile application work properly in notify the parents or carer of any seizures occurrences. Meanwhile, user acceptance test is conducted to demonstrate to users that the application is ready for deployment phase. Any feedback and criticism will be taken into accounts and any improvements will be made immediately after the test been conduct and the new version of prototype will be released. The entire test is done to ensure that the application will satisfy the user need and achieve the objectives of this project.

3.5 Deployment

Next, the procedure will proceed onward to the deployment stage where the application will be created with the selected tools and method. The component focus amid this stage is to ensure it will be easy to understand (user friendly) to the target group. A few reasonable functions will be made and utilized as a part of building up the mobile application. The purpose behind this stage is to transform the necessities into complete and detailed system design specification before moving to the last stage that is the deployment stage. Graphical User Interface (GUI), database and system architecture will be design at this stage.

4. Result and Discussions

A system architecture is a conceptual model that characterizes the structure, behaviours, and more perspectives of a system. An architecture description is a formal depiction and portrayal of a framework, sorted out in a way that backings thinking about the structures and behaviours of the system. In this study, the author uses the below illustration as the conceptual model for the project which comprise the children and parents and their respective device used to detect the seizures and get the notification from information apps. The cloud database will be used to store history data and current collection data of seizures.

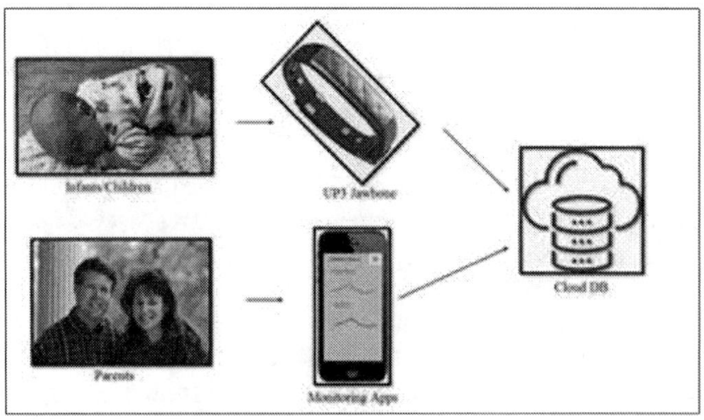

Figure 1. System Architecture

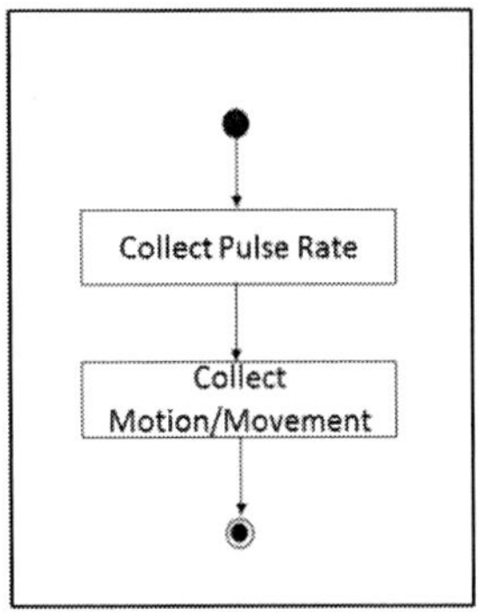

Figure 2. Activity Diagram Patient

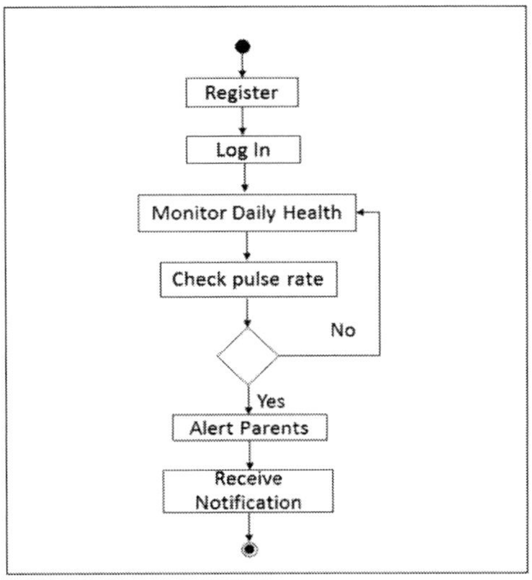

Figure 3. Activity Diagram Parent/Carer

Activity diagram above shows the overall activity in information apps. Two main actors will involve which are patient and carer. Both have their own activity flow for this application. For patients, they will be the source for the information apps to collect the pulse rate and motion. They also able to monitor their daily health through the wristband attached. For carer, they shall register/sign up to use the information apps. Then, they must login to enter to the homepage. Information apps also help them to monitor their children daily health, give them alert and they can receive the notification from the information apps.

5. Conclusion

In a nutshell, the author found that information apps and wristband attached is a potential mobile application and device to be used in Malaysia after had done many research and interview. It is very useful and helpful in assisting the parents to monitor their children's daily health, get the notification and alert them should there is any triggers of seizure detection to their children when they are busy with their work. This will also help their children to freely play outside from their home. It will help millions of victims of seizures around the globe. For future recommendation is to develop the information apps that can alert the carer through detection of many types of seizures. Thus, there will be more user can use the device and an information apps and therefore can saves many lives. Next recommendation is to develop cloud monitoring for the doctors to monitor the seizures. The parents no need to go to the hospital if the doctors got an application or desktop monitoring. He will be able to monitor his patient remotely. It will reduce time for patient's family to go there and can get the results instantly.

References

[1] Riazul Islam, S. M., Daehan Kwak, Humaun Kabir, M., Hossain, M., & Kyung-Sup Kwak. (2015). The Internet of Things for Health Care: A Comprehensive Survey. *IEEE Access* **3** 678–708.
[2] Pang, Z. (2013). Technologies and Architectures of the Internet-of-Things (IoT) for Health and Well-being. KTH-Royal Institute of Technology.
[3] BR, J., & Rajasekaran, M. (2017). Internet of Things in Healthcare Management. Journal of Chemical and Pharmaceutical Sciences ..., (2), 172–175. Retrieved from http://jchps.com/specialissues/2017 Special Issue 2/1221116.pdf
[4] Malarvili, M. B., & Mesbah, M. (2009). Newborn seizure detection based on heart rate variability. *IEEE Trans. Biomedical Eng.* 56(11) 2594–2603.
[5] Mormann, F., Andrzejak, R. G., Elger, C. E., & Lehnertz, K. (2007). Seizure prediction: The long and winding road *Brain* 130(2) 314–333.
[6] Elliott, I. M., Lach, L., & Smith, M. Lou. (2005). I just want to be normal: A qualitative study exploring how children and adolescents view the impact of intractable epilepsy on their quality of life. Epilepsy and Behavior 7(4) 664–678.
[7] Elger, C. E., Widman, G., Andrzejak, R., Arnhold, J., David, P., & Lehnertz, K. (2000). Nonlinear EEG analysis and its potential role in epileptology *Epilepsia* 41(3) 34–S38.
[8] Baumgartner, C., Serles, W., Leutmezer, F., Pataraia, E., Aull, S., Czech, T., Podreka, I. (1998). Preictal SPECT in temporal lobe epilepsy: regional cerebral blood flow is increased prior to electroencephalography-seizure onset *J. Nuclear Medicine : Official Publication, Society of Nuclear Medicine* 39(6) 978–982.
[9] Weinand, M. E., Carter, L. P., El-Saadany, W. F., Sioutos, P. J., Labiner, D. M., & Oommen, K. J. (1996). Cerebral blood flow and temporal lobe epileptogenicity *Neurosurgical Focus* 1(5) 5.
[10] Adelson, P. D., Nemoto, E., Scheuer, M., Painter, M., Morgan, J., & Yonas, H. (1999). Noninvasive continuous monitoring of cerebral oxygenation periictally using near-infrared spectroscopy: a preliminary report *Epilepsia* 40(1 1), 1484–1489.
[11] Federico, P., Abbott, D. F., Briellmann, R. S., Harvey, A. S., & Jackson, G. D. (2005). Functional MRI of the pre-ictal state. Brain, 128(8), 1811–1817.
[12] Novak, V., Reeves, a L., Novak, P., Low, P. a, & Sharbrough, F. W. (1999). Time-frequency mapping of R-R interval during complex partial seizures of temporal lobe origin *J. Autonomic Nervous System* 77(2–3) 195–202.
[13] Litt, B., & Echauz, J. (2002). Prediction of epileptic seizures *Lancet Neurology* 1(1) 22–30.

The Growth Morphology and Yield of Grey Oyster Mushrooms (*Pleurotus sajor-caju*) Subjected to Different Durations of Acoustic Sound Treatment

Mazlin Nur Iman Binti Hasnoel Mazidi [1] ,Roshita Ibrahim[1],Noorulnajwa Diyana Yaacob[1]

[1]Department of Chemical Engineering Technology, Faculty of Engineering , UniCITI Alam Campus, Universiti Malaysia Perli (UniMAP), 02100 Sungai Chuhcuh , Padang Besar, Perlis

Abstract. *Pleurotus sajor-caju* (grey oyster mushroom) is a typical edible fleshy fungi. It has fruiting bodies shape like an oyster that contain high nutrient such as carbohydrates, proteins, fats, minerals and multivitamins. Due to its high nutritional values, the demand of grey oyster mushrooms keeps increasing. In this study, the effect of different durations and intervals of acoustic sound treatment on growth morphology, yield enhancement and the quality of the mushroom were studied. Using sound of fire cracker at constant intensity of 75 dB, four different acoustic sound treatments had been applied on the mushroom bags which were 5 pops at 5-day interval, 15 pops at 5-day interval, 5 pops at 10-day interval and 15 pops at 10-day interval and mushroom bags without any sound treatment acted as control. The results showed that the controls had longer time for mycellium to fill-up the bags compared to sound treated bags. Besides, controls also showed lower yield in terms of number of fruiting bodies and biological efficiencies (%) compared to all acoustic sound treated bags. However, 15 pops at 10-day interval had the shortest time taken to complete all the growth stages (mycellium growth, emergence of pinhead and formation of fruiting bodies). For colour of the mushroom, there were significant differences ($P<0.05$) in L* values for 15 pops at 5-day interval and no significant different ($P>0.05$) was observed for a* (redness) and b* (yellowness) values. Controls had the highest firmness value compared to mushroom with sound treatment. In summary, 15 pops at 10-day interval was the best treatment duration and interval as it took the shortest time for different growth stages with the greatest yield of mushrooms.

1. Introduction

Pleurotus mushroom is a fleshy fungi that its fruiting bodies contains spore and comes under the family of Pleurotaceae [6]. It is commonly known as oyster mushroom as it fruiting bodies open up like a shape of an oyster during its morphogenesis process [7]. It provides various macro- and micronutrients for our health. Besides, it also accommodate a compelling levels of zinc, potassium, iron, calcium, phosphorus, Vitamin C, niacin, folic acid, low in calories, high in vegetable proteins, fibre and vitamins C, D and B-complex.[1]. The demand of mushroom market in Malaysia gradually increasing due to the increment of human population and impetus in mushroom uses whether in culinary and medicine. In order to fulfil the demand, various methods could be used to boost the growth and yield of the oyster mushroom. One of the method is sound treatment. Sound treatment has been verified could change the cell cycle based on different sound acoustics, sound intensities and sound frequencies [2]. Various acoustic sounds have been used as physical stimulant that can enhance the growth and yield of oyster mushrooms such as hard core music, soothing instrumental, thunder storm and quranic recital [4].

2. Methodology

2.1 Substrate Preparations

The sawdust will be mixed with rice bran, calcium carbonate and water in the ratio of 100:10:1 respectively. The substrate was manually mixed with water until the desired texture have been achieved. The wet substrate was then filled into polypropylene bags (29 x 46cm) about 850g in each bag. The bags were closed and tied up into cylindrical shape and closed with cotton-stuffed caps. The bags were then sterilized at 121°C for 30 minutes using autoclave or conventionally using steamer for 10 hrs. After sterilization the bags were cooled overnight and ready to be inoculated with *Pleurotus sp.* spawn. The inoculation process was carried out in a laminar flow.

2.2 Mushroom Cultivation

The inoculated bags were incubated (spawn running) at 28-30°C in the dark room (5% light) with RH of 75%-85% until mycelia had completely covered the bags. All the bags were labelled according to the treatment applied and arranged vertically on the racks in the mushroom cultivation chamber. This is because it can accelerate the process of mycelium colonization due to the pulling force of gravity. It took about 29 to 35 days for the mycelium to fill-up the bags. After completion of spawn running stage, the bags were then placed into a mushroom house where the environmental conditions were maintained at 28-30°C; 80 – 90% RH and about 30% light. Water was sprayed on the floor regularly to maintain the relative humidity up to the desired level in the form of fine mist with the help of a nozzle. The bags were opened for cropping (emergence of pinhead and fruiting body formation). Fruiting bodies that have achieved maturity level were then harvested by pulling from the substrate bags. The bag was then covered with the caps to close of the mushroom bags and left for 7 days for the next fruiting cycle in order to let the tissue of mycelium to accumulate the nutrients and to recover from injury.

2.3 Physical Stimulant Treatments

Acoustic sounds is the physical stimulant treatment that was carried out based on completely randomised design with two main factors which are treatment duration and treatment interval. Acoustic sound treatment was applied on the mushroom bag during spawn running stage and before fruiting body formation. The sound of fire cracker at 75 decibels was imposed on the mushroom bags for durations of 5 and 15 pops at 5- to 10-day intervals.

2.4 Growth Performance Analyses

The growth performance of the *Pleurotus sajor-caju* can be evaluate by determination of mycelium growth rate for each bag. The growth rate of mycelium for different durations and intervals of acoustic sound treatment and control were determined. Growth rate of mycelium treated with acoustic sounds were measured from the neck of the bag until the bottom of the bag when the mycelium has fill-up the bag by using a ruler in terms of unit of cm per day at every five day interval during spawn-running process.

Meanwhile, growth performance of the mushroom was determined by noting the numbers of days taken to reach the various growth stages of the grey oyster mushroom which were mycelium fill-up the bag, pinhead emergence and fruiting bodies formation. Each treatment was represented by 5 mushroom bags.

2.5 Mushroom Yield

Once the fruiting bodies have reached maturity, the mushroom can be harvested from the substrate bags. The yields of mushroom fruiting bodies were for each treatment were determined in terms number of fruiting bodies, fresh mushroom weight and percentage of biological efficiency, all the samples from the first harvesting until the last harvesting cycle were collected. All fruiting bodies harvested. Percentage biological efficiency (BE) was calculated as total fresh weight of harvested mushroom (g)/dry weight of substrate x 100.

2.6 Physico-chemical Analyses

Physico-chemical analyses of the mature fruiting bodies of grey oyster mushroom were established based on pileus diameter (largest and smallest), colour, texture (firmness) and percentage of moisture content.

2.6.1 Pileus Diameter. The largest and smallest sizes of the pileus of the mushrooms that had been harvested from different sound treatments and control were chosen. The mushroom's pileus was measured side to side using a ruler as the diameter measurement.

2.6.2 Colour Determination. Colour Measurement was determined on the mushroom pileus using Minolta Chroma Meter (Model CR200 Trimulus Colour Analyser, Minolta Camera Co. Ltd., Japan.) using CIELAB colour parameters, L* (lightness), a* (greenness (-) to redness (+)), and the (blueness (-) to yellowness (+)) values. A standard white tile was used to calibrate the equipment. The colour measurements was carried out immediately after harvesting.

2.6.3 Texture (firmness) Determination. Texture of fresh mushroom was determined by using texture analyser TA.XTplus (Stable Macro System) using P/2 (2 mm stainless steel probe). The selected pileus was placed on the stage of the texture analyser. Three different points were taken from the same surface of the harvested grey oyster mushroom to analyse the firmness of the pileus. In the penetration test, a 2 mm diameter cylinder probe was used to penetrate the harvested grey oyster mushroom pileus from top to 5mm depth. The firmness of the mushroom was determined by the maximum positive force for each test.

2.6.4 Moisture Content. Moisture analyser AND MX-50 was used to determine the moisture content of the mushroom pileus and stem. Freshly harvested grey oyster mushroom was weighed and placed on the pan at 105°C for about 30-45 minutes to obtain the percentage moisture content of the samples.

2.7 Experimental Design and Statistical Analysis

The experiment was designed based on Completely Randomized Design (CRD) with 1 control and 4 different durations and intervals of acoustic sound treatment. Each treatment was represented by 5 replicates. All the data obtained were analysed using statistical package SAS version 16.0 software. One-way Analysis of variances (ANOVA) was used to analyse the differences among the treatments. The significant means ($P<0.05$) was tested through Tukey's Studentized Range Test.

3. Results and Discussion

3.1 Mycelium Growth Rate

Mycelium growth rate of grey oyster mushroom's inflicted to constant sound intensity of 75dB, 5 or 15 pops of fire cracker explosion at 5- or 10-day interval are showed in Table 1. It showed that all the mushroom bags treated with acoustic sound treatment had significantly higher mycelium growth rate ($P<0.05$) compared to control. The fastest rate of mycelium growth (0.791 cm/day) to fully colonized the bags was from the treatment of 15 fire cracker pops at 10-day interval. However, control bags showed the slowest of mycelium growth rate (0.698 cm/day). This showed that acoustic sound treatment had significant effect on the growth of mycelium of grey oyster mushrooms.

Table 1: The mycelium growth rate (cm/day) of grey oyster mushroom subjected to different conditions of acoustic sound treatments.

Acoustic sound treatments	Mycelium growth rate (cm/day)
Control	0.698 ± 0.0359^{b}
5 pops at 5-day interval	0.761 ± 0.036^{a}
15 pops at 5-day interval	0.766 ± 0.0385^{a}
5 pops at 10-day interval	0.791 ± 0.0148^{a}
15 pops at 10-day interval	0.761 ± 0.0328^{a}

Note: Values are means of 5 replicates. Means (n=5) $\pm$ standard deviation.
a-b : Values bearing the different superscript within the same column are significantly different at 5% level ($P<0.05$).

3.2 The Number of Days for Mycelium to Fill-up the Bag, Pinhead Emergence and Fruiting Body Formation

There were significant differences (P<0.05) in the number of days taken for mycelium to fill-up the mushroom bags among all of the different acoustic sound treatment (Figure 1). It can be seen that the acoustic sound treated bags had shorter time for mycelium to fill-up the substrate bags compared to control. Acoustic sound treatment with 75dB, 15 pops at 10-day interval took the shortest time for mycelium to fill-up the substrate bags which was 29.6 days followed by 15 pops at 5-day interval (30.6 days), 5 pops at 5-day interval and 5 pops at 10-day interval (30.8 days).

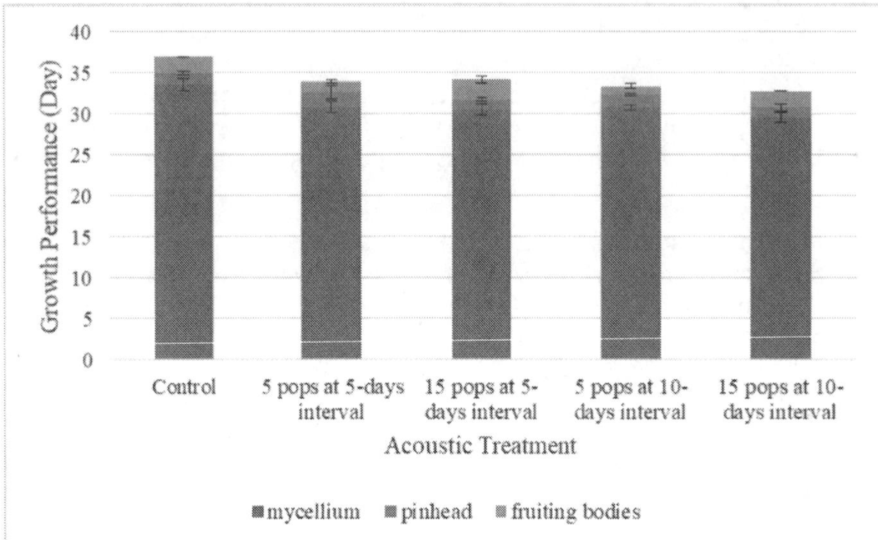

Figure 1: The number of days for mycelium to fill-up the bag, pinhead emergence and fruiting body formation of grey oyster mushroom (*Pleurotus sajor-caju*) subjected to different conditions of acoustic sound treatments. Vertical bars represent standard error.

A study done by [3], had reported that sound wave was able to significantly generate the protective enzymes and endogenous hormones under the suitable intensity, frequency and durations of the acoustic sound treatment.

There were significantly (P<0.05) longer time needed for pinhead emergence for the treatment of 5 pops at 5-day interval compared to other acoustic sound treatments and even control. The shortest time taken was 1.2 days which was shown by the treatment of 15 pops at 5-day interval and 15 pops at 10-day interval. High sound intensity at 75dB with 5 day interval has the shortest time taken for mycelium to fill-up the bag with 23.6 days and the control took the longest time with 40.6 days [5].

The number of days for fruiting bodies formation was counted from the pinhead emergence until the matured mushrooms had been harvested. Treatment of 5 pops at 5-day interval took the shortest time for the fruiting bodies to form which was 1.4 days. However, 15 pops at 5-day interval took the longest compared to all sound treated bags and control which was 2.4 days.

In overall, acoustic sound treatment with durations of 15 pops at 10-day interval took the shortest time for mycelium growth rate and the shortest time taken to complete all the different stages. In this study, the enzymes activity of mushrooms actively stimulated by acoustic sound treatment under suitable durations and intervals thus actively promoted the growth of mushrooms.

3.3 Yield

For the number of fruiting bodies, there was no significant different (P>0.05) among all the different conditions of acoustic sound treatments (Table 2) even though all the acoustic sound treated bags had higher number of fruiting bodies as compared to control.

Table 2: The yield of *Pleurotus sajor-caju* subjected to different conditions of acoustic sound treatments.

Acoustic sound treatments	No. of fruiting bodies	Total weight of fruiting bodies (g)	Biological efficiency (%)
Control	35.000±24.238[a]	333.000±159.515[a]	47.334±23.951[a]
5 pops at 5-day interval	51.000±29.875[a]	456.000±64.265[a]	64.762±9.455[a]
15 pops at 5-day interval	78.000±48.683[a]	329.000±241.076[a]	68.666±13.696[a]
5 pops at 10-day interval	46.000±6.519[a]	428.000±32.519[a]	62.093±8.707[a]
15 pops at 10-day interval	65.000±23.184[a]	538.000±159.044[a]	79.563±23.184[a]

Note: Values are means of 5 replicates. Means (n=5) ± standard deviation.
a : Values bearing the different superscript within the same column are significantly different at 5% level (P<0.05).

The acoustic sound treatment of 15 pops at 5-day interval had significantly (P<0.05) the highest number of fruiting bodies which was 35 followed by 15 pops at 10-day interval (65), 5 pops at 5-day (51), and 5 pops at 10-day interval (46) for the total of 5 harvesting cycles. It is obvious that control had the lowest number of fruiting bodies which was only 35. It is reported that application of acoustic sound treatment at 111dB, had promoted the growth of paddy rice seeds. This is because the sound wave has the ability to change the cell cycle of the paddy rice cells thus increase the rate of reproduction [3]. However, 15 pops at 5-day interval had the lowest total weight of fruiting bodies (329 g) and 15 pops at 10-day interval had the highest total weight of fruiting bodies (538 g). Control obtained a total weight of fruiting bodies of 333 g, 5 pops at 5-day interval had 456 g, and 5 pops at 10-day interval had 428 g.

There was no significant different (P>0.05) among all the treatments. However, Control obtained the lowest biological efficiency as compared to acoustic sound treated bags (Table 2). From the results, it can be seen that control (47.3%) tend to show lower biological efficiency than all the sound treated bags (ranged from 62.093 % to 79.563 %). This showed that all acoustic sound treated bags had a greater yield potentials of mushrooms converted from the substrate.

3.4 Physico-chemical Properties
The pileus size of *Pleurotus sajo-caju* was determined by pileus diameter. There was no significant different (P>0.05) in the largest pileus diameter of harvested mushrooms among all the different durations of acoustic sound treatment (Table 2).

Table 3: The pileus diameter (cm) of *Pleurotus sajor-caju* subjected to different conditions of acoustic sound treatments.

Acoustic sound treatments	Largest Pileus diameter (cm)	Smallest Pileus diameter (cm)
Control	8.600±0.6325[a]	2.700±1.446[a]
5 pops at 5-day interval	11.580±1.300[a]	1.740±0.699[a]
15 pops at 5-day interval	10.960±1.658[a]	3.320±1542[a]
5 pops at 10-day interval	11.760±3.713[a]	2.740±1.757[a]
15 pops at 10-day interval	10.500±1.149[a]	1.900±0.418[a]

Note: Values are means of 5 replicates. Means (n=5) ± standard deviation.
a: Values bearing the different superscript within the same column are significantly different at 5% level (P<0.05).

However, all acoustic sound treated bags produced larger pileus diameter than control (Table 3). The largest pileus diameter for all acoustic sound treated bags were ranged from 10.5 cm to 11.76 cm. However, control had the largest pileus diameter of 8.6 cm only acoustic sound treatment of 5 pops at

10-day interval had the largest pileus diameter of 11.760 cm followed by 5 pops at 5-day interval (11.580 cm), 15 pops at 5-day interval (10.960 cm) and 15 pops at 10 day interval (10.50 cm). This showed that acoustic sound treatment had significant effect on the largest pileus diameter of the mushrooms.

For the smallest pileus diameter of the mushrooms, there was no significant different (P>0.05) observed among all different durations of acoustic sound treatment (Table 2). The range of the small pileus is between 3.320 cm to 1.740 cm. The smallest is pileus 1.740 cm (5 pops at 5-day interval).

The physico-chemical characteristics of grey oyster mushroom in terms of colours were analysed. The colour will be divided into 3 parts, which are L* value, a* value and b* value. L* represents the colour's lightness, which L* 100 was pure white and L* 0 was matte black. a* values was in ranged of -60 (greenness) to +60(redness) and b* values was in ranged of -60 (blueness) to +60 (yellowness).

Table 4: The physico-chemical properties (colours) of *Pleurotus sajor-caju* subjected to different durations of acoustic sound treatment.

Acoustic sound treatments	Color		
	L* value	**a* value**	**b* value**
Control	57.232 ± 4.056^{c}	5.538 ± 0.566^{a}	14.093 ± 0.735^{a}
5 pops at 5-day interval	59.863 ± 1.697^{bc}	5.818 ± 0.799^{a}	17.273 ± 3.768^{a}
15 pops at 5-day interval	64.152 ± 2.440^{a}	5.311 ± 0.701^{a}	15.821 ± 2.000^{ab}
5 pops at 10-day interval	57.944 ± 1.575^{c}	4.850 ± 1.023^{aa}	$14.970 \pm 0..836^{ab}$
15 pops at 10-day interval	62.546 ± 2.672^{ab}	5.301 ± 0.514^{a}	15.179 ± 0.890^{ab}

Note: Values are means of 5 replicates. Means (n=5)±standard deviation.
a-c : Values bearing different superscript within the same column are significantly different at 5% level (P<0.05).

Based on Table 3, in terms of L* values there were significant differences (P<0.05) among all different acoustic sound treatments. All acoustic sound treated grey oyster mushrooms had higher L* values compared to the control. The lightest pileus colour indicated by the highest L* values was 15 pops at 5-day interval which was 64.152, followed by 15 pops at 10-day interval (62.546), 5 pops at 5-day interval (59.863), 5 pops at 10-day interval (57.944) and control (57.232).

In terms of a* values, there was no significant different (P>0.05) and were in range of 4.850 to 5.818 which was in the region of red colour.

There were significant differences (P<0.05) in terms of b* values among all different acoustic sound treatments. All acoustic sound treated mushrooms had higher b* values indicating yellow colour compared to control. The highest b* values was from 5 pops at 5-day interval (17.273) and the lowest was from control (14.093).

Table 5: The physico-chemical properties (firmness,) of *Pleurotus sajor-caju* subjected to different durations of acoustic sound treatment.

Acoustic sound treatments	Firmness (gf)	Pileus's Moisture Content (%)	Stem's Moisture Content (%)
Control	77.520 ± 7.931^{a}	88.216 ± 4.554^{a}	82.560 ± 9.556^{a}
5 pops at 5-day interval	55.555 ± 11.165^{b}	89.880 ± 0.966^{a}	79.940 ± 5.703^{a}
15 pops at 5-day interval	70.088 ± 11.832^{ab}	87.112 ± 4.890^{a}	72.800 ± 7.476^{a}
5 pops at 10-day interval	70.089 ± 11.155^{ab}	72.300 ± 30.969^{a}	78.640 ± 5.892^{a}
15 pops at 10-day interval	61.978 ± 12.466^{b}	88.466 ± 2.823^{a}	75.720 ± 5.042^{a}

Note: Values are means of 5 replicates. Means (n=5)±standard deviation.
a-b : Values bearing different superscript within the same column are significantly different at 5% level (P<0.05).

Based on Table 4, the physico-chemical properties in terms of firmness of grey oyster mushrooms, there were significant differences (P<0.05) among all the treatments. The highest firmness value was from control (77.520 gF), followed by 5 pops at 10-day interval (70.089 gF), 15 pops at 5-day interval (70.088 gF), 15 pops at 10-day interval (61.978 gF) and 5 pops at 5-day interval (55.555 gF). There was no significant different (P>0.05) in terms of moisture content for both pileus and stem of the grey oyster mushroom among all different treatment (Table 4). This shows that the acoustic sound treatment does not have any effect towards the moisture content for all the harvested grey oyster mushrooms.

4. Summary

From this study, the effects of different acoustic sound treatment on the growth performance, yield enhancement and quality of grey oyster mushroom (*Pleurotus sajor-caju*) were investigated. Acoustic sound treated bags had faster time for mycelium growth rate compared to control. 15 pops at 10-day interval took the shortest time for mycelium to fully grown compared to all acoustic sound treatment bags and control (30 days). Besides that, time taken for all acoustic sound treatment has faster time taken to complete the stages of growth (spawn-running process, pinhead emergence and fruiting bodies formation). There were sign of improvement on yield of mushroom especially on number of fruiting bodies. There were significant differences (P<0.05) for L*values and firmness of the grey oyster mushroom harvested from acoustic sound treated bags compared to control. However, there was no significant different (P>0.05) on a* values, b* values and moisture content of pileus and stem of the grey oyster mushroom. In conclusion, acoustic sound treatment with 15 pops at 10 day interval was the best treatment duration especially for the higher growth rate of mycelium and shorter time taken to complete all growth stages.

5. References

[1] Cağlarırmak, N. (2007). The nutrients of exotic mushrooms (Lentinula edodes and Pleurotus species) and an estimated approach to the volatile compound. *Food Chemistry*, 105.

[2] Chuanyun, D., Bochu, W., Huan, Z., Conglin, H., Chuanren, D., Wangqian, L., . . . Sakanishi, A. (2004). Effect of low frequency ultrasonic stimulation on the secretion of siboflavin produced by Ecemothecium Ashbyii. *Colloids and Surfaces B: Biointerfaces, 34*(1), 7-11. doi:https://doi.org/10.1016/j.colsurfb.2003.10.005

[3] Hassanien, R., Hou, T.-z., Li, Y.-f., & Li, B. (2014). Advances in Effects of Sound Waves on Plants. *Journal of Integrative Agriculture, 13*, 335–348. doi:10.1016/S2095-3119(13)60492-X

[4] Ibrahim, R., Amir Mukhlis, A., S. Nur Ain, M., Shi Fern, C., & Zakaria, Z. (2017). Enhancing growth and yield of grey oyster mushroom (Pleurotus sajor-caju) using sound treatment at different intervals (Vol. 1885).

[5] Ibrahim, R., Mohd Aziz, K., Arshad, A. M., & Hasan, S. (2015). Enhancing mushroom production using physical treatments prior to fruiting body formation (Vol. 44).

[6] Ibrahim, R., Mohd Jamil, A. A. I., Hasan, S. M. Z., Mat Arshad, A., & Zakaria, Z. (2017). Enhancing Growth and Yield of Grey Oyster Mushroom (Plearotussajorcaju) Using Different Acoustic Sound Treatments. *MATEC Web Conf., 97*, 01054.

[7] Patil, S. S. (2012). Cultivation of Pleurotus sajor-caju on different agro wastes (Vol. 2).

Acknowledgement

The researchers would like to acknowledge the Ministry of Higher Education for financial support through FRGS Grant, Department of Chemical Engineering Technology, Faculty of Engineering Technology, UniMAP for their facilities and continuous support

IOP Conf. Series: Materials Science and Engineering **767** (2020) 012014 doi:10.1088/1757-899X/767/1/012014

Assessing the Effect of Different Covariates Distributions on Parameter Estimates for Multinomial Logistic Regression (MLR)

Hamzah Abdul Hamid[1,a)], Siti Raudhah Ismail[2,b)], Sahimel Azwal Sulaiman[3,c)] and Nor Azrita Mohd Amin[1,d)]

[1]Institute of Engineering Mathematics, Universiti Malaysia Perlis, Pauh Putra Main Campus, 02600 Arau, Perlis, Malaysia
[2]Kolej Matrikulasi Kedah, Kementerian Pendidikan Malaysia, 06010 Changlun, Kedah Darul Aman, Malaysia
[3]Faculty of Muamalah & Management, Kolej Universiti Islam Perlis, Taman Seberang Jaya, Fasa 3, Seberang Ramai, 02000 Kuala Perlis, Perlis, Malaysia

Corresponding author: hamzahhamid@unimap.edu.my

Abstract. In fitting a multinomial logistic regression model, one of the most important part is estimating the parameter. In Multinomial Logistic Regression (MLR), Maximum Likelihood Estimation (MLE) method is used to estimate the parameters. MLE is the suitable method to be applied to the problems associated with categorical response variables since it has several benefits such as sufficiency, consistency, efficiency and parameterization invariance. This study investigates the different type of continuous distributions (normal, negatively skewed, positively skewed) on parameter estimation via Monte Carlo simulation. From the simulation result, it shows that as the sample size increases, the effect of covariate distribution reduces. The estimated parameter also less affected for model with normal covariate distribution. At sample size 300 and above, the estimated parameter with normal covariate distribution is considered as close to the true parameter value. Interestingly, for the positively skewed, the estimated parameter also obtained unbiased parameter at sample size 300 and above. However, for negatively skewed, it requires a larger sample size to get closer to the true parameter value. The estimated parameters deviate too far from the true parameter at small sample size. As expected, as sample size increases the parameter estimates for all distributions are getting close to the true parameter value. Lastly, the distribution for MLR with more than one covariate give the same effect as the MLR model with only one covariate on parameter estimations.

1. Introduction

The logistic regression is one of the GLMs and it is used to model the relationships between a categorical response variables, *Y* and one or more independent variables, $X=(x_1,x_2,...,x_p)$. The response variables *Y* may have two or more categories and these categories can be nominal or ordinal. If Y has only two categories, it is called binary logistic regression. However, when *Y* has more than two nominal categories, the multinomial logistic regression (MLR) is more relevant to be applied, while in the situation when Y has more than two categories and the categories can be ordered, the ordinal logistic regression is more appropriate to be applied. The independent variables in logistic regression model can be quantitative, qualitative or both (mixed). In logistic regression, the maximum likelihood method is usually applied to estimate the parameter.

The distribution of the covariates may affect the estimation of parameter of statistical model. [1] reported that ANOVA and Kruskal-Wallis tests are affected when the data is not normal. They reported that both tests are affected by the kurtosis of the error distribution, but less affected by the skewness. The simulation results of [2] showed that the presence of kurtosis and small sample size are affected the t-test for simple linear regression model with non-normal errors.

The investigation by [3] shows that the parameter estimates for MLR model with a single covariate affected when data is not normally distributed. The parameter estimates are deviated too far from the true parameter when the distribution of covariate is negatively skewed and required very large sample size to get close to the true parameter value. The parameter estimates for MLR with categorical covariate getting closer to the true parameter when the sample size is 300 and above. However, the effect of distribution on MLR model with more than one covariate and minimum sample size required to minimize the effect of covariate distribution is still vague. Therefore, the objective of this simulation study is to investigate the minimum sample size required to minimize the effect of covariate distributions for MLR model with more than one covariates on parameter estimation.

2. Methodology

2.1. Multinomial Logistic Regression

In multinomial logistic regression, let Y be an outcome variable with c possible value $(0,1,\ldots,c\text{-}1)$ with Y=0 be the reference category. Let $X=(x_1,x_2,\ldots,x_p)$ be the independent variables. Thus, the conditional probabilities of each outcome category can be written as [4]:

$$P(Y=0\mid x)=\frac{1}{1+e^{g_1(x)+\ldots+g_{c-1}(x)}}$$

$$P(Y=1\mid x)=\frac{e^{g_1(x)}}{1+e^{g_1(x)+\ldots+g_{c-1}(x)}}$$

$$P(Y=c-1\mid x)=\frac{e^{g_{c-1}(x)}}{1+e^{g_1(x)+\ldots+g_{c-1}(x)}}$$

It follows that the logit function of category j:

$$g_j(x)=\ln\left[\frac{P(Y=j\mid x)}{P(Y=0\mid x)}\right]=\beta_{j0}+\beta_{j1}x_1+\ldots+\beta_{jp}x_p \quad \text{for } j=1,2,\ldots,c\text{-}1$$

2.2. Simulation Procedures

The simulation study is used to evaluate the effect of distribution for MLR with more than one covariate and the sample size required in order to estimate the parameters. The Monte Carlo simulation is used and the sample sizes of 50, 100, 150, 300, 500, 1000, 1500, 3000 and 5000 are considered in this investigation. The simulation procedure was carried out using R, an open source programming software (version 3.5.1). The data will be generated using the same technique as in [5]. **Table 1** shows the distribution of covariates and true parameter values for the multinomial logistic regression model. At least one of the covariate distributions in this model is N(0,1). This is to avoid

IOP Conf. Series: Materials Science and Engineering **767** (2020) 012014 doi:10.1088/1757-899X/767/1/012014

the imbalanced data set generated for the dependent variables. The coefficient values are chosen to represent an odds ratio for each covariate is greater and lower than 1. Odds ratio greater than 1 indicates exposure associated with higher odds of the outcome while the odds ratio lower than 1 indicates exposure associated with lower odds of outcome.

Table 1. The distribution of covariates and true parameter values

Model	Covariate Distribution	Skewness	Kurtosis	Coefficient
A	N(0,1) and N(0,1)	0.000 and 0.000	2.996 and 2.996	β_{10}= -2.10
				β_{11}= -0.35
				β_{12}= 1.08
				β_{20}= -1.90
				β_{21}= -0.21
				β_{22}= 1.69
B	N(0,1) and Beta (12,1)	0.000 and -1.577	2.996 and 6.108	β_{10}= -2.10
				β_{11}= -0.35
				β_{12}= 1.08
				β_{20}= -1.90
				β_{21}= -0.21
				β_{22}= 1.69
C	N(0,1) and χ^2 (4)	0.000 and 1.405	2.996 and 5.931	β_{10}= -2.10
				β_{11}= -0.35
				β_{12}= 1.08
				β_{20}= -1.90
				β_{21}= -0.21
				β_{22}= 1.69

The simulation steps involved are as follows:

Step 1: The value of x is generated from the stated distribution (normal, positively skewed, negatively skewed).

Step 2: Fit the multinomial logit equation:

$$g_1(x) = \beta_{10} + \beta_{11}x_1 + \beta_{12}x_2$$

$$g_2(x) = \beta_{20} + \beta_{21}x_1 + \beta_{22}x_2$$

Step 3 : Evaluate the multinomial logistic probabilities for each category (π_0, π_1, π_2).

$$\pi_0(x) = \frac{1}{1 + e^{g_1(x)} + e^{g_2(x)}}$$

$$\pi_1(x) = \frac{e^{g_1(x)}}{1 + e^{g_1(x)} + e^{g_2(x)}}$$

$$\pi_2(x) = \frac{e^{g_2(x)}}{1 + e^{g_1(x)} + e^{g_2(x)}}$$

.

Step 4: The value of U is generated by using an independent $U(0,1)$ distribution.

Step 5: Variable (Y) outcome is assigned based on the rule:

i. Y=1 if $u > \pi_0 + \pi_1$

ii. Y=2 if $u < \pi_0 + \pi_1$ and $u > \pi_0$

iii. Y=0 otherwise

Step 6 : Fit multinomial logistic regression model based on the generated x and y and estimate β .

Step 7: Repeat Steps (1) – (6) for 10000 replications to obtain $\hat{\beta}^1, \hat{\beta}^2, ..., \hat{\beta}^{10000}$.

Step 8 : Calculate the mean of $\hat{\beta}$, $\bar{\hat{\beta}} = \sum_{k=1}^{10000} \hat{\beta} \Big/ 10000$

Step 9 : Calculate the 95% Confidence Interval (CI) for $\bar{\hat{\beta}}$.

This simulation study involved three types of continuous distribution. N(0,1) represents normal distribution, Beta(12,1) represents negatively skewed and $\chi^2(4)$ represents positively skewed distribution.

3. Simulation Results

Simulation study is carried out in order to investigate the effect of covariate distribution on parameter estimates for multinomial logistic regression. **Table 2** shows the summary of the parameter estimated and the 95% confidence interval (CI) for model that contains two normal covariate distribution (Model A). At sample size of 300 and above, the estimated parameter can be considered as closed to the true parameter value. Overall, the estimated parameters not deviate too far from the true parameter even the sample size used is small.

Table 2. Parameter Estimation for Model A

Sample size	β_{10}= -2.10	β_{11}= -0.35	β_{12}=1.08	β_{20}= -1.90	β_{21}= -0.21	β_{22}= 1.69
50	-2.068	-0.428	1.301	-2.314	-0.255	2.162
	(-2.080,-2.056)	(-0.441, -0.416)	(1.285 ,1.317)	(-2.408, -2.219)	(-0.282, -0.227)	(2.084,2.239)
100	-2.239	-0.379	1.163	-2.036	-0.230	1.837
	(-2.248,-2.230)	(-0.387,-0.371)	(1.153 ,1.172)	(-2.045, -2.027)	(-0.237, -0.223)	(1.827, 1.847)
150	-2.210	-0.377	1.137	-1.987	-0.221	1.786
	(-2.217, -2.202)	(-0.383,-0.370)	(1.129, 1.145)	(-1.994, -1.980)	(-0.227, -0.216)	(1.778, 1.793)
300	-2.151	-0.358	1.101	-1.946	-0.218	1.738
	(-2.155,-2.146)	(-0.362, -0.353)	(1.096, 1.106)	(-1.951, -1.941)	(-1.951,-1.941)	(1.733, 1.743)
500	-2.130	-0.355	1.096	-1.924	-0.215	1.719
	(-2.134,-2.127)	(-0.358, -0.352)	(1.092,1.100)	(-1.928, -1.921)	(-0.218,-0.212)	(1.715, 1.723)
1000	-2.114	-0.353	1.087	-1.912	-0.213	1.703
	(-2.112,-2.117)	(-0.356, -0.351)	(1.085,1.090)	(-1.914,-1.909)	(-0.215, -0.211)	(1.700,1.705)
1500	-2.109	-0.351	1.084	-1.909	-0.210	1.699
	(-2.107,-2.111)	(-0.353,-0.349)	(1.082, 1.086)	(-1.911,-1.907)	(-0.212,-0.208)	(1.697, 1.701)
3000	-2.105	-0.351	1.083	-1.904	-0.210	1.694
	(-2.103,-2.106)	(-0.353, -0.350)	(1.082, 1.085)	(-1.905,-1.902)	(-0.211,-0.209)	(1.692, 1.695)
5000	-2.103	-0.352	1.082	-1.902	-0.210	1.692
	(-2.102,-2.104)	(-0.353,-0.351)	(1.080,1.083)	(-1.903, -1.901)	(-0.211, -0.209)	(1.691,1.693)

Table 3 shows the summary of parameter estimates and the 95% (CI) for Model B – model with two covariate distributions which are N(0,1) and Beta(12,1). The value of parameter estimates get closer to true parameter value at sample size of 1000 and above. This shows that the model that contains negatively skewed covariate requires larger sample size to get closer to the true parameter value compared to the model without negatively skewed distribution covariate. The estimated parameters become unbiased when the sample size is very large.

Table 3. Parameter Estimation for Model B

Sample size	β_{10}= -2.10	β_{11}= -0.35	β_{12}=1.08	β_{20}= -1.90	β_{21}= -0.21	β_{22}= 1.69
50	-3.454	-0.407	2.436	-2.483	-0.238	2.287
	(-3.609, -3.298)	(-0.418,-0.397)	(2.271,2.602)	(-2.593 ,-2.374)	(-0.246,-0.231)	(2.169,2.404)
100	-2.725	-0.375	1.671	-2.088	-0.223	1.869
	(-2.816,-2.634)	(-0.381,-0.368)	(1.573,1.768)	(-2.156,-2.021)	(-0.228,-0.218)	(1.796,1.942)
150	-2.492	-0.364	1.456	-2.084	-0.217	1.879
	(-2.561, -2.423)	(-0.369,-0.359)	(1.382,1.530)	(-2.137,-2.032)	(-0.220,-0.213)	(1.823, 1.935)
300	-2.278	-0.361	1.249	-1.935	-0.214	1.770
	(-2.232,-2.323)	(-0.364,-0.357)	(1.200,1.298)	(-1.966,-1.900)	(-0.216,-0.211)	(1.732, 1.808)
500	-2.236	-0.353	1.214	-1.930	-0.212	1.720
	(-2.271,-2.201)	(-0.356,-0.351)	(1.252,1.177)	(-1.957,-1.904)	(-0.214,-0.210)	(1.692, 1.749)
1000	-2.138	-0.353	1.113	-1.913	-0.211	1.702
	(-2.161,-2.114)	(-0.355,-0.351)	(1.088,1.138)	(-1.932,-1.894)	(-0.212,-0.210)	(1.682,1.722)
1500	-2.134	-0.350	1.113	-1.924	-0.208	1.714
	(-2.153,-2.115)	(-0.352,-0.349)	(1.092,1.134)	(-1.939,-1.908)	(-0.212,-0.204)	(1.698, 1.731)
3000	-2.118	-0.351	1.096	-1.912	-0.211	1.702
	(-2.131,-2.104)	(-0.352,-0.350)	(1.081,1.110)	(-0.212, -0.204)	(-0.212,-0.204)	(1.690, 1.713)
5000	-2.115	-0.351	1.095	-1.901	-0.210	1.690
	(-2.126,-2.105)	(-0.351, -0.350)	(1.084,1.106)	(-1.909,-1.893)	(-0.211, -0.209)	(1.681,1.699)

Next, **Table 4** shows the summary of the parameter estimates and 95% CI for Model C, which is the model that contains positively skewed covariate. The result shows that the parameter estimates can be considered as close to the true parameter value at sample size of 300 and above. This indicates that the model that contains positively skewed covariate are less affected compared to the model that contains negatively skewed covariate.

Table 4. Parameter Estimation for Model C

Sample size	β_{10}= -2.10	β_{11} = -0.35	β_{12} =1.08	β_{20} = -1.90	β_{21} = -0.21	β_{22} = 1.69
50	-3.305	-0.598	2.239	-3.776	-0.421	3.050
	(-3.459 ,-3.151)	(-0.652,-0.544)	(2.109,2.368)	(-3.930,-3.621)	(-0.474,-0.367)	(2.920,3.180)
100	-2.365	-0.416	1.299	-2.293	-0.256	2.013
	(-2.396,-2.334)	(-0.432,-0.399)	(1.274,1.323)	(-2.323,-2.262)	(-0.272,-0.241)	(1.988,2.037)
150	-2.274	-0.376	1.201	-2.131	-0.236	1.875
	(-2.256,-2.291)	(-0.367,-0.384)	(1.191,1.212)	(-2.147,-2.115)	(-0.243,-0.228)	(1.865,1.885)
300	-2.180	-0.362	1.132	-2.003	-0.220	1.771
	(-2.191,-2.168)	(-0.368,-0.357)	(1.126,1.139)	(-2.014,-1.993)	(-0.224,-0.215)	(1.765,1.777)
500	-2.148	-0.364	1.113	-1.965	-0.220	1.742
	(-2.156,-2.139)	(-0.368,-0.360)	(1.109,1.118)	(-1.973,-1.958)	(-0.224,-0.217)	(1.738,1.747)
1000	-2.118	-0.358	1.094	-1.928	-0.215	1.713
	(-2.124,-2.112)	(-0.361,-0.355)	(1.091,1.097)	(-1.933,-1.922)	(-0.218,-0.213)	(1.710,1.716)
1500	-2.114	-0.353	1.090	-1.921	-0.212	1.706
	(-2.119,-2.109)	(-0.355,-0.350)	(1.087,1.093)	(-1.925,-1.917)	(-0.201,-0.324)	(1.704,1.709)
3000	-2.108	-0.352	1.085	-1.908	-0.212	1.697
	(-2.111,-2.104)	(-0.354, -0.350)	(1.083,1.086)	(-1.911, -1.905)	(-0.213, -0.210)	(1.695,1.699)
5000	-2.105	-0.351	1.083	-1.904	-0.210	1.694
	(-2.107,-2.102)	(-0.353,-0.350)	(1.082,1.085)	(-1.910,-1.905)	(-0.212,-0.210)	(1.694,1.696)

Table 5 shows the summary of the parameter estimates for each model in all three distributions for comparison purposes. Based on this table, it can be seen clearly that Model B (model with negatively skewed covariate) needs larger sample size to get closer to the true parameter value compared to other models.

Table 5. Parameter Estimates for Different Distribution

Sample size	Model	$\beta_{10} = -2.10$	$\beta_{11} = -0.35$	$\beta_{12} = 1.08$	$\beta_{20} = -1.90$	$\beta_{21} = -0.21$	$\beta_{22} = 1.69$
50	A	-2.068	-0.428	1.301	-2.314	-0.255	2.162
	B	-3.454	-0.407	2.436	-2.483	-0.238	2.287
	C	-3.305	-0.598	2.239	-3.776	-0.421	3.050
100	A	-2.239	-0.379	1.163	-2.036	-0.230	1.837
	B	-2.725	-0.375	1.671	-2.088	-0.223	1.869
	C	-2.365	-0.416	1.299	-2.293	-0.256	2.013
150	A	-2.210	-0.377	1.137	-1.987	-0.221	1.786
	B	-2.492	-0.364	1.456	-2.084	-0.217	1.879
	C	-2.274	-0.376	1.201	-2.131	-0.236	1.875
300	A	-2.151	-0.358	1.101	-1.946	-0.218	1.738
	B	-2.278	-0.361	1.249	-1.935	-0.214	1.770
	C	-2.180	-0.362	1.132	-2.003	-0.220	1.771
500	A	-2.130	-0.355	1.096	-1.924	-0.215	1.719
	B	-2.236	-0.353	1.214	-1.930	-0.212	1.720
	C	-2.148	-0.364	1.113	-1.965	-0.220	1.742
1000	A	-2.114	-0.353	1.087	-1.912	-0.213	1.703
	B	-2.138	-0.353	1.113	-1.913	-0.211	1.702
	C	-2.118	-0.358	1.094	-1.928	-0.215	1.713
1500	A	-2.109	-0.351	1.084	-1.909	-0.210	1.699
	B	-2.134	-0.350	1.113	-1.924	-0.208	1.714
	C	-2.114	-0.353	1.090	-1.921	-0.212	1.706
3000	A	-2.105	-0.351	1.083	-1.904	-0.210	1.694
	B	-2.118	-0.351	1.096	-1.912	-0.211	1.702
	C	-2.108	-0.352	1.085	-1.908	-0.212	1.697
5000	A	-2.103	-0.352	1.082	-1.902	-0.210	1.692
	B	-2.115	-0.351	1.095	-1.901	-0.210	1.690
	C	-2.105	-0.351	1.083	-1.904	-0.210	1.694

Model A:Normal (0,1), Model B: Beta (12,1), Model C: $\chi^2(4)$

To measure how close are the fitted parameter estimate values, $\hat{\beta}$ to the true parameter values, the Mean Square Error (MSE) is used. **Table 6** shows the MSE for Model A. The values of MSE for model A (both normal distributed covariate) become smaller (less than 1) at sample size of 100 and above.

Table 6. MSE for Model A

Sample size	β_{10}=-2.10	β_{11}=-0.35	β_{12}=1.08	β_{20}=-1.90	β_{21}=-0.21	β_{22}=1.69
50	0.375	0.425	0.69	23.341	1.985	16.008
100	0.235	0.17	0.26	0.237	0.129	0.275
150	0.157	0.106	0.161	0.136	0.074	0.154
300	0.061	0.047	0.069	0.059	0.034	0.067
500	0.035	0.027	0.04	0.033	0.02	0.038
1000	0.016	0.013	0.019	0.016	0.01	0.018
1500	0.011	0.009	0.013	0.01	0.007	0.012
3000	0.005	0.004	0.006	0.005	0.003	0.006
5000	0.003	0.003	0.004	0.003	0.002	0.003

Table 7 summarizes the MSE result for Model B (contains one negatively skewed covariate). The result shows that the MSE value is very large when the sample size is small. As sample size increases, the MSE value decreases. The model needs larger sample size to get smaller MSE value compared to the model without negatively skewed covariate distribution.

Table 7. MSE for Model B

Sample size	β_{10}=-2.10	β_{11}=-0.35	β_{12}=1.08	β_{20}=-1.90	β_{21}=-0.21	β_{22}=1.69
50	65.037	0.274	73.089	31.665	0.151	36.288
100	22.057	0.110	25.154	11.904	0.061	13.715
150	12.484	0.066	14.36	7.099	0.22	20.904
300	5.366	0.031	6.217	3.243	0.018	3.769
500	3.131	0.018	3.632	1.832	0.011	2.128
1000	1.008	1.710	1.400	55.606	19.490	17.927
1500	0.984	0.006	1.146	0.612	0.003	0.710
3000	0.475	0.003	0.549	0.297	0.002	0.346
5000	0.280	0.002	0.326	0.177	0.001	0.206

The values of MSE for Model C are summarized in **Table 8**. The values of MSE become smaller (less than 1) at sample size of 150 and above.

Table 8. MSE for Model C

Sample size	β_{10}= -2.10	β_{11}= -0.35	β_{12}=1.08	β_{20}= -1.90	β_{21}= -0.21	β_{22}= 1.69
50	63.086	7.654	45.130	65.702	7.475	45.675
100	2.592	0.706	1.665	2.538	0.614	1.695
150	0.843	0.195	0.302	0.743	0.139	0.300
300	0.350	0.082	0.11	0.287	0.057	0.103
500	0.194	0.047	0.06	0.161	0.033	0.057
1000	0.094	0.023	0.028	0.076	0.016	0.026
1500	0.061	0.015	0.018	0.049	0.010	0.016
3000	0.030	0.007	0.009	0.024	0.005	0.008
5000	0.018	0.004	0.005	0.015	0.003	0.005

Table 9 summarizes the MSE values for all three models. The results show that the MSE values for the model that contains negatively skewed covariate are the largest compared to the other models. As expected, for all models, as the sample size is increases, the MSE value also decreases.

Table 9. MSE for Model A, B and C

Sample size	Model	β_{10}= -2.10	β_{11}= -0.35	β_{12}=1.08	β_{20}= -1.90	β_{21}= -0.21	β_{22}= 1.69
50	A	0.375	0.425	0.69	23.341	1.985	16.008
	B	65.037	0.274	73.089	31.665	0.151	36.288
	C	63.086	7.654	45.130	65.702	7.475	45.675
100	A	0.235	0.17	0.26	0.237	0.129	0.275
	B	22.057	0.110	25.154	11.904	0.061	13.715
	C	2.592	0.706	1.665	2.538	0.614	1.695
150	A	0.157	0.106	0.161	0.136	0.074	0.154
	B	12.484	0.066	14.36	7.099	0.220	20.904
	C	0.843	0.195	0.302	0.743	0.139	0.300
300	A	0.061	0.047	0.069	0.059	0.034	0.067
	B	5.366	0.031	6.217	3.243	0.018	3.769
	C	0.350	0.082	0.11	0.287	0.057	0.103
500	A	0.035	0.027	0.04	0.033	0.020	0.038
	B	3.131	0.018	3.632	1.832	0.011	2.128
	C	0.194	0.047	0.060	0.161	0.033	0.057
1000	A	0.016	0.013	0.019	0.016	0.010	0.018
	B	1.008	1.710	1.400	55.606	19.49	17.927
	C	0.094	0.023	0.028	0.076	0.016	0.026
1500	A	0.011	0.009	0.013	0.01	0.007	0.012
	B	0.984	0.006	1.146	0.612	0.003	0.71
	C	0.061	0.015	0.018	0.049	0.010	0.016
3000	A	0.005	0.004	0.006	0.005	0.003	0.006
	B	0.475	0.003	0.549	0.297	0.002	0.346
	C	0.030	0.007	0.009	0.024	0.005	0.008
5000	A	0.003	0.003	0.004	0.003	0.002	0.003
	B	0.280	0.002	0.326	0.177	0.001	0.206
	C	0.018	0.004	0.005	0.015	0.003	0.005

4. Conclusion

MLR is widely applied in many research fields such as medical, business, finance, computer science and environmental research. The aims of this study is to evaluate the effect of different distribution of covariates on parameter estimation for multinomial logistic regression model and to determine the minimum sample size required to minimize the effect of covariate distribution on parameter estimation for multinomial logistic regression model. The simulation result shows that the parameter estimation are affected by the distribution of covariate. A large sample size is required to produce unbiased estimates for the model that contains skewed covariate distribution compared to the model that contains no skewed covariate distribution. The parameter estimation is less affected for model that contains only normal covariate distribution and the model that contains positively skewed covariate distribution compared to the model that contains negatively skewed covariate distribution. For model that contains only normal covariate distribution, and model that contains positively skewed covariate distribution, the parameter estimates can be considered as close to the true parameter value when the sample is 300 and above, while for the model that contains negatively skewed covariate distribution, it needs sample size of 1000 and above. Overall, the results from this simulation study is consistent with the results in Hamid et al. (2016), which stated that the MLR model with only negatively skewed covariate distribution is more affected compared to the model with only normal and model with only positively skewed covariate distribution. Future research can be carried out to investigate the effect of count and categorical data on MLR. More information also can be gathered if the covariates are mixed between continuous, categorical and count data.

References

[1] Khan, A., & Rayner, G. D. (2004). Robustness to non-normality of common tests for the many-sample location problem. Journal of Applied Mathematics and Decision Sciences, 7(4), 187–206.

[2] Jahan, S., & Khan, A. (2012). Power of t-test for simple linear regression model with non-normal error distribution: a quantile function distribution approach. Journal of Scientific Research, 4(3), 609–622.

[3] Hamid, A. H., Yap, B. W., & Xie, X.-J. (2016). Effects of different type of covariates and sample size on parameter estimation for multinomial logistic regression model, Jurnal Teknologi UTM(3), 155–161.

[4] Hosmer Jr, D. W., Lemeshow, S., & Sturdivant, R. X. (2013). Applied logistic regression (Vol. 398). (3rd Edition). (Vol. 398).John Wiley & Sons.

[5] Fagerland, M. W., Hosmer, D. W., & Bofin, A. M. (2008). Multinomial goodness-of-fit tests for logistic regression models. Statistics in Medicine, 27(21), 4238–4253.

[6] Hamid A. H., Yap, B. W., Xie, X.-J., & Hezlin A.R. (2015). Assessing the Effects of Different Types of Covariates for Binary Logistic Regression. AIP Conference Proceedings 1643, 425–430.

[7] Kutner, Michael H., Christopher J. Nachtsheim, John Neter, W. L. (2005). Applied Linear Statistical Models (Fifth Edition). McGraw Hill.

[8] Mccullagh, P., & Nelder, J. A. (2007). Generalized linear models (Vol.37) CRC press.

[9] Xie, X. J., Pendergast, J., & Clarke, W. (2008). Increasing the power: A practical approach to goodness-of-fit test for logistic regression models with continuous predictors. Computational Statistics and Data Analysis, 52(5), 2703–2713.

IOP Conf. Series: Materials Science and Engineering **767** (2020) 012015 doi:10.1088/1757-899X/767/1/012015

Full Reference Image Quality Assessment Algorithm based on Haar Wavelet and Edge Perceptual Similarity

F Mokhtar[1] and R Ngadiran[1]

[1]Embedded, Networks and Advanced Computing Research Cluster (ENAC), School of Computer and Communication Engineering, Universiti Malaysia Perlis, Pauh Putra Campus, 02600 Arau, Perlis, MY.

faizahmokhtar94@gmail.com

Abstract. The research on image quality assessment (IQA) has been become a hot topic in image processing. Many studies show that HVS edge information plays crucial role when human perceive the quality of an image. The proposed metric is called HEPSI. The method from HaarPSI metric is combined with edge structural similarity and a contrast map is added for pooling the structural similarity map, Validation is taken by comparing HEPSI with the well-known state-of-the-art IQA metrics: PSNR, SSIM, MSSIM, FSIM and HaarPSI over the LIVE database. Experiment shows that HEPSI achieved better performance than other 5 IQA metrics.

1. Introduction

Assessing image quality is significant in optimizing image processing performance. Many studies have been performed for the purpose of effectively measure image quality. Digital images often pass through several processing stages such as acquisition, processing, storage and transmission before they reach to the observers [1]. These images are subjected to different kinds of distortions during the stages such as transmission, processing, acquisition and compression. These stages may result in degradation of visual quality of the images.

For example, during the transmission stage, the quality of the received image may decrease because of dropping of some data due to limited bandwidth of the channels. Consecutively, it is significant for image acquisition, communication, processing systems and management to measure the quality of images at each stage. Hence, image quality assessment (IQA) is very important in order to maintain and conserve the quality of the images.

1.1. Image Quality Assessment (IQA)

Image quality measurement (IQA) techniques can be categorized as subjective and objectives. Subjective measurement consists of mean opinion scores (MOS) scored by a number of selected observers that has shown a series of test scenes. Although this measurement will give accurate results, subjective evaluation is usually time-consuming, too inconvenient and cannot be incorporate in real time.

To eliminate the need for expensive subjective studies, numerous efforts have been made to develop objective measurement that can correlate with perceived quality. The goal of objective IQA is to design algorithms that are able to predict the quality of an image automatically and accurately. Objective measurement are automatic algorithms for quality assessment that could analyse images and

IOP Conf. Series: Materials Science and Engineering **767** (2020) 012015 doi:10.1088/1757-899X/767/1/012015

report their quality without human involvement. Such methods could eliminate the need for expensive subjective studies. The most widely used and simplest full reference (FR) IQAs are the peak signal-to-noise ratio (PSNR) and mean square error (MSE), but due to the low correlation between their results with human opinion scores another FR IQAs have been presented [2].

Wang et al. proposed structural similarity index (SSIM) based on the assumption that HVS is highly adapted by extracting structural information from an image [3]. After the huge achievement of SSIM, it gives inspirations and ideas to numerous new IQAs. Besides, Zhang et al. [4] proposed a low-level feature similarity index (FSIM), to assess phase congruency and gradient magnitude with the assumption that the HVS distinguishes images according to low-level features. Recently, Reisenhofer et al. [5] proposed utilizes the coefficients obtained from a Haar wavelet decomposition to assess local similarities between two images, as well as the relative importance of image areas.

2. Proposed quality assessment methods

Given image i is the reference image and image j is the distorted image. Coefficients obtained from a Haar wavelet decomposition is used to calculate the local similarities and the relative importance of image areas between image i and image j. The local similarity measure included chroma channels I and Q from the color images in the YIQ color space. These are calculated according to the method in [5]. This generalization is given by

$$S_{1_{i,j}} = l_\alpha^{-1}\left(\frac{\sum_x \sum_{k=1}^{3} HS_{i,j}^{(k)}[x]\cdot W_{i^Y,j^Y}^{(k)}[x]}{\sum_x \sum_{k=1}^{3} W_{i^Y,j^Y}^{(k)}[x]}\right)^2 \tag{1}$$

where l_α is non-linear mapping to the local similarities obtained from high-frequency Haar wavelet filter response. $HS_{i,j}^{(k)}[x]$ is the local similarity measure that is based on the first two stages of two dimensional Haar wavelet transform where $HS_{i,j}^{(1)}$ and $HS_{i,j}^{(2)}$ differentiates between horizontal and vertical filters and $HS_{i,j}^{(3)}$ is the chroma-sensitive local similarity measure.

$$l_\alpha(x) = \frac{1}{1+e^{-\alpha x}} \tag{2}$$

$$HS_{i,j}^{(1,2)}[x] = l_\alpha\left(\frac{1}{2}\sum_{n=1}^{2} S\left(\left|\left(g_n^{(1,2)} * \mathrm{i}\right)[x]\right|, \left|\left(g_n^{(1,2)} * \mathrm{j}\right)[x]\right|, C\right)\right) \tag{3}$$

$$HS_{i,j}^{(3)}[x] = l_\alpha\left(\frac{1}{2}\left(S(|(m * i^I)[x]|, |(m * j^I)[x]|, C) + S(|(m * i^Q)[x]|, |(m * j^Q)[x]|, C)\right)\right) \tag{4}$$

where S denotes the scalar similarity measure, m is 2×2 mean filter, $*$ is the two-dimensional convolution operator and C is a constant. Meanwhile, $W_{i^Y,j^Y}^{(k)}[x]$ is the weight map that is derived from single low frequency Haar wavelet filter.

$$W_{i,j}^{(k)}[x] = max\left(W_i^{(k)}[x], W_j^{(k)}[x]\right) \tag{5}$$

$$W_{i^Y,j^Y}^{(3)}[x] = \frac{1}{2}\left(W_{i^Y,j^Y}^{(1)}[x] + W_{i^Y,j^Y}^{(2)}[x]\right) \tag{6}$$

Mean square of detail subband (horizontal, vertical & diagonal) used to compute the edge map functions for i and j.

$$i_E(m,n) = \frac{1}{3}(i_H^2(m,n) + i_V^2(m,n) + i_D^2(m,n)) \quad (7)$$

$$j_E(m,n) = \frac{1}{3}(j_H^2(m,n) + j_V^2(m,n) + j_D^2(m,n)) \quad (8)$$

where i_E and j_E represent the edge maps of i and j respectively. (m,n) represents the sample position within the wavelet subbands. i_H, i_V and i_D denote horizontal, vertical and diagonal detail subbands of image i while j_H, j_V and j_D denote detail subbands of image j. The edge SSIM map is calculated between the two images.

$$SSIM_E(i_E, j_E) = \frac{2\sigma_{i_E,j_E}+c}{\sigma_{i_E}^2+\sigma_{j_E}^2+c} \quad (9)$$

where σ_{i_E,j_E} is the cross correlation between image patches i_E and j_E while $\sigma_{i_E}^2$ and $\sigma_{j_E}^2$ are the variances of i_E and j_E respectively. Next, contrast map is formed for weighted pooling of the edge SSIM map. It is well known that the HVS is more sensitive to areas near the edges [6]. The contrast map is calculated within a local Gaussian square window that moves (pixel-by-pixel) over the entire approximation subband i_A and edge map i_E.

$$Contrast\ (i_E, i_A) = \left(\mu_{iE}\ \sigma_{iA}^2\right)^{0.1} \quad (10)$$

$$\sigma_{iA}^2 = \sum_{k=1}^{N} w_k\left(x_{A,k} - \mu_{iA}\right)^2 \quad (11)$$

$$\mu_{iE} = \sum_{k=1}^{N} w_k\, i_{E,k}\ ,\ \mu_{iA} = \sum_{k=1}^{N} w_k\, i_{A,k} \quad (12)$$

$$S_2 = \frac{\sum_{n=1}^{M} Contrast(i_{E,n}|j_{A,n})\ .\ SSIM_E(i_{E,n},j_{E,n})}{\sum_{n=1}^{M} Contrast(i_{E,n},j_{A,n})} \quad (13)$$

The edge similarity scores, local similarities and the relative importance of image areas are combined to obtain overall quality measure between images i and j. The final quality score is calculated using the following formula.

$$HEPSI(i,j) = \gamma S_1 + (1-\gamma)S_2 \quad (14)$$

where γ is a constant. We set $\gamma = 0.99$ in this paper.

3. Evaluation of results

3.1. Image database

All the IQAs are validated on a publicly available image database which is LIVE database. LIVE image quality database contains 29 reference images and 779 distorted images in 24-bpp color BMP format at different image resolutions ranging from 634 ×438 to 768 ×512 pixels. There are five different types of distortions in this database: JPEG compression, JPEG2000 compression, additive Gaussian white noise, Gaussian blurring and JPEG2000 with bit errors via a simulated Rayleigh fading channel.

Each types of distortions were generated at five to six difference amounts of distortions. The scores were collected from 29 different subjects. LIVE database is intended for evaluation of FR IQA metrics. It allows estimating how a metric corresponds to mean human perception.

3.2. Performance evaluation

Four commonly used correlation coefficient (CC) are applied to assess the performance of the proposed technique between objective and subjective score pertaining to image quality. These four performance metrics are the Spearman rank order correlation coefficient (SROCC), Kendall rank-order correlation coefficient (KROCC), Pearson linear correlation coefficient (PLCC), and root-mean-squared error (RMSE). SROCC and KROCC assess the prediction monotonicity while PLCC and RMSE evaluate the prediction accuracy of an IQA [10].

In this research, a non-linear logistic regression function is chosen for non-linear mapping in the scatter plot. The function is recommended by the video quality experts' group (VQEG) and has been widely used by many IQA researchers [7]. The function is defined as following:

$$Q(x) = \beta_1 \left(\frac{1}{2} - \frac{1}{1+e^{\beta_2(x-\beta_3)}}\right) + \beta_4 x + \beta_5 \quad (15)$$

where Q represents the predicted score after regression, x denotes the raw objective scores of a metric and β_1, β_2, β_3, β_4 and β_5 are regression model parameters. The five regression model parameters are calculated by fitting the function to the objective and subjective data. The importance of this function is to analyse the performance in a common space and remove the non-linearity caused in the process of subjective scores [4], [8], [9].

4. Results and Discussion

4.1. Performance Comparison

The performance of HEPSI is compared with selected state-of-the-art IQAs which are SSIM [3], MSSIM [11], FSIM [4], HaarPSI [5] and PSNR models (MATLAB source code of all the methods are obtained from the original authors). The following Table 1 lists the SROCC, PLCC, KROCC and RMSE results of proposed metric and other 5 IQAs on LIVE database. The best one metric produces the highest CC and lowest RMSE is marked in **boldface**. The results show that HEPSI outperform all other metrics in terms of correlation coefficients and root mean square error. Thus, HEPSI performs better than other IQA metrics

Table 1. Performance comparison of IQA metrics on LIVE database.

	PSNR	SSIM	MSSIM	FSIM	HaarPSI	HEPSI
KROCC	0.6865	0.7963	0.8044	0.8337	0.8448	**0.8488**
PLCC	0.8723	0.9449	0.9489	0.9597	0.9695	**0.9709**
SROCC	0.8756	0.9479	0.9513	0.9634	0.9683	**0.9703**
RMSE	13.3597	8.9455	8.6187	7.6780	6.6935	**6.5383**

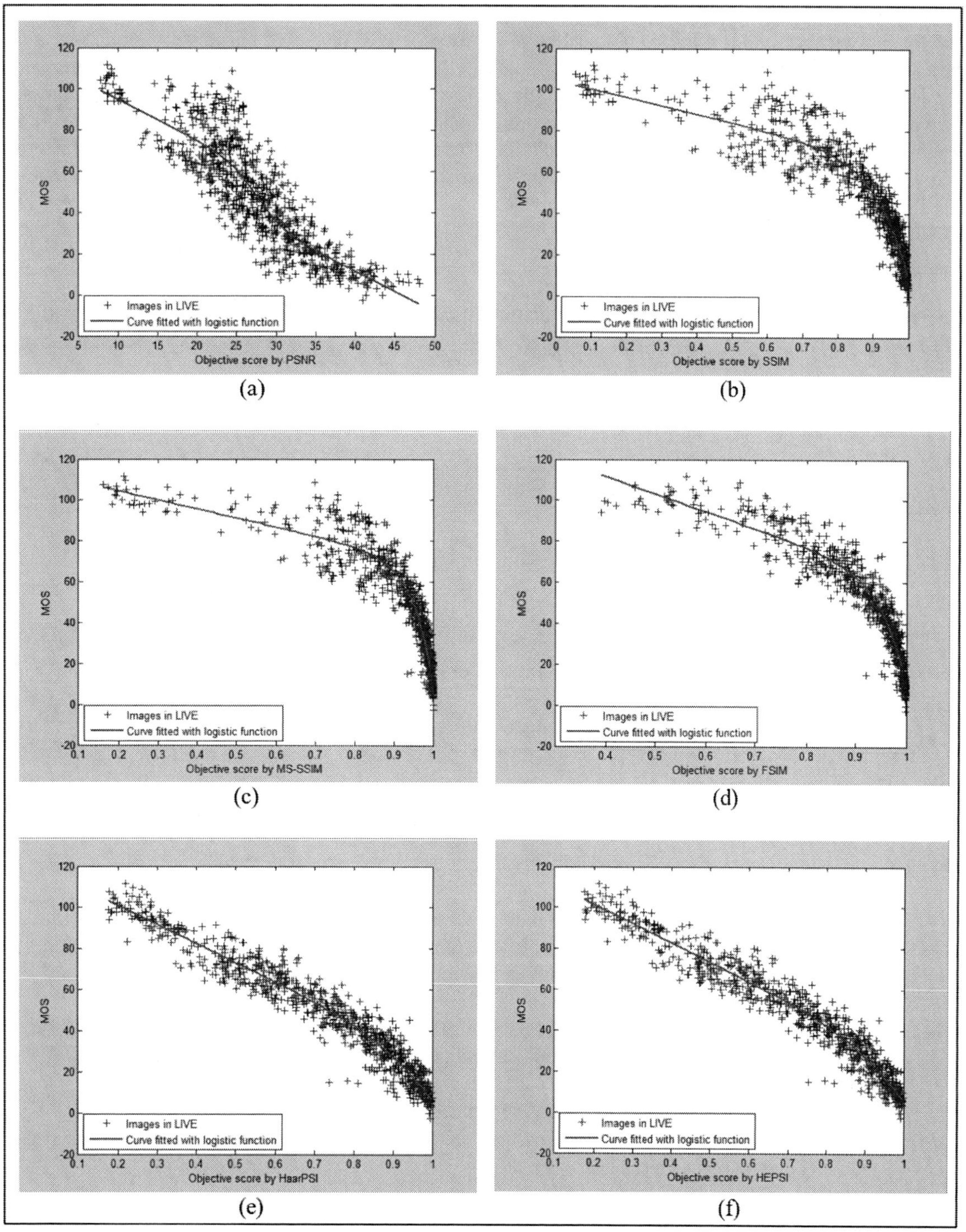

Figure 1 Scatter plots of 6 IQAs on LIVE database (a) PSNR, (b) SSIM, (c) MSSIM, (d) FSIM, (e) HaarPSI and (f) HEPSI

In order to provide a visual comparison of the proposed metric with five other IQAs (SSIM, MSSIM, FSIM, HaarPSI and PSNR), their scatter plots of subjective rating DMOS versus objective scores obtained by the IQAs on LIVE database are shown in Figure 1 above. Each point represents one test image. Generally speaking, if the scatter plots are closer to the fitted curve (red line), the results will be with lower RMSE. Ideally, for an IQA metric, these points should lie on top of each other.

The fitted curve (red line) plotted in each graph is actually the logistic function defined in Equation 1 that is widely used in IQA researches. In comparison with other scatter plots, HEPSI's points (blue +) are closer to each other and fitted curve, which means that HEPSI correlates well with subjective scores. This shows that proposed metric HEPSI performs better than the other 5 IQA metrics which are PSNR, SSIM, MS-SSIM, FSIM and HaarPSI.

5. Conclusion

In this research, we proposed a full reference IQA metric that is called HEPSI. This metric combines the assessment of local similarity and the relative importance of image areas from HaarPSI with the edge similarity between two images. Experiments show that HEPSI achieved good and better performance than other 5 IQA metrics across LIVE database. The work also indicates the high correlation between the image edges and visual perception of image quality. Assessing the edges of reference and distorted image can well measure the image structural distortion and become an efficient IQA metric.

6. References

[1] Abraham Anju V and Priya S 2017 A Study on Different Types of FR IQA 6 pp 46–51

[2] He L, Gao F, Hou W and Hao L 2014 Objective image quality assessment: a survey Int. J. Comput. Math 91 pp 2374–88

[3] Wang Z, Bovik Alan C, Sheikh Hamid R and Simoncelli Eero P 2004 Image quality assessment: From error visibility to structural similarity IEEE Trans. Image Process 13 pp 600–12

[4] Zhang L, Zhang L, Mou X and Zhang D 2011 Correspondence FSIM : A Feature Similarity Index for Image IEEE Trans. Image Process 20 pp 2378–86

[5] Reisenhofer R, Bosse S, Kutyniok G and Wiegand T 2018 A Haar Wavelet-Based Perceptual Similarity Index for Image Quality Assessment Signal Processing: Image Communication 61 pp 33–43

[6] Wang Z and Bovik Alan C 2005 Modern image quality assessment Synth. Lect. Image, Video, Multimed. Process

[7] Rohaly Ann M, et al. 2000 Video Quality Experts Group: current results and future directions Visual Communications and Image Processing pp 742–53

[8] Liu A, Lin W and Narwaria M 2012 Image Quality Assessment Based on Gradient Similarity IEEE Trans. Image Process 21 pp 1500–12

[9] Xue W, Zhang L, Mou X and Bovik Alan C 2014 Gradient magnitude similarity deviation: A highly efficient perceptual image quality index IEEE Trans. Image Process 23

[10] Gupta S, Gore A, Kumar S, Mani S and Srivastava P K 2017 Objective color image quality assessment based on Sobel magnitude Signal, Image Video Processing 11 pp 123–8

[11] Wang Z, Simoncelli Eero P and Bovik Alan C 2003 Multi-Scale Structural Similarity For Image Quality Assessment 2 pp 9–13

Cloud-based Embedded System for Object Detection and Recognition

Z T Olalekan and S A Anwar

School of Computer and Communication Engineering, Universiti Malaysia Perlis (UniMAP), Pauh Putra Campus, 02600 Arau, Perlis, Malaysia

said@unimap.edu.my

Abstract. Object detection and recognition techniques require large image datasets, memory, a workstation with specific graphics processing capability to train the algorithm and might have high power consumption. Embedded platforms on the other hand are characterized by portability, low power consumption and space, and energy resources making the deployment of such algorithms on them difficult. In order to overcome these drawbacks, cloud-based processing embedded system for object detection and recognition is proposed in this work. The system consists of an image acquisition device set up using embedded board and camera to capture, process and send images to the remote computer via cloud storage platform. This cloud platform serves as an interface between the embedded board and the remote computer. The detection algorithm of Faster R-CNN is executed on the remote computer and is trained and validated with 3000 images obtained from ImageNet. The training of the algorithm aims to detect five classes of object. The proposed system was validated off-line and have achieved a mean Average Precision (mAP) of 0.67. The performance of entire system procedure took about 45 seconds and have obtained an average confidence score of 0.86.

1. Introduction

In recent years, computer vision domain has not been left out from thriving due to the endless efforts of researchers and significant advancement in related fields. Computer vision uses digital images to model and emulate human vision using a computer through three major steps. These steps are image acquisition, image processing with the third being image analysis and understanding. As a result of this, applications of computer vision such as pattern recognition, medical imaging, 3D model building, surveillance, object detection and recognition and face detection has been made a reality.

Object detection is basically determining instances of real-world objects in images or videos while object recognition is the identification of target objects in still images or videos. Deep learning is one of the approaches used in object detection and recognition. Although several deep learning models have been proposed for object detection, Convolutional Neural Networks (CNN) is the most popular because it produces state-of-the-art recognition results, eliminates the requirement for manual feature extraction and it can be retrained for a new recognition task [1]. CNN models learn directly from image data and classify the images using pattern thus removing the need for manual feature extraction. They are used to recognize objects, scenes, and faces. CNN can be used to visually detect cancer cell, enable autonomous vehicles to detect objects [2].

The region proposal methods and region-based CNN have shown advances in object detection tasks [3]. However, this R-CNN algorithm has drawbacks where it was found to be slow in execution time due to its heavy computation. It was later overcome by the introduction of Fast R-CNN algorithm

[4]. The algorithm was very recently improved by merging the Region Proposal Network (RPN) with Fast R-CNN into a single network by sharing their convolutional features [5].

Notwithstanding the improvement in terms of object detection accuracy, the implementation of Faster R-CNN still involves a lot of computational intensity. The general-purpose computer with only central processing unit (CPU) becomes heavily loaded and more often cannot achieve the real-time requirements when implementing the algorithm except when equipped with graphics processing unit (GPU) [6]. With the recent trends towards portable and mobile devices, the issue of power supply, memory size and processing capability becomes a major problem if the detection algorithm is about to be implemented such devices. These issues are extended to the use of embedded or single board computer system.

The availability of embedded platforms which are reliable, portable, easy to use, have low energy consumption with lower prices has increased [7]. These merits make the embedded system suitable for the role of acquiring images. However, as mention earlier, due to its limitation in terms of processing capability and memory resource, the implementation of object detection algorithm such as Faster R-CNN is almost impossible. Hence, a new strategy to incorporate the use of embedded board for image acquisition and the benefits of the accuracy of object detection algorithm has been proposed. This involved utilizing high speed of the internet connection and the availability of access to the cloud storage facility.

In this paper, the implementation of cloud-based processing embedded system for object detection and recognition is presented. The front-end system utilizes the commercially available embedded system board Rasberry Pi 3 Model B with the camera module attached for image acquisition and pre-processing. On the other end is a remote computer to execute the Faster R-CNN algorithm and perform the object detection and recognition tasks. Both elements in the system are connected to the internet while the cloud storage service was used as a medium for uploading and downloading the images.

The rest of this paper is organized as follows. In Section 2, related previous works on the detection and recognition of objects with their implementation on embedded platforms were presented. In the next section, the implementation of the system is described and follows by the brief description of Faster R-CNN algorithm. This section also includes the explanation of the datasets and quality metrics used in this work. In Section 4, experimental results and discussion are presented to report the performance of the proposed system. Finally, the conclusion is presented.

2. Related Works

This section presents the related works previously done in terms of the implementation of the deep learning algorithm such as CNN on the embedded platforms or general-purpose computers. These works are aims to solved various problems and applications.

Shiddieqy et al. [8] implemented CNN using Python programming language and TensorFlow on the Raspberry Pi 3 to classify images (Dogs and Cats). They limited the number of ConvNet layers in the neural network to two and five. The model was trained on a computer with i7 7700 Quad-core 3.6GHz, 32GB RAM, 256GB SATA SSD and 8GB Nvidia GTX 1080 GPU. The accuracy of the two layers and five layers network are 55.6% and 77% respectively. The trained model was tested on 100 images on the i7 7700 PC and Raspberry Pi. It is worthy to note that the time recorded was only the time in the CNN model. On the Raspberry Pi, 13.699ms and 31.479ms were gotten for the best and worst run while the duration of the best run and worst run on the i7 7700. Although the layers of the neural network were greatly reduced, the processor usage on the Raspberry Pi was 100% while the PC was 17%. Moreover, a five-layer CNN is not enough for real-time applications as the more the layers of a neural network, the better the accuracy of the network's prediction.

Faster R-CNN algorithm was used, trained and tested on a computer (64-bit, Windows 7, 4GB Nvidia Quadro K2200) using which is used to automatically recognize books and their positions [9]. The database consists of 10 classes and exceed 4000 images in total. About 1000 images which were selected randomly was used to test the detector performance in terms of speed and accuracy. Testing of 1000 images took about 190.85 seconds and the classification accuracy of the network was 98.6%. In this work, the detector was not implemented on an embedded system as all stages (training and testing) was done on a general-purpose computer.

Xu et al. [10] introduced an application of depth learning in commercial video analysis and accelerates the embedded depths of learning algorithm model. They implemented a CNN for face attribute recognition based on a pre-existing model GoogleNet and added the structure of inception (pre-existing CNN model) to increase network's depth and width. They have deployed two independent networks for the identification of age and gender. Adience dataset was used which consists of images that were automatically uploaded from smartphones. The accuracy of 95% and 96% were achieved on age and gender classification respectively. The embedded hardware used was NVIDIA Jetson TX1 GPU alongside TensorRT optimization library to achieve a dual model with a speed of 15 frames per second when analysing commercial videos.

On another application, a real-time vehicle detection using deep learning scheme was proposed [11]. This algorithm was implemented on an embedded system in real-time. The detection system has two classes (vehicle and non-vehicle). The authors modified an already existing Integral Channel Feature (ICF) for the detection. The classifier used was Adaptive boosting (Adaboost) to generate strong detectors by linearly combining weak weighted detectors. In order to reduce the rate of false detection, a model which was generated by deep learning was applied to the result obtained from AdaBoost. The processing time was approximately 70ms on Freescale's I.MX6Q embedded board. The accuracy was 92.3% after the validation phase. The deep learning model was used only to validate.

Lee et al. [12] implemented a built-in system with Nvidia Jetson TX1 board to recognize a vehicle number plate. The plate recognition was based on AlexNet (pre-existing CNN model). AlexNet was trained on 500 vehicle license plate images each having a size of 1392x1040. About 20% of the image along the vertical and horizontal plate was removed with the remainder taken as the region of interest. The numbers and letters were recognized using DIGITS. The system was tested on the embedded platform which was equipped with a CCTV camera installed on the road. 63 images were used at the testing phase and a recognition accuracy realized was 95.24%. Although the system was able to identify the numbers that exist in the region of interest, images with reflective or broken plate number were not recognized. The processing time was not stated.

It can be inferred that in most cases, the models are implemented on general-purpose computers with the GPU acting as accelerators when training the neural networks. For actual implementation on the embedded board, the network layer is reduced which is most likely not enough for real-time applications. In the case of the main algorithm that is running on embedded platform, there are no effort were reported in terms of outsourcing the heavy computational works to the more powerful system through any means of cloud usage.

3. Methodology

In this section, the proposed embedded system with the cloud-based processing ability is presented. It consists of the system overview where the hardware elements and the flow of the processes involved are described in detail. Then, the object detection and recognition algorithm featuring Faster R-CNN is briefly described. Next, the description of standard image dataset and the quality metrics used to evaluate the system are also presented.

3.1. System Implementation

The proposed cloud-based object detection and recognition embedded system are shown on figure 1. It consists of the embedded board Rasberry Pi with the camera attached for image acquisition and preparation at the front-end and the computer at the backend. Both of these elements are connected to the internet and have access to the cloud-storage where the images are kept during the processing.

IOP Conf. Series: Materials Science and Engineering **767** (2020) 012016 doi:10.1088/1757-899X/767/1/012016

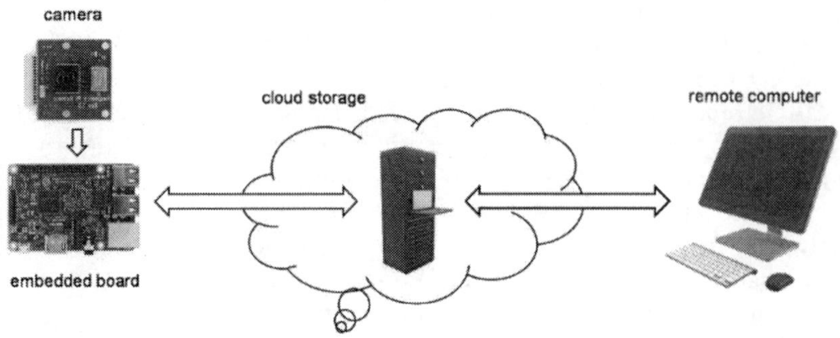

Figure 1. Proposed object detection and recognition system

The specification of the embedded board and computer involves in this system are tabulated in table 1. The Raspberry Pi Camera Module is connected to the embedded board has 5MP native resolution with a sensor capability of 2592x1994x3 pixels. The computer has the technical computing software package MATLAB 2019A installed. Meanwhile, the embedded board is programmed using Phyton programming language. The cloud storage used in this work is Dropbox with the capacity of 5GB.

Table 1. Specification of the system.

Item	Raspberry Pi Model 3B	Remote Computer
Processor	64-bit Quad Core, 1.2GHz	64-bit Core i5, 2.50GHz
Memory	1GB RAM	12GB RAM
Storage	32GB SD	1TB HDD
Connectivity	WiFi on board	WiFi
Operating System	Raspbian Stretch	Windows 10

In general, the proposed work has four processes which are image acquisition, image preparation and transmission, deployment of Faster R-CNN algorithm (briefly explained in section 3.2) and lastly, the display of the detected object. Image acquisition and preparation are executed on the embedded board and the image later is transmitted through internet and stored in the cloud storage. After this process, the image is download by the remote computer and the Faster R-CNN algorithm is performed to detect and recognize the objects in the image. Later, the resulted output image is uploaded back to the cloud storage where it is available for embedded board to download and display them. Figure 2 shows each process implemented in the system and which element is involved.

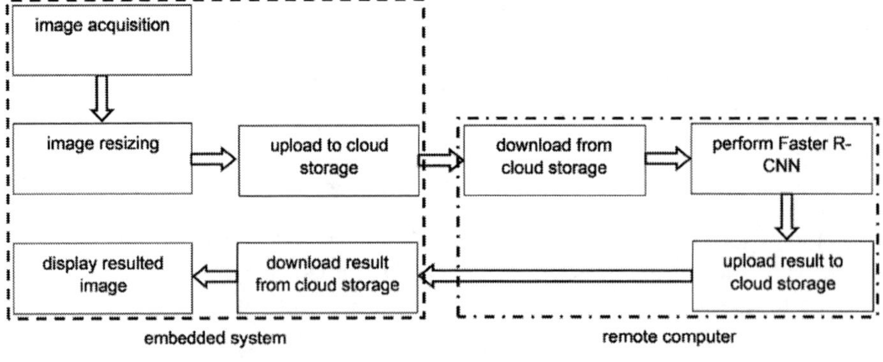

Figure 2. Flow of the process involved.

3.2. Object Detection and Recognition

In this work, the object detection and recognition tasks are performed on the remote computer using Faster Region-based Convolutional Neural Network (Faster R-CNN) [5]. This algorithm is trained on the remote computer by using MATLAB 2019A software package. Training is done by feeding the network myriads of already labelled images as input. The network then learns features of the desired objects in the images which in turn allows it find objects of the same class in unlabelled images. The flowchart in figure 3 shows the implementation of this algorithm.

As can be seen from figure 3, in the Region Proposal Network (RPN), image is fed to CNN for feature extraction and anchor boxes are generated. These input image can be labelled (in the case of training) or unlabelled and must at least be of size 224x224 pixels because it is the minimum size that the base network can accept. The extracted features and anchors are sent to the region proposal layer where the probability of it being a background or foreground is predicted.

Non-maximum suppression is executed on the generated proposals to eliminate duplicate proposals. Based the foreground probability, the top 2000 proposals are sent to the Region of Interest (ROI) pooling layer where the fixed-sized feature maps for each region proposal is extracted in order to classify the proposals into one of the object classes and also readjust the bounding box in accordance to the predicted class.

In this work, options like maximum number of times an entire dataset is passed through a neural network, mini batch size and initial learn rate are configured to 10, 1 and 0.001 respectively. The mini batch size is set to 1 because multiple image regions are processed from one training image on every iteration. Initial learn rate is the amount that the weights are updated during training and if the value is set too low, the training takes longer time to complete and if it is too high, the result might be suboptimal. In addition, a checkpoint path is set to a temporary location thereby saving the partially trained detector during the process.

3.3. Datasets

The dataset containing 2500 images is used for the training of the Faster R-CNN algorithm and was collected from ImageNet (http://imagenet.stanford.edu/). The collected images were of the 8-bit RGB colour scheme and in various sizes ranging from 110x110 to 1036x776 pixels. This dataset contains five classes of objects which are Broom, Fan, Television, Keyboard and Mouse. Each of the classes has 500 images.

Objects in the training images were labelled by drawing bounding boxes over them with an aid of the MATLAB Image Labeller application. The resulting ground truth file which was exported from MATLAB played a crucial role while training the object detection algorithm. For training purposes, all images are resized to 800x600x3 (SVGA resolution) in order to reduce the sampling point hence speeding up the training process.

After training, the algorithm is validated using another 500 independence images from the ImageNet where 100 images are allocated for each of the classes. Lastly, for the testing images, camera on the embedded board was used to capture the real scene. The scenery chosen to be captured as testing image must contains at least one out of five objects described earlier. Similar to the training images, 20 captured images were then resized to 800x600x3 before being uploaded to the cloud storage.

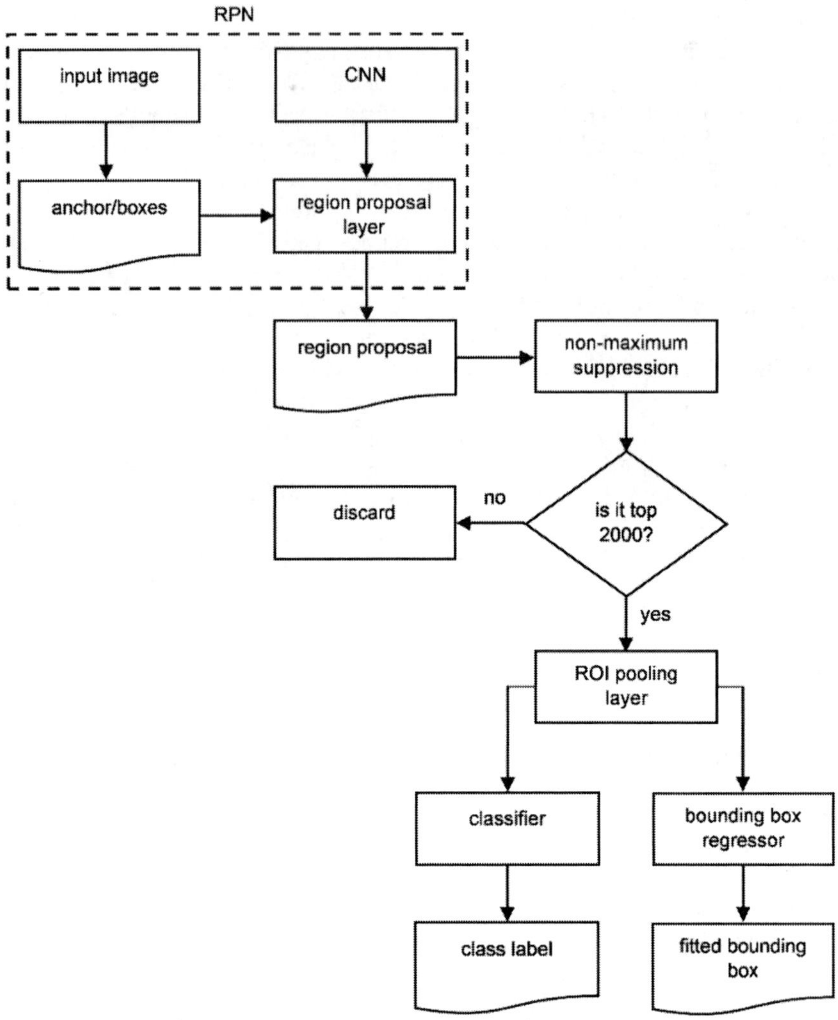

Figure 3. Faster R-CNN process flow.

3.4. Quality Metrics

Mean Average Precision (mAP) is a metric used to measure the accuracy of object detection algorithm. The ability of the detection process to find the relevant objects only is a measures precision of the algorithm. Sometimes this can also be defined as the percentage of correct predictions. In this case, Intersection-over-Union (IoU) is the precision measure of the object detection algorithm used in this work [12] is given below

$$\text{IoU}=\frac{A_o}{A_u} \tag{1}$$

where A_o and A_u is area of overlap and union respectively. It measures the overlap between two bounding boxes (i.e. the predicted box and the ground truth) and a prediction is correct when IoU ≥ 0.5.

In order to compute the mAP for a class, average IoU of all classes is determined and the mean of the IoU averages is calculated. The average precision can also be considered as the area beneath the precision-recall curve. Recall is the ability of algorithm to identify all relevant (i.e. the percentage of predicted true positive out of all positive instances based on the ground truth). Equations (2) and (3) shows the mathematical definition of precision P and recall R.

$$P = \frac{tp}{tp + fp} \tag{2}$$

$$R = \frac{tp}{tp + fn} \tag{3}$$

where tp (true positive) is a correct prediction, fp (false positive) is a wrong detection (i.e. predicted positive when it is negative) and fn (false negative) is when an object is predicted as negative when it actually is positive.

4. Result and discussion

In this section, the experimental results from the methods and procedures described in the above sections are presented. It consists of two parts: offline is where the object detection algorithm is validated and online is when the system is test in real-time.

4.1. Offline validation

For the purpose of evaluating the algorithm, 500 ground truth images were fed to the detector. The result from the detector and the available ground truth was compared. The mAP is computed for each image and the precision-recall graph is derived for each object class as shown in figure 4.

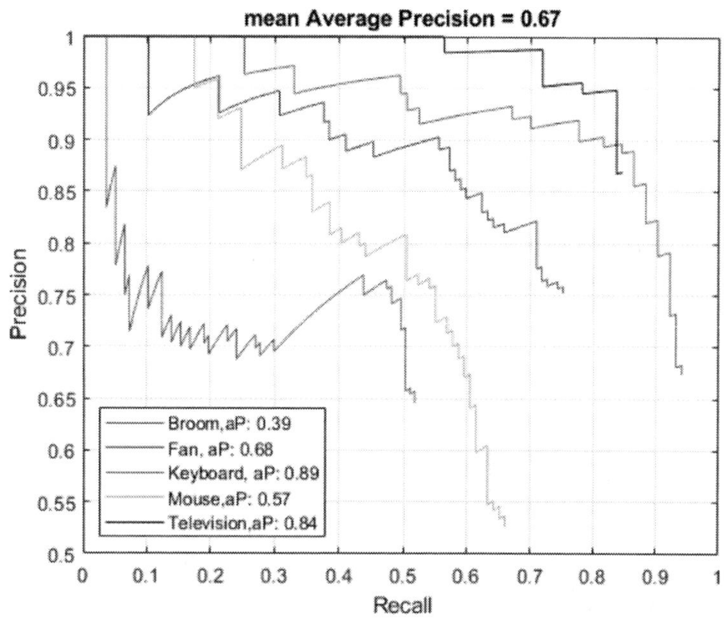

Figure 4. mean Average Precision (mAP) of the object detection algorithm.

As can be seen, the lowest Average Precision (aP) is 0.39 which is for the Broom class. This is translated into only 39% of the data point are relevant out of the total detected object. Meanwhile, the Keyboard class scores the highest aP at 0.89 follows by the Television class at 0.84. These two classes have more than 80% of their detected points in the dataset are relevant. However, upon closer

IOP Conf. Series: Materials Science and Engineering **767** (2020) 012016 doi:10.1088/1757-899X/767/1/012016

examination, the precision for both classes start to reduce at 0.24 and 0.58 respectively. In terms of spans of precision value, the Television class have tighter range between 0.87 to 1 compared to the Keyboard class which is between 0.68 to 1.

Overall, figure 4 shows the mean average precision (mAP) of the multi-class detection. Since the mAP is 0.67, it can be deduced that 67% of objects detected in that dataset is significant. This shows that the detector locates one-third of the objects that are irrelevant as relevant.

4.2. Online testing

In order to test the system online, 20 images some of which contains multiple objects were captured and processed. It is important to note that the program on the embedded board pauses for 30 seconds to prevent error while it waits for the output image to be uploaded to cloud. In addition, the network strength also plays an important role as a weak network signal will result in an increase in the process duration.

Figure 5 shows the first 9 output images from the online testing. The bounding box is drawn by the detection algorithm to highlight the object found as well as the class and the confidence score. The result also shows that the algorithm can detect more than one object in a single image as can be seen in figure 5 (g) and (i).

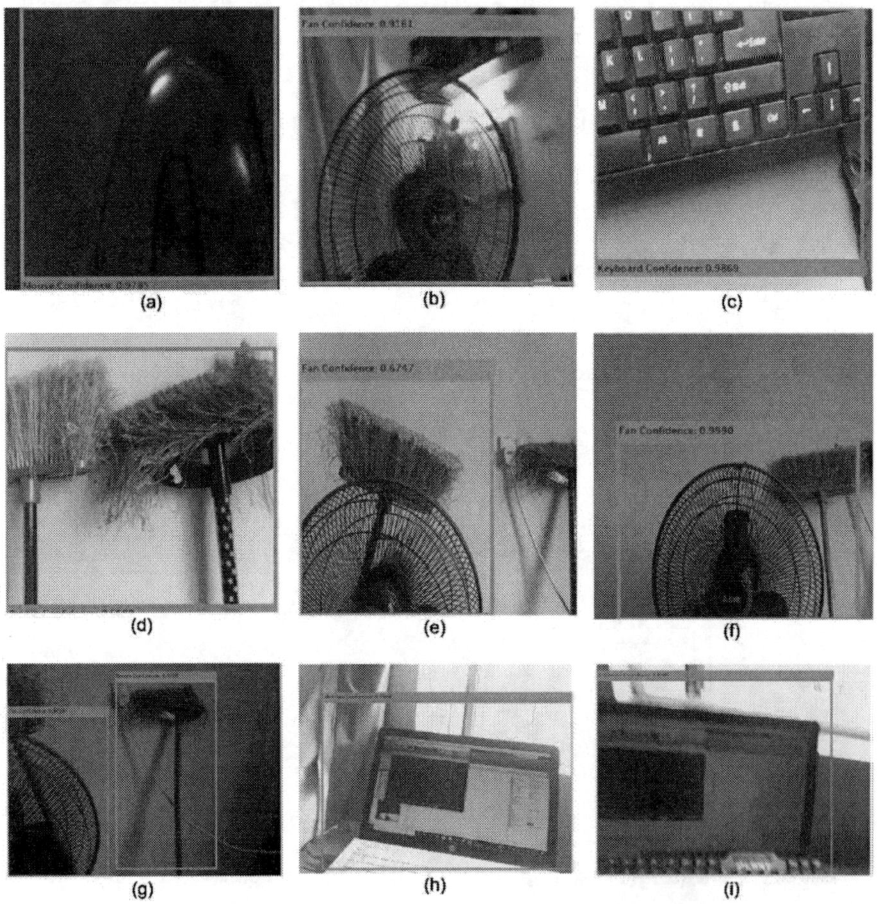

Figure 5. Example of detected objects in real scene.

In addition, table 2 tabulated the confidence score and the execution time of the system starting from the image capturing to the output display for the all testing images. The average time taken to

IOP Conf. Series: Materials Science and Engineering **767** (2020) 012016 doi:10.1088/1757-899X/767/1/012016

process these images is approximately 44.66 seconds with the lowest execution time is 43.45 seconds and the highest is 47.20 seconds. Since the buffering period of 30 seconds is artificially introduced in the software, it can, therefore, be deduced that the major processes the system take about 13 to 17 seconds.

Table 2. System performance evaluation

Image	Detected Object	Score	Time Elapse (seconds)
001.jpg	Mouse	0.9785	45.46
002.jpg	Fan	0.9161	44.04
003.jpg	Keyboard	0.9869	45.48
004.jpg	Broom	0.5681	44.00
005.jpg	Fan	0.6747	46.05
006.jpg	Fan	0.9990	47.20
007.jpg	Broom Fan	0.9538 0.9716	47.13
008.jpg	Television	0.9989	45.13
009.jpg	Television Keyboard	0.9553 0.9205	44.46
010.jpg	Mouse Keyboard	0.6506 0.9873	45.24
011.jpg	Mouse	0.9990	44.04
012.jpg	Television Fan Keyboard	0.9999 0.9878 0.8012	43.72
013.jpg	Broom Television Keyboard	0.9992 0.9995 0.7446	43.70
014.jpg	Keyboard Broom Keyboard	0.8243 0.6488 0.7246	43.45
015.jpg	Keyboard Mouse Keyboard	0.8243 0.7858 0.6637	43.84
016.jpg	Broom Fan	0.9693 0.5844	43.70
017.jpg	Television Keyboard Broom	0.8750 0.9117 0.7818	45.09
018.jpg	Fan	0.9775	44.05
019.jpg	Keyboard Fan	0.9626 0.9884	43.71
020.jpg	Keyboard	0.7246	43.69
	Average	**0.8668**	**44.66**

5. Conclusion

Object detection and recognition are one of the aspects of computer vision that detects objects in still or moving images by placing a bounding box around it. The goal of object detection in this paper is to be able to recognize objects in real-time and classify them accordingly without requiring expensive and high computational power. This is achieved by employing the embedded board with camera as front-end processing elements to capture and prepare the images. Utilizing the internet and the cloud storage facility, the images are processed by remote computer elsewhere. In this work, Faster R-CNN algorithm is implemented, and this multi-class detection algorithm was trained to locate five classes of objects.

The performance of the algorithm was evaluated off-line by validating the detection process on random test images and has yielded a mAP of 67% accuracy. Lastly, the system was tested on-line in real-time by capturing real-scene images and sending them cloud storage, processed by the remote computer and downloaded back into the embedded board. Total of 20 real-scene images were tested and the average confidence score produced is 0.86 with the average processing time of 45 seconds. These results suggest that an embedded system platform can have access to the effectiveness of computationally heavy algorithm running elsewhere with the proposed strategy of utilizing the cloud storage and remote computer as external resources.

References

[1] Simonyan K and Zisserman A 2014 Very deep convolutional networks for large-scale image recognition

[2] Nagaoka Y, Miyazaki T, Sugaya Y and Omachi S 2017 Text detection by Faster R-CNN with multiple region proposal networks *2017 14th IAPR Int. Conf. on Document Analysis and Recognition (ICDAR)* vol 06 pp 15–20

[3] Girshick R, Donahue J, Darrell T and Malik J 2014 Rich feature hierarchies for accurate object detection and semantic segmentation *2014 IEEE Conf. on Computer Vision and Pattern Recognition* pp 580–587

[4] Girshick R 2015 Fast R-CNN *2015 IEEE Int. Conf. on Computer Vision (ICCV)* pp 1440– 1448

[5] Ren S, He K, Girshick R and Sun J 2017 *IEEE Transactions on Pattern Analysis and Machine Intelligence* **39** 1137–1149

[6] Lee S, Son K, Kim H and Park J 2017 Car plate recognition based on CNN using embedded system with GPU *2017 10th Int. Conf. on Human System Interactions (HSI)* pp 239–241

[7] Fuente A M, Castro V G, Robles L F and Guti´errez E A 2015 Design and implementation of an embedded system for image acquisition of inserts in a headtool machine

[8] Shiddieqy H A, Hariadi F I and Adiono T 2017 Implementation of deep-learning based image classification on single board computer *2017 Int. Symposium on Electronics and Smart Devices (ISESD)* pp 133–137

[9] Zhu B, Wu X, Yang L, Shen Y and Wu L 2016 Automatic detection of books based on Faster R-CNN *2016 3rd Int. Conf. on Digital Information Processing, Data Mining, and Wireless Communications (DIPDMWC)* pp 8–12

[10] Xu J, Wang B, Li J, Hu C and Pan J 2017 Deep learning application based on embedded GPU *2017 First Int. Conf. on Electronics Instrumentation Information Systems (EIIS)* pp 1–4

[11] Shin J, Kim U, Lee D, Park S, Oh S and Yun T 2017 Real-time vehicle detection using deep learning scheme on embedded system *2017 9th Int. Conf. on Ubiquitous and Future Networks (ICUFN)* pp 272–274

[12] Rosebrock A 2017 *Deep Learning for Computer Vision with Python* (pyimagesearch)

A Self-Adapting Ant Colony Optimization Algorithm Using Fuzzy Logic (ACOF) for Combinatorial Test Suite Generation

Mohd Zamri Zahir Ahmad, Rozmie Razif Othman, Mohd Shaiful Aziz Rashid Ali and Nuraminah Ramli.
School of Computer and Communication Engineering, Kampus Pauh Putra, Universiti Malaysia Perlis, 02600 Arau, Perlis, Malaysia.

E-mail: zamri.zahir@yahoo.com, rozmie@unimap.edu.my, shaifulaziz@unimap.edu.my, nuraminah.ramli@gmail.com.

Abstract. Software testing is one of most crucial phase in software development life cycle (SDLC). The main function of testing is to cater bugs between interactions of the inputs. It is not possible to eliminate all bugs in one system but by using a suitable testing optimization, it can provide a good enough solution for it. Reducing effort on this phase is not only could lead to numerous bugs between the input interactions, but it also leads to a greater loss such as loss of profits, reputations and even loss of life. Generally, there are three categories of combinatorial testing techniques which is computational, metaheuristic, and hyper heuristic. Ant colony optimization (ACO) is a heuristic technique where its mimic the nature of ants in finding the best route from the nest to the food node and vice versa. Many optimization problems have been solved by using ACO. This paper is to proposed a self-adapting ant colony optimization algorithm using fuzzy logic (ACOF) for combinatorial test suite generation, where it will dynamically determine number of ants and edge selection (i.e. either to explore or to exploit) based on percentage of remaining tuple list and covered test cases.

Keywords: Combinatorial testing, T-way testing, Ant Colony Optimization.

1. Introduction

In development of a software system, there are five main stages that engage throughout of the entire process. Also knows as software development life cycle (SDLC), it consists of software requirement analysis, system design, implementation, testing and evolution. It is reported that, from this five stages, testing phase consume up to 50% of total cost development. This numbers could growth if the system involves with safety-critical applications [1]. Software failure is a disaster situation as it could increase the cost of software development tremendously. In order to minimize the software failure, a rigorous testing needs to be conducted. To ensure the effectiveness of the testing activity, many software testing technique have been developed in literature. Combinatorial testing (or also known as t-way testing) is one of the test design technique that has been used widely. It can be used to generate test suite that can cater bugs due to interaction between input parameters [2]. In generating the t-way test suite, t-way strategy uses either mathematical or computational methods in reducing the test suite size but at the same time still covering all interactions involving specified parameters-values combination for specific strength, t [3,4].

IOP Conf. Series: Materials Science and Engineering **767** (2020) 012017 doi:10.1088/1757-899X/767/1/012017

As the main aim of all combinatorial test suite generation strategy is to produce as small test suite size as possible, many optimization algorithm has been proposed and adopted for the purpose of generating t-way test suite (e.g IPOG, Jenny, GTWay, ITTDG and TVG). At the same time, there are researchers that use metaheuristic optimization technique in generating t-way test suite. Bestoun implements Particle Swarm Optimization technique in generating uniform and variable strength t-way test suite while Alsewari adopted Harmonic Search Algorithm for generating uniform strength t-way test suite. There is also an attempt made by Zamli of using more than one metaheuristic technique (namely hyper-heuristic technique) to generate uniform strength t-way test suite. The capability of metaheuristic techniques in producing smaller size of test suite become the motivation of researchers to implements these techniques in generating t-way test suite.

One of the most used metaheuristic technique in generating t-way test suite is Ant Colony Optimization (ACO) algorithm. In 2004, Shiba successfully implements ACO algorithm in generating uniform strength t-way test suite [5]. Enhancing Shiba works, Chen added local search in his implementation of ACO (called Ant Colony System) for generating variable strength t-way test suite [6]. Recently, Ramli proposed her implementation of ACO algorithm for input-output based t-way test suite [7]. Overall, all implementation of ACO successfully producing the smallest t-way test suite in many system configurations [5,6,8–12].

The success of ACO in producing smaller test suite size is highly depending on number of ants used in the algorithm. Increasing number of ants increase the probability of producing the optimum test suite. Nevertheless, increasing number of ants will also increase the execution time. Although the execution time are also affected by system environment used, optimizing the numbers of ant possibly could optimize the ant colony efficiency. Thus, we propose a study on a self-adapting ACO algorithm using fuzzy logic for combinatorial test suite generation. The fuzzy component will dynamically determine the suitable number of ants that will be used in ACOF and edge selection either to explore or to exploit the search space.

2. Related Works

Particle Swarm Optimization (PSO) is an algorithm for t-way combinatorial testing published in 2010. The algorithm is inspired by swarm animals hunting for their food. Each individual in the swarm moves towards the best individual location and the best global location based on optimal solution calculated from their position and velocity [13]. A few algorithm or tools is developed based on PSO algorithm. The generation of PSO algorithm for t-way testing starts with the development of pairwise PSO [13]. The researchers have developed two different algorithms. One algorithm is OTAT strategy and another one is OPAT. The utilization of PSO algorithm in the previous research has encouraged the development of Particle Swarm Test Generator (PSTG) [9]. Faster convergence rate attitude, requires a few parameters to be controlled, easily apply to any optimization problem and lighter computational load are the reasons PSO has been chosen as PSTG's basic platform [9,10]. PSTG only support uniform strength up to 6 levels. Next, VSPSTG has been developed to improve PSTG by enhancing variable strength interaction [11]. Discrete Particle Swarm Optimization (DPSO) [14] is another algorithm based on PSO and discrete particle swarm as its basis. This new algorithm is designed by adopting S-PSO that uses set-based scheme as its discrete search space. The new DPSO has improves the performance by having two auxiliary strategies; particle reinitialization and additional evaluation of gbest. It also offers guidelines for parameter settings. DPSO is reported to be a promising improvement of PSO. The last algorithm that motivated by PSO between 2010 and 2017 is Swarm Intelligent Test Generator (SITG) [12]. SITG could support 2 to 6 uniform strength and variable strength. SITG outperform PSTG in some cases in uniform strength while performing better results for variable strength. As PSO is a metaheuristic algorithm, its entire variant follows the similar search technique. PSTG, V-PSTG, DPSO and SITG use OTAT strategy approach while Pairwise PSO has both OTAT and OPAT strategy for different type of Pairwise PSO algorithm.

In 2011, GTWay, ITTDG and AURA were introduced as another t-way combinatorial testing algorithms. Those algorithms applied computational search technique to find the best test cases.

GTWay [15] applies backtracking concept. The algorithm needs to map the actual data with symbolic representation. Then, it generates t-way pair possible interactions. After that, the strategy backtracks the uncovered t-way pairs. The t-way pairs are merge if they are combinable and complement each other's missing value and the new merge interaction could cover the most uncovered t-way pairs. If the pairs fail to merge, backtrack algorithm goes back to the first defined values. GTWay is an OTAT strategy that supports uniform strength for t greater than six. ITTDG [8] practices a list of candidate test data to be used in deciding a final test data. The visited tuples are put into a candidate list. Next, parameter that has a value that covers the most uncovered tuples is added to the candidate test data, one parameter at a time. In a situation where tie occurs, the corresponding test data that has the tie values is duplicated and put into the candidates list. Once the candidate list finished forming a complete test data, the strategy calculates weight for every test data. Test data with a highest weight has a value of the most covered tuples. If tie happens, the first round in the candidate list will be selected to be put in the final test suite and removed from the uncovered tuples list. The process repeated until all uncovered tuples are being covered. ITTDG could support all three types of interaction; uniform, variable and IOR.

Similar to ITTDG, AURA [16] has the same ability in supporting all type of interactions. AURA started by exploiting the interactions combination created earlier to generate test suite. Lastly actual data mapping algorithm is used to support symbolic values and actual data output. Only one type of algorithm published in 2012. Harmony Search algorithm is the basis for two other algorithms, HSS and HS-PTSGT. Another algorithm that uses metaheuristic search technique to generate t-way test suite is Harmony Search Strategy (HSS) [4,15]. It is based on Harmony Search (HS) algorithm. Similar to PSO, HS algorithm requires lightweight computation with only few parameters setting. HSS could support uniform and variable interaction strength with high interaction strength, up to 14 and constraints. Besides HSS, HS-PTSGT [6] is a pairwise generator tool that was developed based on HS algorithm.

In 2013 DA-RO and DA-FO [17] emerged. Both algorithms are a computational technique and support all type of interactions. DA-FO starts with an empty test case which contains unfixed factors. Local density for each factor is determined to get the order of coverage requirements. A coverage requirement with highest local density is selected to be fixed factor based on global density. For each combination in the test case where the values of factors are fixed, DA-RO will calculate global density. The combination which holds greatest global density is selected to fix the value of factors. Unlike DA-RO, DA-FO algorithm produces single test case by fixing value in order of factor. Priority number is defined to measure priority of different factors to determine order of factors. Coverage requirements that hold high priorities and local densities will get a great factor density and will be chosen to fix the values. Similar to DA-RO, after fixing values, the local, global and factor densities may change due to re-modification of densities. DA-FO also suffers from tie problem. To break the ties, the similar factors as DA-RO can be used. GVS [18] is the only algorithm published in 2014. The algorithm is inspired by GTWay. Similar to GTway, GVS is a computational technique and improves the interaction by supporting all type of interactions. It generates one tuple at a time before the iteration starts. Each tuple which is generated in the tuple generator is assigned with a don't care value to complete the test data. The test data with the first don't care value, is compared with the uncovered tuples list. If the test data exists, the test data with the first don't care is selected. However, if not, the algorithm changes the next remaining value parameter in the don't care value space. The process of determining the don't care values is repeated until all parameter values finish. Afterward, GVS checks which test cases could cover the most uncovered tuples. The chosen test case is then placed into final test suite and covered tuples are added in the covered tuples list.

In 2015, TCA, Cuckoo Search (CS) and Flower Strategy (FS) were published. They are metaheuristic algorithms. TCA [19] is the algorithm that combines greedy Tabu search and random walk heuristic. Greedy Tabu search is used during initialization of test cases, and then TCA apply heuristic search technique to expand the opportunities in covering the uncovered interactions. By using both techniques, it effects the runtime test suite generation. Even though TCA utilizes combination of search techniques, it is still under metaheuristic category because Tabu search is a metaheuristic algorithm. TCA supports 3-way constraint uniform strength. Cuckoo Search [20] has been explored in t-way combinatorial testing as a capacity that Cuckoo algorithm offers search capabilities using Levi flights [21] to update search space and it only consists of few parameters to be tuned [20]. Cuckoo Search supports uniform, up to 3 interaction strength and also variable strength. Flower Strategy (FS) [22] evolved from the beauty of Flower Pollination Algorithm (FPA). FPA is a simple, flexible and requires lightweight computation. It also offers a balance between exploitation and exploration by employing levi flight. FS is a metaheuristic that supports uniform strength.

In 2016, hyper-heuristic strategy emerged. High Level Hyperheuristic (HHH) [23] is the pioneer in utilizing hyper-heuristic strategy in t-way combinatorial testing. It consists of high level metaheuristic Tabu Search and other four (4) low level metaheuristic algorithms (i.e. Teaching Learning based Optimization, Global Neighbourhood Algorithm, Particle Swarm Optimization, and Cuckoo Search Algorithm). The algorithm uses metaheuristic search technique and could support uniform strength from 2 until 6 level of strength. There are two proposed t-way testing strategies found in the literature. Artificial Bee Colony (ABC) [24] has been proposed to generate an optimum test suite. The strategy is inspired by a group of bees searching for nectar for their hive. ABC was chosen by the researchers as it was proven to be a great strategy for combinatorial field. Another proposed strategy found in 2016 by Ramli is Ant Colony System (ACS) [7]. ACS is a variant of Ant Colony Optimization (ACO) algorithm. It has been successfully solving many combinatorial optimization problems. Ant Colony Algorithm (ACA) [5] and ACS [6] have been built to generate an optimum test suite to cater uniform and variable strength respectively. The proposed ACS strategy is to cater all type of interactions especially IOR. Adaptive TLBO (ATLBO) is a metaheuristic t-way testing that emerged from Teaching Learning-based optimization (TLBO) algorithm in the first quarter of 2017. It was based on Mamdani fuzzy inference system. ATLBO could support uniform and variable strength [25].

3. Problem Definition Model

To demonstrate uniform strength of t-way testing, let's consider a system for scholarship sponsor application. The system is to help filtering application from student to determine which student is eligible to obtain the scholarship. There system will filter the application based on four factor which is cumulative grade point average (CGPA), course, study level and parent salary as shown in Table 1. To ease the discussion, a representation of parameters and its value is shown in Table 2. The CGPA is represented as 'C', course represented as 'K', study level represented as 'S' and parent salary represented as 'P'. For the values, each of the values represented by small letter of the represented filtering factor. For example, c1 referring to input parameter CGPA and the value is less and equal to 3.

Table 1. Requirement for Scholarship Application.

Filtering Factor	CGPA	Course	Study Level	Parent Salary
Values	≤ 3	Engineering	Diploma	$\leq$ RM2000
	> 3	Business	Bachelor Degree	> RM2000

Table 2. Simplified Inputs Parameter for Scholarship Application.

Input Parameter	C	K	S	P
	c1	k1	s1	p1
Values	c2	k2	s2	p2

To minimize the potential bugs in this system, appropriate testing must be made before release the system to end user. Interaction testing is the crucial test that need to be conduct as fault may occurred between this input parameter. One of the technique to test the system is by using exhaustive testing. In exhaustive testing, all possible input combinations need to be tested. This means, all input is interacting with each other where faults are expected to arise between the interactions. As a result, 16 test cases (i.e. $2^4 = 16$) need to be executed. The list of exhaustive test case is shown in Table 3.

IOP Conf. Series: Materials Science and Engineering **767** (2020) 012017 doi:10.1088/1757-899X/767/1/012017

Table 3. Exhaustive Test Suite for Scholarships Application.

C	K	S	P
c1	k1	s1	p1
c1	k1	s1	p2
c1	k1	s2	p1
c1	k1	s2	p2
c1	k2	s1	p1
c1	k2	s1	p2
c1	k2	s2	p1
c1	k2	s2	p2
c2	k1	s1	p1
c2	k1	s1	p2
c2	k1	s2	p1
c2	k1	s2	p2
c2	k2	s1	p1
c2	k2	s1	p2
c2	k2	s2	p1
c2	k2	s2	p2

Table 3 shown exhaustive testing of the system where the strength, t is 4. The interaction strength is possible to be reduce to 2 (t=2) or also known as pairwise testing. This means, only two input parameter are interacting and faults is expected to arise within these interactions as compared to four input parameter in the exhaustive testing. The rest of input parameter that not related to interaction are considered as 'don't care' (DC) value, as it will not affect the input interactions testing. The purpose to reduce the strength is to minimize the generated test case without compromise the interacting input. The possible input interaction for 2-way testing is CK, CS, CP, KS, KP and SP. As the result, the generated test case now become smaller where the repeated test case is removed from the list as shown in Figure 1. Noted that, it almost achieves 44 percent reduction of test cases as compared to exhaustive testing. The reduction may not significant for a small and simple application, but it is significantly reduced for a larger and complex applications environment. For example, system with 40 input with 2 possible value each will resulted 1099511627776 (i.e. 2^{40}) test cases need to be executed. It is impractical to do the exhaustive testing as it will consume a lot of time as well as many software testers to be part in, therefore the overall cost of testing will increase enormously. Hence, by implementing t-way testing technique, it is proven could reduce the test cases and save a lot of resources with an acceptable fault detection.

CK

C	K	S	P
c1	k1	s1	p1
c1	k2	s2	p2
c2	k1	s2	p1
c2	k2	s1	p2

CS

C	K	S	P
c1	k1	s1	p2
c1	k2	s2	p1
~~c2~~	~~k2~~	~~s1~~	~~p2~~
~~c2~~	~~k1~~	~~s2~~	~~p1~~

CP

C	K	S	P
~~c1~~	~~k1~~	~~s1~~	~~p1~~
~~c1~~	~~k2~~	~~s2~~	~~p2~~
~~c2~~	~~k1~~	~~s2~~	~~p1~~
~~c2~~	~~k2~~	~~s1~~	~~p2~~

KS

C	K	S	P
~~c1~~	~~k1~~	~~s1~~	~~p2~~
~~c2~~	~~k1~~	~~s2~~	~~p1~~
~~c2~~	~~k2~~	~~s1~~	~~p2~~
~~c1~~	~~k2~~	~~s2~~	~~p2~~

KP

C	K	S	P
c1	k1	s2	p1
~~c1~~	~~k1~~	~~s1~~	~~p2~~
c2	k2	s1	p1
~~c2~~	~~k2~~	~~s1~~	~~p2~~

SP

C	K	S	P
c1	k2	s1	p1
~~c1~~	~~k1~~	~~s1~~	~~p2~~
~~c2~~	~~k1~~	~~s2~~	~~p1~~
~~c1~~	~~k2~~	~~s2~~	~~p2~~

Final Test Suite

C	K	S	P
c1	k1	s1	p1
c1	k1	s1	p2
c1	k1	s2	p1
c1	k2	s1	p1
c1	k2	s2	p1
c1	k2	s2	p2
c2	k1	s2	p1
c2	k2	s1	p1
c2	k2	s1	p2

Figure 1. Test Suite for 2-Way Testing.

4. Overview of ACO

Ant algorithm is a metaheuristic approach where it is inspired by nature life behavior of ant colony in finding shortest path to foods from their nest. Some ants' species are completely blind and they can't see anything, yet they have a very powerful searching system where they can find shortest path to the foods source. Ants sense the path by a special chemical deposited by them while travelling the path called as pheromone. The more pheromone deposited to the trail increases the probability for the ants to choose that path.

There were several algorithms have been proposed inspired from the nature behavior of ant colony. In early nineties, an Ant System (AS) was the first proposed by Dorigo et al. [26] for travelling salesman problem (TSP). Ant system consist of three main techniques which is ant-cycle, ant-density, and ant-quantity. Among of this three techniques, ant-cycle is the most performed compared the other two hence it was later established as the Ant System [27]. This algorithm was then being enhance producing several more algorithm based on this Ant System. Among of the rise algorithm are Ant System Elitist, rank-based Ant System (ASrank), MAX–MIN Ant System (MMAS), and Ant Colony System (ACS). ACS is the tuned version of AS where several modifications been made to enhance the searching effectiveness. It is reported that ACS give the most competitive performance compared with others AS strategies because it goods in the exploitation and exploration in searching of the best solutions [6,28]. To be comprehensive, ACS is improved because of;

a) Edge selection

The edge selection rule provides, which called as random-proportional rule; a through probability method to balance searching solutions, between exploration of the new path or exploitation of the path with previous information. A random variable of q and user-define q_0 is used to define the probability of the ants in selecting the edges. Variable q is uniformly distributed in $[0,1]$ and variable q_0 $(0 \leq q_0 \leq 1)$ is used to determine either the ant to do the exploration or exploitation.

- Exploration; if $(q_0 > q)$

$$p_{i,j}(t) = \frac{[\tau_{i,j}(t)]^{\alpha}[\eta_{i,j}(t)]^{\beta}}{\sum_{h=1}^{l_i}[\tau_{i,h}(t)]^{\alpha}[\eta_{i,h}(t)]^{\beta}} \tag{1}$$

- Exploitation; if $(q_0 \leq q)$

$$Argmax_{1 \leq h \leq l_i}\left\{[\tau_{i,h}(t)]^{\alpha}[\eta_{i,h}(t)]^{\beta}\right\} \tag{2}$$

where;

- $[t_{i,j}(t)]$ is the amount of pheromone values of edges $e_{i,j}$,
- $\eta_{i,j}(t)$ is the heuristic values value of edge $e_{i,j}$, and
- $\alpha(\alpha \leq 0)$ and $\beta(\beta > 0)$ are the factor that determine relative influence of pheromone and heuristic value on the edge decision of the ant.

b) The pheromone update

- Global pheromone update

Ants that travel along the path will construct the path based on probability of the edge chosen. Eventually there is only one best ant identified as the single best ant (or globally best ant) where it constructs the shortest path compare to the others. This single best ant is the only ant that allowed to deposit pheromone to the trail. Meanwhile, pheromone evaporation occurs in order to avoid the algorithm stuck in a local minima solution. This is process done after all the ants have successfully complete construct the path in the current iteration.

$$\tau_{i,j}(t+1) = \begin{cases} (1-\rho)\tau_{i,j} + \rho\Delta\tau_{i,j}^{bs} & if\ e_{i,j} \in test^{bs} \\ \tau_{i,j}(t) & otherwise \end{cases} \tag{3}$$

where;

- ρ[0,1] is the pheromone evaporation rate and,
- Δτ_(i,j)^bs is the fitness function.

- The local pheromone update

  The main intention of local pheromone update is to make the search performed by following ants during the iteration is diversified. While travelling in trail path, each ant is allowed to deposit their pheromone at the last edges. Meanwhile, pheromone evaporation occurs as to make sure subsequent ant explore the different edges. Ants will update the pheromone level by applying local update rule;

$$\tau_{i,j}(t) = (1 - \varphi)\tau_{i,j}(t) + \varphi\tau_0 \quad (4)$$

where;

- τ_0 is the initial value of pheromone and
- φ $(0 < \varphi < 1)$ is the evaporation of the pheromone.

5. Proposed Strategy

The aim of t-way testing test is to find faults between interaction input parameter. In previous section, the algorithm of ant and variant of ant colony optimization is explained. However, there is room for improvement that still can be made to the algorithm, where number of ants is still not optimized. As ACOF will adapt OTAT searching technique, number of ant's agent can be optimized based on remaining uncovered tuple using fuzzy logic algorithm. Besides, the probability in selecting the edge to travers also need to be tuned with exploration in early iteration and exploitation in middle or near end. Figure 2 shows the main component function of ACOF. It consists of four main components that drive towards the effectiveness of ACOF.

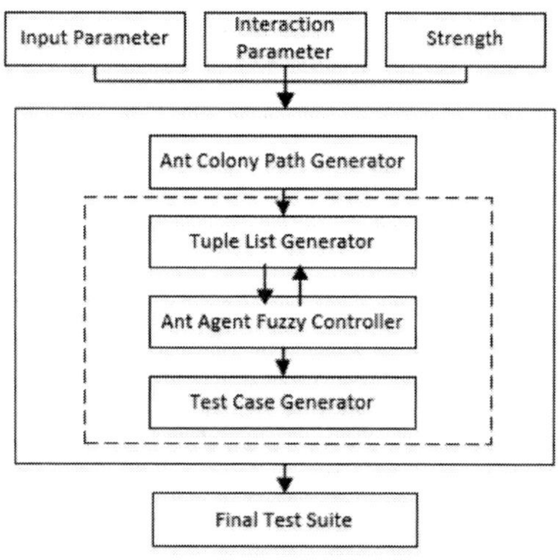

Figure 2. Main Framework of ACOF

5.1. Ant Colony Path Generator (ACPG)

This component function is to generate path of the ant based on configuration setting set by software tester. It consists of virtual nodes where it represents the inputs and branch for each node where it represents value of each input. This path will be used by ant agents to travel from each input node, in finding the optimum path. This will help in generating the optimum test case.

5.2. Tuple List Generator (TLG)

This function is to generate a full list of interaction tuples. This functions retrieve inputs parameter (i.e. number of input and its values), interaction parameters as well as interaction strength prior generating the tuple list. The tuple list then will be used by test case generator to determine which test case covered the most tuple. Tuple list will eventually will be empty as every covered tuple by test cases will be removed from the list.

5.3. Ant Agent Fuzzy Controller (AAFC)

Different input configuration tends to have different amount of ant agent to be released. This function is to dynamically assign number of ant agent without regarding the input setting. Different from others ACO algorithm, number of ant agent will be optimized automatically using fuzzy logic algorithm. It is based on the percentage of uncovered tuples in the tuple list and percentage of covered test case of every iteration. For the early iteration, less number of ant will be assigned as there are still many uncovered tuples in the list. While the iterations reach to half or near end, more ant will be assigned as to help test case generator to find the most uncovered tuple.

5.4. Test Case Generator (TCG)

TCG is the main component to generate test case. As ACOF will adapt OTAT, one best test case will be generated in one process iteration. Test case generator uses ACPG as the path for ants to travers in constructing the test case. In traveling from node to node, ant will have to make decision in which branch they will choose. The decision is based on heuristic value, pheromone, probability and fitness function. However, as mentioned in AAFC earlier, the probability need to be tune as to control the exploration and exploitation of the path travel by the ants. In finding the best test case for early iteration, ant will explore the search space while in middle or near end of the process, ants will exploit path from the last best test case. This is since, at beginning of the iteration, there are still a lot of uncovered tuple remain in the tuple list. Thus it is to avoid the algorithm stuck in local minima solution, where there might be other best solution in other space. Eventually, ant will do the exploitation as remaining of uncovered tuple now become fewer to empty.

5.5. Flowchart of ACOF

An overview of how the strategy works is presented in Figure 3. The flowchart presents the process in general. The strategy accepts input parameters and its values as input to the strategy and produces test suite. Firstly, TLG generates a tuple list. In the meantime, ACPG starts to generate path to be travelled by ant's agent based on input parameters and its values. After both TLG and ACPG finished their tasks, TCG starts to generate a single best test case.

Next, the generated single best test case will construct its own tuples. Each tuple then is compared to the tuples list. Any similar tuples will be removed from the tuple list. The similar tuples mean that the single best test case covers at least one combination of input parameters. The single best test case is then stored in the Test Suite. However, if none of the constructed tuple is similar to the tuples in the tuple list, the TCG continues to generate another single best test. Throughout the process, the tuple list is becoming smaller as it has been removed whenever it is covered by a single best test. The process iterates until the tuple list is empty.

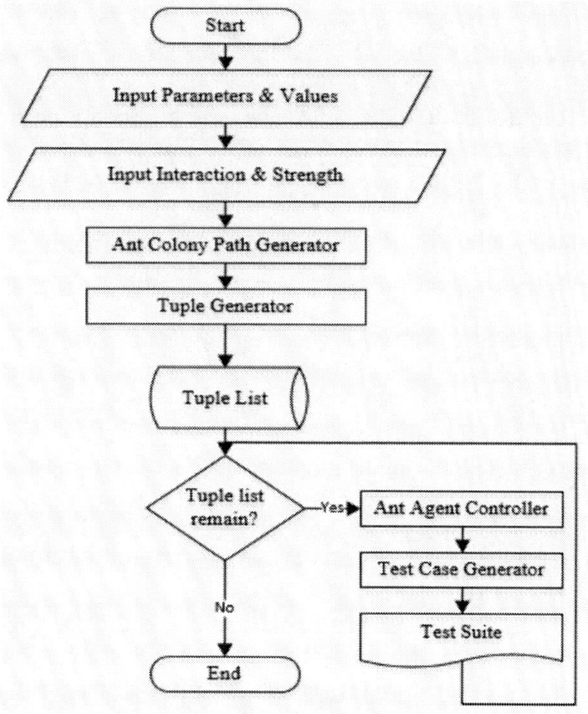

Figure 3. Flowchart of ACOF

6. Conclusion and Future Works

In this paper, we proposed a t-way strategy that based on ant colony optimization with fuzzy logic. The framework of ACOF is consist of four main components which is Tuple Generator, Ant Colony Path Generator, Ant Agent Controller and Test Case Generator. Next, the algorithm will be implemented and the performance of the ACOF (i.e. test suite size) will be evaluated and compare against other strategies with several benchmarking experiment that have been published.

Acknowledgement

The author would like to acknowledge the support from Fundamental Research Grant Scheme (FRGS) under a grant number of FRGS/1/2018/ICT01/UNIMAP/02/1 from the Ministry of Education Malaysia.

References

[1] Ammann P and Offutt J 2008 *Introduction to Software Engineering* (Cambridge University Press)

[2] Ahmad M Z Z, Othman R R and Rashid Ali M S A 2016 Sequence Covering Array Generator (SCAT) for Sequence Based Combinatorial Testing *Int. J. Appl. Eng. Res.* **11** 5984–91

[3] Kuhn D R, Kacker R N and Lei Y 2009 Combinatorial Software Testing *Computers* **42** 94–6

[4] Alsewari A R A and Zamli K Z 2011 Interaction Test Data Generation Using Harmony Search Algorithm *2011 IEEE Symp. Ind. Electron. Appl. ISIEA 2011* 559–64

[5] Shiba T, Tsuchiya T and Kikuno T 2004 Using Artificial Life Techniques to Generate Test Cases for Combinatorial Testing *Proc. 28th Annu. Int. Comput. Softw. Appl. Conf. 2004. COMPSAC 2004.* 72–7

[6] Chen X, Gu Q, Li A and Chen D 2009 Variable Strength Interaction Testing with an Ant Colony System Approach *Proc. - Asia-Pacific Softw. Eng. Conf. APSEC* 160–7

[7] Ramli N, Othman R R and Ali M S A R 2017 Optimizing Combinatorial Input-Output Based Relations Testing Using Ant Colony Algorithm *2016 3rd Int. Conf. Electron. Des. ICED 2016*

586–90

[8] Othman R R and Zamli K Z 2011 ITTDG: Integrated T-way Test Data Generation Strategy for Interaction Testing *Sci. Res. Essays* **6** 3638–48

[9] Ahmed B S and Zamli K Z 2010 PSTG: A t-way strategy adopting particle Swarm Optimization *AMS2010 Asia Model. Symp. 2010 - 4th Int. Conf. Math. Model. Comput. Simul.* 1–5

[10] Ahmed B S, Zamli K Z and Lim C P 2012 Application of Particle Swarm Optimization to uniform and variable strength covering array construction *Appl. Soft Comput. J.* **12** 1330–47

[11] Ahmed B S and Zamli K Z 2011 A Variable Strength Interaction Test Suites Generation Strategy Using Particle Swarm Optimization *J. Syst. Softw.* **84** 2171–85

[12] Rabbi K, Mamun Q and Islam R 2015 An Efficient Particle Swarm Intelligence Based Strategy to Generate Optimum Test Data in T-way Testing *2015 IEEE 10th Conf. Ind. Electron. Appl.* 123–8

[13] Chen X, Gu Q, Qi J and Chen D 2010 Applying Particle Swarm Optimization to Pairwise Testing *Proc. - Int. Comput. Softw. Appl. Conf.* 107–16

[14] Wu H, Nie C, Kuo F C, Leung H and Colbourn C J 2015 A Discrete Particle Swarm Optimization for Covering Array Generation *IEEE Trans. Evol. Comput.* **19** 575–91

[15] Zamli K Z, Klaib M F J, Younis M I, Isa N A M and Abdullah R 2011 Design And Implementation Of A T-Way Test Data Generation Strategy With Automated Execution Tool Support *Inf. Sci. (Ny).* **181** 1741–58

[16] H Y O and Kamal Z Z 2016 Development of interaction test suite generation strategy with input-output mapping supports *Sci. Res. Essays* **6** 3418–30

[17] Wang Z and He H 2013 Generating variable strength covering array for combinatorial software testing with greedy strategy *J. Softw.* **8** 3173–81

[18] Othman R R, Khamis N and Zamli K Z 2014 Variable Strength T-Way Test Suite Generator With Constraints Support *Malaysian J. Comput. Sci.* **27** 204–17

[19] Lin J, Luo C, Cai S, Su K, Hao D and Zhang L 2016 TCA: An Efficient Two-Mode Meta-Heuristic Algorithm for Combinatorial Test Generation Jinkun *Proc. - 2015 30th IEEE/ACM Int. Conf. Autom. Softw. Eng. ASE 2015* 494–505

[20] Ahmed B S, Abdulsamad T S and Potrus M Y 2015 Achievement of Minimized Combinatorial Test Suite for Configuration-aware Software Functional Testing Using The Cuckoo Search algorithm *Inf. Softw. Technol.* **66** 13–29

[21] Yang X S and Deb S 2009 Cuckoo search via Lévy flights *2009 World Congr. Nat. Biol. Inspired Comput. NABIC 2009 - Proc.* 210–4

[22] Nasser A B, Sariera Y A, Alsewari A R A and Zamli K Z 2016 Assessing Optimization Based Strategies for t-way Test Suite Generation: The Case for Flower-based Strategy *Proc. - 5th IEEE Int. Conf. Control Syst. Comput. Eng. ICCSCE 2015* 150–5

[23] Zamli K Z, Alkazemi B Y and Kendall G 2016 A Tabu Search hyper-heuristic strategy for t-way test suite generation *Appl. Soft Comput. J.* **44** 57–74

[24] Ali M S A R, Othman R R, Yahya Z R, Ahmad M Z Z and Ramli N 2017 Implementation of Artificial Bee Colony Algorithm for T-Way Testing *2016 3rd International Conference on Electronic Design, ICED 2016* (Institute of Electrical and Electronics Engineers Inc.) pp 591–4

[25] Zamli K Z, Din F, Baharom S and Ahmed B S 2017 Fuzzy Adaptive Teaching Learning-Based Optimization Strategy for The Problem of Generating Mixed Strength T-Way Test Suites *Eng. Appl. Artif. Intell.* **59** 35–50

[26] Dorigo M and Di Caro G 1999 Ant Colony Optimization: A New Meta-Heuristic *Proc. 1999 Congr. Evol. Comput. CEC 1999* **2** 1470–7

[27] Dorigo M and Stutzle T 2010 *Ant Colony Optimization: Overview and Recent Advances* vol 146 (Springer International Publishing AG)

[28] Dorigo M 1997 Ant Colony System: A Cooperative Learning Approach to the Traveling Salesman Problem *IEEE Trans. Evol. Comput.* **1** 53

Adaptive DCO-FBMC in Visible Light Communication

Abdullah Ali Qasim[1,2], M F L Abdullah[1], R Talib[1]

[1] Dept. Communication Engineering University Tun Hussein Onn Malaysia Parit Raja, 83000 Batu pahat, Johor, Malaysia.
[2] Eng. Technical College-Najaf, Al-Furat Al-Awsat Technical University, Iraq.

alzubydea@atu.edu.iq

Abstract. Filter Bank multicarrier (FBMC) has been modified for VLC. FBMC is the main filter suitable for high-speed data transfer using multicarrier modulation (MCM). Therefore, the FBMC is an appropriate alternative to the OFDM system which suffers from several weaknesses like limited bandwidth due to the presence of the cyclic prefix (CP). FBMC has been adapted to IM / DD compliant using Hermitian Symmetry to make the real signal, but this method increases system complexity and power consumption. In this paper, FBMC was adapted using a different technique than the above which consisted of generating an original complex FBMC signal. The signal was then dismantled into two real and imaginary parts (juxtaposing the real and imaginary parts of the complex signal in the time domain), thus the imaginary signal turned real. The proposed technique saved 50% in the number of the FFT/IFFT (Inverse Fast Fourier Transformer/ Fast Fourier Transformer) operations, which resulted in a great decrease in power consumption and the occupied chip area.

1. Introduction

The significant development in solid-state light-emitting diodes (LED or LD) began when it was used in luminance at a large level inside buildings and outdoors [1] due to its high luminance efficiency, low cost, low power consumption compared to other types of illumination and long life. The speed of switching that has been used in digital communication as it is harmless to the human eye [2]. Optical communications have attracted the attention of many researchers, especially the VLC system because of the use of visible light, which has an unlicensed spectrum with a wide range of (400nm-800nm) in addition to this type of communication. The VLC system with a large unregulated spectrum is an excellent alternative to the limited spectrum of RF [3].

Orthogonal frequency division multiplexing (OFDM) is popularly employed in high data rate communications because of its capacity to combat inter-symbol interferences (ISI) [4]. However, the signal from OFDM cannot be used in intensity modulation / direct detection (IM / DD) systems because it is complex and bipolar, and the electrical signal is modulated on the transmitter. The Hermitian Symmetry technique is used to obtain a real signal in the system where the incoming symbols are restricted to inverse Fast Fourier transform (IFFT) at which the imaginary signal is excluded [5]. The signal emerging to be compatible with the IM / DD system must be positive and non-negative. This is done by applying DC-Bias technology to make sure it is not negative i.e. unipolar. This condition of the system working within IM / DD is identical (the signal is real and positive) [6].

OFDM converts a frequency-selective channel into a set of parallel frequency-flat channels. Provided the cyclic-prefix (CP) is longer than the maximum delay introduced by the channel, the data at each sub-carrier can be equalized independently using a simple one-tap equalizer [7]. However, the insertion of CP will lead to reduced spectral efficiency [8]. To increase spectral efficiency, a specially designed filter bank was used instead of CP using an offset quadrature amplitude modulation / OFDM (OQAM / OFDM) system, which is an appropriate alternative to the OFDM system [9].

The biggest challenge facing the VLC system is the limited bandwidth modulation, thus, the modification of FBMC as the solution to this challenge was modified in [2] and [10] to be compatible with the IM / DD real and positive signal.

The techniques used to obtain the real signal in the FBMC system are complicated as in [2],[9] and [10]. The most prominent technique used was Hermitian Symmetry. FFT/IFFT are two of the major components in the system FBMC, however, the result from Hermitian Symmetry symbols frequency was double for the size of the components IFFT / FFT and frequency modulation codes N needed to 2N-point IFFT / FFT [9]. This shows that large size require high precision FFT and thus consume more energy and take up more space in the chip. This causes the greatest complexity.

This paper, to the best of our knowledge, is the first time FBMC modulation will be modified by generating a real signal without using Hermitian symmetry or any complicated technique. This method depends on generating a regular complex FBMC signal and juxtaposing the real and imaginary parts in the time domain to achieve a real FBMC signal. Resulting from this, only N-point IFFT/FFT transforms are demanded to generate a 2N-point real-time OFDM signal.

2. DC-OFDM in VLC

The block diagram in figure 1 on the DCO-OFDM within the VLC system shows where the input data stream was using QAM technology, followed by IFFT using sub-band carrier N to convert the signal from the frequency domain to time domain.

The IFFT algorithm outputs the discrete OFDM symbol vector $\boldsymbol{x}$ in the time-domain

$$x_m = \frac{1}{N}\sum_{k=0}^{N-1} X_k\, e^{j\frac{2\pi km}{N}} \qquad for\ 0 \le m \le N-1 \tag{1}$$

where N is the size of IFFT and X_k is the k_{th} subcarrier symbol in the frequency domain.

The output signal in (1) is complex (meaning (a+jb) a is real and jb is the imaginary) which means that it is incompatible with IM / DD based on VLC. Hermitian Symmetry is used to be the IFFT signal to be real and IM / DD compliant (figure 2 illustrates the imaginary signal after the complex signal is inserted through the Hermitian Symmetry). This is a transpose-conjugate copy of the active subcarriers, which is added to the other half of the IFFT frame (acts as a mirror). New elements of the IFFT input vector $\boldsymbol{X_H}$ are:

$$X_H = [X_0, X_1, X_2, X_3, \ldots . X_{N-1}, X_N, X_{N-1}^*, \ldots . X_3^*, X_2^*, X_1^*] \tag{2}$$

and the DC component, $X_0 = X_N = 0$ results in a 2N-point IFFT output of OFDM, therefore, the equation (1) must be modified to

$$x_m = \frac{1}{N}\sum_{h=0}^{2N-1} X_{H,h}\, e^{j\frac{2\pi hm}{N}} \qquad for\ 0 \le m \le 2N-1 \tag{3}$$

where h is the hth -subcarrier symbol of X_H.

The advantage of OFDM is that it is able to overcome this ISI problem by using the CP listed at the beginning of the OFDM framework. To ensure the unipolar signal meets IM/DD requirements, a DC-offset is added in the DC-coupled optical OFDM (DCO-OFDM) scheme. A DC-offset was implemented in practice as a dc-bias current to drive the LED transmitter. It was added to ensure unipolarity. After it became a real signal and the non-negative being sent was received through the photodetector, the DC-Bias was removed and the CP was also excluded. The signal was converted from analogue to digital (A / D) and the conversion to recover back to the parallel signal was demodulated using 2N -point FFT, thus restoring the bitstream by performing QAM [11].

Figure 2 illustrates the imaginary signal coming out of IFFT after the Hermitian Symmetry mode where N-point was used and needed to double the number to 2N-point. This in turn increases the complexity of the system and increases power consumption.

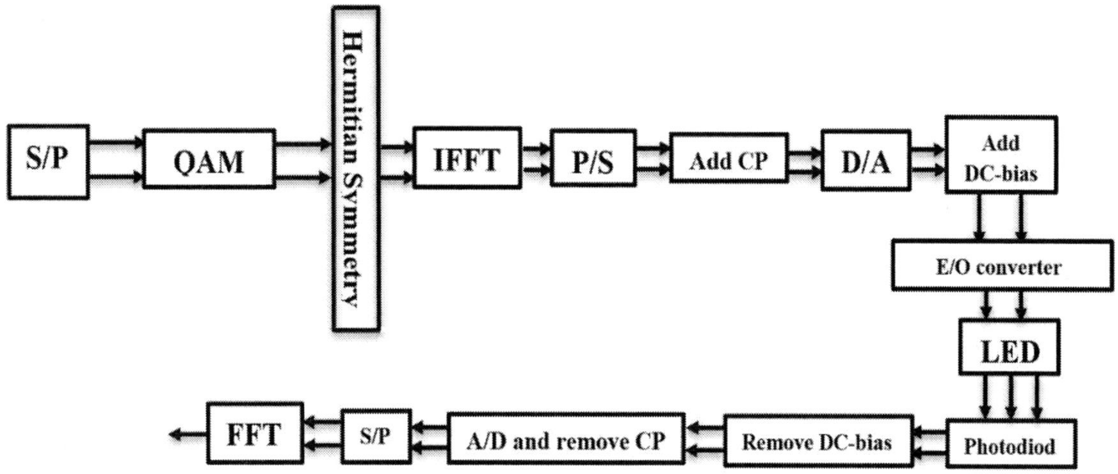

Figure 1. DCO-OFDM Block diagram

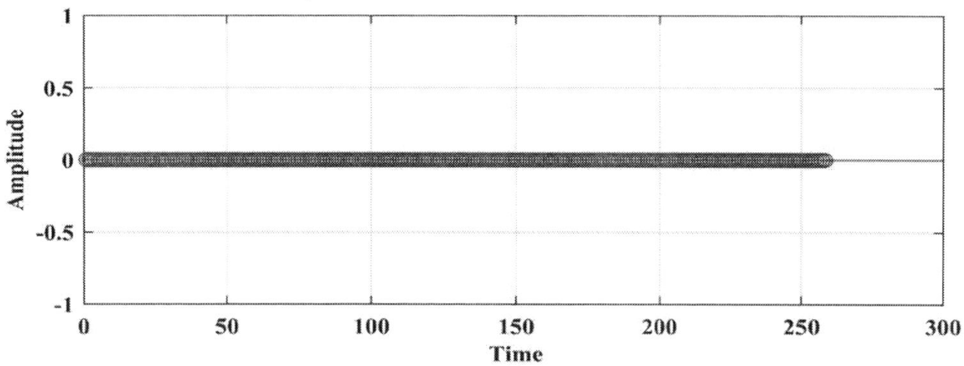

Figure 2. Output IFFT imaginary signal after used Hermitian Symmetry.

3. Conventional DCO-FBMC in VLC

FBMC has become promising to replace the conventional OFDM in 5G technology which has contributed to improve spectral efficiency because it uses CP and overcomes all the weaknesses of OFDM modulation or any other modulation technique [12]. To make FBMC within a VLC system, it must be IM / DD compliant, i.e. the signal is real and positive. To ensure this, the conventional Hermitian Symmetry method was used to obtain a real signal as shown in figure 3, which shows the main structure of DCO-FBMC.

In equation 4, the signal format is shown for the traditional FBMC system where the signal is complex. Therefore, traditional methods must be used to obtain a real signal from this system.

$$x(n) = \sum_{n=-\infty}^{+\infty} \sum_{m=0}^{N-1} a_{m,n} g_{m,n}(t) \tag{4}$$

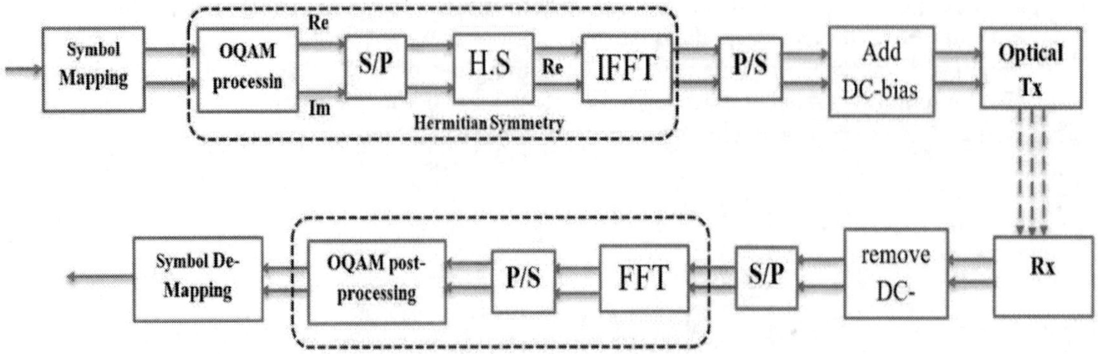

Figure 3. Conventional block diagram DCO-FBMC used Hermitian Symmetry.

Figure 3 illustrates the itinerary signal from entering and processing in the Offset QAM (OQAM), where the signal emerges as a complex signal. Tto make a real signal, it must pass through the Hermitian Symmetry and after converting the frequency domain to a time domain, the signal from the IFFT will be real but bipolar. Equation 4 should be modified as shown in equation 5 as in DCO-OFDM.

$$x(n) = \sum_{n=-\infty}^{+\infty} \sum_{m=0}^{2N-1} a_{m,n} g_{m,n}(t) \quad (5)$$

where *m* is the time index; and *n* is the subcarrier index; $a_{m,n}$ is the symbol (massage) being transmitted; *N* is the number of subcarriers; and *g(t)* is the synthesis function which maps $\boldsymbol{a_{m,n}}$ into the signal space; $g_{m,n}(t)$ is the shifted version in time and frequency as (6).

$$g_{m,n}(t) = h(n - mT_0)e^{j2\pi/Nk\left(n-\frac{L_p-1}{2}\right)}e^{j\Phi_{m,n}} \quad (6)$$

where, τ_0 is the symbol spacing in time; *Lp* is the length of the pulse shaping filter (*LP=K*N*); K is the overlapping factor as shown in Table 1, *h(n)* is the pulse shaping filter (known as the prototype filter). *k* is the subcarrier index.

$$h(n) = 1 + 2\sum_{k=1}^{K-1} (-1)^k H_k \cos\left(2\pi \frac{kn}{L_P}\right) \quad (7)$$

Hk is the set of coefficients related to *K*.

The main factor that defines and distinguishes the performance of the prototype of the candidate FBMC system is an overlapping factor *K*, where the *K* value greatly affects the BER, the filter shape and sidelobe. Figure 4 gives different filter windows because of the changing value of *K*.

Table 1. Overlap factor K for FBMC

coefficients	K2	K3	K4	K5	K6	K7	K8
h_0	1	1	1	1	1	1	1
h_1	0.707107	0.911438	0.971960	0.99184131	0.99818572	0.99938080	0.99932588
h_2		0.411438	0.707107	0.86541624	0.94838678	0.97838560	0.98203168
h_3			0.235147	0.50105361	0.70710678	0.84390076	0.89425129
h_4				0.12747868	0.31711593	0.53649931	0.70710678
h_5					0.06021021	0.20678881	0.44756522
h_6						0.03518546	0.18871614
h_7							0.03671221

As a result of the use of Hermitian symmetry and the properties of the Inverse Fast Fourier Transformer (IFFT), the imaginary signal was excluded and the outgoing signal was real as shown in figure 5. The complexity that the Hermitian Symmetry adds to the system was observed because it used twice the value of the IFFT, i.e. $2N$. These are simulated as figure 4 at $N = 128$ and $K = 2$.

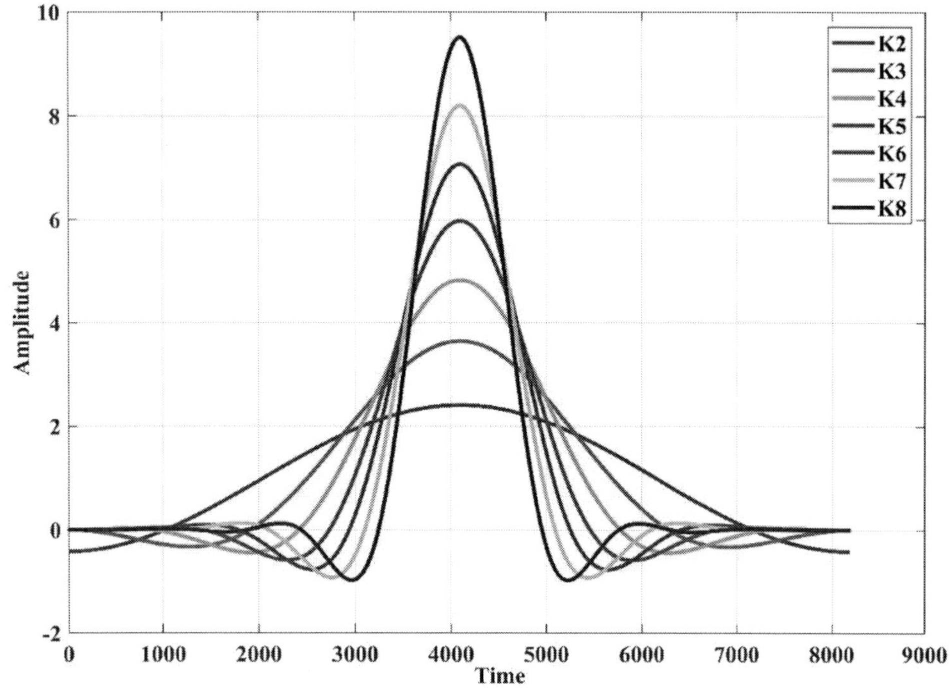

Figure 4. Time domain function of the proposed filter with different K values.

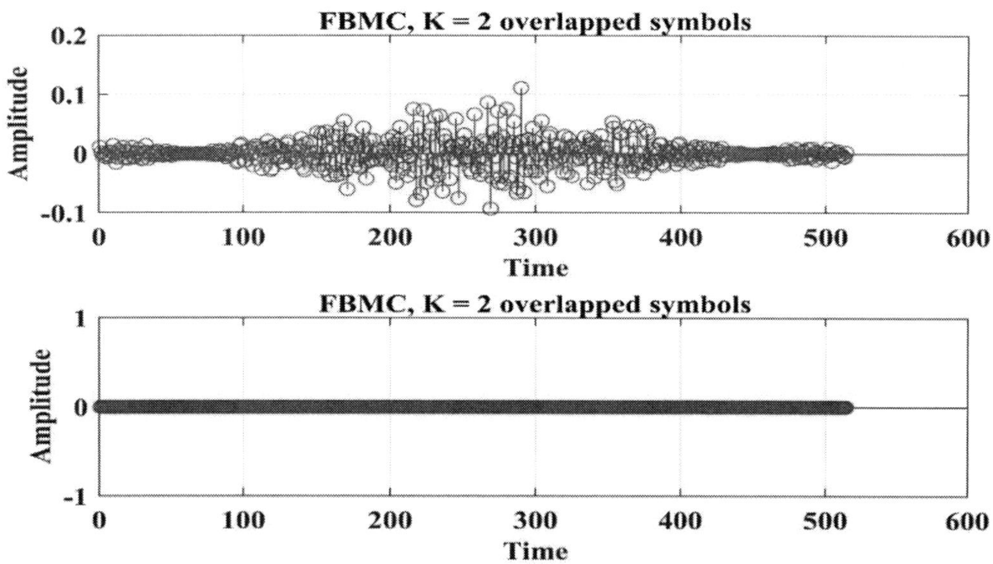

Figure 5. The shape of the signal after passing through IFFT denoted that Hermitian Symmetry was successfully used.

4. Juxtaposed optical DCO-FBMC

For the FBMC modulation system to be IM / DD compliant, the signal must be real. It was found that this technique was characterized by complexity because it leads to a doubling of the size of IFFT / FFT. This reason and others motivated the researchers to find a way to perform a real signal without complication and not use the Hermitian Symmetry. Therefore, the modulation of FBMC had to be improved to be within VLC. It was found that this technique was characterized by complexity because they lead to a doubling of the size of IFFT / FFT.

4.1. *Design Juxtaposed optical DCO-FBMC transmitter.*

Figure 6 illustrates the main structure of the DCO-FBMC enhanced design where it is noted that this design focuses mainly on the time domain. In other words, the signal was after IFFT and passing block parallel to serial transform (P/S). This signal is complex (consisting of a real part and an imaginary part) and is similar to the signal emerging from the conventional FBMC system, which is incompatible with IM / DD because as it is not real. Therefore, the imaginary signal should be converted into a real signal as will be explained below. To get a real signal by separating the complex signal into a real and imaginary signal, then converting the imaginary signal to real, and then juxtaposing the real and imaginary respectively, the process is as shown in figure 6.

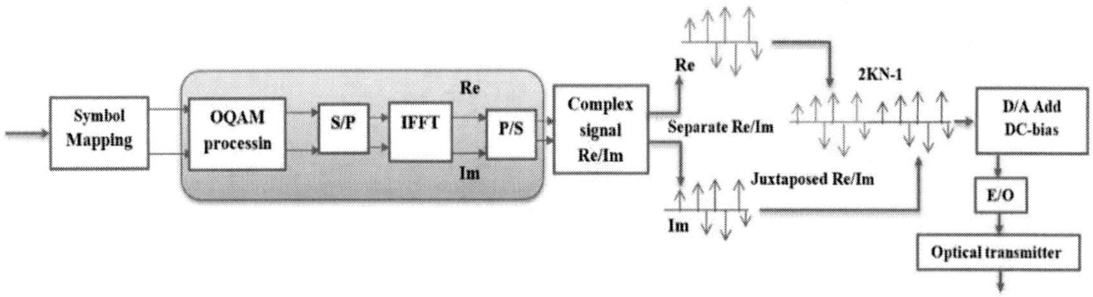

Figure 6. Main structure Juxtaposed real and imaginary DCO-FBMC without using Hermitian Symmetry.

In (4) represents the complex main signal coming out of the conventional FBMC system, hence the improvement on the system as explained above. The complex time signal can be represented exactly in (8), and this method is employed after the P/S block.

$$x(n) = a(n) + jb(n), \qquad n = 1,2,3, \ldots . K * (N-1) \tag{8}$$

where ***a*** and ***b*** are the real and imaginary parts of ***x(n)*** respectively. A real signal is generated using the above method in (9).

$$x_{2KN} = \begin{cases} a(n) & n = 0, \ldots\ldots, K(N-1) \\ b(n-KN) & n = KN, \ldots\ldots, 2K(N-1) \end{cases} \tag{9}$$

Figure 3 shows the real signal and the imaginary signal after their separation before applying the new technique. After that, the imaginary juxtaposed real signal in the new method is employed as in figure 7, thus, the denoted signal length is $2KN$.

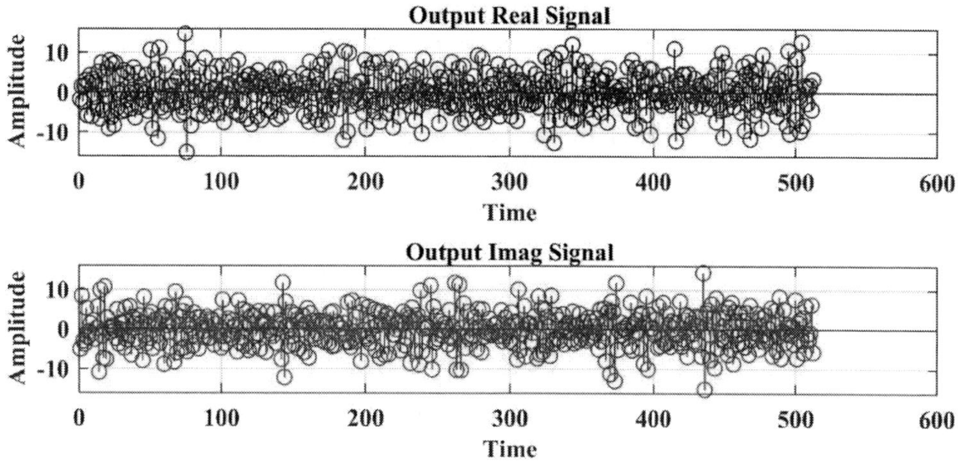

Figure 7. Separate the complex signal into a real and imaginary signal at N=256 and K=2.

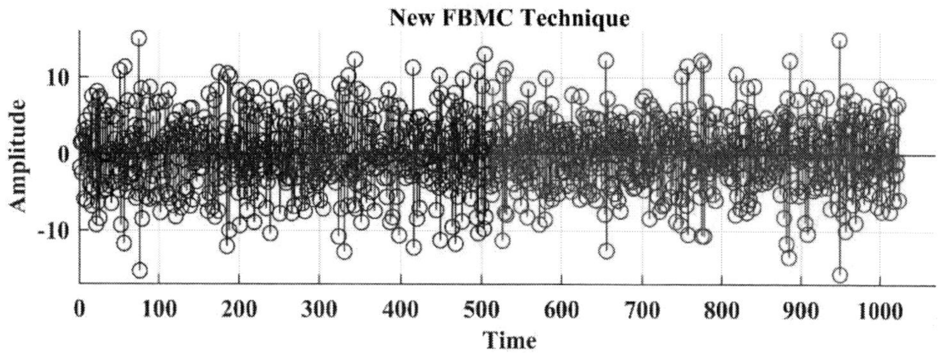

Figure 8. FBMC Juxtaposed signal at length $2KN$.

The signal shown in figure 8 is real but bipolar, which means that it is incompatible with IM / DD because one of the two conditions is incomplete. To resolve this problem, DC-bias technology was added to convert the bipolar to unipolar, and the second condition achieved a real positive signal which is therefore compatible with the IM / DD.

4.2. *Design Juxtaposed optical DCO-FBMC receiver.*

The optical signal is received by the photodetector and is converted into an electrical signal, as shown in figure 9, where the first step is to remove the DC-bias effect. Then, the signal is separated into two real and imaginary parts as described in (10).

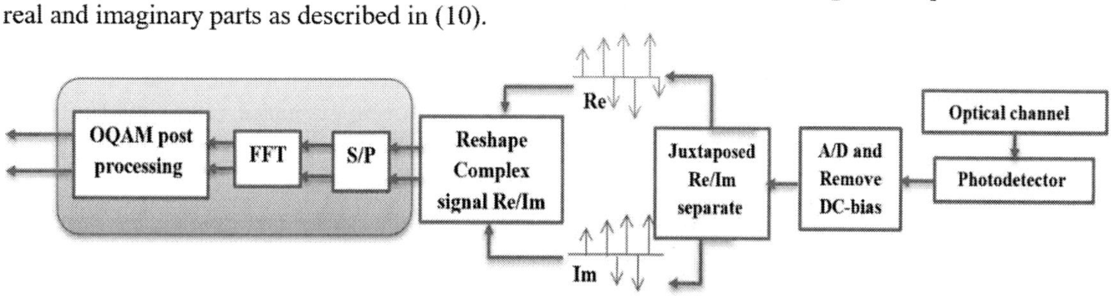

Figure 9. Design new optical DCO-FBMC receiver

$$y(n) = \begin{cases} a_r(n) = y_{2KN}(n), & n = 0,.., K(N-1) \\ b_r(n) = y_{2KN}(KN+n) & n = 0,.., K(N-1) \end{cases} \quad (10)$$

where a_r and b_r are respectively the real and imaginary components of received signal *y(n)* which then reshaped the complex to pass the signal to the analysis filter bank demodulation as the conventional FBMC system.

5. BER performance

Figure 10 shows the results compared to the simulated bit error rate (BER) technology DCO-OFDM, DCO-FBMC traditional (utilising Hermitian Symmetry), and finally, DCO-FBMC technology which uses the new method as a signal-noise ratio (SNR) in dB. This simulation was done at 1000 different M-QAM codes in AWGN from 16 QAM to 256QAM at N = 1024. The simulation results were great as the DCO-FBMC performance in both new and traditional methods was better and BER was lower than DCO-OFDM and they are similar at the same time.

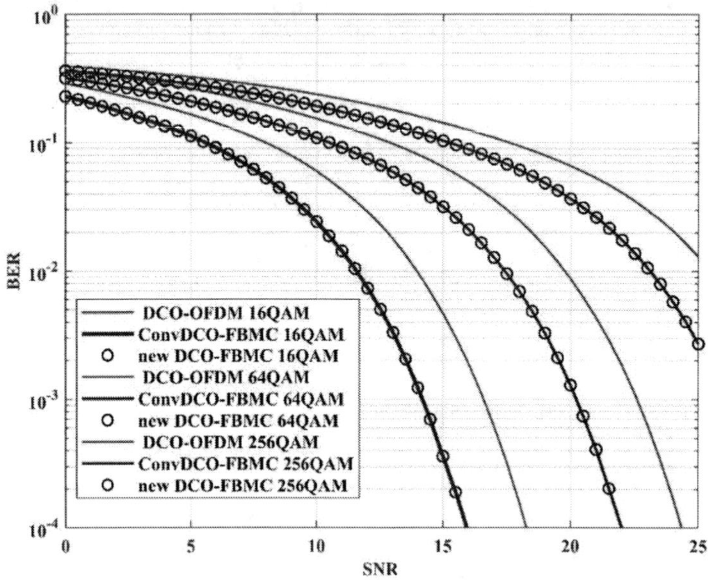

Figure 10. The BER as function of SNR of the conventional (Conv) DCO-FBMC, DCO-OFDM the proposed (new) techniques DCO-FBMC.

6. Complexity and power consumption

Using Hermitian Symmetry in the traditional DCO-FBMC system to get a real signal requires additional carrier and additional complexity. This new method to get the real signal does not require any complexity, is very simple and uncomplicated, and does not occupy a place. The complexity was calculated as shown in figure 11.

$$A_{anew} = 3N\left(log_2\left(\frac{N}{2}\right) - 1\right) + 8 + 4(NK - N + 1) \quad (11)$$

$$A_{acon} = 6N(log_2(N) - 1) + 8 + 4(NK - N + 1) \quad (12)$$

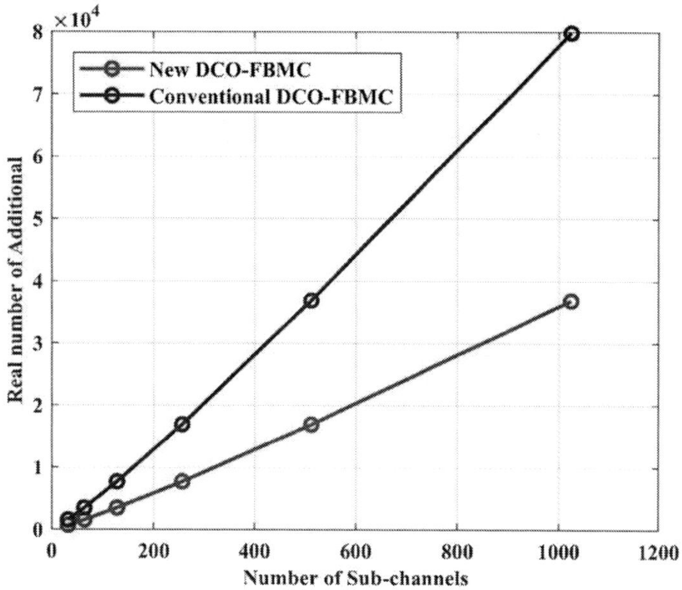

Figure 11. Illustrates the differences in complexity between the new method and the conventional method.

Figure 11 above shows that the conventional method is more complex than the new method by 50% and this shows an increase in the size of IFFT / FFT which increases power consumption

7. Conclusion

The new FBMC modulation is compatible with IM / DD without using Hermitian Symmetry as a conventional system to operate within the VLC system. The new method of this technique relies on the complex signal in the time domain where the real signal is separated from the imaginary signal and then the imaginary signal is converted to the real signal. This method increases spectral efficiency, reduces complexity, and reduces energy consumption, which denotes better performance.

8. References

[1] Mabrouk, Wafi A, M.F.L Abdullah, M.S.M Gismalla, B Das M I N 2019 Configuration model in visible light communication 2019 *Int. Conf. Inf. Sci. Commun. Technol.* 1–6

[2] Lin B, Tang X, Ghassemlooy Z, Fang X, Lin C, Li Y and Zhang S 2016 Experimental Demonstration of OFDM/OQAM Transmission for Visible Light Communications *IEEE Photonics J.* **8**

[3] Afgani M Z, Haas H, Elgala H and Knipp D 2006 Visible Light Communication Using OFDM *2nd Int. Conf. Testbeds Res. Infrastructures Dev. Networks Communities, 2006. TRIDENTCOM 2006.* 129–134

[4] Singh S, Mishra S, Kumar S and Asthana P 2017 Review of OFDM Modulation Scheme for Indoor Visible Light Communication System *Int. J. Tech. Res. Appl.* **42** 92–95

[5] Barrami F, Guennec Y Le, Novakov E, Duchamp J, Busson P, Crolles S, Monnet J and Neel L 2013 A novel FFT / IFFT size efficient technique to generate real time optical OFDM signals compatible with IM / DD systems *Proc. 43rd Eur. Microw. Conf.* 1247–50

[6] Dissanayake S D, Member S, Armstrong J and Member S 2013 Comparison of ACO-OFDM , DCO-OFDM and ADO-OFDM in IM / DD Systems *J. Light. Technol.* **31** 1063–72

[7] Saju S C and George A J 2015 Comparison of ACO-OFDM and DCO-OFDM in IM / DD Systems *Int. J. Eng. Res. Technol.* **4** 1315–9

[8] Qasim A A, Abdullah M F L, Talib R, Muwafaq H, Omar K A and Abdulrahman A M 2019 Visible Light Communication the next Future Generation System 2019 *Int. Conf. Inf. Sci. Commun. Technol.* 1–7

[9] Tabach M El, Tortelier P, Pyndiah R and Bouchet O 2008 Diffuse Infrared Personal Optical Wireless Based on Modified OFDM / OQAM *2008 6th Int. Symp. Commun. Syst. Networks Digit. Signal Process.* 161–164

[10] Chen R, Park K H, Shen C, Ng T K, Ooi B S and Alouini M S Visible light communication using DC-biased optical filter bank multi-carrier modulation *Glob. LIFI Congr. GLC 2018* Jan 1–6

[11] Wu L, Zhang Z, Dang J and Liu H 2015 Adaptive Modulation Schemes for Visible Light Communications *Liang J. Light. Technol.* **33** 117–25

[12] Viholainen A, Ihalainen T, Stitz T H, Renfors M and Bellanger M 2009 Prototype filter design for filter bank based multicarrier transmission *Eur. Signal Process. Conf.* 1359–1363

Designing One Degree of Freedom Nursing Robot for Stroke Rehabilitation

Nurul Muthmainnah Mohd Noor, Muhammad Haziq Suhaimi

Fakulti Kejuruteraan Mekanikal, Universiti Teknologi MARA, Cawangan Pulau Pinang, Kampus Permatang Pauh, 13500 Permatang Pauh, Pulau Pinang, Malaysia.

nurul.muth@gmail.com

Abstract. Rehabilitation training is a method of post-treatment for physically injured part of body especially the stroke patients. The training task such as massage and repetition exercises are needed to help them to have a normal life as before by a physiotherapist. The physiotherapist or physical therapy is a person who has experienced to treat disease or injury by physical methods. Therefore, a main objective of this paper is to develop a nursing-care robot system that can assist the stroke patient in doing daily activities. A one-degree of freedom nursing robot was designed based on the right-hand bicep using CATIA software. This robot was associated with the electromyography (EMG) sensors in order to collect the data from the muscle contractions. Meanwhile the EMG sensor is a muscle sensor that detect an electrical potential between the two-contraction flexion or extension muscle using the polymer Ag/AgCl coated electrodes. Then, the data from the muscle was sent to the Arduino microcontroller to control the servo motor. The different angle of servo motor can give the direction of movement of the nursing-care robot. The nursing robot was activated by electrical potential from the contraction of the muscle. The value of servo angle from 0 to 100 degrees angle made the robot will move up and down. The movement of robot in up and down direction is likely same as the therapist assistance doing to their patient. The size of robot is about 600 mm X 100 mm length and width and then was fabricated using 3D printer with one degree of freedom (DOF). The limitation of this project is focusing only the hand right hand bicep brachii muscle physiotherapy exercise.

1. Introduction

In industrial revolution 4.0, using robotics and automation is a method in solving the various problems especially in rehabilitation purposes. Most of these problems are relatively related to the repetitive and time-consuming tasks such as the muscle rehabilitation or physiotherapy exercises [1, 2]. The patient needs to regain their fully functional part of body that was injured by assisted with the rehabilitation training. Therefore, it requires the patient to meet a physiotherapist in few times either weekly or monthly [3]. This may give some trouble to some patient who has limited time to meet the schedule of appointment for rehabilitation exercises [4]. Other problems also need to be considered is the patient's health condition, attention and effort in an intensity and task-orientation of the training. Intense repetitions of coordinated motor activities constitute a significant burden for the therapists assisting patients [5]. The idea to solve these problems is designing an automated robot that can assist the rehabilitation exercises without any physiotherapist and can be done by himself/herself.

Robotic devices with the ability to perform repetitive tasks on patients are advanced devices and called as rehabilitation robot. Most of these robotic devices are already used in the clinical practice as well as the clinical evaluation. It can support and enhance clinicians' productivity and effectiveness as they try to facilitate the individual's recovery. The demand for caregivers and rehabilitation services are growing apace with the aging population. Therefore, a new approach is needed to improve the effectiveness and efficiency of rehabilitation to assist in the recovery process. There are two types of rehabilitation robot from the mechanical design point of view: end-effector based robots, and exoskeleton-type robots. One of examples of end-effector based robots is MIT-MANUS. This robot is will interact with subjects at the end of robot arm, which is based on different size body [6]. While exoskeleton-type robots can resemble human anatomy and apply torque to specific joints that can be used for rehabilitation training. Most of prosthesis robot and robot-aided rehabilitation are used the EMG signals to trigger and control the movement of robots. The EMG signals reflect the activities of the muscles and generate the formation of muscle force. These robots were designed like human's joints and could be worn by the human operators as an assistive device.

Therefore, in this study, the myo-electrically controller of robotic system based was designed and evaluated its effect during rehabilitation training. This project hopefully can help the stroke patient can regain his/her injured back to the normal life.

2. Electromyography (EMG) Biopotential Sensor

EMG is a method of biopotential sensor for measuring muscle activities in human [7]. This method is an active topic to be improved for accurate results. One of the different with EEG signal is EMG signal is harvested from the muscle voluntary contraction, meanwhile EEG signal from the brain [8]. A vast analysis of EMG and its information can provide an accurate picture of the function of individual muscles during voluntary movement. By using EMG, the mechanical function of a muscle or the muscle kinematics is determined by the electrical potential of the muscle, not by its mechanical events of a contraction [9]. The EMG signal can be read in amplitude information that sent via electrodes attached on the target muscle. Therefore, the amplitude information then can be manipulated to vary device control [10]. Using EMG signals, the system can predict the intention movement of certain movement in order to perform a task [11].

In this project, a MYOWARE muscle sensor from Advancer Technologies was used. The MYOWARE board acts by measuring the filtered and rectified electrical activity of a muscle with output 0 - 5 Volts depending on the amount of activity in the targeted muscles [12]. Figure 1 shows the position of 24 mm diameter biomedical pads (electrodes) that attached to the MYOWRAE board. The electrodes are coated with polymer silver chloride electrode (Ag/AgCl) to read the electrical potential difference [13]. The advantage of using MYOWARE muscle sensor is that the output signal gained was in filtered and rectified signal. Hence, reducing the time to convert raw signal into filtered signal. In addition, the position and orientation of pin A and pin B need to be aligned with the muscle fibres' orientation. Incorrect positioning of electrodes will reduce the strength and quality of the sensor's signal. The summary of the position of electrodes can be seen in Table 1.

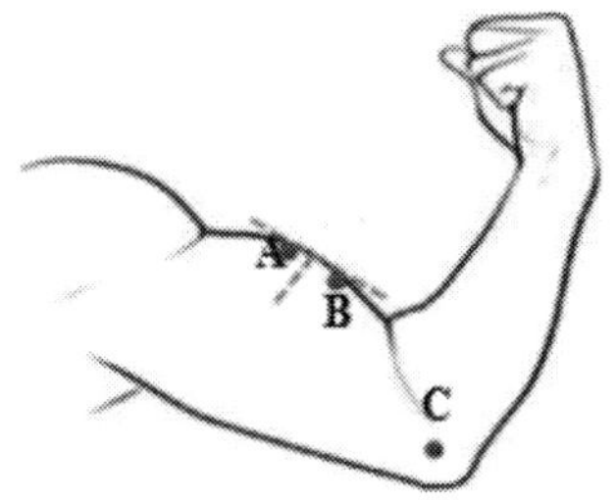

Figure 1: Position of the electrodes [8].

Table 1: Position of electrodes and its pin.

Point	Pin	Position
A	Middle	Above the midline of the biceps brancii
B	End	Below the midline of the biceps brancii
C	Reference	Elbow side

3. Electromyography Data Acquisition

3.1 EMG data acquisition from Arduino Serial plotter

In this experiment, the Arduino serial plotter was used to acquire the muscle signal. The movement of flexion and extension of the elbow joint were recorded until the subject reached full range of motion (ROM) for each movement. The experiment setup for this project as shown in figure 2, where the electrodes were placed on the user's bicep brancii. Then the electrodes were connected to the pins on MYOWARE sensor board. The sensor board then connected to the Arduino NANO microcontroller.

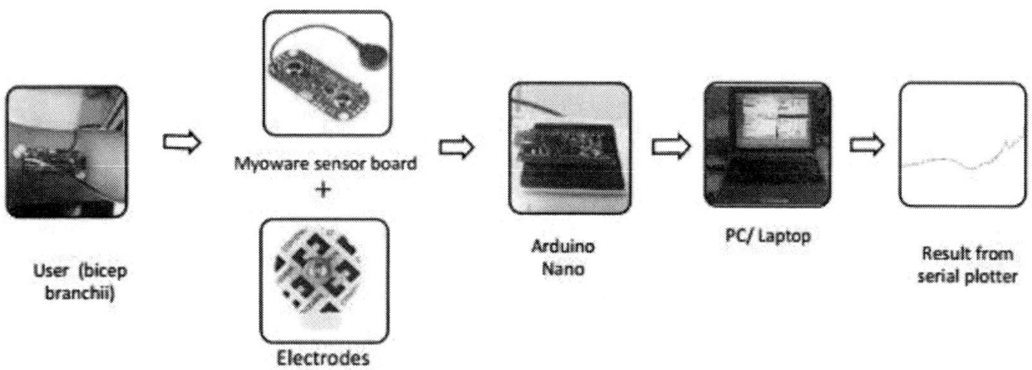

Figure 2: The experimental setup.

From the data acquired based on EMG signals towards the movement of the user's arm, the pattern of signal was classified. There are only two movements involved which are flexion and extension of muscle. The movements of the arm were needed to be actuated separately. Therefore, an algorithm was developed by using the two different conditions based on the movement of the arm either in extension or flexion. These both conditions were making the system can be repeated the process for a few times same as the manual training exercise. The flow of algorithm can be seen in figure 3.

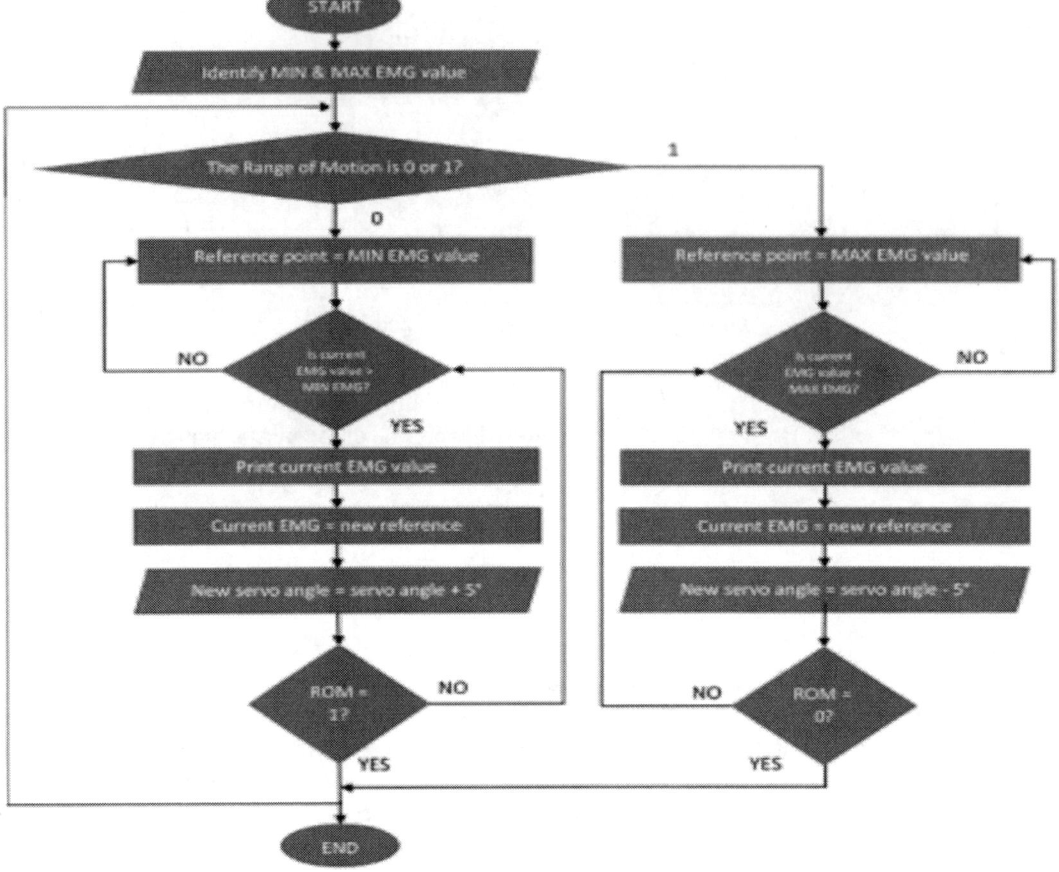

Figure 3: Flowchart of algorithm.

3.2 Design exoskeleton arm using CATIA software

In designing the exoskeleton of nursing robot, a few factors are considered such as weight, size and strength as well as the ergonomics. The exoskeleton dimensions were designed based on the 50th percentile male manikin according to right-hand side (RHS) arm measurements which includes the average size of a human arm. Therefore, the exoskeleton can be fitted for average human population. This exoskeleton was designed using CATIA software. In this software, the strength of exoskeleton can be found by doing the simulation. Figure 4 shows the illustration of exoskeleton that attached to human by CATIA software.

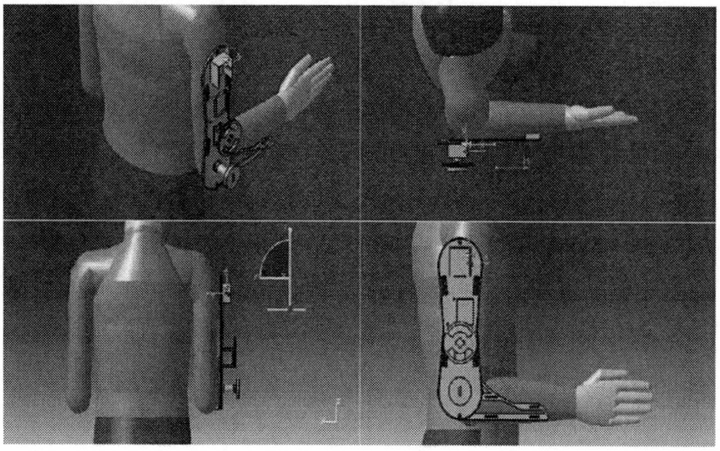

Figure 4: The exoskeleton designed attached to 50th percentile male manikin.

Meanwhile, figure 5 shows the full assembly the exoskeleton of nursing robot. It consists of four major parts. The major parts of exoskeleton structure are upper arm, lower arm with forearm placeholder and two gears. The upper arm part was designed to put the mechanical and electrical components such as Arduino NANO. The lower arm part with forearm placeholder was designed to assist the forearm of the user and act as a supporter for the arm. There are two gears – driver and driven gears were attached to the servo motor and the lower arm part respectively. The gears were linked using 6 mm width timing belt. The mechanism of single degree of freedom (SDOF) was located at the elbow joint, which the robot will rotate. The size of the exoskeleton was 600 mm X 100 mm length and width respectively.

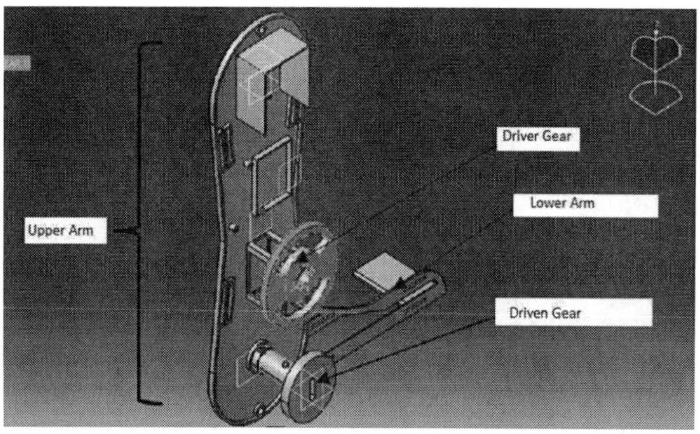

Figure 5: The assembly of the exoskeleton.

3.3 Fabrication of exoskeleton

In figure 6 shows the fabricated exoskeleton of nursing robot. The material used was Acrylonitrile Butadiene Styrene (ABS) plastic and fabricated using 3D printer. The product is twenty-percent infill with the total weight of 167 grams. Table 2 shows the description of the parts consist in the nursing robot.

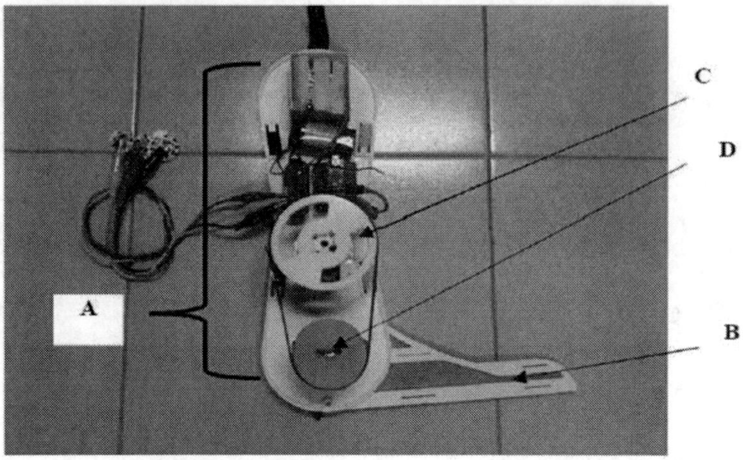

Figure 6: The part of fabrication of nursing robot.

Table 2: Description of the exoskeleton parts.

Label	Description
A	The upper arm
B	The lower arm with forearm placeholder
C	Driver gear (100 teeth)
D	Driven gear (75 teeth)

4. Results and Analysis

4.1 Exoskeleton Finite Element Analysis (FEA)

In designing the nursing robot, a few analyses need to be considered in order to achieve the objective. One of analyses in this project is to identify the maximum yield strength of the product design. Using CATIA software, the method of finite element analysis (FEA) was conducted. In this simulation, the load needs to be chosen as well as the direction of force. In this analysis, the most concerned problem was whether the exoskeleton parts can support the weight of the user's arm. Referred at anthropometric data adapted in Principle of Biomechanics in 2009, an average value for 50th percentile male upper limb weighs around 3.70kg while 95th percentile weigh about 4.33kg. Therefore, to exceed any additional weight, 10kg weight or 100 N force was applied at vital points that seemed to be vulnerable to bear the load. The result of analysis can be seen in figure 7 and figure 8.

IOP Conf. Series: Materials Science and Engineering **767** (2020) 012019 doi:10.1088/1757-899X/767/1/012019

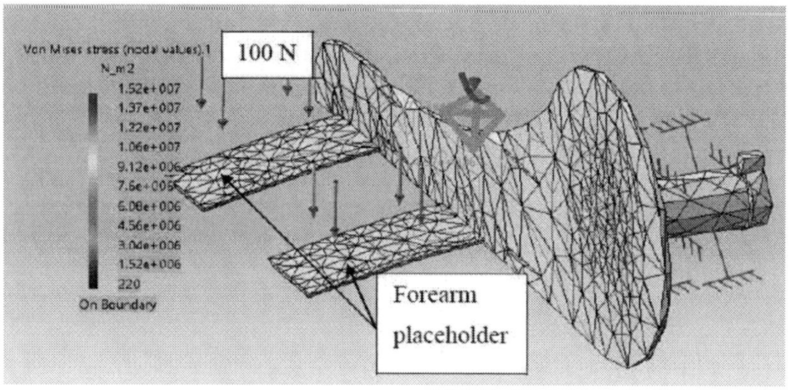

Figure 7: The lower part of exoskeleton.

In figure 7, it shows a simple 100 N force analysis exerted at the forearm placeholder. This analysis simulate force exerted from the weight of the forearm of the user. The Von Mises Stress shows the maximum stress was 15.2 MPa. While in figure 8 shows the maximum stress was 3.43 MPa.

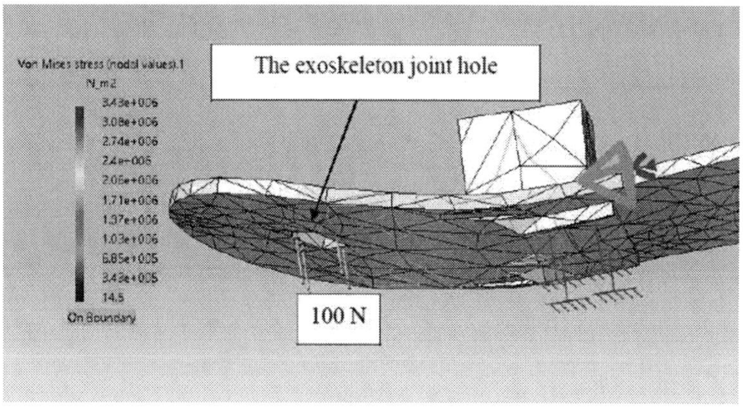

Figure 8: Von Mises Stress on upper part of exoskeleton.

Based on Table 3, the data was recorded based on the Von Mises value. The yield strength of common Acrylonitrile Butadiene Styrene (ABS) is about 40 MPa. Meanwhile, the maximum stress for lower part and upper part were 15.2 MPa and 3.43 MPa respectively. Both maximum values did not reach the ABS yield strength. Therefore, the design should be able to withstand any force applied on the exoskeleton such as the forearm mass. The summary has been recorded in Table 3.

Table 3: The data recorded from FEA.

Item	Lower part	Upper part
Force (N)	100	100
Point of force exerted	Forearm placeholder	The exoskeleton joint hole
Maximum yield stress (Pa)	15.2×10^6	3.43×10^6
Minimum yield stress (Pa)	220	14.5

4.2 Electromyography data acquisition results

The data of EMG is collected on the bicep muscle by using Arduino software serial plotter. The subject was asked to extend the arm during sitting on the chair. This condition is called relax condition. Meanwhile, the subject was needed slowly raises his forearm in the flexion position until reach the maximum flexion and then slowly lowering the forearm to its original position. The value of EMG can be seen in figure 9.

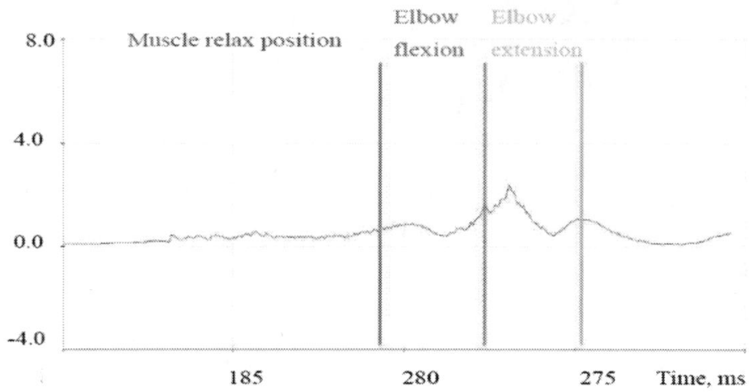

Figure 9: The data EMG signal from using serial plotter.

Figure 9 shows the EMG signal plotted in voltage value. When the subject flexed his forearm, the muscle value or EMG value increases but not directly proportional to the elbow angle. This shows that the EMG value only concern about the contraction strength when flexing or relaxing the target muscle. The overall reading was satisfied showing that the muscle sensor can be used in the project.

In this project, there are three subjects involved, which are two males aged 24 and 28 while the other female aged 23. Figure 10 shows the signal response by the serial plotter of the nursing robot for the three sets of movement (extension – flexion - extension). The blue graph indicates the muscle value was needed to manipulate the servo angle (green graph). The EMG value was varied for every subject. This is because everyone has their own muscle values. The sudden raised of signal gained between point 80 and point 100 was caused by the subject touched the electrode while the system was tested. The mean value was gained by adding the maximum and minimum EMG values then divided by 2.

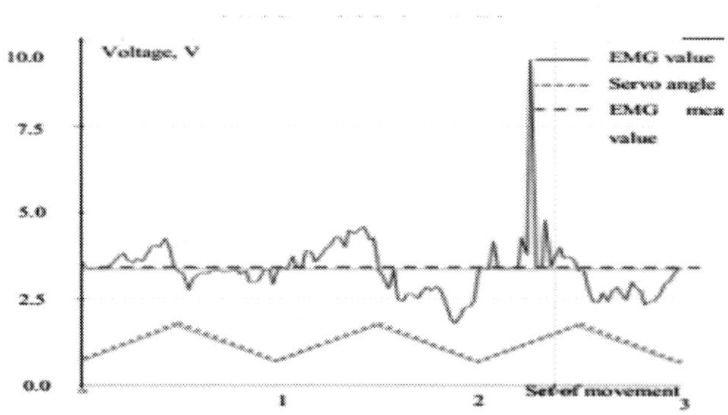

Figure 10: The EMG values with servo angle.

Table 4 shows the results of the EMG values for each subject and the servo angle actuated based on their EMG value. The EMG value acquired can be divided into the maximum, the minimum and mean EMG values. Based on the results recorded, the highest value among the subject's average mean EMG value was 3.16 V and the lowest was 1.774 V.

Table 4: The subjects' details

Details	Subject 1	Subject 2	Subject 3
Age	23	24	28
Gender	Female	Male	Male
Max EMG value (V)	248	435	264
Min EMG value (V)	88	240	192
Mean EMG value (V)	168	337.5	228
ROM flexion, degree	100	100	100
ROM extension, degree	0	0	0

Meanwhile Table 5 shows the sequences of input and output in the both flexion and extension of the elbow joint that obtained from the test. In flexion movement, the servo motor rotated from 0° to 100° degree when the muscle value was higher than the mean value and when the muscle value lower than mean value, the servo motor maintains in the position. However, in the extension movement, the servo motor rotated from 100° to 0° angle when the muscle value was lower than mean value and motor stopped when the value of angle is more than 100°.

Table 5: The array of nursing robot system.

Movement	**Input (muscle value)**	**Output (servo actuation)**
Flexion	1	1
	0	0
Extension	1	0
	0	1

5. Conclusion and Recommendations

In this project, the Arduino microcontroller-based nursing robot for rehabilitation purpose was developed and fabricated using the 3D printer respectively. Based on finite element analysis using CATIA software, the nursing robot was designed according to human kinematics motion especially the movement at the elbow joint. Therefore, from the result, it can conclude that the nursing robot is successfully fabricated and suitable to be used for both male and female. It can replace the human job by using this robot. The exoskeleton design also meets the desired yield stress within the forearm mass. The limitation of the nursing robot is that it is applicable only to be used at the right-hand side of the upper limb. This is because, the servo angle needs to be calibrated as the left and right arm are on the opposite side. Several improvements can be considered in the future development. For example, the product can be used on both sides of arms. Graphical user interface (GUI) can also be added to increase the user interaction with the product and track the user's exercise routine.

References

[1] S. Bhattacharyya, and D. N. Tibarewala, "Motor imagery, P300 and error-related EEG-based robot arm movement control for rehabilitation purpose," *Med Biol Eng Comput*, p. 1007–1017, 2014.

[2] H. Bastiaens, G. Alders, P. Feys,S.Notelaers, K. Coninx, L.Kerkhofs, V. Truyens, R. Geers, and A. Goedhart, *"Facilitating robot-assisted training in MS patients with arm paresis: a procedure to individually determine gravity compensation," in Proceedings of the 12th IEEE International Conference on Rehabilitation Robotics (ICORR '11),* pp. 1100–1105,

Zurich, Switzerland, July 2011.

[3] P. Maciejasz, J. Eschweiler, K. Gerlach-Hahn, A. Jansen-Troy, and S. Leonhardt, *"A survey on robotic devices for upper limb rehabilitation"*, Journal of Neuro-Engineering and Rehabilitation, **vol 11(3)**, pp: 1-29, 2014.

[4] J. R. Octavia, P. Feys, and K. Coninx, "Development of Activity Related Muscle Fatigue during Robot-Mediated Upper Limb Rehabilitation Training in Persons with Multiple Sclerosis: A Pilot Trial," Multiple Sclerosis International, **vol. 2015**, p. 11, 2015.

[5] H. I. Krebs, and B. T. Volpe, "Rehabilitation robotics", *Hand ClinNeurol*, **vol: 110**, pp: 283-294, 2013

[6] R. Song, K. Tong, X. Hu, and W. Zhou, *"Myoelectrically controlled wrist robot for stroke rehabilitation,* "Journal Neuroeng Rehabil, **Vol: 52**, 2013.

[7] R. G. Whittaker, "The fundamentals of electromyography," *Practical Neurology*, **vol. 12**, p. 187–194, 2012.

[8] A. T. Poyil, F. Amirabdollahian, and V. Steuber, *"Study of Gross Muscle Fatigue During Human-Robot Interactions,"* The Tenth International Conference on Advances in Computer-Human Interactions, pp: 187 – 192, 2017.

[9] A. Basteris, S. M. Nijenhuis, A. H. Stienen, J. H. Buurke, G. B. Prange, and F. Amirabdollahian, "Training modalities in robot-mediated upper limb rehabilitation in stroke: a framework for classification based on a systematic review," *Journal of Neuro-Engineering and Rehabilitation*, **vol. 11**, no. 1, 2014.

[10] L. Grover, A. Arcelus, R. Wang, R. Huq, K. Zabjek, and D. Hebert, *"Investigation of EMG Fatigue Patterns While Using an Upper Limb Rehabilitation Robotic Device,"* in Proceedings of the RESNA 2013 Annual Conference, 2013.

[11] M, Yamik, B. Purohit, and K. George, *"Electromyography (EMG) sensor controlled assistive orthotic robotic arm for forearm movement,"* Sensors Applications Symposium (SAS), pp. 1-4, 2017.

[12] S. Didier, D. F. Stegeman, and J. H. van Dieën, "Redundancy or heterogeneity in the electrical activity of the biceps brachii muscle? Added value of PCA-processed multi-channel EMG muscle activation estimates in a parallel-fibered muscle," *Journal of Electromyography and Kinesiology*, **vol. 23**, no. 4, pp. 892-898, 2013.

Acknowledgement

This research was supported by Universiti Teknologi MARA, Cawangan Pulau Pinang. I also would to thank my final year student who involved this project by providing his expertise, knowledge and time.

IOP Conf. Series: Materials Science and Engineering **767** (2020) 012020 doi:10.1088/1757-899X/767/1/012020

A Modified Artificial Bee Colony Based Test Suite Generation Strategy for Uniform T-Way Testing

Mohd Shaiful Aziz Rashid Ali, Rozmie Razif Othman, Zainor Ridzuan Yahya and Mohd Zamri Zahir

School of Computer and Communication Engineering, Kampus Pauh Putra, Universiti Malaysia Perlis, 02600 Arau, Perlis, Malaysia.

E-mail: shaifulaziz@unimap.edu.my, rozmie@unimap.edu.my, zainoryahya@unimap.edu.my, zamri.zahir@yahoo.com.

Abstract. Today, t-way testing has been widely known with the ability to reduce test suite size compared to exhaustive testing. At the same time, it has been proven by many researchers to provide maximum bug detection capability. Thus, various t-way strategies were developed since the past three decades. The paper proposed a new test generation strategy, named Modified Artificial Bee Colony T-Way Test Suite Generation (MABCTS). It supports uniform strength t-way testing. Experimentation results are compared with present strategies and produced comparable results. Since t-way testing is considered an NP-hard problem, there are no strategies that can be demanded to produce the best results.
Keywords: Artificial bee colony, uniform strength, test suite, t-way testing.

1. Introduction

Every day, we are dealing with software, whether it is directly or indirectly. For example, when we use the washing machine, we are indirectly using the software imbedded inside the machine by pressing the buttons on the panel. This appliance is controlled by the software in the controller board. To ensure the software interact with the inputs and outputs of the machine, the test engineer must test the software correctly. Exhaustive testing is the best solution where all of the possible combinations of configurations are tested to detect any faulty interactions. Unfortunately, this testing is impossible to be completed due to the time and resources constraint especially the manufacturing cost. Consider a device with 20 inputs with two possible values. If the whole testing process is applying exhaustive testing, there are 220 (1,048,576) test cases to execute.

As a consequence, there are several t-way generation test suite strategies that have been recommended by researchers. It will aid in generating a minimal test suite to cover all intended interaction strengths [1]. Strategies such as Automatic Efficient Test Generator (AETG) [2], Myra Implementation of AETG (mAETG) [3], In-Parameter Order (IPO) [4], In-Parameter-Order General (IPOG) [5] and Jenny [6] generate test cases for uniform strength t-way and pure computational based. [1, 7]

At the same time, several researchers have attempted to use the metaheuristic technique in generating t-way test suite. Shiva began the work by applying the Genetic Algorithm (GA) [8, 9] and Ant Colony Optimization (ACO) [10] to produce uniform strength t-way test suite. It was then continued by Simulated Annealing (SA) [11, 12], Particle Swarm Optimization (PSTG) [13], Harmony Search Strategy (HSS) [14-17], Cuckoo Search (CS) [18] and others until recently it has been occupied by

Hybrid Hyper Heuristic (HHH) [19]. From the literature review, test suite size can be improved when the metaheuristic in t-way strategies [7, 19] is applied.

These results suggest that there is a room further improvement, this paper manage to produce a solution to generate an optimum test suite by adopting an artificial bee colony as a searching algorithm. The original artificial bee colony was modified to suit with the T-way problem as well as to act as a balance between exploration and exploitation.

This paper is structured as follows. Section 2 describes the general idea of T-way testing, while Section 3 explains the proposed strategy. It is followed by the presentation of results and discussion in Section 4. Lastly, in section 5, conclusions were stated.

2. The general idea of t-way testing

In order to further explain the idea of t-way, try to imagine an embedded system consisting of 4 toggle switches as inputs. We name all switches with the letters A, B, C and D, as typically known, by toggling the switch it gives two possible values ("High" and "Low"). Table 1, represents all possible values with symbolic representation of the switches.

To ensure that the system is working accurately, exhaustive testing can be applied to test all possible combination of inputs. There are 16 tests cases (2^4=16) that can be generated from the exhaustive testing where it can be referred to in table 2.

Table 1. Symbolic representation for input switches of embedded system

Input Parameter	A	B	C	D
	a1	b1	c1	d1
Symbolic Representation	a2	b2	c2	d2

To minimize potential bugs in this system, ample testing must be made before releasing the system to the end-user. Interaction testing is a crucial test that needs to be conducted as faultiness may occur between the input parameter. One of the technique to test the system is by using exhaustive testing. In exhaustive testing, all possible input combinations need to be tested. Thus all input will interact with each other, this is where the defects are expected to arise between the interactions. As a result, 16 test cases (i.e. $2^4 = 16$) need to be executed. The list of the exhaustive test case is shown in Table 3.

Table 2. Test suite of exhaustive testing

Test Case	A	B	C	D
1	a1	b1	c1	d1
2	a1	b1	c1	d2
3	a1	b1	c2	d1
4	a1	b1	c2	d2
5	a1	b2	c1	d1
6	a1	b2	c1	d2
7	a1	b2	c2	d1
8	a1	b2	c2	d2
9	a2	b1	c1	d1
10	a2	b1	c1	d2
11	a2	b1	c2	d1
12	a2	b1	c2	d2
13	a2	b2	c1	d1
14	a2	b2	c1	d2
15	a2	b2	c2	d1
16	a2	b2	c2	d2

As an alternative method, T-way testing can be applied to test all possible combination inputs. In t-way testing, the interaction strength (t) must be decided first before the test suite is generated. The number of test cases can be decreased when lower interaction strength is picked. The 't' can range from 2 until maximum number of input parameters is reached. In this example, we will choose two as the interaction strength. The 2-way test suite will be generated, it is commonly known as pairwise. AB, BC, CD, AC, AD and BD are the combination of input parameters that have produced tuples that was charted in Figure 2. The result produced in figure 2 is considered as only tuples which was covered by each combination. By applied t-way testing method, 25 per cent of test cases can be reduced compared to exhaustive testing. Only 9 test cases are required to test the complete system. The most glaring advantage of using the t-way testing is it will reduce the number of test cases will severely save time and countless resources.

The covering array notation is used to represent the final test suite (F) as in Equation 1. Therefore for the system in Figure 1, the final test suite is written as CA (9, 2, 2^4).

$$F = CA(N; t, C) \quad (1)$$

Where,

- N = the number of test data inside the final test suite.
- t = the interaction strength
- C = value configuration can be represented as following:
- $v_1^{p1}, v_2^{p2}, \ldots, v_n^{pn}$ which indicates that there are p_1 parameters with v_1 values, p_2 parameters with v_2 values and continue

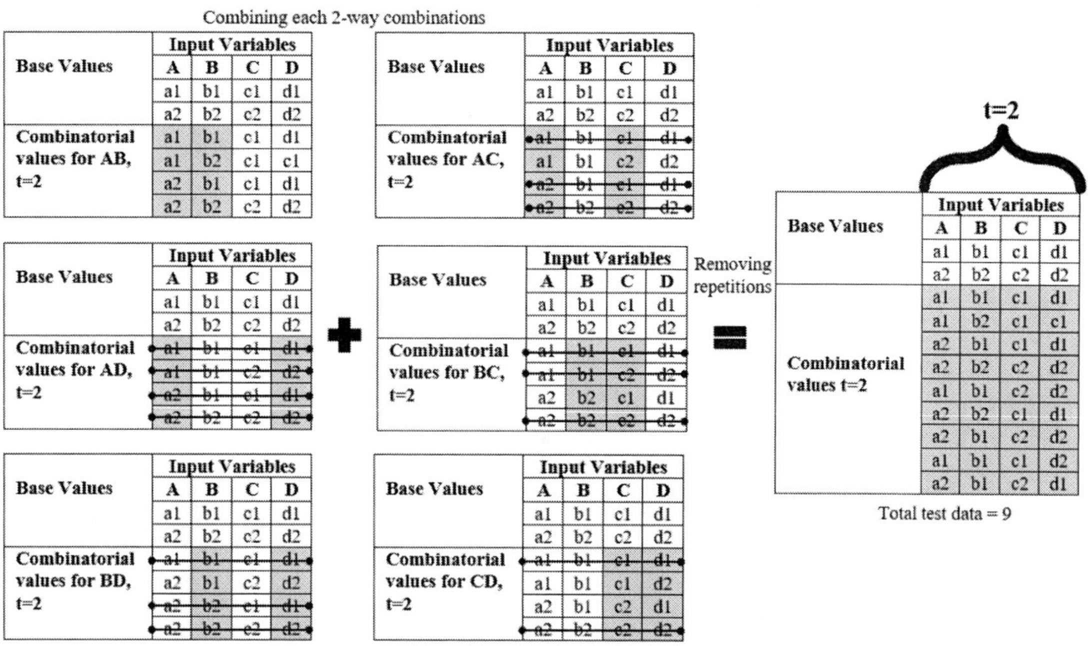

Figure 1. Uniform Strength with two interaction strength

3. Proposed strategy

Karaboga has introduced the artificial Bee Colony (ABC) algorithm in 2005[21]. Since then, it has been recommended in solving problems which involve non-linear optimization[22]. Fundamentally, the idea of adopting ABC is by mimicking the behaviour of bees on how they find a food source for their hive. The grouping of bees according to ABC is characterized into three different characters as below:

- Employer bees are the workers who find the food source and bring the information to the onlooker bees.

- Onlooker bees are in charge to evaluate the quality of the food source. The evaluation process is determined by researching the Waggle dance move via the employer bees.
- Scout bees are bees that randomly search for a new food sources. The employer bees on the hand can transform into scout bees if the current food source is exhausted.

As explained earlier on the matter of exploitation and exploration of the bees' behaviour, this algorithm focuses more on exploitation compared to exploration. The restriction on the exploration behaviour is that it only occurs once in the scout bee phase compared to the exploitation behaviour which happened twice in the employer and onlooker bee phase. This fact is the main reason why we are choosing the ABC as the core of our searching algorithm in generating the optimum test suite.

By relating the ABC algorithm with our problem, tuples produced from exhaustive testing can become one of the potential solutions or in ABC it is known as a food source. Any tuples are randomly chosen from the search space. The number of food sources is equal to the number of employer bees. Afterwards each food sources will be assigned to one employer bee.

$$x_{ij} = x_{j\,min} + rand\ (0,1).\left(x_{j\,max} - x_{j\,min}\right) \tag{2}$$

Where:

- x_{ij} : Initialization possible solutions i with parameter j
- $x_{j\,min}$: The value of the smallest possible solutions based on parameters j.
- $x_{j\,max}$: Greatest value possible solutions based on parameters j
- $rand\ (0,1)$: A random value between 0 and 1
- $i = 1 \ldots$ SN, SN is the number of possible solutions (food source).
- $j = 1 \ldots$ D, D is the number of parameters

These employer bees will exploit their position in finding food source hence they will find the best local optimum source in their area.

$$v_{ij} = x_{ij} + \phi_{ij}.\left(x_{ij} - x_{kj}\right) \tag{3}$$

Where:

- v_{ij}: New food source
- x_{ij}: the value of the possible solution i with parameter j
- $i = 1 \ldots$ SN, SN is the number of possible solutions (food source)
- $j = 1 \ldots$ D, D is the number of parameters
- $k = 1 \ldots$ SN, SN is the number of parameters used
- ϕ_{ij}: Random real number between [-1, 1]

Then the amount of nectar for each food source will be evaluated by using the fitness function. The function has been modified to suit our problem, fn was given by:

$$f_n = \sum_{k=0}^{Total\ numbers\ of\ tuples\ pair} w_k \tag{4}$$

Where:

- w_k is the number of uncovered tuples for kth uniform strength.

The best food source has the highest amount of nectar which is determined by the onlooker bees. This indicates the best solution or the best tuples pair which can cover the highest number of test cases.

If a solution is not improved further through a predetermined number of cycles, then that food source is assumed to be exhausted. This exhausted food source will be replaced by a scout bee. A standard searching method by a scout bee is by randomly searching. The method has a high potential to be trapped in a local minima or a cycle of repeated old solutions. The second reason is that there is too much exploitation in the employee and onlooker bee phases hence it might not cover the whole solution space.

For these reasons, a mutation is applied to replace the random search. As a result, a new food source will be explored in a new region from the search space. This modification will overcome the problem of being trapped in the local minima and avoid the possibility of returning back the same location.

As an example, we select test case 2 [1, 1, 1, 2] in Table 2. We take the first parameter to be flipped as below.

Step 1: Finding the number of bit binary to represent. The formula is:

$$N = \log_2(UB + 1) \tag{5}$$

Where:

- N = number of bits
- UB = upper bound

In this example UB = 2, so

$$N = \log_2(2 + 1)$$
$$=1.58 \text{ (round it up to next integer)}$$
$$=2$$

Step 2: Convert the first parameter into a binary and then flip all bits. Finish, return to decimal. Now the new value is two and a new food source [2, 1, 1, 2]. By referring to Table 2, the position has jumped to number 10. This proves that the embedded mutation in the scout bee phase has better exploration process.

Two scenarios are in need of a normalized process. In the first scenario, the mutation result is above the upper bound. In this case, the value from mutation is subtracted with the upper bound. In another scenario, mutation result is less than the lower bound. Therefore it needs to be added with the lower bound. Either one is repeated until the result obtained is between the ranges.

Finally, the ending criteria of MABCTS will be checked. The generation process will be stopped when there is no further uncovered tuples. Otherwise, the searching algorithm will be repeated. The flow chart in figure 2 summarizes the proposed method.

IOP Conf. Series: Materials Science and Engineering **767** (2020) 012020 doi:10.1088/1757-899X/767/1/012020

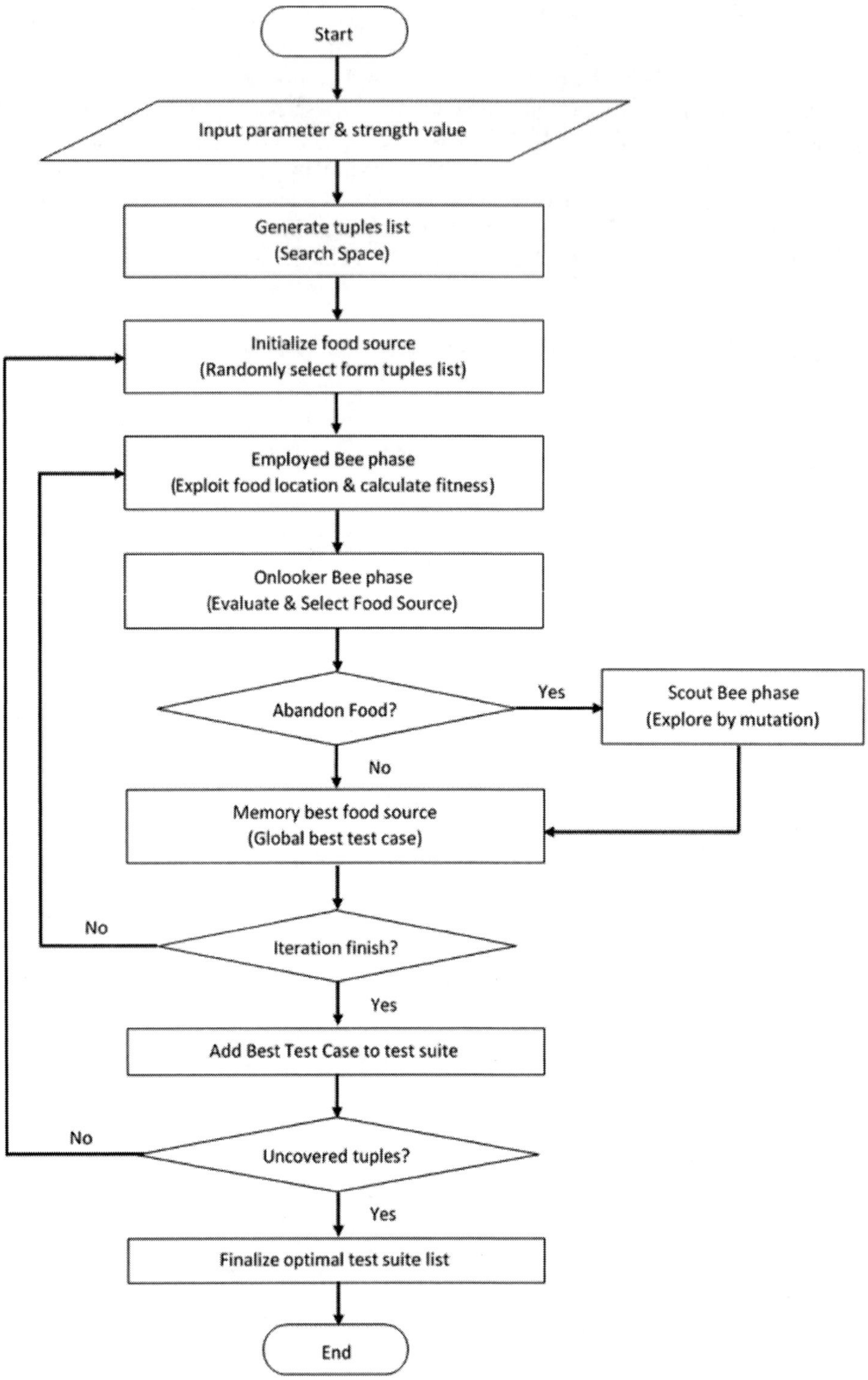

Figure 2. Proposed MABCTS flow chart

IOP Conf. Series: Materials Science and Engineering **767** (2020) 012020 doi:10.1088/1757-899X/767/1/012020

4. Results

Table 3.Benchmarking CA configurations. [19]

System configuration	Meta-heuristic-based strategies								General Computational-based Strategies				
	MABCTS	HHH	HSS	PSTG	CS	SA	GA	ACO	mAETG	AETG	IPOG	Jenny	TGV
CA(N;2,3^4^)	**9**	**9**	**9**	**9**	**9**	**9**	**9**	**9**	**9**	**9**	**9**	10	11
CA(N;2,3^13^)	18	17	18	17	20	16	17	17	17	**15**	20	20	19
CA(N;2,5^10^)	**42**	**42**	43	45	NA	NA	NA	NA	NA	NA	50	45	51
CA(N;3,3^6^)	39	**33**	39	42	43	**33**	**33**	**33**	38	47	53	51	49
CA(N;3,4^6^)	96	**64**	70	102	105	**64**	**64**	**64**	77	105	**64**	112	123

*Bold value specifies the most minimum results (i.e. optimal)

Table 4. CA (N; t, 2^{10}) with t varied from 2 to 6.[19]

t	Meta-heuristic-based strategies					General Computational-based Strategies				
	MABCTS	HHH	HSS	PSTG	CS	IPOG	ITCH	Jenny	TConfig	TVG
2	8	8	7	8	8	10	**6**	10	9	10
3	**16**	**16**	**16**	17	**16**	19	18	18	20	17
4	**36**	**36**	37	37	**36**	49	58	39	45	41
5	**78**	79	81	82	79	128	NA	87	95	84
6	**153**	**153**	158	158	157	352	NA	169	183	168

*Bold value specifies the most minimum results (i.e. optimal)

Table 5. CA (N; t, 5^{10}) with t varied from 2 to 6. [19]

t	Meta-heuristic-based strategies					General Computational-based Strategies							
	MABCTS	HHH	HSS	PSTG	CS	IPOG	ITCH	Jenny	PICT	TConfig	TVG	GTWay	MIPOG
2	**42**	**42**	43	45	45	50	45	45	47	48	50	46	45
3	**271**	280	276	287	297	313	225	290	310	312	342	293	281
4	**1623**	1638	1624	1716	1731	1965	1750	1719	1812	1878	1971	1714	1643
5	**8524**	8704	8866	9425	9616	11009	NA	9437	9706	NA	NA	9487	8169
6	**42886**	47800	47550	50350	50489	57290	NA	NA	47978	NA	NA	44884	45168

*Bold value specifies the most minimum results (i.e. optimal)

Table 6. CA (N; 4, v^{10}) with v varied from 2 to 7[19]

v	Meta-heuristic-based strategies					General Computational-based Strategies							
	MABCTS	HHH	HSS	PSTG	CS	IPOG	ITCH	Jenny	PICT	TConfig	TVG	GTWay	MIPOG
2	36	36	37	34	**28**	49	58	39	43	45	40	46	43
3	208	**207**	211	213	211	241	336	221	231	235	228	224	217
4	660	668	691	685	698	707	704	703	742	718	782	**621**	637
5	**1623**	1635	1624	1716	1731	1965	1750	1719	1812	1878	1917	1714	1643
6	**3375**	3405	3475	3880	3894	3935	NA	3519	3735	NA	4159	3514	3657
7	**6215**	6412	6398	NA	NA	7061	NA	6462	NA	NA	7854	6459	5927

*Bold value specifies the most minimum results (i.e. optimal)

To validate the performance of the strategy, MABCTS is compared with existing strategies which is by grouping it in two, meta-heuristic-based strategies and general Computational-based Strategies. The configuration is set up to meet the standard benchmarking [19]. Only the final number of the test suite is evaluated. Table 3, Table 4, Table 5 and Table 6 represent those results.

Table 3 shows results for the interaction strength $t \leq 3$, which is the low interaction strength. At this interaction strength, MABCTS's performance is average.

At covering arrays CA(N; 2, 3^4) and CA(N; 2, 5^{10}) are best configurations for MABCTS compare with others. SA strategy can be assumed as the best for most configurations.

For Table 4 and 5, the covering arrays CA (N; t, 2^{10}) and CA (N; t, 5^{10}) with t varied from 2 to 6. From the result, MABCTS is showing the best result with other strategies. For the last result in Table 6, MABCTS has improved the results when more input values are configured.

5. Conclusion and Future Works

In this study, a new strategy for t-way test suite generation has been proposed based on the Modified ABC algorithm. As mentioned earlier, problems generated by the test suite is considered NP-hard problem. It can be seen in the results where none of the strategy's best result can be used for all configuration and interaction strength. Therefore, this can be improved upon as a niche for future studies and the MABCTS strategy can further be optimized for other types of t-way strength.

Acknowledgments

The author would like to acknowledge the support from the Fundamental Research Grant Scheme (FRGS) under a grant number of FRGS/1/2018/ICT01/UNIMAP/02/1 from the Ministry of Education Malaysia.

References

[1] Othman RR, Zamli KZ. *T-way strategies and its applications for combinatorial testing*. International Journal of New Computer Architectures and their Applications. 2011;1(2):459-73.

[2] Cohen DM, Dalal SR, Kajla A, Patton GC. *The automatic efficient test generator (AETG) system.* InProceedings of 1994 IEEE International Symposium on Software Reliability Engineering 1994 Nov 6 (pp. 303-309). IEEE.

[3] Cohen DM, Dalal SR, Fredman ML, Patton GC. *The AETG system: An approach to testing based on combinatorial design.* IEEE Transactions on Software Engineering. 1997 Jul;23(7):437-44.

[4] Lei Y, Tai KC. *In-parameter-order: A test generation strategy for pairwise testing.* InProceedings Third IEEE International High-Assurance Systems Engineering Symposium (Cat. No. 98EX231) 1998 Nov 13 (pp. 254-261). IEEE.

[5] Lei Y, Kacker R, Kuhn DR, Okun V, Lawrence J. *IPOG: A general strategy for t-way software testing.* In14th Annual IEEE International Conference and Workshops on the Engineering of Computer-Based Systems (ECBS'07) 2007 Mar 26 (pp. 549-556). IEEE.

[6] Jenkins RJ. *Jenny: a pairwise testing tool* [Internet]. 2005 Available from: http://burtleburtle.net/bob/math/jenny.html [Accessed 16 Nov. 2019].

[7] Al-Sewari AA, Zamli KZ. *An orchestrated survey on t-way test case generation strategies based on optimization algorithms.* InThe 8th International Conference on Robotic, Vision, Signal Processing & Power Applications 2014 (pp. 255-263). Springer, Singapore.

[8] Shiba T, Tsuchiya T, Kikuno T. *Using artificial life techniques to generate test cases for combinatorial testing.* InProceedings of the 28th Annual International Computer Software and Applications Conference, 2004. COMPSAC 2004. 2004 Sep 28 (pp. 72-77). IEEE.

[9] McCaffrey JD. *An empirical study of pairwise test set generation using a genetic algorithm.* In2010 Seventh International Conference on Information Technology: New Generations 2010 Apr 12 (pp. 992-997). IEEE.

[10] Chen X, Gu Q, Li A, Chen D. *Variable strength interaction testing with an ant colony system*

approach. In2009 16th Asia-Pacific Software Engineering Conference 2009 Dec 1 (pp. 160-167). IEEE.
[11] Stardom J. *Metaheuristics and the search for covering and packing arrays.* Burnaby: Simon Fraser University; 2001 May 16.
[12] Colbourn CJ, Ling AC. *Constructing strength three covering arrays with augmented annealing.* Discrete Mathematics. 2008 Jul 6;308(13):2709-22.
[13] Ahmed BS, Zamli KZ, Lim CP. *Constructing a t-way interaction test suite using the particle swarm optimization approach.* International Journal of Innovative Computing, Information and Control. 2012 Jan 1;8(1):431-52.
[14] Alsewari AA, Younis MI, Zamli KZ. *Generation of pairwise test sets using a harmony search algorithm.* Comput Sci Lett. 2011 Apr 2;3(1).
[15] Alsewari AA, Zamli KZ. *Interaction test data generation using harmony search algorithm.* In2011 IEEE Symposium on Industrial Electronics and Applications 2011 Sep 25 (pp. 559-564). IEEE.
[16] Alsewari AR, Zamli KZ. *Design and implementation of a harmony-search-based variable-strength t-way testing strategy with constraints support.* Information and Software Technology. 2012 Jun 1;54(6):553-68.
[17] Alsewari AR, Zamli KZ. *A harmony search based pairwise sampling strategy for combinatorial testing.* International Journal of Physical Sciences. 2012 Feb 9;7(7):1062-72.
[18] Yang XS, Deb S. *Cuckoo search via Lévy flights.* In2009 World Congress on Nature & Biologically Inspired Computing (NaBIC) 2009 Dec 9 (pp. 210-214). IEEE.
[19] Zamli KZ, Alkazemi BY, Kendall G. *A tabu search hyper-heuristic strategy for t-way test suite generation.* Applied Soft Computing. 2016 Jul 1;44:57-74.
[20] Ali MS, Othman RR, Yahya ZR, Ahmad MZ, Ramli N. *Implementation of artificial bee colony algorithm for T-way testing.* In2016 3rd International Conference on Electronic Design (ICED) 2016 Aug 11 (pp. 591-594). IEEE.
[21] Karaboga D, Akay B. *A comparative study of artificial bee colony algorithm.* Applied mathematics and computation. 2009 Aug 1;214(1):108-32.
[22] Lindfield G, Penny J. *Introduction to Nature-Inspired Optimization.* Academic Press; 2017 Aug 10.

IOP Conf. Series: Materials Science and Engineering **767** (2020) 012021 doi:10.1088/1757-899X/767/1/012021

Prediction of carbon concentration profile within carburised-carbon steel 1024 using Alternating Direction Implicit (ADI) method

A H Norzilah[1] and N Rusli[2]

[1]Centre of Excellence of Frontier Materials Research, School of Materials Engineering, Universiti Malaysia Perlis (UniMAP), Jalan Kangar-Alor Star, Kampung Seriab, 01000 Kangar, Perlis, Malaysia.
[2]Institute of Engineering Mathematics, Universiti Malaysia Perlis (UniMAP), Pauh Putra Campus, 02600 Arau, Perlis, Malaysia.

norzilah@unimap.edu.my

Abstract. Pack carburisation is recognised to improve the carbon surface content of carburised-carbon steel 1024 that enhance the surface hardness and further can be used in automobiles, form implements, machines, gears and springs. The transient two- dimensional diffusion equation that represents the carbon concentration profile within the carburised- carbon steel 1024 is discretised using Alternating Direction Implicit (ADI) technique. A semi- analytical method is used to verify the ADI method and the numerical algorithm is developed in MATLAB software. To materialise the pack carburisation process, the boundary condition, initial condition and step time are encoded to signify the experiment conditions from the literature. The simulation results successfully predicted the carbon concentration profile within the carburised-carbon steel 1024 with variation of carburisation time (2 hours to 60 hours) and temperature (900°C - 1000°C). The effect of carburisation time and temperature are evaluated and the relation of both effects on the carbon development is explained. Simulation results show that carburisation extended to 60 hours only increases the carbon content at the core to 0.3wt % from 0.24 wt% of its original content. The small increment is satisfactory to maintain its ductility by hardening the surface.

1. Introduction

Pack carburisation is categorised as a heat treatment for surface hardening of low carbon steel 1024. According to American Iron and Steel Institute/Society of Automotive Engineers (AISI/SAE) steel grade, -10 represents the categorisation of plain carbon steel and 24-denotes the weight percentage of carbon content which is 0.24wt%. Because of its low carbon content, surface hardening is necessary to increase the carbon content hence to be used in applications such as arbours, hinges, link components, driving pins, camshafts and cardan joints. These applications require only the surface to be hardened by maintaining the core ductility.

The carbon steel is packed with a mixture of solid carburiser, such as charcoal, activated carbon or graphite powder, along with energiser in the carburisation box. Then the prepared box is exposed at high temperature ranges from 900 to 1000°C for 2 to 12 hours. The process involves 4 stages - the oxidation between the air with the packed carbon, the decomposition of carbon monoxide gas, the deposition of carbon on the steel surface and finally the carbon diffusion within the steel structure at high temperature.

Improvement of surface hardness of low carbon steel is theoretically instigated by the rapid cooling of carburised layer resulted from the carburisation. Nevertheless, often found in the literature the surface hardness reading is presented without demonstration of carbon concentration profile at corresponding depths [1-3]. In practice, carbon content at each depth is measured using spark emission spectroscopy once pack carburisation process completed. However, the measurement is restricted to a certain depth. On top of that, it is difficult and time-consuming to control the desired carbon concentration on the surface of low carbon steel experimentally.

It is established that the carburisation process is dominated by diffusion. Thus, understanding about the carbon diffusion during carburisation is crucial in order to determine the effective case depth and this is the reason why the carbon profile concentration is essential. The carbon concentration profile is resulted from the carbon diffusion equation recognised as Fick's Second Law. The Fick's Second Law is a parabolic partial differential equation that represents the relationship between the rate of carbon concentration with the diffusivity and the carbon concentration gradient.

One of the techniques from finite difference method to solve the two-dimensional transient or unsteady-state of Fick's Second Law is Alternating Direction Implicit (ADI). The technique is basically initiates with breaking the 2-D problems into the 1-D problem and solves each problem implicitly. It is unconditionally stable and also overcomes the drawbacks of the normal implicit finite difference method which is very time-consuming and less efficient. The technique has proven to be more efficient and stable in solving parabolic and elliptic differential equation compared to the implicit scheme alone [4,5]. The heat or mass equation is closely represented the physical or chemical processes occurred in engineering. Determination of dependent variables such as temperature and concentration within the solid or fluid in multi-dimension are essential in order to analyse the process in details prior to design purposes.

For example, detailed studies of heat and mass transfer during the drying process of a square cylinder for confined flow are visualised by the usage of ADI-based software [6]. The 2-dimensional numerical analysis is reported comparable with the experimental data. A more complicated process from coupling the heat equation with the continuity equation and carbon dioxide (CO_2) properties during tubing and annulus injection also solved by ADI method. As a result, Zhao et al., successfully estimated the wellbore temperature and pressure profile at the axial and radius direction [7]. Aslib et al., used the ADI method to resolve the transient heat transfer problem of an anisotropic cylindrical pin fin with heat sink. The numerical results stated as well-validated with the analytical solution [8]. The ADI method also exhibited good agreement with the experimental data in more complex heat transfer problem involving Vapour Phase Soldering (VPS) for electronic components [5]. The findings also highlight the potential of the ADI method in advanced manufacturing engineering application.

Therefore, this paper presents the application of ADI method in solving transient 2-dimensional diffusion equation to visualise the carbon concentration growth within the carbon steel 1024 during pack carburisation. The discretised algorithm is validated with semi-analytical method and then used to evaluate the carbon concentration at various carburisation temperature and time.

2. Methodology

2.1 Discretisation of transient two-dimensional diffusion equation

The transient two-dimensional diffusion equation is given by equation (1):

$$\frac{\partial C}{\partial t} = D\left(\frac{\partial^2 C}{\partial x^2} + \frac{\partial^2 C}{\partial y^2}\right) \tag{1}$$

Using ADI formulation, Equation (1) is discretised to:

$$\frac{C_{i,j}^{n+\frac{1}{2}} - C_{i,j}^{n}}{\left(\frac{\Delta t}{2}\right)} = D\left[\frac{C_{i+1,j}^{n+\frac{1}{2}} - 2C_{i,j}^{n+\frac{1}{2}} + C_{i-1,j}^{n+\frac{1}{2}}}{(\Delta x)^2} + \frac{C_{i,j+1}^{n} - 2C_{i,j}^{n} + C_{i,j-1}^{n}}{(\Delta y)^2}\right] \tag{2}$$

and

$$\frac{C_{i,j}^{n} - C_{i,j}^{n+\frac{1}{2}}}{\left(\frac{\Delta t}{2}\right)} = D\left[\frac{C_{i+1,j}^{n+\frac{1}{2}} - 2C_{i,j}^{n+\frac{1}{2}} + C_{i-1,j}^{n+\frac{1}{2}}}{(\Delta x)^2} + \frac{C_{i,j+1}^{n+1} - 2C_{i,j}^{n+1} + C_{i,j-1}^{n+1}}{(\Delta y)^2}\right] \tag{3}$$

Where Δt = time step size; Δx = x step size; $\Delta y = y$ step size; n = time level; i and j = Cartesian coordinates x and y .

Equation (2) and (3) are written in the tridiagonal form as:

$$-d_1 C_{i-1,j}^{n+\frac{1}{2}} + (1+2d_1) C_{i,j}^{n+\frac{1}{2}} - d_1 C_{i+1,j}^{n+\frac{1}{2}} = d_2 C_{i,j+1}^{n} + (1-2d_2) C_{i,j}^{n} + d_2 C_{i,j-1}^{n} \tag{4}$$

and

$$-d_2 C_{i,j-1}^{n+1} + (1+2d_2) C_{i,j}^{n+1} - d_2 C_{i,j+1}^{n+1} = d_1 C_{i+1,j}^{n+\frac{1}{2}} + (1-2d_1) C_{i,j}^{n+\frac{1}{2}} + d_1 C_{i-1,j}^{n+\frac{1}{2}} \tag{5}$$

where $d_1 = \frac{1}{2} d_x = \frac{1}{2}\frac{D\Delta t}{(\Delta x)^2}$ and $d_2 = \frac{1}{2} d_y = \frac{1}{2}\frac{D\Delta t}{(\Delta y)^2}$

The solution procedure starts with the solution of tridiagonal system in equation (4). The formulation of this equation is implicit in the x-direction and explicit in the y-direction. Thus, the solution at this stage is referred as the x-sweep. Solving the tridiagonal system of equation (4) provides necessary data for the right-hand side of equation (5) to solve the tridiagonal system of equation (5). In this equation, the finite difference equation is implicit in the y-direction and explicit in the x-direction and referred as y-sweep. Numerical algorithm is then developed based on equation (4) and (5).

2.2 Development of numerical algorithm and implementation in MATLAB

Numerical algorithm of ADI are established in matrix form [9]. For x-sweep:

$$\begin{pmatrix} b_1 & c_1 & 0 \\ a_1 & b_1 & c_1 \\ a_1 & b_1 & c_1 \\ 0 & a_1 & b_1 \end{pmatrix} \begin{pmatrix} C_{2,2} \\ C_{3,2} \\ C_{IN-1,2} \end{pmatrix} = \begin{pmatrix} D_1 - a_1 C_{1,2}^{n+\frac{1}{2}} \\ D_1 \\ D_1 \\ D_1 - c_1 C_{IN,2}^{n+\frac{1}{2}} \end{pmatrix} \tag{6}$$

For y-sweep:

$$\begin{pmatrix} b_2 & c_2 & 0 \\ c_2 & b_2 & a_2 \\ c_2 & b_2 & a_2 \\ 0 & c_2 & b_2 \end{pmatrix} \begin{pmatrix} C_{2,2} \\ C_{2,3} \\ C_{2,4} \end{pmatrix} = \begin{pmatrix} D_{2,j=2} - c_2 C_{2,1}^{n+1} \\ D_{2,j=2} \\ D_{2,j=3} \\ D_{2,j=IN-1} - a_2 C_{2,IN}^{n+1} \end{pmatrix} \tag{7}$$

Where

$a_1 = -d_1; b_1 = (1+2d_1); c_1 = -d_1;\ a_2 = -d_2; b_2 = (1+2d_2); c_2 = -d_2;$

$D_1 = d_2 C^n_{i,j+1} + (1-2d_2) C^n_{i,j} + d_2 C^n_{i,j-1};\ D_2 = d_1 C^{n+\frac{1}{2}}_{i+1,j} + (1-2d_1) C^{n+\frac{1}{2}}_{i,j} + d_1 C^{n+\frac{1}{2}}_{i-1,j};$ $IN =$ Number of *i*, and $JN =$ Number of j .

The generated solution will be verified the solution from semi-analytical method.

2.3 Verification with the semi-analytical method

Due to the mathematical similarity between the heat conduction and diffusion equation, the numerical algorithm is verified by semi-analytical method for two dimensional heat conduction equation [10]. Therefore, the variable is written as T which is the temperature to be calculated for the square domain of $0 \le x \le 3.0$ m, $0 \le y \le 3.0$ m, with $k_x = k_y = 1.25$ Btu/(hr.m.°F), $\Delta x = \Delta y = 0.3$ m, at $t = 1.2$ hr. The equation is subjected to the following boundary and initial conditions:

$$T(0,y,t) = T(x,y,t) = T(L_x,y,t) = T(x,L_y,t) = 0 \quad (8)$$

$$T(x,y,0) = 30 \quad (9)$$

Where L_x and L_y are the lengths of the solution domain in the x and y directions, respectively. The semi-analytical solution for this problem is:

$$T(x,y,t) = \sum_{n=1}^{\infty}\sum_{j=1}^{\infty} A_n \sin\frac{n\pi x}{L_x} \sin\frac{j\pi y}{L_y} \exp\left[-\left(\frac{k_x n^2\pi^2}{L_x^2} + \frac{k_x j^2\pi^2}{L_y^2}\right)t\right] \quad (10)$$

Where $A_n = \frac{4(30)}{nj\pi^2}\left[(-1)^n - 1\right]\left[(-1)^j - 1\right]$

The verified numerical method is then used to predict the carbon concentration profile.

2.4 Prediction of carbon concentration profile during pack carburisation

2.4.1 Process description of pack carburisation This numerical studies is based on experimental work conducted by Ali et al. which provide detail information of the carburising condition and the carbon steel dimension [1]. Carbon steel 1024 with carbon content of 0.24wt% is pack carburised with charcoal using barium carbonate ($BaCO_3$) as carburiser. The stainless steel container is filled with the mixture of charcoal powder and $BaCO_3$ surrounding the carbon steel as illustrated in figure 1. The stainless steel container (2) contained with the charcoal, $BaCO_3$ (3) and the carbon steel (4) is pack carburised in the muffle furnace (1) that already being heated before that to reach carburisation temperature of 900-950 °C. The dimension of carbon steel is 0.0508m × 0.0127m. The carburisation occurred for 2, 3, 5 and 9 hours before taken out from the furnace and quenched at room temperature. Heat from the muffle furnace is transferred to the carbon atoms from the surface of carbon steel 1024 to its case.

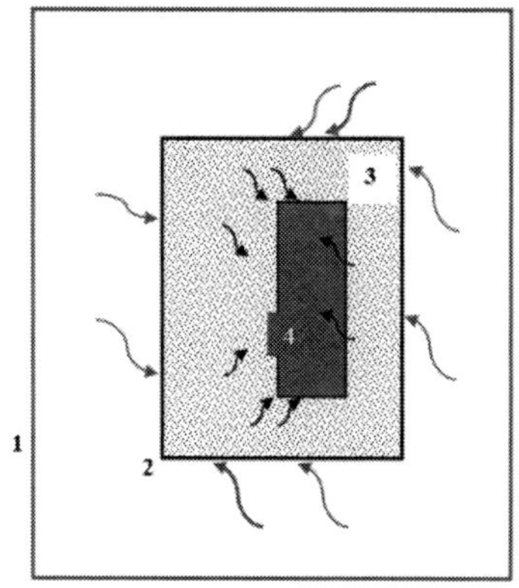

Figure 1. Schematic diagram of pack carburising process in this study
1- Muffle furnace; 2- Stainless steel container; 3 - Mixture of charcoal and barium carbonate; 4 - Carbon steel plate; ➙ : Heat conducted from 1 to 2; ➙ Carbon diffused from 3 to 4.

2.4.2 Mathematical model of pack carburisation The pack carburisation process for carbon diffusion is represented by equation (1). The diffusion coefficient of carbon, D to diffuse into the γ-Fe phase is calculated by equation (11) [11].

$$D(T) = 0.0052\exp\left(-\frac{127950}{R_J T}\right) \tag{11}$$

Where R_J = Gas constant in 8.314 J/mol.K and T = Temperature in Kelvin (K).

2.4.3 Initial and boundary conditions The initial condition is written as:

$$C_o(x,y,0) = 0.24 \tag{12}$$

C_o is the initial carbon composition before carburised. The boundary condition for each side of the carbon steel, C_b is written as:

$$C_b(0:0.0508,0,t) = 1 \tag{13}$$
$$C_b(0:0.0508,0.0127,t) = 1 \tag{14}$$
$$C_b(0,0:0.0127,t) = 1 \tag{15}$$
$$C_b(0.0508,0:0.0127,t) = 1 \tag{16}$$

2.4.4 Computational procedure A computer code is developed in MATLAB. The spatial step size for x and y is 0.0001m respectively and the time step used is 10s. The time step used is restricted to the stability condition as mentioned in equation (17):

$$\left[\frac{D\Delta t}{(\Delta x)^2} + \frac{D\Delta t}{(\Delta y)^2}\right] \leq \frac{1}{2} \tag{17}$$

IOP Conf. Series: Materials Science and Engineering **767** (2020) 012021 doi:10.1088/1757-899X/767/1/012021

3. Results and discussion

3.1 Verification of ADI method

The comparison of the ADI method with the semi-analytical method is illustrated in figure 2. The ADI approximation agrees well with the semi-analytical solution. Next, the verified ADI method is implemented to estimate the carbon concentration profile within the carbon steel 1024 during carburisation process.

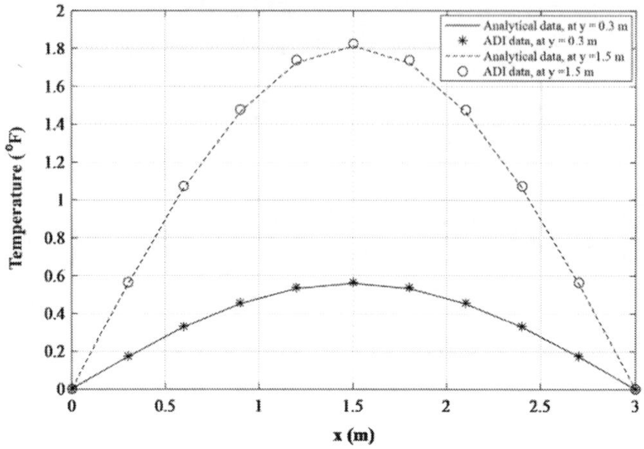

Figure 2. Temperature distribution for the heat conduction problem along x-axis at $y = 0.3$ m and $y = 1.5$ m for 1.2 hr.

3.2 Prediction of carbon concentration profile within the carbon steel 1024 structure during pack carburisation

The contour plot of the carbon concentration profile for carburised- carbon steel after 3 hours of reaction in figure 3(a) displays a very small changes near the sample surface. This is common in carburisation process, where the main focus of the treatment is to increase the carbon content within the case or certain depth below the steel surface, not the core of the steel. In each direction, carbon concentration progressively slow as depicted in figure 3(b). Hence, for clearer image, the discussion is focused on the optimum depth of 0.0015m as shown in figure 3(c).

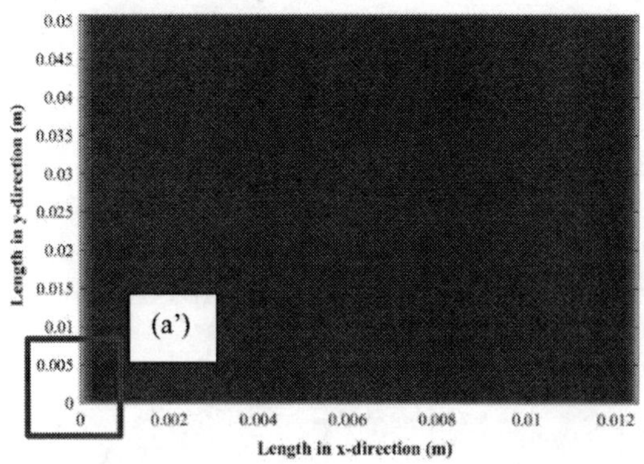

(a)

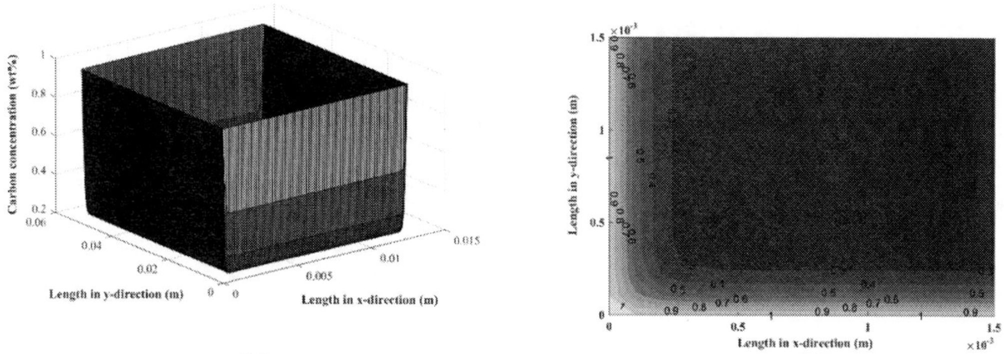

(b) (c)

Figure 3. Simulation results of carburisation at 900°C for 2 hours (a) Overall contour plot of carbon steel; (b) Surface plot of carbon concentration profile; (c) Enlarged image of contour plot (a') at the corner of the carbon steel in x and y direction.

3.3 *Effect of carburisation time*

The carburisation time is varied from 2 hours to 60 hours. The development of carbon concentration within the carbon steel are illustrated by contour plots in igure 5. The core of the carbon steel constantly having 0.3 wt% of carbon content with its surface slowly penetrated by the diffused carbon over time.

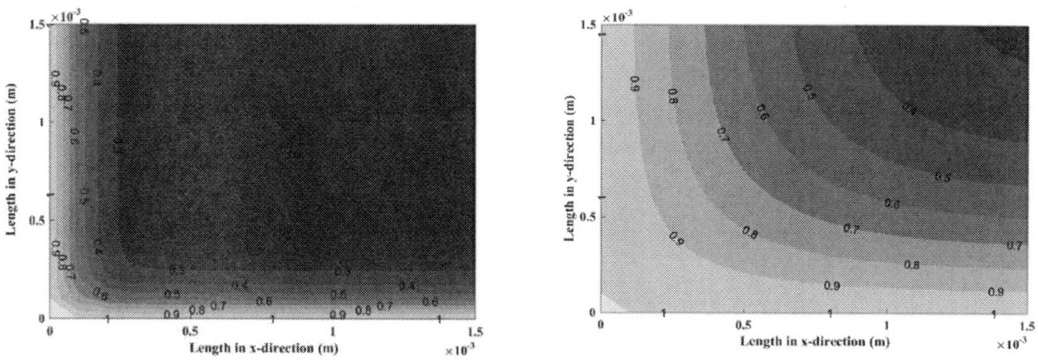

(a) (b)

Figure 4. Contour plot of carbon concentration within the carbon steel at various carburising time; (a) 2 hours; 60 hours.

Figure 5 shows a closer look on the dynamic of carbon diffusion at 0.0015 m, the optimum depth from each dimension surface. It is interesting to note that the simulated result for carbon concentration of carburisation for 3 hours can explain the case depth of carbon steel conducted by Ali et al., 2016 [1]. The measured case depth is 0.468 × 10^{-3} m which comparable with the simulated carbon concentration profile in this study.

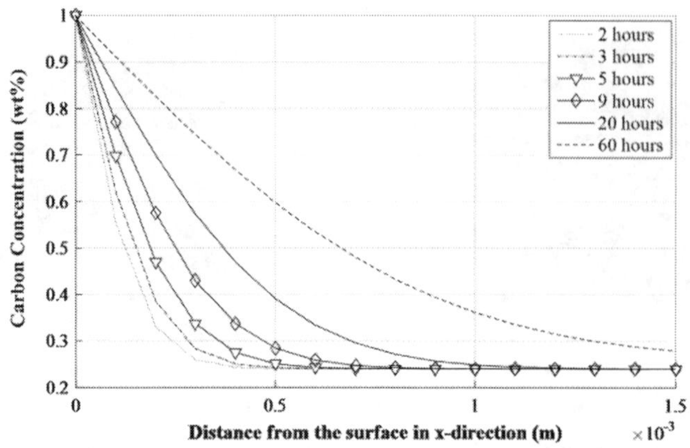

Figure 5. Carbon concentration profile at optimum depth of 0.0015 m from the carbon steel surface carburised at 2 hours, 3 hours, 5 hours, 9 hours, 20 hours and 60 hours in x direction along $y = 0.0015$ m.

Increasing the carburising time will ensure the increment of diffused carbon for both directions. This is due to the continuous decomposition of CO_2 gas and $BaCO_3$. At 2 hours of reaction, more carbon atoms enter the structure of steel. Thus, the rate of diffusion is also increases. But after some time, less carbon atoms remained in the pack. Therefore, the carbon diffusion eventually slow [12].

3.4 Effect of carburisation temperature

The carbon steel is simulated with three different carburisation temperature of 900°C, 950°C and 1000°C for 3 hours. Figure 6 shows contour plots of carbon concentration within the carburised- carbon steel. The carbon content is increasing with carburisation temperature.

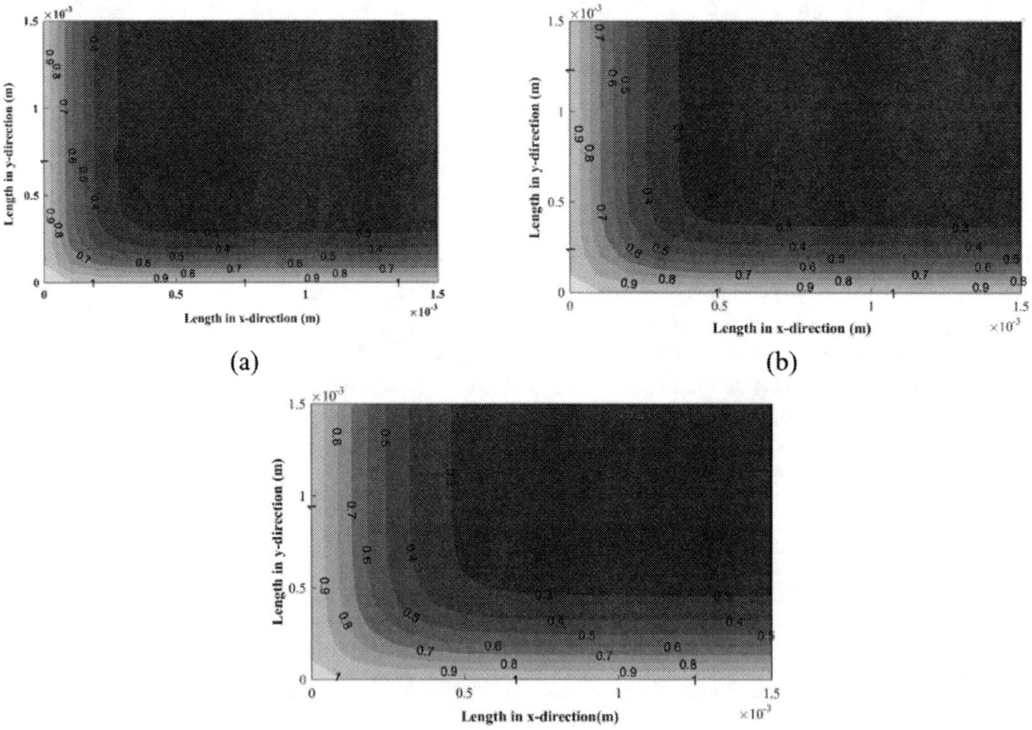

(c)

Figure 6. Contour plots of carbon concentration within 3 hours carburised- steel at carburising temperature of (a) 900°C; (b) 950°C; (c) 1000°C

The region contains 0.3 wt% becoming smaller as carbon diffuses within the structure. This is because of increases in the flux of carbon atoms due to the positive influence of the carbon diffusion coefficient D with the temperature of carbon steel. Higher temperatures provide sufficient thermal energy to the diffusing carbons that overcome the activation energy barrier and facilitate their movement to new sites in atomic arrangements. The graphical representation of the carbon concentration at the optimum depth of 0.0015 m below the surface of x and y dimension as illustrated in figure 7 shows that the rapid changes can be made by extending the carburisation duration time and increasing the temperature.

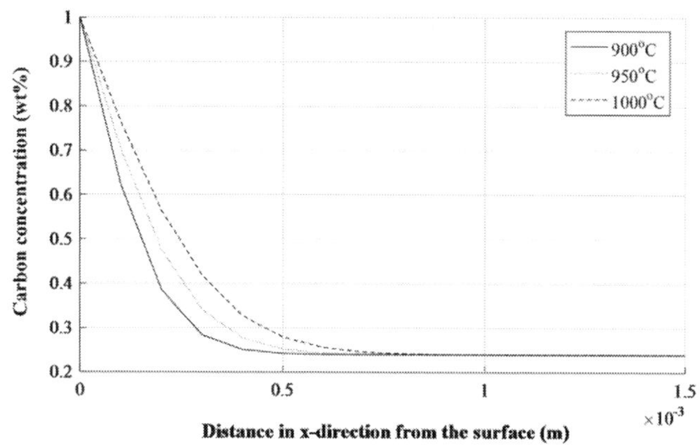

Figure 7. Carbon concentration profile at optimum depth of 0.0015 m from the carbon steel surface carburised at 900°C, 950°C and 1000°C in x -direction along y=0.0015 m.

4. Conclusion

The transient two-dimensional diffusion equation has been discretised using ADI method and the numerical algorithm in MATLAB has been successfully verified analytically. As a results, prediction of carbon concentration profile within the carburised-carbon steel is efficaciously visualised using various representations. Simulation results show that carburisation extended to 60 hours only increases the carbon content at the core to 0.3wt % from its original content which is 0.24 wt%. The small increment is required to maintain its ductility even though the surface is hardened. The developed ADI method can be used to solve engineering process involve transient two dimensional parabolic equation such as mass diffusion and heat conduction.

References

[1] Ali, M., Husnain, R., Ayman, A., Rafiq, A., Inam, A., & Ali, F. (2016). Carburising of Low-Carbon Steel Using Carbon Black Nanoparticles. *Arab J Sci Eng*, **41**, 4661–4667.

[2] Darmo, S., Soenoko, R., Siswanto, E., & Widodo, T. D. (2018). Study pack carburizing for subsoil plow chisel with alternative carburizer media corn cob charcoal–pictada maxima shell powder. *International Journal of Mechanical Engineering and Technology*, **9**(6), 443–449.

[3] Izanaty, H. (2014). The effects of carburization on hardness and wear properties of the mild steel samples. *International Journal of Innovation and Applied Studies*, **6**(4), 995–1001.

[4] Chen, H., Da, X., Cao, J., & Zhou, J. (2019). A formally second order BDF ADI difference scheme for the three-dimensional time-fractional heat equation. *International Journal of Computer Mathematics, 7160.*

[5] Bozsóki, I., Géczy, A., and Illés, B. (2019). Component level modelling of heat transfer during

vapour phase soldering with finite difference ADI approach. *International Journal of Heat and Mass Transfer*, **128**, 562–569.

[6] Korukcu, M. O. (2017). Numerical modeling of heat and mass transfer characteristics during the forced convection drying of a square cylinder under strong blockage. *Numerical Heat Transfer, Part A: Applications*, **72**(2), 171–184.

[7] Zhao, J., Huang, S., Gong, L., and Huang, Z. (2016). Numerical study and optimizing on micro square pin-fin heat sink for electronic cooling. *Applied Thermal Engineering*, *93*, 1347–1359.

[8] Aslib, I., Hamza, H., Hanchi, N., Lahjomri, J., & Oubarra, A. (2019). Numerical Study of Transient and Steady-State Anisotropic Cylindrical Pin Fin Thermal Behavior. *Journal of Thermal Science and Engineering Applications*, **11**(June), 1–12.

[9] Klaus A. and Hoffman, S. T. C. (2000). *Computational Fluid Dynamics:* Vol. I (4th ed.). Engineering Education System.

[10] Bruch, J. C., and Zyvoloski, G. (1974). Transient two-dimensional heat conduction problems solved by the finite element method. *International Journal for Numerical Methods in Engineering*, **8**(3), 481–494.

[11] Lee, S., Matlock, D. K., and Van Tyne, C J.(2011). An Empirical Model for Carbon Diffusion in Austenite Incorporating Alloying Element Effects. *ISIJ International*, **51**(11), 1903–1911.

[12] Reza, S., Hosseini, E., Khosravi, H., Sohrabi, R., and Hosseini, Z. (2013). Correlation between Hardenability Curves and Microstructure of a Cementation Steel Carburized in the Presence of Na_2CO_3 as an Energizer Material. *ISIJ International,* **53**(12), 2213–2217.

IOP Conf. Series: Materials Science and Engineering **767** (2020) 012022 doi:10.1088/1757-899X/767/1/012022

A Review on Methods to Improve and Balance the Assembly Line

M H Kharuddin[1] and M F Ramli[2]

[1,2]Institute of Mathematics Engineering, University Malaysia Perlis, Perlis, Malaysia

khazrinhus@yahoo.com

Abstract. Assembly line is well known as an important part in manufacturing process. It consists of several workstations that linked together to create a complete product. Each of workstation is doing their own tasks. However, during the process of item moving from workstation to another workstation to become a finished product, the unbalance of workload among workstation will create a bottleneck effect. This problem will cause the performance of assembly line is not optimal. Therefore, to improve the performance of assembly line, many methods are applied and used to solve assembly line balancing problem. In this paper, different methods which can be used to improve and balance the assembly line are discussed.

1. Introduction

The assembly lines have a big influence on industrialization. It was first introducing by Henry Ford in 1913 to manufacture a new Ford T model. The mass automobile production was using assembly line and it able to reduce the processing time to build a car.

In assembly line, the complete product will be divided into several different parts and each part is assigned to the workstation. However, which part to process first must follow the precedence flow strictly. The part that should process first must not process second and so on because the item from previous process is needed to begin the next process. These parts are added together until the finished product is produced. Each workstation is linked together but their tasks are different and they are doing their tasks.

Even though the assembly line able to reduce the time it took to complete a product, the bottleneck effect was happened during the process of manufacturing. It happens because of unbalance distribution of workload among workstations and time each workstation required to do their task are different. Therefore, the assembly line is not in best condition and its performance is not optimal.

The term of assembly line balancing (ALB) is normally utilized to represent a procedure of allocating the tasks which their aim is to allocate the tasks into suitable workstations so that each workstation has approximately equal processing time to complete their tasks and precedence constraint is respected. The tasks are operations which required converting raw material into finished product.

There are some reviews and surveys on assembly line which can be found in current literature. They are presented in the paper whose first were studied by Baybars [1] and publish in year 1986, followed by Ghosh and Gagnon [2] in year 1989, Erel and Sarin [3] in year 1998, Rekiek, Delchambre,

Dolgui and Bratcu [4] in year 2002, Becker and Scholl [5] in year 2006, Boysen, Fliedner and Scholl [6] and Kriengkorakot and Pianthong [7] in year 2007, Rashid, Hutabarat and Tiwari in year 2012 [8], Kumar and Mahto [9] and Battaia and Dolgui in year 2013 [10], Saif, Guan, Wang, Mirza and Huang [11], Pachghare and Dalu [12] and Sivasankaran and Shahabudeen in year 2014 [13], Jusop and Ab Rashid [14], Rane, Sunnapwar, Sudhakar and Rane [15] and Hazir, Delorme and Dolgui [16] in year 2015, Dolgui and Gafarov [17] in year 2017, Mohammed, Ramli and Zakaria [18] and Bakar, Ramli, Sin and Masran [19] in year 2019.

This paper is especially focused on methods that can be applied to improve and balance the assembly line. Numerous methods to optimize the assembly line can be found used in many available research papers. Their aims mainly to solve different kind of problems related to the assembly line and improve its performance. A few methods to improve ALB like heuristic, meta-heuristic, simulation approach and queuing concept are going to be discussed later.

2. Methods to improve the Assembly Line

Nowadays several methods have been used to tackle the assembly line balancing problem (ALBP) and improving the performance of the assembly line. The ALBP can be categorised as the NP-hard problem because of its nature. However, a few exact procedures have been proposed to tackle and find it optimum solution. Even though by using those methods guaranteeing an optimal solution, the size of tasks become it limitation. Its unable to achieve an optimal solution for problem of a large number of tasks.

Therefore, new methods have been proposed and developed to discover the best solution to a problem with such limitation and thus, able to obtain optimum result which is closely to an optimal solution. That was how approximate methods such as heuristic and meta-heuristic methods come to exist. Besides that, the simulation approach is quite popular among researcher and also be able to measure, evaluate and improve the assembly line. Queuing concept is also one of the methods that can be used to analyze the assembly line.

2.1. Exact Methods

Usually, to aid decision makers and designer to have a clear understanding of assembly system, assembly line balancing problem is described by using integer linear programming model. However, by considering the real world scaled problems, most often the standard solvers were not efficient and cannot produce the optimal solution. Hence, the exact procedures to solve ALBP that can be found in literature is based on 1) dynamic programming (DP) and 2) branch-and-bound (B&B) procedures.

1) DP procedure is a method that breaks the problem into small sub-problems and changes it into a multi-stage decision process. The small sub-problems are solved to find it optimal solution and then in order to build the optimum solutions of the real problem, the optimal solution of sub-problem is used.

2) B&B procedure is an enumeration method, where the aim was to determine the optimum solution by discovering the possible solution subsets. A branching process is process of branch the solution space to create the Sub-regions. For a bounding process, by using different searching strategies, it is used to find lower or upper bounds of the optimum result within each sub-region.

2.2. Heuristic Methods

A heuristic method normally used to solve ALB is the greedy approach. The solution is selected on the basis of a given requirement until a final solution is produced at each step of the procedure. The simplest procedure will randomly produce the solutions, analyzes every of them and retains the best possible solution. Fundamentally, constructive methods are based on priority rules where certain element must be considering such as the number of predecessor's number, the successors number and the task processing time. The Ranked Positional Weight (RPW) was the first proposed heuristic method to solve ALBP. By arranging the tasks in descending order of the positional weight, where the positional weight is summation of the task processing time and all its successors processing times. The biggest values of positional weight will be ranked at top whereas the smallest value at the bottom. Then, the process of allocating the tasks into their specific workstation can be done. However, the precedence flow must be strictly followed.

Besides that, there are several other heuristic methods based on priority rules such as large candidate rule, Kilbridge and Wester's method, longest processing time of task, maximum total successors number, minimum earliest and latest workstation and Gray-Kidd algorithm or known also as minimum slack. By combining several priority rules some other heuristics are created. The method based on priority rule generates the assignable tasks rank list. If the summation of task time and the current time of workstation less or equal to the cycle time and all of its predecessors have already been assigned to workstation then that task is assignable. The process of select and assign the tasks to the workstations was done by following one of the two strategies.

The first strategy is Station-oriented. This strategy is started with one workstation at a time, and then others are sequentially considered one by one. From the ranked list obtained from each iteration process, tasks are picked and allocated to the current workstation. After completely loaded the current workstation, start to open a new workstation to assign the remaining tasks and so on.

The second strategy is task-oriented. The task with the highest priority, which is the first task in the rank list is chosen and allocated to the earliest workstation that can be assigned the task to it. This procedure is further separated into immediate-update-first-fit and general-first-fit procedures. These two procedures can be distinguished based on whether the rank list of tasks available is updated instantly after allocating a task or after allocating all the tasks currently available in the rank list, respectively.

2.3. Meta-heuristic Methods

The main advantage of classical heuristic methods is dropping in a local optimum. In opposition to this, meta-heuristics have been developed in the past years to overwhelm such a limitation. Glover (1986) is the first person that has introduced the term of meta-heuristic. Generally, the procedures in heuristic methods are based on constructive methods where an initial solution (a population of initial solutions) is needed to be found and local search algorithms is required to move to an improved neighbour solution. In contrast to local search approaches, meta-heuristics do not stop when no improving neighbour solutions can be found. Furthermore, the movement to worsening solutions is allowed in the meta-heuristic system as to avoid premature convergence to a local optimum solution. Various concepts derived from artificial intelligence and evolutionary algorithms inspired by mechanisms of natural evolution are used in meta-heuristics.

GRASP (Greedy Randomized Adaptive Search Procedure) is an iterative process in which each iteration contains of two stages. The first one is the construction stage where an initial solution is produced. The initial solution is generated by probabilistically selecting the next element to be incorporated in a partial solution from a restricted candidate list (RCL). Secondly, is the improving stage. In this stage, a local optimization procedure is used in finding a local optimum.

Tabu search (TS) is a meta-heuristic that focused on memory structures which prevent solution from returning and keeping trap in a local optimum solution. It is allowed to move from a local optimum to worse solutions to escape and obtained a better solution. To prevent the solution from moving back to recently visited solutions, a tabu list is used. The size of the set and main parameters can determine the number of iterations during which a given solution to reoccurring can be avoided. The procedure stops when several search movements have been carried out and no further progress has been achieved.

Ant colonies algorithms (ACO) is a model inspired by looking at the behavior of ants. For example, in the process of finding the food, the ants connect two different locations by searching for an optimal path. The path selection is stochastic. It is affected by both the quantity of pheromone that other ants have placed on a path and the local objective function values that can be determined if the path is chosen.

A method inspired by an analogy with the physical annealing of solids is called Simulated Annealing (SA). It models how at high temperatures disorder the molecular structure of metals, and at low temperatures, the molecular structure of metals is ordered and crystalline. To formulate a problem instance, developed it in such a way that it resembles disordered material. Gradually, the temperature is reduced in such a manner that the ordered states correspond to good problem solutions. By allowing uphill moves based on a model of the annealing process in the physical world, SA technique able to prevent getting trap in a local optimum and lead to a much better solution.

Genetic algorithms (GA) are related to imitating biological evolution as they map programs and data into structures similar to DNA and convey some fitness notion. GA method uses a set of initial solutions, i.e. individuals, each point in the search space represents one feasible solution to a particular problem. A certain number of people conform to a population of feasible solutions. By applying crossover and mutation operators along with an objective function (i.e. the function of fitness) that determines how likely individuals are to be reproduced, the population is evolved.

2.4. Simulation Approach

The simulation concept has adapted by many enterprises for improving the system of production. The possibility of the simulation causes enterprises to evaluate the outcomes for different alternatives. Hence, unnecessary errors and costs can be reduced. Besides, numerical and logical models based on real-world problems are also generated through simulation, and different scenarios using computers are imitated to find solutions for problems. An effective tool provided by computer simulation technology help in planning for solving, analysing and evaluating different alternatives especially for complex problems with high risks or impossible for real-world testing.

The continuous improvement and updating of computer hardware and software cause computer software simulation to be demanded by the people due to high computation efficiency and accuracy. Simulation can observe the different operational patterns quickly and choose the appropriate plan to solve problems. Commonly, a set of techniques for computers to imitate, or simulate the operations of various kinds of real-world facilities or processes is known as simulation. A real phenomenon is imitated in this process using a set of mathematical formulas. The numerical model of reality is generated using a computer for the determination of describing complex interaction among components of a system. Furthermore, other functions of the simulation are to see the behaviour of the process under different conditions and to test new theories. After creating a theory of causal relationships, the theorist can codify the relationships in the form of a computer program. If the program then behaves in the same way as the real process, there is a good chance that the proposed relationships are correct. Nevertheless, the simulation result cannot solve the problem but may display the problem clearly.

A modelling technique where only changes in states of a system are defined as Discrete Event Simulation (DES). Basically, a queue of events is created via DES has affected the system state. The arrangement of these events is based on their timings. The simulation then moves via these events and the changes on the system is applied without modelling the time between any two events. The arrival of a part, the initial and end of cycle times on machines and the occurrence of breakdowns are examples of such events in a typical manufacturing system. Therefore, it is a dynamic simulation technique where changes in the system are represented over time. The reader is referred to look for more details on DES as it has been applied successfully in a wide range of business and manufacturing applications. In fact, it is the most popular technique to model manufacturing systems.

The modelling variability in statistical or empirical distributions and rapid modelling by providing built-in modules that accelerate the modelling process are the main features including in typical DES software. Moreover, a visual interactive simulation where changes in the system are animated and users can interact during the simulation are enabled in typical DES software. Last but not least, a better understanding of the model can be gained from a visual interactive simulation by visualizing, interactive experimentation, improved communications to all stakeholders and the facilitation of model verification.

2.5. Queuing Concept

The assembly line consists of workstation that linked together and they are performed their own tasks with aim at the end of the production to produce a finished product. Raw materials are moving from one workstation to another workstation. The combination of a few queuing systems is also recognised as the assembly line.

To model the assembly line with the principles of queuing theory, first we must adapt the production system components to the components of a queuing system. Regarding to this modelling, products of assembly line are considered as customers of queuing systems and workstations are assumed as servers in the queuing system.

This method can be used to solve ALB however there are so few researchers who's interested in this method. Hence lead to lack of literature.

3. Research gaps

According to the thorough literature survey, it is concluded that no buffer, no breakdown, no material shortage, fixed demand and many more are the assumptions that most of the papers referred on. However, these cases do happen in a real-world situation. The difficulties in the collection of data are the effect of the limitation in using the statistical tools, as the real-world does not permit data collection due to competition and confidentiality.

Cost restriction has been paid very little attention due to confidentiality. Trends and shifts are needed to develop and advance algorithms such as GA, ACO and SA are required to be compared. To the best of our knowledge, there is a lack of attention on the combinatorial effect of feeder lines, automation, failure, worker's skill and number, absenteeism and material shortage (delay and rejection).

There are lacks literature related to queuing system for assembly line balancing compared with other methods. Many researcher tendencies to choose other method than queuing concept even though this method can improve the assembly line just same as other method and can be much better for certain cases.

4. Suggestion and future research

Various categories of ALBPs and state-of-the-art research in each category can be understood through the ALBP review. Several number of the researchers have developed many meta-heuristics, namely a genetic algorithm, simulated annealing algorithm, tabu search algorithm, ant calonies algorithm, greedy randomized adaptive search procedure and many more for multi-modelling of ALBPs.

Different crossover methods may be proposed by the researchers are compared in terms of their effectiveness in the accuracy of the solution during the development of a genetic algorithm. Similarly, for the development of a SA algorithm, an efficient seed generation algorithm is required in designing in its first stage, which will result in a greater effect on the solution accuracy in the second stage of the SA algorithm. Moreover, the hybrid development method is encouraged to approach as, two or more basic heuristic(s) or meta- heuristic(s) are to be combined to bring a synergetic effect on the solution accuracy. On the other hand, the researchers may research finding the best combination of the basic heuristic(s) or meta-heuristic(s) in terms of greater solution accuracy for a selected problem to serve as a hybrid algorithm.

Besides that, a specific crossover method is used by the researchers in genetic algorithm. Alternatively, a crossover method may be selected randomly by them for each pair of chromosomes from a pool of crossover methods, while performing the crossover operation in each of the iterations. Consequently, this will result in an opportunity to form offspring with improved fitness function values.

In most of the meta-heuristics, such as SA algorithm, GA algorithm, ACO algorithm, and many more, the parameters are either fixed through trial-and-error method using an experiment or based on past published works, which may not guarantee the best performance. Hence, the parameter of these algorithms should be optimized by the researchers using Taguchi's method before executing each of these algorithms. As a result, the accuracy of the solution of the problems that are solved using that algorithm can be enhanced.

In all the categories, statistical comparisons of the performance of new approaches with that of existing approaches are not comprehensive in terms of applying design of experiment to test the effects of the factors or parameters of the problems on the response variable(s). So, the comparison of the algorithms must be done using a complete factorial experiment by considering relevant factors with interaction effects, in which "Algorithm" is a factor to check the superiority of the proposed algorithm. It is observed that very less work is done on multi-model ALB categories and the categories with U-type ALBPs.

To sum up, the development of meta-heuristics with single or multi-objective can be carried out in these categories for future studies. Also, it is found that less work is carried out on the SALB-2 problem. Meanwhile, it is also considered as a significant problem in ALB and more research efforts should be made on this problem.

5. Conclusion

Many kinds of literature included the past and recent research papers have been mentioned on the numerous balancing methods of the assembly line. The exact method, Heuristic method and Meta-heuristic method are such examples. These methods are used to obtain an optimal solution and improve the assembly line. However, there are still some gaps in research that can be thinking about and explore. By doing so will provide some help to other researchers to develop a new idea and contribute to existing knowledge.

References

[1] Baybars I 1986 A survey of exact algorithms for the simple assembly line balancing problem *Management Science* **30** 909-32

[2] Ghosh S and Gagron R J 1989 A comprehensive literature review and analysis of the design, balancing and scheduling of assembly systems *International Journal of Production Research* **27** 637-70

[3] Erel E and Sarin S C 1998 A survey of the assembly line balancing procedures *Production Planning and Control* **9** 414-34

[4] Rekiek B, Delchambre A, Dolgui A and Bratcu A 2002 Assembly line design: survey *IFAC Proceedings Volumes* **35** 155-66

[5] Berker C and Scholl A 2006 A survey on problems and methods in generalized assembly line balancing *European Journal of Operational Research* **168** 694-715

[6] Boysen N, Fliedner M and Scholl A 2007 A classification of assembly line balancing problems *European Journal of Operational Research* **183** 674-93
[7] Kriengkorakot N and Pianthong N 2007 The assembly line balancing problem: review articles *KKU Engineering Journal* **34** 133-40
[8] Rashid M F F, Hutabarat W and Tiwari A 2012 A review on assembly sequence planning and assembly line balancing optimisation using soft computing approaches *Int J Adv Manuf Technol* **59** 335-49
[9] Kumar N and Mahto D 2013 Assembly line balancing: a review of developments and trends in approach to industrial application *Global Journal of Researches in Engineering Industrial Engineering* **13** 29-50
[10] Battaia O and Dolgui A 2013 A taxonomy of line balancing problems and their solution approaches *International Journal of Production Economics* **142** 259-77
[11] Saif U, Guan Z, Wang B, Mirza J and Huang S 2014 A survey on assembly lines and its types *Front. Mech. Eng.* **9** 95-105
[12] Pachghare V and Dalu R S 2014 Assembly line balancing - a review *International Journal of Science and Research* **3** 807-11
[13] Sivasankaran P and Shahabudeen P 2014 Literature review of assembly line balancing problems *Int J Adv Manuf Technol* **73** 1665-94
[14] Jusop M and Ab Rashid M F F 2015 A review on simple assembly line balancing type-e problem *IOP Conf. Ser.: Mater. Sci. Eng.* **100** 012005
[15] Rane A B, Sunnapwar V K, Sudhakar D S S and Rane S 2015 Improving the performance of assembly line: review with case study *International Conference on Nascent Technologies in the Engineering Field (ICNTE 2015)*
[16] Hazir O, Delorme X and Dolgui A 2015 A review of cost and profit oriented line design and balancing problems and solution approaches *Annual Reviews in Control* **40** 14–24
[17] Dolgui A and Gafarov E 2017 Some new ideas for assembly line balancing research *IFAC PapersOnLine* **50** 2255-59
[18] Mohammed F D, Ramli M F and Zakaria M Z 2019 A review on hybrid metaheuristics in solving assembly line balancing problem *AIP Conference Proceedings* 2138 040017
[19] Bakar N A, Ramli M F, Sin T C and Masran H 2019 A review on robotic assembly line balancing and metaheuristic in manufacturing industry *AIP Conference Proceedings* 2138 040005

IOP Conf. Series: Materials Science and Engineering **767** (2020) 012023 doi:10.1088/1757-899X/767/1/012023

Effect of RTV coating on the electrical performance of porcelain insulator string under dry condition

M Othman[1], M Isa[2], Z C M Kasa[2], M N Mazlee[3], M A M Piah[4]

[1]Faculty of Engineering Technology, Universiti Malaysia Perlis, Malaysia
[2]School of Electrical Systems Engineering, Universiti Malaysia Perlis, Malaysia
[3]School of Materials Engineering, Universiti Malaysia Perlis, Malaysia
[4]School of Electrical Engineering, Universiti Teknologi Malaysia, Malaysia

mardianaliza@unimap.edu.my

Abstract. Pollution flashover is one of the primary reasons cause to power system failure in Malaysia. The presence of accumulation on the porcelain insulator surface reduces the electrical performance and also causes the failure of insulation in transmission and distribution lines. This research presents a study on the effect of Room Temperature Vulcanize (RTV) coating on the performances of a string porcelain insulator and their impact on environmental contamination. An attempt has been made to simulate the electric field distribution of 33 kV insulators using the finite element method (FEM). From the simulation results, the maximum electric field distribution is obtained at the pin insulator of a clean insulator, while the cap insulator indicated the minimum electric field. With the existence of the contamination layer, the electric field distribution of the porcelain surface results in a reduction trend.

1. Introduction

Nowadays, the consumption of electric energy in developing countries increases dramatically due to fast population growth and rapid development in economic activities. Malaysia is the most developing country in the Southeast Asia region, with an annual average growth rate under Eleventh Malaysia Plan is 6.2%, as referred to the Malaysian National Development Strategy. Electricity consumption of commercial sectors and residential had recorded 7559 ktoe in 2015 and contributed around 53.6% of the total electricity consumption. Moreover, the electricity demand for both sectors has increased up to 53.7% for five years, which is from 2000 to 2015[1]. Hence, the power systems industries keep updating the information to develop long-distance transmission and high voltage [2]–[5]. For distribution and safe transmission of electric power, the insulator is the most critical part of power systems to prevent the current flow from the wire to the earth through ground supporting towers.

Porcelain insulator was first made in China in the year 1909 and are used until today in electrical power application, for instance, distribution and power transmission. They also provide excellent insulation from the power plant to distribution lines. The selection of porcelain insulators compared to other types of insulators could consider as a safe decision due to material that made from raw material such as clay, fillers and feldspar, which is naturally available, prevent from industrial processes materials and cheaper. The essential material for a porcelain insulator is clays that contain ball clay and kaolin, which have different chemical properties and physical depending on the geological and geophysical environment. Traditional porcelain formulations for hard porcelain consists of 50% clay, 25 % silica and 25 % flux[6]–[8]. While, soft porcelain usually consists of 25 % plastic component clay, 25 % filler and 50 % flux. Other than that, porcelain insulators offer some advantages such as high electrical stability, high mechanical strength and corrosion resistance, especially in humid or wet conditions.

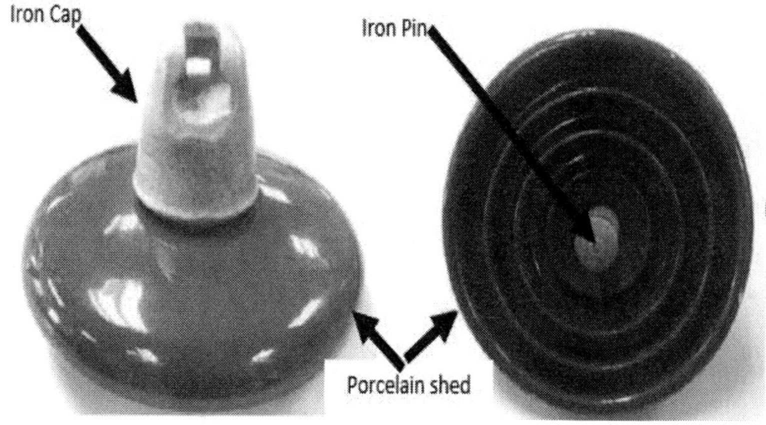

Figure 1. Porcelain insulator.

From previous studies, most of the researchers found that outdoor environmental exposure, for instance, ultraviolet radiation (UV), contamination (salt, dust and bird secretions), rain and humidity contribute to porcelain insulator degradation due to chemical changes from dry band arcing and environmental especially in wet conditions. These contaminants allow electric field distribution of the insulator to become unstable, which acts as a nonlinear or highly variable resistor[9]–[11]. Thus, a path of leakage current is created and heat is generated until a flashover is caused. Room Temperature Vulcanize (RTV) coating has been chosen as the most alternative way to recover electrical power insulation performances in ceramic insulator due to excellent hydrophobicity, excellent electrical insulation properties, excellent self-cleaning and water resistance and excellent arc resistance for preventing insulator surface from contamination. It has been proven that the coating reduces the number of flashover voltages and extends the life of ceramic insulators. The life of the insulator can be as long as 15 years, depending on the material being manufactured[12]. However, for the type of porcelain insulator string, there is still a minor approach from previous researches. Thus, an attempt of Room Temperature Vulcanize coating on a porcelain insulator string is essential to solve the accumulation of contaminants on the surface insulator and to prevent flashover as a safety step. Evaluation of the withstand voltage capability and electrical performance of the insulator string is compulsory to determine the effectiveness of transmission lines and to prevent the power systems engineering from fire hazard, which in turn of low carbon economy[13]–[15]. This paper aims to study the effects of an RTV coating on a porcelain insulator string in order to increase electrical insulation under clean and contaminated conditions[16]–[18].

2. Methodology

Flashover phenomenon is one of the most serious problems detected in high voltage outdoor insulators[19]–[23]. This phenomenon exists by many factors, for instance, the accumulation of contaminants in different environments such as dry and wet conditions and non-uniform contamination distribution on the insulator surface. In this paper, simulation work was used to investigate the performance of a insulator string with the effect of RTV coating under the dry condition on 33kV high voltage. The 2-D CAD model of four-unit suspension porcelain insulator has imported into the finite element method (FEM) software, COMSOL multiphysics. The dimensions of each insulator are tabulated in Table 1.

Moreover, the cap insulator at the top of the insulator is assigned 0 kV, which is referred to as ground and the pin insulator at the bottom of the insulator is assigned 33 kV as high voltage input, while the rectangular shape represented as the air surrounding the insulators. Ceramic insulators divided into three main parts, such as pin & cap, ceramic disk, and cement. The material properties consist of relative permittivity and conductivity, as tabulated in Table 2. RTV coating was applied on the upper surface for each porcelain insulator with a 0.2 mm dimension. The electric field is simulated under electrostatic mode. In this case, the operating frequency of the porcelain insulator is 50 Hz, assuming electric and magnetic fields change slowly. In addition, the induced current of electric field and magnetic field are neglected.

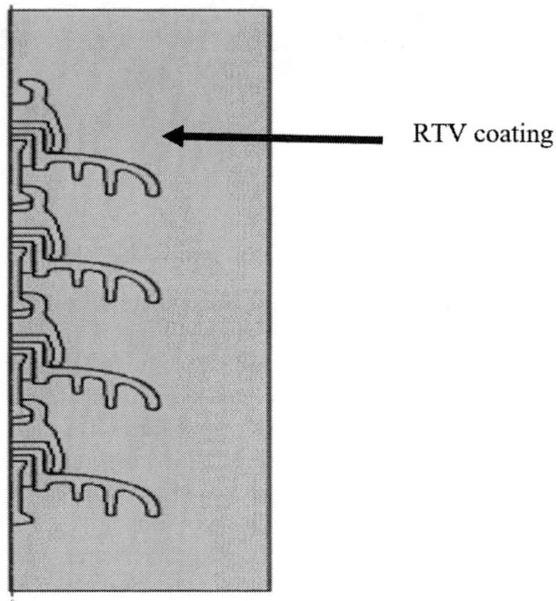

Figure 2. Geometrical model of string porcelain insulator.

Table 1. Dimensions of coated insulator.

Parameters	**Dimensions (mm)**
Maximum diameter	250
Height	145
Creepage distance	320
Pin diameter	15
Coating thickness	2

Table 2. Material properties.

Material	**Relative permittivity**	**Conductivity (S/m)**
Cement	2.1	1×10^{-14}
Porcelain	6	1×10^{-14}

IOP Conf. Series: Materials Science and Engineering **767** (2020) 012023 doi:10.1088/1757-899X/767/1/012023

RTV coating	2.9	1 x 10^{-13}
Air	1	1 x 10^{-15}

3. Results and discussion

Figure 2 shows the effect of RTV coating on the electric field distribution without contamination. From the simulation, it shows that the electric field close to the pin or high voltage input is the highest under dry condition compared to other surfaces of the insulator. The bottom insulator displays in red color represent "hot" temperature due to resistance heating along the insulator surface. This was for the same reason as explained in [24], that is, the maximum electric field is obtained close to the pin of some dielectric mediums such as air, pin and silicone. The second upper insulator is "cold" as shown in a greenish-blue color because of low resistance through the insulator surface. Lastly, the upper insulator is the best insulator represented in blue color, it has the lowest surface heating as a result of surface leakage currents due to weak electric field.

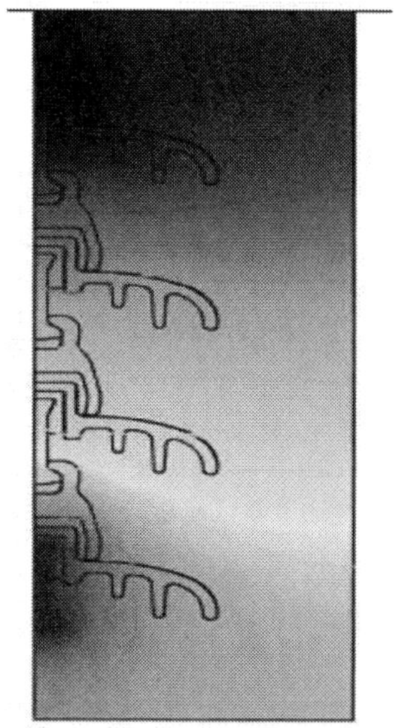

Figure 3. Electric field distribution of clean insulator under dry condition.

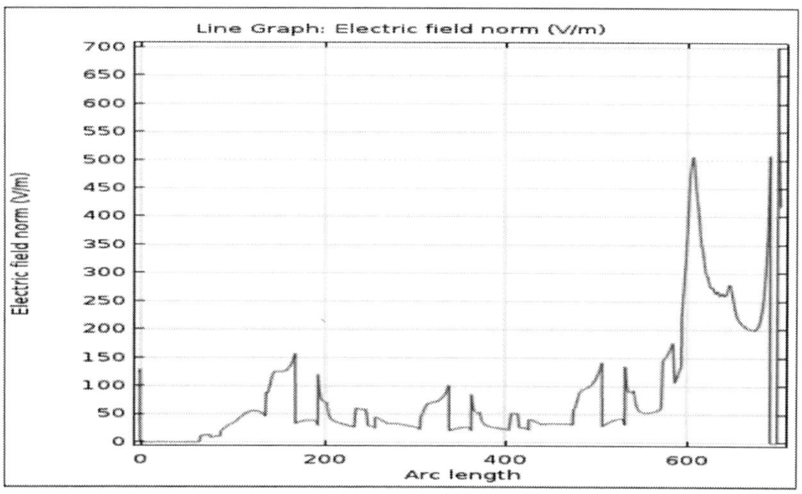

Figure 4. Electric field distribution of contaminated insulator during dry condition.

Figure 4 shows the electric field distribution with non-uniform contamination applied to the RTV coating. The presence of contamination layer decreases electric field at the overall insulator surface. By increasing the contamination conductivity, the electric field will be decreased especially for both areas of pin and cap insulator.

4. Conclusion

In the present work, the application of the Room Temperature Vulcanize (RTV) coating was proven to be an effective method in high-voltage insulation and can solve the accumulation of contamination on the porcelain insulator. The simulation results summarized that the maximum electric field distribution is obtained at the pin insulator of a clean insulator, while the cap insulator indicated the minimum electric field. The presence of the contaminated layer leads to a decrease in the distribution of electrical field on the porcelain surface.

REFERENCES

[1] Q. J. Kwong, J. E. Lim, and M. S. Hasim, "Miscellaneous electric loads in Malaysian buildings - Energy management opportunities and regulatory requirements," *Energy Strateg. Rev.*, vol. 21, no. March, pp. 35–49, 2018.

[2] S. Samuel, T. Sayo, O. Esther, and U. Kingsley, "Influence of waste glasses on the physico-mechanical behavior of porcelain ceramics," *Boletín la Soc. Española Cerámica y Vidr.*, vol. 58, no. 2, pp. 77–84, 2018.

[3] A. Merga, H. C. A. Murthy, E. Amare, K. Ahmed, and E. Bekele, "Heliyon Fabrication of electrical porcelain insulator from ceramic raw materials of Oromia region, Ethiopia," *Heliyon*, vol. 5, no. August, p. e02327, 2019.

[4] N. Singh, P. Kumar, P. Tripathi, R. Pyare, and M. R. Majhi, "Influence of alumina and silica addition on the physico-mechanical and dielectric behavior of ceramic porcelain insulator at high sintering temperature," *Boletín la Soc. Española Cerámica y Vidr.*, vol. 57, no. 4, pp. 151–159, 2017.

[5] F. A. Jamaludin, M. Izadi, N. Azis, J. Jasni, and M. S. A. Rahman, "Considering the Effects of a RTV Coating to Improve Electrical Insulation against Lightning," *2016 33rd Int. Conf. Light. Prot.*, pp. 1–5, 2014.

[6] D. P. Lagunas, "Anestesia en el paciente séptico con distrés respiratorio," no. 2. pp. 47–60, 1992.

[7] M. T. Gençoğlu and M. Cebeci, "The pollution flashover on high voltage insulators," *Electr. Power Syst. Res.*, 2008.

[8] Q. Wang, J. Liao, H. Li, X. Wei, and Z. Peng, "Voltage sharing characteristics of V-type suspension ceramic insulator strings for UHV AC substation," *Proc. IEEE Int. Conf. Prop. Appl. Dielectr. Mater.*, vol. 2015-Octob, pp. 580–583, 2015.

[9] H. Alfiadi, D. Fauziah, Rachmawati, and S. Suwarno, "Study on leakage current characteristics of naturally aged silicon rubber coated ceramic insulator," *ICECOS 2017 - Proceeding 2017 Int. Conf. Electr. Eng. Comput. Sci. Sustain. Cult. Herit. Towar. Smart Environ. Better Futur.*, pp. 259–263, 2017.

[10] K. Marimuthu, S. Vynatheya, N. Vasudev, and P. Raja, "Quality Analysis of Ceramic Insulators Under Steep Front Impulse Voltage," *2019 Int. Conf. High Volt. Eng. Technol.*, pp. 1–5, 2019.

[11] Y. Chao, F. Huang, S. Zhao, C. Wang, F. Wang, and Y. Yue, "Study on natural pollution accumulating characteristics of cap and pin suspension ceramic insulator with composite shed of DC 500kV transmission line in central China," *ICHVE 2016 - 2016 IEEE Int. Conf. High Volt. Eng. Appl.*, pp. 1–4, 2016.

[12] F. A. Jamaludin, M. Z. A. Ab-Kadir, M. Izadi, N. Azis, J. Jasni, and M. S. Abd Rahman, "Considering the effects of a RTV coating to improve electrical insulation against lightning," in *2016 33rd International Conference on Lightning Protection, ICLP 2016*, 2016.

[13] D. Zhang, Z. Zhang, X. Jiang, W. Zhang, J. Zhao, and M. Bi, "Influence of fan-shaped non-uniform pollution on the electrical property of typical type HVDC insulator and insulation selection," pp. 3555–3562, 2016.

[14] R. Divigalpitiya, "Effect of Insulative Debris on the Contact Resistance of Anisotropic Conductive Film (ACF) Adhesives : A Comparison Between a Solid Conducting Particle and a Coated Insulative Particle," vol. 31, no. 1, pp. 222–228, 2008.

[15] T. Koga, A. Itose, H. Prasetia, M. Kozako, and M. Hikita, "Degradation Diagnosis of Electrical Insulation Performance of 6 . 6 kV Class Epoxy Resin Insulator Contaminated with Iron Powder and Moisture," pp. 786–789, 2016.

[16] N. R. M. Putra, N. Sartika, Rachmawati, and Suwarno, "The study on leakage current waveform characteristics and computer simulation of ceramic insulator under artificial tropical condition," *Proc. IEEE Int. Conf. Prop. Appl. Dielectr. Mater.*, vol. 2018-May, pp. 320–323, 2018.

[17] P. Taklaja, I. Kiitam, J. Niitsoo, J. Klüss, and P. Hyvönen, "Electric field distribution in glass and porcelain pin insulators," 2015.

[18] T. Deng, Y. Yao, X. Liang, T. Jiang, J. Zhou, and H. Gao, "Application of pre-coated RTV insulators in UHV OHLs and steep-front impulse voltage test," *ICHVE 2016 - 2016 IEEE Int. Conf. High Volt. Eng. Appl.*, pp. 1–5, 2016.

[19] N. A. Othman, M. A. M. Piah, Z. Adzis, H. Ahmad, and N. A. Ahmad, "Simulation of voltage and electric-field distribution for contaminated glass insulator," in *Proceeding - 2013 IEEE Student Conference on Research and Development, SCOReD 2013*, 2015.

[20] Q. U. Kai, D. Zhen, F. A. N. Qiang, L. I. U. Guanghui, and C. H. E. Bin, "The Influence of Low and Zero Resistance Insulators on the Voltage Distribution of Insulator Strings," *2018 13th IEEE Conf. Ind. Electron. Appl.*, pp. 2089–2094, 2018.

[21] U. D. A. Y. A. Kumar and L. S. A. T. I. Sh, "Potential and electric field distribution in a ceramic disc insulator string with faulty insulator ' "," *2012 IEEE 10th Int. Conf. Prop. Appl. Dielectr. Mater. July 24-28, 2012, Bangalore, India*, pp. 10–13, 2012.

[22] D. Stefanini and J. M. Seifert, "Three Dimensional FEM Electrical Field Calculations for EHV Composite Insulator Strings," *2010 IEEE Int. Power Modul. High Volt. Conf.*, pp. 238–242, 2010.

[23] M. P. Lalitha and M. Ieee, "Modeling and Simulation of Porcelain Disc Insulators using Finite Element Method in OPERA Software," *2014 Annu. Int. Conf. Emerg. Res. Areas Magn. Mach. Drives*, pp. 1–7, 2014.

[24] T. Sörqvist and A. E. Vlastós, "Performance and ageing of polymeric insulators," *IEEE Trans. Power Deliv.*, vol. 12, no. 4, pp. 1657–1665, 1997.

Acknowledgment
The authors would like to thank the Faculty of Engineering Technology, School of Electrical Systems Engineering for providing laboratory facilities and valuable assistance in conducting the simulation and to acknowledge the support from the Fundamental Research Grant Scheme (FRGS) under a grant number of FRGS/2/2014/ICT06/UNIMAP/02/3 from the Ministry of Education Malaysia.

Investigation on the Effect of Deflected Ground Structure in Multi-Direction Steerable Antenna for WSN

N A F M Zainudin, M N Osman, T Sabapathy and M Jusoh

Bioelectromagnetics Research Group (BioEM), School of Computer and Communication Engineering, Universiti Malaysia Perlis, Pauh Putra Main Campus, 02600 Arau, Perlis.

mohamed.nasrun@gmail.com

Abstract. In this paper, a multi-direction steerable radiation pattern antenna is designed and investigated, which operates at 2.45 GHz frequency. The radiation pattern reconfigurable is achieved by locating four parasitic elements surrounding the driving element at the centre. The artificial switches, are currently used to replace the ideal diode for proof of the concept, are inserted at the specific position on the parasitic element. By controlling the state ON/OFF of the switches, the parasitic is grounded via shorting pin, thus realizing the beam steering to be directed into four direction angles; right, left, upwards and downwards. In this research, an investigation on the effect of deflected ground plane on the steering angles has been carried out. With minimum numbers of switches deployed, the proposed antenna is a potential candidate to serve as a transmitting element in WSN applications.

Keywords: beam steering, deflected ground plane, parasitic patch, WSN.

1. Introduction

The rapid growth and evolution of the telecommunication technology has leads to a change of the system conditions to meet the current trends and demands from the end-user such as; lower in cost, compact in size and light in weight, with the enhanced performance. Conventional antenna may face restriction to meet the requirements and adapt to new conditions due to inflexible characteristics. One solution to overcome this issue is the use of beam-forming antennas. Beamforming can also be known as spatial selectively where the signal at the particular angles may be constructive and destructive thus the signal can be transmitted to a particular direction [1]. Beamforming works by combining elements in a phased array at particular angles that experience constructive and destructive interference with the gains in one directional, attenuations, signal to noise ratio is significantly increased at the receiver [2]. However, beamforming has its disadvantage where it requires complex algorithms and high-end embedded system such as digital signal processor. In such a scenario, radiation pattern reconfigurable (RPR) antenna [3] could be used to serve the similar purpose with less design complexity.

One of the critical components and key in wireless communication for WSN is the antenna. WSN is recently creating new and exciting applications that can be used in many areas for remote sensing, tracking, detecting, and monitoring. In WSN the most desirable features are flexibility, high sensing fidelity, low cost, and fault tolerance. However, to design an optimal WSN to satisfy the performance objective such as extended operation periods, battery lifetime and maximum sensing coverage, antenna is best candidate to be optimized. There are many types of antenna that have a certain advantages and disadvantages to WSN. Thus, the antenna that will be used in WSN are carefully studied in terms of its function and the design complexity.

The RPR antenna could be an interesting candidate in WSN as it requires antenna with capability of beam steering. However, various works have shown that RPR antenna requires more switching elements to perform the radiation pattern reconfiguration to multiple directions [4][5]. This would be a problem where more switches will drain the power faster. An RPR antenna design with minimum number of switches with comparable beam directions needs to be developed. The use of RPR antenna with minimum number of elements degrades the beam steering performance of the antenna. The maximum tilt angle that could be obtained with 3-element RPR antenna is usually about 25° [4]. It should be noted that the antenna design in [4] used Taconic substrate which has low loss tangent. The tilt angle is expected to reduce with the use of high loss substrate such as FR-4. Therefore, in this paper, the effect of deflected ground structure (DGS) towards the tilt angles is investigated. DGS technique has been applied in [6] and has proven its ability to enhance the performance of the antenna. The aim of the research is to improve the maximum tilt angle of DGS introduced RPR antenna as compared to RPR antenna with full ground plane. High degree of beam tilting can improve the sensing coverage in WSN. Similar design approach has been proposed in[7]–[9], however the tilt angle steering value is not more than 30°. The aim of this work is to attain at least 30° tilt angle at xz plane even if the antenna is fabricated using FR4. Simulation results are compared and analysed, and important parameter results for optimization is presented.

2. Antenna design and reconfigurations.

Figure 1 shows the initial structure of the proposed multi-element RPR antenna. It consists of one driven element at the centre, and four parasitic elements located surrounding the driven element. This antenna is designed on FR-4 substrate with the dimension of 150 mm x 100 mm, height of 1.6 mm and the permittivity of ε_r=4.3. Rectangular patch shape is adopted for the driven and parasitic element design due to its simplicity in optimization and easy integration with switches. The patch of the driven element (L_p x W_p) and the all the parasitic elements are placed on the top of the substrate. At the other side of the substrate, the DGS with a dimension of $((W_s \times L_s)/2)$ is introduced. The patch and the ground are made by copper with a thickness of H_t = 0.035 mm. The dimension of the driven and the parasitic elements are almost identical (a=0.9 and b=0.998), and all the parasitic are positioned with the gap of 3 mm from the driven element. The switches (currently represented by the shorting pins for proof of the concept) are placed between the parasitic elements and the ground plane. The shorting pins are positioned at the specific location of the parasitic patch. The direction pattern of the beam steering is changed to the left, right, up or down depending on the switching configurations as tabulated in Table 1. The material of the shorting pin set as copper indicates the switch is ON, whereas it is represented with vacuum for OFF state. When the switch is ON, the parasitic is grounded via shorting pin, thus it acts as a reflector to reflect the energy of the radiation pattern towards the driven element. Figure 2 shows the ground plane with DGS and optimized ground plane structure, which the parameter W_s is varied to obtain maximum tilt angle as compared to the RPR with full ground plane. The dimensions of the optimized design are as follows (unit in milimeters): W_p=37.58, L_p=27.4, W_g=38, L_g=2, W_s=150, L_s=100, and W_{DGS} = 37.5.

Table 1. Switching configurations of the proposed antenna.

Configuration	Tilt direction	Switch 1	Switch 2	Switch 3	Switch 4
I	Left	ON	OFF	OFF	OFF
II	Right	OFF	ON	OFF	OFF
III	Upwards	OFF	OFF	ON	OFF
IV	Downdwards	OFF	OFF	OFF	ON

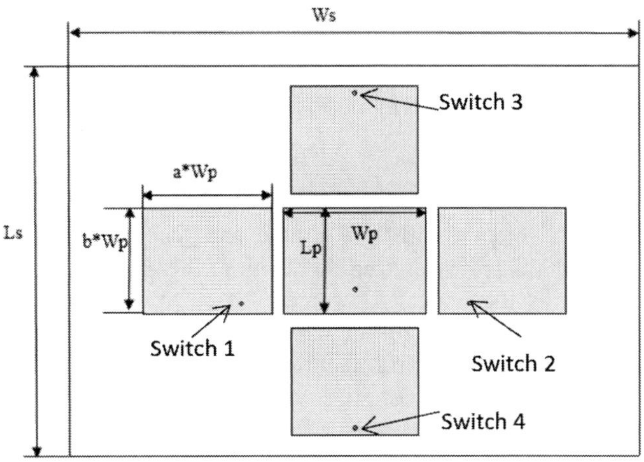

(a)

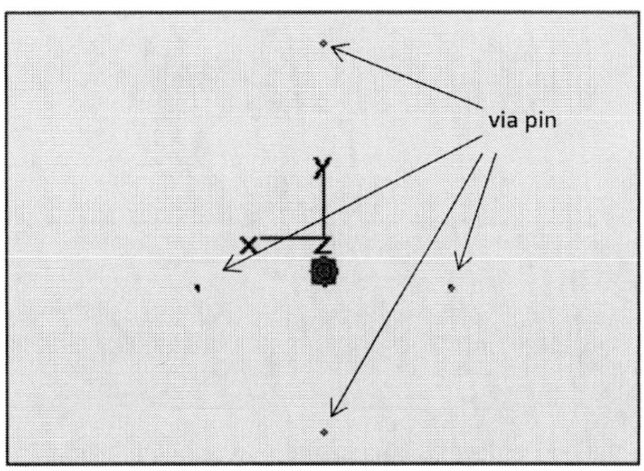

(b)

Figure 1. Geometry of the initial design of the multi-element RPR antenna (a) front view (b) back view.

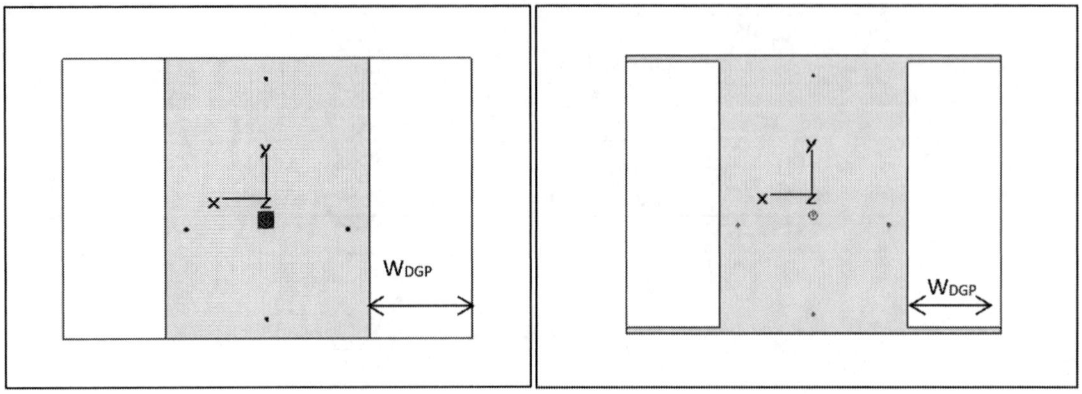

Figure 2. The back view (a) with DGS and (b) with optimized DGS.

3. Result and discussion

The preliminary works has been done by using ideal diodes for proof of the concept. The antenna was designed as microstrip patch antenna with full ground and feed at the middle of the driven patch. Parameter sweep has been conducted by cutting the width of the ground plane the beam steering of the antenna. The effect of reducing the width of parameter W_{DGS} towards the beam steering is shown in Figure 3. It can be seen that the value of tilt angle increases significantly when the width is reduced. Figure 4 shows the simulated reflection coefficient result of the proposed antenna with and without the defected ground structure. The DGS does not give significant effect towards reflection coefficient. This is understood since the matching is only take place for driven element.

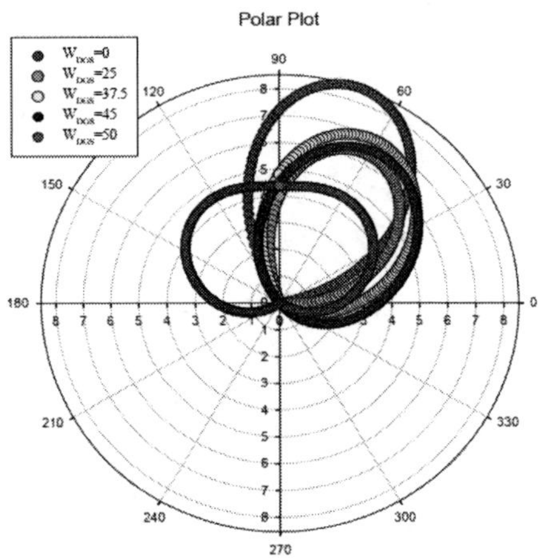

Figure 3. Simulation results of the parameters sweep with DGS for switching condition I.

IOP Conf. Series: Materials Science and Engineering **767** (2020) 012024 doi:10.1088/1757-899X/767/1/012024

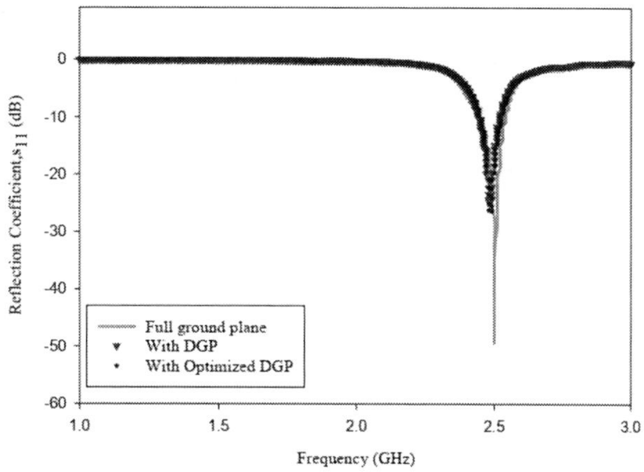

Figure 4. Simulated reflection coefficient, S_{11} for the proposed antenna.

The tilt angles of the radiation pattern seem to be changed when the parasitic is placed only on one side of the patch. First, the parasitic are placed on the left side (Configuration I) of the patch and the results simulations show the changes in the direction of the radiation pattern. The process has been repeated for the remaining three configurations. The simulated 3D radiation pattern and polar plot for three type of ground plane; full ground plane, with DGS and optimized DGS, for all configurations is tabulated in Table 2 and Table 3, respectively. When the parasitic is grounded via copper pin, the direction is changes accordingly to the opposite of the parasitic, acts as a reflector. The tilt angle results are summarized in Table 4. From this table, the DGS technique has improved the tilt angle for the left and right configurations, whilst reduced the tilt angle for upwards and downwards configurations. This is due to the antenna cannot radiates properly because the ground was cut.

Table 2. Simulated 3D radiation pattern for three type of ground plane.

Configuration	Full ground plane	With DGP	With optimized DGP
I	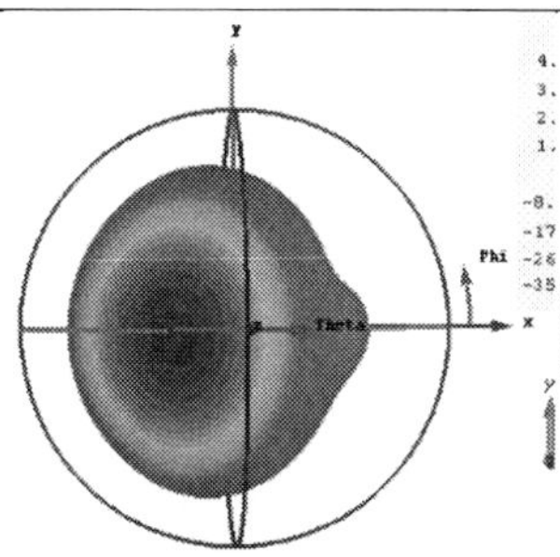	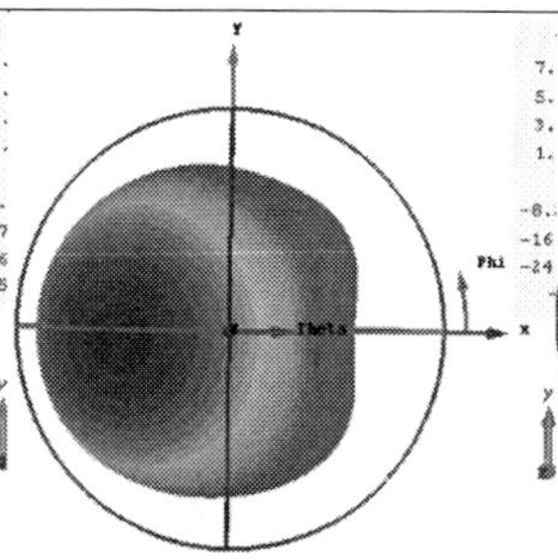	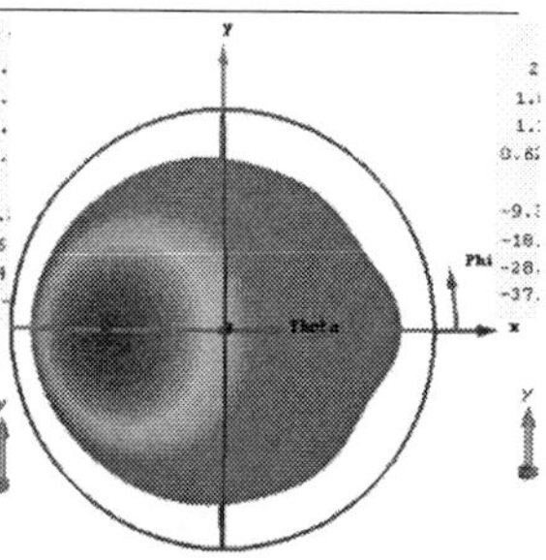
II	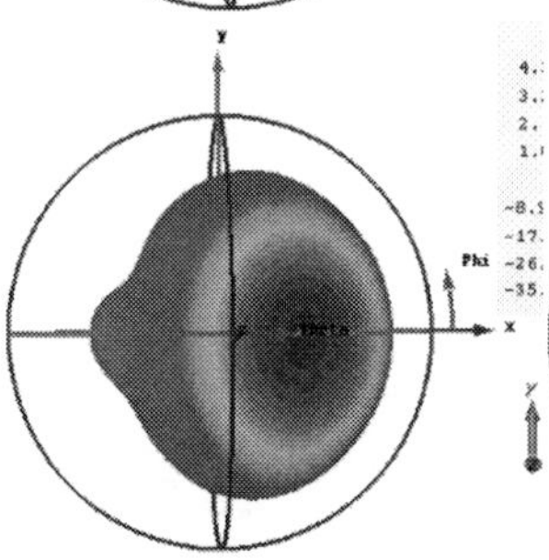	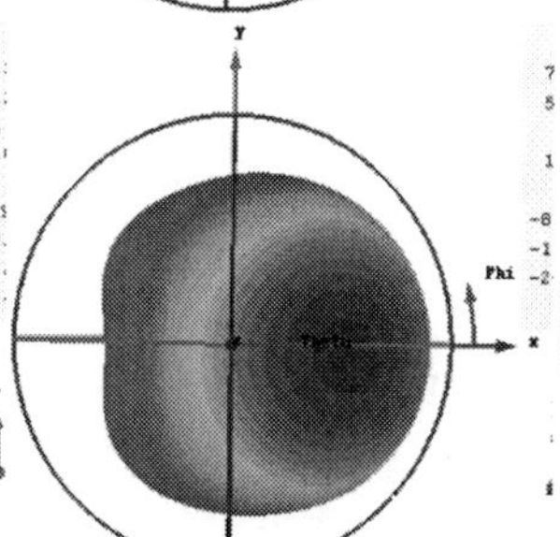	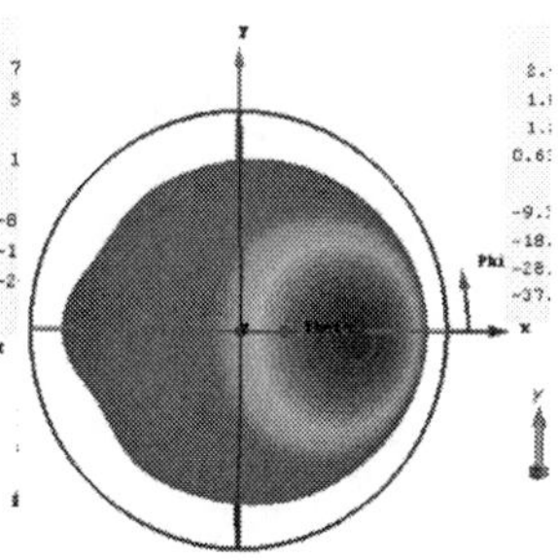

IOP Conf. Series: Materials Science and Engineering **767** (2020) 012024 doi:10.1088/1757-899X/767/1/012024

III	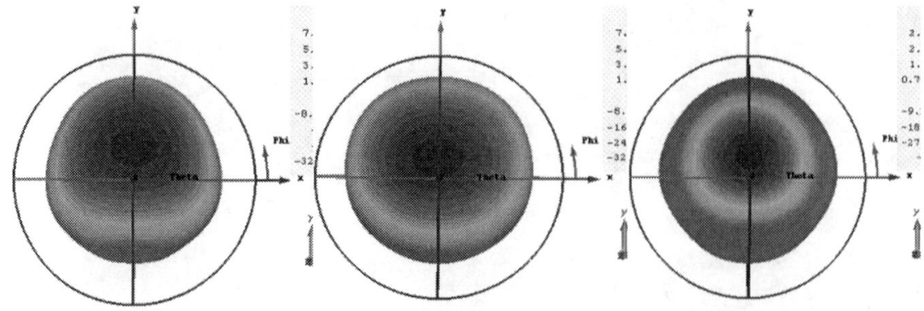
IV	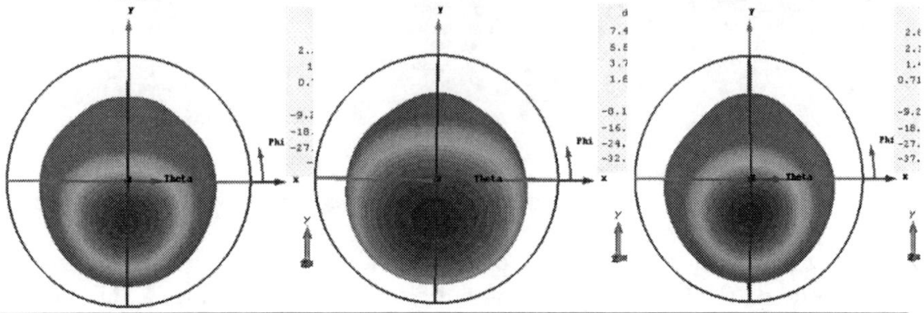

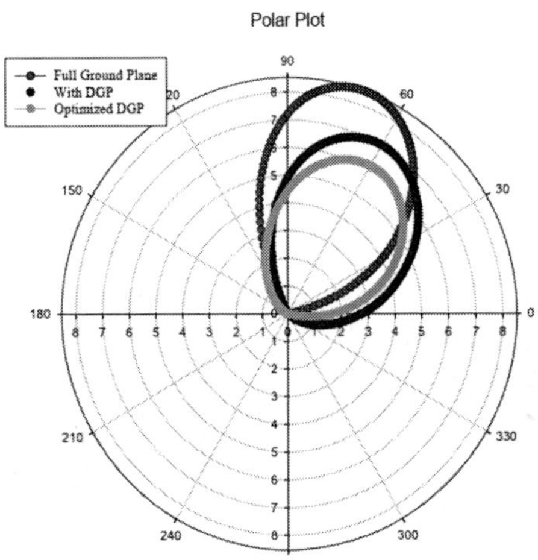

(a)

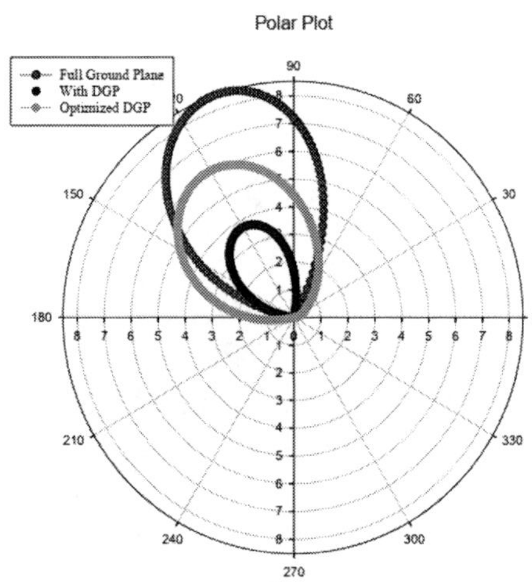

(b)

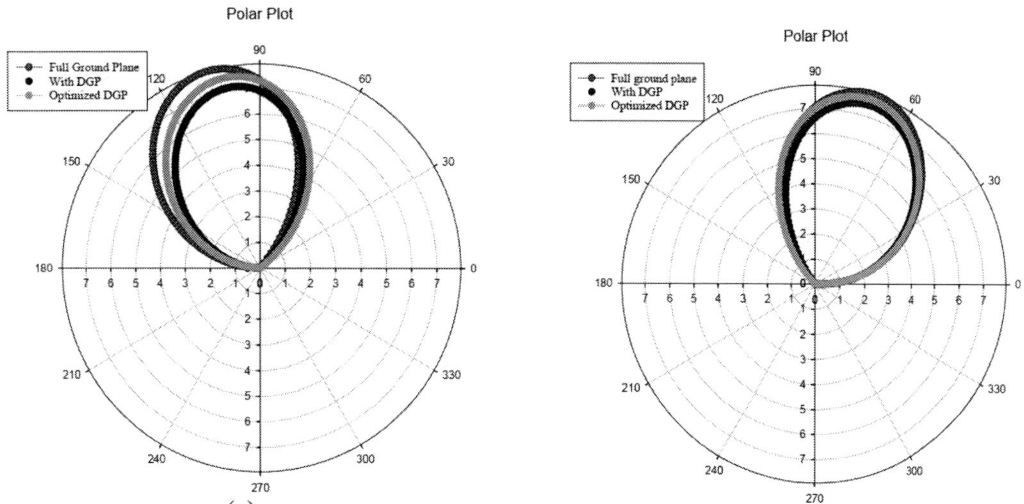

(c) (d)

Figure 5. The polar plot for three type of ground plane (a) configuration left (b) configuration right (c) configuration upwards (d) configuration downwards.

Table 3. Summarized simulated results of the proposed antenna.

	Degree of tilt angle			
Type of ground plane	Config. I	Config II	ConfigIII	ConfigIII
With full ground plane	19°	19°	15°	18°
With DGP	30°	30°	10°	18°
With optimized DGP	32°	32°	10°	17°

4. Conclusion

This paper proposed a design of multi element RPR beam steering antenna with optimized ground plane by using DGS technique. To achieve the steering capability, the parasitic array technique is applied on the proposed design, by placing almost identical parasitic patches next to the driven patch. The beam steering antenna is realized by grounded the shorting pin of the parasitic. The direction of the beam pattern is capable to be changed in four directions. The tilt angle is more than 30 degree when the DGS technique is integrated to the structure.

5. References

[1] J. Feng *et al.*, "Energy-efficient transmission for beamforming in wireless sensor networks," in *SECON 2010 - 2010 7th Annual IEEE Communications Society Conference on Sensor, Mesh and Ad Hoc Communications and Networks*, 2010.

[2] G. Charis and N. Showme, "Beamforming in Wireless Communication Standards: A Survey," *Indian J. Sci. Technol.*, 2017.

[3] A. Iqbal *et al.*, "Frequency and Pattern Reconfigurable Antenna for Emerging Wireless Communication Systems," *Electronics*, vol. 8, no. 4, p. 407, 2019.

[4] T. Sabapathy, M. F. Bin Jamlos, R. B. Ahmad, M. Jusoh, M. I. Jais, and M. R. Kamarudin, "Electronically Reconfigurable Beam Steering Antenna Using Embedded Rf Pin Based Parasitic Arrays (Erppa)," *Prog. Electromagn. Res.*, vol. 140, no. April, pp. 241–261, 2014.
[5] B. Ban, Z. Cheng, and S. L. Chen, "Center-Fed Shorting-Via-Loaded Circular Patch Antenna with Reconfigurable Polarization and Switchable Beam," *2018 12th Int. Symp. Antennas, Propag. EM Theory, ISAPE 2018 - Proc.*, pp. 1–3, 2019.
[6] T. Sabapathy, M. Jusoh, R. B. Ahmad, M. R. Kamarudin, and P. J. Soh, "A Ground-Plane-Truncated, Broadly Steerable Yagi-Uda Patch Array Antenna," *IEEE Antennas Wirel. Propag. Lett.*, vol. 15, pp. 1069–1072, 2016.
[7] Z. Li, E. Ahmed, A. M. Eltawil, and B. A. Cetiner, "A beam-steering reconfigurable antenna for WLAN applications," *IEEE Trans. Antennas Propag.*, vol. 63, no. 1, pp. 24–32, 2015.
[8] R. Florencio Díaz, R. Rodríguez Boix, F. Carrasco Yépez, J. Encinar Garcinuño, M. Barba Gea, and G. Pérez Palomino, "Broadband reflectarrays made of cells with three coplanar parallel dipoles," *Microw. Opt. Technol. Lett.*, vol. 56, no. 3, pp. 748–753, 2014.
[9] V. V. Khairnar, B. V. Kadam, C. K. Ramesha, and L. J. Gudino, "A reconfigurable parasitic antenna with continuous beam scanning capability in H-plane," *AEU - Int. J. Electron. Commun.*, vol. 88, no. February, pp. 78–86, 2018.

Acknowledgements
Authors wishing to acknowledge the support from the Fundamental Research Grant Scheme (FRGS) under a grant number of FRGS/1/2017/ICT01/UNIMAP/02/1 from the Ministry of Higher Education.

Determination of soft starter firing angle performance to mitigate motor high inrush current using current limitation method

Mohamad Shukor Abdul Rahim[1], Muhammad Mokhzaini Azizan[2],Mohd Irwan Yusoff[1] , Mohd Hafizuddin Mat[1], Nur Irwany Binti Ahmad[3], Noor Fazliana Binti Fadzail[1], Suhaireza Binti Md Esa[1].

[1]Centre for Diploma Studies, UniCITI Campus, Universiti Malaysia Perlis, 02100, Pdg Besar, Perlis, Malaysia.
[2]School of Electrical System Engineering, Pauh Putra Campus, Universiti Malaysia Perlis (UniMAP), Pauh, 02000,Arau, Perlis, Malaysia.
[3]Faculty of Engineering Technology, UniCITI Campus, Universiti Malaysia Perlis, 02100, Pdg Besar, Perlis, Malaysia.

mshukorrahim@unimap.edu.my

Abstract. Inrush current in the simplest form can also be determined as current drawn by an induction motor during startup period. This starting current will shoot up about 5 to 7 times the rated current. However, this high current usually occur in the starting period only. To overcome this, several techniques can be implemented to reduce the high current. The configuration of soft starter just involving some power semiconductor device act as switches that control the current flow from power source to the motor. The switches is in form of thyristor and are connected back-to-back because the system conduct in AC system. The current output can be controlled by varying the firing angle. This changing of firing angle will be managed by a firing angle control circuit. This soft starter was connected between power source and motor. The thyristors that built in soft starter act like a gate to control the voltage applied to the motor. The firing angle for current limitation soft starter was changed to several angle and what can be concluded that the high current succeed to mitigate with increasing the firing angle. The current drawn for this type of starter is steadily constant. The lower current during starting took longer time for motor to reach its rated speed. This type of starter successfully reduces inrush current about 42 percent. Finally what can be concluded is that the soft starter was proven to mitigate inrush current. Type of soft starter that going to implement is depending on the application of motor. When the application need to control the torque is more suitable to use current limitation soft starter because the current is steadily control.

1. Introduction

Inrush current can easily be explained as the current drawn by electrical appliances which consist of mechanical parts. This starting current will shoot up about 5 to 7 times the rated current. However, this high current usually occurs in the starting period only. These temporary conditions might cause difficulties to both the power network and motor. [1]–[3] . Usually the inrush current caused by change in the flux linking with a coil where electro motive force is induced in the coil. Basically, when the current drawn is high means that the torque exerted also high. This starting current will severely disrupt the voltage in power supply until it quickly drops and disturb all other running devices that use the same power network. This loads coincidently become the major causes and the major victims of power quality problems. Due to their non-linearity, all these loads create disturbances in voltage waveform[4], [5]. This disturbance will bring harm to motor such as overheating[6]. Three phase induction motor has broad application in industry control and electrical drive. Assorted motor starters were created to overcome the problem. There are several types of motor which divided to conventional starters and electronic drives. Conventional starter are star-delta, autotransformer, and direct on line starter [7]. However the existing method still create current spike when the starter is being bypassed after the starting period. They also have disadvantages such as high installation cost, low efficiency and can be applied to certain type of motor only[8]–[10]. In order to overcome this improper method, the determination of soft starter firing angle can be performed to mitigate high inrush current. The angle that been determined will be controlled by firing control circuit. This control circuit operates by controlling the current flow through the circuit. Power electronic device are more reliable because it only consists several power semiconductor switches and controller[11], [12]. Nowadays, power electronic drives soft starter are broadly used in industry because it is capable in providing low inrush to prevent breakdown and finally minimize the maintenance cost.

2. Experimental method

2.1. Soft starter firing angle circuit system

Figure 1 demonstrates the block diagram of the determination of soft starter firing angle. The entire system contains inserting supply, managing the signal, and keeping track of the output. Initially the direct on line motor starting method will be implement and the fluctuation of output current will be noted. Then, the output will be compare to the soft starting method. The circuit is controlled by a firing control circuit. Then, thyristor firing control circuit get signal of firing angle from the firing angle control circuit.

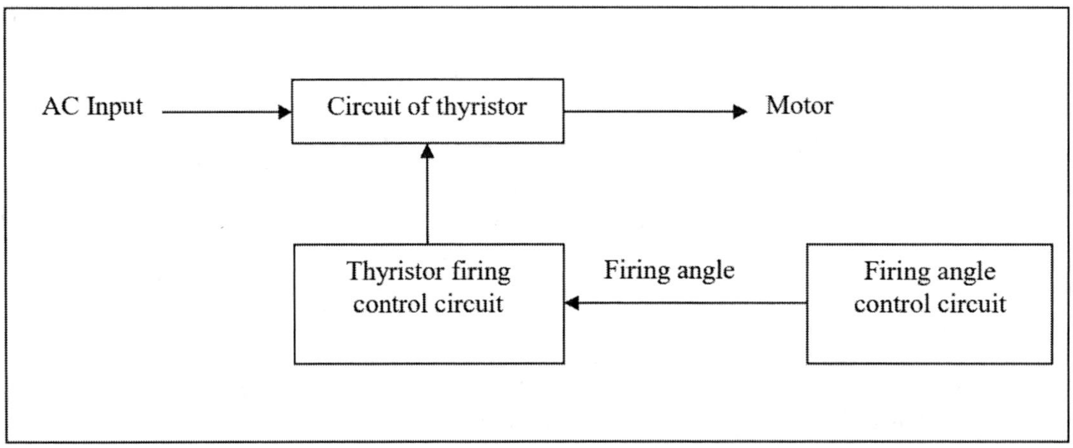

Figure 1. Soft starter firing angle block diagram.

The firing angle was diversified to manipulate the voltage and current. To be able to alter the firing angle provided to circuit of the thyristor, the reference signal is determined to intersect the saw tooth voltage.

2.2. Calculation for Firing Angle

The most familiar type of SCR power control is phase control. Through this setup of operation, the SCR stayed in the open condition for a period of half cycle and then is triggered into short condition in the half cycle which is decided by the control circuit. One SCR itself can only manage one half cycle of the waveform. When applying AC full wave, two SCR are needed in opposite direction. The period when the SCR is running known as the conduction angle. This signifies that the SCR operate by governing the voltage. The power source and load determine the circuit current. The power curve shows that when conduction angle of 30° supply only 3% of full power and 150° conduction angle supply 97% of full power. Therefore, it may become senseless in any situation to attain conduction angles more than 150° and smaller than 30°. Fundamentally when varying firing angle, the adjusting value only take place within the reference signal whilst the sawtooth remains to be the same. The reference signal have been adjusted that it is proportional with 360°. The time of conduction period is relies on firing angle.

$$Triggering\ Time = \frac{Time\ for\ one\ period}{360} \times Firing\ Angle \tag{1}$$

Right after deciding the triggering moment, next the conduction current could be researched with various firing angle. Notice that the thyristor current flow only can be controlled in form turn-on, therefore the current need to be zero again to turn off.

2.3. Direct-on-line Starting Method

A direct-on-line starter first was developed and conducted through PSCAD/EMTDC as shown in Figure 2. Motor parameters on W, S and TL were neglected. W at the motor represent speed input, S represent selection and TL represent torque input. The current then was recorded for further differentiation to the soft-starter current. This direct-on-line starting up technique typically will spike about 6 times the rated current.

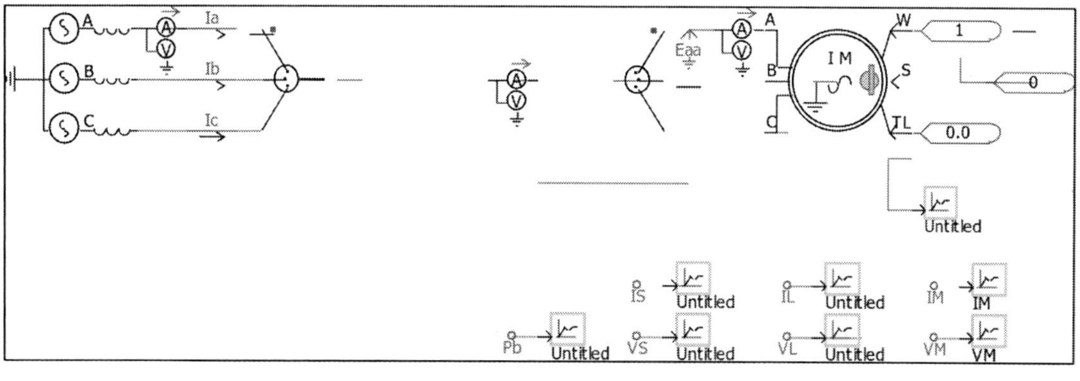

Figure 2. Direct-on-line Simulation.

In addition, this configuration is the most famous type of starter due to its simplest connection which did not require any additional device. However, when the inrush current occur during startup, the fuse or circuit breaker will detect as fault then trip. This will create breakdown to the system. The current will increase from 5 to 7 times from the normal current. The normal rated current is 50A. Whilst, the line current for the motor could be determined by Equation 2:

$$I_L = \frac{S}{\sqrt{3} \times V_{LL}} \tag{2}$$

S is complex power and V_{LL} is line to line voltage. With the S = 100kV and V_{LL} = 0.69kV, the value of the line current can be calculated as 83.67A.

2.4. Simulation of Current Limitation Soft Starter

Figure 3 shows how the motor and power source was interrupted by antiparallel thyristor in the middle which use breaker symbol. This thyristor was connected to each three phase. The placement of soft starter can be subjective and can be located anywhere in between.

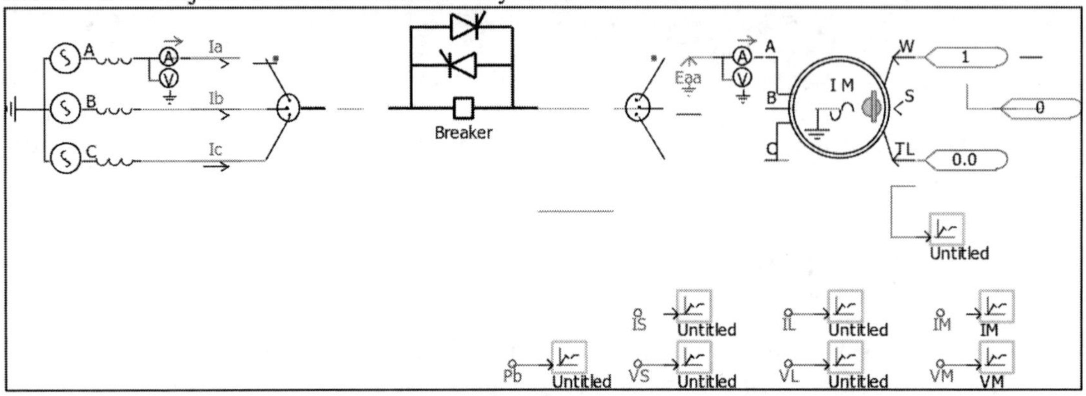

Figure 3. Soft Starter Simulation Circuit.

The breaker shown in Figure 3 then was zoomed in and look like in the Figure 4. There are two set of breakers was connected in the circuit. The thyristor breaker was closed short at starting and the main breaker was open because the soft starter operation took place. When the motor has reaches its rated speed, the source was straightly supplied to the motor. Hence the thyristor breaker open the circuit and main breaker directly supply the source.

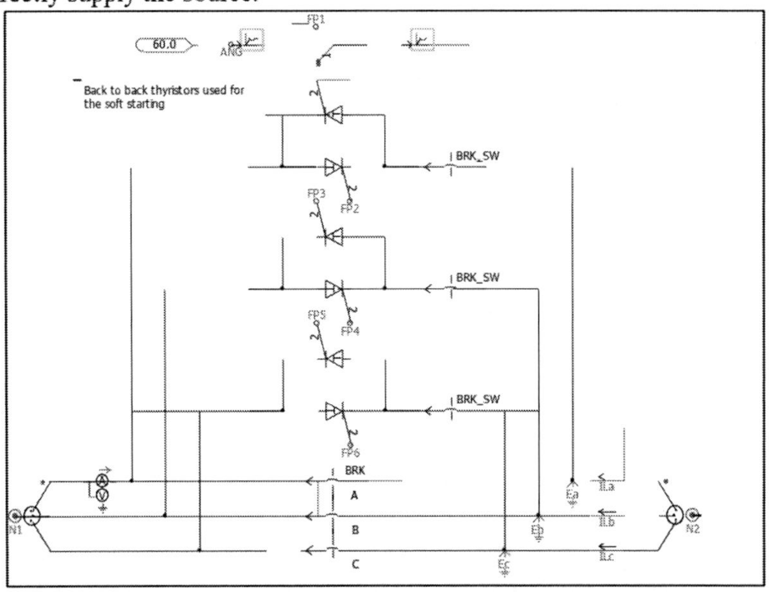

Figure 4. Breaker Simulation Circuit.

Breaker shown in Figure 4 along with the thyristor firing angle circuit is the additional part which is different from the direct-on-line starting. The parameter of power supply and motor used in both simulation is the same. The firing angle then was varied to several increment so that it is suitable with desired torque which is depends on the load and application itself.

2.5. Firing Circuit for Current Limitation Soft Starter

Firing control circuit is the main heart of whole operation in order to perform firing angle in the middle between the source and motor. Figure 5 shows how the circuit was implemented and using a certain device to generate the pulses for the thyristor. The devices used are phase locked loop, interpolated firing pulse and summing junction.

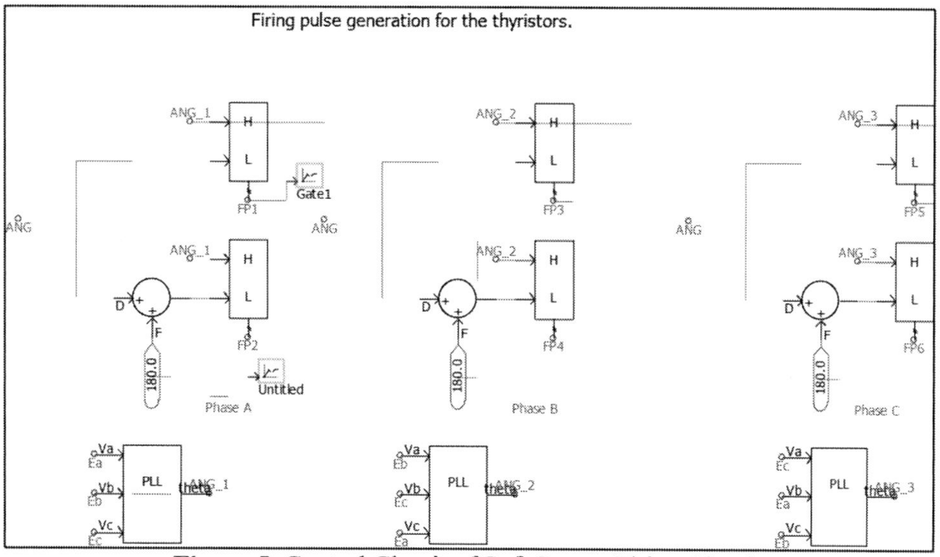

Figure 5. Control Circuit of Soft Starter Firing Circuit.

Firstly, the fundamental product required is *sawtooth* signal and reference voltage. The intersection between this two will generate the pulse. The sawtooth was generated by phase locked loop. The parameter of the sawtooth then was adjusted so that the peak is 360 as shown in Figure 6, which is easier to calculate the reference voltage when varying the firing angle. The frequency of this device also need to be alike with line frequency because it will be more tally and neat.
The sawtooth signal then is going to be connect with the input of interpolated firing pulse to ensure the device can function, which its purpose is to compare and contrast both the sawtooth signal and reference signal and then finally generates pulses.

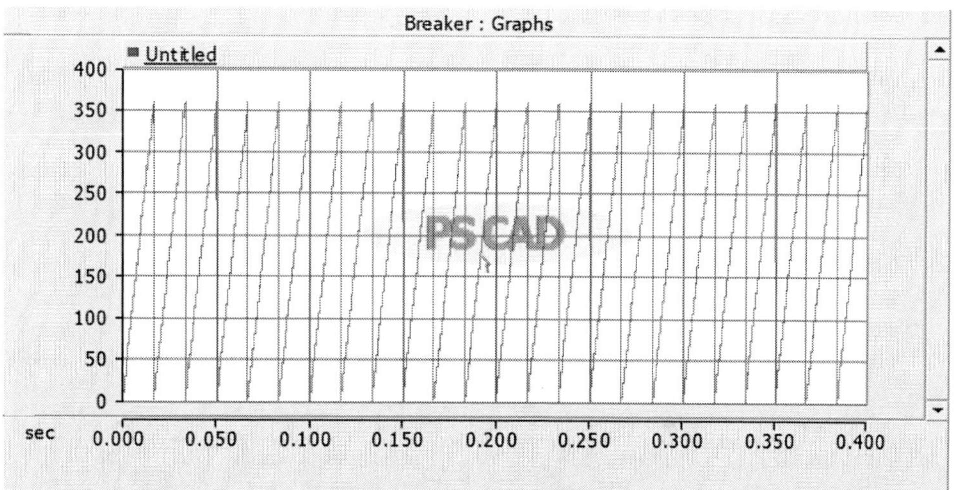

Figure 6. Generated Sawtooth.

The thyristor can only function in one half of the cycle only, so in this situation there are two SCRs which connected in parallel back-to-back in order to perform both cycle conduction period. To perform this operation, summing junction device was used as mathematical constant which will total up from first reference signal. The constant which is to be use is 180 due to the 180° delay from the primary triggering angle. Out of this parameter, the pulse will efficiently be produced.

3. Results and Discussion

3.1. Direct-on-Line Motor Starting

Direct-on-line may be the most straightforward form configuration of all the starters. It entails quite simple devices to apply this particular starter. However, the inrush current occurrence in the mean starting period annoyed the motor. The current rise can be as high as 0.6 kA such shown in Figure 7.

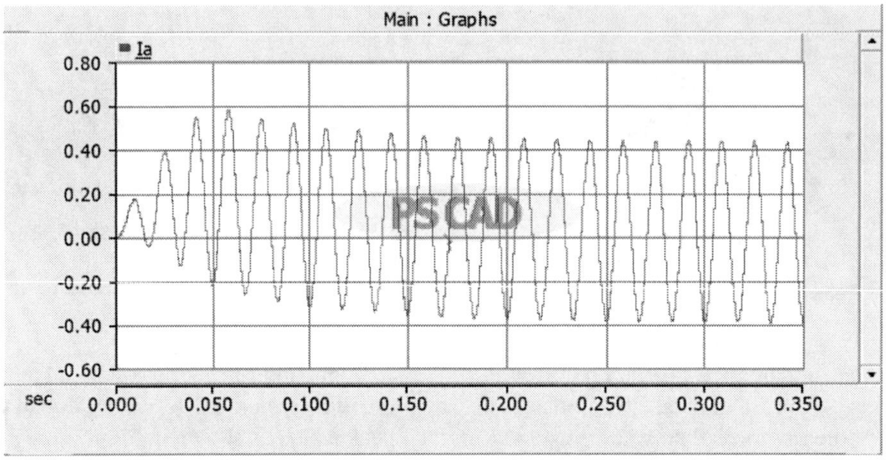

Figure 7. Highest Magnitude of Inrush Current.

Besides of the high magnitude in inrush current, the time taken for the motor to reach rated speed also being affected. The high current shown in Figure 8 delay for a very long time until 1.75s can harm the motor from several aspect. The motor winding will likely be harm by continuous excessive current caused by power loss in heating form.

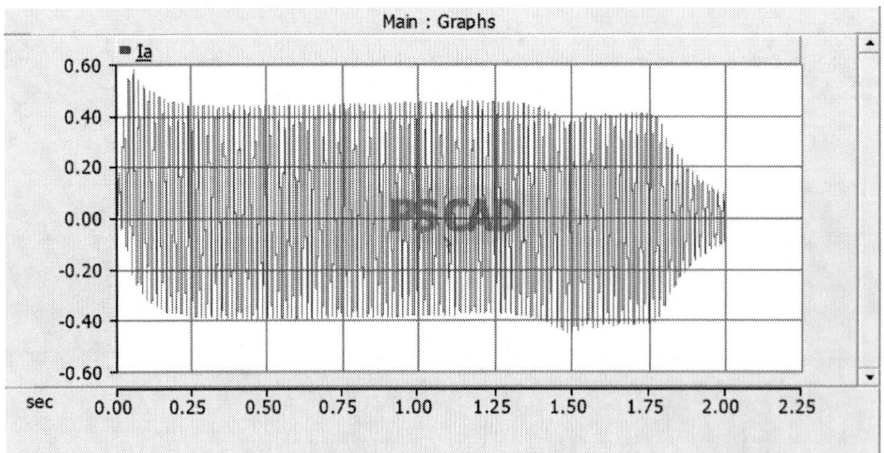

Figure 8. Time Taken for DOL Starter to Reach Rated Speed.

3.2. Current Limitation Soft Starter

Thyristor works by controlling the voltage by the gate signal. Should there be pulse existing, the voltage begin to supply and current is going to be lagging caused by inductive load as demonstrated in Figure 9.

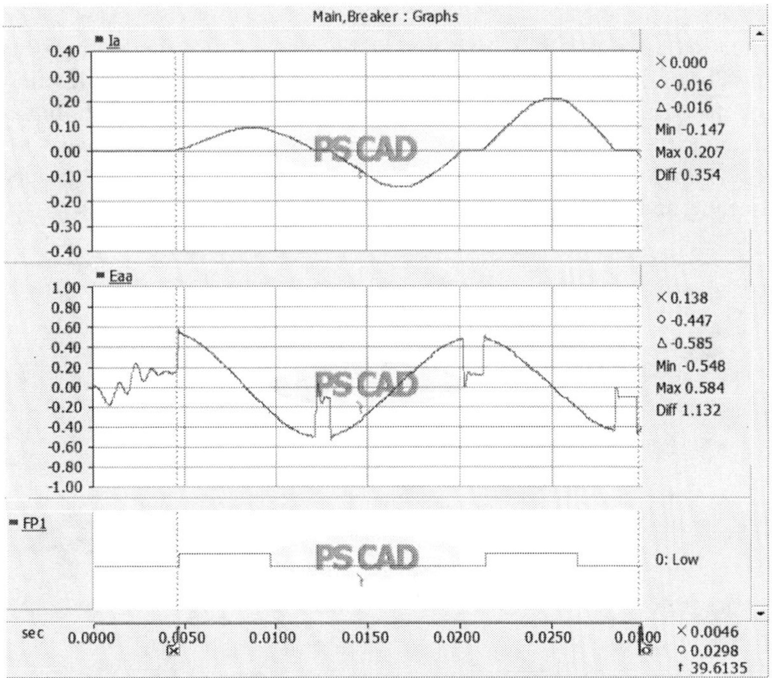

Figure 9. Triggering Time and Pulse for Thyristor.

In conclusion, what can be relate from Table 1 is that the main function of this type of soft starter is set limit to protect the motor from breakdown. The increasing of firing angle also means that the conduction angle is decreasing. The main idea to decrease the conduction angle is to decrease the voltage applied to the motor. When the average voltage applied is decreasing, the current to the motor also decreasing. Despite the goodness of current reduction, the time taken by the motor to reach its rated speed escalating as the current minimize due to the torque. The connection amongst torque and current is the fact that torque of motor deviates the square of the current.

Table 1. Current Limitation.

Firing Angle	Current Limit(kA)
40	0.45
50	0.445
60	0.44
70	0.435
80	0.42
90	0.4
100	0.36
110	0.31

120	0.26
130	0.19
140	0.12

From the fluctuation of every firing angle that being change, what the Figure 10 try to show is that when the current is being limit or reduce, the motor torque also reduce that affecting the time for the motor to reach its rated speed. This type of starter is very much suitable when the amount of current need to be limit and the time taken for the motor to reach the rated speed or the torque needed is not an issue. The current level is selected by the user to provide a suitable balance between load on the electrical system, starting torque, and acceleration time.

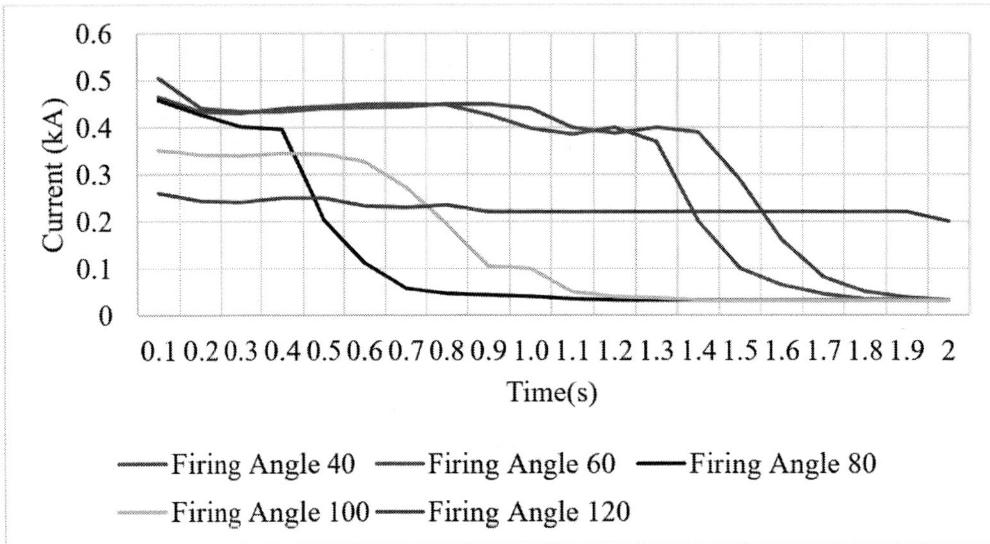

Figure 10. Graph of Current against Time.

4. Conclusion

An induction motor is well known for drawing high current during startup. The maximum spike up of the current can goes from 5 to 7 times higher than the rated current. This high current could take up to 30 to 40 cycles or higher until the current normalize again. It can be conclude that the firing angle set on 100° is the best because the inrush current can be limit to 0.36kA only and the time frame taken up to arrive at rated speed may be the fastest which is 1.1s. This type of starter is very good in reducing the current depends on application. When a certain motor which demands for large amount of torque, then the setting for current limit must be increase in order to fulfill the demand. It also very reliable to implement because the tripping of circuit breaker can be avoid when there is no transient current rise occur on the system.

References

[1] R. S. Girgis and E. G. TeNyenhuis, "Characteristics of inrush current of present designs of power transformers," in *2007 IEEE Power Engineering Society General Meeting, PES*, 2007.

[2] K. Pillay, M. Nour, K. H. Yang, D. N. D. Harun, and L. K. Haw, "Assessment and comparison of conventional motor starters and modern power electronic drives for induction motor starting characteristics," *2009 IEEE Symp. Ind. Electron. Appl. ISIEA 2009 - Proc.*, vol. 2, no. Isiea, pp. 584–589, 2009.

[3] W. X. Li, J. G. Lu, M. S. Liu, and J. Zhao, "Design of intelligent soft-start controller for induction motor," *Proc. 2004 Int. Conf. Mach. Learn. Cybern.*, vol. 2, no. August, pp. 908–912, 2004.

[4] M. G. Mehebub Alam, "Power Quality Problems and Solutions: An Overview," *Int. J. Sci. Res.*, vol. 3, no. 10, p. —, 2014.

[5] E. Asizehi, "Effect of Switching angle on Magnetizing flux and Inrush current of a Transformer," *IOSR J. Electr. Electron. Eng.*, vol. 8, no. 3, pp. 20–22, 2013.

[6] Y. Youxin, X. Zezhong, W. Yalan, Y. Peigang, and P. Wanquan, "A Soft Starter of Three Phase A synchronous Motor," pp. 563–567, 2007.

[7] H. H. Goh, M. S. Looi, and B. C. Kok, "Comparison between Direct-On-Line, Star-Delta and Auto-transformer Induction Motor Starting Method in terms of Power Quality," *Lect. Notes Eng. Comput. Sci.*, vol. 2175, no. 1, pp. 1558–1563, 2009.

[8] C. C. Yeh and N. A. O. Demerdash, "Fault-tolerant soft starter control of induction motors with reduced transient torque pulsations," *IEEE Trans. Energy Convers.*, vol. 24, no. 4, pp. 848–859, 2009.

[9] T. Sen and P. K. Bhattacharjee, "Design and Implementation of Firing Circuit for Single-Phase Converter," *Int. J. Comput. Electr. Eng.*, vol. 3, no. 3, pp. 368–374, 2011.

[10] J. Larabee, B. Pellegrino, and B. Flick, "Induction motor starting methods and issues," *Rec. Conf. Pap. - Annu. Pet. Chem. Ind. Conf.*, pp. 217–222, 2005.

[11] G. L. Arsov and S. Mircevski, "Quo vadis, thyristor?," *Proc. EPE-PEMC 2010 - 14th Int. Power Electron. Motion Control Conf.*, pp. T2-152-T2-157, 2010.

[12] J. De, A. Nied, and R. Pinho, "Study on Energy Efficiency of Induction Motor Soft-Starting with Torque Control," in *Advances in Motor Torque Control*, InTech, 2011.

Design and on-body evaluations of a low-cost wearable PIFA for on-body applications

Azadeh Ahmadihaji[1], Hadi Aliakbarian[1] , Nordiana Mohamad Saaid[2] and Ping Jack Soh[2]

[1] The Wireless Terminals Measurement Lab, K.N.Toosi University of Technology, Iran.

[2] School of Computer and Communication Engineering, Universiti Malaysia Perlis, 02600 Arau, Perlis, Malaysia.

dianams@unimap.edu.my

Abstract. A flexible low profile textile antenna based on planar inverted-F antenna (PIFA) structure is presented in this paper. This topology has been chosen due to its design simplicity, relatively large bandwidth and the existence of a full rear ground plane, which shields against the effects of the body. The antenna topology which is designed on a 4 mm-thick felt substrate is presented. Conductive textile (Aaronia-shield) is applied as the antenna's radiator and ground plane due to its light weight, low surface resistance and good conductivity. The antenna performance in free space and bent conditions in the proximity of human body are studied, followed by a numerical investigation on the influence of relative humidity on the textile antenna performance. The fabricated PIFA is assessed experimentally and is found to be in good agreements with the simulation results. Finally, the antenna is attached to a compact RF module and is evaluated on body in a practical environment. The wireless link quality is also evaluated in an indoor laboratory environment.

Keywords: Textile antennas, planar inverted-F antennas (PIFA), humidity effect , specific absorption rate, link budget.

1. Introduction

In recent years, wearable technologies are increasingly popular and are potentially useful for applications such as in security services (police or military services), rescue applications (firefighters), sports and entertainment. For instance, they can be applied to athletes for health and performance monitoring purposes. The wearable technologies can also be extended to improve the quality of life for people with long term and chronic diseases. Depending on the type of application and activities they perform, various wearable antennas have been presented with different requirements and features. However, they should generally be low-profile, lightweight, simple to fabricate and cost effective. Furthermore, they must feature simple installation and easy maintenance.

There are many types of wearable antennas designed based on conventional topologies such as planar dipoles antenna [1], monopoles [2], planar inverted-Fs (PIFAs) [3], [4], and microstrip patches [5]. These are popular antenna types due to their low cost and ease of fabrication. In this paper, a planar inverted-F antenna (PIFA) is chosen due to their small size and simple design. The availability of a rear ground plane also enables the operation in close proximity to human body. This antenna is a suitable prototype that fulfills all the mentioned characteristics. This article mainly aims at exploring PIFA from its challenges such as body effect, specific absorption rate (SAR) and humidity effect on the antenna. Thereafter, a fabricated PIFA and its measurement analysis are presented. Due to very limited works discussing antennas link quality at the end, a communication link evaluation with a base station using the proposed is performed.

This article is organized as follows. In Section 1, the design of a PIFA as a wearable antennas will be introduced. Section 2 discusses the sensitivity analysis and the PIFA performance in different conditions, such as in bent situation or in proximity of human body. The assessment of the humidity effects on antenna performance is investigated in the following section, prior to the SAR evaluations. In Section 5, the fabrication of the PIFA and the measurements of its reflection coefficients and radiation patterns is presented. Finally, the antenna is integrated with a wireless module and worn on-body to evaluate its wireless link quality in Section 6, prior to the conclusions.

2. Wearable textile PIFA design

The PIFA's structure consists of a ground plane, a radiator, a shorting wall and a feeding probe. This antenna is simulated based on [6] to achieve a large bandwidth. Its ground plane prevents serious detuning in operation in the proximity of the human body. The design calculation is based on an empirical formula in [7]. This formula, shown in equation (1), is appropriate for antennas smaller than $S_h + R_w + R_L < \lambda$. In (1) W_f is the width of the feed. The horizontal and vertical distance of a 50 Ω SMA feeding probe from the edge of the shorting plate are f_h and f_v, respectively. Other dimensions are marked in Fig.1.

$$f_c = \frac{c}{3R_w + 5.6R_L + 3.7S_h - 3W_f - 4.3\sqrt{f_h^{\,2} + f_v^{\,2}}} \qquad (1)$$

The antenna structure is a conventional simple PIFA shown in Figure 1. A 0.17 mm-thick textile, namely ShieldIt is applied as antenna's conductive part. A 4 mm-thick felt, with a relative permittivity of $\varepsilon_r = 1.45$ and $\tan\delta = 0.044$, is used as the substrate. In order to simplify the analysis, the conductive textile is considered as a good metal (with a conductivity, $\sigma = 5.9\times10^7$ S/m). The antenna dimensions are given in Table 1.

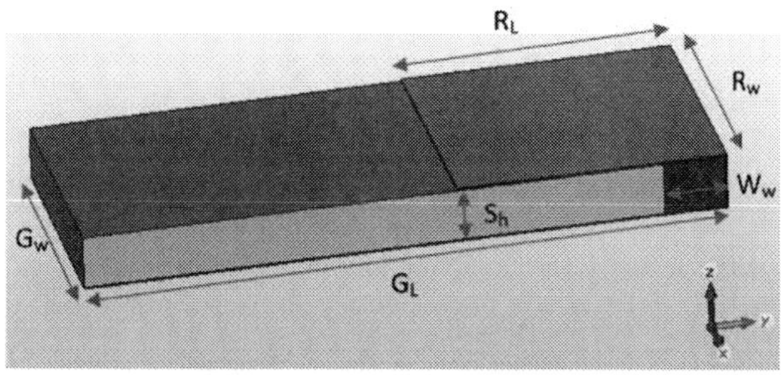

Figure 1. Antenna structure [6].

Table 1. Summary of the dimension of proposed PIFA.

Parameter	W_w	f_v	f_h	R_W	R_L	G_W	G_L
Size (mm)	5.0	9.5	8.5	19	21	19	50

3. Sensitivity analysis

The first analysis considers the effects of the body tissue presence on the antenna proximity, which is the most common factor affecting the antenna performance. For this analysis, it is assumed that the antenna has been mounted on an arm model. Simulations were performed using CST Microwave Studio (MWS). A four-layered arm model is used in the analysis: bone, muscle, fat and skin (listed from the inner to outer layer). A 15 mm gap between the antenna and human tissue is considered. The thickness of each layer is listed in Table 2 and the model is shown in Figure 4(a). The permittivity, conductivity and density of layers are obtained from CST MWS material library. The length of tissue model is assumed to be infinite due to the use of the open boundary condition.

The human body is an electrically lossy medium with different complex dielectric constants for each tissue. In Figure 2, the simulation result of the reflection coefficient (S_{11}) of the cases with and without the tissues in flat conditions are compared. A minor difference between the two cases is shown, where the presence of the high dielectric constant of the tissue resulting in an extra resonance at 2.26 GHz and shifted the center frequency of 2.45 GHz to about 2.62 GHz. The -10 dB impedance bandwidth of the antenna, however, increased from 700 MHz to 750 MHz approximately. The radiation patterns of antenna with and without human body are compared in Figure 3, where it is observed that the radiation pattern is directed away from the body due to the presence of body tissues.

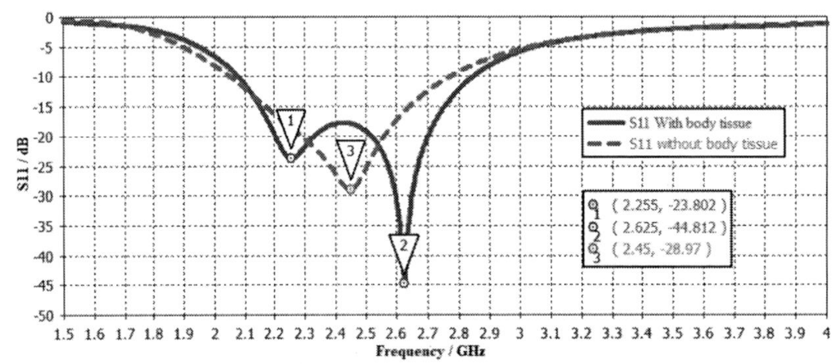

Figure 2. Reflection coefficients of the PIFA.

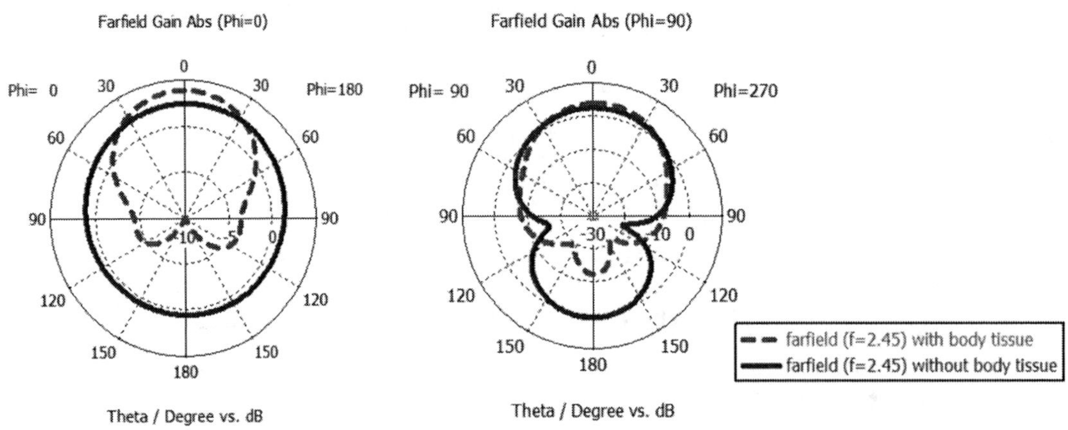

Figure 3. Radiation patterns of antenna at the $\varphi = 0°$ and $\varphi = 90°$ cuts.

The antenna is then placed bended around an arm using radii given in Table 2, as shown in Figure 4(b). The reflection coefficients and radiation patterns of the bended antenna are depicted in Figures 5 and 6. As seen from Figure 5, the changes in the radius of the arm did not result in considerable resonant frequency shift, but instead, having using the antenna around a smaller arm radius reduces the antenna bandwidth. Comparisons are then made between the flat case and radii of 67.5 mm and 40.5 mm, which are near to realistic radii of the arms of normal adults. On a 67.5 mm arm radius, the antenna resonates at 2.35 GHz, with bandwidth of 750 MHz while bending on a 40.5 mm arm radius resulted in a slightly lower resonant frequency of 2.32 GHz and a bandwidth of 650 MHz. Note that a 750 MHz bandwidth for the antenna operating in flat condition is observed previously. The dual resonance characteristic remains for 67.5 mm arm radius, but disappeared in the case of the 40.5 mm arm. The antenna's maximum gain is 3.187 dB and 0.804 dB for larger and smaller arm radii, respectively. This is mainly due to the degradation in radiation efficiency. In general, the bigger the arm radius, the better the antenna performance, as shown in Figure 6.

Table 2. Multilayered human model [8].

Layer	Air Gap	Skin	Fat	Muscle	Bone
Thickness (mm)	15	2	10	28	12.5

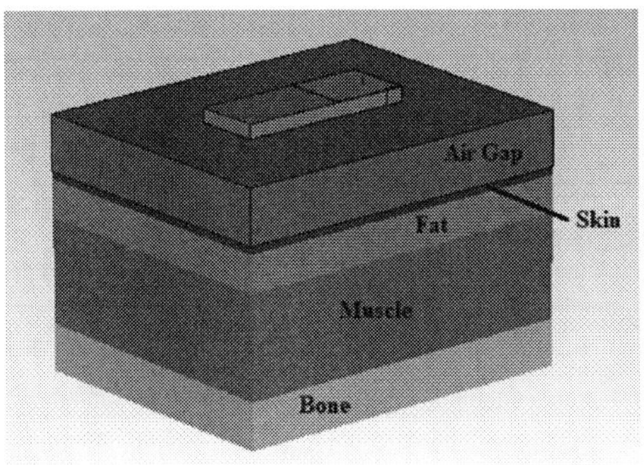

(a)

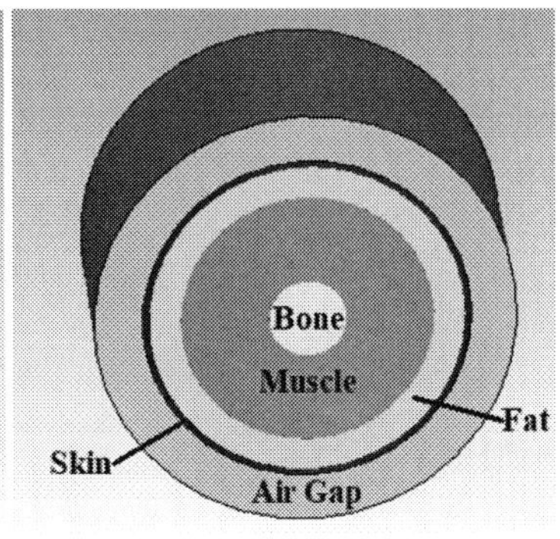

(b)

Figure 4. 3D view of the two phantom models: (a) flat (b) cylindrical arm.

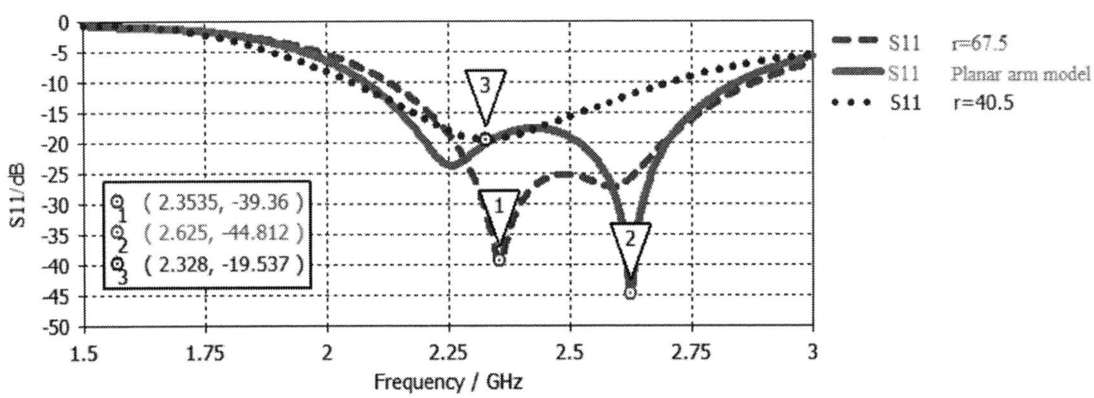

Figure 5. Comparison of the reflection coefficient (S_{11}) simulations of the bent PIFA antenna on the human arm with radii of R = 67.5 mm and 40.5 mm in comparison with planar condition.

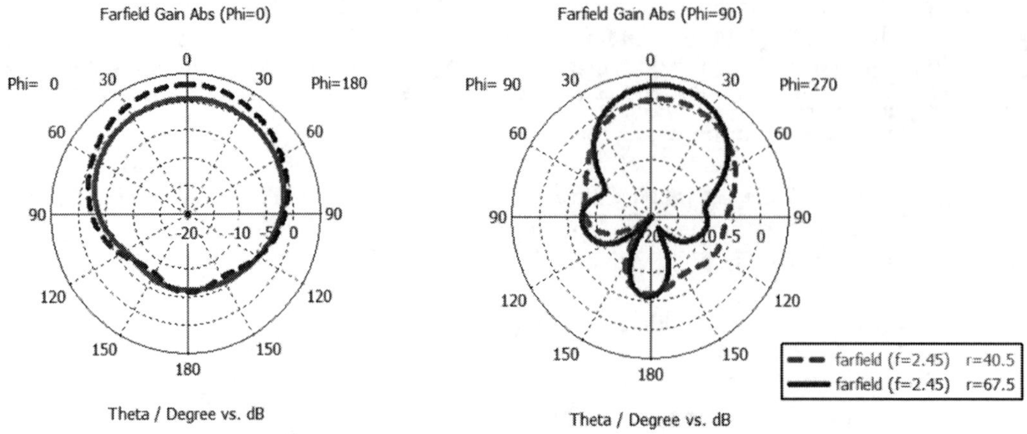

Figure 6. Simulations of radiation pattern of the bent antenna in the presence of human body with radii of 67.5 and 40.5 mm at: a) phi=0° cut; and b) phi=90° cut.

The reflection coefficients for the antenna in flat and bent conditions evaluated in free space is shown in Figure 7. In comparison to evaluating on human tissue model in Figure 6, these two results illustrate inconsistencies in terms of resonant frequencies. However, a clear trend in terms of gain can be seen in Figure 8. The antenna resonates at 2.54GHZ with bandwidth of about 730 MHz and the gain of 2.465 dB when bent with 67.5 mm of radius. On the other hand, for 40.5 mm radius, the resonant frequency, bandwidth and gain is about 2.3 GHz, 690 MHz and 2.389 dB, respectively. It is seen from these results that the reduction in the bending radii decreases the antenna gain.

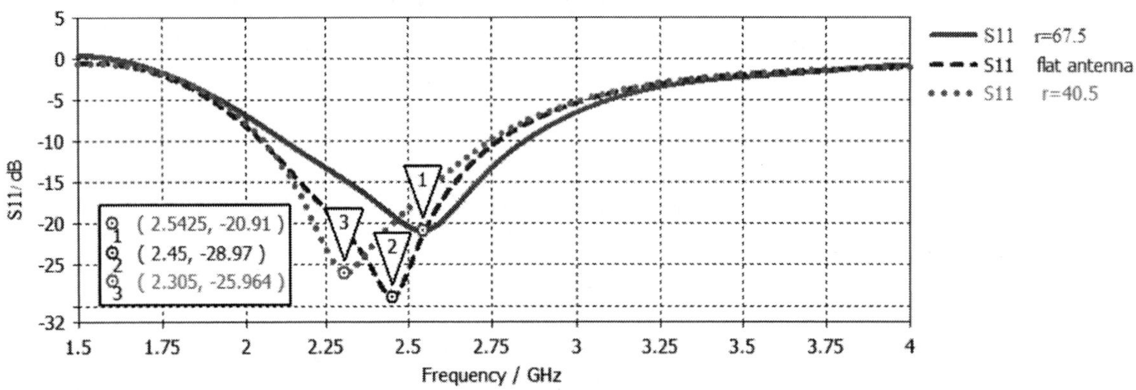

Figure 7. Reflection coefficients of the antenna in bent condition in free space (without human body).

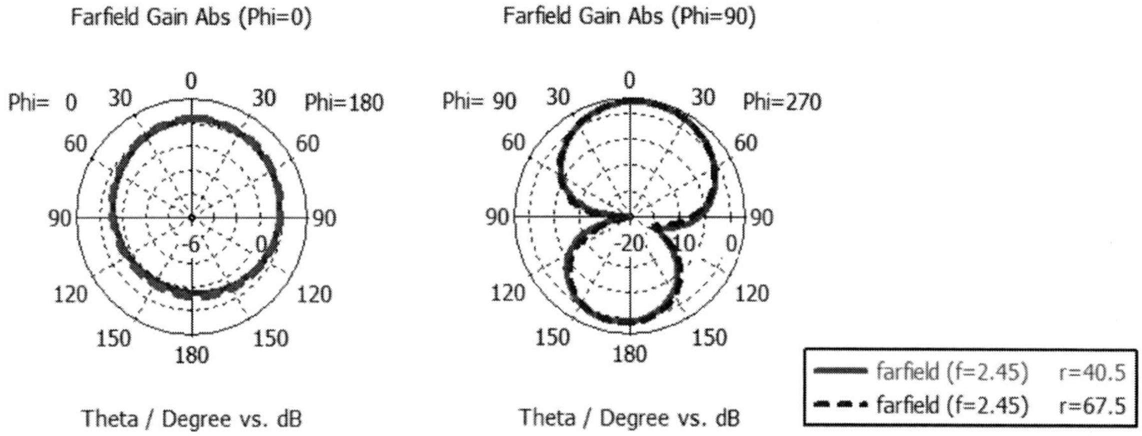

Figure 8. Radiation patterns of the bent antennas in free space and with radii of 67.5 mm and 40.5 mm at: a) phi=0° cut; and b) phi=90° cut.

Next, SAR levels for the bent PIFAs are illustrated in Figure 9(a). In this simulation, an input power of 250 mW is used, resulting in a SAR value of 1.97 W/kg. This does not exceed the ICNIRP standard threshold of 2W/kg averaged over 10g of tissues. Moreover, the 250 mW of input power also meets the FCC standard threshold for power transfer in the 2.43 GHz band, which is also limited to 1 W. In [9], it was reported that the measured SAR for the slotted PIFA (SPIFA) when placed at distance of 10 mm from the body in flat condition using an input power of 100 mW at 2.45 GHz is about 0.5 W/kg. Similarly, simulated SAR for the proposed PIFA under bent condition is 0.807 W/kg when evaluated under the same conditions, indicating a reasonable agreement. The small discrepancies may be due to the slight difference in the evaluation conditions, in which in [9], it is evaluated in flat condition, but in this work, it is evaluated in bend conditions.

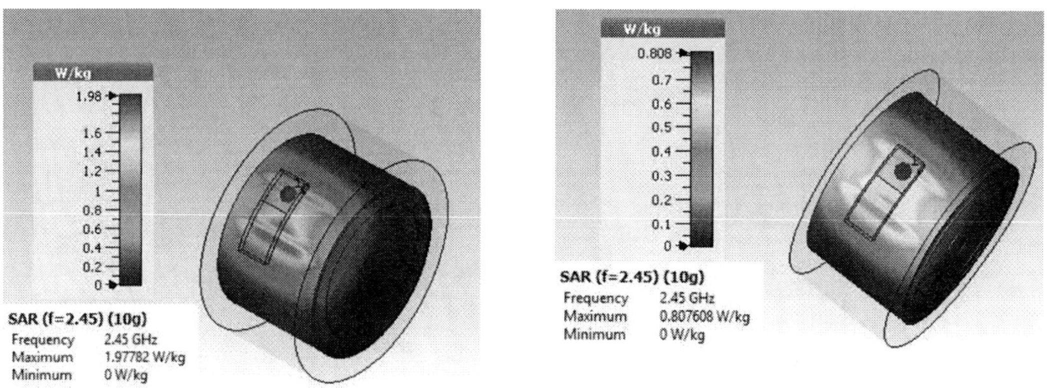

(a) (b)

Figure 9. Simulated SAR of the bent PIFA with (a) input power=250mW (b) input power=100mW

4. Effects of Humidity on the Antenna Performance

Generally, humidity or moisture will potentially affects textile antenna performance. Due to the high permittivity of water ($\varepsilon_r = 78$ at 2.45 GHz and 25°C), if it penetrates into a substrate, it may lead to an increase in permittivity. Therefore, to assess the severity of the humidity effect towards antenna performance, a simulation with a varied relative permittivity is performed. It is also assumed that all other properties of the substrate remain unchanged. Note that the inital relative permittivity of this substrate is $\varepsilon_r = 1.45$. Figure 10 shows that as the relative permittivity increases, the resonant frequency shifts towards lower frequencies. On the other hand, as the permittivity increases, the $|S_{11}|$ performance is gradually deteriorated such that with a permittivity of about 2, the reflection coefficient degraded to less than -10 dB. This moisture effect simulation results are in good agreement with the measurements performed in [10]. Therefore, the humidity effect on the substrate loss tangent of textile materials can be further investigated in order to estimate this effect more accurately.

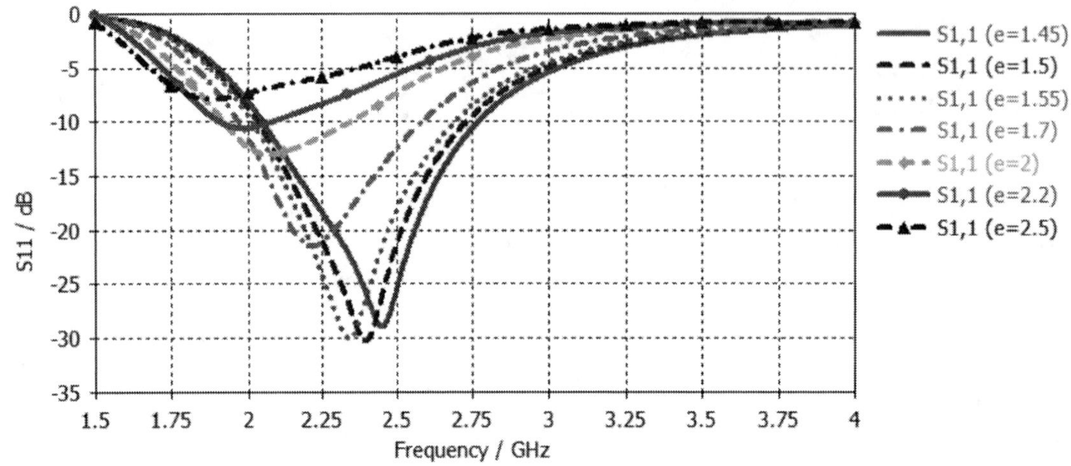

Figure 10. Effect of permittivity change (humidity) on reflection coefficient of PIFA.

5. Fabrication and measurements

To fabricate the antenna, felt is chosen as its substrate, while a conductive textile called Aaronia-shield is used for the conductive part. The latter is used as the antenna's radiator, ground plane and shorting wall, as it is light weight (about 34g/m^2), low surface resistance (lower than 0.1Ω/sq) and good surface conductivity. The dimensions of the PIFAs are as listed in Table 1. Both the substrate and conductive textiles are sewn to connect them. A SMA connector is used to feed the antenna by soldering to the ground plane. Figure 11 shows the fabricated PIFA. A comparison between simulated and measured reflection coefficients and radiation pattern of the antenna are shown in Figure 12 and 13 respectively.

(a) (b)

Figure 11. (a) Upper and (b) front view of antenna prototype.

As depicted in Figure 12, the antenna operation is centered at 2.5 GHz with a good bandwidth of 750 MHz from 2.25 GHz to 3 GHz. However, in Figure 13, the radiation patterns show disagreement between simulations and measurements. This could be due to slight damage to the substrate during the soldering process, and inaccurate values of the electrical properties of the substrate (permittivity and loss tangent). It was assumed initially that these properties are as listed in [6], but are found to be different in practice. Moreover, the sewing process also causes inaccuracy towards the measurement data.

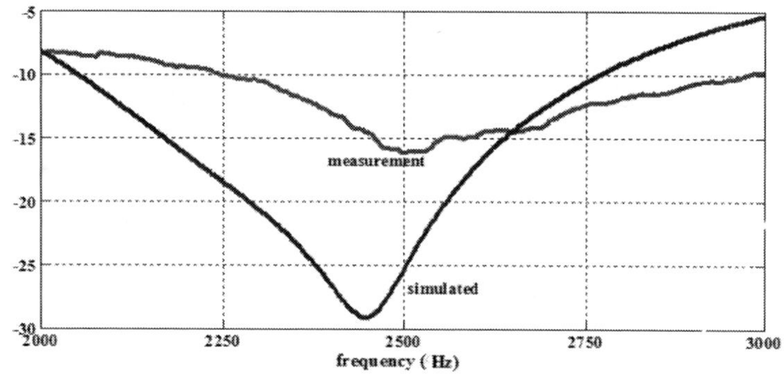

Figure 12. Comparison between simulated and measured reflection coefficients of the antenna.

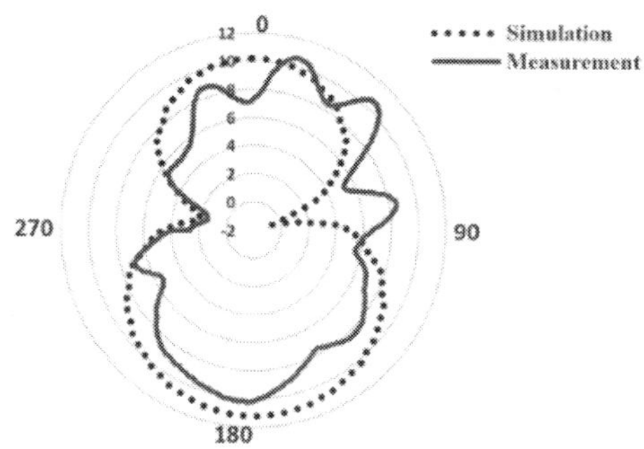

Figure 13. Comparison between simulated and measured normalized radiation patterns of the PIFA.

6. Wireless Link Evaluation

From the previous section, it is validated that the antenna is operated from 2.25 GHz to about 3 GHz with less than -10 dB of reflection coefficient. To further evaluate the performance of the antenna in practice, a low-cost NRF24L01 module from Nordic Semiconductor with a nominal transmission range of 1800 m is used. This module is used to collect health data using a series of small integrated thermal sensors, heart beat rate sensors and accelerometers to communicate with the access point located at a distance away in an indoor environment. To obtain real-time test data in a personal computer, a converter was used to connect the RF-antenna system to the computer via its USB port. The fabricated PIFA is then connected to the NRF24L01 module and the USB converter board as presented in Figure 14. The base station (in the form of a laptop) is connected to a monopole antenna to communicate with the wearable system as shown in Figure 15(a). To assess the quality of the link, the wearable system is mounted on a human body to transmit repeated simple bits. The user of this wearable system then moves and the wireless link quality is tracked until it is totally disconnected. The measurement is performed in an indoor laboratory environment and the quality of the wearable system is compared with the case of a simple monopole replacing PIFA when worn on body. It is observed that quality of the link for both are similar with none of them being superior. Next, the wireless link quality is also tested when the base station is moved. The connectivity range of the link is then decreased to around 100 meters due to severe multipath of the indoor environment. The communication between the on-body wearable system to the base station is illustrated in Figure 15(b). The ultimate final goal of this research is to be able to fully integrate the wearable module with the antenna and sensors to track human health, which is aimed to be prototyped in the near future.

IOP Conf. Series: Materials Science and Engineering **767** (2020) 012026 doi:10.1088/1757-899X/767/1/012026

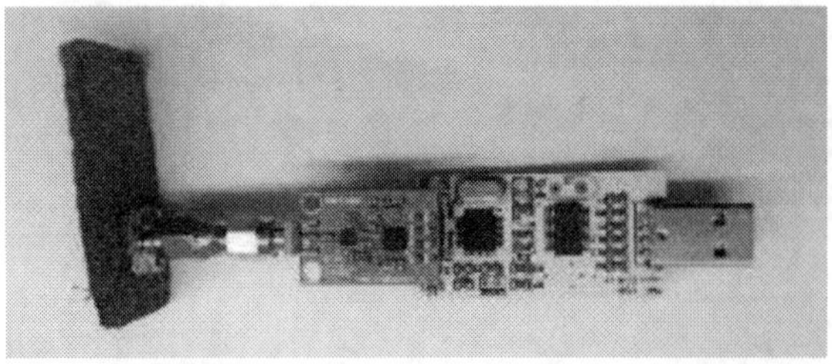

Figure 14. Wearable PIFA connected to a NRF24L01 and driver board.

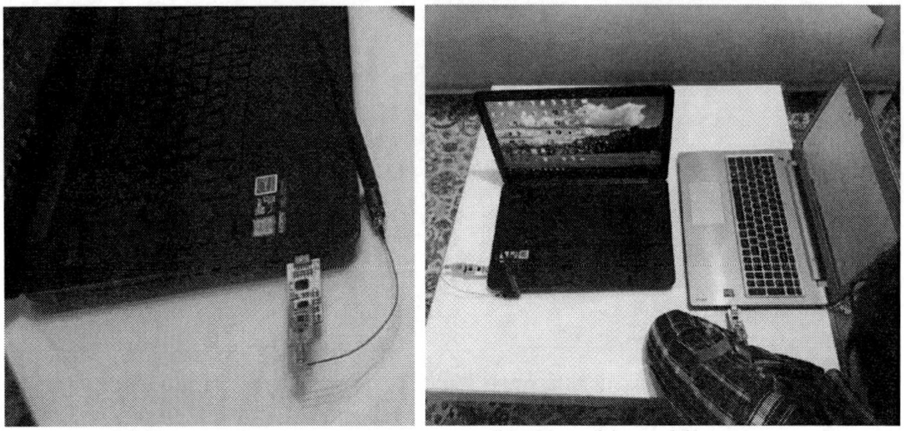

(a) Base station (b) Wearable mobile system communicating with the base station

Figure 15. (a) The fixed base station for the test and (b) The on-body wearable mobile system while connected to a laptop to track communication quality.

7. Conclusion

A planar inverted-F antenna designed using textile materials is presented in this paper. The design procedure and method of estimating the initial dimensions of this PIFA is discussed, followed by its evaluation in planar and bent conditions, both in free space and on-body. To ensure the safety level of the proposed antenna, specific absorption rate of this antenna is calculated when placed in the proximity of two types of human phantoms, a three-layered rectangular body tissue model, and a similar three layered cylindrical model replicating an arm. For the latter, two cylindrical radii were used, 67.5 and 40.5 mm to assess the effects of the severity of the bending on the antenna performance. Next, the antenna performance was also evaluated with various humidity levels of the substrate. This is done by changing the relative permittivity of the substrate and assessing the reflection coefficient of the antenna. The optimized antenna was then fabricated and measured, and then compared with simulations performance, in terms of reflection coefficient and radiation patterns. Finally, the proposed antenna is integrated with a wireless module to evaluate its performance in practice when worn on the human body. The quality of the wireless link performed in an indoor laboratory environment is also compared with the performance from a monopole antenna, indicating similarity in performance.

IOP Conf. Series: Materials Science and Engineering **767** (2020) 012026 doi:10.1088/1757-899X/767/1/012026

References

[1] J. Roh, Y. Chi, J. Lee, Y. Tak, S. Nam and T. J. Kang, "Embroidered Wearable Multiresonant Folded Dipole Antenna for FM Reception," in *IEEE Antennas and Wireless Propagation Letters*, vol. **9**, pp. 803-806, 2010.

[2] A Rahim, Hasliza & AbdulMalek, MohamedFareq & Adam, I & Ahmad, Sahadah & Mohd Hashim, Nur Baya & S Hall, P. (2012). Design and Simulation of a Wearable Textile Monopole Antenna for Body Centric Wireless Communications.

[3] P. J. Soh, G. A. E. Vandenbosch, S. L. Ooi and N. H. M. Rais, "Design of a Broadband All-Textile Slotted PIFA," in *IEEE Transactions on Antennas and Propagation*, vol. **60**, no. 1, pp. 379-384, Jan. 2012.

[4] Il Kwak, Sang, et al. "Design of PIFA with metamaterials for body-SAR reduction in wearable applications." *IEEE transactions on electromagnetic compatibility* 59.1 (2016): 297-300.

[5] C. Hertleer, A. Tronquo, H. Rogier, and L. Van Langenhove, "The use of textile materials to design wearable microstrip patch antennas," *Textile Research Journal*, vol. **78**, no. 8, pp. 651–658, August 2008.

[6] Soh, Ping Jack, et al. "Characterization of a plain broadband textile PIFA."*Radioengineering* 20.4 (2011): 718-725.

[7] Chattha, H.T., Huang, Yi., Xu, Z., Yang, L. An empirical equation for predicting the resonant frequency of planar inverted-F antennas. *IEEE Antennas and Wireless Propagation Letters*, 2009 vol. **8**, p. 856 - 860.

[8] Hebelka, Vladimir. "Planar antenna in proximity of human body models." In *Antennas and Propagation* (EuCAP), 2013 7th European Conference on, pp. 3309-3311. IEEE, 2013.

[9] Soh, Ping Jack, et al. "Specific absorption rate (SAR) evaluation of textile antennas." *IEEE Antennas and Propagation Magazine* 57.2 (2015): 229-240.

[10] Hertleer, Carla, et al. "Influence of relative humidity on textile antenna performance." *Textile Research Journal* (2009).

IOP Conf. Series: Materials Science and Engineering **767** (2020) 012027 doi:10.1088/1757-899X/767/1/012027

Digital photonics parity bit error detection system

M. Rakib Uddin and F. K. Law

Eelectrical and Electronic Engineering Programme Area, Faculty of Engineering
Universiti Teknologi Brunei

rakib.uddin@utb.edu.bn

Abstract. The optical communication medium is emerging to be one of the potential replacements to the current electronic systems widely used today. This is due to its much higher data transmission with extremely low loss properties, not to mention that it is almost invulnerable to electromagnetic interference. In this paper we demonstrated our proposed design of a digital photonic parity bit error detection system, one of the fundamental digital electronic circuits used during long-range digital signal transmission. Silicon photonic micro-ring resonators are used as its core component, where each of the rings operates as a digital logic XOR mode, by taking advantage of the resonance shifting properties of the silicon waveguide. Dynamic response characterization has been carried out by simulating the proposed circuits at the data rate of 1Gbps, with the data sampling rate of 1.6THz, and clear timing waveform was generated to show that the proposed circuit operates as the parity bit error detection system.

1. Introduction

Digital electronic circuits have become one of the most important aspects of any CPU builds. However, with the exponential rise in technological development, many believed that electronic circuits have reached their limit, as the data transmission inside the circuit is limited to the speed of electrical signals traveling to the physical material. Optical channels, however, allows for the data transmissions at higher data rate (at the speed of Gbps), with lower channel transmission loss and relatively small crosstalk as well as electromagnetic interference [1, 2]. Photonic integrated circuit (PIC) is a circuit in which photonic devices used are built by taking advantage of the currently available CMOS fabrication technology, thus making it available for design and extensive research to be carried out. It also enables the circuit to be built based on different kinds of materials, in which Lithium Niobate (LiNbO3), as well as Galium Arsenide (GaAs), have been demonstrated [3, 4].

Amongst many of the digital logic circuits, combinational logic circuits are a group of logic circuits that is time-independent and can produce logical output operation instantaneously, with no feedbacks needed. Combinational logic circuits are commonly classified into three categories, arithmetic, and logical functions, data transmission and code converters. Amongst many combinational logic circuits, parity bit generator, as well as checker, are the two logic circuits used widely in binary serial transmissions. A parity bit is generated and embedded together with the data bits and sent to the receiving node. At the receiving end, number of bits are counted to detect errors in transmission, depending on the system, whether it is odd or even parity bit, errors in transmission will be detected. This has been researched and demonstrated in quantum dot applications for low power nanocommunications [5-7].

Several breakthroughs have been demonstrated in silicon photonics, where digital logic mode operations have been achieved by using the single mode Fabry-Perot laser diode (FP-LD) [8, 9], however, such designs cannot be incorporated into on-chip CMOS level circuitry, thus micro-scale photonic micro-ring resonators are being implemented into digital logic circuits. We have also shown that one silicon micro-ring resonator can be used and configured to operate as digital logic AND, NAND, OR as well as NOR modes [10]. Several logic gates and circuits have also been demonstrated, where two symmetric micro-ring resonators are used to design directed OR/NOR and AND/NAND logic circuits [11], micro-ring resonators-based D type flip-flop [12], half adder [13], bit magnitude comparator and digital encoders [14, 15]. Micro-ring resonators have also been used in communication circuits such as WDM systems [16].

This paper presents the principal design of the digital photonic even parity bit checker circuit, by using silicon micro-ring resonator as its core element. The ring waveguide uses electro-optic effect based on free carrier plasma diffusion PIN diode setup. This work also details the time-varying simulation for both circuits at the data rate of 1Gbps in order to demonstrate that the proposed bit checker is working as desired.

2. Design Principle of the Proposed Error Detection Circuit

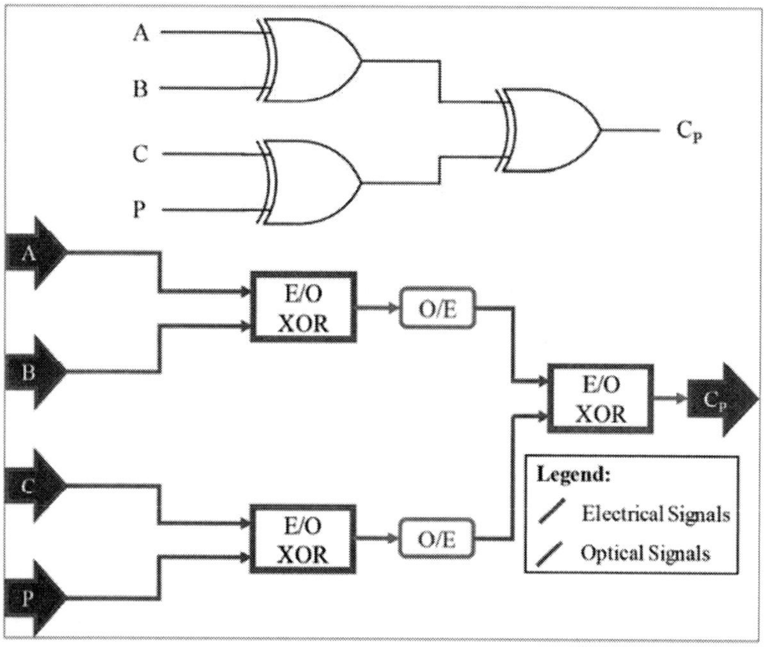

Figure 1. A logic circuit (on top) and proposed design (on bottom) of the bit checker.

A 3-bit even parity error detection circuit is basically a digital parity bit checker that has a total of three digital logic XOR gates, where the first two XOR gates output is then fed into the third XOR gate, to generate a parity check (C_P) bit. In our proposed bit generator, we replaced the digital logic gates with silicon micro-ring resonators operating as digital logic XOR mode as shown in Figure 1. The equation representing the operation of the bit checker output is $C_P = (A \oplus B) \oplus (C \oplus P)$ and the entire operation bit by bit for the circuit is as shown in Table 1, where C_P is only at digital logic state '1' when the total number of logic state '1's for the received four bits is an odd number, indicating that the received information contains error.

Table 1. Even Parity Bit Checker Truth Table.

4-bits Message Received				Parity Check
A	**B**	**C**	**P**	**C_P**
0	0	0	0	0
0	0	0	1	1
0	0	1	0	1
0	0	1	1	0
0	1	0	0	1
0	1	0	1	0
0	1	1	0	0
0	1	1	1	1
1	0	0	0	1
1	0	0	1	0
1	0	1	0	0
1	0	1	1	1
1	1	0	0	0
1	1	0	1	1
1	1	1	0	1
1	1	1	1	0

Notice that in Figure 1, components labeled O/E is shown, which is the optical to electrical converter. This circuit converts the optical output power of 5dBm to electrical signal of 0.2V linearly, which can be achieved using PIN photodetector connected to electrical attenuator. PIN photodetector converts the optical signal to electrical signal linearly and electrical attenuator lowers the voltage to the desired level, which in this case is 0.2V.

3. Design Architecture of the Silicon Micro-Ring Resonator

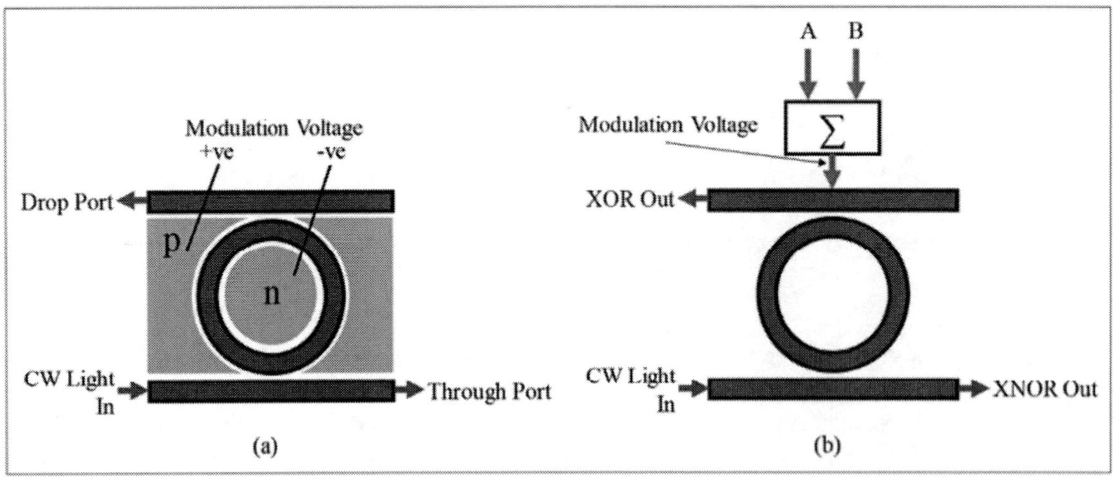

Figure 2. (a) Architecture of silicon micro-ring resonator. (b) Schematic setup for XOR and XNOR mode operation.

The silicon micro-ring resonator we used is the PIN diode structure ring waveguide to facilitate the shift in resonance properties of the silicon material [17]. The ring is constructed with a ring length of 31.8um, with its initial effective index of 2.77 and its group index of 3.961. The coupling coefficient is found to be 0.5. The ring waveguide is doped accordingly so that free carriers can be injected into the waveguide's core, thus effectively changing its effective index. Initial simulation for the ring shows that the ring propagation loss is 5.93dB/cm. The modulation voltage polarity is set up so that the positive terminal is connected to the p-doped side of the ring waveguide, while the electrical negative terminal is connected to the other n-doped side of the ring waveguide, as shown in Figure 2(a). The change in effective index versus the modulation voltage applied to the ring is as shown in Figure 3, where there is no major change in effective index from 0V up to 0.7V, but there is a linear change in effective index from 0.8V up to 1.4V [18]. This change in effective index profile can then be used to design the ring so that it operates as digital logic XOR mode.

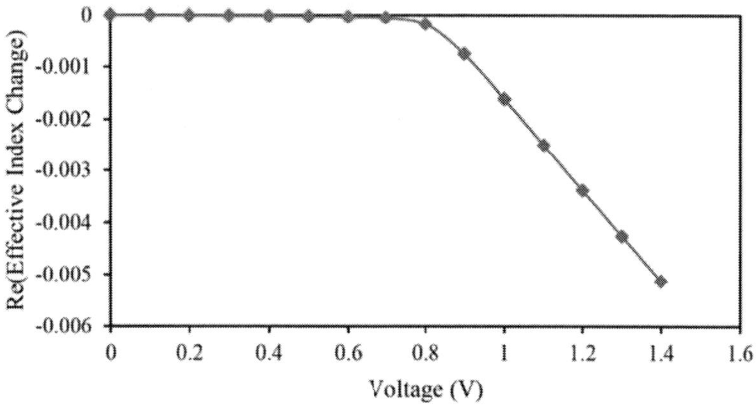

Figure 3. Change in effective index versus modulation voltage applied to the ring.

However, since there are two electrical input signals to be fed into the ring resonator, we used electrical voltage adders to add the two electrical signals and finally fed the signal to the ring, as shown in Figure 2(b). Based on our previous research, we can obtain XOR mode operation by taking the drop port output and applying three voltage levels, 0.9V, 1.1V and 1.3V so that the resonance spectra for the drop port looks like shown in Figure 4. The working wavelength we selected is the resonance wavelength when the voltage applied is 1.1V, which is 1550.3nm. When the voltage applied is 0.9V, the optical output power at working wavelength is at logic state '0' (optical power = -11.7dB). When the voltage applied is 1.1V, the optical power at working wavelength is at logic state '1' (optical power = -2.15dB), and when the voltage is 1.3V, the resonance shifted even more, thus at working wavelength, the output is at logic state '0' again (optical power = -11.2dB). With these voltages, together with a constant voltage source of 0.9V, and each of the input shown in Figure 3(b) is at amplitude of 0.2V, we can achieve the XOR mode operation with a single silicon micro-ring resonator.

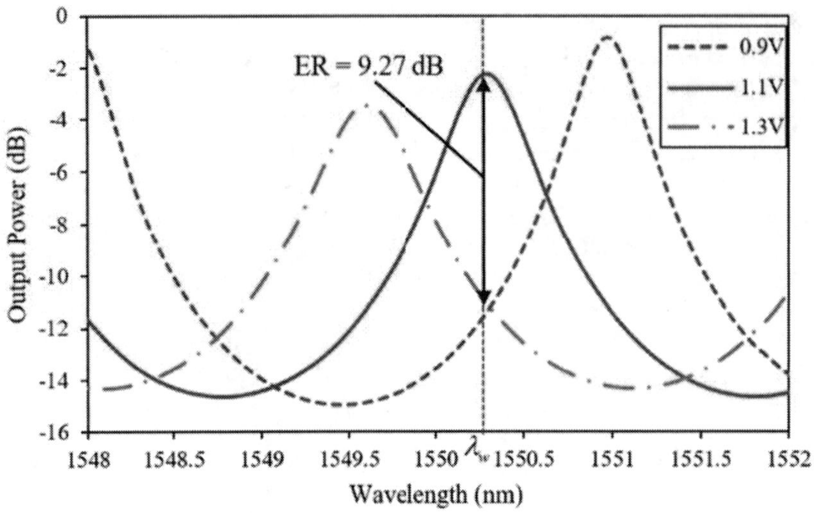

Figure 4. Resonance spectra at the drop port of the ring resonator at specific modulation voltage.

4. Dynamic Response Test

The proposed circuits were then tested in one of the commercially available photonic simulation software, Lumerical INTERCONNECT. The ring resonators are injected with a single wavelength light source of 1550.3nm with the optical power of 5dBm. Four digital bit signal generators are used to generate predetermined bit information with the amplitude of 0.2V, where the input A received the repeating $(0000000011111111)_2$, while input B received a repeating $(0000111100001111)_2$ with input C is injected with a repeating information signal of $(0011001100110011)_2$ and finally the parity bit input signal is generated with repeating information of $(0101010101010101)_2$. A single PIN photodetector is used to convert the optical output signal P to an electrical signal, in which 1dBm of optical light power is linearly converted into 1V of electrical voltage signal. The entire simulation setup for the bit checker circuit is as shown in Figure 5.

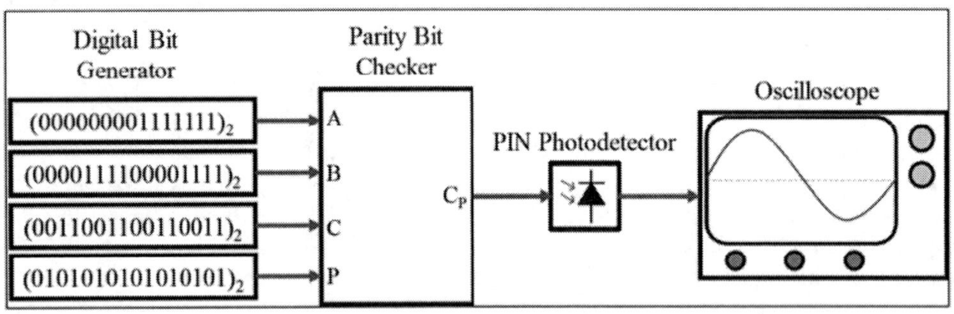

Figure 5. Simulation block setup for the bit checker.

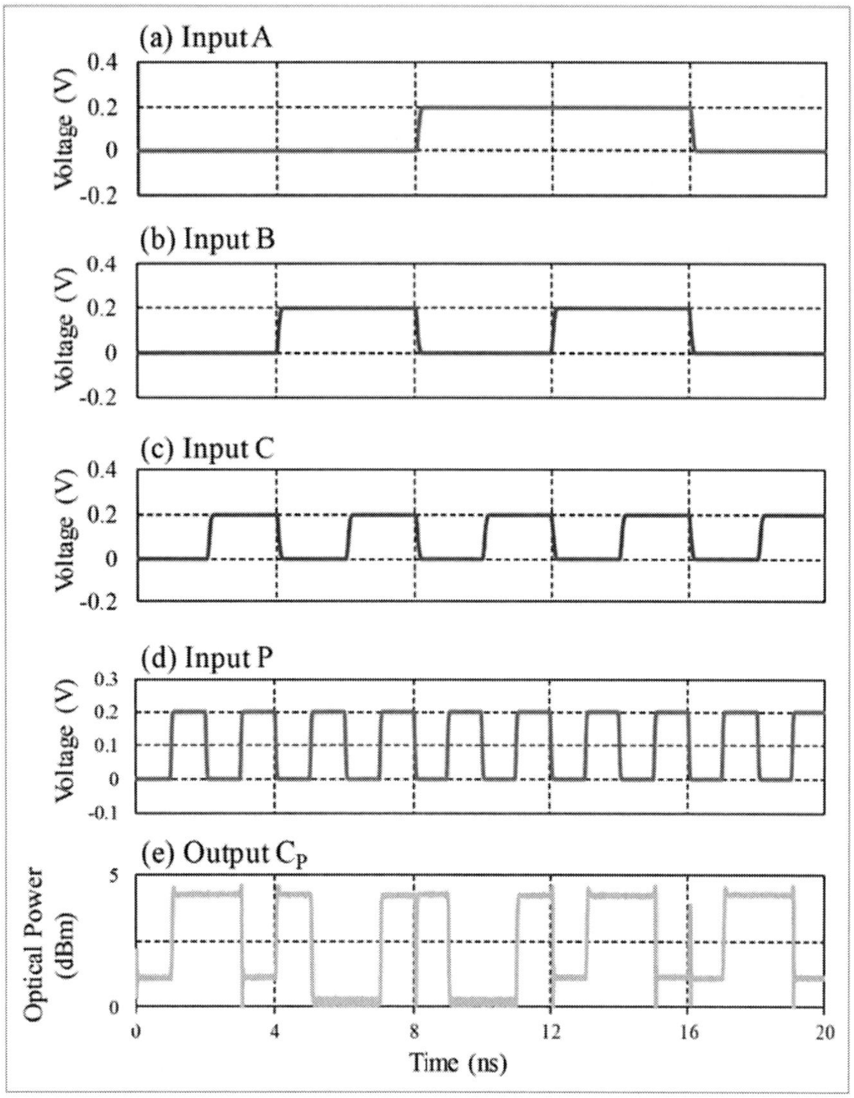

Figure 6. Timing waveform for the proposed bit checker.

Figure 6 shows the timing waveforms for all the input as well as output signals received from the simulation that have been carried out with the setup in Figure 5. The time window for the simulation is 20ns with the data injection rate of 1Gbps in order to demonstrate our proposed circuit operation. Predetermined electrical inputs are as shown in (a), (b), (c) and (d), while the parity bit error signal was generated as shown in (e), where each data iteration is in accordance with the truth table shown in Table 1. Based on the generated output, we measured the highest logic '1' power level to be 4.3dBm while the highest logic '0' power level to be 1.3dBm. If we are to consider the logic '1' power level to be within 4dBm to 5dBm and logic '0' power level to be within 0dBm up to 1.3dBm, then the generated result is as desired for our target of designing the parity bit checker error detection circuit.

5. Conclusion

This paper has presented the circuit design of a digital photonic parity bit error detection system with silicon micro-ring resonator as its base element. We have shown that the micro-ring resonator used is following the free carrier plasma diffusion electro-optic effect, where no change in the effective index when the voltage applied to it is within 0V up to 0.7V, and a linear change is observed from 0.8V up to 1.4V. We have also detailed the use of the micro-ring resonator to operate as a digital logic XOR mode by using electrical adder as an extra essential component. This XOR mode operated micro-rings were then used and replaced the logic gates in the digital bit checker. Dynamic response characterization has also been carried out with the predetermined digital input signals at the data rate of 1Gbps in order to demonstrate that our proposed circuit operates as digital parity bit checker, which yields the clear timing waveform results following the operation summarized by the digital bit checker truth table.

References

[1] H. J. Caulfield and S. Dolev, "Why future supercomputing requires optics," *Nature Photonics*, vol. 4, no. 5, p. 261, 2010.

[2] J. Hardy and J. Shamir, "Optics inspired logic architecture," *Optics Express*, vol. 15, no. 1, pp. 150-165, 2007.

[3] M. J. Kobrinsky *et al.*, "On-Chip Optical Interconnects," *Intel Technology Journal*, vol. 8, no. 2, 2004.

[4] D. A. Miller, "The role of optics in computing," *Nature Photonics*, vol. 4, no. 7, pp. 406-406, 2010.

[5] F. Ahmad, P. Z. Ahmad, and G. M. din Bhat, "Design and analysis of odd-and even-parity generators and checkers using Quantum-dot Cellular Automata (QCA)," in *2015 2nd International Conference on Computing for Sustainable Global Development (INDIACom)*, 2015, pp. 187-194: IEEE.

[6] J. C. Das and D. De, "Quantum-dot cellular automata based reversible low power parity generator and parity checker design for nanocommunication," *Frontiers of Information Technology & Electronic Engineering*, vol. 17, no. 3, pp. 224-236, 2016.

[7] M. Mustafa and M. Beigh, "Design and implementation of quantum cellular automata based novel parity generator and checker circuits with minimum complexity and cell count," 2013.

[8] B. Nakarmi, M. Rakib-Uddin, T. Q. Hoai, and Y. H. Won, "A simple controlled all-optical on/off switch using gain modulation in single mode FP-LD," *IEEE Photonics Technology Letters*, vol. 23, no. 4, pp. 212-214, 2010.

[9] M. R. Uddin, J. Lim, Y. Jeong, and Y. Won, "All-optical digital logic gates using single-mode Fabry–Perot laser diode," *IEEE Photonics Technology Letters*, vol. 21, no. 19, pp. 1468-1470, 2009.

[10] F. Law, M. R. Uddin, H. Hashim, and Z. Hamid, "Simulation and demonstration of electro-optic digital logic gates based on a single microring resonator," *Optical and Quantum Electronics*, vol. 49, no. 12, p. 413, 2017.

[11] Y. Tian *et al.*, "Proof of concept of directed OR/NOR and AND/NAND logic circuit consisting of two parallel microring resonators," *Optics letters*, vol. 36, no. 9, pp. 1650-1652, 2011.

[12] F. K. Law, M. R. Uddin, and H. Hashim, "Photonic D-type flip flop based on micro-ring resonator," *Optical and Quantum Electronics*, vol. 50, no. 3, p. 119, 2018.

[13] F. Law, M. R. Uddin, and N. M. Masir, "Micro-Ring Resonator based Digital Photonic Half Adder Demonstration," in *Laser Science*, 2018, p. JTu2A. 80: Optical Society of America.

[14] F. Law, M. R. Uddin, and N. Musyiirah, "Design, Simulation and Analysis of a 4 to 2 Digital Electro-Optic Encoder," in *2018 IEEE Student Conference on Research and Development (SCOReD)*, 2018, pp. 1-4: IEEE.

[15] F. K. Law, M. R. Uddin, H. Hashim, and Y. H. Won, "Demonstration of photonic micro-ring resonator based digital bit magnitude comparator," *Optical and Quantum Electronics*, vol. 51, no. 1, p. 1, 2019.

[16] N. M. H. Masri, M. R. Uddin, and L. F. Kui, "WDM system based on radius variation of photonic microring resonators," in *Research and Development (SCOReD), 2017 IEEE 15th Student Conference on*, 2017, pp. 243-246: IEEE.
[17] F. K. Law, M. R. Uddin, and N. M. H. Masri, "Simulation and Analysis of a Silicon PIN Waveguide Based on Electro-Optic Carrier-Depletion Effect," *International Journal of Nanoelectronics & Materials,* vol. 11, no. 4, pp. 499-515, 2018.
[18] L. F. Kui and M. R. Uddin, "Photonic microring resonator modulated resonance response analysis," *Optical and Quantum Electronics,* vol. 49, no. 8, p. 275, 2017.

IOP Conf. Series: Materials Science and Engineering **767** (2020) 012028 doi:10.1088/1757-899X/767/1/012028

Characterizations of Cascaded Micro-ring Resonators

M Rakib Uddin[1*], M Yazid[1], F K Law[1] and C B M Rashidi

[1]Eelectrical and Electronic Engineering Programme Area, Faculty of Engineering Universiti Teknologi Brunei.
[2]Advanced Communication Engineering, Centre of Excellence, School of Computer and Communication Engineering, Universiti Malaysia Perlis, Malaysia

rakib.uddin@utb.edu.bn

Abstract. The effects of cascading the ring resonator to form wavelength filters is a major focus of this paper, the results tested between the two methods of cascading the rings are the trends between the two types of cascaded rings for the ON-OFF ratio where the serially cascaded method was found to make better filters, -3dB bandwidth which is the width of the pass-band of through-port and drop-port where both the two cascading methods have the same effect as the numbers of rings increases. The delay times of the output for use in telecommunication are also simulated illustrating that only the parallel cascaded method provides this property.

1. Introduction

Silicon photonics provide faster ways of transmitting signals when compared to standard ways of transmitting data. In silicon photonic chips, these signals being transmitted are in the form of light pulses; this means data travels faster in silicon photonics when compared to traditional electrical signals. There have been researches done in order to design digital logic gates in optical domain using Fabry Perot Laser Diode [1, 2], but these designs cannot be incorporated and fabricated at CMOS-level circuits, thus electro-optic circuits have been proposed using one of the widely researched silicon photonic devices, the micro-ring resonator [3-5]. It is possible to connect multiple micro-ring resonators to create a cascaded effect. Cascaded micro-ring resonators have multiple benefits; cascaded micro-ring resonators are an important part of wave division multiplexing networks. The benefits include their small-volume, lightweight, tunability, re-configurability and electromagnetic immunity [6].

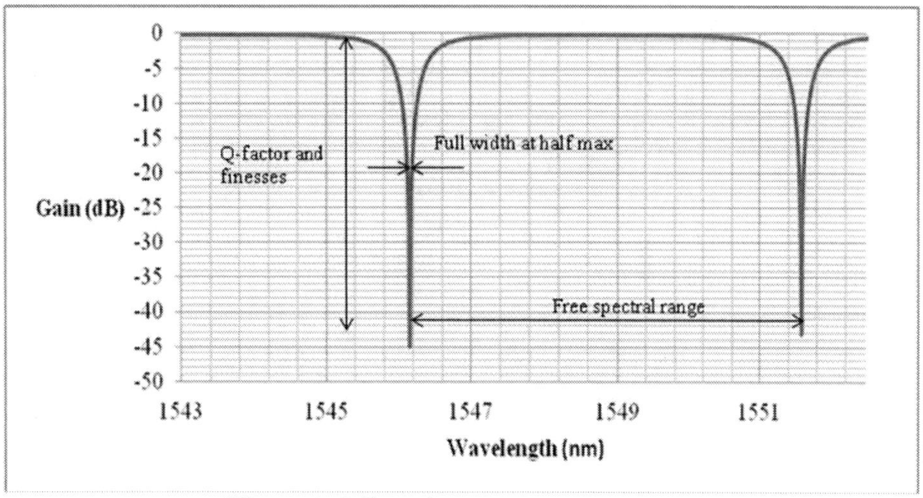

Figure 1. Micro-ring resonator parameters.

Micro-rings are very important, as the world is quickly moving towards replacing electronic communication systems with optical communication systems. Figure 1 shows the output of a typical micro-ring resonator and its parameters These parameters include the free spectral range (FSR), full width half max, finesse and Q-facto [7]. In a paper done done in 2002 titled "Box-like filter response and expansion of FSR by vertically triple couples microring resonator filter" illustrates the design and the nature of ring resonators when they are coupled and when multiple resonators are set adjacent to each other in a vertical manner. The output filter response at the dropping port of micro-ring resonator filters (series-coupled) is found by the product of filter responses of each individual single micro-ring resonators. Due to this, a sharper roll-off and lower crosstalk can be expected. The FSR is also expanded [8]. From the two previous works mentioned, it is clear that higher order filter designs yield better filter like responses. A paper done by Fengnian Xia, Mike Rooks, Lidija Sekaric and Yuri Vlasov compares the outputs of a 5th order ring resonator optical filters to a 3rd order filter. The result from this paper show that increasing the number of rings in a filter structure will provide better filters and will change aspects of the output of the micro-ring resonator.

2. Methodology

This paper will set about to characterize the effects of cascading the rings between two different methods of cascading, namely the parallel cascading and the serially cascading method. For a parallel and series cascaded micro-ring are designed to be analyzed, the topography of these two cascaded micro-ring configurations differentiated by the positions of the consecutive micro-ring resonator within the system.

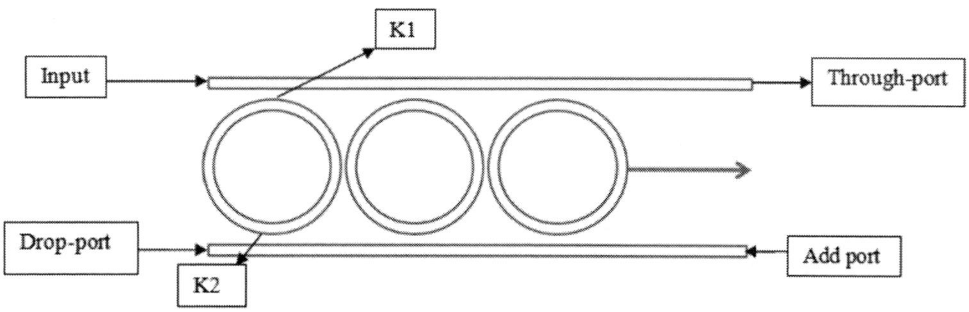

Figure 2. Parallel cascaded ring resonator.

The block diagram in Figure 2 of parallel micro-ring shows that the micro-rings are adjacent to one another in between two straight waveguides, for this configuration, the incoming signal at the input will be coupled into the first ring with a coupling coefficient of K1 as the remaining signal propagates through the straight waveguide to the through port, the part of the signal that propagates in the ring will be coupled into the waveguide below the ring with a coupling coefficient of K2 to travel to the drop port of the micro-ring configuration [10]. This propagation action continues for all the rings. This process is reversed for signals from the add port to the through port of this parallel combination.

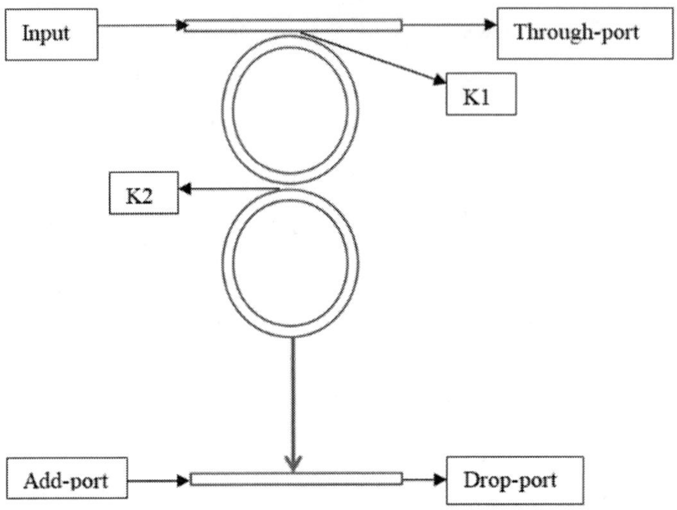

Figure 3. Serially cascaded ring resonator.

Figure 3 shows a serially cascaded ring resonator filter. In serially cascaded micro-ring resonator filters, a part of the signal from the input port will be coupled into the first ring with a coupling coefficient of K1 while the remaining part of the signal will travel through the straight bus waveguide to the through-port, a part of the signal coupled into the first ring from the top waveguide will be coupled into the second ring, a distance of half the ring length will be travelled by the signal before it is coupled to the consecutive rings with their respective coupling coefficients until the signal reaches the drop port of the lower waveguide.

In parallel configuration, the arrangement is formed in such a way that the rings do not interact between one another, the interaction between these consecutive rings are indirect, however, via the waveguides between the rings, because of this, the light propagates in a unique way and there is no counter propagation of the wave.

In series combination, the arrangement is formed in such a way that the rings interact between one another, the interaction between the rings are direct in the form of propagation between the rings themselves, in this configuration, only 2 rings will interact with the straight waveguides, the top and bottom most rings will be the rings that interact with the waveguides leading to the through-port and drop-port.

For the simulation of these cascading ring structures, the number of rings will be increased from one ring to 5 rings, this corresponds to the order of the rings. The parameters for the rings will all be set to the same value. As only the effects of cascading are being tested for results, the coupling coefficient will be set of 0.1 for all rings and the length will be set of 40 um for all tests except when stated otherwise.

The parameters to be tested are:

1. Widening effect on the free spectral range both cascading methods
2. Rejection/ON-OFF ratio

3. 3dB width at drop-port and through-port
4. Propagation delay.

3. Simulation Results

The simulation procedure and the results will be gone through in this section. For all the rings, the parameters will be the same [11] as shown in Table 1. This is to ensure that the effects of cascading are isolated and no other parameters will disrupt the results.

Table 1. Cascaded ring resonator parameters.

Parameter	Settings
Configuration	Bidirectional
Resonant wavelength	1550 nm
Length	40 μm
Loss	3 dB/m
Effective index	2.5
Group index	5
Coupling Coefficient	0.1

3.1. Widening effect on Free Spectral Range

An important quality that's brought about by cascading is its ability to increase the free spectral range of the entire micro-ring resonator system. This can be simulated and tested. To test this, two rings are cascaded with different lengths to find the effects of the free spectral range of the micro-ring resonator.

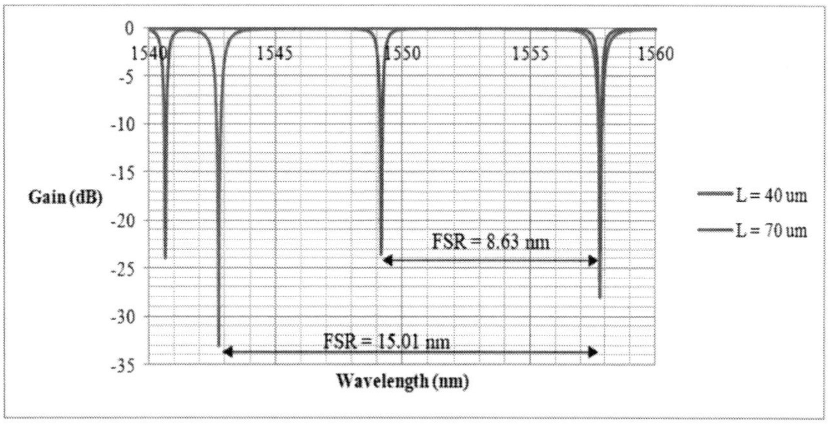

Figure 4. Superimposed output of 70um and 40um ring resonator.

In simulation testing, the FSR outputs of two micro-ring resonator configurations need to be known. The lengths used in this simulation are a 70 um ring resonator and a 40um ring resonator cascaded according to Figure 3 and 4 to achieve a 2nd order ring resonator structure. Other parameters for these two ring resonator parameters will be the same for both rings.

The output of the 2nd order rings to be tested is expected to be an interaction between the two spectral outputs shown in Figure 4 of the 70 and 40 um ring resonators.

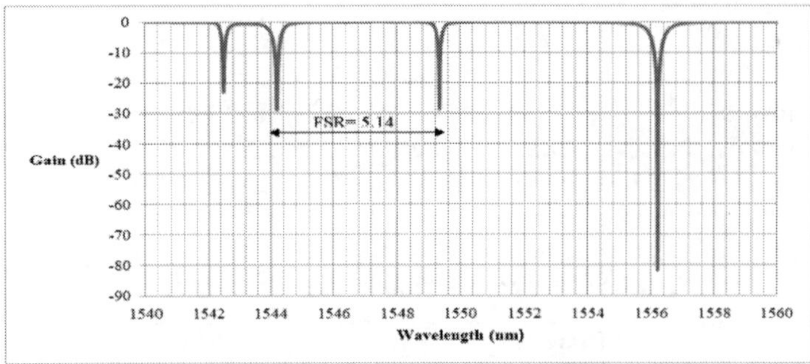

Figure 5. Spectral output and FSR widening effect of parallel cascaded ring resonator.

The two values of length chosen to test the effects of cascading the rings on the FSR are 40um and 70um. A 40um micro-ring resonator has 2 minimas in the spectral output while the 70um has 3 minimas. This is shown in Figure 5.

From these simulation results, it is seen that to achieve a high free spectral range value by cascading the rings a serial configuration is suitable as the output of a cascaded ring structure in this fashion will yield the desired result as shown in Figure 6, for parallel cascaded ring resonator however this is not achieved, the free spectral range value does change however it does not become wider as the minimas from each individual ring appears in the overall spectral output of the cascaded ring resonator structure.

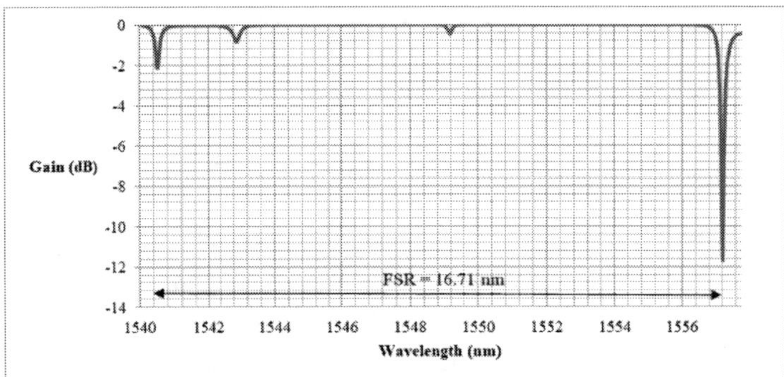

Figure 6. Spectral output and FSR widening effect of serially cascaded ring resonator.

In conclusion, a serially cascaded ring resonator will result in a wider FSR but a parallel cascaded ring resonator will only add the gain values of the spectral output if the spectral dips are aligned with each other.

3.2. ON-OFF ratio

The performance of the device can be determined by the ON-OFF ratio this is determined at the drop-port response of the micro-ring filter configuration shown in Figure 7.

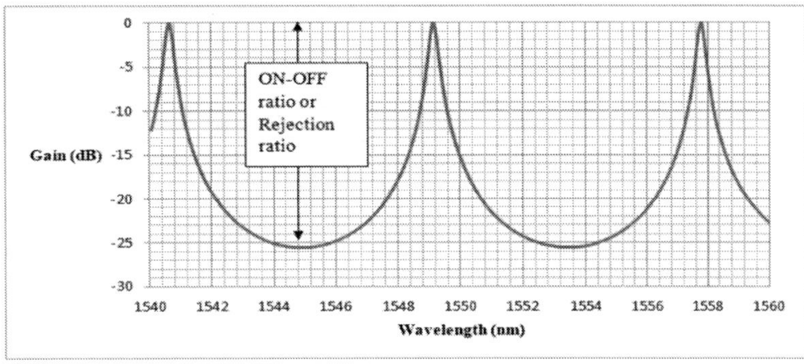

Figure 7. ON-OFF ratio measurement.

For the results of this parameter, the ON-OFF ratio is measured for the cascaded ring structures from its 1st order to its 5th order for both cascading methods as shown in Table 2.
It is illustrated in Table 2 that the value of the ON-OFF ratio has different trends for the different cascading methods.

Table 2. ON-OFF Ratio.

Order	Parallel Cascaded	Serially Cascaded
1st	-25.57 dB	-25.57 dB
2nd	-19.58 dB	-41.36 dB
3rd	-16.10 dB	-58.82 dB
4th	-13.65 dB	-74.62 dB
5th	-11.79 dB	-90.41 dB

3.3. 3dB bandwidth at through-port an drop-port

The gain at the through-port and drop of the ring resonators structures can also be simulated at the through-port and drop-port; the outputs can then be compared between the 1st order and the 5th order. This simulation comparison will be done with the 3dB bandwidth [Figure 8] of the response. The 3dB bandwidth comparison is shown in Table 3.

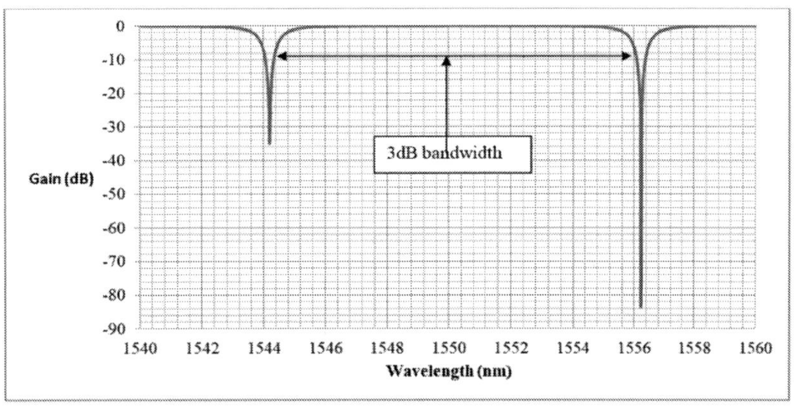

Figure 8. 3dB bandwidth measurement.

Table 3. 3dB width for cascaded ring resonators.

Order	Parallel	Serial
1st	11.61 nm	11.61 nm
2nd	11.21 nm	10.61 nm
3rd	10.79 nm	10.35 nm
4th	10.38 nm	10.06 nm
5th	9.940 nm	9.910 nm

The drop-port has a different affect however, the output of the drop-port will have the opposite effect of the through-port and this is because the output results for the same wavelength band from the input, and the summation of the output at the drop-port and the through-port will result in the input signal. Table 4 shows the 3dB width of the drop-port of the parallel ring resonator [Figure 9] increasing as the number of rings increases, vice versa to the 3dB width at the through-port.

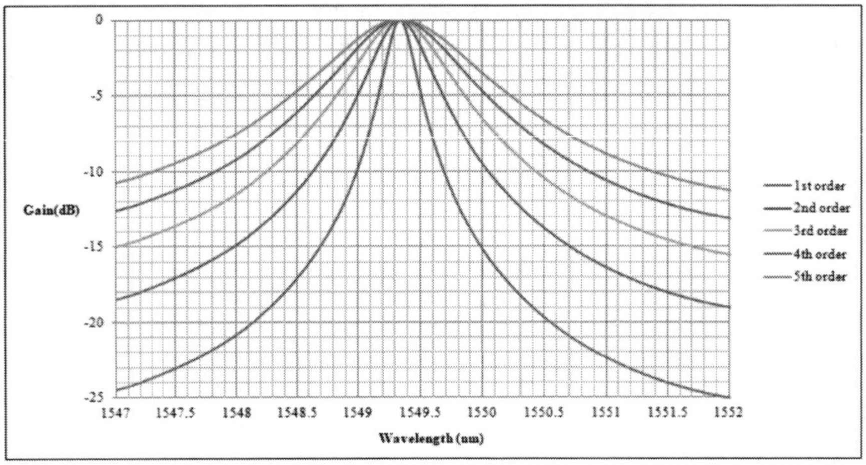

Figure 9. Drop-port output for parallel cascaded ring resonator.

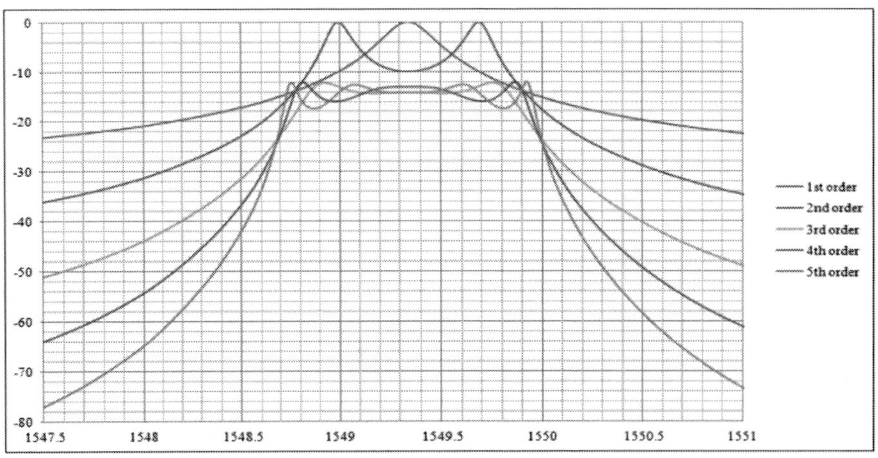

Figure 10. Drop-port of serially cascaded ring resonator.

From Figure 10 it is illustrated that the output of the drop-port at the 3rd order has a gain of less than -10 dB, this is a value that is too low, receivers will not be able to detect this signal. The reason for this is because of the topography of the serially cascaded ring resonator, when light propagates from the input port to the drop-port the signal has to propagate between many rings that traps some of that signal.

Table 4. 3dB width for drop-port of parallel cascaded ring resonator.

Order	**Parallel ring resonator**
1st	0.23 nm
2nd	0.43 nm
3rd	0.69 nm
4th	0.92 nm
5th	1.17 nm

3.4. Propagation delay

An important characteristic also tested in cascaded simulation is the delay [as shown in Table 5] Fcaused by cascading the rings. Delays in micro-ring resonators are used in Optical time division multiplexing networks in foe the demultiplexing of the channels as well as in antenna beam forming networks in telecommunications.

Coupled micro-ring resonators can be used for this purpose, as the number of micro-ring resonators increase, the signal will need time (in picoseconds) to complete propagate from the input port to the through-port (output) due to this mechanism, delays can be achieved.peak delay of each micro-ring resonator configuration can be simulated to achieve a value of time for the delay of transmission for the through-port, this value is tabulated for the rings, and this value is shown to increase as the value of the rings increases.

Table 5. Propagation delay

Order	**Parallel**	**Serial**
1st	0.01765 ps	0.01765 ps
2nd	0.03350 ps	0.01805 ps
3rd	0.05230 ps	0.01805 ps
4th	0.06930 ps	0.01805 ps
5th	0.15030 ps	0.01805 ps

4. Conclusion

This paper simulates the cascading of ring resonators in two different methods; this includes the serially cascaded and the parallel cascaded ring resonator structure.

The results measured are the widening effects of the FSR, ON-OFF ratio, the 3dB width of the through-port and the drop-port and the propagation delay caused by cascading. The actual parameters of the ring resonator are kept constant so that the effects of cascading on the output can be isolated from the effects of the other parameters of the ring.

It is seen that the serial and parallel cascaded rings differ slightly, in terms of widening the FSR, the serial cascaded ring is the only structure that achieves this effect. The ON-OFF ratio which is an indicator of the performance of the ring resonator shows that the serially cascaded ring is a better filter than the parallel cascaded. For the 3dB width for the through and drop-port, both of the ring structures have the same effect occur at the output however, the output at the drop-port of the serial cascaded ring resonator is no viable for use as the gain of the signal is too small. Moving on the propagation, it is illustrated that only the parallel cascaded ring resonator filter provides this property so for synchronization in telecommunication components that implement micro-ring resonators, this structure can be used for synchronization.

IOP Conf. Series: Materials Science and Engineering **767** (2020) 012028 doi:10.1088/1757-899X/767/1/012028

References

[1] B. Nakarmi, M. Rakib-Uddin, T. Q. Hoai, and Y. H. Won, "A simple controlled all-optical on/off switch using gain modulation in single mode FP-LD," *IEEE Photonics Technology Letters,* vol. 23, no. 4, pp. 212-214, 2010.
[2] M. R. Uddin, J. Lim, Y. Jeong, and Y. Won, "All-optical digital logic gates using single-mode Fabry–Perot laser diode," *IEEE Photonics Technology Letters,* vol. 21, no. 19, pp. 1468-1470, 2009.
[3] F. K. Law, M. R. Uddin, and H. Hashim, "Photonic D-type flip flop based on micro-ring resonator," *Optical and Quantum Electronics,* vol. 50, no. 3, p. 119, 2018.
[4] F. K. Law, M. R. Uddin, H. Hashim, and Y. H. Won, "Demonstration of photonic micro-ring resonator based digital bit magnitude comparator," *Optical and Quantum Electronics,* vol. 51, no. 1, p. 1, 2019.
[5] W. Bogaerts, P. Heyn, T. Vaerenbergh, K. Vos, S. Selvaraja, T. Claes, P. Dumon, P. Bienstman, D. Thourhout and R. Baets, "Silicon microring resonators", Academia.edu, J. Clerk Maxwell, A Treatise on Electricity and Magnetism.
[6] Parallel-Cascaded Semiconductor Microring Resonators for High-Order and Wide-FSR Filters R. Grover, Student Member, IEEE, V. Van, Member, IEEE, T. A. Ibrahim, P. P. Absil, L. C. Calhoun, F. G. Johnson, J. V. Hryniewicz, and P.-T. Ho.
[7] Q. Xu, B. Schmidt, J. Shakya and M. Lipson, "Cascaded silicon micro-ring modulators for WDM optical interconnection", Optics Express, 2006.
[8] Y. Yanagase, S. Suzuki, Y. Kokubun and Sai Tak Chu, "Box-like filter response and expansion of FSR by a vertically triple coupled microring resonator filter", Journal of Lightwave Technology, 2002.
[9] F. Xia, M. Rooks, L. Sekaric and Y. Vlasov, "Ultra-compact high order ring resonator filters using submicron silicon photonic wires for on-chip optical interconnects", Optics Express, 2007.
[10] R. Katti and S. Prince, "Analysis of serial and parallel cascaded microring resonators: An FDTD approach", Optik, vol. 152, pp. 36-48, 2018.R. Nicole.
[11] F. Law, M. R. Uddin, N. M. Masri, and Y. H. Won, "Digital Photonic Even Parity Bit Generator," in *2018 IEEE Photonics Conference (IPC)*, 2018, pp. 1-2: IEEE.

Beam Steering Monopole Antenna for Low Power Wireless Communication in WSN

N A F Mahdir, M N Osman, T Sabapathy, M Jusoh and N A F M Zainudin

Bioelectromagnetics Research Group (BioEM), School of Computer and Communication Engineering, Universiti Malaysia Perlis, Pauh Putra Main Campus, 02600 Arau, Perlis.

mohamed.nasrun@gmail.com

Abstract. Antenna for wireless sensor networks (WSN) application need to satisfy a number of additional properties like antenna efficiency. Since there are many WSN nodes, the use of low cost and high efficient antenna is highly important. Therefore, in this paper, a beam steering antenna resonate at 2.4 GHz with monopole element by using parasitic array technique is proposed. The parasitic array element is introduced to achieve a better gain and acts as a reflector to steer the beam to the desired degree of angles. The structure of the proposed design consists of monopole antenna as the radiating element, and to parasitic patches with the same dimension of monopole are placed next to the radiating patch. Two switches (PIN diode) are used and embedded at the back of the structure. By activating and deactivating of the switches (ON or OFF), the parasitic is either grounded or disconnected from ground. Consequently, depending on the switches configuration, the beam pattern can be steered two direction angles which ±25° left or right. The simulated and measured results are presented to demonstrate the performance of the proposed antenna.

1. Introduction

Nowadays, wireless networks are one of the most promising technologies that current been used in the industries and one of the technologies that always fits well with all the wireless network type is wireless sensor network (WSN). WSN are distributed system compromising of a number of nodes and each node consists of sensors, battery, microprocessor, radio chip and antenna [1]. There are several requirements in designing the antenna for WSN which are low cost, flexible, high efficiency and low power. In reliability aspect, there are challenges in making sure the system is sufficiently reliable and in the same time increase the lifetime of the sensor nodes.

However, high energy efficiency of WSN being a big problematic issue that leads to low performance of WSN [2], [3]. Life span of WSN decreasing when it is cannot consume enough power. The radiation signals of WSN's antenna on sensor node become a waste of power because it radiated all directions not only toward the user [4]. Moreover, the power radiated in all directions will be having interference by other users. In order for the sensor nodes to consume enough power for the continuous life span, it needs to exchange the battery regularly which is really not practical and causing high cost. Not enough energy will also lead to sensor failure which is will reduce to low accuracy of the data. Improvisation of the data transfer on the sensor is said can be a solution to recover this problem. By using beam steering antenna, the direction of the beam can be changed to the desired position, making more effective communication link due to higher gain and interference reduction [5].

Therefore, the aim of the project is to design the beam steering antenna with monopole as a driven element and parasitic array technique for tilting. The switching of the beam pattern will be accomplished by inserting two switches and manipulating the parasitic array which acts as a reflector of the designed antenna. The same concept of reconfigurability has been proposed in [6],[7], but this works focusing on using omni directional type of antenna as a driven element and less complexity.

2. Antenna design and reconfigurations.

The geometry of the monopole antenna with parasitic element is shown in Figure 1. This antenna is designed on FR-4 substrates with a dimension of 105mm x 90mm, height, h= 1.6mm and permittivity, ε_r=4.3. On the top of the substrate, the patch of the driven element (L_p x W_p) and both parasitic elements are placed, and the partial ground plane (dimension of k x W_g) is located at the back of the design. The dimension of feed line is w_fx L_f mm. The dimension of the driven and parasitic element is almost identical, and parasitic is positioned at a gap of g_p from the driven. Patch and ground plane are made up by the copper which is the thickness, t = 0.035mm. Two PIN diode switches are placed in between the two shorting copper pin behind the both parasitic elements and the ground plane. The shorting copper pin is inserted at the centre bottom of the parasitic patch. For the activation of the PIN diode, the antenna structure is modified and integrates with biasing circuit, includes biasing line, biasing cable and RF components. Three RF inductors with a value of 27 nH are placed as a bridge and provide blocking of the dc current to flow through dc biasing wire. The cathodes of the diodes are connected to the dc biasing wire (positive terminal), while the anodes are soldered to the partial ground plane (connected with negative terminal). The direction of the pattern is changed to the left or right depending on the switch configurations. When the switch is ON, one of the parasitic patches is grounded via shorting pin and switch, and acts as reflector to reflect the energy and pattern to the opposite side, and vice versa.

A photograph of the prototype antenna using PIN diode is shown in Figure 2. The dimensions of the proposed antenna are as follows (unit in millimeters): L_p= 21.5, W_p= 31, k= 47, W_g= 105, w_f= 3, L_f= 48, g_p= 3, r_p= 13.6.

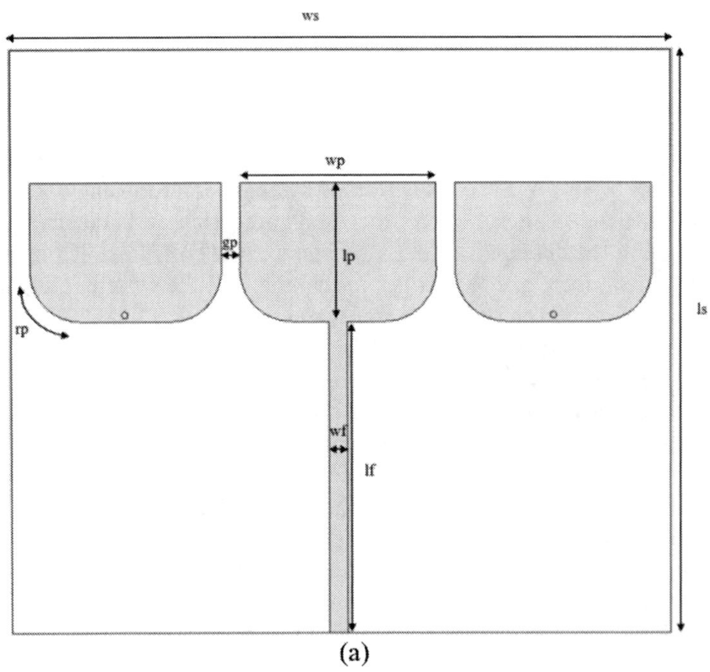

(a)

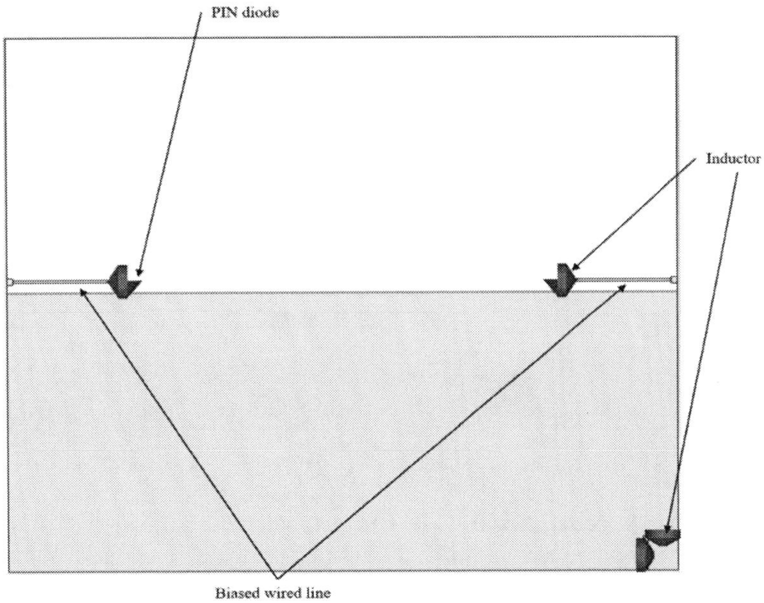

(b)

Figure 1. Geometry of the proposed antenna (a) front view (b) back view

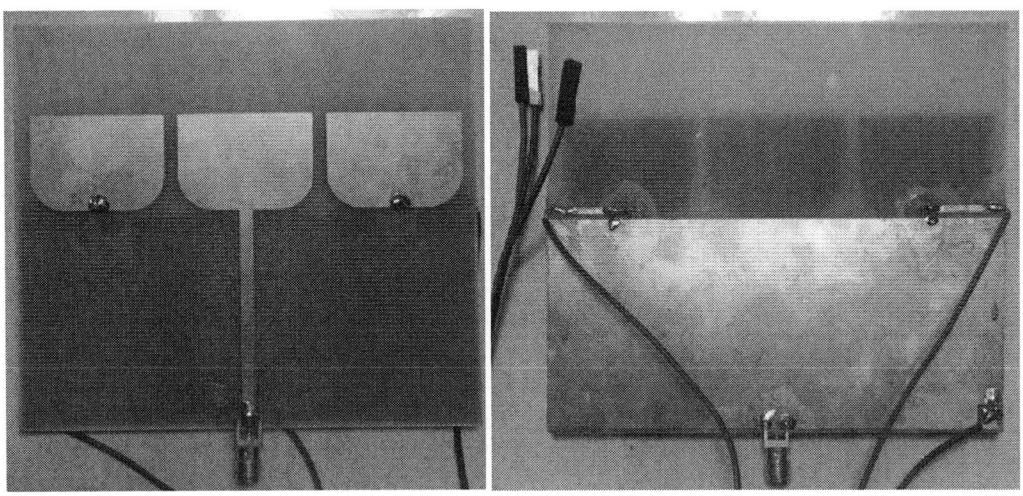

Figure 2. View of front (left) and back (right) of fabricated proposed antenna

3.Result and discussion

The preliminary works has been done by using ideal diodes for proof of the concept. Before the parasitic plane has been placed on the antenna, the antenna was design as monopole antenna with single patch and feed at the front and half ground plane at the back of the antenna. Figure 2(a) and (b) shows the simulated reflection coefficient result of the monopole antenna without and with parasitics, respectively.

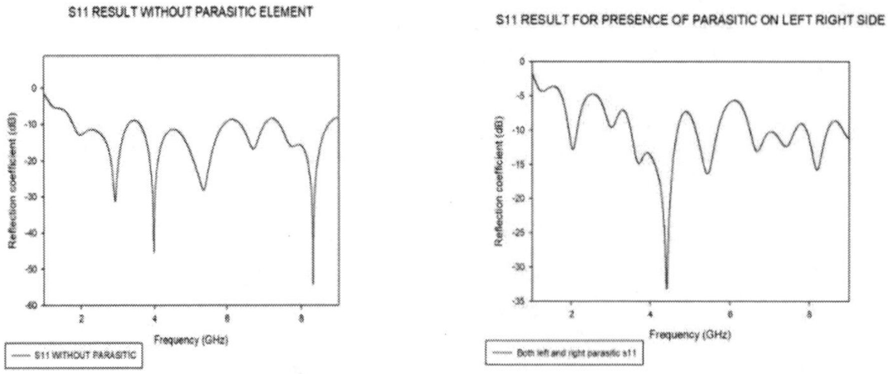

(a) (b)

Figure 3. Simulated S11 for the monopole antenna (a) without parasitics and (b) with parasitics.

The direction of the radiation pattern seems to be changed when the parasitic is just placed only on one sided. First, the parasitic was placed on the right side of the patch and the result of the simulation shows the changes in the direction of the radiation pattern, as shown in Figure 4(a). It goes direct toward the left of the antenna, about 25° to the left. On the other hand, the direction of the pattern in steered on the opposite direction when the parasitic patch was placed on the left side, as demonstrates in Figure 4(b). When both of the diodes is either ON or OFF state, the 'butterfly shaped' pattern is achieved, as presented in Figure 4 (c) and (d). This type of shape is very important to indicate the function of parasitic in coupling the energy. The switching configuration with simulated tilt angle is summarized in Table 1.

Table 1. Switching configuration of proposed antenna

State of the switch		**Angle tilt (°)**	**Realized Gain (dBi)**
Right (Switch 1)	**Left (Switch 2)**		
ON	ON	0°	1.23
OFF	ON	-25°	2.51
ON	OFF	+25°	2.41

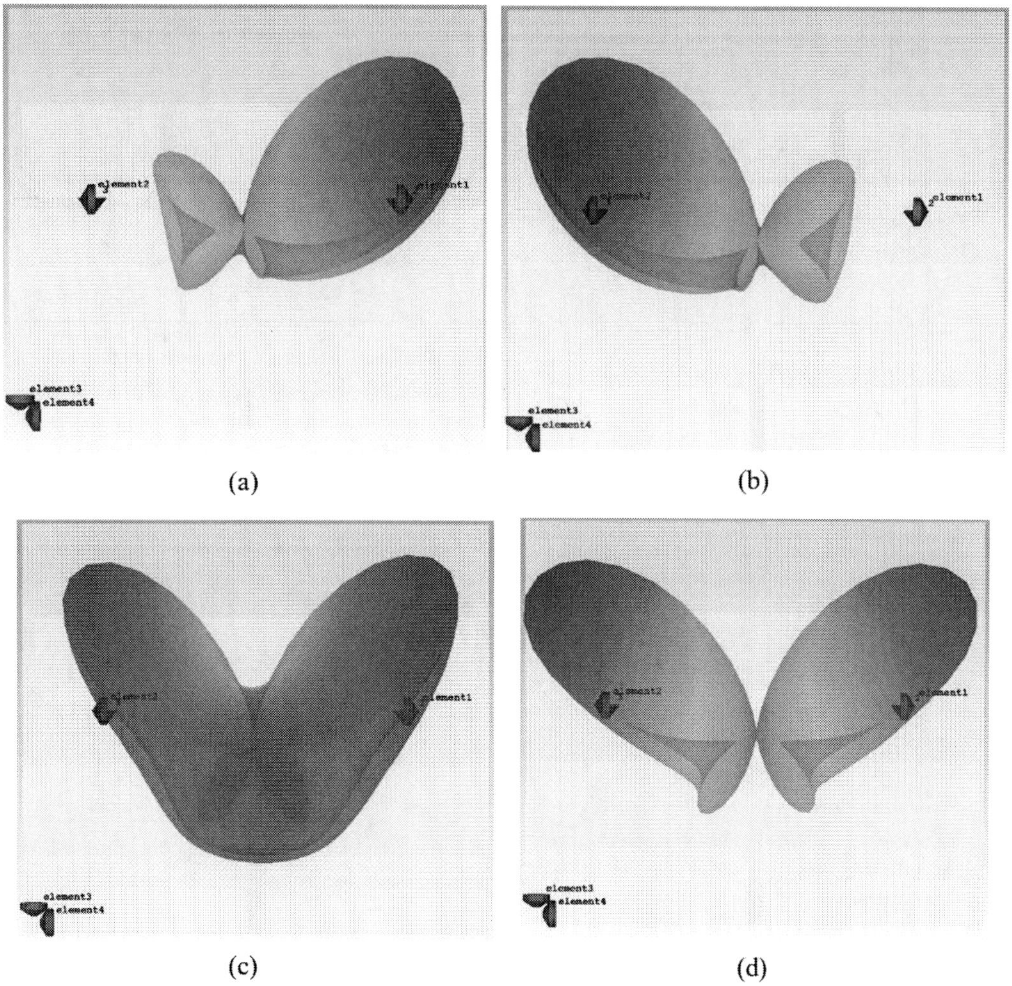

Figure 4. Simulated of the beam steering antenna when (a) switch 1 is ON, (b) switch 2 is ON, (c) both switches are ON and (d) both switches are OFF

The fabricated antenna is then measured using the Rohde & Schwarz ZVL network analyzer (VNA) to measure the S – parameter of the antenna. Figure 5 shows the comparison results of simulation and measurement for all configurations. Based on this figure, it clearly can be seen that the measured antenna is slightly different with the simulated result. This is occurred maybe due to the fabrication tolerance that might affect the results. In addition, via copper used to connect the parasitic patch and the pin diode need to be embedded properly with exactly the same dimension of the simulation. However, due to the limitation of size of hole saw of the drill in our lab, the hole dimension is not exactly match with the size in the simulation. This will disturbs the dimension of the overall patch, and consequently affect the reflection coefficient result.

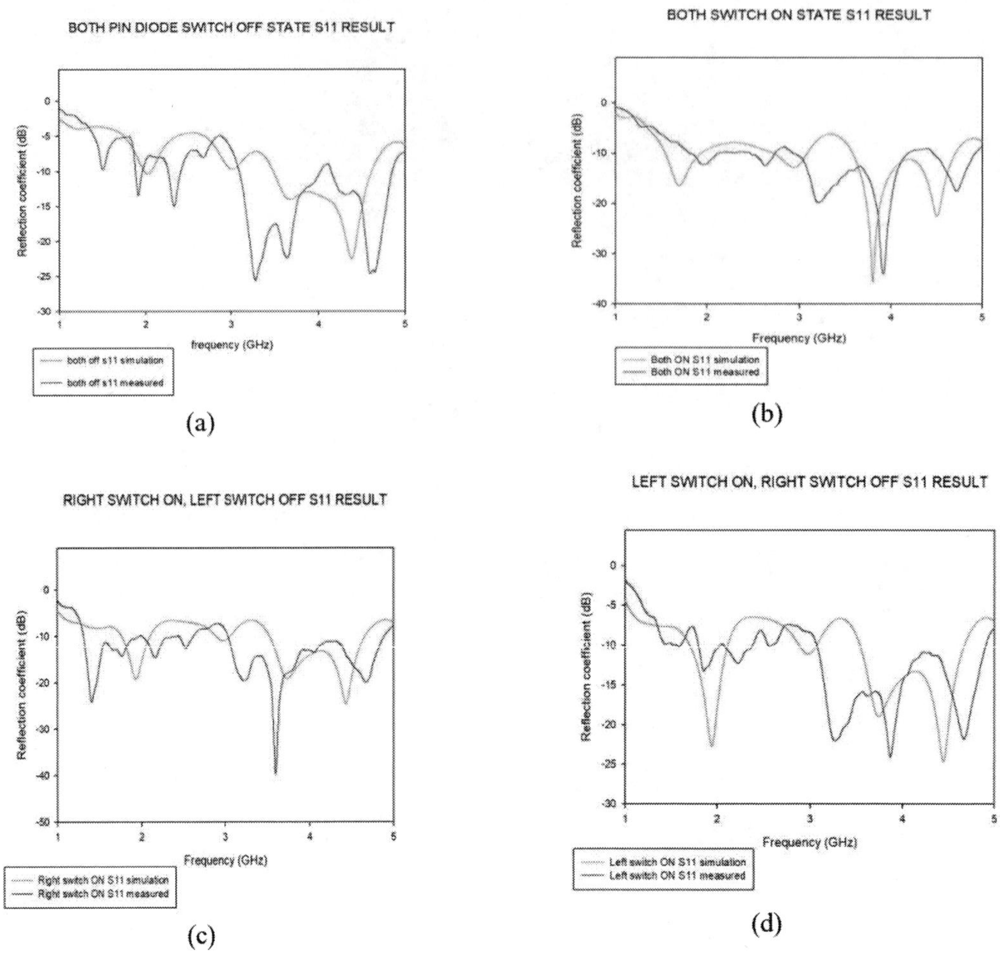

Figure 5. Reflection coefficient comparison of simulation and measurement result for (a) both switches OFF, (b) both switches ON, (c) right switch ON and (d) left switch ON

4.Conclusion

This paper proposed a design of beam steering antenna for low power wireless communication in WSN application. The beam steerable antenna is realized by using two PIN diodes. The parasitic array technique is applied on the proposed design, by placing two identical parasitic patches next to the radiating element. The antenna is successfully steering the beam with presence of the parasitic element. The direction of the beam pattern is capable to be changed in two directions which are in a degree of ±25°. The antenna is fabricated on the FR4 substrate and the simulation and measurement results is compared and analyzed.

5.References

[1] A. Kundu and S. Roy, "Performance of Smart Antennas in Wireless Sensor Networks," *Int. J. Comput. Appl.*, vol. 957–888, no. 1, pp. 3–6, 2012.

[2] E. Network, A. Computing, P. Putra, and M. Campus, "Power Consumption Optimization Analysis Based on Berkeley-Mac Protocol Using," vol. 88, no. 2, pp. 295–302, 2016.

[3] M. Elshaikh, M. N. B. M. Warip, O. B. Lynn, R. B. Ahmad, P. Ehkan, F. F. Zakaria, and F. A. A. Fuad, "Energy consumption optimization with Ichi Taguchi method for Wireless Sensor Networks," *2014 2nd Int. Conf. Electron. Des.*, pp. 493–498, 2014.

[4] L. Catarinucci, S. Guglielmi, L. Patrono, and L. Tarricone, "Switched-Beam Antenna for

Wireless Sensor Network Nodes," *Prog. Electromagn. Res. C*, vol. 39, no. April, pp. 193–207, 2014.

[5] J. Ahn, K. Kim, Y. Yoon, and K. Hwang, "Pattern reconfigurable antenna for wireless sensor network system," *Electron. Lett.*, vol. 48, no. 16, pp. 984–985, 2012.

[6] T. Sabapathy, M. F. Jamlos, R. B. Ahmad, M. Jusoh, M. I. Jais, and M. R. Kamarudin, "Electronically reconfigurable beam steering antenna using embedded RF PIN based parasitic arrays (ERPPA)," *Prog. Electromagn. Res.*, vol. 140, pp. 241–261, 2013.

[7] M. Jusoh, T. Sabapathy, M. F. Jamlus and M. R . Kamarudin,"Reconfigurable four-parasitic-element patch antenna for high beam switching application, *IEEE Antennas and Wireless Propagation Letters*, vol. 13, pp. 79-82, 2014.

Acknowledgments

The author would like to acknowledge the support from the Fundamental Research Grant Scheme (FRGS) under a grant number of 9003-00608 from the Ministry of Higher Education.

Implementation of Particle Swarm Optimization and Genetic Algorithms to Tackle the PAPR Problem of OFDM System

A Abdalmunam[1], MS Anuar[1], MN Junta[1], NM Nawawi[1], A Noori[1]

[1]School of Computer and Communication,University Malaysia Perlis, Malaysia

Corresponding author: 1988aymen7@gmail.com

Abstract. A multi-carrier modulation technique, which represented in this paper as orthogonal-frequency-division-multiplexing (OFDM), ensured wireless high-speed data transmission. The transmission of modulated symbols uses a large number of subcarriers in the OFDM system. Consequently, the OFDM signals have an extended dynamic range, or a high output power peak envelope fluctuation or high PAPR. To mitigate the PAPR, in this paper, we implement two algorithms to reduce the output power envelope fluctuation of the OFDM system, namely PSO and GA. Also, the PTS method and PAPR in OFDM systems difficulty described briefly. We present an OFDM system through the use of conventional PTS based on PSO and GA. The simulation result shows that both evolutionary approaches outperform the conventional PTS OFDM in-terms of reducing the Peak-to-Average-Power-Ratio (PAPR). Furthermore, the performance of the PSO algorithm is found to be better than GA in-terms of its simplicity and the time execution. On the other hand, the GA algorithm outperforms the PSO and the conventional OFDM, in terms of the PAPR reduction.

1. Introduction

The digital world of today filled with wireless devices that facilitate humanity's daily life. The convenience run by wireless technology leads to a rapid increase in the number of people who are simultaneously aware of and actively using the technology. These increasing numbers of active users result in high demand for quality services and wireless broadband services[1, 2] such as High-definition television, high-speed internet, mobile video, and video conferencing. Orthogonal-frequency-division-multiplexing (OFDM) is one of the standard transmission techniques for multi-carriers, is a simple and efficient wideband communication solution, because unlike single carrier transmission systems, the channel divided into subchannels for parallel data transmission with longer symbol durations. which transforms the selective frequency fading channel into a flat fading channel and, further can encapsulate the dilemma of inter-symbol interference (ISI) estimation [3, 4]

The high dynamic range or high output power envelope fluctuation which is represented as a peak-to-average-power-ratio (PAPR) is the most common OFDM restricted problem that degrade the efficiency of OFDM system to overcome this drawback need to transmit high power amplifier (HPA) to cover non-linear signal which causes signal distorted that negatively effects on battery lifetime, to solve the above mentioned problem, several method had been proposed such as, clipping and filtering[5], iterative clipping[6], companding [7], coding[8], interleaved[9], tone-reservation[10], tone-injection [11], active-constellation-extension[12], selective-mapping(SLM)[13], partial-transmit-sequence (PTS) [14]. All these methods have a trade-off between the PAPR, BER, and computational complexity. In the OFDM system, PTS is the most effective technique for reducing PAPR with the lowest distortion.

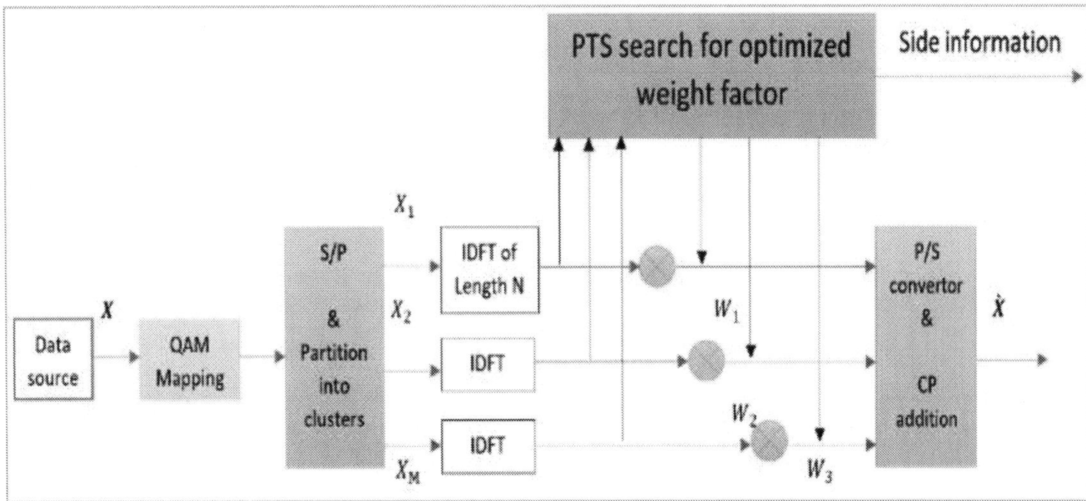

Figure1. the transmitter of PTS-OFDM

PTS is a non-distortion arrangement in which the data block inputs divided into several disjointed sub-blocks, and the IFFT performed. These IFFT sub-block outputs are weighted or scrambled by several rotation factors, which are totaled to create different candidate signals. Ultimately, the minimum PAPR signal is chosen for transmission[15], as in fig(1). As long as a PTS needs to implemented, complexity must taken into account which is a substantial parameter in the transmitter because it raises exponentially with the quantity of subsequence, for this reason, the vector's rotation should be limited to a set of finite number of element PTS is weakened by high computational complexity [16] that results from thoroughly searching for candidate signals and additionally needs data regarding rotation factors to be sent to the receiver as a side information.

In the literature PSO-PTS of OFDM system proposed in [17] by Wen, Horng et al. uses heuristics to search for the optimum integration of low-complex phase factors, got lessen of computational complexity but PAPR slightly high. The authors in [18] Presented an OFDM system that uses a sub-optimal PTS method anchor on PSO for finding optimal phase weighting factors got quite a good result in term of PAPR and complexity but a low number of iteration. Also, in [19], they worked on PTS-OFDM method, which Presented a new approach to overcoming computational complexity based on PSO, the output of approached near to be accomplished but with the trade of PAPR. According to the combination of GA-PTS with a partheno-crossover operator (PCGA), the least PAPR got, but the complexity load still a bit high[20]. Further the GA and PSO algorithms in PTS-OFDM compared and conclude that the GA is giving PAPR reduction but with the expense of computational complexity and vice versa in PSO [21, 22]. There is another algorithm called fireworks algorithm (FWA) that outperformed the above two algorithms the reader may read [23] for more information. It proposed that many evolutionary PTS-based optimization algorithms for reducing search numbers, which are particle swarm optimization (PSO) and genetic algorithm (GA)[21]. In this paper, the comparison of these two algorithms illustrated clearly especially of two notorious parameters of OFDM that are PAPR and computational complexity. Also a numnerical comparison for which method resulted in better PAPR

The rest of this paper organized as follows: firstly, the introduction of OFDM and its drawback is explained, and the next described the PAPR of the OFDM-PTS signal, then implementing PSO in PTS-OFDM is discussed, also illustrate the GA. Finally, before the conclusion, we explained the simulation result of PAPR reduction.

2. PAPR of the OFDM-PTS signal

The high-rate binary input data sequence on the transmitter of the OFDM technique is mapped to complex symbols using a modulation structure such as QAM, which simulated in this paper. To produce a block vector with complex symbols valued, $X = \{X_k\}_{k=0}^{N-1} = [0\ N-1 = [X0, X1, \ldots, X_{N-1}]^T$ The symbols are divided and simultaneously transmitted to low-rate with N subcarriers, where each symbol modulates one set of the subcarriers. Then each length N grouped then fed into the IFFT module, the signal of an OFDM gotten by adding all the autonomously modulated sub-carriers at the IFFT block output. The signal of OFDM transmitted using a discrete-time complex envelope defined as in (1).

$$x[n] = \frac{1}{\sqrt{N}} \sum_{k=0}^{N-1} x_k e^{\frac{j2\pi nk}{LN}}, 0 \le N \le LN-1 \quad (1)$$

Where N represents a subcarrier number (IFFT size), x_k is the nth complex symbol carried and conveyed by the kth subcarrier, the factor of oversampling represents as L that equal four (L=4) in OFDM signals to get an efficient, accurate value of estimating PAPR in discrete-time compared to the continuous-time. The large subcarrier in the time-domain causes significant output power envelope fluctuation, which know as the peak-to-average-power-ratio (PAPR) parameter. PAPR described as the ratio between the supreme power of the transmitter and its average that happened after the IFFT block executed. It can define as

$$P_{X(N)}(dB) = PAPR = \frac{P_{peak}}{P_{avarage}} = 10\log_{10} \frac{\max\left[(x_n)^2\right]}{E\left[(x_n)^2\right]} \quad (2)$$

The PAPR reduction efficiency is calculated through Complementary Cumulative Distribution Function (CCDF) [24]:

$$p_r(PAPR > z) = 1 - p_r(PAPR \le z) = 1 - (1 - e^{-z})^k \quad (3)$$

3. PSO Based PTS-OFDM

Kennedy and Eberhart established an evolutionary computing technique in [25]that is a PSO algorithm by simulated a random movement of a bird's bunch. We implement the PSO algorithm to mitigate high peak power fluctuation with lower computational-complexity in the PTS technique of an OFDM system. The phase factor of large space represents every value in a dimension as a factor. A factor's set can be a point or place in space [19]. A sequence of the least PAPR represented by the best place. PSO algorithm initializes a particle, which is an incipient population of solutions. Every particle is a point in the space of M dimensions (M is the sub-block number of PTS). Any particle has phase vector M, which is a vector element chosen from phase vectors. Particles fly in space to search on the best place, and the position of the particle modified depends on the relationship between each other[26]. Moreover, the points occur meanwhile searching on the local and global optimization. Eventually, all particles coverage to optimal location, which is the ability to discover causes particles to find the sequence of phase vectors. Briefly, The PTS technique can be applied by altering the optimization for block W appropriately. The first goal of PTS that comes by a weighted combination of N sub-block written as

$$X' = \sum_{i=1}^{M} W_i X_{i,} W_i = e^{j\theta i} \quad (4)$$

The factor of phase weighted limited within $[0,2\pi]$ where W_i, i=1, 2,…..,M. Generally, the selection of phase weighting factors (W) is limited to a finite set of elements to reduce the complexity search. The second important goal is to reduce the PAPR of X' by chose $W = W_1,\ W_2 \ldots, W_i$ and the optimum of OFDM symbol parameter

$$\hat{W} = \arg W_{\min} \left\{ \max | \sum_{i=1}^{M} W_i X_i | \right\} \tag{5}$$

The population, in this context, pointing to swarm and the individuals pointing to particle, resembling the swarm social behavior, which clearly explained in[27, 28]. The PSO procedures optimization based on a populace of particles flying with dynamic velocity in the solution space according to their own experience of flying and the best experience of flying among the swarm. Each possible solution pointed as swarm with a position vector x, during the process of PSO, which is referred respectively as a phase weighting factor (W) and velocity (V). Therefore, the K-dimensional optimization and the ith a particle of position and velocity respectively represented as $W_i = (W_{i,1}, W_{i,2}, \ldots, W_{i,K})$ and $V_i = (V_{i,1}, V_{i,2}, \ldots, V_{i,K})$. Every particle has its best position $W_i^P = (W_{i,1}, W_{i,2}, \ldots, W_{i,K})$. Corresponds to the best individual objective value obtained in time t, referred to as $pbest$ and $W^G = (W_{g,1}, W_{g,2}, \ldots, W_{g,K})$ denoted as a global best particle ($gbest$), which represents the optimal particle detected at a time t in entire swarm. The equation below had been structured to update the velocity for particle i.

$$V_i(t+1) = wv_i + c_1 r_1 V_i(t+1) = (W_i^P(t) - W_i(t)) + c_2 r_2 (W^G(t) - W_i(t)) \tag{6}$$

Further, the new position of a particle calculated according to the subsequent equation based on the updated velocities

$$W_i(t+1) = W_i(t) + V_i(t+1) \tag{7}$$

Where the initial velocity of particle i at time t represented in (6) as $v_i(t)$. The new velocity is associated with the weight w of the old velocity and also with the particle's position and that of the global best velocity by acceleration factors of $c_1$and c_2. Therefore those acceleration factors are denoted as cognitive and social relates respectively; these weighting factors are pulling individual particle toward the $pbest$ and $gbest$ positions. The inertial weight w is used to manipulate the impact of the previous velocity history on the current velocity. Generally, the suitable value of $w(t)$ provides the desired balance between the globe and the swarm's ability of local exploration. The particle populations then moved by the new velocity and locations calculated by (6,7) Tend from different directions to cluster together. Thus, the assessment of each associated fitness of the new particle population begins again. Further, in this paper, we set the particle swarm size is 30 with 1000 iterations, and the acceleration coefficient (c1, c2) is (1,3), respectively. The flow chart of PSO portrayed below.

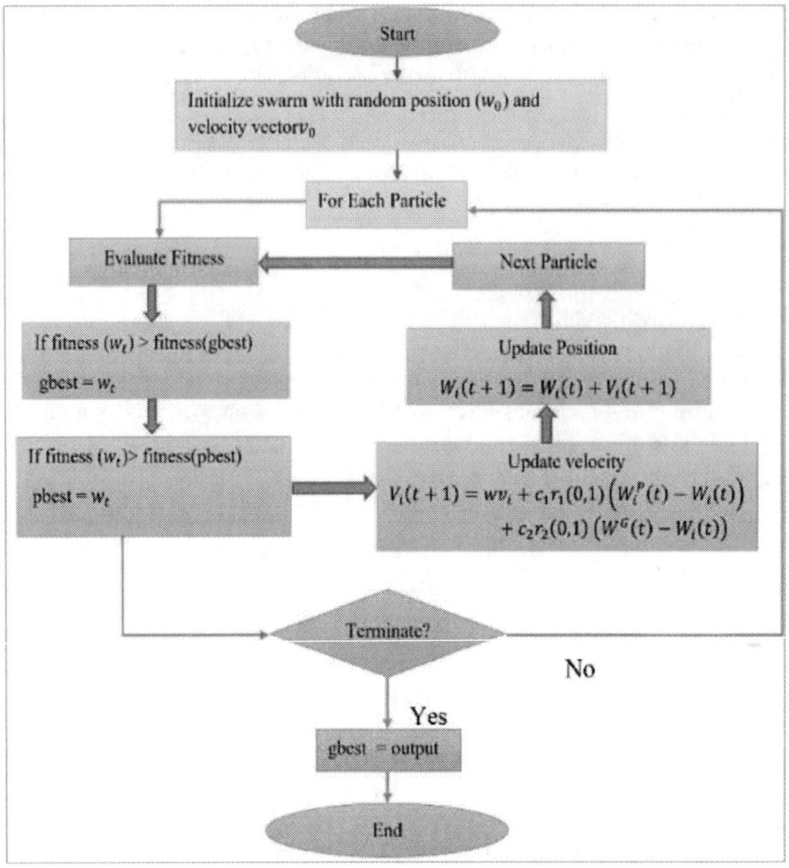

Figure2. Block diagram of the PSO algorithm

4. GA Based PTS-OFDM

John Holland introduced the Genetic Algorithm in the United States in the 1970s. The fundamental ideas of the algorithm are natural selection and genetic inheritance[29]. GA is Focusing on finding a suitable phase factor to diminish PAPR. Hence, the evaluation function defined in order to increase the value of the evaluation function as the PAPR decreases. This algorithm initially picks the random population defined as a chromosome, which is then the chosen value multiplied by a set of phase factors leading to the calculation of PAPR. Each chromosome's fitness value can be calculated by

$$F(ya(t)) = \frac{1}{10\log_{10} PAPR(ya(t))} \tag{8}$$

Candidates are chosen for the generation of additional chromosomes for the subsequent population according to the calculated fitness value of each chromosome[30]. The crossover procedure is carried out by selecting a crossover point for the chromosomes of the parent. For both chromosomes, the whole process is repetitive and adding new offspring to the population. Some of their genes mutated in some new offspring formed, i.e., Some of the bits inverted in the bitstream. The best fit the chromosome selected as the desired object among the current survivors. Further we set the simulation of GA by giving the population size is 30, and consequently, the mutation and cross over are 0.001, single point, respectively. The process step of GA portrayed below.

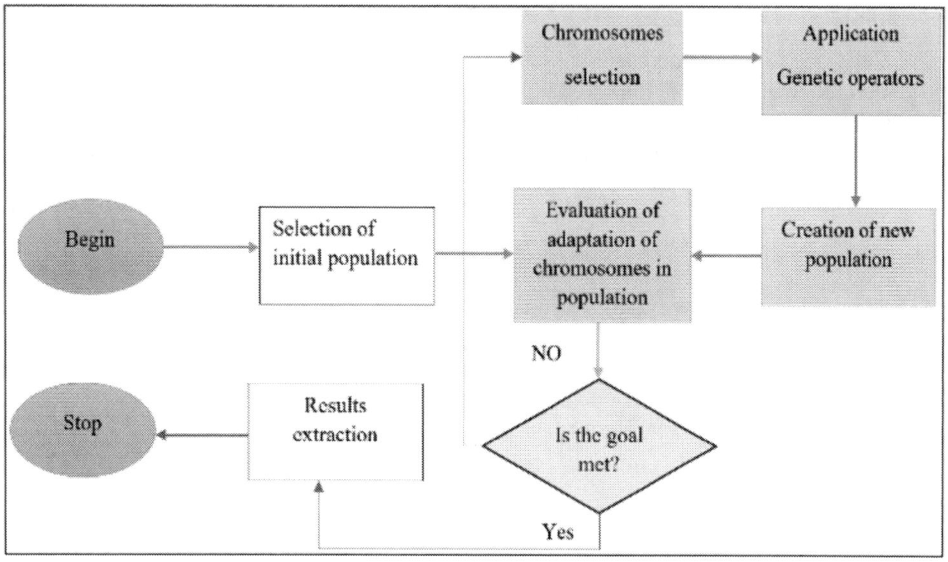

Figure3. Flowchart of genetic algorithm steps

Table 1. Previous studies

Ref.	OFDM technique	Brief description	Advantage	Drawback
[17]	PTS	Heuristics search of optimum combination used as a proposed PSO scheme to get low complex phase factors	Got lower complexity than the optimum method	PAPR slightly high
[21]	Comparis on of GA and PSO OFDM	Adaptive subcarrier and bit allocations were applied to GA and PSO to minimize the overall transmit power of an OFDM multi-user system.	PSO better than GA in term of coding capability, execution time, simplicity and computational resource	PSO takes higher power while GA got better PAPR reduction
[18]	PTS-OFDM	Presented an OFDM system that uses a sub-optimal PTS method based on particle swarm optimization (PSO) to find optimal phase weighting factors	Got power reduction same as the optimal PTS with low complexity	Computation complexity is direct proportionally with iterations
[19]	PTS-OFDM	Presents a new approach to overcoming computational complexity based on PSO	Got lower complexity	The PAPR of the proposed method is higher than compared exhaustive search

[23]	FWA-PTS in OFDM	Suggested a reduced computational complexity in fireworks algorithm (FWA) PTS-OFDM bases to mitigate PAPR	Got minimum complexity	PAPR Does not reduce
[22]	GA-PTS, PSO PTS, ABC-PTS A, and BBO-PTS	Described the PAPR reduction technique for reducing various computational, time, and space complexity factors, thereby making the system more optimized.	GA outperformed PSO, and another optimization in lessening PAPR and PSO is easier to implement and has fewer components	Got higher complexity compared to another mentioned optimization
[20]	GA-PTS using PCGA	uses a two-stage crossover operator and a mutation operator to generate the offspring and selects the appropriate offspring candidates for the new population	Got least PAPR than conventional GA-PTS	Still complexity load slightly high

5. Simulation Result

In this section, partial swarm optimization and the genetic algorithm have been studied extensively along with conventional PTS-OFDM features. The channel has assumed here is an additive white Gaussian noise (AWGN). Computer simulations conducted for 2048 as a number of a subchannel to run a 32-QAM PTS-OFDM system with N = 64 and Number of Monte-Carlo Iterations=100 iterations. The simulations have done with different numbers of initial population and swarms. Further, every algorithm iterated 100 times. Performance metrics for conducting a comparative evaluation between GA-PTS and PSO-PTS are considered PAPR and computational complexity as metrics; the simulation shows that GA needs more time to converge here than PSO in all cases. Also, the computational complexity GA more than PSO but it is outperformed PSO in terms of PAPR reduction as shown in fig 4. the complementary cumulative distribution function (CCDF) at 0.1 dB almost got the same PAPR but at 0.001dB GA got more power to lessen. Moreover, those two algorithms are remarkably outperforming the original PTS-OFDM.

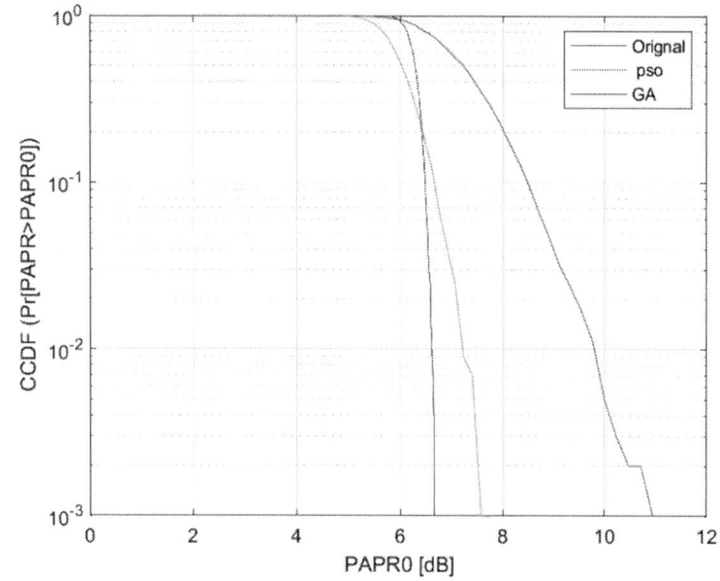

Figure4. comparison of PAPR using two optimizations and the original PTS-OFDM

6. Conclusion

The PAPR reduction performance of the PTS technique has been analyzed in this paper. The PTS search for phase rotation factors was simulated by a comprehensive search to obtain the optimal combination of phase factors, we have seen, whenever the number of sub-blocks increased the searching for complexity exponential increased as well. To overcome this challenge, we used GA and PSO algorithms to get optimum phase rotation factor with low complexity. The simulation result of implemented algorithms gives a practical approach to the trade-off between PAPR reduction and computational complexity. Further, GA remarkably gives higher performance in-term of PAPR lessen than PSO and conventional PTS, but the computational complexity of PSO notably lower than GA and the execution time as well.

7. Recommendation

Both of the evolutionary approaches outperform the conventional PTS. We can develop these algorithms to get better results by integrating or hybrid upcoming algorithms with these approaches. Further, the improvement of the PAPR lessen of GA and PSO-PTS in OFDM is still active research.

8. References

[1] D. Tse and P. Viswanath, Fundamentals of wireless communication: Cambridge university press, 2005.

[2] R. Prasad, OFDM for wireless communications systems: Artech House, 2004.

[3] I. G. Muhammad, K. E. Tepe, and E. Abdel-Raheem, "QAM equalization and symbol detection in OFDM systems using extreme learning machine," Neural Computing and Applications, vol. 22, pp. 491-500, 2013.

[4] N. Taşpınar and Ş. Şimşir, "Pilot tones design using particle swarm optimization for OFDM–IDMA system," Neural Computing and Applications, pp. 1-10, 2018.

[5] V.-N. Tran, "Hybrid scheme using modified tone reservation and clipping-and-filtering methods for peak-to-average power ratio reduction of OFDM signals," Signal Processing, vol. 158, pp. 166-175, 2019.

[6] X. Zhu, W. Pan, H. Li, and Y. Tang, "Simplified approach to optimized iterative clipping and filtering for PAPR reduction of OFDM signals," IEEE Transactions on Communications, vol. 61, pp. 1891-1901, 2013.

[7] S. Jiang, H. Deng, W. Xiao, and J. Du, "New nonlinear companding technique to decrease peak to average power ratio of the optical orthogonal frequency division multiplexing system based on Hartley transform," Optical Engineering, vol. 54, p. 106112, 2015.
[8] M. Peng, M. Chen, H. Zhou, Q. Wan, L. Jiang, L. Yang, et al., "Hybrid PAPR reduction scheme with Huffman coding and DFT-spread technique for direct-detection optical OFDM systems," Optical Fiber Technology, vol. 40, pp. 1-7, 2018.
[9] S.-y. Zhang and B. Shahrrava, "A Selected Mapping Technique Using Interleavers for PAPR Reduction in OFDM Systems," Wireless Personal Communications, vol. 99, pp. 329-338, 2018.
[10] T. Jiang, C. Ni, C. Ye, Y. Wu, and K. Luo, "A novel multi-block tone reservation scheme for PAPR reduction in OQAM-OFDM systems," IEEE Transactions on Broadcasting, vol. 61, pp. 717-722, 2015.
[11] W. Wang, M. Hu, Y. Li, and H. Zhang, "A low-complexity tone injection scheme based on distortion signals for PAPR reduction in OFDM systems," IEEE Transactions on Broadcasting, vol. 62, pp. 948-956, 2016.
[12] S.-H. Wang, W.-L. Lin, B.-R. Huang, and C.-P. Li, "PAPR reduction in OFDM systems using active constellation extension and subcarrier grouping techniques," IEEE Communications Letters, vol. 20, pp. 2378-2381, 2016.
[13] D. Mestdagh, J. G. Monsalve, and J.-M. Brossier, "GreenOFDM: a new selected mapping method for OFDM PAPR reduction," Electronics Letters, vol. 54, pp. 449-450, 2018.
[14] M.-J. Hao and P.-H. Huang, "Pilot assignment for PTS-OFDM with channel estimation for PAPR reduction," in Intelligent Signal Processing and Communication Systems (ISPACS), 2016 International Symposium on, 2016, pp. 1-6.
[15] Y. Rahmatallah and S. Mohan, "Peak-to-average power ratio reduction in OFDM systems: A survey and taxonomy," IEEE communications surveys & tutorials, vol. 15, pp. 1567-1592, 2013.
[16] Y. Polyanskiy and Y. Wu, "Peak-to-average power ratio of good codes for Gaussian channel," IEEE Transactions on Information Theory, vol. 60, pp. 7655-7660, 2014.
[17] J.-H. Wen, S.-H. Lee, Y.-F. Huang, and H.-L. Hung, "A suboptimal PTS algorithm based on particle swarm optimization technique for PAPR reduction in OFDM systems," EURASIP Journal on wireless communications and networking, vol. 2008, p. 14, 2008.
[18] H.-L. Hung, Y.-F. Huang, C.-M. Yeh, and T.-H. Tan, "Performance of particle swarm optimization techniques on PAPR reduction for OFDM systems," in 2008 IEEE International Conference on Systems, Man and Cybernetics, 2008, pp. 2390-2395.
[19] M. H. Aghdam and A. A. Sharifi, "PAPR reduction in OFDM systems: An efficient PTS approach based on particle swarm optimization," ICT Express, 2018.
[20] H. Liang, Y.-R. Chen, Y.-F. Huang, and C.-H. Cheng, "A modified genetic algorithm PTS technique for PAPR reduction in OFDM systems," in 2009 15th Asia-Pacific Conference on Communications, 2009, pp. 182-185.
[21] I. Ahmed and S. P. Majumder, "Adaptive resource allocation based on modified genetic algorithm and particle swarm optimization for multiuser OFDM systems," in 2008 International Conference on Electrical and Computer Engineering, 2008, pp. 211-216.
[22] A. Joshi, A. Garg, E. Garg, and N. Garg, "PAPR Reduction Analysis of OFDM Systems Using GA-, PSO-, and ABC-Based PTS Techniques," in Advances in Signal Processing and Communication, ed: Springer, 2019, pp. 145-155.
[23] A. Lahcen, H. Mustapha, E. Ali, A. Saida, and A. Adel, "Peak-to-Average Power Ratio Reduction Using New Swarm Intelligence Algorithm in OFDM Systems," Procedia Manufacturing, vol. 32, pp. 831-839, 2019.
[24] F. Sandoval, G. Poitau, and F. Gagnon, "Hybrid peak-to-average power ratio reduction techniques: Review and performance comparison," IEEE Access, vol. 5, pp. 27145-27161, 2017.
[25] R. Eberhart and J. Kennedy, "A new optimizer using particle swarm theory," in MHS'95. Proceedings of the Sixth International Symposium on Micro Machine and Human Science, 1995, pp. 39-43.
[26] M. Miyatake, M. Veerachary, F. Toriumi, N. Fujii, and H. Ko, "Maximum power point tracking of

multiple photovoltaic arrays: A PSO approach," IEEE Transactions on Aerospace and Electronic Systems, vol. 47, pp. 367-380, 2011.
[27] P. A. Abdulla, S. Aronis, M. F. Atig, B. Jonsson, C. Leonardsson, and K. Sagonas, "Stateless model checking for TSO and PSO," Acta Informatica, vol. 54, pp. 789-818, 2017.
[28] F. Marini and B. Walczak, "Particle swarm optimization (PSO). A tutorial," Chemometrics and Intelligent Laboratory Systems, vol. 149, pp. 153-165, 2015.
[29] Y. Kohama, T. Takada, N. Kozawa, and A. Miyamura, "Collapse analysis of rigid frame by genetic algorithm," WIT Transactions on The Built Environment, vol. 31, 1970.
[30] S.-S. Kim, M.-J. Kim, and T. A. Gulliver, "A new PTS for PAPR reduction by local search in GA," in The 2006 IEEE International Joint Conference on Neural Network Proceedings, 2006, pp. 2370-2373.

Acknowledgments
The author would like to acknowledge the support from the Fundamental Research Grant Scheme (FRGS) under a grant number of ***FRGS/1/2017/ICT03/UNIMAP/02/3*** from the Ministry of Higher Education Malaysia.

A Mini-Review of Coupled Convection-Diffusion Equations in a Fixed-Bed Adsorption

Nurhusnina Mohd Supian[1], **Norzilah Abdul Halif**[2] **and Nursalasawati Rusli**[1]

[1] Institute of Engineering Mathematics, Universiti Malaysia Perlis (UniMAP), Pauh Putra Campus, 02600 Arau, Perlis, Malaysia.
[2] Centre of Excellence of Frontier Materials Research, School of Materials Engineering, Universiti Malaysia Perlis (UniMAP), Jalan Kangar-Alor Star, Kampung Seriab, 01000 Kangar, Perlis, Malaysia.

nurhusnina.supian@gmail.com

Abstract. Environmental contamination triggered by dyes has gained global attention. Industrial effluent of dye in high concentration into the rivers must be reduced to minimise the harmful effect on the quality of water that threatens human health. The removal of dye from wastewater is a significant step in addressing the problem of dye emission which can be achieved through adsorption separation technologies. In the most adsorption process, the adsorbent is in contact with fluid in a fixed bed. The performance of adsorptive separation of dye can be predicted through a fixed bed mathematical model which consists of coupled partial differential equations for optimizing the design and operating conditions. This paper presents a mini-review of mathematical modelling of fixed-bed adsorption of dye. The discussions are limited to the used of single component-dye separation in, convection-diffusion equations coupled with Langmuir isotherm. Besides, the numerical methods used for solving the convection-diffusion equations are also discussed in this paper.

1. Introduction

Pollution triggered by industrial wastewater has become a regular problem for many countries [1]. Dyes are the most common water pollutants that broadly exist in industrial waste streams [2]. The most significant sources of dyes released are from industries of cosmetics, textile, printing, plastics, paper production, leather and food industries [3]. Wastewater containing dyes has much impact on environmental protection that causes harm to aquatic life and humans [4]. A primary concern of increasing dye concentration and its effect on water quality has caused major environmental agencies and organisations effort including the United States Environmental Protection Agency (US EPA) to control and reduce the highly toxic and hazardous pollutants [5]. Therefore, it is essential to eliminate dyes from wastewater emitted by industries to avoid excess dyes discharge [4]. Various methods of treatment, for example chemical oxidation, electrochemical oxidation, coagulation, degradation, biological process and adsorption are increasingly been applied to reduce the presence of dye in wastewater [6].

Currently, the most extensively used technology for the elimination of dye from wastewater is adsorption [7] since it is taken into account as the best and most universal equilibrium separation method [8]. Besides, it was reported as an effective, efficient and economical method [7]. Generally, two types of adsorption can be categorised as static adsorption and dynamic adsorption. Static adsorption is known as batch adsorption that takes place in a closed system that only is used to treat the low volume of wastewater [9]. However, this process is not suitable in the case of column-mode operations [10]. Hence, fixed-bed adsorption is found to be more convenient over batch technique according to an industrial point of view as the adsorbent continuously remains in contact with fresh sorbate solution [8]. Fixed-bed adsorption is normally used for gas and liquid pollution control in flowing mode [9]. It is designed in such a way that the flowing polluted fluid encounters a fixed amount of adsorbent thereby creating room for treatment of a large volume of effluent fluid with less monitoring requirement [9]. Moreover, the system is easy to operate, low price and can easily be scaled up [9]. High selectivity and working capacity, as well as strong adsorption capability, is the key to choose a suitable adsorbent in the design of the dye adsorption process. As a result, a variety of adsorbents, such as alumina [11], silica gel [12], zeolites [13] and activated carbon [14] have been examined in recent years for this purpose. Many studies have chosen activated carbons in the elimination of different types of dyes since they are good materials but however, their use is limited and increase in cost [15]. Besides, it will become finished and are no longer be able to further adsorb the dyes. This has caused in attempts to prepare alternative adsorbents that are a low cost [16] from industrial waste, natural material, or agricultural by-products as these materials do not require any expensive additional pre-treatment step [17].

Determine the breakthrough curve is the main information needed in the design of appropriate column adsorption [18]. To gain the breakthrough curve of a given column adsorption system, two approaches that are widely used are direct experimentation or mathematical modelling [18]. A direct and brief breakthrough curve of a given system can be constructed from the experimental method. Nevertheless, it is regularly a time-consuming and economically is not always a desirable process, mostly for the trace pollutants and long amount of time that adsorbed molecules of dyes spend on the surface. Also, it greatly relies on the experimental condition, such as air temperature surrounding a component and modelling of residence time [18]. Comparatively, mathematical modelling is easy [18] and required in design and optimisation adsorption columns for representing and correlating the breakthrough curve data of the experiment [19]. Additionally, the number of experiments associated with new operating conditions can be minimised once mathematical modeling has been validated [20]. Attention among the researchers in the mathematical modelling process has increasingly attracted due to the reasons stated before.[18]. Therefore, the primary aim of this paper is to review recent research on the convection-diffusion equation and the numerical study used to simulate the behavior of the dye column adsorption system. Numerous methods for solving the mathematical equation for dye column-adsorption as well as different descriptions of the mass transfer mechanisms within the particle of adsorbent are reviewed.

2. Description of the Mathematical Equation

2.1. Convection-Diffusion Equation Coupled with Source Term

The fixed-bed adsorption system can be analysed and understood through the formulation and mathematical model. Mathematical model development plays a vital role in the prediction of breakthrough curve behavior [18]. The mathematical equations describing the fixed-bed adsorption are partial differential equations that can be derived based on the mass conservation principle for fluid. The general form of a mass balance equation can be written as [21]

$$Rate\ of\ material\ in + Rate\ of\ material\ out = Rate\ of\ accumulation\ of\ material + Rate\ of\ loss\ by\ adsorption$$

The modelling of dye column adsorption has been divided into four basic parts: (1) liquid phase mass transfer including convective mass transfer and molecular diffusion, (2) interface diffusion between bulk solution and external surface of adsorbents and (3) intraparticle mass transfer including pore diffusion and surface diffusion. Liquid phase mass transfer is where the movement of molecules or ions in the column are in both axial and radial directions and usually is considered by other researchers in modelling of the fixed-bed adsorption process [6] [20] [22] [23] [24]. It is general to assume that homogeneous for all cross-sections and the movement is in axial as for simplification. Thus, radial movement can be neglected. Therefore, the equation for a mass balance that is acquired to represent the dynamic behavior of dye through fixed-bed is the convection-diffusion equation coupled with source term as shown in Equation (1) [6] [20] [22] [23] [23] [25]. This equation includes axial dispersion term, convection flow term, accumulation in the fluid phase and adsorption term.

$$-D_L \frac{\partial^2 c}{\partial z^2} + \frac{\partial}{\partial z}(vc) + \frac{\partial c}{\partial t} + \frac{(1-\varepsilon)}{\varepsilon}\frac{\partial \bar{q}}{\partial t} = 0 \quad (1)$$

Here, c is the concentration of the bulk solution, D_L is coefficient of axial dispersion, v is the superficial velocity, t is time, ε is the bed porosity, z is the distance to inlet and $\bar{q}$ is the average adsorbed phase dye concentration [22]. Equation (1) is derived considering the assumptions as follows : (1) the system is in condition of isothermal, (2) no occurrence of chemical reaction in column, (3) radial dispersion is negligible, (4) the filling material is made of particles that are porous and uniform in size, (5) the flow rate is constant and (6) the equilibrium of adsorption is described by Langmuir isotherm The next step is the interface diffusion between bulk solution and outer surface of adsorbent [18]. This is the first step in the adsorption process where the dye is transported from the fluid to the exterior surface of adsorbent across the boundary layer [6] [20] [22] [23] [24], and it is known as the film diffusion. It predominates the total uptake rate to some level and in some cases turn out to be the rate control step. The rate of mass transfer is stated as Equation (2) below;

$$\frac{\partial \bar{q}}{\partial t} = k_f a(c - c_s) \quad (2)$$

where a is volumetric surface area, c_s is the dye concentration at the exterior surface of adsorbent and k_f is the coefficient of film diffusion. The second step in column adsorption is intrapellet diffusion.

The intraparticle or internal particle diffusion is a complicated step and plays a key role in the consideration of modelling column adsorption [6] [20] [22] [23] [24]. The diffusion of the dye occurs in the liquid within the pores of the adsorbent (pore diffusion) or along the pore walls (surface diffusion) are involved in intrapellet transfer simultaneously. Normally, Fick's law is used to describe the intraparticle diffusion of the dye inside the pores of adsorbent particles. However, to present the mass transfer into the pellets, some of the researchers [20] [22] [23] [24] apply a linear driving force to simplify the solution of the differential equations. It considers that the rate of mass transfer at a given point of the bed is proportional to the difference between the current of solute concentration on the adsorbent at a local level and that corresponding when the equilibrium with the bulk phase is achieved. The rate of uptake expression based on a linear driving force model is as shown [20]

$$\frac{\partial \bar{q}}{\partial t} = k_e\left(q - \bar{q}\right) \tag{3}$$

Where q is the loading of the dye at the exterior surface of the adsorbent and it is obtained from the Langmuir isotherm and k_e is the kinetic coefficient for intraparticle diffusion in the adsorbent. The process of dye flow through a packed bed based on Equation (1) is represented in Figure 1 [21].

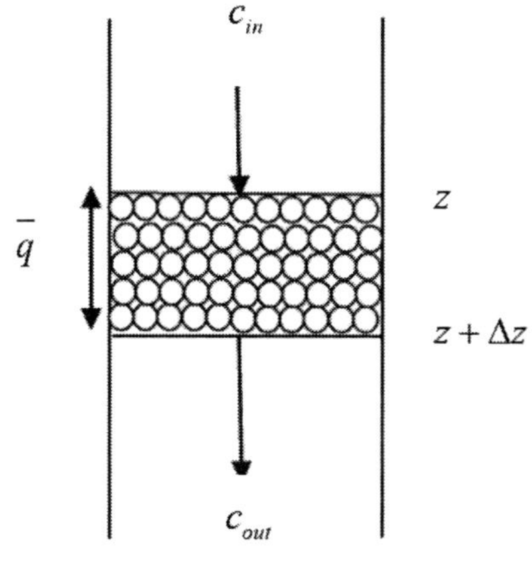

Figure 1. Flowing of dye through a packed-bed

Another relationship is required to formally describe the fourth term on the left-hand side of Equation (1) as to relate the concentration of dye in the bulk with the adsorbed amount at the interface. This relationship is vital for the last step of adsorption process itself onto the surface and the relationship is given by the Langmuir isotherm as following [1] [16] [26] [27] [28];

$$q = \frac{q_m K c_s}{1 + K c_s} \tag{4}$$

where q is the equilibrium concentration in the solid phase, q_m is the maximum adsorption capacity of the adsorbent, K is the Langmuir isotherm constant and c_s is the concentration of solute in the liquid phase inside the pores or equilibrium concentration in the liquid phase [18]. Choosing the isotherm style is a prerequisite in the modelling of the dynamic adsorption as it will directly affect the effect of mathematical modelling. The Langmuir isotherm assumes that the adsorption process occurs as a monolayer, homogeneous for the surface of the adsorbent and no interactions between the adsorbed molecules. No additional adsorption can occur at that site when once a site is occupied with dye molecule, it has been conclusively shown that the adsorption process is a monolayer in nature. Langmuir isotherm is frequently chosen by the researchers [6] [20] [22] [23] [24] due to its simple form and good-fitting performance [18]. Therefore, Langmuir isotherm has turn-into one of the most famous models in adsorption studies. Thus, Equation (1) along with Equation (4) gives a coupled system of differential equations due to the temperature-dependent in concentration and the adsorption rate [30]. Two dependent variables which make Equation (1) to be coupled are c and $\bar{q}$.

Proposing a general model is a difficult task but significant when considering the different components of the adsorption systems (dye and adsorbent), a variation of operation conditions and specific demands of accuracy and calculative simplicity since most models are derived from different assumptions. The differences in the mathematical models of adsorption processes arise from the different expressions of the mass transfer inside and outside the adsorbent particle and the equilibrium isotherm considered in convection-diffusion equation for the dye in a fixed-bed column [23]. Table 1 summarizes the main assumptions and considerations about the convection-diffusion equation for modelling the fixed-bed adsorption of dye reported in the literature.

Table 1. Summary of mathematical models based on convection-diffusion equations for fixed-bed adsorption of dye

Adsorbate	Adsorbent	Assumptions	Equation	Reference
Maxilon Goldgelb GL EC 400 % (MG-400) and Maxilon Schwarz FBL-01 300 % (MS-300).	Granular activated carbon and natural zeolite	• Isothermal adsorption of a single component. • Negligible concentration gradient in the radial direction. • The Linear Driving Force (LDF) model is used. • The cross-section is constant and the properties of the adsorbent bed throughout the column are uniform.	Governing equation: $v\frac{\partial c}{\partial z}+\frac{\partial c}{\partial t}+\frac{1-\varepsilon}{\varepsilon}\frac{d}{dt}\left(\varepsilon_p \bar{c}+\bar{q}\right)=D_L\frac{\partial c}{\partial z^2}$ Rate of mass transfer of dye from bulk fluid to the external surface of adsorbent: $m=\frac{d}{dt}\left(\varepsilon_p \bar{c}+\bar{q}\right)$ In terms of external mass transfer, $m=\frac{3k_f}{R_p}(c-c_s)$ Rate of uptake equation based on the linear driving force (LDF) model for solid diffusion: $m=\frac{15D_s}{R_p^2}\left(q-\bar{q}\right)$ Langmuir equation:	[23]

IOP Conf. Series: Materials Science and Engineering **767** (2020) 012031 doi:10.1088/1757-899X/767/1/012031

			$q=\frac{q_m K c_s}{1+K c_s}$	
Dye	Natural Zeolite	• Isothermal adsorption of a single component. • Negligible concentration gradient in the radial direction. • The Linear Driving Force (LDF) model is used. • Constant cross-section and uniform properties of adsorbent bed throughout the column.	Fluid phase mass balance equation: $-D_L\frac{\partial^2 c}{\partial z^2}+\frac{\partial}{\partial z}(vc)+\frac{\partial c}{\partial t}+\left(\frac{1-\varepsilon}{\varepsilon}\right)\frac{\partial \bar{q}}{\partial t}=0$ Mass transfer rate of dye from bulk fluid to external adsorbent surface: $m=\frac{d\bar{q}}{dt}$ External mass transfer phase, $m=\frac{3k_f}{R_p}(c-c_s)$ Rate of uptake equation based on the LDF model for solid diffusion: $m=\frac{15D_s}{R_p^2}\left(q-\bar{q}\right)$ Langmuir equation: $q=\frac{q_m K c_s}{1+K c_s}$	[22]
Basic Green 4	Activated carbon	• The superficial velocity inside the column is constant. • The flow moves in the axial direction. • Isothermal system. • No chemical reaction occurs on the bed. • Dispersion in the radial direction is negligible.	Conservation equation for dye: $\frac{\partial c}{\partial t}+\left(\frac{1-\varepsilon}{\varepsilon}\right)\frac{\partial \bar{q}}{\partial t}+v\frac{\partial c}{\partial z}-D_L\frac{\partial^2 c}{\partial z^2}=0$ LDF model is applied to describe the adsorption rate process of the dye for the fluid phase of the adsorbent in terms of the global mass transfer of the fluid phase is shown below: $\frac{\partial \bar{q}}{\partial t}=k_f a\left(q-\bar{q}\right)$ Langmuir isotherm: $q=\frac{q_m K c_s}{1+K c_s}$	[20]

IOP Conf. Series: Materials Science and Engineering **767** (2020) 012031 doi:10.1088/1757-899X/767/1/012031

		• The porosity of the bed inside the column is constant.		
C.I Acid Blue 92	Exfoliated Graphite	• Implementing nonlinear isotherm. • Linear driving force mass transfer. • Axial dispersion	Applying mass conservation for dyestuff in the liquid phase: $\left(\frac{\rho_b}{\varepsilon}\right)\frac{\partial \bar{q}}{\partial t}+\frac{\partial c}{\partial t}++u\frac{\partial c}{\partial z}-D_L\frac{\partial^2 c}{\partial z^2}=0$ Mass transfer resistance in the liquid phase is described by the LDF approximation: $\frac{\partial \bar{q}}{\partial t}=k_f\left(c-c_s\right)$ Langmuir isotherm: $q=\frac{q_m K c_s}{1+K c_s}$	[6]
Reactive blue 5G dye	Dowex Optipore SD-2	• Axial dispersion model. • Homogeneously distributed adsorption particles. • Isothermal and isobaric process. • Constant physical properties of the bed and liquid phases.	**Model 1** Mass balance in the liquid phase: $\frac{\partial c}{\partial t}+v\frac{\partial c}{\partial z}+\left(\frac{\rho_b}{\varepsilon}\right)\frac{\partial \bar{q}}{\partial t}-D_L\frac{\partial^2 c}{\partial z^2}=0$ External mass transfer resistance: $\frac{\partial \bar{q}}{\partial t}=\frac{k_f \varepsilon}{\rho_B}\left(c-c_s\right)$ Langmuir isotherm: $q=\frac{q_m K c_s}{1+K c_s}$ **Model 2** Mass balance in the liquid phase: $\frac{\partial c}{\partial t}+v\frac{\partial c}{\partial z}+\left(\frac{\rho_b}{\varepsilon}\right)\frac{\partial \bar{q}}{\partial t}-D_L\frac{\partial^2 c}{\partial z^2}=0$ LDF model is used for internal mass transfer resistance $\frac{\partial \bar{q}}{\partial t}=-k_f\left(\bar{q}-q\right)$	[24]

Langmuir isotherm:

$$q = \frac{q_{\mathrm{m}} K c_s}{1 + K c_s}$$

Model 3

Mass balance in the liquid phase:

$$\frac{\partial c}{\partial t} + v\frac{\partial c}{\partial z} + \left(\frac{\rho_b}{\varepsilon}\right)\frac{\partial \bar{q}}{\partial t} - D_L\frac{\partial^2 c}{\partial z^2} = 0$$

Adsorption in the adsorbent sites described by pseudo second-order reaction rate law:

$$\frac{\partial \bar{q}}{\partial t} = k_a\left(q_{\mathrm{m}} - \bar{q}\right)c - k_d\bar{q}$$

Reduces to Langmuir isotherm at equilibrium:

$$\frac{\partial \bar{q}}{\partial t} = k_a\left(q_{\mathrm{m}} - \bar{q}\right)c - \frac{k_a}{K}\bar{q}$$

As can be seen from Table 1 (above), a number of researchers have proposed a few mathematical models to describe the dye column adsorption process, which can commonly be categorised into adsorption reaction models [24] and diffusion models [6] [20] [22] [23]. The kinetic process of adsorption can be described based on both models. However, both models are quite different. It has been noted that the construction of adsorption diffusion models are usually based on three different stages in a fixed-bed adsorption process such as external, internal mass transfer and adsorption in adsorbent sites where the main equation involved is the convection-diffusion equation, Equation (1) together with Equation (2), Equation (3) and Equation (4). However, Marin, Borba, Módenes, Espinoza-Quiñones, Oliveira and Kroumov [24] have used the adsorption reaction model in their research. This model originating from chemical reaction kinetics that is based on the entire process of adsorption without considering stages in the fixed-bed adsorption process. They describe the adsorption in the adsorbent site by using a pseudo-second-order reaction rate model. Generally speaking, this model [24] cannot offer valuable information to obtain a deep understanding of the adsorption mechanism and to design a system of fixed-bed and are therefore not representative of the real course of adsorption [30]. On the contrary, a real adsorption course can be represented more reasonably through the adsorption diffusion models that are based upon basic steps together with the convection-diffusion equation [30]. The mathematical models [6] [20] [23] [24] develop have considered the convection-diffusion equation together with mass transfers are shown to be suitable for the prediction of concentration of dye in the fixed-bed adsorption since they can represent a significant tool for the textile industry used. To solve the mathematical formulation, the numerical method is needed for the simulation of fixed-bed dynamics.

2.2. Numerical Method

The solution of the coupled convection-diffusion equation and their associated initial and boundary conditions play a significant role in the modelling of the breakthrough curve for a fixed-bed adsorption system of dye. Accurate prediction of the breakthrough curve is crucial to determine the operation and the dynamic behaviour of column adsorption. Very few partial differential equations have the analytical or exact solution due to the complexity and nonlinearity of mathematical. Even in linear and extremely simplified cases, analytical solutions are scarce and generally slowly converging. Hence, numerical methods are the only possible alternative to find an approximate solution of the convection-diffusion equation to simulate the breakthrough curve of the fixed-bed column [31].

Several sophisticated models have been successfully developed by researchers to solve equations by implementing numerical methods that are finite difference method (FDM) [6] [22] [23] and finite volume method (FVM) [20]. FDM is the oldest method used to replace the derivatives in the differential equations by finite difference approximations which give an algebraic system of equations to be solved in place of the differential equation [32]. The major advantages of FDM are that it is simple, efficient in solving parabolic equations and ease of varying initial and boundary conditions [31]. However, this method cannot directly be applied in curvilinear coordinates, without transforming into a Cartesian coordinate system [33]. Markovska, Meshko, Noveski and Marinkovski [23] have transformed the coupled convection-diffusion equation into a set of an ordinary differential equation using FDM and then solve using SPEEDUP software for adsorption of Maxilon Goldgelb GL EC 400 % (MG-400) and Maxilon Schwarz FBL-01 300 % (MS-300) onto granular activated carbon and natural zeolite. The result appeared a satisfying agreement between the experimental and the simulation of breakthrough curves. However, the model is not quite fulfilled for these experiments as some deviations are found between the experimental and predicted data. This denotes especially to the assumption of a constant coefficient of solid diffusion. The values of the solid diffusion coefficient, the coefficient of axial dispersion and the coefficient of external mass transfer are required to be manipulated to obtain satisfactory agreement between the simulated and the experimental breakthrough curves.

Babu and Gupta also [22] considered solving the equations simultaneously using the explicit finite difference method and simulating the solution using MATLAB. The simulations are carried out to understand the influence of axial dispersion, external film resistance and solid diffusion resistance on adsorption of dye on natural zeolite. However, they found the axial dispersion, coefficient of solid diffusion and external mass transfer need to be adjusted to give the best-fit curve. Goshadrou and Moheb [6] also applied FDM in discretization and simulate the behavior by ODE integrator subroutine of MATLAB for C.I Acid Blue 92 dye adsorption on exfoliated graphite. The breakthrough curves obtained from modeling and experimental work are compared showing a very good agreement.

FVM used a discretization technique for partial differential equations based on the integral form of the conservation laws, rather than their differential form [34]. This method is particularly useful for direct discretization in the physical space [33]. Thus, there are no problems with any kind of transformation between the physical and the computational coordinate system. Souza, Peruzzo and de Souza [20] obtain a discretized form of the coupled convection-diffusion equation and showed a good agreement with the experimental data. Model [20] based on the Basic Green 4 dye on granular activated carbons shown to be convenient for the prediction of the mass transfer process. Downhill simplex optimization method is used to solve the coupled convection-diffusion equations. The mass transfer limiting step of reactive Blue 5G dye is identified through the simulation result using Maple software for removal process in a fixed-bed column packed with Dowex Optipore SD-2 adsorbent.

3. Conclusion

Mathematical modelling and simulation processes are the essential tools in the design and understanding of the fixed-bed adsorption system performance under various conditions to improve operation. A review of mathematical modelling of fixed-bed adsorption of dye conducted by various researchers is provided in this study. Most of the model reviewed here include mass transfer rate, resulting in a theoretical represent that more closely approximates an actual process. Recently, researchers have shown an increased interest in solving a convection-diffusion equation together with mass transfer rate and Langmuir isotherm for modelling and design of a column adsorption process. So far, however, there has been little discussion about solving the convection-diffusion equation for dye fixed-bed column adsorption. The solution of the convection-diffusion equation requires a numerical method to solve the equation. The convection-diffusion equation has been solved by many researchers mostly using FDM and FVM. There are two main obstacles may be faced in the simulation of convection-diffusion equations using numerical methods. This is due to the coupling between convection-diffusion equations with the adsorption isotherm. The used of a reduced mathematical model is recommended in getting the simulation results since it can aid in the identification of mass transfer mechanism, interpretation and analysis of experimental data, prediction of responses due to changes in the operational condition and process optimisation [24]. Simplify the representation of mass transfer phenomena within the adsorbent particles is one of the alternatives in facilitating the fixed-bed adsorption calculation. To represent numerical data realistically, it is suggested to solve the convection-diffusion equation with the combination of other numerical methods or improved methods [31] to accurately represent the dye adsorption onto a fixed-bed column and thus, several experiments associated with new operating conditions can be reduced.

Acknowledgements

The authors would like to acknowledge the financial support from Fundamental Research Grant 2019 (FRGS/1/2019/STG06/UNIMAP/02/6) for the development of this research work.

References

[1] Hamdaoui O 2006 Batch study of liquid-phase adsorption of methylene blue using cedar sawdust and crushed brick *Journal of Hazardous Materials* **135** 264-73
[2] Mahmoodi NM, Taghizadeh M and Taghizadeh A 2019 Activated carbon/metal-organic framework composite as a bio-based novel green adsorbent: Preparation and mathematical pollutant removal modeling *Journal of Molecular Liquids* **277** 310-22
[3] Ma D, Zhu B, Cao B, Wang J and Zhang J 2017 Fabrication of the novel hydrogel based on waste corn stalk for removal of methylene blue dye from aqueous solution. *Applied Surface Science* **422** 944-52
[4] Wu L, Sun J and Wu M 2017 Modified cellulose membrane prepared from corn stalk for adsorption of methyl blue *Cellulose* **24** 5625-38
[5] Vikrant K, Giri BS, Raza N, Roy K, Kim KH, Rai BN and Singh RS 2018 Recent advancements in bioremediation of dye: current status and challenges *Bioresource Technology* **253** 355-67
[6] Goshadrou A and Moheb A 2011 Continuous fixed bed adsorption of CI Acid Blue 92 by exfoliated graphite: An experimental and modeling study *Desalination* **269** 170-6
[7] Sarode S, Upadhyay P, Khosa MA, Mak T, Shakir A, Song S and Ullah A 2018 Overview of wastewater treatment methods with special focus on biopolymer chitin-chitosan. *International Journal of Biological Macromolecules* **121** 1086-100
[8] Kukučka M, Kukučka N and Kukučka A 2016 A novel approach to adsorption kinetics calculation

Microporous and Mesoporous Materials **228** 123-31

[9] Auta M and Hameed BH 2014 Chitosan–clay composite as highly effective and low-cost adsorbent for batch and fixed-bed adsorption of methylene blue. *Chemical Engineering Journal* **237** 352-61

[10] Foroughi-dahr M, Esmaieli M, Abolghasemi H, Shojamoradi A and Sadeghi Pouya E 2016 Continuous adsorption study of congo red using tea waste in a fixed-bed colum *Desalination and Water Treatment* **57** 8437-46

[11] Adak A and Pal A 2006 Removal kinetics and mechanism for crystal violet uptake by surfactant-modified alumina *Journal of Environmental Science and Health* **41** 2283-97

[12] Murathatunyaluk S, Srichanvichit K, Anantpinijwatna A and Kitchaiya P 2018 Modeling of non-isothermal adsorption process in a silica gel desiccant packed bed *Computer Aided Chemical Engineering* **43** 49-54

[13] Zhang Y, Jin F, Shen Z, Wang F, Lynch R and Al-Tabbaa A 2019 Adsorption of methyl tert-butyl ether (MTBE) onto ZSM-5 zeolite: Fixed-bed column tests, breakthrough curve modelling and regeneration *Chemosphere* **220** 422-31

[14] Tan IA, Ahmad AL and Hameed BH 2008 Adsorption of basic dye using activated carbon prepared from oil palm shell: batch and fixed bed studies *Desalination* **225** 13-28

[15] Gupta VK 2009 Application of low-cost adsorbents for dye removal–a review *Journal of Environmental Management* **90** 2313-42

[16] Wen X, Yan C, Sun N, Luo T, Zhou S and Luo W 2018 A biomass cationic adsorbent prepared from corn stalk: low-cost material and high adsorption capacity *Journal of Polymers and the Environment* **26** 1642-51

[17] Han R, Wang Y, Zhao X, Wang Y, Xie F, Cheng J and Tang M 2009 Adsorption of methylene blue by phoenix tree leaf powder in a fixed-bed column: experiments and prediction of breakthrough curves *Desalination* **245** 284-97

[18] Xu Z, Cai JG and Pan BC 2013 Mathematically modeling fixed-bed adsorption in aqueous systems *Journal of Zhejiang University SCIENCE A* **14** 155-76

[19] Ahmed MJ and Hameed BH 2018 Removal of emerging pharmaceutical contaminants by adsorption in a fixed-bed column: a review. *Ecotoxicology and Environmental Safety* **149** 257-66

[20] De Souza SM, Peruzzo LC and de Souza AA 2008 Numerical study of the adsorption of dyes from textile effluents *Applied Mathematical Modelling* **32** 1711-8

[21] Yusuff AS, Popoola LT, Omitola OO, Adeodu AO and Daniyan IA 2013 Mathematical modelling of fixed bed adsorption column for liquid phase solute: effect of operating variables *International Journal of Scientific & Engineering Research* **4** 811-22

[22] Babu BV and Gupta S 2004 Modeling and simulation for dynamics of packed bed adsorption *Proceedings of international symposium & 57th annual session of IIChE in association with AIChE (CHEMCON-2004)*

[23] Markovska L, Meshko V, Noveski V and Marinkovski M 2001 Solid diffusion control of the adsorption of basic dyes onto granular activated carbon and natural zeolite in fixed bed *Journal of the Serbian Chemical Society* **66** 463-75

[24] Marin P, Borba CE, Módenes AN, Espinoza-Quiñones FR, de Oliveira SP and Kroumov AD 2014 Determination of the mass transfer limiting step of dye adsorption onto commercial adsorbent by using mathematical models *Environmental Technology* **35** 2356-64

[25] Selvaraju N and Pushpavanam S 2009 Adsorption characteristics on sand and brick beds *Chemical Engineering Journal* **147**(2-3) 130-8

[26] Blanco SP, Scheufele FB, Módenes AN, Espinoza-Quiñones FR, Marin P, Kroumov AD and Borba CE 2017 Kinetic, equilibrium and thermodynamic phenomenological modeling of reactive dye adsorption onto polymeric adsorbent *Chemical Engineering Journal* **307** 466-75

[27] Srisorrachatr S, Kri-arb P, Sukyang S and Jumruen C 2017 Removal of basic dyes from solution

using coconut shell charcoal *InMATEC Web of Conferences* **119** 01019

[28] Elmoubarki R, Moufti A, Tounsadi H, Mahjoubi FZ, Farnane M, Machrouhi A, Elhalil A, Abdennouri M, Zouhri A and Barka N 2016 Kinetics and thermodynamics study of methylene blue adsorption onto Aleppo pine cones *Journals of Materials and Environmental Science* **7** 2869-79

[29] Shafeeyan MS, Daud WM and Shamiri A 2014 A review of mathematical modeling of fixed-bed columns for carbon dioxide adsorption *Chemical Engineering Research and Design* **92** 961-88

[30] Qiu H, Lv L, Pan BC, Zhang QJ, Zhang WM and Zhang QX 2009 Critical review in adsorption kinetic models *Journal of Zhejiang University-Science A* **10** 716-24

[31] Sun LM and Meunier F 1991 An improved finite difference method for fixed-bed multicomponent sorption *AIChE Journal* **37** 244-54

[32] LeVeque RJ 2007 *Finite difference methods for ordinary and partial differential equations: steady-state and time-dependent problems* (Washington: Society for Industrial and Applied Mathematics) chapter 1 p 3

[33] Blazek J 2005 *Computational fluid dynamics: principles and applications (*Butterworth: Elsevier) pp 37-38

[34] Eymard R, Gallouët T and Herbin R 2000 *Finite volume methods Handbook of numerical analysis* vol 7 (Elsevier) pp 713-1018

IOP Conf. Series: Materials Science and Engineering **767** (2020) 012032 doi:10.1088/1757-899X/767/1/012032

Design of C-band Low-noise Amplifier (LNA) Using E-pHEMT Device for Satellite Communication System

F Kamsaini[1], M S Razalli[2], S Z Ibrahim[3] and M Z Ilyas[4]

[1,2,3,4] Embedded, Networks and Advanced Computing Research Cluster (ENAC), School of Computer and Communication Engineering (SCCE), Universiti Malaysia Perlis (UniMAP), Pauh Putra Campus, 02600 Arau, Perlis, MY.

faaiqahkamsaini@gmail.com[1]

Abstract. A low-noise amplifier (LNA) plays an indispensable role in a communication system for amplification purposes at the transceiver. LNA design in radio frequency (RF) circuits involve various key attributes, namely noise figure (NF), gain, and power consumption. This paper focuses on the design of a C-band LNA for satellite communication system with a centre frequency of 6 GHz using ATF-55143 enhancement-mode pseudomorphic high-electron-mobility transistor (E-pHEMT) technology. The LNA performance is augmented by adding inductors to the drain and gate of the ATF-55143 transistor. Smith chart impedance matching technique is implemented to foster a more precise matching for input and output of the LNA. In this work, the C-band LNA is biased at V_{DS} of 2.7 V and I_{DS} of 10 mA. Electromagnetic (EM) software is used to design and simulate the performance of the LNA circuit layout. Simulation results indicate NF of 2.66 dB and power gain (S_{21}) of 12.29 dB. The LNA consumes 27 mW from a 3 V DC supply.

1. Introduction

Low-noise amplifier (LNA) is commonly the most crucial element in a radio frequency (RF) transceiver system. At the receiver section, LNA is used to amplify weak RF signals at microwave frequency with minimal distortion prior to passing them to the later stages of the system [1-2]. Critical parameters which determine the LNA performance are input and output impedance matching, noise figure (NF), gain, and linearity [3-5]. The LNA performance, in turn, has a dominant influence on the gain, NF, signal bandwidth, and power consumption of the overall reception chain [6-8].

The standard configuration of an LNA includes an input matching network, a transistor amplifier, and an output matching network [9]. Numerous LNA designs have been stated to apply high-electron-mobility transistors (HEMTs) or pseudomorphic high-electron-mobility transistors (pHEMTs) process due to their low-noise intrinsic traits [10-11]. However, pHEMT technology is favoured in LNA transceiver systems as it offers low noise, high gain, high reliability, high linearity, and high transconductance [12-13]. The C-band LNAs are also widely implemented for wireless local area networks (WLANs) and satellite communications since they offer low-noise performance, high data rate, and high speed with less power dissipation [14].

RF devices are built to function at the diverse frequency spectrum of the RF bands. The proposed LNA is initiated to operate at 6 GHz which is in the C-band frequency range. This particular band is appropriate for both wireless and satellite communications, and radiolocation applications [15].

Communication satellites operating in the C-band employ uplink/downlink frequencies of 6/4 GHz. The C-band is the first portion of the radio spectrum to be applied for satellite communications and is still widely used due to its usability in heavy rain and high accessibility of components.

Iyer and Shanmuganantham designed an LNA for C band applications with a power gain (S_{21}) of 13 dB and NF of 2.2 dB at 6 GHz [7]. In 2017, the design of LNA for satellite uplink proposed by Iyer and Shanmuganantham obtained an S_{21} of 8.96 dB and NF of 2.56 dB at 6 GHz [16]. A wideband p-HEMT LNA designed by Kumar and Pathak revealed an S_{21} of 9 dB and NF of 1.5 dB over the frequency range of 5-6 GHz [17]. Parkavi and Ravi has introduced an RF low noise and high gain amplifier for wireless communication in 2016 with S_{21} of 13.2 dB and NF of 0.8 dB at 6 GHz [18].

In this paper, an effort to design a C-band LNA operating at 6 GHz for satellite communication system using enhancement-mode pseudomorphic high-electron-mobility transistor (E-pHEMT) technology is initiated. Single stub matching network on both sides (input and output) of the transistor amplifier are designed and included in the LNA circuit to attain an accurately matched impedances for better S-parameter performance. Additionally, they also provide better bandwidth while ensuring maximum output power and maximum possible gain supply by the transistor. DC biasing is performed across the LNA system which permits DC power flow while simultaneously impedes the coupling of microwave signals to the DC supply. The LNA is designed by using electromagnetic (EM) simulator. Simulated results of the proposed C-band LNA are presented.

2. C-band LNA circuit design

The LNA design procedure is as listed below:

- Selecting a proper active device i.e. transistor.
- Inspecting the stability of the device using S-parameters.
- Employing appropriate techniques to stabilise the device if it is found unstable at the chosen operating frequency.
- Exerting proper biasing based on the operating goal desired and design purpose.
- Optimising the input and output matching circuit design.

The general LNA structure consists of an active device represented by its S-parameters [19]. When designing for an LNA, it is essential for the active device to have the least intrinsic noise. E-pHEMT is a semiconductor technology optimised for wireless applications, operating from a single positive voltage source. E-pHEMT devices are non-conductive and only sink current when both the drain and gate are biased, irrespective of the sequence. Therefore, negative voltage supply, drain switch, and sequencing circuitry are not necessary. This technology solves system-integration problems and allows pHEMT circuits to be easily applied for assorted applications [20].

Stability is determined based on the selected S-parameters using Rollet's stability factor given by (1) and (2). The transistor amplifier is unconditionally stable if K is greater than unity ($K > 1$) whereas it is conditionally stable if the K is less than unity ($K < 1$). Plotting stability circle on the Smith chart is crucial to establish knowledge in which area the LNA is not stable to avoid it from oscillating. Rollet's stability factor defined in [21] is given by:

$$K = \frac{1-|S_{11}|^2-|S_{22}|^2+|\Delta|^2}{2|S_{12}S_{21}|} \geq 1 \tag{1}$$

where

$$\Delta = S_{11}S_{22} - S_{12}S_{21} \tag{2}$$

In this work, the stability factor, K is 1.04. Thus, the device is unconditionally stable for any arrangement of source and load impedances.

Figure 1 shows the simulated K-factor of the E-pHEMT device (ATF-55143).

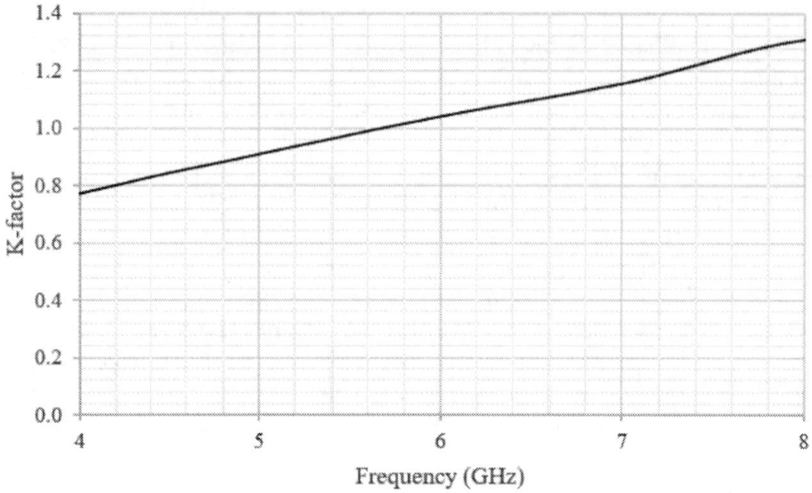

Figure 1. Simulated K-factor of ATF-55143.

If the transistor amplifier is conditionally stable ($\square < 1$), its gain becomes maximum stable gain (MSG) which is given by

$$\square\square\square = \frac{|\square_{21}|}{|\square_{12}|} \tag{3}$$

However, if the device is unconditionally stable ($\square > 1$), the gain is denoted as

$$\square\square\square = \frac{|\square_{21}|}{|\square_{12}|}(\square \pm \sqrt{\square^{2} - 1} \tag{4}$$

Figure 2 shows the simulated maximum allowable gain (MAG) of ATF-55143.

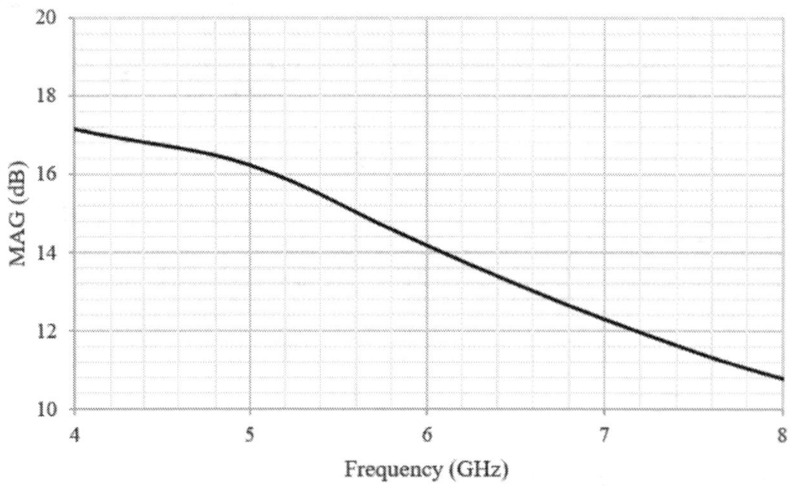

Figure 2. Simulated MAG of ATF-55143.

The E-pHEMT process is selected for the design as it reaches the requirements of the proposed C-band LNA. With reference to the stability analysis, biasing conditions for the LNA are determined. Operating bias points of ATF-55143 are designed by adhering to the datasheet with $V_{DS} = 2.7$ V and $I_{DS} = 10$ mA. The bias circuit includes resistors between the drain and gate of the transistor amplifier, and DC power supply, which represent a voltage divider network. The resistors are coupled with a series capacitor ($\square = 1\ \square\square$) in a feedback loop to inhibit feedback effect on the DC bias. The structure of the DC bias circuit is constructed by an EM simulator. Resistor values are calculated as $\square_1 = 4460\,\Omega$, $\square_2 = 28.6\,\Omega$, $\square_3 = 10\,\square\Omega$, and $\square_4 = 940\,\square$ to obtain the bias points mentioned. The bias circuit, powered with a 3 V DC supply, is shown in figure 3.

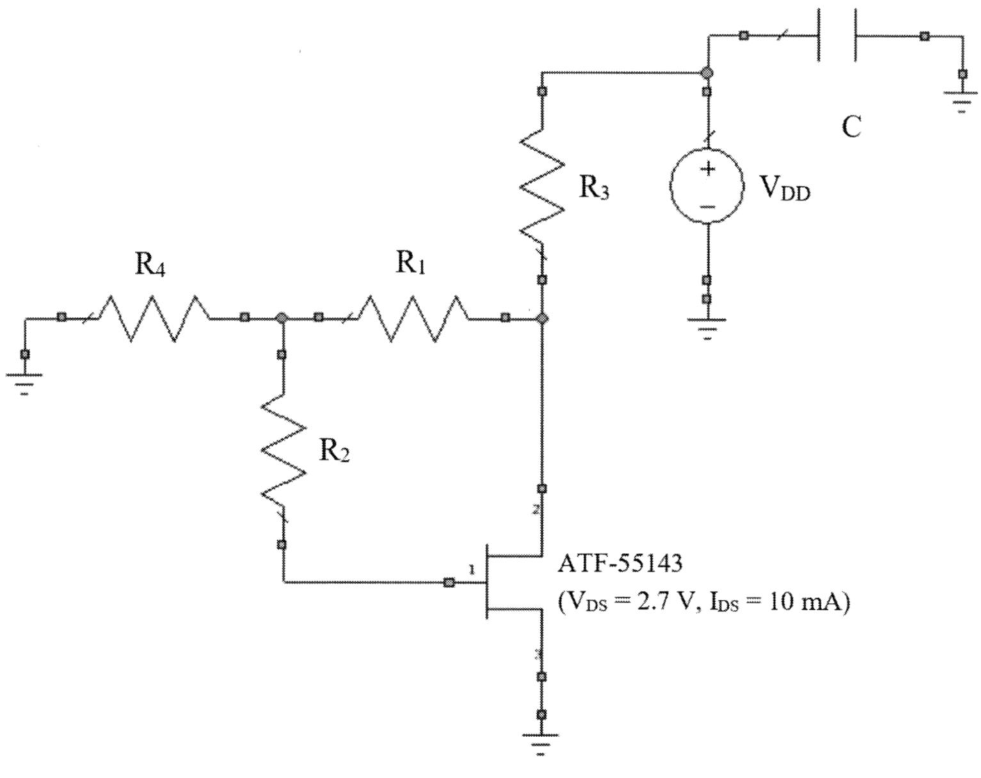

Figure 3. DC bias circuit with $V_{DS} = 2.7$ V and $I_{DS} = 10$ mA.

From the ATF-55143 datasheet, the S-parameter S_{12} (reverse-transmission coefficient) is 0.094 at 6 GHz which is considered insignificant ($S_{12} = 0$). Therefore, unilateral amplifier design is assumed while designing the impedance matching circuits of the LNA. This suggests that the transistor network is guaranteed to have almost zero internal feedback. According to the unilateral amplifier design rule, when $S_{12} = 0$, input reflection coefficient, $\Gamma_{\square\square} = \square_{11}$ and output reflection coefficient, $\Gamma_{\square\square\square} = \square_{22}$ [21]. 50 Ω load terminations are utilised to calculate S-parameters of the network. The Smith chart matching impedances are drawn and calculated to obtain the length and width of the transmission lines and stubs for both input and output matching impedances.

From the plotted Smith chart, the transmission line and stub lengths at the input matching network are $l_1 = 7.9$ mm and $l_2 = 8.1$ mm. For the output matching network, the transmission line and stub lengths are $l_3 = 11.6$ mm, $l_4 = 17.6$ mm, $l_5 = 11.4$ mm, and $l_6 = 9.0$ mm respectively. All transmission line and stub widths are the same ($w = 3.0$ mm) since they are designed at 50 Ω terminations.

The LNA circuit is designed on FR-4 with specifications of $\varepsilon_{□}$= 4.6, H = 1.6 mm, T = 0.035 mm, and tan δ = 0.01 according to [22]. A complete schematic diagram of the proposed C-band LNA is shown in figure 4.

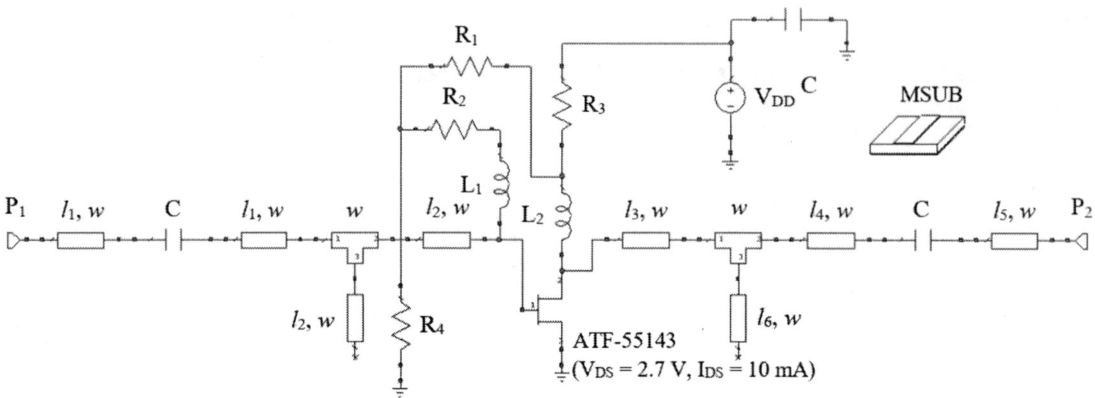

Figure 4. Schematic diagram of the proposed C-band LNA design.

3. Results and discussion

The C-band LNA circuit is designed and simulated by using an EM simulator. The DC power supply, V_{DD} is set at 3 V with bias points V_{DS} = 2.7 V and I_{DS} = 10 mA. Inductors ($□_1$ = $□_2$ = 1 μH) are incorporated into the DC bias network to boost the LNA gain. When a high-frequency signal is fed into port 1 of the LNA, the impedances of L_1 and L_2 escalated up to 37.7 kΩ at 6 GHz. This provides large input impedance for the DC bias circuit.

At 6 GHz signal frequency, impedances of L_1 and L_2 are increased and the DC bias network i.e. R_1, R_2, R_3, and R_4 are totally separated from the matching impedance network (transmission lines and stubs). This provides a purely matched input and output for the source and load of the LNA. In other words, as the 6 GHz signal enters the transistor amplifier, both L_1 and L_2 act as chokes where they block the AC signal from entering the DC power supply, V_{DD}.

The LNA circuit incorporated with inductors to improve the gain response is shown in figure 5.

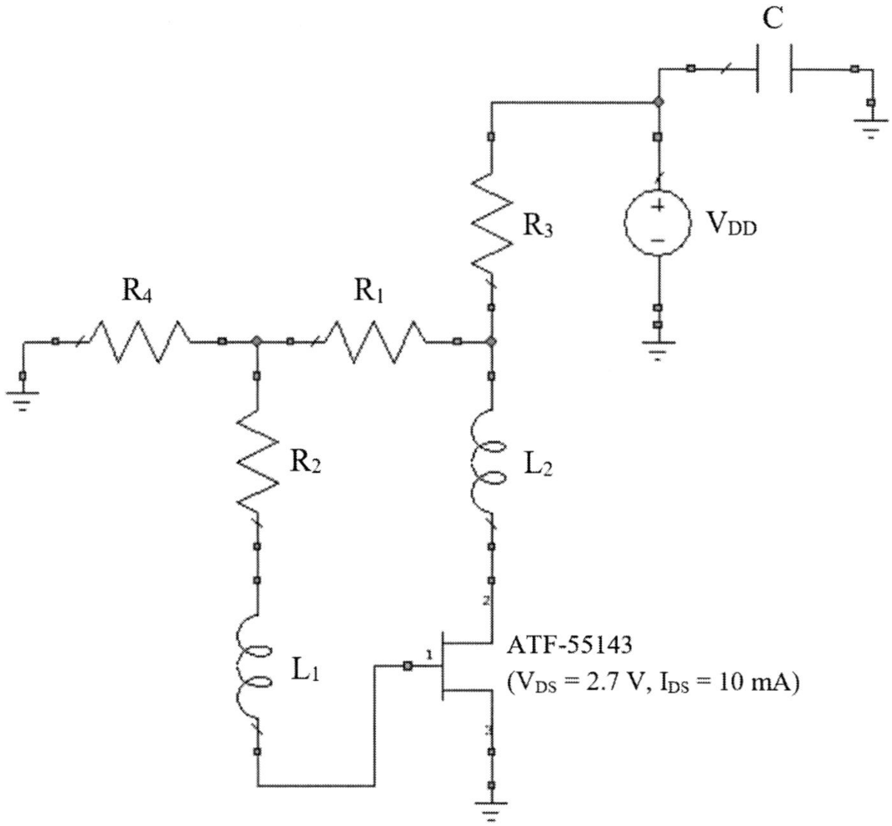

Figure 5. Addition of inductors to the LNA circuit.

Assimilating inductors to the C-band LNA system offers high-frequency isolation which upgrades the LNA performance and increases the power gain delivered at the output of the LNA up to 12.29 dB.

Figure 6 shows the simulated S-parameter results of the C-band LNA. The input and output reflection coefficients (S_{11} and S_{22}) shown in figure 6(a) are -25.21 dB and -20.12 dB respectively at 6 GHz.

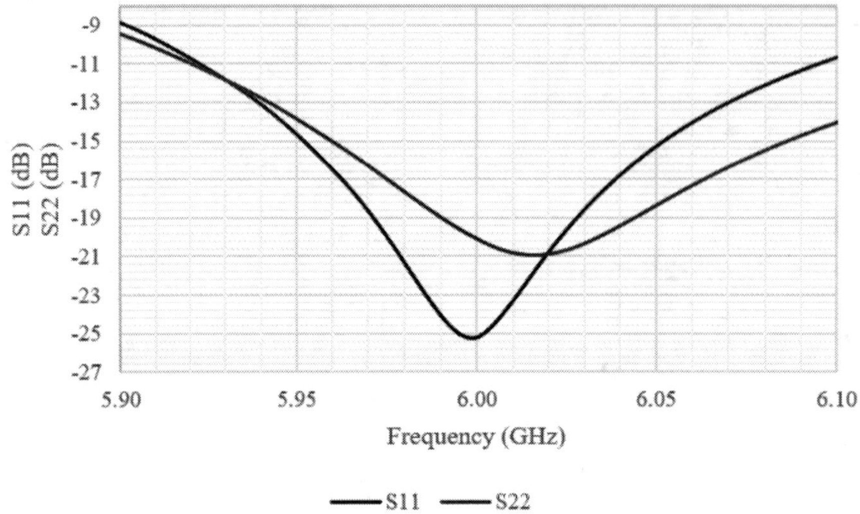

6(a)

Meanwhile, the LNA power gain (S_{21}) shown in figure 6(b) has a peak value of 12.29 dB at the designated frequency operation.

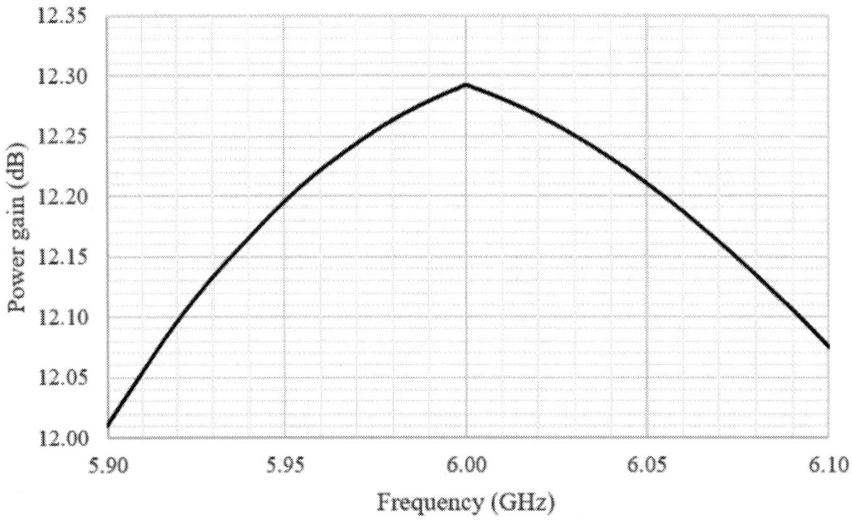

6(b)

Figure 6. (a) Input and output reflection coefficient (S_{11} and S_{22}). (b) Power gain (S_{21}).

Figure 7 shows the simulated NF. At 6 GHz, the NF obtained is 2.66 dB.

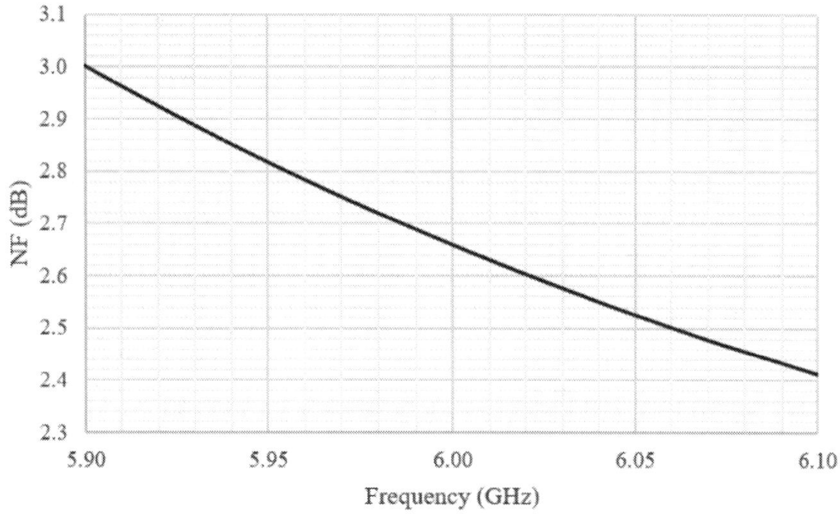

Figure 7. Simulated NF.

Table 1 refers to figure 6 and figure 7 which illustrate the simulation results of the proposed LNA circuit design.

Table 1. Simulated results of the proposed LNA.

Parameters	Simulated results (dB)
Input reflection coefficient (S_{11})	-25.12
Output reflection coefficient (S_{22})	-20.12
Power gain (S_{21})	12.29
Noise figure (NF)	2.66

4. LNA performance comparison

Table 2 represents the performance comparison of the designed LNA with other similar, published studies operating at close frequency ranges (6 GHz). The reviewed characteristics of the LNAs include input reflection coefficient (S_{11}), output reflection coefficient (S_{22}), power gain (S_{21}), NF, and power consumption.

Table 2. Comparative simulated results.

Parameters	Simulated results				
	Proposed design	[7]	[16]	[17]	[18]
Input reflection coefficient (S_{11})	-25.12 dB	-48.00 dB	-49.32 dB	-12.50 dB	-12.91 dB
Output reflection coefficient (S_{22})	-20.12 dB	-40.00 dB	-44.94 dB	N.A.	-14.61 dB

Power gain (S_{21})	12.29 dB	13.00 dB	8.96 dB	9.00 dB	13.20 dB
Noise figure (NF)	2.66 dB	2.20 dB	2.56 dB	1.50 dB	0.80 dB
Power consumption	27.00 mW	120.00 mW	30.00 mW	19.25 mW	60.00 mW

5. Conclusion

A C-band LNA for satellite communication system application operating at 6 GHz has been presented in this paper. Both input and output matching networks are designed by employing a single open-circuited stub. Smith charts are used to obtain the matched input and output impedance networks. The LNA DC bias is set as such V_{DS} = 2.7 V and I_{DS} = 10 mA. Inductors are added to the DC bias circuit to isolate it from the matching impedance layout circuit which instigated a better LNA power gain. EM software is utilised to design and simulate the proposed LNA circuit layout. The simulated input and output reflection coefficients (S_{11} and S_{22}) are -25.21 dB and -20.12 dB respectively at 6 GHz. A power gain (S_{21}) of 12.29 dB is attained from the LNA simulation, with the NF of 2.66 dB. The LNA consumes 27 mW of power from a 3 V DC supply.

References

[1] Kalamani C 2019 Design of differential LNA and double balanced mixer using 180 nm CMOS technology *Microproc. Microsyst.* **71** 102850
[2] Haji Konang S 2015 Design of low noise amplifier for radar application (Melaka: Universiti Teknikal Malaysia Melaka)
[3] Chen K, Lu J, Chen B et al. 2007 An ultra-wide-band 0.4-10-GHz LNA in 0.18-µm CMOS *IEEE Trans. Cirs. Syst. II: Express Briefs* **54** 217–221
[4] Wang R, Chen S, Huang C et al. 2008 2-6GHz current-reused LNA with transformer-type inductors *Asia-Pacific Microwave Conf.* (Macau: IEEE)
[5] Kushwah A and Katare S 2017 An ultra low voltage, wideband low noise amplifier design technique *Int. Res. J. Eng. Technol. (IRJET)* **4** 805–8
[6] Diaz-Chinea D, Garcia-Vazquez H, Montesdeoca M, Khemchandani S and Pino J del 2015 A wide-band noise-cancelling CMOS LNA based on current conveyors *Conf. on Design of Circuits and Integrated Systems (DCIS)* (Estoril: IEEE)
[7] Iyer M and Shanmuganantham T 2017 Design of LNA for C band applications *Int. Conf. on Circuits and Systems (ICCS)* 211–214
[8] Kumar M and Deolia V 2019 Performance analysis of low power LNA using particle swarm optimization for wide band application *Int. J. Electron. Commun.* **111** 152897
[9] Pozar D 2008 *Microwave Engineering* ed Kulek P and Aiello G (New Jersey: Wiley) chapter 11 540-541
[10] Wu C, Pao C, Yau W et al. 1995 Pseudomorphic HEMT manufacturing technology for multifunctional Ka-band MMIC applications *IEEE Trans. Microw. Theory Tech.* **3** 257–266
[11] Mimura T 2002 The early history of the high electron mobility transistor (HEMT) *IEEE Trans. Microw. Theory Tech.* **50** 780–2
[12] Sharma S and Sharma S 2014 Design and simulation of three stages pHEMT LNA at C-band *Int. J. Res. Appl. Sci. Eng. Technol.* **2** 358–361
[13] Khan M, Zhang H, He P, Shahzad S and Ullah R 2015 A pHEMT based wideband LNA for wireless applications *China Commun.* **12** 109
[14] Larson L 1996 *RF and Microwave Circuit Design for Wireless Communications* ed Larson L E (Norwood: Artech House Inc.)
[15] U.S. Department of Commerce National Telecommunications and Information Administration, United States Frequency Allocations [Online] Available: https://www.ntia.doc.gov/files/ntia/publications/january_2016_spectrum_wall_chart.pdf
[16] Iyer M and Shanmuganantham T 2017 Design of LNA for satellite uplink *Int. Conf. on Intelligent Computing Systems (ICICS)* 255–264

[17] Kumar A and Pathak N 2014 Design and characterization of a wideband p-HEMT low noise amplifier *Int. Conf. on Advances in Computing, Communications and Informatics (ICACCI)* 785-788
[18] Parkavi N and Ravi T 2016 Design and analysis of RF low noise and high gain amplifier for wireless communication *ARPN J. Eng. App. Scis.* **11** 11584-8
[19] Lahsaini M, Zenkouar L and Bri S 2013 Design of broadband low noise amplifier based on HEMT transistors in the X-band *Int. J. Eng. Technol.* **5** 468–476
[20] Kong S, Lee H, Kim C and Park B 2017 A 21.9-dB Gain 18.9–35.9-GHz low noise amplifier using InGaAs E-mode 0.15-µm pHEMT technology *IEEE Asia Pacific Microwave Conf. (APMC)* 1203–1206
[21] Yip P 1990 *High-Frequency Circuit Design and Measurements* (London: Chapman & Hall) chapter 6 101-102
[22] Turkish Electric Power *FR-4 data sheet* [Online] Available: \ http://www.turkishelectricpower.com/wp-content/uploads/2015/11/FR4-_used-for-PCB_-Technical-Specifications.pdf

An Identification of Software Requirement for Web Based Judging System

N.A.A.Zahari, R.A.JM.Gining, S.S.M.Fauzi*, T.R.Razak and M.N.F.Jamaluddin

Faculty of Computer and Mathematical Sciences, Universiti Teknologi MARA, Perlis Branch, Malaysia.

shukorsanim@perlis.uitm.edu.my

Abstract. The paper aims to identify the requirement for the judging system. The requirement identification of the system involves two phases; requirement gathering and requirement analysis. Several activities are done in each of the phases and the result is discussed. This study translated the requirement gathered from the interview and document analysis into several diagrams; use case diagram, activity diagram and entity relationship diagram.

1. Introduction

Every product in an innovation competition needs to be evaluated. The assignation of judges to the products or booths is done a few weeks before the competition. During the day of the innovation competition, the judges need to identify themselves to the secretariat, and they will be briefed and provided with evaluation forms for their assigned product. After the briefing, all judges start to conduct the judging processes. by visiting their assigned booth and filling in the evaluation form. Once the judging process is completed, the judges need to return the evaluation form to the secretariat. The secretariat then enters the suggested scores by the judges into a selected platform - usually Microsoft Excel. Finally, all the scores are sorted to identify the winners.

The above-mentioned process will lead to several issues and drawbacks. Firstly, during the judging process, judges had to carry several evaluation forms on hand and sorting them would be hard, as they need to go through all of the forms to assign scores. Sometimes, the manually written scores are misunderstood by the secretariat as the judges would have varied writing styles. There is also a possibility that the evaluation form will be misplaced, damage and loss. Secondly, other than needs more time and resources, manual data entry is also burdensome [1]. Moreover, the problem of missing data from its records has been proven as a main problem for this approach [2-5]. The data entry from physical to electronic undertaken by a human is also a slow. If this situation continues, a massive delay in announcing the winners in innovation competition surely will occur.

Therefore, to improve the judging process, a web-based judging system is developed. The objective of the paper is to identify the software requirements for the judging system. The software requirement identification is a crucial process to ensure the success of the system. The requirements are presented in a form of diagrams as it will be used as the main source for reference in the next phase - the development phase.

This paper is alienated into several sections. The first section introduces the issues and motivation of the study. Followed by the second section which discusses the methodology used. Results and discussion are presented in the third section and the paper is concluded in section 4.

2. Methodology

The methodology used in the system requirement identification involves two phases which are requirement gathering and requirement analysis. Each phase consists of activities that lead to the requirement collection and affirmation before the development process.

2.1. Requirement Gathering

The first phase focuses on the interview activity, which conducted among several experienced innovation competition committee and judges. An information-gathering interview is a dialogue with a specific goal that uses a question-and-answer format. A recorded formal or informal interview can be used for study purposes. The interview content may comprise of a semi-structured and open-ended question which to be answered by the respondent. The purpose of conducting the interview is to gather information about the current judging system used in an innovation competition. The judges for innovation competition who are responsible for giving scores to the participants were interviewed to gain an in-depth understanding of the current system used in the judging process. These involve the flow of the judging process and the problems that judges frequently face. A face to face conversation is preferred in this activity since it allows the respondent to answer the question spontaneously and any unclear information could be clarified immediately.

Other than the interview, the activity of analysing existing documentation has also been done in a way of obtaining an evaluation form used by the respondents in an innovation competition. Documentation assists the requirement identification process in term of understanding the overall flow of the judging process. Information required such as the product, judges, and the evaluation criteria are also analysed and extracted from the available documentation.

2.2. Requirement Analysis

The second phase of the method is the Requirement Analysis phase. This phase utilises several modelling tools and diagram such as Use Case Diagram, Activity Diagram, and Entity Relationship Diagram. These diagrams are used to visualise the user requirement compiled in the first phase for better understanding. Specifically, it is developed to aid the developers to visualize, specify, document and construct the artefacts of the system to be developed.

2.2.1. Use Case Diagram. A use case is a methodology used in to recognize, elucidate, and organize system requirements [6]. The use case consists of a set of possible interactions flows between users and systems in a specific environment and associated with a specific goal. It is composed of three main elements; actors, use case and association.

2.2.2. Activity Diagram. Activity Diagram purpose is to model the step-by-step actions taken in a portion of a larger activity [7]. Activity Diagram is essentially an improved version of the flow chart that demonstrate the flow from one activity to another activity. By explaining the flow of actions in a process, it also indirectly gives a view of the system behaviour. An Activity Diagram is denoted by shapes that are linked by arrows that illustrate the sequential order of executed activities.

2.2.3. Entity Relationship Diagram (ERD). Entity Relationship Diagram (ERD) is also known as an entity-relationship model, is a graphical representation of an information system that portrays the relationships among objects, people, concepts, places, or events within that system [8]. In Database Management System (DBMS) terminology, an entity is a table or attribute of a table in a database, thus by displaying the relationship between tables and their attributes, ERD illustrates the complete logical structure of a database. ERD also can be used as the foundation for a relational database, which is the type of database that will be used in the development process.

3. Result and Discussion

The results of the interview sessions indicate most of the judging approach used by the respondents in their participated innovation competitions are still using the manual approach - using paper-based evaluation form to record scores and then the responsible judging committee would transfer the score to their selected platform - usually to Microsoft Excel. This approach could increase the tendency for errors, mostly on the committee and such as mistype or inserting data in the wrong column, and they might misunderstand the score given by the judges as a manually written score sometimes lack clarity. Not only that, the possibility that the evaluation form to be misplaced exists, in this case, the judges may need to re-evaluate the innovation product and the result may be different from the score given previously.

Therefore, the respondents prefer an easy alternative for judging which could reduce their workload on evaluating and calculating the score for each participant or booth. The use of an information system is highlighted in some of the interviews as it could help to store the needed data and calculate the score automatically.

Through the review of the input gathered from the interview sessions and document analysis, the requirement for the judging system has been determined. The requirements are analysed and translated into modelling tools diagrams; Unified Modelling Language (UML) diagram, Entity Relationship Diagram (ERD) and sitemap. The UML diagram presents both the Use Case Diagram and Activity Diagram.

The requirement analysis phase classified the users' requirements into two; functional and non-functional requirements. Listed below are the requirements for the system. Each requirement is assigned with an ID and in the priority column, the following abbreviations are used:

M – Mandatory requirements (something the system must do)
D – Desirable requirements (something the system preferably should do)
O – Optional requirements (something the system may do)

Table 1. Requirement of the judging system.

A. FUNCTIONAL REQUIREMENTS			
No.	**Requirement ID**	**Requirement Description**	**Priority**
	JS_01	**Login into the system (Judge and Admin)**	
1.	JS _01_01	Judge and admin must enter email and password to login.	M
2.	JS _01_02	Message will be prompted if email and password are not match.	D
3.	JS _01_03	Could click on Forgot Your Password if user forgot their password.	O
	JS _02	**View All Group (Admin)**	
4.	JS _02_01	Admin could display the list of groups that participate in the innovation competition.	M
5.	JS _02_02	Admin could edit the group name, product name, product type and group member.	D
6.	JS _02_03	Admin could add group member by clicking on Add Group Member button.	M
7.	JS _02_04	Admin could select product type by using drop down list.	D
8.	JS _02_05	Message will be prompted when the number of group member exceed five.	O

9.	JS _02_06	Admin could add new group by clicking Add Group.	M
10.	JS _02_07	Alert message will be prompted when blank mandatory field detected (Group name, Product type and Product name)	O
11.	JS _02_08	Admin can delete group record by clicking Delete Group button.	M
	JS _03	**View All Booth (Admin)**	
12.	JS _03_01	Admin could display the list of booth number as well as the product assigned to it.	M
13.	JS_03_02	Admin could edit the booth number and product name.	D
14.	JS _03_03	Admin could add booth by clicking on Assign Booth.	M
15.	JS _03_04	Alert message will be prompted when blank mandatory field detected (Booth number, Product Name).	O
16.	JS _03_05	Admin can delete booth record by clicking Delete Booth button.	M
	JS _04	**View Assigned Judge to Booth (Admin)**	
17.	JS _04_01	Admin could display the booth number along with the list of assigned judges and group name.	M
18.	JS _04_02	Admin could edit the assigned judge.	D
19.	JS _04_03	Admin could add judges to booth.	M
20.	JS _04_04	Admin could select judges by using drop down list.	D
21.	JS _04_05	Admin could view the group details during assigning the judge.	D
22.	JS _04_06	Admin could delete the assigned judge.	M
	JS _05	**Register New User (Admin)**	
23.	JS _05_01	Admin could register new user (Admin and Judge).	M
24.	JS_05_02	Display message for successful registration.	D
	JS _06	**View Assigned Booth (Judge)**	
25.	JS _06_01	Judge could view the assigned booths.	M
26.	JS _06_02	Judge could access the evaluation form for each group.	M
27.	JS _06_03	Judge could view the rubric during the evaluation process.	M
28.	JS _06_04	Judge could view the scores they gave to each group everytime they access the evaluation form.	D
29.	JS _06_05	Judge could edit the score they gave in the evaluation form.	D
	JS _07	**Report (Admin)**	
30.	JS _07_01	Admin could view the scores obtain by all groups.	M
31.	JS _07_02	Admin could view the list of groups that already sorted by higher score to lower.	M
		B. NON-FUNCTIONAL REQUIREMENTS	
No.	**Requirement ID**	**Requirement Description**	**Priority**
	JS_08	**Usability Issues**	
32.	JS_08_01	Easy to use and does not need training.	M
	JS_09	**Efficiency Issues**	
33.	JS_09_01	Performance.	D
34.	JS_09_02	Ability to save and retrieve data.	D
35.	JS_09_03	Support the system with large database.	D

The requirement ID is used for identification based on the activities done in the system. JS in the ID is the abbreviation of Judging System and the numbering of the ID is provided as in number increment. The priority classification of each requirement is determined from the result of the interview. The non-functional requirements centre around usability and efficiency but mostly it is just preferred features which are not mandatory to the system.

A use case is a methodology used in to identify, interpret, and organize system requirements [7]. There are two actors involved in this system; Admin and Judge. Admin is referring to the secretariat of the innovation competition. The Admin actor is associated with six use cases; 1) login into the system, 2) register new user, 3) manage the innovation booths, 4) manage group, 5) manage judge and 6) view all scores obtained by all groups. On the other hand, the Judge actor is connected to four use cases; 1) login into the system, 2) fill in the evaluation form, 3) view rubric and 4) view list of assigned booths. The figure below shows the Use Case Diagram of the judging system.

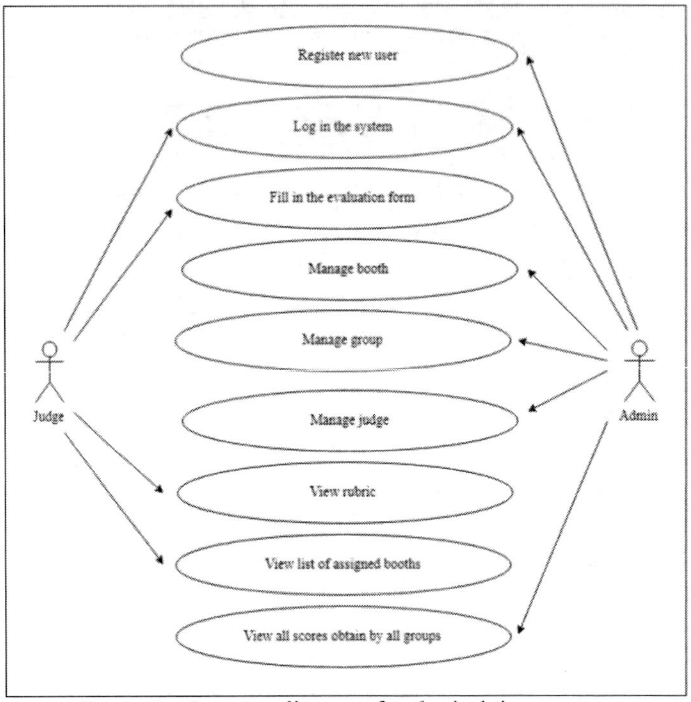

Figure 1. Use case diagram for the judging system.

The Activity Diagram shows the detailed process for both actors (Admin and Judge) interacting with the system. For the sake of simplicity, only two Activity Diagram is included in this paper - Login and Adding New Group diagram.

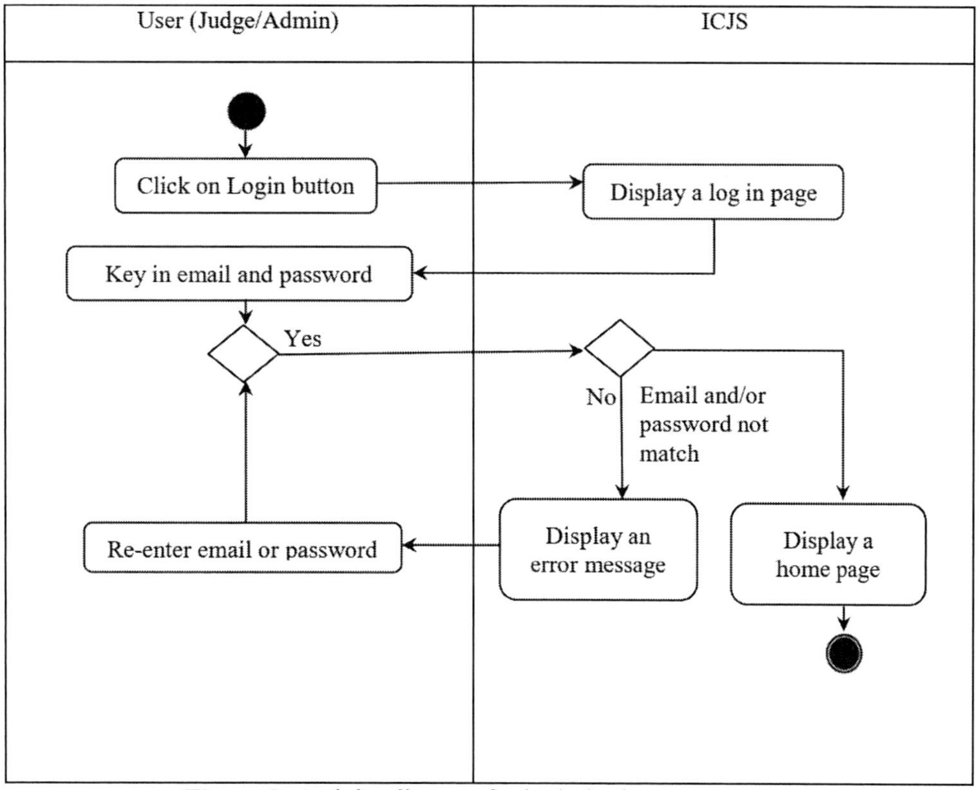

Figure 2. Activity diagram for login in the system.

The figure above illustrates the process occurs when both actors login into the system. After the actors click on the login button, the page for the user to enter their email and password will appear, which is the Login Page. The actors need to enter a correct combination of email and password to be directed to the Home Page. If any of the information provided is incorrect, an error message will appear, and user needs to re-enter the correct email and password.

The figure below shows the processes occurs during the creation of a new group or participant in the innovation competition. This process is done by the admin. When Admin clicks on Add Group menu, the corresponding form will be displayed. Several data needed to be entered such as the group name, product name, product type and the group members before submitting the form. There is a restriction for the numbers of group member - which is limited to five members, adding more and an error message will appear. After the Admin submits the form by clicking on the Add Group button, the system will check for validation, if all the data is valid, it will be saved in the storage and the Admin will be shown a 'successful' pop-up message, otherwise, the Admin need to re-enter any invalid data.

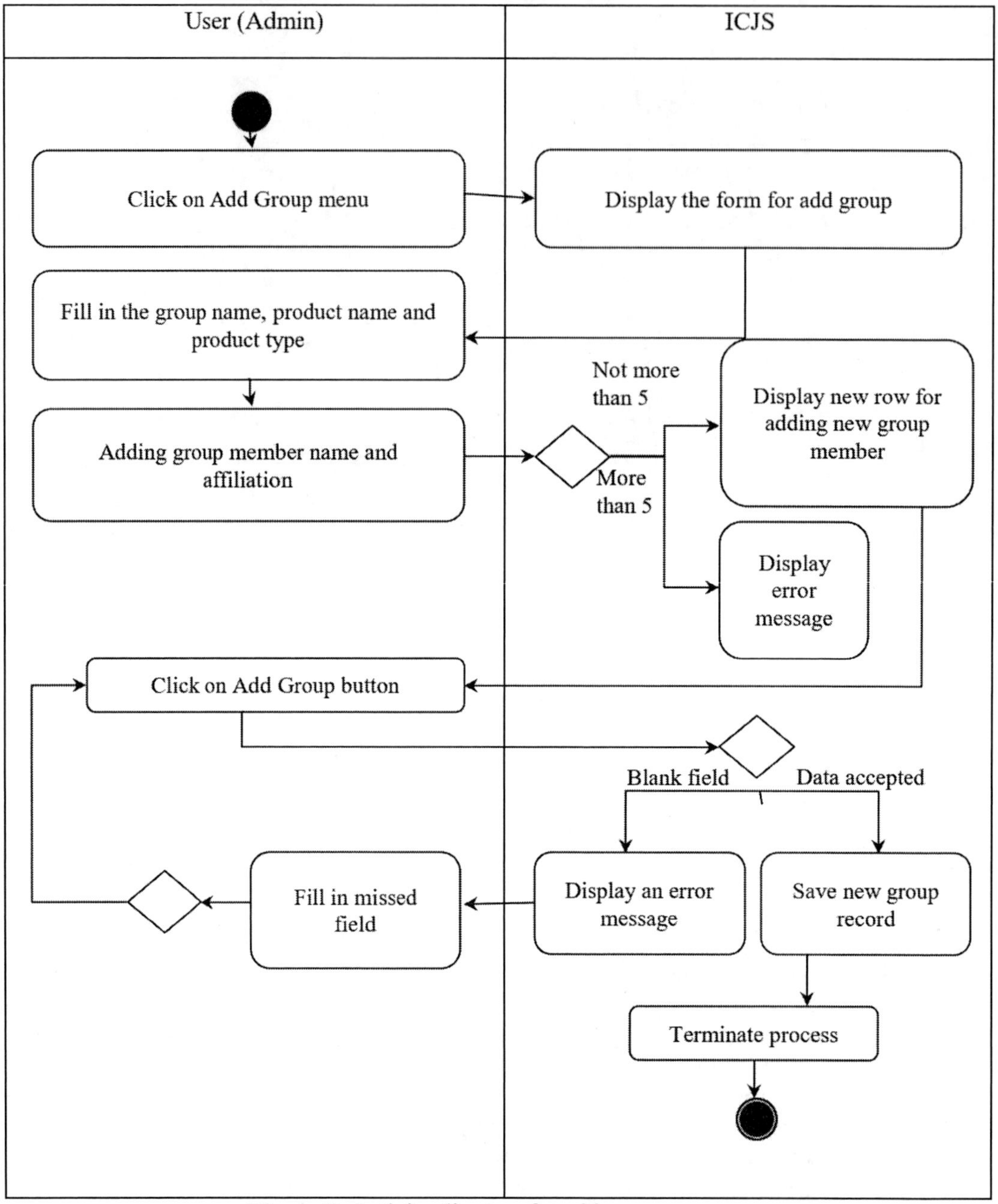

Figure 3. Activity diagram for add group.

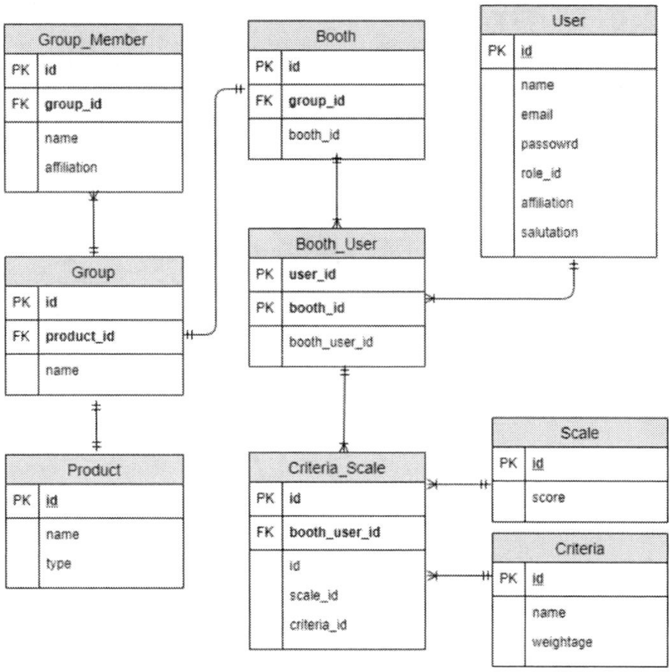

Figure 4. Entity-Relationship Diagram (ERD) for the judging system.

The figure above shows the Entity-Relationship Diagram (ERD) of the judging system. It consists of nine entities and two of these are a bridge entity. The Booth_User table is the bridge between Booth and User table while Criteria_Scale table is the bridge between Criteria and Scale table. Both of these tables become a pivot table or bridge because of a many-to-many relationship. Every table has its own primary key which is id except for Booth_User table that have two primary keys, user_id and booth_id.

The relationship between the Group and Group_Member is one-to-many because one group can have many members and one member can only have one group. Group and Product have a one-to-one relationship because one group can own one product and one product can only belong to one group. Group and Booth also are connected with one to one relationship. This is due to one group can be assigned to one booth and one booth can only have one group. The relationship between Booth and User is many to many therefore there is a pivot table named Booth_User. This is because one booth can be assigned to many judges and one judge can be assigned to many booths. Criteria and Scale are connected with a many-to-many relationship, therefore there is a pivot table named Criteria_Scale. This is due to one scale can be assigned to many criteria and one criterion can have many scales. Booth_User and Criteria_Scale have a one-to-many relationship because one judge can insert scores into many evaluation forms while one evaluation form can only be filled by one judge.

These requirements are the main foundation of the judging system that will be developed in the future as it acts as a communication tools and references among the developer and designer of the system. The success and failure of the development process rely on the requirement collected during the requirement identification process.

4. Conclusion

The main purpose of this study is to identify the software requirements for the web based judging system. 31 functional and 4 non-functional requirements were identified. The identified requirements were classified into 9 use cases, and 2 actors. The proposed ERD consists of 9 entities, which two of them are bridge entities. Future works would be the development of the proposed system based on the identified requirements.

References

[1] Vlada, M., Babiy, I., & Ivanescu, O. (2010). ABBYY recognition technologies – ideal alternative to manual data entry . Automating processing of exam tests . *Star*, *3*(1), 3–8.
[2] Tufo, H. M., & Speidel, J. J. (1971). Problems with medical records. Medical care, 509-517.
[3] Dawes, K. S. (1972). General Practice Observed: Survey of General Practice Records. Br Med J, 3(5820), 219-223.
[4] Zuckerman, A. E., Starfield, B., Hochreiter, C., & Kovasznay, B. (1975). Validating the content of pediatric outpatient medical records by means of tape-recording doctor-patient encounters. Pediatrics, 56(3), 407-411.
[5] Romm, F. J., & Putnam, S. M. (1981). The validity of the medical record. Medical care, 310-315.
[6] Rouse, M. (2007, April). *use case.* Retrieved from https://searchsoftwarequality.techtarget.com/definition/use-case
[7] Bell, D. (2003). UML basics Part II: The activity diagram. IBM Global Services, Rational Software.
[8] Rouse, M. (2018, July). *entity relationship diagram (ERD).* Retrieved from https://searchdatamanagement.techtarget.com/definition/entity-relationship-diagram-ERD

Design and Development of GreedLea Routing Protocol for Internet of Vehicle (IoV)

Normaliza Omar, Naimah Yaakob, Zulkifli Husin and Mohamed Elshaikh

School of Computer and Communication Engineering (SCCE), c/o School of Manufacturing Engineering Complex, Universiti Malaysia Perlis (UniMAP), First Floor, Pauh Putra Campus, 02600 Arau, Perlis, Malaysia

normaliza@unimap.edu.my

Abstract. In Internet of Vehicle (IoV), each vehicle uses a routing protocol to find a path for sending its messages to the last destination. Nowadays, the studies of IoV routing protocols and their impact on the performances of network with different network scenarios has significantly developed a precise understanding of the requirements and goals for designing an IoV routing protocol. In IoV, topology of network diverse promptly which leads to the fragmentation of network, frequent route breakage, and packet loss. This paper discusses on the development of an integrated routing protocol for IoV scenario. Greedy Perimeter Stateless Routing (GPSR) and Reinforcement Learning (RL) is integrate to determine a route based on demand. Then, the mobility model has been designed to reduce road collision. Lastly, traffic management also been focused to deal with the loss, mobility and network delay to meet the application demands.

1. Background

IoV is a superset of Vehicle Ad-Hoc Network (VANET). IoV enhanced VANET's applications, scale and structure. Its aims is to preserve and enhance travel comfort and to provide real time traffic information to vehicle drivers smoothly. Generally, IOV is use in both urban and rural traffic environment to allow network access to traffic management, vehicles and drivers. The motivation behind improving conventional transportation system is due to traffic congestion and accident which leads to low road efficiency. Growing number of vehicles in the traffic system may composed to the increasing number of traffic accidents. In the IoV, communication is generally known as Vehicle to Infrastructure (V2I), Vehicle to Everything (V2X) and Vehicle to Vehicle (V2V). Figure 1 shows the communication in IoV.

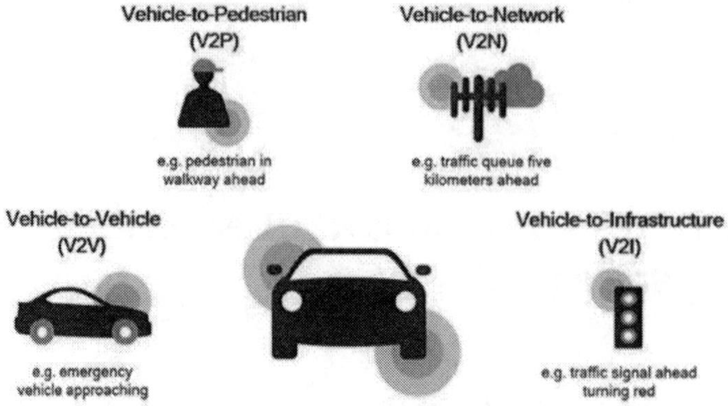

Figure 1. Communication in IoV

2. Internet of Vehicles (IoV) Concept

Nowadays, number of vehicles on the road has increased significantly. It can be concluded that one fourth of the vehicles will be connected to internet and by the year of 2020, the worldwide vehicular traffic is believed to reach 300 000 Exabyte. IoV concept typically refers to vehicles and infrastructure which are linked to all IP-based infrastructure, able to transfer information immediately and capable to solve some problems, in order to produce more safe, efficient and new vehicle system. The development of an IoV environment must meet some requirement such as:

- A scenario-driven method through a comprehensive analysis for several scenarios expected to form part of internet-connected vehicles.
- Autonomic management systems expanding the internet-connected vehicles operation:
  - Generating outputs based on a per-scenario to enhance the internet connected vehicles performance
  - Vehicle, driver and environment modelling for assuming personalized information to be used in context handling
  - A comprehensive efficient and system architecture for IoV, based on current standards and providing expansion when needed.

3. IoV Routing Protocols Classification

IoV's routing protocols classification has been a focus topic for several works recently [1], [2], [11], [3]–[10]. IoV's routing protocols can be classified into three categories; geographical, topological and bio-inspired divided in network routing protocols. The first aspect is with regard to geographical routing protocols which uses the advantages of positioning systems in vehicles to find a path by utilizing the location information of nodes. The second aspect relates to the criteria used to determine topological routing which calculate the path by distributing topology information between vehicles. Moreover, topological algorithms are divided into reactive, proactive and hybrid protocols. Further geographical protocols divided into greedy and stateless, street aware, connectivity aware and infrastructure assisted protocols. Furthermore, the classification are classified into revolutionary and Swarm intelligent protocols added into a new paradigm of routing protocol based on bio-inspired algorithm. Figure 2 depicts the taxonomy of this classification. In this research, GPSR routing protocol has been used and integrated with reinforcement learning method to propose a novel robust routing protocol for IoV environment. GPSR is a stateless protocol that allows nodes to figure out who is its immediate neighbors which is near to the information destination. This implies that it does not attempt to preserve routes between sender and receiver, permitting to each node independently forward packets to the best neighbor. The statelessness of GPSR makes it highly scalable, and the protocol is convenient to networks with highly dynamic changing topologies like VANET. GPSR expends Link State (LS), Path Vector and Distance Vectors (DV) routing algorithms.

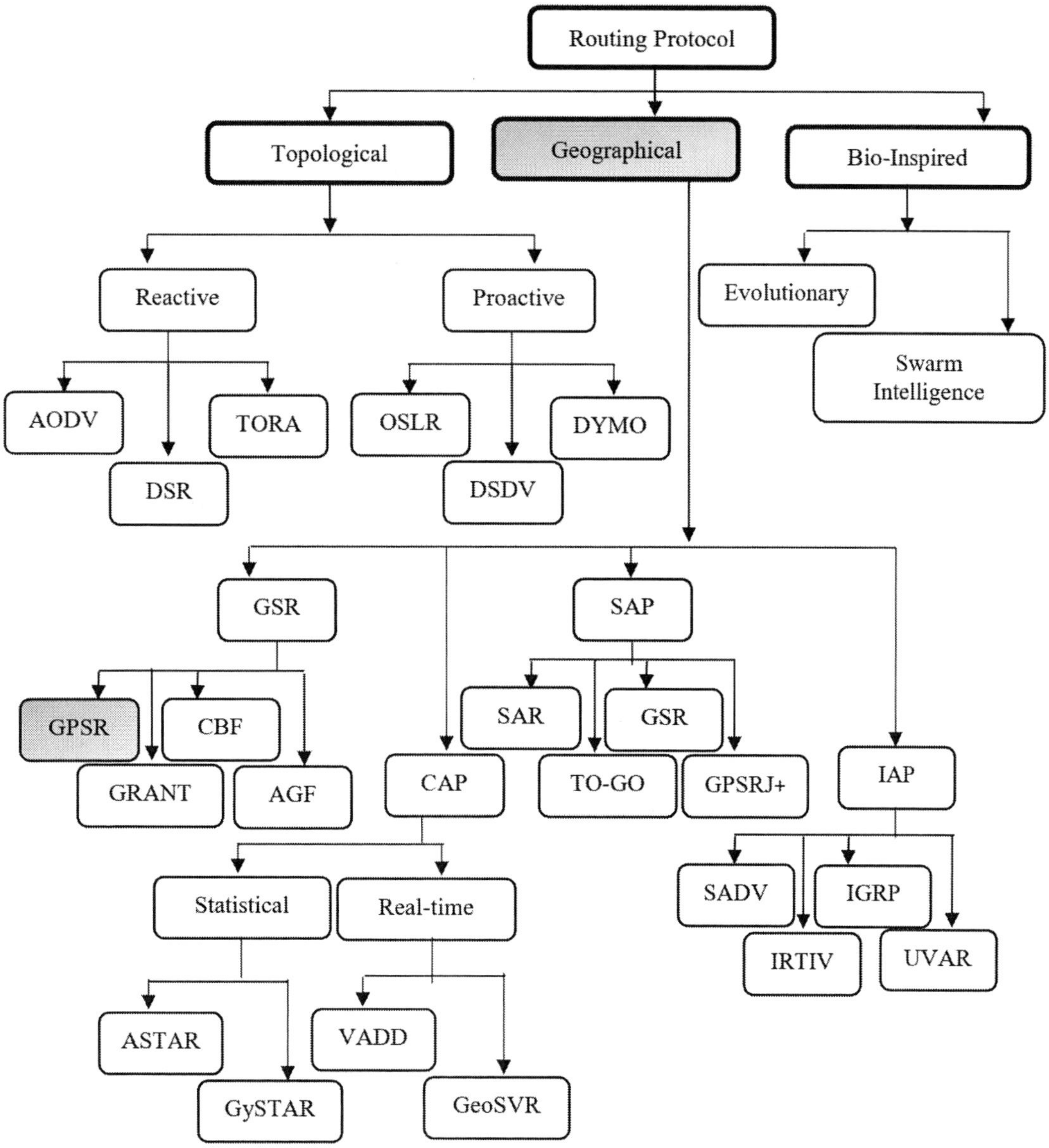

Figure 2. Taxonomy of network Routing Protocols Classification.

4. Reinforcement Learning (RL)

RL algorithm is a process of learning that connect through the environment by generating action and determine error or reward. The most significant features of RL are trial and error search and delayed reward [12][12]. This process permits software agents and machines to automatically regulate the ideal behavior of network environment in a particular context in order to increase its performance [13]. Markov Decision Process (MDP) is the example of RL. This research explores the idea of utilizing the vehicles in an attempt to solve routing in an IoV environment. The idea of dynamic rewards within the environment for the vehicles are incorporated based on analysis of many of the latest advances in traffic routing strategies, reinforcement learning and multi-agent interaction. RL algorithm is the most effective way in the unknown environment because the priori knowledge of environment is not required with this method. Therefore, the vehicle has the ability to select the optimal location from the experienced location sequence in the process of Markov Decision. The vehicles adjusts its behavior through trial-and-error interactions in the dynamic environment without the model of its IoV environment.

5. Related Work

This section provides a general review of different elements related to routing protocols for network. The main goal is however to present the evolution and classification of network routing mechanisms including bio-inspired, geographical and topological routing protocols. A summary of the reviewed work is presented in table 1.

Table 1. Comparison of the reviewed network Routing Protocols.

Protocol	Class	Routing Algorithm	Routing Metric	Conclusion
Greedy Perimeter Coordinator Routing (GPCR) [14]	Greedy and stateless routing protocols (GSR)	beaconing	Street & junction	Map based
Intersection-Based Geographical Routing (IGRP) [14]	Infrastructure assisted protocols (IAP)	Infrastructure	Delay, bandwidth and error rate	
Geographic Stateless VANET Routing (GeoSVR) [15]	Connectivity Aware Protocols (CAP)	Dijkstra	Channel condition	
Multicast with Ant Colony Optimization based on MAODV (MAV-AODV) [16]	Ant Colony Optimization (ACO)	Multicasting	Link life time, hops	
Best-Energy Aware OLSR (BEA-OLSR) [17]	ACO	Proactive	Energy	Reduces PDR 8%
Intelligent Optimized Link State Routing (IOLSR) [18]	ACO	Proactive	Routing Parameter	
Trust dependent Ant Colony Routing (TACR) [19]	ACO	Cluster	Pheromones	
Parallel Particle Swarm Optimization - Ad Hoc On-demand Distance Vector (pPSO-AODV) [20]	ACO	Reactive	Routing Parameter	

Consequently, a summary of the reviewed work presented to address the advantages and disadvantages with respect to each category; thus highlighting the goals, merits, research gap and problems in each class of routing protocols. In recent years, the analysis of IoV routing protocols and their impact on the performances of network with different network scenarios has significantly developed a precise understanding of the requirements and goals for designing an IoV routing protocol. Further, in the literature many routing protocol mechanisms has been propose to deal with IoV's requirements. Nonetheless, proposed routing mechanisms in the literature considered a single network scenario in IoV. However, vehicles or moving nodes in IoVs are tend to travel in long distances, which implies their engagement in multiple network scenarios and topologies. The adhered behavior of VANET's nodes results in a need for a routing mechanism that addresses the requirement of more than one network scenarios and topologies. This problem is less considered in the literature. The IoV routing protocol must capably to manage a fragmented network and prompt change of topology. Nonetheless, existing well-known routing protocols unable to fully address these particular requirements.

6. Research Design and Procedure

RL algorithm is a process of learning that connect through the environment by generating action and determine error or reward. This process is integrated with Greedy perimeter Stateless Routing (GPSR) protocols. The integration of RL algorithm and GPSR routing known as GreedLea routing. Each protocol is firstly design for two different IoV scenarios (City and Highway) in-terms of its routing parameters to improve a certain performance metric. Further, the selected protocols are modified to automatically change their routing parameters values according to the current network scenario. Three metrics (throughput, delay, packet delivery ratio) are selected as a target for improving the network performances in this research. The research process is depicted in figure 3. The process consists of three stages:

- Routing parameter selection
- GreedLea-routing design and implementation
- Results and Evaluation

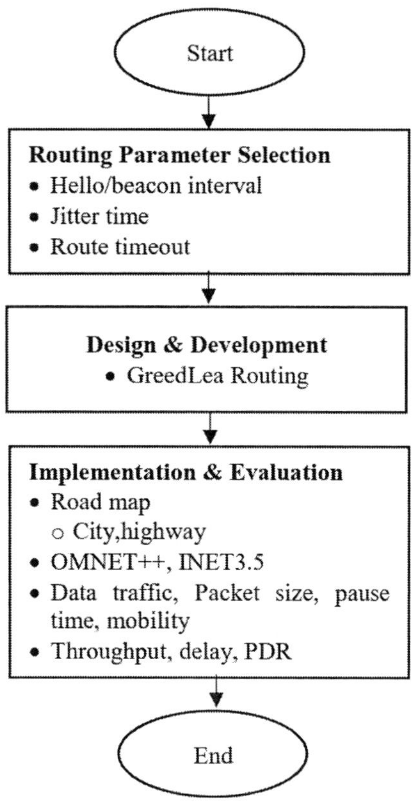

Figure 3. Research methodology flow diagram.

GPSR routing is designed for self-configured, mobile and scalable networks. GPSR routing consisted of basic routing operations such Route discovery and route maintenance. Thus, to accomplish these operations a set of inner-parameters are used to guard these operations from being dragged to unexpected situations such as: count for infinity with the use of sequence number and maintaining a route freshness in the routing table.

The first phase will focus on identifying the potential protocols parameters to be fine-tuned in the integration process. The selection of routing parameters is based-on the protocols tunable. With the parameters selected, each protocol will go through a simulation performance analysis to identify the effect of each parameters in the network performances. This stage resulted in a list of three inner-parameter to be fine-tuned in the integration stage.

The second stage emphasis on the design of the GreedLea routing that is aimed at obtaining the best performance of a routing protocol when a network device moves from one network scenario to another. The flow of the proposed routing model is start with the data transmission using GreedLea routing. In the GreedLea routing, data are being transmitted using two method. The transferred data from RouteSel process will be goes to RouteBea process before being process as a profile mapped to the desired IoV environment. Further, the TraMan process select the best profile for the current running.

The third stage focused on the implementation, evaluation and analyzation. The discrete event simulator OMNET++ 5.0 and the INET 4 framework is used for these purposes. The proposed GreedLea routing is evaluated and tested for IoV scenarios specifically the highway and city. Vehicles mobility and placement will be achieved by utilizing the mobility modules.

7. Proposed GreedLea Routing Model

This section discusses on the proposed new routing protocol named GreedLea Routing. GreedLea Routing is the integration of GPSR routing with RL method that proposed to manage a fragmented network and prompt change of topology. The research design and procedure is one of the most important parts of this research, it consists of three steps including routing algorithm, mobility model, and traffic management. An efficient IoVs routing protocol is designed using geographical information to handle highly dynamic topology, different communications environments (e.g., in city, highways etc), and time-constraints applications. In the next step, mobility model is developed by using detailed synthetic and actual mobility traces in large obstacle environment and high mobility. In the third step, a traffic management is studied to deal with the loss, mobility and network delay to meet the application demands. Figure 4 shows the flow diagram of the GreedLea Routing Method. Further, the RL method is integrated with the GPSR routing as profiles mapped to the desired IoV scenario.

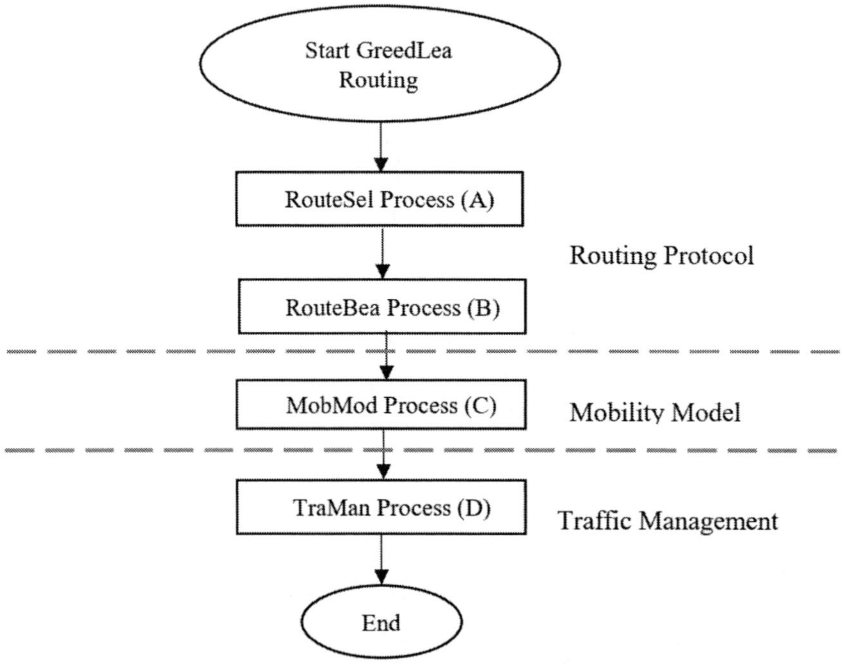

Figure 4. GreedLea routing flow diagram.

7.1. RouteSel Process (A)

In this research, an IoVs routing protocol is developed to adapt network situation by integrating real time traffic information, path stability and geographical forwarding. An integrated routing protocol is designed to discover routes. GPSR routing is used for the development of GreedLea Routing Protocol. This protocol implements two modes of operations (greedy and perimeter) to forward data. Greedy mode starts by calculating the distance between the position of the last destination and the current node, and finds the shortest path from a neighbor's table. Perimeter forwarding mode utilizes neighbor's table to construct a planar graph for the network topology. Data in GPSR are preferred to be forwarded in greedy mode. However, greedy mode might fail to find the shortest path to a destination. Hence, the perimeter is used when greedy mode is failed.

7.2. RouteBea Process (B)
The second type of message implemented by GreedLea Routing is the Beacon message. Beacon is used for the beaconing mechanism in GreedLea protocol. GreedLea routing requires each vehicle to periodically send a beacon contains nodes identification and location to its neighbors. This mechanism is used by GreedLea routing to build and update its neighbor's table. Once a beacon is received, a receiving node firstly check if the sender is in its table or not. In case the sender is in the table, the node updates the table with time of receiving the beacon, and if not in the table, the receiving node adds the sender to its table. This step is followed by recalculation of the network planar graph. Moreover, beaconing mechanism removes a neighbor node from the table if no beacon or packet is received from that node for a certain duration of time called Neighbor Validity Time.

7.3. MobMod Process (C)
Mobility model is an important element in designing networking protocols for Vehicular Networks. A practical mobility model for urban areas is particularly important to predict the network capacity, efficiency and infrastructure requirements. In this research, Mobility Model Generated for Vehicular Network (MOVE) is used for mobility model. MOVE comprise of two modules, which is the vehicle movement editor and road map editor. MOVE is develop using RL method. A RL table is establish between each state of the environment and each potential location while the vehicle keep exploring in the IoV environment. The proposed models allows vehicles to accelerate at maximum safe speed to make sure there is no collision with the prior vehicles. Since city and highway scenarios are targeted, proposed mobility model will consider traffic light system at the road intersection, multi lane and bidirectional traffic. The large-scale network simulations is performed using detailed mobility traces together with an accurate urban map that includes street direction and lane description, stop signs, traffic lights and their timing, building shape and height.

7.4. TraMan Process (D)
GreedLea routing implements the afore-mentioned mechanisms to process the routing profile. The last part of the development is the traffic management (TraMan). In TraMan process, each vehicle in the network will selects the closest Roadside Units (RSUs) as its parent RSU, and then acquires its routing profile by sending a profile request message (Profile REQ) to its parent RSU. GreedLea routing selects parent RSU based-on the geographical distance, where the closest RSU to the running vehicle is selected as its parent RSU. Once an RSU receives a profile request from a vehicle, it replies with the newest routing profile in a response message (Profile REP).

8. IoV Simulation Scenario
IoV scenarios that is implemented in this project are designed based-on the roads and traffic in city and highway scenarios. The geographical placement of the highway crossing the city is suitable for the purposes of evaluating the proposed mechanisms in a combined scenario. Table 2 shows the key simulation parameters of IoV scenarios.

Table 2. IoV Scenarios Key Simulation Parameters.

Parameter	City	Highway
Simulation area	1000m x 1000m	50m x 2000m
Number of Vehicles	30 ~ 60	20 ~ 40
Number of RSU	5	5
Vehicle's Speed	40 ~ 60km/h	80 ~ 110km/h
Simulation time	NILL	NILL
Data packet size	512 ~ 1024 bytes	512 ~ 1024 bytes
Vehicle Mobility Module	Road trajectory	Linear

9. Issue of Reliability and Validity

Some data in simulation (e.g., obstacles) might not be in real time and will be delay-tolerant. There is no limitations of storage data. Data is sent to the node by the access points in the transmission range. It is impossible to deliver data to more than one destinations.

10. Network Performance Evaluation

The goal of evaluation is to measure the effectiveness of the proposed methods in the performances of IoV network. To accomplish this goal, three performance metrics will be used (throughput, delay and Packet Delivery Ratio (PDR)). Further, these metrics will be evaluated for four design metrics (mobility, traffic generation, packet size and simulation pause time).

Figure 5 shows the PDR observations of GreedLea, GPSR, DSDV and DYMO routing for twenty (20) nodes in multi network scenarios. GreedLea routing achieved almost 100% PDR in multi scenario, and GPSR routing obtained PDR is between 91% and 93%. DSDV routing maintains its PDR in an average of 87% in both single scenario and multi scenarios which obtained PDR in varies between 86% and 88%. The observed PDR for DYMO routing in multi scenario increases compared to single scenario. From the observation, it shown that the PDR for all routing are increase as the number of node increase. This is because each node interact with each other to deliver the packet from source to the desired destination.

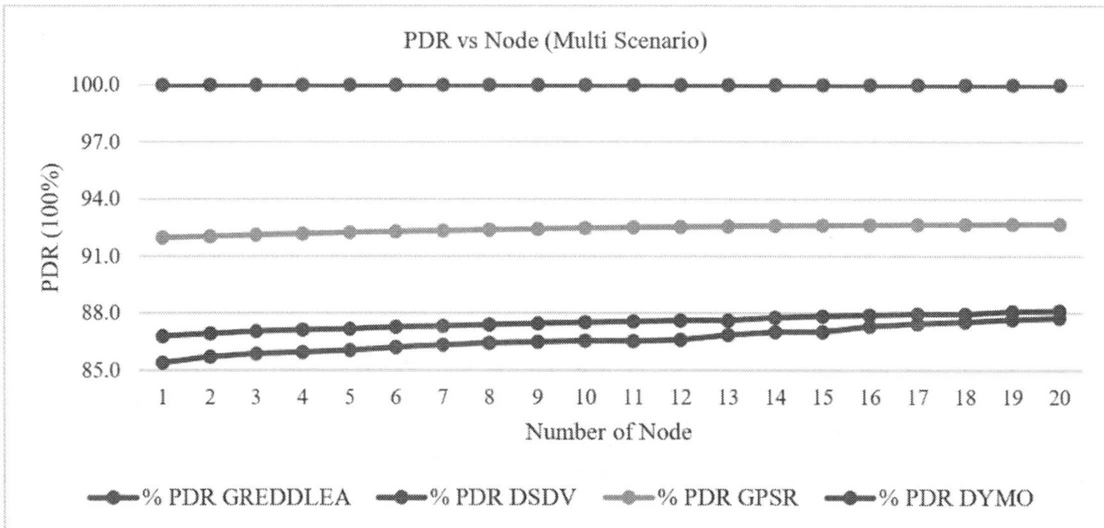

Figure 5. PDR vs Node for multi scenario

11. Conclusion

With increasingly growing number of vehicles, road traffic has become major problem of urban area. The integration of IoV can solve various traffic problem effectively. It plays an important role in enhancing road utilization, reducing energy consumption, reducing pollution and improving road safety. IoV is an integrated network that is capable to deliver vehicle smart control, smart dynamic information service and smart traffic control. This paper focuses on designing and developing an integrated routing protocol to manage hugely dynamic topology, packet loss, sparse connectivity and intermittent connectivity. Second, the development of IoV mobility model for highway and city scenario to establish realistic communication between vehicles in highway and cities. Lastly, the development of a novel IoV traffic management solution to overcome the large number of nodes participating in a transportation network. From the simulation result, it shows that the proposed GreedLea routing provided better PDR percentage compared to other existing routing protocol.

References

[1] S. Bitam, A. Mellouk, and S. Zeadally, “Bio-inspired routing algorithms survey for vehicular ad

hoc networks," *IEEE Commun. Surv. Tutorials*, **vol. 17**, no. 2, pp. 843–867, 2015.

[2] D. S. Gaikwad and M. Zaveri, "VANET Routing Protocols and Mobility Models :," pp. 334–342, 2011.

[3] S. M. Tornell, C. T. Calafate, J. C. Cano, and P. Manzoni, "DTN protocols for vehicular networks: An application oriented overview," *IEEE Commun. Surv. Tutorials*, **vol. 17**, no. 2, pp. 868–887, 2015.

[4] R. Hajlaoui, H. Guyennet, and T. Moulahi, "A Survey on Heuristic-Based Routing Methods in Vehicular Ad-Hoc Network: Technical Challenges and Future Trends," *IEEE Sens. J.*, **vol. 16**, no. 17, pp. 6782–6792, 2016.

[5] H. Houda and M. Salah, "A survey of trajectory based data forwarding schemes for vehicular ad-hoc networks," *Proc. 2015 IEEE Int. Conf. Commun. Softw. Networks, ICCSN 2015*, pp. 399–404, 2015.

[6] M. Zhang and R. Wolff, "Routing protocols for vehicular Ad Hoc networks in rural areas," *IEEE Commun. Mag.*, **vol. 46,** no. 11, pp. 126–131, 2008.

[7] L. Kristiana, C. Schmitt, and B. Stiller, "Survey of angle-based forwarding methods in VANET communications," *IFIP Wirel. Days*, **vol. 2016-April**, pp. 1–3, 2016.

[8] S. Kumar and A. K. Verma, "Position Based Routing Protocols in VANET: A Survey," *Wirel. Pers. Commun.*, **vol. 83**, no. 4, pp. 2747–2772, 2015.

[9] M. Royer, A. Pirovano, and F. Garcia, "Survey on Context-Aware Publish / Subscribe Systems for VANET," *Nets4Cars/Nets4Trains*, pp. 46–58, 2013.

[10] S. Singh and S. Agrawal, "VANET routing protocols: Issues and challenges," *2014 Recent Adv. Eng. Comput. Sci. RAECS 2014*, pp. 6–8, 2014.

[11] M. H. Tareque, M. S. Hossain, and M. Atiquzzaman, "On the Routing in Flying Ad hoc Networks," **vol. 5**, pp. 1–9, 2015.

[12] M. Van Otterlo and M. Wiering, "Reinforcement Learning and Markov Decision Processes," *Reinf. Learn.*, pp. 3–42, 2012.

[13] L. Pack Kaelbling, M. L. Littman, and A. W. Moore, "Reinforcement Learning: A Survey," *J. Artif. Intell. Res.*, pp. 237–285, 1996.

[14] H. Saleet, R. Langar, K. Naik, R. Boutaba, A. Nayak, and N. Goel, "Intersection-based geographical routing protocol for VANETs: A proposal and analysis," *IEEE Trans. Veh. Technol.*, **vol. 60**, no. 9, pp. 4560–4574, 2011.

[15] X. Wang, W. Shi, Y. Xiang, and J. Li, "Policy Space Analysis," vol. 24, no. 5, pp. 1–13, 2015.

[16] A. B. Souza, J. Celestino, F. a. Xavier, F. D. Oliveira, A. Patel, and M. Latifi, "Stable multicast trees based on Ant Colony optimization for vehicular Ad Hoc networks," *Int. Conf. Inf. Netw. 2013*, pp. 101–106, 2013.

[17] J. Toutouh, S. Nesmachnow, and E. Alba, "Fast energy-aware OLSR routing in VANETs by means of a parallel evolutionary algorithm," *Cluster Comput.*, **vol. 16**, no. 3, pp. 435–450, 2013.

[18] J. Toutouh, J. García-Nieto, and E. Alba, "Intelligent OLSR routing protocol optimization for VANETs," *IEEE Trans. Veh. Technol.*, **vol. 61**, no. 4, pp. 1884–1894, 2012.

[19] R. R. Sahoo, R. Panda, D. K. Behera, and M. K. Naskar, "A trust based clustering with Ant Colony Routing in VANET," *2012 3rd Int. Conf. Comput. Commun. Netw. Technol. ICCCNT 2012*, no. July, 2012.

[20] J. Toutouh and E. Alba, "Parallel swarm intelligence for VANETs optimization," *Proc. - 2012 7th Int. Conf. P2P, Parallel, Grid, Cloud Internet Comput. 3PGCIC 2012*, pp. 285–290, 2012.

IOP Conf. Series: Materials Science and Engineering **767** (2020) 012035 doi:10.1088/1757-899X/767/1/012035

A New Modulation Technique to Improve Received Power Under Turbulence Effects For Free Space Optical Communication

A K Rahman[1, a], A L Tom[2, a], N.A.A Mohtadzar[3,a], YMY Buswig[4,a], Nurdiani Zamhari[5, a], D N A Zaidel[6, a], S K Sahari[7, a], N Julai[8, a]

[a]Department Of Electrical & Electronic, Universiti Malaysia Sarawak (UNIMAS), Sarawak, Malaysia

[1]karahman@unimas.my,[2]amlionyy@gmail.com, [3]mnaathirah@unimas.my
[4]byonis@unimas.my, [5]znurdiani@unimas.my, [6]azdnorkhairunnisa@unimas.my,
[7]sskudnie@unimas.my,[8]jnorhuza@unimas.my

Abstract. This paper focus on new modulation technique to improve the performance of the conventional modulation that uses intensity modulation/direct detection (IM/DD) for On-Off Keying (OOK) modulation. This new modulation technique that is Dual Diffuser Modulation (DDM) can create superior modulation with able to reduce scintillation index, enhance power received and threshold signal level. The analysis result shows that the free space optical (FSO) can have good performance of power received under strong turbulence. The DDM are better performance as compare with conventional IM/DD-OOK and IM/DD-OOK with diffuser. This can help FSO system to combat with severe turbulence effect for optimum operation.

1. Introduction

The FSO communication is strongly influenced by the atmospheric channel effect which can cause the beam signal fading and wander. There are two main effects that can deteriorate the quality signal of FSO, namely the atmospheric attenuation and atmospheric turbulence as reported in [1]–[4]. Atmospheric attenuation will cause the signal scattering and absorption, therefore modification is needed on an FSO system design in order to transmit higher power without exceeding the safety limits or has to reduce the propagation link [5], [6]. Meanwhile, atmospheric scintillation is caused by the atmospheric temperature inhomogeneity as presented in [7], [8]. The scintillation effect will cause the signal fading due to the constructive and destructive interference of the optical beam traversing the atmosphere. Typical scintillation fading margins for short propagation link is around 2 to 5 dB which is less than the margins for atmospheric attenuation [8], making scintillation insignificant for short range FSO systems. Nevertheless, if the range is exceeding more than 1 km the scintillation will impairs the FSO link availability and degrade the FSO performance significantly [2], [4], [7], [9].

The OOK modulation technique is a simple modulation technique to be implemented and it has become the most popular for commercially modulation technique available for terrestrial FSO systems currently [4], [10]. The major problem with OOK is it suffers from the threshold signal level. The threshold level or point in the decision circuitry is used to distinguish between bits zero and one is always fixed midway between the expected levels of data bits one and zero. This will produce an optimum less error performance in non-fading channels. However when there is presence of turbulence, the received signal level will fluctuate and at the same time the threshold detector has to track this fluctuation in order to determine the optimum decision point [11], [12]. As a result, this will need a great design challenge, as the channel noise and fading will have to be continuously tracked for the OOK in FSO to perform optimally. Ignoring the signal fluctuation and letting the OOK FSO system to operate with a fixed threshold level will cause the increment of detection error [13].

By using a phase screen diffuser which is mounted at the laser exit of transmitter, it can reduce the scintillation index and enhanced the FSO performance. Nevertheless, the use of the phase screen diffuser will attenuate the power transmit due to the expended of beam divergence caused by the diffuser effect. Consequently, less power will be received at the receiver detector. The effect of the diffuser will also become less effective when increased to strong turbulence. This condition will not give the advantage over FSO to improve the overall performance. Therefore, the new DDM overcomes this limitation of conventional IM/DD-OOK where it optimizes the diffuser effects, increases the power received and fix the zero threshold level signal to create robust modulation against atmospheric turbulence effect.

2. Effect Scintillation in Laser Beam

The most practical implementation for FSO communication involves the use of an intensity modulation/direct detection (IM/DD) system. The transmitted data is on-off keying (OOK) intensity modulated and goes through an atmospheric channel to the receiver. The receiver aperture collects the received optical signal and focuses it onto a photodetector, which converts the instantaneous optical power into electrical current for the detection process. In the presence of atmospheric turbulence between transmitter and receiver, the received signal exhibits random intensity fluctuations.

The instantaneous received signal power P_R as expressed in Equation (1) is a random quantity. The observed quantity is now the averaged or mean received signal power $\langle P_R \rangle$ given by

$$\langle P_R \rangle \cong \frac{\pi}{8} D^2 \langle I(0,L) \rangle = \frac{P_R}{1+1.33\sigma^2 \Lambda^{5/6}} \tag{1}$$

Furthermore, it follows that the mean signal current is $\langle i_s \rangle = R\langle P_R \rangle$, where R is the photodetector responsivity. The output current from the detector $i = i_S + i_N$ in the case is a random variable, which has the mean value $\langle i_s \rangle$ and the variance $\sigma_i^2 = \sigma_S^2 + \sigma_N^2$, , where σ_S^2 represents fluctuations in the signal that become a contributor to the detector noise and related to the normalized intensity variance σ_I^2 by

$$\sigma_S^2 = \langle i_S^2 \rangle - \langle i_s \rangle^2 = \langle i_s \rangle^2 \sigma_I^2 \tag{2}$$

Using the relations given in Equation 1 and Equation 2 and the averaged SNR $\langle \Gamma_0 \rangle$ at the output of the detector assumes the form

$$\langle \Gamma_0 \rangle = \frac{\langle i_S \rangle^2}{\sigma_i^2} = \frac{\Gamma_0}{\left(1+1.33\sigma_I^2 \Lambda^{5/6}\right)^2 + \Gamma_0 \sigma_I^2} \tag{3}$$

The BER can be expressed as

$$P_r(e)_{OOK} = \frac{1}{2} erfc\left(\frac{i_s}{2\sqrt{2}\sigma_N}\right) = \frac{1}{2} erfc\left(\frac{1}{2\sqrt{2}}\sqrt{\Gamma_0}\right) \tag{4}$$

In the presence of optical turbulence, the expression in Equation (1) must be modified to incorporate the effects of signal fluctuations. In this case, the threshold level is now set to half the average signal corresponding to a received pulse ($i_T = \langle i_S \rangle / 2$).

The false alarm probability P_f does not depend on the random received signal and can be written as

$$P_f = P_R|(1|0) = \frac{1}{2} erfc\left(\frac{1}{2\sqrt{2}}\sqrt{\Gamma_0}\right) \tag{5}$$

However, the miss probability p_m is now,

$$P_m = P_R|(0|1) = \frac{1}{2} erfc\left[\left(\frac{2i_s}{\langle i_s \rangle} - 1\right)\frac{1}{2\sqrt{2}}\sqrt{\Gamma_0}\right] \tag{6}$$

where i_s is the random detector signal corresponding to the instantaneous received intensity. To compute the average BER $\langle P_R(e)\rangle$, these questions must average over the intensity fluctuation spectrum. This gives:

$$\langle P_r(e)\rangle = \frac{1}{4}\left\{erfc\left(\frac{1}{2\sqrt{2}}\sqrt{\Gamma_0}\right) + \int_0^\infty P_I(i_s)\left[\left(\frac{2i_s}{\langle i_s \rangle} - 1\right)\frac{1}{2\sqrt{2}}\sqrt{\Gamma_0}\right] di_s\right\} \tag{7}$$

where $P_I(i)$ is the PDF of the intensity fluctuations.

The limiting performance of average BER can be achieved by assuming the threshold level is dynamically set at half instantaneous received signal level i_t, which lead to expression,

$$\langle P_r(e)\rangle_L = \frac{1}{2}\int_0^\infty P_I(i_s) erfc\left(\frac{i_s}{2\sqrt{2\langle i_s \rangle}}\sqrt{\langle \Gamma_0 \rangle}\right) di_s \tag{8}$$

This shows that the turbulence will deteriorate the signal performance FSO and the most problem OOK signalling is relating threshold ability.

3. Results and Discussion

The parameters used for modulation comparison performance are wavelength λ= 1550 nm, distance $L =$ 2000 m, electron charge $e = 1.6 \times 10^{-19}$ C, plank constant $h = 6.6 \times 10^{-34}$ Js, speed of light $c = 3 \times 10^8 ms^{-1}$, photodetector efficiency $\eta = 0.7$, power transmit $P_o = 0\ dBm$, receiver aperture diameter $l_c = 0.1$, radius of curvature $F_O = \infty$, spot beam at transmitter $(z = 0)$ $W_o = 0.025$m.

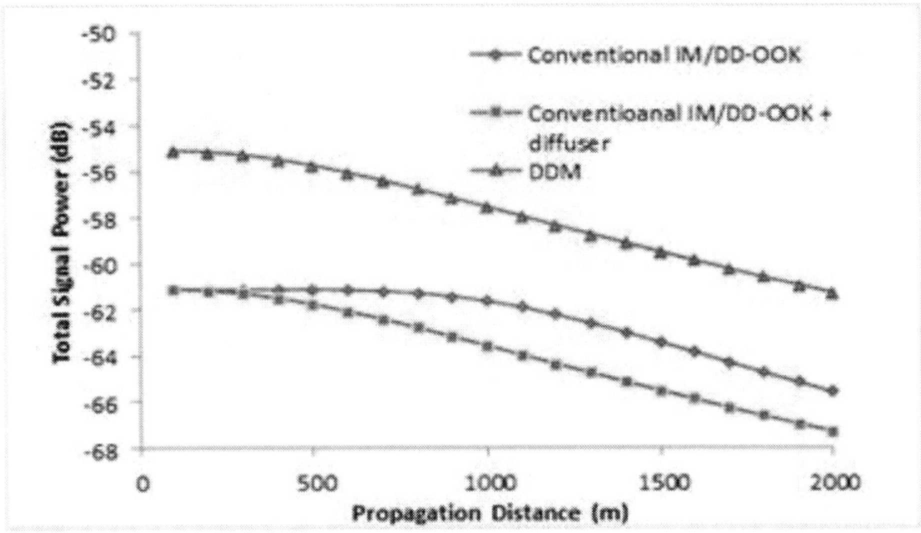

Figure 1. Performance of total signal power versus propagation distance in without presence of turbulence

The Figure 1 shows the corresponding of the total signal power over propagation distance in the absence of turbulence. From the graphs, the signal power under diffuser effect experiencing decline condition due to diffraction beam as shown by the curve conventional IM/DD-OOK with diffuser. Nonetheless, DDM technique has a better signal power due to double magnitude power yield after subtracting process. When there is presence of turbulence, the intensity of signal begins to become unstable and the diffuser helps to reduce the scintillation index which can produce a better signal power. Figure 2 and Figure 3 present the intensity signal power when experiencing the turbulence effect.

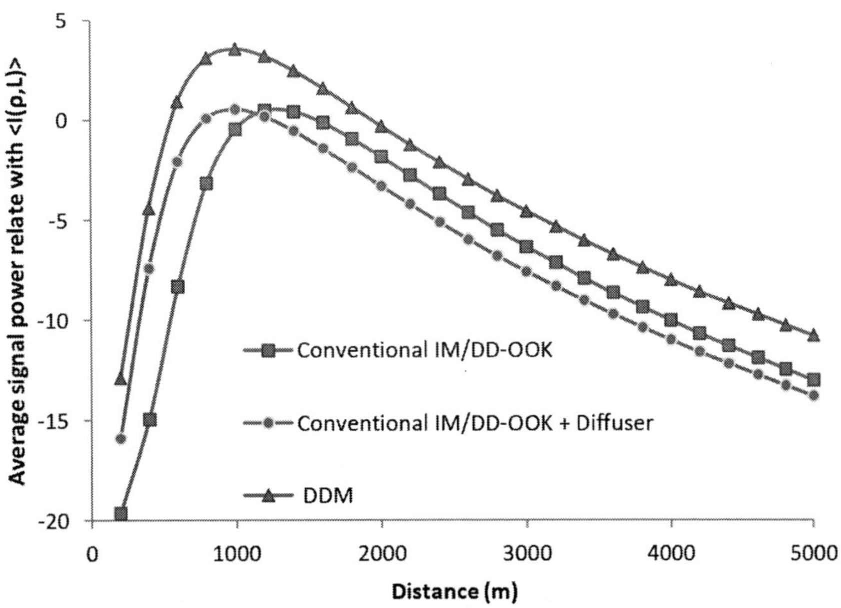

Figure 2. Performance of mean intensity signal versus distance under strong turbulence

The Figure 2 shows the mean intensity signal power correspond to propagation distance for strong turbulence. From the graphs, the DDM increases the mean intensity signal if compares with the conventional IM/DD-OOK and conventional IM/DD-OOK with diffuser. However, all the performances of mean intensity decreases as the propagation distance increases.

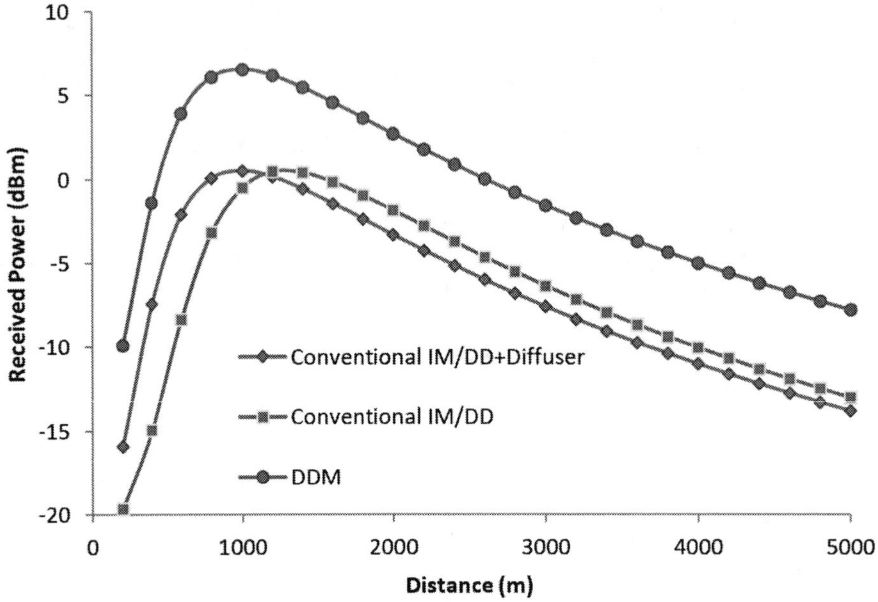

Figure 3. Performance of receiving power versus distance under strong turbulence

Meanwhile Figure 3 shows the power received versus the distance corresponds to mean intensity signal. Under strong turbulence, DDM technique is able to support high power received compared to the conventional IM/DD-OOK and conventional IM/DD-OOK with diffuser. It shows that the capability of this new technique of modulation is able to detect weak signal and maintain the better level of power received.

4. Conclusion

This DDM technique improves the ability of received power in free space optic transmission. The results show that under strong turbulence the DDM technique can have a better power received as compared with conventional IM/DD-OOK and IM/DD-OOK with diffuser. The phase screen diffuser effect enhances the impact of DDM. Therefore, it is possible to be operated on long distance transmission laser beam.

References

[1] M. A. A. Ali, "Atmospheric Turbulence Effect on Free Space Optical Communications," *Int. J. Emerg. Technol. Comput. Appl. Sci.*, vol. 5, no. August 2013, pp. 345–351, 2014.

[2] A. Malik and P. Singh, "Free Space Optics : Current Applications and Future Challenges," *Int. J. Opt.*, vol. 2015, pp. 1–7, 2015.

[3] J. A. Akinwumi and J. O. Bandele, "Free Space Optical Communication : Review Paper," *Int. J. Eng. Sci. Invent.*, vol. 7, no. 10, pp. 58–67, 2018.

[4] H. Kaushal, G. Kaddoum, and C. Engineering, "Optical Communication in Space : Challenges and Mitigation Techniques," *IEEE Commun. Surv. Tutorials*, vol. 19, no. 1, pp. 57–96, 2016.

[5] F. Wang, X. Liu, and Y. Cai, "Propagation of Partially Coherent Beam in Turbulent Atmosphere :

A Review," *Prog. Electromagn. Res.*, vol. 150, pp. 123–143, 2015.

[6] A. A. Farid and S. Hranilovic, "Outage Capacity Optimization for Free-Space Optical Links With Pointing Errors," *J. Light. Technol.*, vol. 25, no. 7, pp. 1702–1710, 2007.

[7] K. S. Altowij, A. Alkholidi, and H. Hamam, "Effect of clear atmospheric turbulence on quality of free space optical communications in Yemen," *Front. Optoelectron. China*, vol. 3, no. 4, pp. 423–428, 2010.

[8] A. C. Motlagh, V. Ahmadi, Z. Ghassemlooy, and K. Abedi, "The Effect of Atmospheric Turbulence on the Performance of the Free Space Optical Communications," in *2008 6th International Symposium on Communication Systems, Networks and Digital Signal Processing*, 2008, pp. 540–543.

[9] I. E. Lee, Z. Ghassemlooy, W. P. Ng, and M.-A. Khalighi, "Reducing Pointing Errors in Free-Space Optical Communication Links over Turbulences with a Partially Coherent Gaussian Beam," in *2016 IEEE International Conference on Communications Workshops (ICC)*, 2016.

[10] G. Kaur, H. Singh, and A. S. Sappal, "Free Space Optical Using Different Modulation Techniques – A Review," *Int. J. Eng. Trends Technol. (IJETT*, vol. 43, no. 2, pp. 109–115, 2017.

[11] B. Barua, "Comparison the Performance of Free-Space Optical Communication with OOK and BPSK Modulation under Atmospheric Turbulence," *Int. J. Comput. Eng. Sci. Technol.*, vol. 3, no. 5, pp. 4391–4399, 2015.

[12] P. Bhardwaj and Sawhil, "Effect of Atmospheric Turbulence and Pointing Error on OOK in Free Space Optics," *Int. J. Eng. Trends Technol.*, vol. 59, no. 3, pp. 122–126, 2018.

[13] E. J. Lee and V. W. S. Chan, "Part 1 : Optical Communication Over the Clear Turbulent Atmospheric Channel Using Diversity," *IEEE J. Sel. Areas Commun.*, vol. 22, no. 9, pp. 1896–1906, 2004.

Acknowledgements
This research is funded under Fundamental Research Grant Scheme (FRGS), MOHE.

IOP Conf. Series: Materials Science and Engineering **767** (2020) 012036 doi:10.1088/1757-899X/767/1/012036

Investigation to Reduce Flux Variance Effects Using Partial Coherent Beam in Free Space Optical Communication

A K Rahman, A L Tom, N A A Mohtadzar, YMY Buswig, Nurdiani Zamhari, M R M Sharip and S K Sahari

Department of Electrical & Electronic, Universiti Malaysia Sarawak (UNIMAS), Sarawak, Malaysia

karahman@unimas.my

Abstract. This paper focus on the employment of partial coherent beam (PCB) in free space optical communication to reduce the flux variance effects in the system. The implementation of PCB in the free space optical system is significantly able to reduce the flux variance effect. The reduction of the flux variance effect by using PCB in free space optical communication by using Gamma-Gamma distribution is presented in this paper. The analysis result shows that the FSO system with PCB have an improved in performance compares with FSO system without PCB implemented on it as it capable to decrease the flux variance and scintillation index. This can help FSO system to encounter with severe turbulence effect for optimum operation.

1. Introduction

The FSO communication is strongly influenced by the atmospheric channel effect which can cause the beam signal fading and wander. The study in the spatially Partial Coherent Beam (PCB) has an attractive number of researchers due to the capability for improving the performance of laser communication systems was indicated recently in a number of publications [1]–[4]. In this paper, the theoretical model for PCB focus on the scintillation index calculation in turbulence. This is because the scintillation index is the most important statistic for practical application FSO systems.

In this research study, the Kolmogorov spectrum model will be used in the case of weak atmospheric turbulence. By using the theory of Andrews [5], [6] the weak scintillation index of a PCB can be used in all atmospheric conditions. Therefore, the scintillation index depends on the strength of the diffuser and the strength of the atmospheric turbulence.

2. Partially Coherent Beam

The figure 1 shows the diffuser model for PCB with propagation channel. For the transmitted beam wave of the diffuser is characterized by TEM_{00} Gaussian beam wave parameters.

$$\Lambda_0 = \frac{2L}{kW_0^2} \tag{1}$$

$$\Theta_0 = 1 - \frac{L}{F_0} \tag{2}$$

where $\Lambda_0(L)$ represent the initial Fresnel ratio and the $\Theta_0(L)$ represent the initial phase curvature. The $k = \frac{2\pi}{\lambda}$ is the laser wave number, λ (in meters) is wavelength, L (in meters) is propagation distance to the collecting (Gaussian) lens, F_0 (in meters) is the phase front radius of curvature, and W_0 (in meters) is the laser exit aperture radius. From the Andrews and Phillips [5] the parameters Θ_1 and Λ_1 for the beam incident on the collecting lens can be written as:

$$\Lambda_1 = \frac{2L}{kW_1^2} = \frac{\Lambda_0}{\Theta_0^2 + \Lambda_0^2} \quad (3)$$

$$\Theta_1 = 1 + \frac{L}{F_1} = \frac{\Theta_o}{\Theta_0^2 + \Lambda_0^2} \quad (4)$$

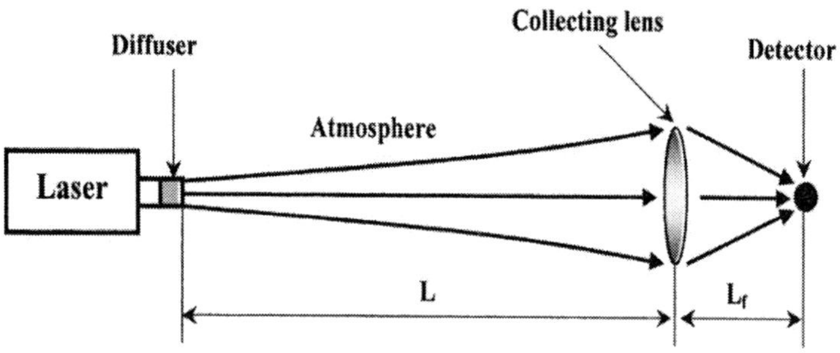

Figure 1. PCB's model and its configuration.

The Gaussian lens at the receiver is assumed has radius W_G and phase front radius of curvature F_G. After propagating through the lens to the detector located in the image plane at distance L_f behind the lens, the beam has radius W_2 and radius of curvature F_2, which are characterized by beam parameters:

$$\Lambda_2 = \frac{2L_f}{kW_2^2} = \frac{L}{L_f}\left[\frac{\lambda_1 + \Omega_G}{\left(L/L_f - L/F_G + \overline{\Theta}_1\right)^2 + (\Lambda_1 + \Omega_G)^2}\right] \quad (5)$$

$$\Theta_2 = 1 + \frac{L_f}{F_2} = \frac{L}{L_f}\left[\frac{L/L_f - L/F_G + \overline{\Theta}_1}{\left(L/L_f - L/F_G + \overline{\Theta}_1\right)^2 + (\Lambda_1 + \Omega_G)^2}\right] \quad (6)$$

where $\overline{\Theta}_1 = 1 - \Theta_1$ and the nondimensional parameter Ω_G is defined by

$$\Omega_G = \frac{2L}{kW_G^2} \quad (7)$$

The model the diffuser in front of the laser transmitter by a thin random phase screen diffuser can be characterized by a single-scale Gaussian spectrum model [7], [8]:

$$\Phi_s(k) = \frac{\langle n_1^2 \rangle l_c^3}{8\pi\sqrt{\pi}} exp\left(-\frac{1}{4} l_c^2 k^2\right) \quad (8)$$

where k is wave number, and l_c is directly related to the variance σ_g^2 to describe the partial coherence properties of the source [1]:

$$l_c^2 = 2\sigma_g^2 \quad (9)$$

The parameter $\langle n_1^2 \rangle$ is the fluctuation in the index of refraction induced by the diffuser and the nondimensional parameter analogous that relating the strength of diffuser is defined as:

$$q_c = L/kl_c^2 \tag{10}$$

Therefore the 'new' Gaussian beam from the diffuser effect can be characterized by an effective set of beam that are Θ_e, Λ_e and N_s represent the number of speckles cells.

$$N_s = 1 + \frac{2W_0^2}{l_c^2} = 1 + \frac{4q_c}{\Lambda_0} \tag{11}$$

$$\Lambda_e = \frac{\Lambda_0 N_s}{\Theta_0^2 + \Lambda_0^2 N_s} \tag{12}$$

$$\Theta_e = \frac{\Theta_0}{\Theta_0^2 + \Lambda_0^2 N_s} \tag{13}$$

For the effective spot beam $W_\zeta(L)$ for PCB is described by [9] to consider the global coherent parameter ζ. It can be written as

$$W_\zeta(L) = w_o \left(\Theta_o^2 + \xi \left(\frac{2L}{k{w_o}^2} \right)^2 \right)^{1/2} \tag{14}$$

$$\zeta = \zeta_s + \frac{2w_o^2}{{\rho_o}^2} \tag{15}$$

where the $\zeta_s = 1 + \frac{w_o^2}{\sigma_\mu^2}$ represent the source of the coherence parameter of the laser beam emitted by the transmitter, $\rho_0 = [0.545 C_n^2 k^2 L]^{-3/5}$ is the coherence length through turbulence atmosphere and σ_μ^2 is the variance of the Gaussian describing the ensemble average of the random phases which relate to diffuser strength. If the $\zeta_s = 1$, the beam is categorized as a perfectly coherent beam and the beam is categorized as a partially coherent beam if the $\zeta_s > 1$ [10]. If without diffuser effect the beam spot radius at the receiver can be expressed as [11].

$$W(L) = W_o(\Theta_o + \Lambda_o)^{1/2} \tag{16}$$

where the Θ_o and Λ_o are the initial phase curvature and initial Fresnel ratio respectively as shows in Equation (1) and Equation (2).

3. Flux Variance

The reduction of the scintillation index due to PCB alone would not be sufficient for the high-quality data transfer calculated in terms of the bit error rate (BER) in communication systems. However, the combination of partial coherence with large enough collecting lens (aperture averaging effect) can provide the required BER level. The flux variance of the intensity fluctuations at the detector plane that calculated for the collecting lens with normalized radius $\mathbf{\Omega_G}$ is valid for all atmospheric conditions as given in [6] :

$$\sigma_{flux}^2(L + L_f, \Omega_G) = exp\left[\sigma_{\ln x}^2(L + L_f, \Omega_G) + \sigma_{\ln y}^2(L + L_f, \Omega_G)\right] - 1 \tag{17}$$

where $\sigma_{\ln x}^2$ is the flux variance associated with large-scale fluctuations [12].

The effect of aperture averaging can be described as when the receiving aperture is larger than a spatial scale size that produces the intensity fluctuations, the receiver will average the fluctuations over the aperture and the scintillation will be less compared to scintillation measured with a point receiver [13], [14]. It has been shown that aperture averaging causes a shift of the relative spatial frequency content of the intensity spectrum toward lower frequencies, since the fastest fluctuations caused by small-scale sizes average out [15]. Hence, the scintillation measured by a receiving aperture is produced by scale sizes larger than the aperture. The most significant advantage of increasing the aperture size is in collecting more light. However, as the aperture size exceeds multiple times the coherence length of incident light on it, the scintillation index gets reduced, this phenomenon is known as aperture averaging [16], [17].

Under strong turbulence, there is two-scale behaviour in the intensity flux variance obtained by a finite size receiving aperture. A sharp decrease in scintillation is observed as the aperture sizes increasing up to the coherence scale, ρ_o, after which there is a levelling effect followed by a second decrease in the intensity flux variance when the aperture exceeds the scattering disk $L/_{\kappa\rho_0}$ [13], [15]. The levelling effect arises since medium-scale sizes are inefficient in producing scintillation in strong turbulence. Hence, if a receiving aperture is larger than ρ_o in strong turbulence, the small-scale scintillation is mostly averaged out. This would affect the PDF of the intensity fluctuations since existing models developed for point receivers are obtained by modulating the small and large scale distributions. Therefore in this paper we use the Gamma-Gamma distribution and it is suitable for all fluctuation regime [18].

The log intensity due to large scale eddies is given as:

$$\sigma_{\ln x}^2\left(L+L_f,\Omega_G\right)=0.49\sigma_1^2\left(\frac{\Omega_G-\Lambda_e}{\Omega_G+\Lambda_e}\right)^2\times R\left[\frac{\eta_x}{1+0.4\eta_x(1+\Theta_e)/(\Lambda_e+\Omega_G)}\right]^{7/6} \quad (18)$$

where, $R=1/3-(1/2)(1-\Theta_e)(1/5)(1-\Theta_e)^2$

The quantity η_x in Equation (18) is the normalized large-scale cutoff frequency determined by the asymptotic behaviour of $\sigma_{\ln x}^2$ in weak turbulence and saturation regime [12], [19]:

$$\eta_x=\frac{R^{-6/7}(\sigma_B/\sigma_1)^{12/7}}{1-0.56\sigma_B^{12/5}} \quad (19)$$

Meanwhile, the log intensity due to small scale eddies is given as;

$$\sigma_{\ln y}^2\left(L+L_f,\Omega_G\right)=\frac{1.27\sigma_1^2\eta_y^{-5/6}}{1+0.4\eta_y/(\Lambda_1+\Omega_G)} \quad (20)$$

where the corresponding cutoff frequency is

$$\eta_y=3(\sigma_1/\sigma_B)^{12/5}\left(1+0.69\sigma_B^{12/5}\right) \quad (21)$$

4. Results and Discussion

In terms of the effect of diffuser strength relating aperture averaging effect, the figure 2 shows the effect normalize collecting lens, Ω_G versus the several partially coherent beams with q_c = 0.1, 1, and 10, and coherent beam for the weak turbulent condition. The propagation distance $L = 1000$m, wavelength, λ

= 1μm, Ω_G = 0.01m and $C_n^2 = 10^{-15}\text{m}^{-2/3}$. As we can see, the flux variance value is low when under this region turbulent with the highest flux is at 0.136 for coherent beam. Meanwhile the lowest flux is at 0.0054 for $q_c = 10$. This indicates the flux variance reduce when the q_c increase which relating to the strength of the diffuser.

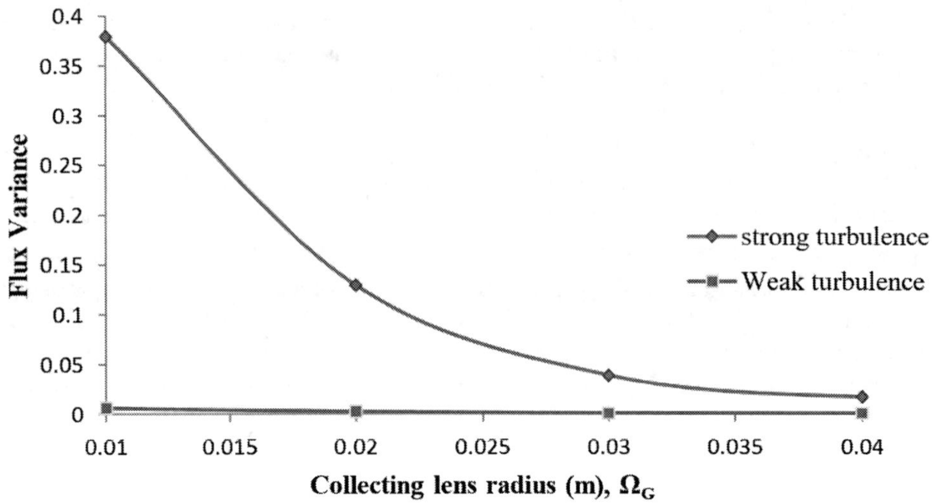

Figure 2. Comparison of flux variance $\sigma_{flux}^2(L + L_f, \Omega_G)$ versus the collecting lens radius, Ω_G between weak and strong turbulence condition for partially coherent beam (diffuser effect).

Similar curves are generated in weak turbulence regime and strong turbulence regime in figure 3 and figure 4, respectively. In strong turbulence regime, the effect of the diffuser is declining as shown in figure 3. Consequently the flux variance value is high and can lead to increase of signal to noise ratio (SNR). To be compared with figure 4, the highest value of flux is at 0.09 for coherent beam and the lowest flux is produced for q_c = 10 with value 0.052. This indicates that the reduction of flux is reduced even the increasing of Ω_G. The flux variance effect decrease with the increase of the collecting aperture size.

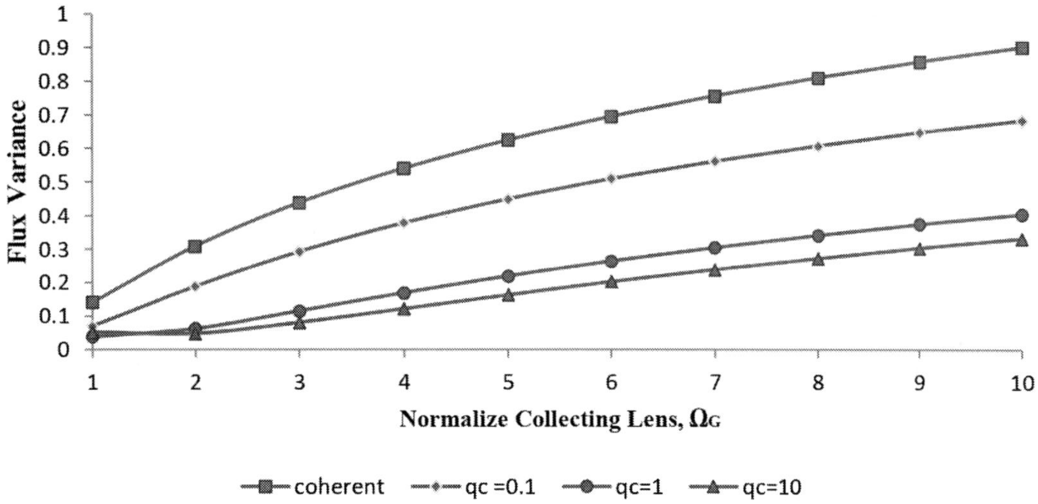

Figure 3. Effect of normalize collecting lens, Ω_G vs. flux variance for $q_c = 0.1, 1$, and 10 and coherent beam in strong turbulence regime.

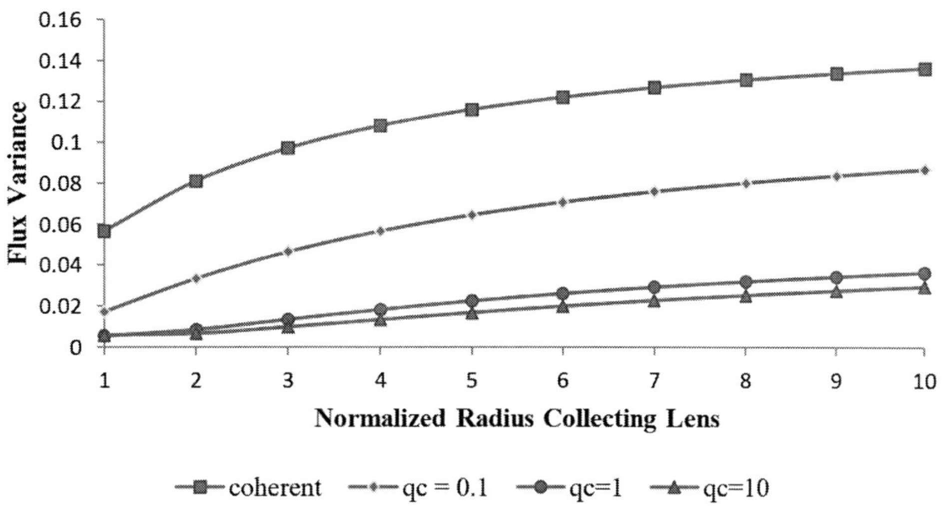

Figure 4. Effect of normalizing collecting lens, Ω_G vs. flux variance for $q_c = 0.1, 1$, and 10, and coherent beam in weak turbulence regime.

The figure 5 shows the comparison of flux variance versus the various of collecting lens radius, W_G between weak and strong turbulent condition. The propagation distance $L = 2000$m and for weak and strong turbulent strength $C_n^2 = 10^{-15}\text{m}^{-2/3}$ and $C_n^2 = 10^{-13}\text{m}^{-2/3}$ respectively which leading to Rytov variance $\sigma_I^2 = 0.07$ and $\sigma_I^2 = 7.09$. The significant reduction of the flux variance occurs in weak turbulent the flux varies from 0 to 0.05 but for the strong turbulence the flux value varies from 0 to 0.4. This illustrated that the effect of flux variance is greater when the strength of turbulence increase and the reduction of flux become less. From the graph also shows that the effect of flux variance can be reduced by increase the collecting lens radius. As the collecting lens radius increase the value of flux variance will decrease.

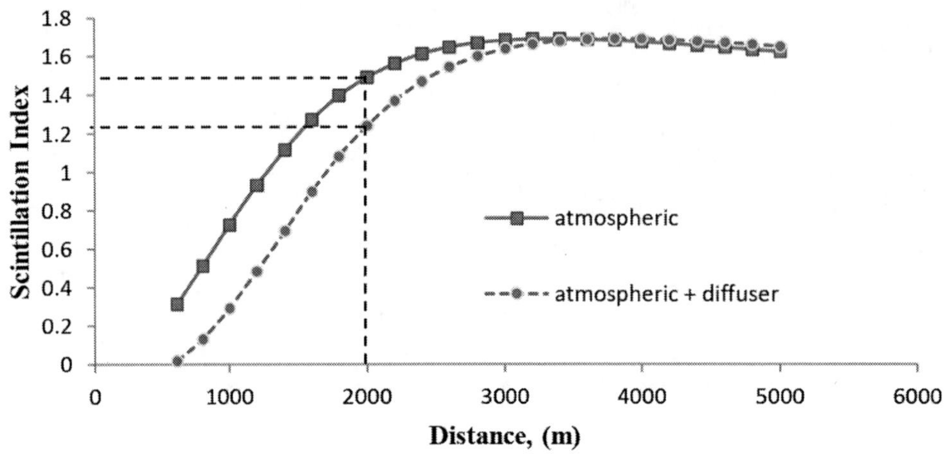

Figure 5. Comparison scintillation index due to strong turbulence between with and without diffuser

5. Conclusion

The effect of atmospheric turbulence from flux variance can be reduce by using partial coherent beam. Without combining it with diffuser, it can be seen at distance of $L = 2000$m that the scintillation index is at high value. However, by using the diffuser, the scintillation index can be reduced and the reduction of scintillation index where estimated approximately 19 percent.

References

[1] J. C. Ricklin and F. M. Davidson, "Atmospheric turbulence effects on a partially coherent Gaussian beam: implications for free-space laser communication," *J. Opt. Soc. Am. A* **19** 1794–1802.

[2] G. Gbur and E. Wolf, "Spreading of partially coherent beams in random media," *J. Opt. Soc. Am. A* **19** 1592–1598.

[3] V. Sidorovich, V. Ragulsky, M. Vasil'ev, A. Leshchev, and M. Sadovnikov, "Mitigation of aberration in a beam-shaping telescope and optical inhomogeinity in a free-space optical path using an extended light source coupled to the telescope," *Proc. SPIE - Int. Soc. Opt. Eng.*

[4] T. Shirai, A. Dogariu, and E. Wolf, "Directionality of Gaussian Schell-model beams propagating in atmospheric turbulence," *Opt. Lett.* **28** 610–612.

[5] L. C. Andrews, R. L. Phillips, and A. R. Weeks, "Propagation of a Gaussian-beam wave through a random phase screen," *Waves in Random Media* **7** 229–244.

[6] C. Y. H. Larry C. Andrews, Ronald L. Phillips, *Laser Beam Scintillation with Application.* Bellingham, WA: SPIE Press.

[7] L. Andrews and R. Phillips, "Laser Beam Propagation Through Random Media".

[8] J. W. Goodman, *Statistical Optics, 1st Editio. New York: Wiley-Interscience.*

[9] Y. Li, Z. Wu, and R. Wu, "Characteristics of the Partially Coherent Gaussian Schell-model Beam Propagating in Atmospheric Turbulence" pp. 0–3.

[10] S. Wang, C. Ouyang, and M. Plonus, "Optical beam propagation of a partially coherent source in the turbulent atmosphere," in 1979 *Antennas and Propagation Society International Symposium* **17** 319–322.

[11] J. Ricklin, S. Bucaille, and F. Davidson, "Performance loss factors for optical communication through clear air turbulence," *Proc. SPIE - Int. Soc. Opt. Eng.* **5160.**

[12] O. Korotkova, L. C. Andrews, and R. L. Philips, "Phase Diffuser at the Transmitter for Lasercom Link: Effect of Partially Coherent Beam on the Bit-Error Rate," *Proc. SPIE* **4976** 70–77.
[13] J. H. Churnside, "Aperture averaging of optical scintillations in the turbulent atmosphere," *Appl. Opt.* **30** 1982–1994.
[14] F. S. Vetelino, C. Young, L. Andrews, and J. Recolons, "Aperture averaging effects on the probability density of irradiance fluctuations in moderate-to-strong turbulence," *Appl. Opt.* **46** 2099–2108.
[15] L. C. Andrews, R. L. Philips, and C. H. Hopen, "Aperture averaging of optical scintillations: power fluctuations and the temporal spectrum, Waves in Random Media," *Waves in Random and Complex Media* **10** 53–70.
[16] D. L. Fried, "Aperture Averaging of Scintillation," *J. Opt. Soc. Am.* **57** 169–175.
[17] R. Kerr, "Experiments on Turbulence Characteristics and Multiwavelength Scintillation Phenomena," *J. Opt. Soc. Am.* **62** 91040–1049.
[18] L. Andrews, M. Al-Habash, C. Y. Hopen, and R. Phillips, "Theory of optical scintillation: Gaussian-beam wave model," *Waves in Random and Complex Media* **11** 271–291.
[19] L. C. Andrews, Special Functions of Mathematics for Engineers, Second Edi. Bellingham, Washington: *SPIE Optical Engineering Press.*

Acknowledgements
This research is funded under Fundamental Research Grant Scheme (FRGS), MOHE.

IOP Conf. Series: Materials Science and Engineering **767** (2020) 012037 doi:10.1088/1757-899X/767/1/012037

Optimizing Speedup on Multicore Platform with OpenMP Schedule Clause and Chunk Size

Z I Abdul Khalib[1,2] , H Q Ng [1] , M Elshaikh[1,2] and M N Othman[1,2]

[1] Embedded, Network and Advanced Computing Cluster, School of Computer and Communication, Kampus Pauh Putra, Unversiti Malaysia Perlis 02600 Arau Perlis
[2] Bioelectromagnetics Research Group, Kampus Pauh Putra, Unversiti Malaysia Perlis 02600 Arau Perlis

zahereel@unimap.edu.my

Abstract. Despite the description of schedule type which can be easily found at any OpenMP reference material, little is known about the effect of different schedule type and chunk size on the parallel performance of shared memory multicore processor. Literature shows that performance analysis on different multicore platform overlooked the effect of different schedule type and chunk size, where often it was not explicitly specified. Hence, default assignment of the loop iterations among threads is assumed. By default, static schedule is used and size of chunk which is the ratio of total number of iteration to the number of threads is implemented. This research analyses the effect of different schedule type and chunk size on speedup achieved of different shared memory multicore platform under regular workload. Apart from that, the performance gain obtained after turning on/off certain multicore technologies and after turning on/off selected number of active cores per processor is also analysed. Results shows that different multicore technology exhibit different speedup value under different combination of schedule type and chunk size. Apart from that, it also observe that different multicore platform is better than the other in terms of speedup as the number of cores are increased.

1. Introduction

Current trend of increasing the number of cores on a single die increases the demand of parallel programming, so as to fully utilize all the cores at the same time. As opposed to traditional single-threaded application which only uses one of the cores during execution time, multithreaded application has the tendency to exploit multicore processor to its full capabilities. This can be accomplished by generating multiple threads, assigning each thread to one of the cores, and running all the threads in different cores concurrently.

Nevertheless, not all applications support multithreading by default. With regards to OpenMP , to make a serial application becomes parallel, a programmer needs to hunt for the optimum spots in their code and insert parallel construct in between lines of the code. Sometimes, a particular code is required to be rewritten so that the parallel construct can be fit into it. As parallel programming is becoming popular, OpenMP API has become the de-facto standard for high level shared memory parallel programming [1]. OpenMP API is a C/C++ and Fortran compiler extension which gives programmer the ability to add parallelism into a source code without the need to significantly rewrite the code. To date, OpenMP Architecture Review Board (ARB) had introduced an ample amount of parallel directives which offer different patterns of parallelism, in order to optimize CPU utilization [2].

In enhancing parallel performance, part the of compiler directives – schedule type and chunk size [3] are of paramount importance. With respect to multicore technology, some work has been done to evaluate the performance of different parallel programming language, construct, tools, and libraries [4].

However, with respect to literature review done, the performance of a parallel code with different OpenMP schedule types and varied chunk sizes on shared memory multicore processor is barely known. The lack of guideline on assigning appropriate schedule type and chunk size could cause an OpenMP user to leave the setting as default. Thus, the higher parallel performance gain which might be produced by other schedule option is indirectly being neglected. Since schedule type decides how the workload is divided among threads, whereas chunk size defines the granularity of work, there is a need to determine the suitable schedule type along with appropriate chunk size of a parallel code. Nonetheless, how much contribution does these directives gives in terms of performance enhancement of different multicore processors is yet to be identified.

2. Related work

Please To date, there are very little research that has been done on investigating the effect changing schedule clause and chunk size. Since there is no explicitly specified schedule by [5], it is assumed that the authors were using default setting. In the paper, the authors compared the parallel performance of Intel Core i7-2600 and AMD Opteron Processor 8347. They obtained speedup which is close to the number of physical cores when the matrix size is large enough at both platform. Also, they found that increasing the number of cores reduces the parallel efficiency.

Running dense matrix multiplication algorithm at a server which has two Intel Xeon 5650 processors, static schedule without specified chunk size gave fastest execution among than other schedule type with chunk sizes varied from 1 to 32 [4]. The study showed the effect of small size of chunk and compared the performance to a loop scheduling policy they created.

By using transpose then multiply approach, the highest speedup achieved by [6] is slightly higher than the total number of logical cores detected. And, the experimental platform was a quad-core processor with hyper-threads. In certain cases, hyper-threading could be beneficial to improving performance of parallelism.

With parallel extension of Microsoft .NET Framework version 4.0, 3.5× speedup could be achieved at Intel Core i5-2400, which is a multicore platform with 4 physical cores [7]. Though different programming language and parallel programming model were used, a speedup value which is close to the ideal case can be obtained when the cores have their own duplicated CPU resource.

[8] used Pthread with SIMD instruction, he obtained 8.35× speedup in running native matrix-matrix multiplication algorithm. However, the experimental platform is not clarified. Nonetheless, the author did mention that 8 threads were used and the AVX SIMD instruction implemented was of 256 bit.

By running a parallel matrix-matrix multiplication at a processor with MIC architecture, the parallel speedup of 1.83× was achieved [9]. The rather low speedup is because the problem size is not large enough, as the serial execution of the algorithm is just 312.83s. Higher speedup could be expected as the problem size grows. Running the same algorithm, 8.58× speedup was achieved by 12 cores (24 threads in total) at Intel Xeon processor [10].

[11] compared the SIMD performance of the matrix multiplication algorithm obtained through inline assembly language and intrinsic function. Besides, they also compared the speedup gained between Intel and MSVC++ compiler. They found that Intel compiler generates code which requires fewer number of cycles, hence it is a better choice in running a code with SIMD instructions. Nonetheless, the average speedup achieved was about 2.1× and 2.08× when the intrinsic function and inline assembly were implemented with Intel and MSVC++ compiler respectively.

IOP Conf. Series: Materials Science and Engineering **767** (2020) 012037 doi:10.1088/1757-899X/767/1/012037

3. Experimental Methods

The methodology is adopted to observe the performance improvement provided by multicore architecture of three processors, specifically Intel Core i5-2410M and AMD A12-9700P. The research is conducted by first preparing the Square matrix multiplication algorithm which involves regular workload per loop iteration.

Next the serial code is parallelize using OpenMP directives. Then, the parallel code is verified for correctness against known parallel problem like race condition. Next, the code is executed repetitively for sampling, where certain multicore technologies are enabled/disabled by issuing command at the Linux terminal.

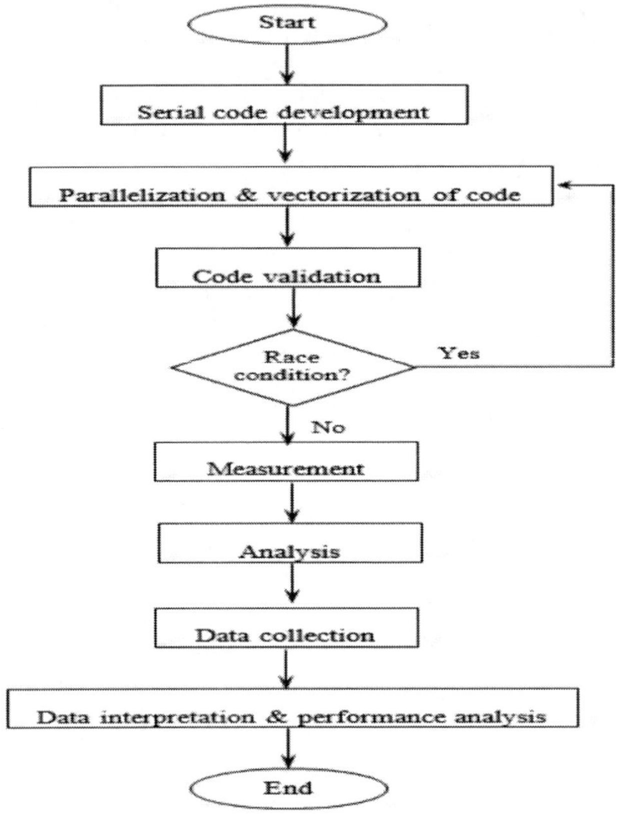

Figure 1. Workflow Methodology

4. Result and discussion

This section focuses on evaluating the parallel performance of CPU + memory bound code with regular workload per iteration under different OpenMP schedule type and chunk size for the three different types of multicore processor. The assigned chunk size divides the total outermost loop iteration among threads available.

Refering to figure 2, If the directive combination of (static, 250) is used, and there is N×N×N iteration to be run in parallel, each thread would be assigned 250×N×N iteration in a round-robin fashion and execute it until all the N×N×N iteration is finished.

```
#pragma omp parallel shared (A,B,C) private (i,j,k)
#pragma omp for schedule (static,250)
for i = 0 to N - 1 do
  for j = 0 to N - 1 do
    for k = 0 to N - 1 do
      begin
        A[i, j] += B[i, k] × C[k, j];
    endfor
  endfor
endfor
```

Figure 2. Matrix multiplication algorithm

Figure 3 compares the speedup achieved when different schedule type and chunk sizes are used in parallelizing matrix multiplication algorithm, on Intel Core i5-2410M platform, with different number of active cores. The maximum speedup achieved is approximately 2.0×, instead of 4×, which is the ideal case when there are four cores. This is due to the fact that Intel multicore processor comes with two terms, which are “physical core” and “logical core”. Physically, there are two physical cores on a processor. Because of HT, one physical core can run two threads simultaneously. Hence, there are 4 logical cores detected by the OS. Also, this logical core could not give equivalent performance as to physical core in this case, because the code bounded by CPU resources. Thus, the lower speedup than the ideal case. This fact is further proven by the graph of C0C1 and C0C2. Both C0C1 and C0C2 are viewed as two logical cores by OS. By right, C0C1 consists of two physical cores whereas C0C2 consists of one physical core and one logical core. The higher speedup between the two tells that physical cores are in demand in running this particular CPU- and memory-bound parallel code. By comparing the speedup between C0C1C2C3 (two physical cores with HT) and C0C1 (two physical cores), the former is of higher value. By comparing the parallel execution time of C0C1 and C0C1C2C3, it has been found that HT gives 1.2× ~ 1.3× faster execution in this case.

Turbo boost technology is a dynamic frequency scaling mechanism. The clock frequency of each core could be increased to a certain limit, if current workload is operating under power and temperature limit, and under limit of the Thermal Design Power (TDP) of that particular platform [12]. Since this technology is a dynamic feature, the clock frequency of processor will change dynamically, depending on the limiting factors. Thus, it is uncertain about the operating frequency would be. The faster parallel execution time of the code, and hence higher speedup, in any number of active cores might be due to the activation of this feature.

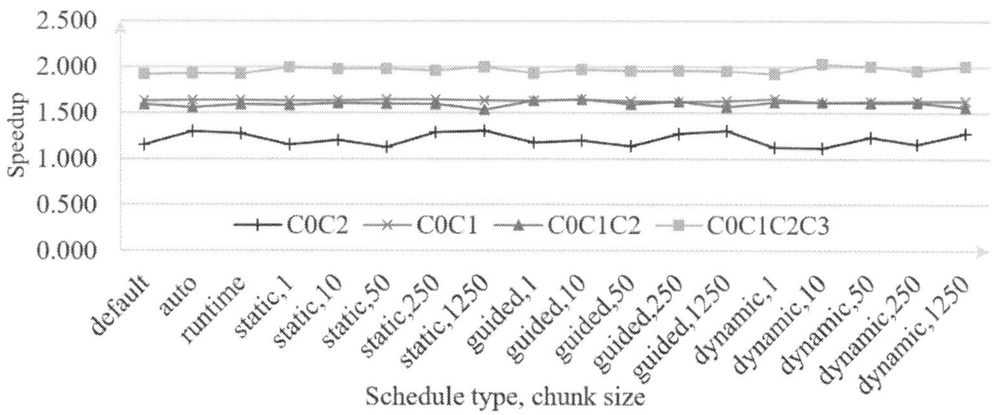

(a)

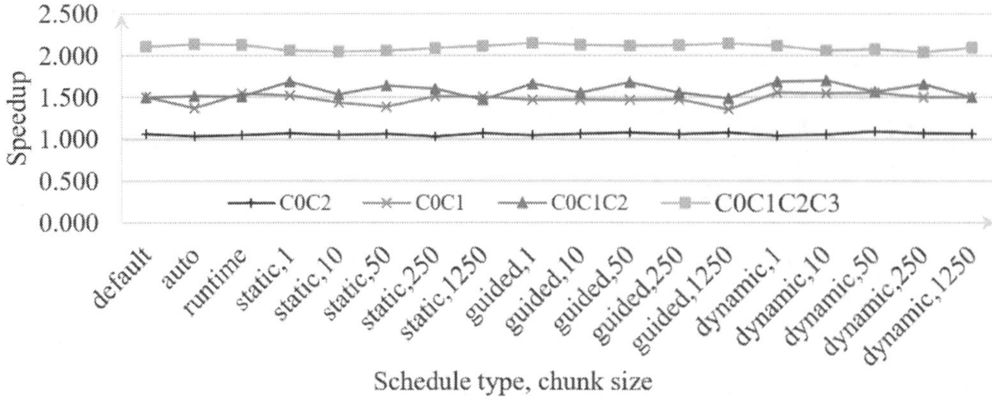

(b)

Figure 3. Graph of speedup against different schedule type and chunk size of parallel matrix multiplication algorithm at Intel Core i5-2410M (a) Turbo Boost Enabled (b) Turbo Boost Disabled

Figure 4 compares the speedup obtained when different OpenMP directives are used in parallelizing matrix multiplication algorithm, on AMD A12-9700P platform, with varied number of active cores. Similar to figure 3, dual-core is capable of producing comparable speedup as to tri-core, just like two non-HT cores (in Intel Core i5-2410M) and one module with two cores (in AMD A12-9700P). It can be deduced that the two multicore platforms are meant to perform better in even-numbered cores with each core having duplicated CPU resources.

As shown in figure 4(a), the speedup of dual- and tri- cores give higher speedup than quad-core. In terms of execution time, quad-core is still the winner among them. The serial execution time of quad-core is much lower than that of dual- and tri- cores. Although the parallel execution time of quad-core is lower than the other two, it differs not much (about ~60 seconds). Hence, when applied to the speedup formula, quad-core gives lower speedup than dual- and tri- cores, due to the small gap of time difference between serial and parallel execution time on quad-core and large gap of time difference between serial and parallel execution time on dual- and tri- cores.

When Turbo is disabled, both serial and parallel execution time decreases with the increment of number of active cores. Despite the shorter serial execution of tri-core as compared to that of dual-core, these two cases possess similar parallel execution time, hence the higher speedup produced by dual-core, as illustrated in figure 4(b).

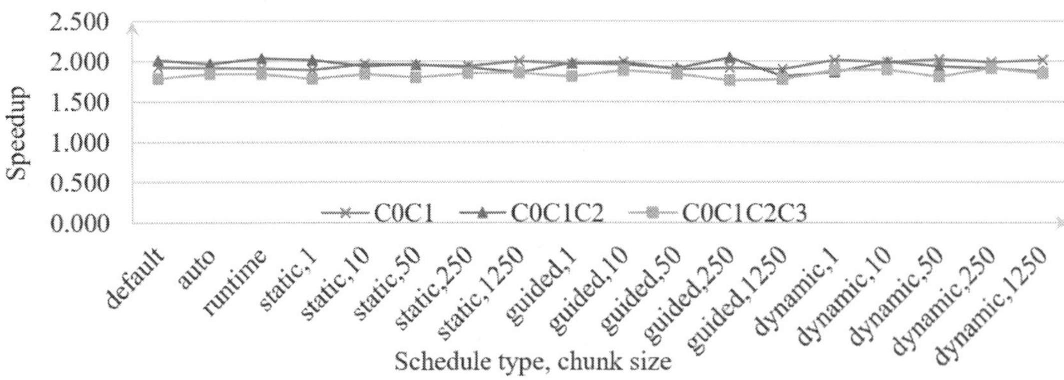

(a)

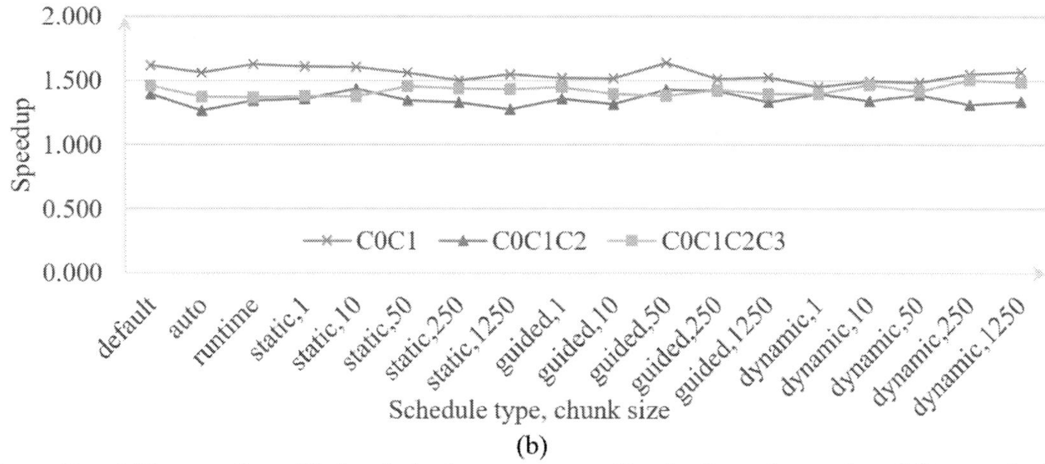

(b)

Figure Error! No text of specified style in document.. Graph of speedup against different schedule type and chunk size of parallel matrix multiplication algorithm at AMD A12-9700P (a) Turbo Technology Enabled (b) Turbo Technology Disabled

Figure 5. illustrates the speedup delivered when different OpenMP directives are used in parallelizing matrix multiplication algorithm, on ARM Cortex-A53 platform, with different number of active cores. ARM Cortex-A53 is a quad-core processor with each core having individual CPU resources. The two former multicore platforms which have shared and duplicated resources give the maximum speedup which is slightly more than 2, even when all cores are active, due to CPU resource restriction.

As is observed from figure 5, when number of active cores increases, the speedup increases. The graph of speedup of all the multicore platforms shows the range of performance improvement that could be obtained from parallelism. In running parallel matrix multiplication algorithm with regular workload per iteration, the workload is enough to make the scheduling overhead negligible, regardless of the combination of OpenMP directives used. Since all of the size of chunk specified is divisible by the problem size and number of threads, the workload is equally divided among threads in all kinds of directive combination. The size of chunk decides how many times the scheduling of task occur. The bigger the chunk, the lesser the scheduling overhead, as there is less request of chunk from thread. Yet, the overhead is negligible in all cases.

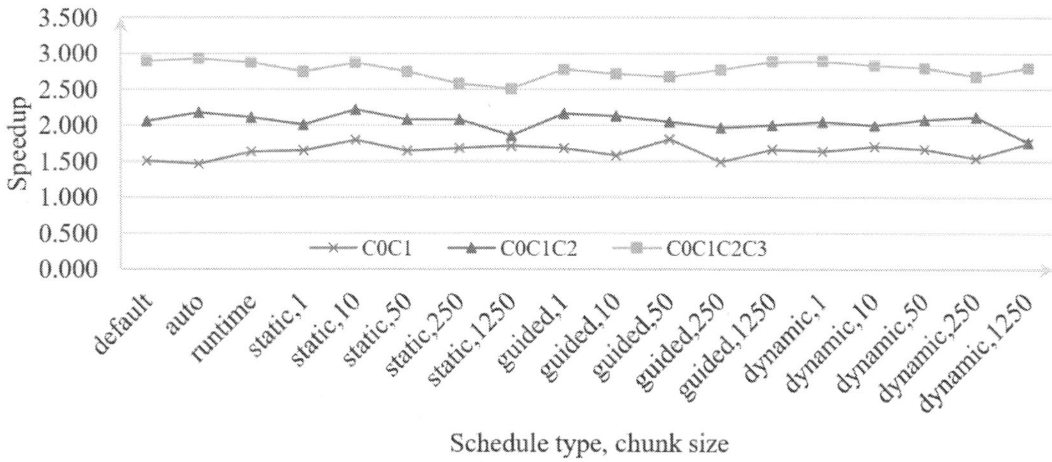

Figure 5. Graph of speedup against different schedule type and chunk size of parallel matrix multiplication algorithm at ARM Cortex-A53

5. Conclusion

In a nutshell, it is not always easy to identify the appropriate schedule and value for chunk size in running a parallel algorithm. It may depend (among other things) not only on the code in the loop, but also on the specific problem size and the number of threads used (Chapman, Jost, & Pas, 2008). The observations produced in this chapter may vary when the algorithm and/or problem size is different. The performance when a total of 4 logical cores detected is summarized in Table 1. The Turbo technology contribution for parallelism is the ratio of execution time of parallel algorithm when Turbo is disabled and when Turbo is enabled. The speedup of 4 logical cores as compared to 2 logical cores is the ratio of execution time of parallel algorithm when C0C1 is active and when C0C1C2C3 is active.

6. References

[1] Cownie J 30 Nov 2016 Retrieved December 1, 2016, from OpenMP: http://www.openmp.org

[2] ARB O November, 2015. *OpenMP Application Programming Interface.* Retrieved November 3, 2016, from OpenMP: http://www.openmp.org/specifications/

[3] *OpenMP Loop Scheduling.* 4, September 2012. Retrieved November 6, 2016, from Intel Developer Zone: https://software.intel.com/en-us/articles/openmp-loop-scheduling

[4] Ajkunic E, Fatkic H, Omerovic E, Talic K and Nosovic N 2012 A Comparison of Five Parallel Programming Models for C++ *Proc. of the 35th International Convention MIPRO* Opatija, Croatia IEEE p1780-84

[5] Anchev N, Gusev M, Ristov S and Atanasovski B 20-24 May 2013 Intel vs AMD: Matrix multiplication performance. *36th International Convention on Information and Communication Technology, Electronics and Microelectronics (MIPRO)*, p 182-7

[6] Tomikj N and Gusev M 21-25 May 2018 Parallel Matrix Multiplication. *41st International Convention on Information and Communication Technology, Electronics and Microelectronics (MIPRO)* p 204-9

[7] Delić M, Omerović E, Sokolovic A and Nosović N 2013 Performance Analysis of the Parallelized Matrix Multiplication Using C# Threading. *36th International Convention on Information and Communication Technology, Electronics and Microelectronics (MIPRO)* p. 1285-88 Opatija IEEE

[8] Pradyumna S 2017 Performance Comparison of Matrix Multiplication. *International Conference on Innovative Mechanisms for Industry Applications (ICIMIA 2017)* p 461-6 Bangalore IEEE

[9] Wang E, Zhang Q, Shen B, Zhang G, Lu X, Wu Q and Wang Y 2014 High-Performance Computing on the Intel® Xeon Phi™: How to Fully Exploit MIC Architectures. In *MIC Optimization Example: Matrix Multiplication* p. 231-55 Haidian Beijing China Springer International Publishing Switzerland.

[10] Thouti K and Sathe S R 2012 Comparison of OpenMP & OpenCL Parallel Processing Technologies. *International Journal of Advanced Computer Science and Applications (IJACSA)* **3** 56-61.

[11] Hassan S A, Hemeida A M and M M Mahmoud, M. 2016 Performance Evaluation of Matrix-Matrix Multiplications Using Intel's Advanced Vector Extensions (AVX). *Microprocessors and Microsystems* **47** 369-374.

[12] *Intel.* 1 November, 2016 Retrieved November 11, 2016, from Frequently Asked Questions about Intel® Turbo Boost Technology: http://www.intel.com/content/www/us/en/support/processors/000005641.html?wapkw=turbo+boost+2.0

Development and Performance Analysis of the Rogowski coil sensor for Arcing Fault Measurement

A N Nanyan, M Isa, M N K H Rohani, H A Hamid, A S Rosmi, M.Othman, A. Z. Abdullah, Z.C.M Kasa and A.Nazifah Abdullah

School of Electrical System Engineering, Universiti Malaysia Perlis (UniMAP), Pauh Putra Kampus, 02600 Arau Perlis Malaysia.

ayob.nazmy@gmail.com

Abstract. Arcing fault is an overvoltage that usually occurs in the power system network, and it is necessary to monitor this phenomenon in electrical equipment. This paper presents the Rogowski coil (RC) as an arcing fault sensor. Finite element method (FEM) software used for analysing the RC magnetic flux density with two types of RC with the different turns and size demonstrated. The sizing of RC based on the 240 mm^2 cross-link polyethene (XLPE) 11 kV underground cable. The prototypes of the RC sensors were designed using Solidworks software and three-dimensional (3D) printer for fabrication purposes. The main objective of the experiment is to investigate the RC effect with high magnetic flux density by implementing in real measurement. An experimental setup for real arcing fault with various voltages (up to 15 kV) was conducted to verify the RC performance, such as sensitivity and bandwidth range. The result has shown that the bandwidth of RC 1 is higher than RC 2 in all measurement by 22.6%.

1. Introduction

Overvoltage phenomenon is the most common problem that always occurs in the power system such as lightning, switching, arc fault or flash and others which can lead to an ignition of the combustible near the failure point and causing a fire or even explosion. Meanwhile, arc fault overvoltage may produce arc flash, which can interrupt the power system network. However, the incipient faults last a few milliseconds and some of the conventional protective schemes cannot detect it and also hard to detect by circuit protection system because this fault current is lower than usual current [1]. Generally, some overvoltage detection sensors have been discussed by other researchers, such as in [2]–[5]. Mostly, overvoltage protection is well developed and applied in commercial and domestic user, but there are not many researchers focusing on the early detection of overvoltage. Theoretically, arc fault overvoltage is the result of the current that flows through the air between conductors in phase-to-phase, single phase-to-ground or multiple phase-to-ground configurations [6][7] and includes both high-frequency components which ranging from several kilohertz up to tens of megahertz (MHz) [8]. Commonly, arc fault caused by insulation ageing, broken or continuous contacting, human error, uncorrected electrical fault, loose jointing at the terminal, moisture present and others [6], [7], [9]. The underground cable widely used in the power system network and always bare to wet conditions and quickly get serious moisture problem which may penetrate the cable's splice and results in a voltage arc. Furthermore, the pre-fault in the cable can't be detected by a human; thus, an early stage monitoring sensor is needed.

In this paper, geometrical analysis of the RC for arcing fault overvoltage detection sensor is presented, which is very helpful especially for the sensor fabrication purposes. The arcing fault experiment was set up to verify the performance of the sensor prototype by injecting voltage range between 3kV to 15kV. Each of the RC geometry has a different characteristic and has been studied in [10]. The advantages of the RC have been discussed in [11]–[13].

2. Rogowski Coil As Arcing Fault Sensor

Generally, the RC is a family of an inductive coil group which was introduced by Walter Rogowski in 1912. The RC consists of wire wound on a non-magnetic core and the required current to be measured will flow through the conductor or cable conductor and for measuring current purposes, the RC will be placed around the cable or conductor. Theoretically, a current-carrying wire produces the magnetic field around it and measured in Weber. This field will convert into a signal according to Faraday's induction law, which is proportional to the current or proportional to the rate of change of the current flowing in the conductor[14][15]. The RC detection method based on the high-frequency current pulse that passes through the conductor such as during arcing fault. The RC equivalent circuit and also known as lumped parameter circuit is shown in figure 1 which consists of lumped inductance, Ll, lumped resistance, Rl, lumped capacitance, Cl and terminating impedance, Zo.

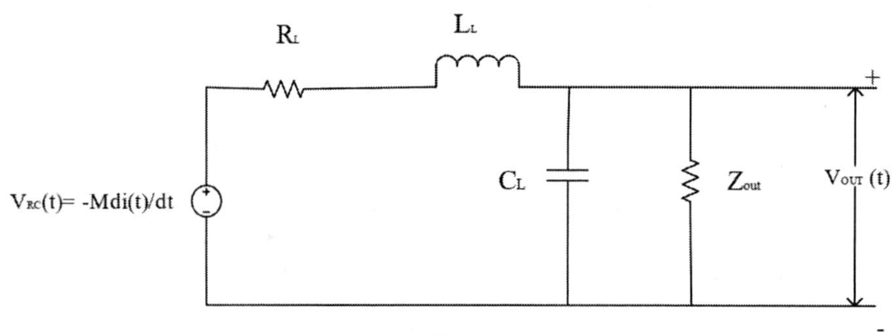

Figure 1. The equivalent of RC Sensor (Lumped parameter model).

3. Rogowski Coil Design

In this works, the finite element method (FEM) was used for analysing the magnetic flux density, B of the sensor. Theoretically, magnetic flux, Φ defined as the group of magnetic field lines emitted outward from the north pole of a magnet, and generally, a current-carrying wire produces the magnetic field around it, and it measured in Weber. The magnetic field intensity, H and the magnetic flux density, B at some defined point are in a reciprocal relation to the distance from the wire and the related between magnetic field intensity and magnetic flux density as expressed in Equation 1. The μ_o is the absolute permeability of the medium and the value is $4\pi \times 10^{-7}$ H/m.

$$B = \mu_o . H = \frac{\mu_o . I}{2. r} \tag{1}$$

Another parameter to be considered when designing the RC sensor is mutual inductance, M and it is determined by mathematical calculation. Mutual inductance calculation is a conventional method to determine the sensitivity of the coil sensor. The values of the parameter of RC (R_c, Cc and Lc) are slightly different among others and depend on the coil geometry [10]. The mathematical formula of mutual inductance is widely discussed in [15][16] .The calculation of mutual inductance for rectangular cross-section is expressed in Equation 2. From equation 2, N is representing the number of turns, h is the height of core and μ_o is equal to $4\pi \times 10^{-7}$. Based on the previous researchers reviewed and discussion

[16], the rectangular cross-section had high sensitivity; thus, this cross-section was selected for arcing fault RC sensor development.

$$M = \frac{\mu_o}{2\pi} Nh \ln\left(\frac{R_{out}}{R_{in}}\right) \quad (2)$$

1.1. RC Sensor Geometry

Firstly, two (2) types of the rectangular cross-section; RC 1 and RC 2 were designed and simulated using FEM software. The FEM analysis was applied for the electromagnetic flux density, B analysis of RC before fabrication is made (the lowest B is the best for manufacture). The geometrical sizing of the RC sensor is based on three cores 240 mm^2 cross-link polyethene (XLPE) 11 kV unarmoured underground cable. The basic rectangular cross-section design and calibration of the Rogowski coil have widely discussed by [17]. Figure 2 shows the RC cross-section, which is consist of height, h, inner radius, R_{in} and outer radius, R_{out}. The constructed parameters of the RC 1 and RC 2 is shown in table 1.

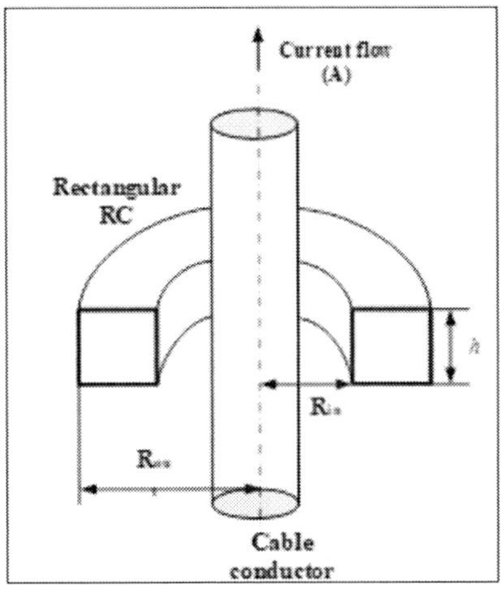

Figure 2. The rectangular cross-section of RC

Table 1: The parameters of constructed RC.

Coil Parameters	RC1	RC 2
Inner radius, R_{in} (mm)	44	44
Outer radius, R_{out} (mm)	54	54
D_{rc} ($R_{out} - R_{in}$)	10	10
Height of core, h	20	20
Number of turns, N	30	50
Mutual inductance, M (μH)	24.58	40.96
Magnetic Flux Density, B (mTesla)	1.47	3.68
Coil Inductance, L_c (μH)	4.21	5.76
Coil Capacitance, C_c (pF)	104.72	172.27
Coil Resistance, R_c (Ω)	0.17	0.26

For fabrication purposes, the RC was designed by using 3D Solidworks tool software which is compatible with 3D printer function. A filament material, Acrylonitrile Butadiene Styrene (ABS) was used as the casing of prototype and this material was proposed in [18]. The ABS is categorised as a non-conducting attribute and has low conductivity and classified as an insulator[19]. The advantages of ABS are widely discussed in [20]–[22].

1.2. RC Design Considerations

Rogowski coil consists of of a copper helical coil wire that has two winding techniques; return loop and return wire. The purpose of these techniques is to cancel the electromagnetic field from outside the coil. In this work, the RC was designed by wounding the copper wire with a diameter of 0.7 mm around the ABS rectangular shape and loop back through the centre of the coil to the end as a return wire to give less affected to the electromagnetic field.

The number of turns will give the effect to the coil sensitivity and the coil inductance, L_C. The L_C increased when the core height increased or adding a higher number of turn (the copper wire becomes more longer in this case). The value of inductance should be applicable to result in elevated sensitivity while preserving the high-frequency band sensitivity. Meanwhile, by increasing the capacitance value, it will result in reducing the reaction of the upper frequency while maintaining the sensitivity. The capacitance must be held as low as possible to prevent high-frequency fall.

Therefore, the prototype of the RC's was built using Solidworks and 3D printer based on the design consideration and feature above. Figure 3(a) and (b) show the prototype of RC's that fabricated in the lab. The coil resistance, R_c and coil inductance, L_c were measured using Agilent U1731C Handheld LCR Meter, and 1kHz frequency was selected.

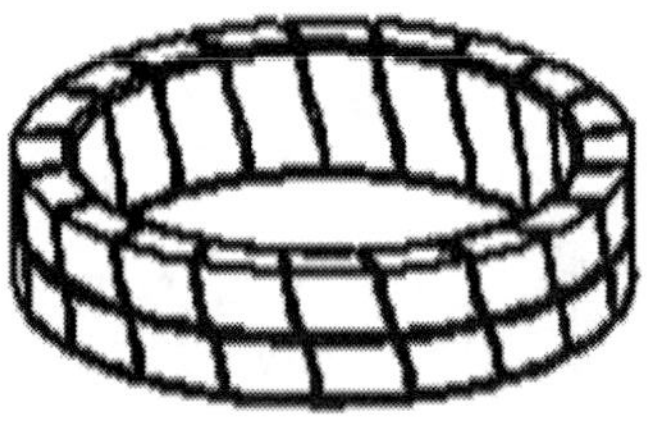

(a) RC designed using Solidworks

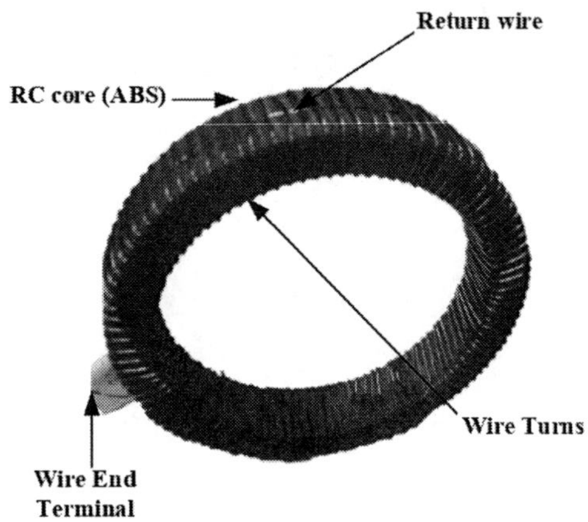

(b) ABS RC prototype with N turns

Figure 3. The prototype of RC's sensor (rectangular cross-section).

4. Experimental Setup

The experiment set up of the arcing fault signal was conducted in high voltage lab. An arc fault forms when the air becomes conductive due to a high electrical potential between two conductors (above the breakdown voltage). Theoretically, the average breakdown voltage ranges from 25 kV / cm to 30 kV /cm and depends on the insulating material, the shape of the tip of the conductor, and the gap between the two conductors [23]. However, it is found that the temperature also influenced the breakdown voltage. Arc generator model was set up by using two (2) copper rode; primary and secondary with 1.0-meter length each. These copper rods represent as the power cable which arcing fault signal travels until the end. The RC sensors are clamped and placed at secondary copper and measuring the travel signal. Figure 4 shows the arcing fault measurement setup which is consists of the BAUR PGK 70HB (for high voltage injection signal) and 1 kΩ resistor as load. The digital oscilloscope, Keysight DSO-X 3024A was used to detect the output signal that captured from the RC.

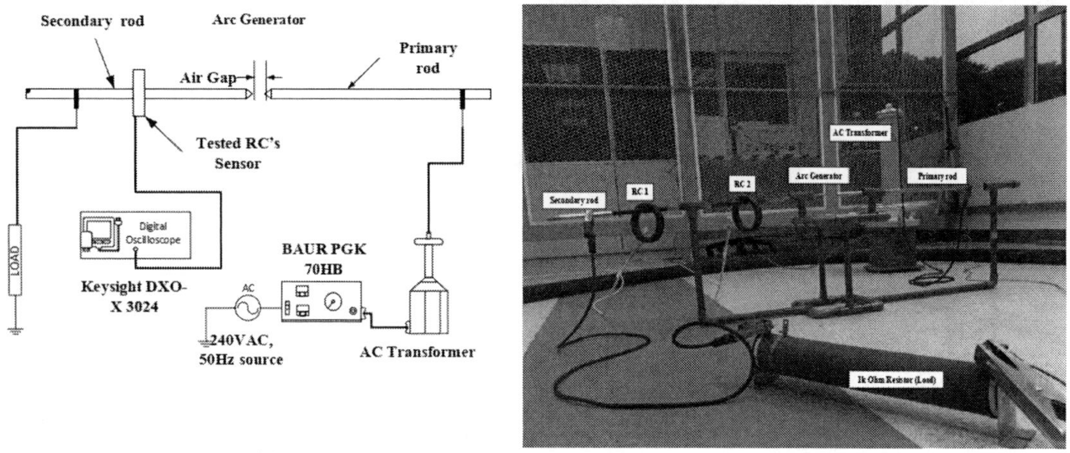

(a) (b)

Figure 4. Experimental setup for Arcing Fault measurement (a) Schematic diagram for experimental setup, (b) Laboratory experimental setup.

5. Simulation And Experimental Result For Arcing Fault Measurement

In this measurement, RC 1 and RC 2 were clamped on the secondary copper rode, and the voltage from 3 kV to 15 kV was injected at primary copper which results in voltage breakdown at air gap rod area (arc spark). The Fast Fourier Transform (FFT) analysis was carried out by importing the oscilloscope captured data signal and analysed by using MATLAB. Table 2 presents the arc fault measurement result of the RC 1 and RC 2. Figure 5 and figure 6 show the measurement waveform result of RC1 and RC 2, which was analysed using MATLAB at 3 kV. As can be seen, the amplitude of arc fault signal at 3 kV captured by RC 1 and RC 2 are 92.5 V and 123.0 V respectively. Similar trends have been observed in frequency resonance of RC 1 and RC 2, which are 7.14 MHz and 5.51 MHz. It shows that the RC with a more significant turn area has a significantly lower bandwidth. As seen in table 1, the more substantial number of turns gradually increased coil inductance. Thus, this observation indicates that the sensitivity of the RC 2 sensor decreased as it has a higher inductance than RC 1. In this case, RC 2 coil wire became longer (having 30 turns extra) than RC 1. Given the results obtained, the percentage difference of frequency resonance between RC 1 and RC 2 is 22.96%.

Table 2. The Arc Fault Measurement Result of RC 1 and RC 2 at 3 kV and 15 Kv.

Sensor ID	Injected voltage (kV)	Resonance frequency (MHz)	Amplitude (V)
RC 1	3	7.14	92.5
RC 2	3	5.51	123.0
RC 1	15	7.03	250.0
RC 2	15	5.47	260.0

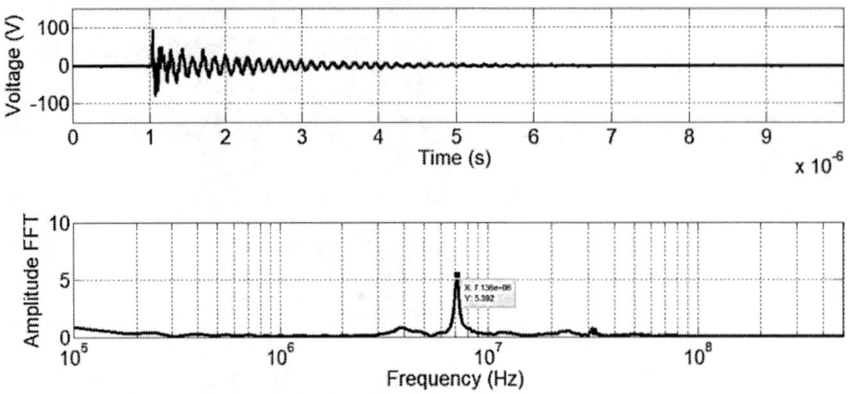

Figure 5. Arc Fault Detected by RC 1 at 3 kV breakdown voltage.

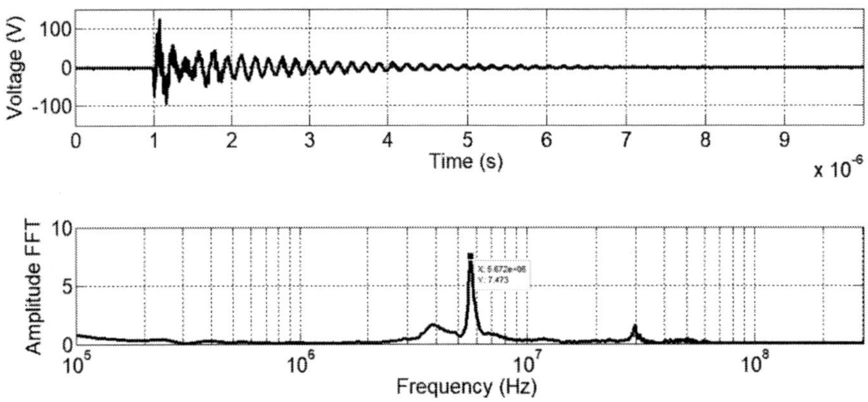

Figure 6. Arc Fault Detected by RC 2 at 3 kV breakdown voltage.

The measurement of sensor performance was verified on a 15 kV breakdown voltage as well, and the results are shown in figure 7 and figure 8. In this voltage level, the resonance frequency of RC 1 was 7.03 MHz while 5.47 MHz was captured by RC 2. It can be seen that, at this voltage, small changes of the resonance frequency has been measured and similar to 3kV measurement signal. Similarly, the voltage amplitude increased when the arc fault signal increased. However, the amplitude of RC 1 was 250 V and 260 V captured by RC 2. As can be seen, the result indicates that the RC with a lower number of turns has better resonance frequency compared to the higher number of turns.

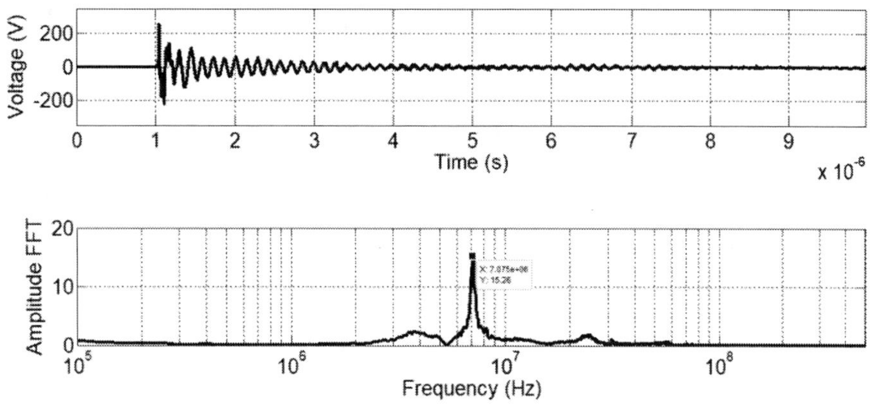

Figure 7. Arc Fault Detected by RC 1 at 15kV breakdown voltage.

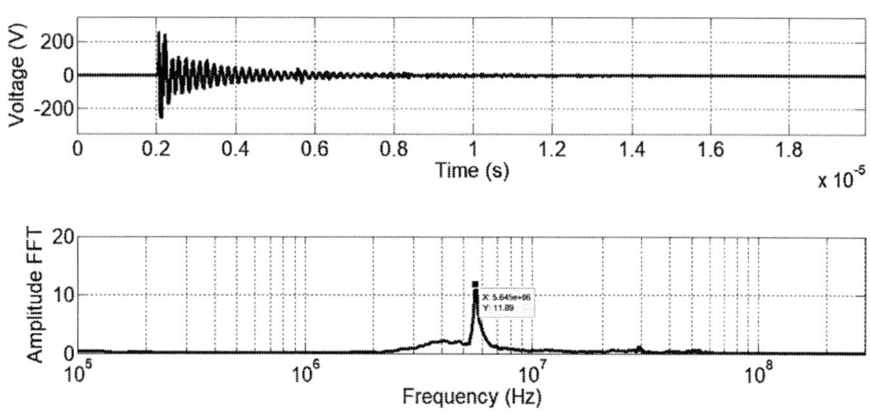

Figure 8. Arc Fault Detected by RC 2 at 15kV breakdown voltage.

6. Conclusion

In this paper, the measurement of arcing fault signal using RC as a sensor was successfully done. The performance in terms of sensitivity of each RCs design has been discussed. This real measurement has proved that the higher magnetic flux density and mutual inductance does not produce a high-frequency response. At 3kV and 15kV, the lower number of turns results in better resonance frequency at 7.14 MHz and 7.03 MHz for arcing fault detection. Also, the RC with low coil capacitance value has improved the upper-frequency limitation. The proposed ABS as sensor material is suitable and successfully measured the high voltage arcing fault.

References

[1] F. Naseri, T. Ghanbari, and E. Farjah 2015 Incipient fault monitoring of medium voltage UD-EPR power cable using Rogowski coil *International Congress on Electric Industry Automation* , pp. 33–36.

[2] R. Han, Q. Yang, W. Sima, T. Yuan, H. Dong, and S. Sun 2014 A non-contact electro-optic sensor for lightning overvoltage measurement *International Conference on Lightning Protection* (ICLP), pp. 1147–1150.

[3] T. Wang, Z. Wang, Q. Zhang, and L. Li 2013 Measurement method of transient overvoltage distribution in transformer windings *Annual Report Conference on Electrical Insulation and*

Dielectric Phenomena, pp. 1093–1096.

[4] D. Chatrefou, M. Pristchepa, and D. Uhde 2000 Application of optical sensors for measurement of high frequency overvoltages in power transformers *Power Engineering Society Winter Meeting IEEE*, **vol. 3**. pp. 2257–2268

[5] S. Wenrong 2011 Study on a new method for overvoltage measurement using CVT *East*, **No. 9**, pp. 321–324.

[6] R. Mendler, R. Hasan, J. B. Trolle, and N. L. Mina 2014 Hazardous Rated Electrical Equipment and the Arc Flash Hazard *IEEE Petroleum and Chemical Industry Technical Conference (PCIC)*. pp. 45–51.

[7] T. B. Dugan 2007 Reducing the flash hazard *IEEE Ind. Appl. Mag.*, **vol. 13**, no. 3, pp. 51–58.

[8] S. Chen, L. Zhu, S. Ji, and X. Liu 2017 Detection of series DC arc fault using rogowski coil *IEEE Conference on Electrical Insulation and Dielectric Phenomenon (CEIDP)*, pp. 401–404.

[9] Yuan Wu, Zhengxiang Song, and Xue Li 2013 A method for arc fault detection based on the analysis of signal's characteristic frequency band with wavelet transform *2nd International Conference on Electric Power Equipment - Switching Technology (ICEPE-ST)*, pp. 1–4.

[10] M. Shafiq, G. A. Hussain, L. Kütt, and M. Lehtonen 2014 Effect of geometrical parameters on high frequency performance of Rogowski coil for partial discharge measurements *Meas. J.*, **vol. 49**, pp. 126–137, Mar..

[11] M. H. Samimi, A. Mahari, M. A. Farahnakian, and H. Mohseni 2015 The Rogowski Coil Principles and Applications: A Review *IEEE Sens. J.*, **vol. 15**, no. 2, pp. 651–658

[12] A. H. Abdulwahid and Shaorong Wang 2016 A busbar differential protection based on fuzzy reasoning system and Rogowski-coil current sensor for microgrid *IEEE PES Asia-Pacific Power and Energy Engineering Conference (APPEEC)*,pp. 194–199.

[13] I. A. Metwally 2013 Performance Improvement of Slow-Wave Rogowski Coils for High Impulse Current Measurement *IEEE Sens. J.*, **vol. 13**, no. 2, pp. 538–547.

[14] L. A. Kojovic 2007 Comparative Performance Characteristics of Current Transformers and Rogowski Coils used for Protective Relaying Purposes *IEEE Power Engineering Society General Meeting*, pp. 1–6.

[15] A. N. Nanyan, M.Isa, H. A. Hamid, M. N. K. Hafizi Rohani, and B. Ismail 2018 The Rogowski Coil Sensor in High Current Application: A Review *IOP Conf. Ser. Mater. Sci. Eng.*, **vol. 318**, no. 1, p. 012054.

[16] M. N. K. H. Rohani *et al* 2016 Evaluation of Rogowski coil sensor performance using EMTP-ATP software *3rd International Conference on Electronic Design (ICED)*, pp. 446–451.

[17] R. Styblíková and K. Draxler 2011 Design and Calibration of Rogowski Coils *Power Energy Work.*, p. 100000,.

[18] M. N. K. H. Rohani *et al* 2016 Geometrical Shapes Impact on the Performance of ABS-Based Coreless Inductive Sensors for PD Measurement in HV Power Cables *IEEE Sens. J.*, **vol. 16**, no. 17, pp. 6625–6632,.

[19] S. Mohd Alauddin, I. Ismail, F. Shafiq Zaili, N. Farahanis Ilias, and N. Fadhilah Kamalul Aripin 2018 Electrical and Mechanical Properties of Acrylonitrile Butadiene Styrene/Graphene Platelet Nanocomposite *Mater. Today Proc.*, **vol. 5**, pp. S125–S129.

[20] Manish, D. Gurjar, S. Sharma, and Akash 2018 A Review on testing methods of recycled Acrylonitrile Butadiene-Styrene *Mater. Today Proc.*, **vol. 5**, no. 14, pp. 28296–28304.

[21] Y. Li *et al* 2017 Reparation of recycled acrylonitrile- butadiene-styrene by pyromellitic dianhydride: Reparation performance evaluation and property analysis *Polymer (Guildf).*, **vol. 124**, pp. 41–47.

[22] R. Goutham, T. R. Veena, Babagowda, and K. R. S. Prasad, 2018 Study on mechanical properties of recycled Acrylonitrile Butadiene Styrene (ABS) blended with virgin Acrylonitrile Butadiene Styrene (ABS) using Taguchi method *Mater. Today Proc.*, **vol. 5**, no. 11, pp. 24836–24845.

[23] J. C. Kim, D. O. Neacsu, B. Lehman, and R. Ball 2019 Series AC Arc Fault Detection Using Only Voltage Waveforms *IEEE Applied Power Electronics Conference and Exposition (APEC)*, pp. 2385–2389.

Acknowledgement
The authors would like to acknowledge the support form the Fundamental Research Grant Scheme (FRGS) under a grant number of FRGS /1/2019/TK04/UNIMAP/02/9 from Ministry of Education Malaysia.

IOP Conf. Series: Materials Science and Engineering **767** (2020) 012039 doi:10.1088/1757-899X/767/1/012039

Critical Review: Adaptive Pole Assignment PID Controller on DC-DC Converters

A W N Husna [1], M H Mat [2], M N M Yasin [3], M Jusoh[4], Y M Irwan[5], M S Abdul Rahim[5] and S Md Esa[5]

[1]School of Electrical System Engineering, Universiti Malaysia Perlis, Malaysia
[2]Bioelectromagnetics Research Group (BioEM), Centre for Diploma Studies, Universiti Malaysia Perlis, Malaysia
[3]Bioelectromagnetics Research Group (BioEM), School of Computer and Communication Engineering, Universiti Malaysia Perlis, Malaysia
[4]Bioelectromagnetics Research Group (BioEM), School of Microelectronic Engineering, Universiti Malaysia Perlis, Malaysia
[5]Centre for Diploma Studies, Universiti Malaysia Perlis, Malaysia

nurulhusna.awb@gmail.com

Abstract. Recently, control usages of DC-DC converters have been widely investigated particularly in renewable energy; as the primary sources. Discovering the most appropriate control method to be applied in DC-DC converter topologies is the most significant interest of research and development in this field. Thus, the review is carried out in selecting a control method that capable to improve the functioning of the converters as well as reducing the effect of disturbances and load variances

1. Overview of DC-DC converters

Recently, control applications of DC-DC converters have been widely investigated particularly in renewable energy; as the primary sources. The most important interest of research and development in this area is constantly to discover the very appropriate control technique to be employed in DC-DC converter topologies.

This paper reviews past work done for DC-DC converters controlled by an adaptive PID controller with the application of pole assignment method which is required on demand to achieve better performance of the overall system. Its rising importance in industrial applications and green technology is due to the need to improve the system so that it can be adapted with existing technology of renewable energy and become more efficient in manufacturing. All reviews come from different sources which consist of various methods, in order to obtain a comprehensive view about the latest technique and technology of controller for DC-DC converters. Thus, a comparative study can be done amongst converters and controllers involved so that a better performance can be obtained. Thus, the most important goal of this review is to study a recent method of adaptive pole assignment PID to control DC-DC converters.

2. Critical Review

PID controller has been designed in 1890s and since that there is a lot of improvement has been done on it thus more approaches were developed to adapt to the latest technology. From conventional method, nowadays PID was improved to the digital controllers that widely implemented with microcontroller, field-programmable gate array (FPGAs), or digital signal processor (DSP). This method has successfully developed through the design of DC-DC converters that controlled by DSP based on the implementation of a digital controller. Beginning by a DC-DC buck converter and a given set of functioning specifications, a digital PID controller is employed in TMS320F2812 DSP. In this work, the effectiveness of the design is established to analyze the steady state and dynamic response performance of the controller. A digital based PID control approach for a DC–DC buck converter has been introduced along with a digital implementation of the controller using DSP. At the end of this work, the experimental performance shows that; steady-state accuracy and settling time are consistent with the simulation results [1].

The classic technique for tuning a PID loop has become even more popular with the beginning of controllers capable of tuning themselves. The tuning techniques namely as Ziegler-Nichols, are still used nowadays even it was published in 1942. As referred to [2], specified that John "Zeke" Ziegler and Nathaniel Nichols may not have developed the proportional-integral-derivative (PID) controller, even though the PID system is the most common of entirely feedback control approaches used in industrial applications, but the well-known loop tuning methods designed by John "Zeke" and Nathaniel Nichols facilitated a lot in creating the PID controller as an outstanding controller. In fact, by tweaking a PID loop the responses of the controller to faults between the determined process variable besides desired set point can be tweaked. Doubt the controlled procedure happens to be fairly sluggish, the design of PID process will take instantaneous and remarkable actions whenever a random disturbance shifts the method variable, or an operator shifts the set point. These reviews stated that an adaptive technique that applied in DC-DC converters show an improvement in transient response besides it produced a superior performance in terms of system improved stability, disorder elimination and yield control. Nevertheless, all this can be achieved using PID controller. Furthermore, ringing and oscillation occurrence in transient response can be averted by smoothly transitioning between steady-state operation as well as dynamic transient operation. The proposed controller; PID has proven that the operation of DC-DC converters is not depending on any further hypothesis of the transient magnitudes or kinds.

In past a decade, the appliance of an adaptive PID controller has been attracting other researchers [3]. An immediate pole assignment control approach is announced during this workplace yet because the execution of a discrete- time controller within the model of a DC-DC buck converter has been used. the answer involves a feed forward component within the control strategy so as to remove steady- state inaccuracies. The worth of the feedforward gain that's eliminates the steady- state error depends upon the gain of the plant, which cannot be known precisely. In their model, the feedforward gain is resolved adaptively to drive the steady state inaccuracy to zero. This method allows precise placement of the closed-loop poles as anticipated. so as for the steady-state error to be exactly zero using this feedforward technique, the feedback gain must be determined exactly. As this can be impossible, they adapt the feedback gain using least-mean-square (LMS) techniques, so the control error is forced back to zero. The direct pole-placement could be a possible approach for DC-DC converter model which is permitting the choice of an easy complex conjugate pair of closed-loop poles, yet on establish a completely unique technique to achieve a zero steady-state error. With this work, the performance of a prototype compared very favorably with standard design methods. An adaptation of the feedback gain using LMS techniques was proven to be effective in driving the steady-state control error to zero.

A straightforward auto-tuning procedure for digitally dictated DC-DC synchronous buck converters was introduced by [4]. In this effort, the intended method is based on the relay feedback technique as well as the authors were presenting perturbations on the output voltage throughout converter soft start. Via utilizing an iterative method, the fine-tuning of PID parameters is achieved precisely by incorporating the controller in the relay feedback and by tweaking the controller factors based on control loop bandwidth along with the phase margin. A nice property of the planned solution is that output voltage perturbations are presented as preserving the closed loop control of the digitally controlled converters. The suggested algorithm is minimal, involves minor tuning times and it is compatible with the complexity constraint of integrated digital ICs. Experimental investigation has been performed using discrete components, employing the digital control in an FPGA. Simulation and experimental results of a 1.5V–5A synchronous buck converter confirms the effectiveness of the proposed solution.

H. and Ran, Z. have been discovered the study centered on the integrated circuit version for DC-DC converters [5]. In this work, the planned new controller is an altered single-neuron self-adaptive PID controller invented by applying the 'Hebbian Learning Rule' (HLR). Moreover, mimicking the PID control implementation as well as vigorous responses designed for the given DC-DC converter than evaluated by using Simulink. To assess the control implementations, an investigational on buck converter has been applied. discover out a rapid, simple and effective design method for the controller of DC-DC converters are well attained are the purpose of this work. Constructed on the aim, the integrated circuit model is utilized to develop new conventional prototypes for PWM DC-DC converters. Additionally, a straightforward single-neuron controller has been designed by utilizing the planned control scheme. The modeling and investigational outcomes are attained to demonstrate that the new controller has improved operations competed to the conventional PID device and the control approach might reduce to bare bones of the inventing method of standard smart controllers.

The application of fuzzy logic and PID controller has been explored by [6]. The state-space averaged prototype, is an appraised prototype, is utilized toward produce PWM DC-DC converters. As an Alternative, a process built on Orthogonal-Function Approach (OFA) merely relating arithmetic calculation is recommended in this work to exactly resolve the irregular vigorous calculations of the PWM DC to DC converters. The optimum fuzzy-immune-PID controller model challenge for a group of PWM DC to DC converters is changed into a constant optimizing parameters challenge that is characterized by algebraic equivalents based on the OFA. Instead, a PWM DC to DC converter, a fuzzy-immune-PID controller by merging the conventional PID controller, the fuzzy logic theory as well as the immune feedback law are considered to be a self-adaptive controller in their work to adapt the load variation. But then the Hybrid Taguchi-Genetic Algorithm (HTGA) is used for the static optimization challenge to discover the best possible parameters of the fuzzy immunizing PID controllers used by the PWM DC to DC converters. The recommended integrative technique, fusing the OFA together with the HTGA, is non-differential, non-integral, straightforward and well-adapted to computer application. The outputs from this proof that the recommended method yields an effective approach to synthesizing the best parameters of the PWM DC to DC converters ' fuzzy-immune-PID controllers. An algorithm engaged in this study, based on the OFA, is merely algebraic calculation to accurately resolve the uneven dynamic calculations of the DC to DC PWM converters. In addition, the algorithm presented is combined with the HTGA to synthesize the optimal parameters of the fuzzy immune PID controllers under the minimisation of an integral quadratic cost functional. Since the proposed method requires mere algebraic computation, the design methods of the ideal fuzzy-immune-PID controllers for the PWM DC to DC converters can be either completely reduced or significantly streamlined appropriately. It shows at the conclusion of this work that the proposed methodology is effective in synthesizing the PWM DC to DC converters optimum fuzzy immune PID controllers.

Another alternative approach has been discovered by [7], introduces a form of enhanced fuzzy-PID control technique which approves linear quadratic regulator (LQR) methodology. There is some model of a system is unspecified or complex in most of industrial control systems. Due to this problem, PID parameters are rather difficult to find hence the authors take this opportunity designed the advanced controller with LQR in improving the control performance. An achievement of better performance has been found even according to the fact that the output voltage is always heavy harmonic pollution in DC-DC converter using a traditional PID controller. In addition, it be able to alter the PID parameters online and be considerably improved than standard PID on steady state as well as anti-jamming; consequently it proves the system is achievable, effective and practical. This work indicates that the combination of LQR algorithm and the fuzzy reasoning control can significantly improve the final control effect. Likewise, the output voltage of the optimized DC-DC buck converter has a better load-disturbance rejection capability than the traditional PID control or LQR method. The overshoot at start-up and the steady state error are eliminated using the new approach. Consequently, it has faster and dynamic responded and smaller overshoot under the load change condition.

Jen-Ta, S. discuss a fully digital controller for interleaving DC-DC converter [8]. Adaptive control architecture was presented based on a PID controller with pole-zero cancellation method to optimize the transient response under different load conditions as well to improve the dynamic performance in rapid load current slew rate requirement. According to PID controller, the control architecture with gain scheduling which is based on pole-zero cancellation method is used to construct the parameters of the PID in optimizing the transient response. The proposed digital controller is tested on a 12V input with 1.2V, 30A buck converter by FPGA based. Furthermore, the control method has verified in an FPGA board and the experimental results demonstrate a substantial improvement in transient response.

Although adaptive control mostly applied to the PID controller, there are other works which introduced this method collaborated with another controller as presented by [9]. As a DC-DC converter is strongly nonlinear, a fuzzy self-adaptive adjusting of PID controller is presented in this work, which is a control mode which employs fuzzy control rules to modify the PID parameters online. The equivalent small signal circuit of the phase-shifted full-bridge converter is derived based on a small signal model of buck converter. The performance of the converter has been proved to be perfect by using the program of fuzzy self-adaptive adjusting of PID control using MATLAB simulation. Due to the self-adaptive adjusting of the PID control parameters, the system can steadily quickly in the condition of perturbation.

Another type of adaptive control such as adaptive voltage positioning (AVP) is also taken into account which is operated inside DC-DC switch over energy converters used for driving unified circuits exactly to this one improvements of applying permissible production energy acceptance whilst reducing the production screen capacitance requisite , combined with attaining a more rapidly transient response. To reduce the right half plane (RHP) 0 consequence the rapid Boost DC to DC Converter is constructed through AVP method [10]. A common model theory in former works with AVP design is the concept of steady production impedance . Nevertheless, achievement of constant output impedance complicated by the "RHP zero" exists in controller to production transfer function of the Boost Converter (BS). To attain comparative continuous production impedance as well as achieve the AVP specific inside BS. The AC altering skill suggested in this article is showed. throughout the load transient contrasted with every control scheme introduced in the past. The boost converter model along with the AVP procedure offers reduced voltage overshoots or undershoots and far quicker response time. Once the load current alterations from 100mA to 400mA. The replication outcomes are shown to demonstrate exceptional performing of faster than the 4μ s response time then the 40mV voltage drop. In this study, the proposed AC modification method relieves this constraint from the well-known RHP zero which occurs in the transfer function to output control. The presence of the RHP zeroes slows the energy supplied to the output and causes great fluctuations in voltage during a transient load. The output voltage will settle down within a few micro-seconds using the AVP control technique and have a much smaller drop in voltage than other control schemes. The results verify the proposed technique successfully and obtain an excellent result of decreasing voltage by 40mV when the load current changes from 100mA to 400mA.

Giustina della, D. And Liberali, V. analyzed the sensitivity of the adaptive filter used to control pulse width modulation DC-DC converters to the discrete and continuous parameters [11]. Measuring and digitizing input variables of the structures are selected as the optimum transition method to compensate the power cell. A digitally programmable switching capacitor circuit is used for its synthesis. It discusses the effects that can lead to an incorrect estimation of the operating point and miss the external feedback dynamic behavior of the target. To investigate limitations in the synthesis of the transfer function, an adaptive switched condenser filter using multiplying digital-to-analog converters (MDACs) was analysed. The results of both quantization and inconsistency cause poles and zeros to change from their theoretical position.

Although PID and Fuzzy controller become the priority controller until nowadays, sliding mode still gives a great performance on DC-DC converters. Sliding mode PID approach has discovered by [12], to control DC- DC converter. The integrated vigorous design of DC to DC buck converter has been set up along with 2 slipping method controllers are presented so-called the traditional "sliding mode (SM) controller" The constancy is examined for the DC to DC converter system controlled by the SM PID as well as the optimal SM PID limits are clarified . The system simulation is employed plus simulation findings demonstrate that SM PID controlled DC to DC buck converter requires improved vigorous combined with constant performing compare with the procedure operated by the traditional SM controller. An integrated dynamic design of DC to DC buck converter is initially configured later their work is applied to developing the SM PID optimal considerations when the system is packed various opposition. To inhibit output voltage disruption as well as the merging of lengthy time SM at the instantaneous of the rapidly changeable load, hence the altered SM PID controller is invented.

Additionally, there are some impressive reviews written by [13], have improving the vigorous performing of DC to DC energy converter an method aimed at an adaptive PID device. The control adjustable namely the proportionate constant (Kp) and integral constant (Ki) vary such as a purpose of the uncompensated inaccuracy voltage in improving the dynamic transient response of the energy converter,. It shows that by comparing the conventional PID there is a decline during undershoot or overshoot as well as settling time from their experimental outcomes. The effectiveness of adaptive PID control method in a reduction in the settling time and overshoot/undershoot of the power converter output voltage, dynamically and adaptively variable the proportional and integral constants as a function of the error signal has been proved. Furthermore, any ringing and oscillations have been averted by the smoothly transitioning between steady-state operation and dynamic transient operation. The idea is continued by presenting an adaptive digital proportional–integral–derivative (AD-PID) controller scheme to enhance the vigorous implementation of energy converters in closed-loop structure from their previous work [14]. The controller smartly modifies the Kp and Ki subsequent a latest control rule. The control rule is a purpose of the magnitude adjustment in the inaccuracy signal as well as that one highest rate through vigorous transients. Same as their previous controller, the proposed AD-PID controller smartly identifies the highest rate of the inaccuracy signal, which is a purpose of the magnitude and transient nature and utilizes it in the control rule as a consequence of the smart-process such that no fluctuations are produced. Consequently, there are reduction occurred in the settling time of the output voltage and vigorous output of the voltage divergence. The concept and architecture of the anticipated AD-PID are stated and its intended control rule is considered besides validated with investigational findings. By applying the AD-PID, it has been proved that the vigorous outcome voltage variation coupled with the settling time are diminished with a straightforward plan then minus the must to know every extra info. The AD-PID controller utilizes a system that ensures the switch among the steady-state plus transient process approaches is easy and unrestricted of extra fluctuations as well as reverberating.

The model of the strong Digital PID (DPID) controller for H-bridge soft-switching boost converter was proposed by [15]. In this paper, a strong DPID controller is suggested for the H-bridge soft-switching boost converter (HSBC). The DPID device remains intended to certify load voltage regulation and give a vigorous execution along with step loads plus source decrease. The statistical simulations of the H-bridge boost converter are articulated applying the method identification apparatus in addition to use in DPID model.

A smart prediction error filter (PEF) in the control system response loop a are introduced by [16]. This study explains the concept of the use of an adaptive PEF and statistical study of core principles, such as the Auto-Regressive (AR) cycle generator and the Moving Average (MA) identifier. In addition, authors also submit the direct relation between an adaptive PEF and a digital Proportional-Derivative (PD) controller. This leads to the development of the adaptive control structure Proportional-Derivative + Integral (PD+I) which is an important replacement for a traditional PID controller. In this work, the relation between the PEF and a PD controller are described, including its suitability for usage in an adaptive PD+I control structure. The superior performance of this controller over a classical PID approach; in terms of system disturbance rejection, improved stability, and output regulation proved in this work. The experimental validation has proven that the approach results in a fast-adaptive controller with self-loop compensation.

The design and structure of a fuzzy logic based on a PID controller for the control of a PWM of DC-DC buck converter working in continuous conduction mode (CCM) was introduced by [17]. In this work, the converter is set to use at a swapping frequency of 100KHz. Computer replications are complete to fuzzily the responses plus association tasks, therefore rapid the corollary regulations combining the output and input and variables to de-fuzzify the outcome limit. No linearized prototype of the buck converter is needed this methodology is straightforward. As of the simulation findings, the fuzzy-logic constructed on PID control method offers tough control for non-linear power electronic flexible shifting assembly .

In some cases, negative impedance instability compensation in buck converter using state space pole placement control has been taken as one of the control methods in improving the satisfactory operation of the converter [18]. In atmosphere energy systems, numerous voltage-power-systems and power-converters work together along with each one in addition be able to be resulting wobbly process. It demonstrates volatility once a DC to DC converter is inserted by a continuous power load (PL),. This is as of the undesirable impedance specific of the continuous PL [19]. Several poles of the converter with a continual PL are in the RHP. The work represents a reimbursement procedure used for buck converter utilizing SSP positioning control. Through a SSP position control, the unpredictable poles have transferred to the steady section. The traditional PI controlled buck converter is volatile once it is packed with continuous PL. Buck converter packed with continuous PL is remunerated with pole positioning control.

Though, fast transient (FT) control in conjunction with the Adaptive Phase Margin (APM) to achieve a high transient response in current DC-DC buck converters under different load conditions was uncovered by [20]. The over-shoot / undershoot voltage and the transient recovery time are reduced efficiently. The APM control enables the system phase margin to be continuously maintained at an appropriate value below distinct load circumstance. That is, the pole-zero compensation pair is modified to load current to broaden the bandwidth of the system and get an adequate phase margin. Research findings indicate that the overshoot / undershoot voltage is lower than 60mV (3%) and the transient period is lower than 12μs as the load current changes abruptly from 100mA to 500mA, or vice versa. The undershoot voltage and recovery time are increased by 45 percent and 85 percent compared with traditional models without any fast-transient method

Yushan, L. analyzed how exactly the current mode control delivers dynamic damping for improve converters by continual power loads [21]. Small-signal control-to-output transfer functions are obtained for peak or valley current mode-controlled boost converter with a downstream regulated converter developed as continuous power load. Moreover, this one is demonstrated how load current feedforward grants an efficient way to enhance energy capacity transient responses. Developing coupled with model methodologies are proven with examine circuit imitations, exhibiting balanced processes utilizing current-mode control in steady energy loads. The specified that John "Zeke" Ziegler and Nathaniel Nichols may not have developed the proportional-integral-derivative (PID) controller, even though the PID system is the most common of entirely feedback control approaches used in industrial applications, but the well-known loop tuning methods designed by John "Zeke" and Nathaniel Nichols facilitated a lot in creating the PID controller as an outstanding controller. In fact, by tweaking a PID loop the responses of the controller to faults between the determined process variable besides desired set point can be tweaked. Doubt the controlled procedure happens to be fairly sluggish, the design of PID process will take instantaneous and remarkable actions whenever a random disturbance shifts the method variable, or an operator shifts the set point.

3. Summary

Since DC-DC converters are strongly nonlinear, a conventional PID controller is simply to adapt with 'small-signal control-to-output' transfer function through a compensation technique using state space pole position control. This control procedure will be applied in this work and may lead in improving the satisfactory performance of the converters. The unstable poles will be repositioned to the stable region through a SSP position control. The pole position controller imprints the anticipated outcome relatively good if adjusting the converters boundary to converge with load current remains constructive,

These reviews stated that an adaptive technique that applied in DC-DC converters show an improvement in transient response besides it produced a superior performance in terms of system improved stability, disorder elimination and yield control. Nevertheless, all this can be achieved using PID controller. Furthermore, ringing and oscillation occurrence in transient response can be averted by smoothly transitioning between steady-state operation as well as dynamic transient operation. The proposed controller; PID has proven that the operation of DC-DC converters is not depending on any further hypothesis of the transient magnitudes or kinds. That One can adapt its process for any transient magnitude and kind, as it continuously utilizes it in its control principle such as per a product of the adaptive process and finds the greatest value of the voltage error signal.

References

[1] Tajuddin M F N bin, Rahim N A and Daut I 2009 Design and Implementation of a DSP Based Digital Controller for a DC-DC Converter *Second International Conference on Computer and Electrical Engineering* (IEEE) 209–13.

[2] Anon Control Engineering 2016 Understanding PID control and loop tuning fundamentals.

[3] Kelly A and Rinne K 2005 Control of DC-DC converters by direct pole placement and adaptive feedforward gain adjustment *Conference Proceedings - IEEE Applied Power Electronics Conference and Exposition - APEC.*

[4] Stefanutti W, Mattavelli P, Saggini S and Ghioni M 2005 Autotuning of digitally controlled buck converters based on relay feedback *PESC Record - IEEE Annual Power Electronics Specialists Conference.*

[5] Hui J and Zhao R 2007 Single-neuron self-adaptive PID control based on unified circuit model of DC/DC converters *ICIEA 2007: Second IEEE Conference on Industrial Electronics and Applications* 1994–7.

[6] Hsieh C H 2007 Optimal fuzzy-immune-PID controllers design of PWM DC-DC converters *Annual Conference of the North American Fuzzy Information Processing Society - NAFIPS.*

[7] Wei K, Sun Q, Liang B and Du M 2008 The research of adaptive fuzzy PID control algorithm based on LQR approach in DC-DC converter *Proceedings Pacific-Asia Workshop on Computational Intelligence and Industrial Application.*

[8] Su J-T, Liu D-M, Liu C-W and Hung C-W 2009 An adaptive control method for two-phase DC/DC converter *International Conference on Power Electronics and Drive Systems (PEDS)* (IEEE) pp 288–93.

[9] Jiang Z and Liu Y 2009 Design of DC/DC Converter Based on Fuzzy Self-Adaptive Adjusting of PID Control *International Conference on Information Engineering and Computer Science* (IEEE) 1–4.

[10] Liao J Y, Huang H H and Chen K H 2010 Minimized right-half plane zero effect on fast boost DC-DC converter achieved by adaptive voltage positioning technique *ISCAS 2010 - 2010 IEEE International Symposium on Circuits and Systems: Nano-Bio Circuit Fabrics and Systems.*

[11] Della Giustina D and Liberali V 2010 Practical design considerations on adaptive controllers for PWM DC/DC converters *2010 International Conference on Microelectronics* (IEEE) pp 212–5.

[12] Li H and Ye X 2010 Sliding-mode PID control of DC-DC converter *Proceedings of the 2010 5th IEEE Conference on Industrial Electronics and Applications, ICIEA 2010.*

[13] Arikatla V and Qahouq J A A 2011 DC-DC Power Converter with digital PID controller *Twenty-Sixth Annual IEEE Applied Power Electronics Conference and Exposition (APEC)* (IEEE) 327–30.

[14] Arikatla V P and Abu Qahouq J A 2012 Adaptive digital proportional-integral-derivative

controller for power converters *IET Power Electron.*

[15] Mummadi V 2011 Design of robust digital PID controller for H-bridge soft-switching boost converter *IEEE Trans. Ind. Electron.*

[16] Algreer M, Armstrong M and Giaouris D 2012 Predictive PID controller for DC-DC converters using an adaptive prediction error filter *IET Conference Publications.*

[17] Rabbani M, Maruf H M M, Ahmed T, Kabir M A and Mahbub U Fuzzy 2012 logic driven adaptive PID controller for PWM based buck converter *International Conference on Informatics, Electronics and Vision.*

[18] Kim S and Williamson S S 2012 Negative impedance instability compensation in buck converter using state space pole placement control *25th IEEE Canadian Conference on Electrical and Computer Engineering: Vision for a Greener Future, CCECE 2012.*

[19] Husna A W N, Siraj S F and Mat M H 2011 Effect of load variations in DC-DC converter *Proceedings - CIMSim 2011: 3rd International Conference on Computational Intelligence, Modelling and Simulation.*

[20] Lee Y H, Huang S C, Wang S W and Chen K H 2012 Fast transient (FT) technique with adaptive phase margin (APM) for current mode DC-DC buck converters *IEEE Trans. Very Large Scale Integr. Syst.*

[21] Li Y, Vannorsdel K R, Zirger A J, Norris M and Maksimovic D 2012 Current Mode Control for Boost Converters With Constant Power Loads *IEEE Trans. Circuits Syst. I Regul. Pap.* **59** 198–206.

An Overview of Enhance Bandwidth Management in Cellular Network

H M Isam[1], B L Ong[1], R B Ahmad[1], M Elshaikh[1] and I Al-Mejibli[2]

[1] School of Computer and Communication Engineering Universiti Malaysia Perlis (UniMAP)Perlis, Malaysia
[2]University of Information Technology and Communication, Baghdad, Iraq

heba.moh.1990@gmail.com

Abstract. cellular networks have changed our life. However, current cellular network does not have an efficient bandwidth management during congestion. Some of these techniques was used have a few drawbacks such as the performance of the handoff mechanism depends on the channel quality, QoS, bandwidth, delay and speed of the cellular node, the handoff optimizing tend to increase the ping-pong handoff rate, frequency and bandwidth. The paper indicates the promising factors of the handoff mechanisms as decrease delay, enhance speed, saving energy and control bandwidth to achieve better throughput in wireless networks as compared to predecessor networks. How limited bandwidth are segregated among nodes is discussed in details in this paper. We believe that having implemented the proposed enhanced bandwidth management, congestion in current cellular network can perform better.
Index Terms— Bandwidth, Cellular Network, CBR, VBR, UBR, Handoff.

1. Introduction

Since the advancement of technology nowadays especially in cellular networks, cellular communications have produced enormous evolution in the past decade, with an expected growth in the next decade. Cells size should be minimized in order to utilize the limited frequency spectrum allocation to accommodate more subscribers. On the other hand, challenges of fundamental issues in cellular networks are considered as well. Among the main problems in cellular networks is location management, which deal with tracking subscribers during mobility [1]. Cellular networks can be differentiated and categorized using the word "generation", as we have "first-generation", "second-generation", and "third generation", etc. This differentiation is very important as a big generation difference between each of these technologies [2] as shown in Figure 1.

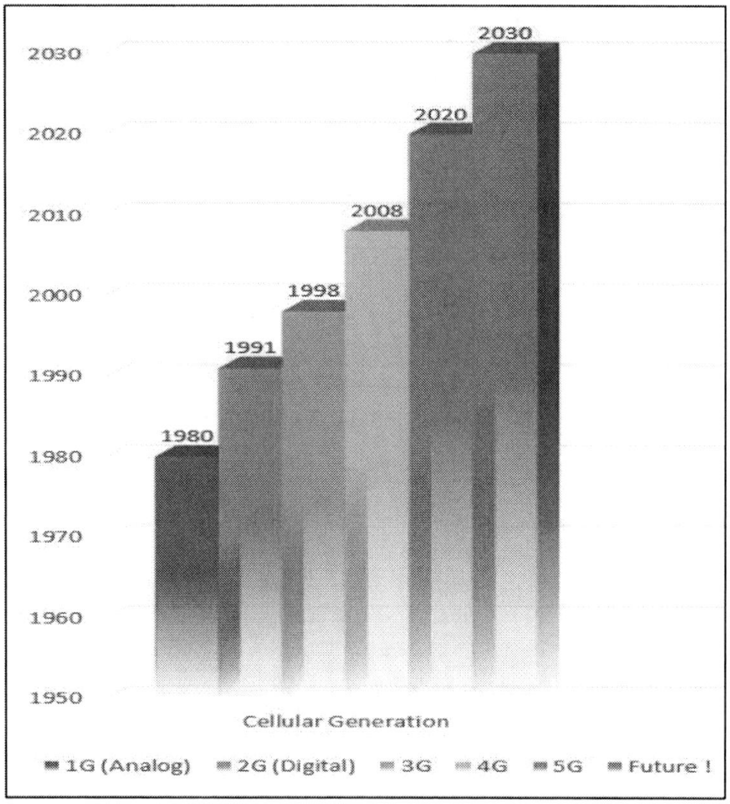

Figure 1. Cellular network generations during the years

Mobility is considered as the most vital advantage of a wireless cellular communication system [3], [4]. This requirement is required to support a continuing service for an individual which is attained by featuring handoff from one cell to some other. Handoff can be explained as the procedure of transferring data program of a continuing call in one access indicate another in a homogeneous or heterogeneous cellular network [5]. Handoff is referred as Handover [5]. While a contact is happening, the handoff process is definitely changing the channel (frequency, bandwidth, period slot, spreading code, or mix of them) which can be linked to the current contact connection, on which it is initiated by either crossing a cell boundary or by a deterioration in the standard of the signal in today's channel [4].

The term "cellular" is a widely used and it became the part of our life [5]. It refers to movement or, in general, it is used for cellular devices. Cellular devices became an essential requirement of day to day life which is unavoidable. There are several factors that can reduce the appearance of Handoff and improve the cellular connection with the lower cost, lower complexity, high QoS and so on. Among these performance metrics is (handoff speed, bandwidth, handoff delay, packet losses, energy, data rate and etc.). In this article different performance metrics that used to improve handoff technique in cellular network are discussed.

2. Cellular Network with Bandwidth

As we mentioned previously that the cellular is a more used and it be the part of the current life of people. Cellular devices are a main requirement of day to day life which is inevitable. Bandwidth management is important to reduce the appearance of Handoff and improve the cellular connection with the lower cost, lower complexity, high QoS and so on. Selecting Bandwidth management is an essential requirement in cellular networks. An efficient and effective bandwidth management scheme must perform the QoS provisioning of cellular connections [6]. To handle the network source fluctuations, adaptive bandwidth allocation can change the allocated bandwidth of ongoing calls dynamically, which is now very attractive in cellular communications [7] as demonstrated in table 1.

Table 1. Evaloation Bandwidth of Wireless Technology (Gupta & Jha, 2015)

Generation	Quantity
1G	30KHZ
2G	GSM 200 MHz CDMA 1.25 MHz
3G	WCDM 5 MHz CDMA 1.25 MHz
4G	LTE (1.4-20) MHz WIMAX mobile (3.5,7 ,5 and 10) MHz
5G	60 GHz

When a cellular user moves in one cell to another, to be able to provide QoS guarantee the required bandwidth ought to be allocated in the brand-new cell. This element needs to be resolved in the handoff scheme. The primary aim of our function is to supply QoS assurance while optimizing the effective utilization of bandwidth by monitoring the user flexibility and reserving needed bandwidth in places to where the consumer can move. This might bring about reducing premature call service and terminations denials. Here affective usage of bandwidth identifies the percentage of bandwidth in fact used by the phone calls to the bandwidth allocated for communication, as shown in Figure 2.

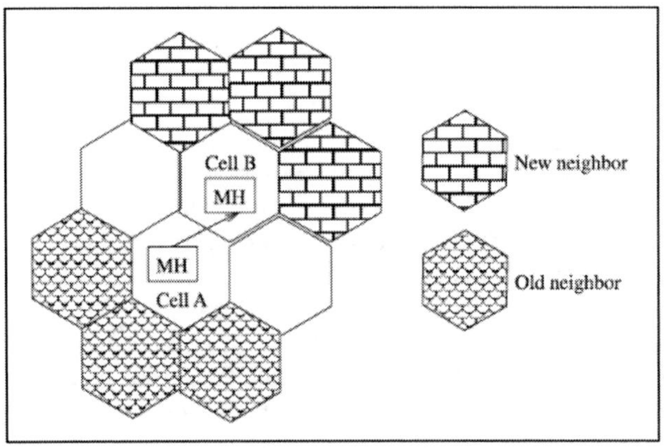

Figure 2. Bandwidth reservation with handoff technique [25]

3. Fair Bandwidth

Congestion control is critically very important to real-time applications because a number of these applications consume significant bandwidth network assets severely [26]. The users have to adapt to the obtainable bandwidth of the network, which fluctuates as time passes. Recently, multi-application and bandwidth-intensive services such as for example online interactive gaming, high-definition internet protocol tv, and video-on-demand have obtained a lot of attention among clients [27]. Moreover, because of the rapid development of portable and wise devices, which need high data prices per customer, a robust gain access to network, with the capacity of allowing an end-to-end quality-of-support (QoS) provisioning and a high-speed data transmission is necessary.

The fairness problem which is how to allocate bandwidth to connection beyond their minimum required bandwidth. We assume that if the maximum required bandwidth is not available, then the connection is not accepted by the network. The fairness issue is only interesting when the utility of an application strictly increases when allocating more bandwidth than its minimum required bandwidth. Connections with on/off utility functions are thus ignored in allocating extra bandwidth once they receive.

4. Call Priority

CoS (class of service) since traffic is classed into groups such as high, moderate, and low and the low the priority, the more "drop eligible" is a packet. E-mail and Website traffic is placed in the cheapest categories often. When the network gets occupied, packets from the lowest categories are dropped 1st.

The issue with network concern schemes is definitely that lower-priority traffic may be held up indefinitely when visitors is weighty unless there is enough bandwidth to manage the best load levels. High-priority visitors may be organized under extreme traffic loads Even. One solution can be to overprovision network bandwidth.

As the traffic loads increase, router buffers begin to fill, which adds to delay. thus, call or data priority depend on amount of bandwidth that available, when the bandwidth is limited thus the first priority for data that is in real time as (video, video conferencing) those called (CBR), second priority for data non in real time as (imaging nrt, data nrt and audio) these called(VBR) and at last for data that very less priority as (file transfer, email, fax and remote login) all of these called (UBR).

5. Discussion of Different Performance Metrics That Was Used to Improve Handoff technique in different Types of Wireless Networks

The researcher Rabe A. utilized a velocity-based handoff control in two-tier cellular networks [9]; in this article, the impact on the handoff (HO) amount is ignored whereas network densification is assumed as an important solution to overcome the expanding capacity demand. HO delays may defuse or even deny the advantages introduced by network densification in dense 5G networks [9]. To improve the average rate for mobile users, the paper suggested user-aware HO skipping schemes for two tier cellular network. In order to study the impact of HO delay on the user amount, the authors designed an analytical paradigm to frame the suggested cooperative HO performance of skipping schemes. The designed mathematical framework is built on stochastic geometry and verified using Monte Carlo simulations [9], QoS, HO delay, and mobile users average rate is improved in this article paper.

In next research paper, a proposal of an adaptive bandwidth borrowing-based call admission management approach for multi-class service is introduced [6]; In this proposal, an adaptive Call Admission Control (CAC) approach for multi-class service wireless cellular networks is presented. The suggested approach utilizes a thorough sharing scheme of all traffic classes for the available bandwidth. The suggested solution is achieved by borrowing of call bandwidth and call preemption methods according to the priorities of the traffic classes. The QoS is achieved in each class by mechanisms to degrade call bandwidth, and call bandwidth upgrading predicated on Max-Min and Min-Max rules for fair resource deallocation and allocation. The strength and performance of the proposed strategy were demonstrated on the simulation outcomes in comparison to other approaches [6]. The proposed results showed a reduction of the dropping probability and blocking probability, and bandwidth utilization enhancement. Moreover, the simulation results showed the performance of the scheme can be enhanced if some parameters, like the utilization threshold and the allowed ratios of traffic classes are carefully tuned to regulate the amount of the provided QoS.

Routing and Bandwidth optimization in wireless cellular networks with relays is presented in another article by [14]; The purpose was to quantify the improvement in performance because of utilizing the set relays in the uplink of the cellular network, where addressing an orthogonal frequency division multiplex (OFDM) based cellular network, because each cell includes a bottom station, multiple cellular users, several relays [14]. This proposal makes enhancement in OFDM optimization of networks with relays. The researchers addressed the simultaneous routing, frequency planning, and power allocation issue with fixed relay infrastructure. They conclude that making the system more equitable while extending coverage is the main advantage relays.

Another research paper is GreenBag: Energy-efficient Bandwidth Aggregation for Real-time Streaming in Heterogeneous Mobile phone Wireless Systems [20]; a promising opportunity for assisting bandwidth-intensive applications, such as high-definition video streaming emerges on recent cellular devices because they are armed with multiple network interfaces such as for example 3G/LTE and WiFi through the use of bandwidth aggregation over LTE and WiFi links. Nevertheless, several challenges eventually accomplish effective bandwidth aggregation in cellular environments such as deployment, hyperlink heterogeneity, network fluctuation, and energy usage [20]. The researcher created an energy-efficient bandwidth aggregation middleware known as GreenBag which will not require adjustments to the prevailing Internet infrastructure and servers; this middleware facilitates real-time data-streaming solutions over asymmetric wireless links. To overcome the pointed-out difficulties, GreenBag utilizes several strategies such as for example medium load balancing, effective segment administration, and energy-aware setting control. A prototype of GreenBag is usually implemented on Android-based mobile devices which hosts the 1st LTE-allowed bandwidth aggregation prototype for energy conserving real-time video streaming [20]. The experiment outcomes acquired for emulated and real-world working space demonstrated that GreenBag achieves great bandwidth aggregation to supply QoS in bandwidth-scarce applications, and energy saving in cellular devices also. Moreover, GreenBag can lower video lag, on a single time consuming just 14-25% reduced energy much better than the non-energy-conscious counterpart in real-globe examinations [20].

The other research paper is Utility-Maximization Bandwidth Adaptation for Multi-Class Traffic QoS Provisioning in Wireless Systems [7]; To handle network source fluctuations, adaptive bandwidth allocation is now promising in cellular communications since it can dynamically change the allocated bandwidth of ongoing calls [7]. In this ongoing work, the experts suggested an utility-centered bandwidth adaptation technique used for multi-class visitors QoS provisioning in wireless networks. The objective of the utility technique is definitely that the accomplished utility of each specific cell can be maximized by assigning each contact a computer program function and based on the network load the bandwidth of ongoing phone calls are improved or degraded. The authors built-in adaptation penalty in to the utility function to handle the unwanted effects of bandwidth adaptation. To judge the overall performance of the proposed technique, simulation experiments were carried out, and results demonstrated that the technique works well in both raising cell utility and reducing the decision blocking and handoff dropping probabilities of cellular networks.

Other research paper is usually performance evaluation of bandwidth-based handoff algorithm for 4G heterogeneous wi-fi networks predicated on WDHOP [22], In this paper, bandwidth-based handoff algorithm for multiple heterogeneous wireless networks are developed [22]. Usually upper coating applications are more delicate on QoS parameters such as for example bandwidth, delay etc.

6. Performance Metrics That was Used to improve Handoff Technique in Different of Wireless Networks

This section contains a comparison between different networks environments, approach with contributions of the papers, as show in table 2.

IOP Conf. Series: Materials Science and Engineering **767** (2020) 012040 doi:10.1088/1757-899X/767/1/012040

Table 2. Comparison between wireless networks preformance meteics

Ye.	Ref.	Environments	Approach	Contribution
2018	A Novel Handoff Necessity Estimation Approach Based on Travelling Distance	Cellular Network & WLAN	Handoff Necessity Estimation algorithm (by MATLAB Simulator).	• QoS. • Reducing the unnecessary HO.
2017	Performance Analysis of Bandwidth Based Handoff Algorithm for 4G Heterogeneous Wireless Networks Based on WDHOP.	Heterogeneous Wireless Network	Bandwidth Based Handoff Algorithm.	• QoS. • Traffic Load. • Flexibility & Connectivity for Network.
2017	Virtual Soft-Handoff for Cellular Heterogeneous Networks	Cellular HetNet.	VSHO Technique	• Lower Complexity. • Better Throughput for users
2013	GreenBag: Energy-efficient Bandwidth Aggregation for Real-time Streaming in Heterogeneous Mobile Wireless Networks.	Mobile Network	GreenBag Architecture.	• Saves Energy. • BW Aggregation. • QoS.
2011	An Adaptive Bandwidth Borrowing-Based Call Admission Control Scheme for Multi-Class Service Wireless Cellular Networks.	Wireless Cellular Network	Call Admission Control (CAC)	• BW Upgrading. • QoS.
2005	A Capable Location Prediction and Bandwidth Reservation Scheme for Multimedia in Mobile Cellular Networks.	Mobile Network	mobility predictive bandwidth reservation scheme.	• QoS. • Throughput.
1998	An Adaptive Bandwidth Reservation Scheme	Wireless Network	Adaptive Bandwidth Reservation Scheme.	• BW. • QoS.

	for High-Speed Multimedia Wireless Networks.

7. Pie Chart

This section shows the classification of performance metrics that used over the years, Figure 3.

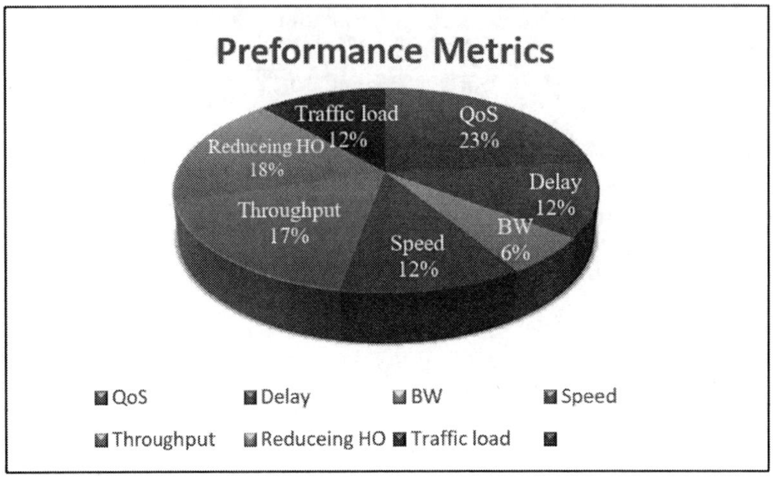

Figure 3. Preformance Metrics Pie Chart

8. Proposed solution

After the studied of the cellular network and see the performances matrices affecting them, noticed there is a problem in the speed of accept new calls and handoff calls where this can cause loss of calls but for each problem there is a solution. There are two major factors which have to be considered for the nice reservation of bandwidth seriously. Initial is, how bandwidth ought to be assigned at first to each cell and just how many quantities of users is there in each cell to keep up the QoS through the entire connection life. Our objective is usually to find the free obtainable bandwidth in virtually any cell and offer it compared to that cell where it needs, keeping the QoS for the incoming contact. This concern distribution of bandwidth is principally divided in three parts:

(1) Constant Bit Rate (CBR): Jobs that can tolerate no delay are assigned the CBR priority. These jobs are provided same number of bits every frame. For example, viewing a video reel definitely requires some blocks in every frame [28]- [30].

(2) Variable Bit Rate (VBR): Jobs that may produce different sized packets at different times are assigned VBR priority. They are provided with a variable number of bits varying between a maximum and a minimum in different frames. e.g. a document may be compressed differently by different machines. Transmitting it will be a variable transmission [28]- [30].

(3) Unspecify Bit Rate (UBR): These jobs are the least priority jobs. The network does not promise anything but simply tries its best to transmit it. From then on, we maximize the capability utilization by assigning any available free bandwidth to the prevailing cell according to user demand. If the bandwidth and frequency of the calls is controlled correctly, it helps to avoid of these problems [28]- [30].

This section focuses on the propose algorithm of enhance bandwidth of handoff technique. Proposed technique includes two algorithms:

1. Sufficient bandwidth algorithm.
2. Insufficient bandwidth algorithm includes:
    a. Sub algorithm 1.

b. Sub algorithm 2.
c. Sub algorithm 3.

Thus, when the bandwidth request is equal to available bandwidth all calls will accepted without any problem but if the available bandwidth is not enough to cover all bandwidth requested from a call, the bandwidth available will divided depend on priority of calls. Thus, that will help to accept all calls without any call's loss and delay.

The proposed algorithm function taking into consideration the good bandwidth which we've initially assigned to each cell, meaning that, we are enhancing the at first assigned bandwidth by using bandwidth reserve further. This algorithm provides high velocity, lower hold up and high QoS for fresh calls and handoff phone calls by allocating reasonable bandwidth and the number of users.

The next Figure 4 illustrates the proposed algorithm described above fully

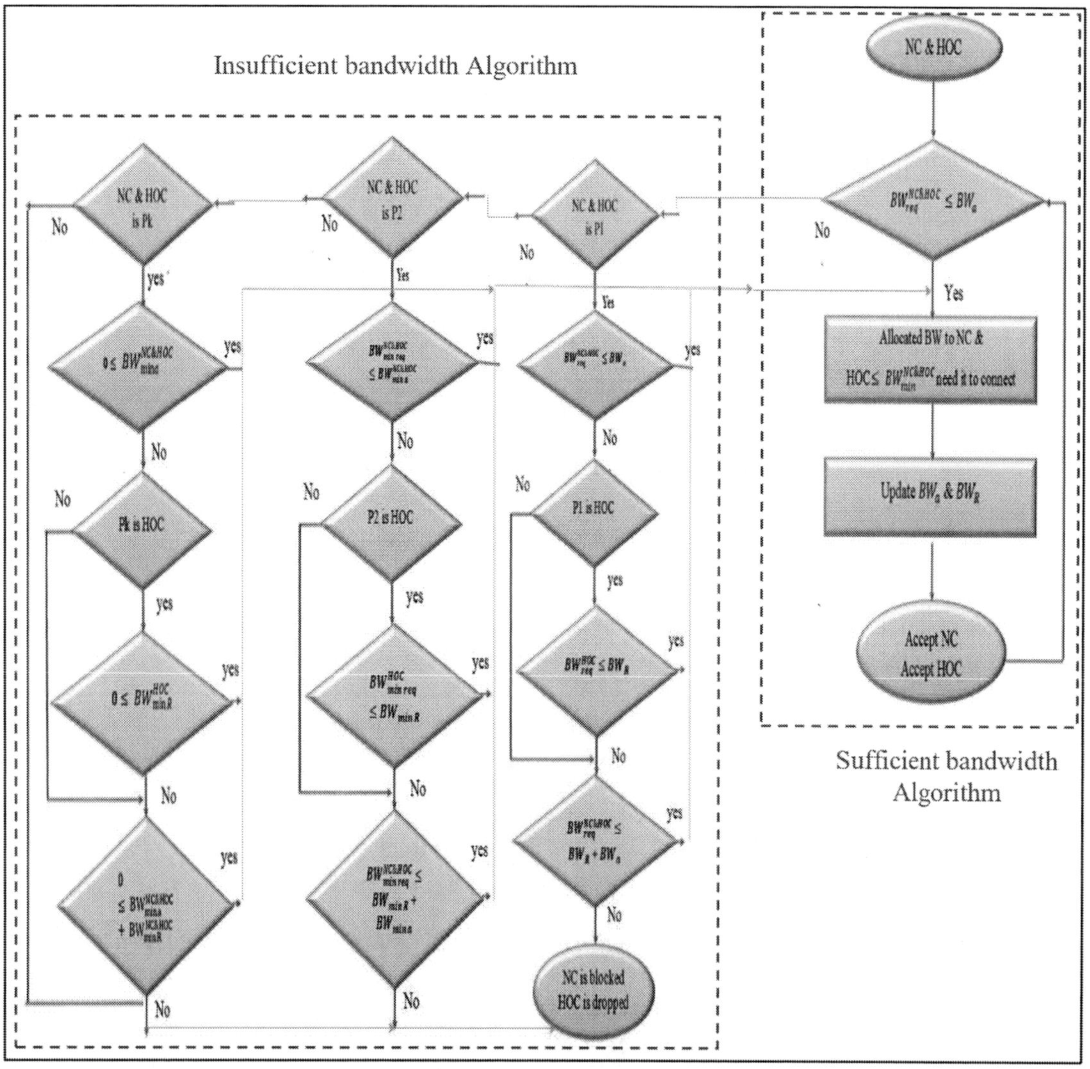

Figure 4. Proposed algorithm

9. Conclusion

In this paper more and different performance matrices in different type of wireless networks are discussed. Handoff technique may have drawbacks as the performance of handoff mechanism depends on the QoS, Speed of cellular nodes, bandwidth, Energy and delay, the handoff optimizing depend on to increase the bandwidth, ping-pong handoff rate and handoff prediction techniques. Therefore, solution to these issues is discussed in details in this paper. The implementation of the proposed mechanism is believed to improve the performance of cellular network during congestion.

10. References

[1] Das P and Chandr A 2011 Location management in wireless networks: A survey. In *World Congress on Information and Communication Technologies* pp. 576-80.

[2] Mishra A R 2004 Fundamentals of cellular network planning and optimisation *Evolution to Wiley-Interscience.*

[3] Zekri M, Jouaber B and Zeghlache D 2012 A review on mobility management and vertical handover solutions over heterogeneous wireless networks *Computer Communications* ***35*** 17 p 2055-68.

[4] Stojmenovic I 2002 Handbook of wireless networks and mobile computing p 1–26.

[5] Jose P G, Chatterjee S, Patodia M, Kabra S and Nath A 2016 Hash and Salt based Steganographic Approach with Modified LSB Encoding *International Journal of Innovative Research in Computer and Communication Engineering* **4** 6 pp 10599-610.

[6] Al Khanjari S, Arafeh B, Day K and Alzeidi N 2011 An adaptive bandwidth borrowing-based call admission control scheme for multi-class service wireless cellular networks. In *IEEE International Conference on Innovations in Information Technology* p. 375-80 IEEE.

[7] Lu N and Bigham J 2005, October Utility-maximization bandwidth adaptation for multi-class traffic QoS provisioning in wireless networks In *ACM Proceedings of the 1st ACM international workshop on Quality of service & security in wireless and mobile networks* p 136-43.

[8] Balachandran K, Kang J H, Karakayali K and Rege K M 2016 Virtual soft-handoff for cellular heterogeneous networks *IEEE Transactions on Vehicular Technology* **66** 4 p 3306-18.

[9] Arshad R, ElSawy H, Sorour S, Al-Naffouri T Y and Alouini M S 2017 Velocity-aware handover management in two-tier cellular networks *IEEE Transactions on Wireless Communications* **16** 3 p 1851-67

[10] Barua S and Braun R 2016 A novel approach of mobility management for the D2D communications in 5G mobile cellular network system. In *IEEE 18th Asia-Pacific Network Operations and Management Symposium (APNOMS)* p 1-4).

[11] Peng C, Lee S B, Lu S and Luo H 2014 GreenBSN: Enabling energy-proportional cellular base station networks *IEEE Transactions on Mobile Computing* **13** 11 p 2537-51.

[12] Vu T T, Decreusefond L and Martins P 2014 An analytical model for evaluating outage and handover probability of cellular wireless networks *Wireless personal communications* **74** 4 p 1117-27.

[13] Xiaofeng L U, Pan H U I and Lio P 2013 Offloading mobile data from cellular networks through peer-to-peer WiFi communication: A subscribe-and-send architecture. *China Communications* **10** 6 p 35-46.

[14] Ji J and Yu W 2009 Bandwidth and routing optimization in wireless cellular networks with relays. *IEEE* in *7th International Symposium on Modeling and Optimization in Mobile, Ad Hoc, and Wireless Networks* p 1-5.

[15] Lao L and Cui J H 2005 Reducing multicast traffic load for cellular networks using ad hoc networks *IEEE in Second International Conference on Quality of Service in Heterogeneous Wired/Wireless Networks (QSHINE'05)* p 10.

[16] Sanabani M, Shamala S, Othman M and Desa J 2005 A capable location prediction and bandwidth reservation scheme for multimedia in mobile cellular networks. *IEEE in Asia-*

Pacific Conference on Applied Electromagnetics p 5.

[17] Elhabib E A, Amin B A, Mustafa N and jabar A 2014 A Adaptive Multistreaming Over Network Via SCTP **4** 12 p 30–4.

[18] Boudko S, Leister W and Gjessing S 2013 Heterogeneous wireless network selection: Load balancing and multicast scenario *International Journal on Advances in Networks and Services* **6** 3&4.

[19] Kamble D P and Terdal S P 2013 Traffic Rank Based QoS Routing in Wireless Mesh Network *International Journal of Computer Trends and Technology* **4** 8 p 2899–904.

[20] Bui D H, Lee K, Oh S, Shin I, Shin H, Woo H and Ban D 2013 Greenbag: Energy-efficient bandwidth aggregation for real-time streaming in heterogeneous mobile wireless networks. In *IEEE 34th Real-Time Systems Symposium* p 57-67.

[21] ZhenTao S and Ning L 2010 Improving throughput by On_Demand Bandwidth Allocation for WiMAX In *IEEE 2nd International Conference on Computer Engineering and Technology* **1** pp 565-570.

[22] Neeraja S and Rao G S 2017 Performance Analysis of Bandwidth Based Handoff Algorithm for 4G Heterogeneous Wireless Networks Based On WDHOP 12(1) pp 96–106.

[23] Kashmar N and Atieh M 2018 Mining frequent patterns to identify vertical handover parameters in cellular networks *Journal of Ambient Intelligence and Humanized Computing* **9** 1 p 31-42.

[24] Madaan J and Kashyap I 2018 A novel handoff necessity estimation approach based on travelling distance *International Journal of Intelligent Systems and Applications* **11** 1 p 46.

IOP Conf. Series: Materials Science and Engineering **767** (2020) 012041 doi:10.1088/1757-899X/767/1/012041

Design and Implementation of Lifting Wavelet Transform Using Field Programmable Gate Arrays

T B Taha[1], R Ngadiran[2]and P L Ehkan[2]
[1]IT Department, Faculty of Sciences, Tishk International University, Erbil, Iraq
[2]School of Computer and Communication Engineering, University Malaysia Perlis (UniMAP), Perlis, Malaysia

phd.taha@gmail.com

Abstract. Lifting Wavelet transform (LWT) has an extensive usage in different image processing applications as image compression and information hiding. LWT is considered a good solution for hardware designs as it relies only on integer calculations while applying the wavelet transform. In this paper, an FPGA design and implementation of LWT is presented, the implementation is achieved using VHDL coding without importing off-shelf components which make the proposed design applicable to a wide range of devices. The design is based on parallel execution to perform LWT implementation with real time response. The design utilized 421 logic registers of DE2 Cyclone II (EP2C35F672C6) FPGA device with 151.91MHz frequency.

1. Introduction

Lifting Wavelet Transform (LWT), was first introduced by Sweldens is considered to be the second generation of wavelets. The main difference that distinguishes LWT from classic wavelet construction is that Fourier transform is not used in the signal transform calculations [1]. In compare to conventional DWT structure, LWT has better computational efficiency in terms of using a lower number of mathematical operations, which results in a smaller implementation area, less power consumption, and lower design complexity. Due to its "in place calculation" where the produced coefficients replaces the old image pixels without extra memory utilization, and its integer-to-integer transformation abilities, LWT can be easily implemented by hardware [2].

The implementation of LWT in hardware is necessary for different applications that need the transformation for implementing digital circuits in real time manner, as image compression and perceptual mapping. The hardware implementation of digital circuits is required because it accelerates the execution speed from hours to several minutes or seconds [3]. However, in order to achieve high speed and real time execution, hardware design needs to be concentrate since the resources as logical components and frequency are limited. Generally to apply digital circuits on hardware, different technologies had been used as application specific integrated circuits (ASIC) and field programmable gate arrays (FPGAs). Cost reduction played a major role in the widespread use of FPGAs since there was a reduction of 100% in the cost of building blocks from 1990 to 2003 [4]. At the same time, the continuous increase in the cost of application specific integrated circuits ASICs compared to advances in semiconductor manufacturing made the FPGA a better alternative for embedded systems, as the cost of making incremental changes to FPGA designs is negligible when compared to the large expenses involved in the re-manufacturing of ASICs [5][6]. FPGA is also considered as one a successful technology for developing systems whose require real time operation, as it has parallel execution nature where several processes can be executed concurrently which produce instance response [7].

Accordingly, the implementation of wavelet transform using FPGA devices will lead to excel the execution of different wavelet-based image processing applications.

A three-level DWT was implemented by Karmani, Djemal & Tourki [8] for a watermarking scheme for images and videos. A pipelined 2D-scan-based architecture was used. The DWT was also used by Darji, Lad, Merchant & Chandorkar [9] by proposing a pipelined blind FPGA-based watermark approach using a quantization method. A greyscale image was used as the host image and a binary image was used as the watermark. In this study, not all the steps of the embedding process were implemented using FPGA, since the DWT was performed using Matlab.

DWT which was used in mentioned studies and several other studies as [10-11], has higher computational complexity than LWT, hence its FPGA implementation used to be resource and time consuming. The LWT was utilized instead, as the model by [12], where the analysis the computational features of 2D DWT algorithm and propose a parallel architecture called Fragment-based Interleaving Dual Pipelines (FIDP). Another lifting based implementation is achieved by [2], in which the 5/3 lifting-based wavelet transform is modeled and simulated using Matlab.

From literature, it's found that there is a need to design wavelet transform using simple calculations, i.e LWT, and the implementation of LWT needs to be designed and simulated codes without importing off-shelf components, to make the design applicable to wide range of applications and devices. In this paper, an optimized LWT implementation is presented using FPGA devices, and all the implementation is achieved using only VHDL codes without relying on Matlab and other external functions or applications.

The paper is organized as follow, in section two a brief description of LWT equations is presented. Section three describes the methodology of LWT implementation on 2D image. The implementation results are shown in section 4. The paper is concluded in section 5.

2. Lifting Wavelet Transform

The filter bank method is used to apply the Discrete Wavelet Transform (DWT) on digital signals is depend on high pass filter and low pass filers to create the detail and approximation bands respectively as shown in figure 1.

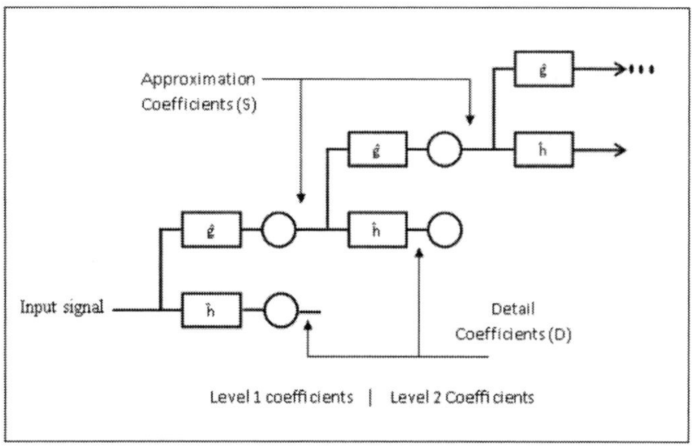

Figure 1. Wavelet transform using filters.

The equivalent processes in LWT is the usage of three steps, Splitting, Prediction and Update. In the Split phase, the signal is divided into odd and even samples. To apply the prediction, which is equivalent to the high pass filter ($\hat{h}$) in the filter banks, equation (1) is used, which is the difference between the odd element X_{2i+1} and the average of its two even neighbors X_{2i}, X_{2i+2}. The output will then be the detail band (Di).

$$D_i = X_{2i+1} - ((X_{2i} + X_{2i+2})/2) \tag{1}$$

For the update, which is equivalent to the low pass filter ĝ in the filter banks, equation (2) is used. In each update application, the value of the even sample is added to the summation of the previously calculated detail coefficient and the recently created one, and the result is divided by four. The output will be the approximation band (Si).

$$S_i = X_{2i} + ((D_i + D_{i-1})/4) \tag{2}$$

For two dimensional signals as digital images, the LWT is applied on rows first followed by columns application. The implementation of the equation above on FPGA device is described in the next section.

3. FPGA Implementations

In this section, a detailed description of the design flow of LWT implementation is presented. As shown in **Figure** 2, input signal X is fed to the LWT and the produced output is approximation band Sn and detail band Dn. The implementation will be described in details for one dimensional signal in the following sub-section followed by two-dimensional implementation.

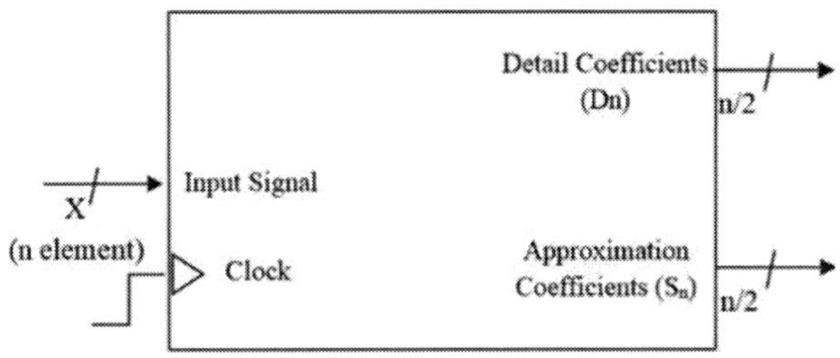

Figure 2. Block diagram of LWT system

A one-dimensional signal of 16 elements is considered as a test input to the design. Values of test signal input a X= 6, 5, 1, 9, 5, 11, 4, 3, 5, 0, 6, 4, 9, 6, 5, 7 that are taken from memory initialization file (mif) that is loaded to M4K RAM which consist of 8 locations each of 16 bits. Each pixel is represented by 8 bits, since each grey scale level pixel has value between 0 to 255. Test signal is shown in figure 3

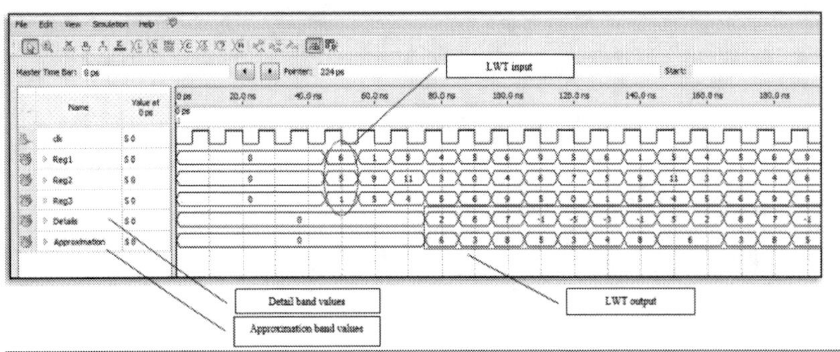

Figure 3. Simulation results of one dimensional LWT application

According to Equations (1) and (2), to obtain one output of each of D (Detail) and S (Approximation) band, three elements from the input are required. Practically, three new elements are required in the first clock followed by two new elements and one from previous reading in subsequent clocks. For instance, the first data inputs according to X signal in the above example must be, 6,5,1 followed by 1,9,5, 5,11,4, and so on.

Three values will be carried by three signals, and implemented using three registers, Reg1sig, Reg2sig, and Reg3sig as shown in figure 4, which depicts the RTL (Register Transfer Level) of the first part of design that responsible of reading the pixels in the appropriate sequence and deliver them to the LWT logic. Registers' values will be fed to LWT procedure for calculating D and S coefficients. LWT, as integer-based transformation, is one clock process because all the LWT-related calculations are performed by variables within a process. The RTL of LWT implementation of LWT is shown in Figure 5.

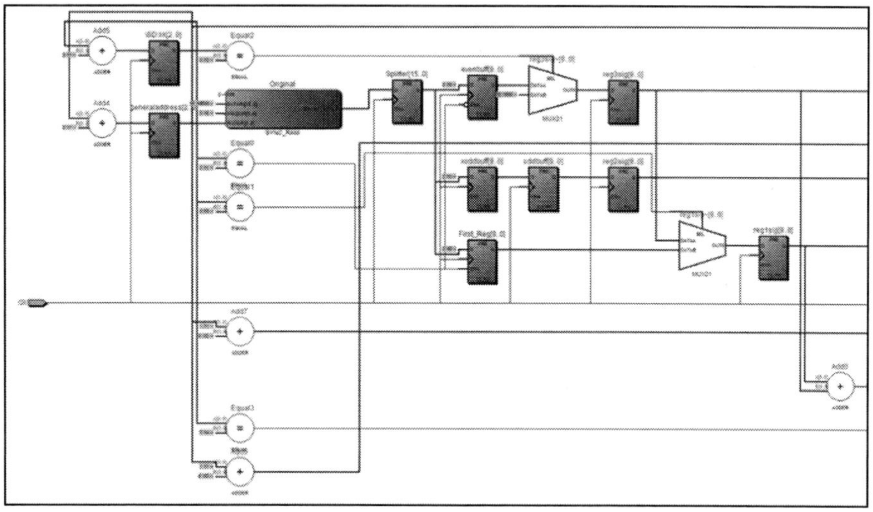

Figure 4. RTL for one level LWT
(Part1: input registers initialization)

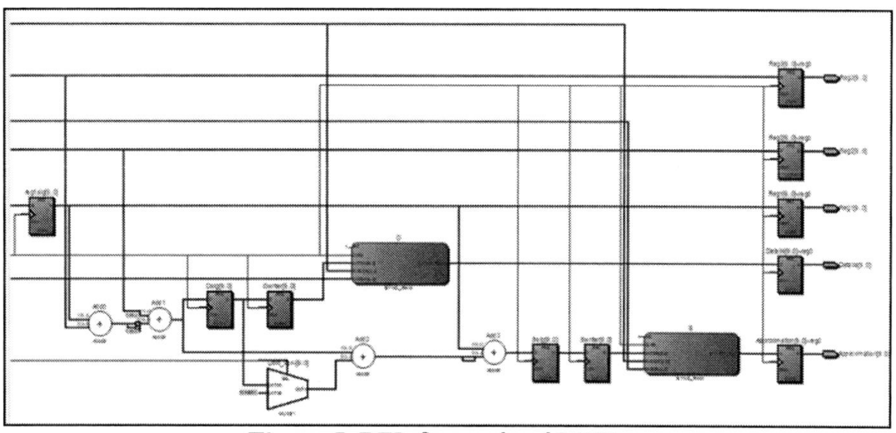

Figure 5. RTL for one level LWT
(Part2: LWT implementation on input data)

In two dimensional signals, further LWT implementation is applied again on each of detailed band D, and approximation band S individually as the same way it was applied on original sequence. S and D are further divided into S1, D1 and S2, D2, respectively. This operation is applied on digital images by considering them as two-dimensional arrays. In the process of applying LWT on an image of size (n

× n), each row is considered as one-dimensional signal, the result is detail and approximation bands with width n/2 and length n. Another LWT operation is applied on each produced band by column wise to obtain four n/2 × n/2 bands as shown in figure 6.

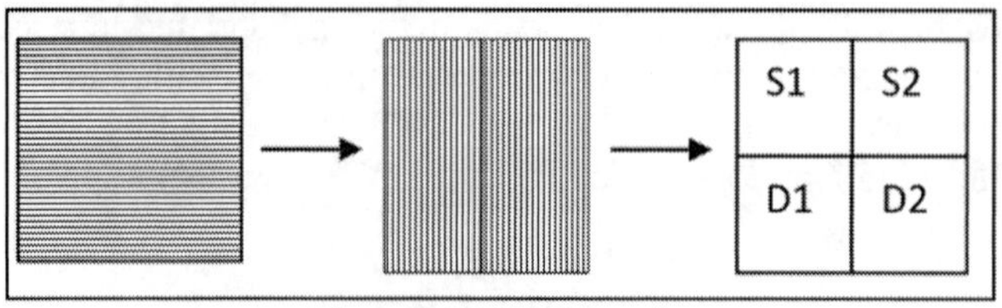

Figure 6. LWT bands.

In order speed up the calculations, concurrent transpose processes were initiated for both bands at the same time, another parallel computation was achieved by creating S1, D1 from S concurrently with creating S2, D2 from D. Figure 7 shows two dimensional LWT process where operations on the same column were processed simultaneously.

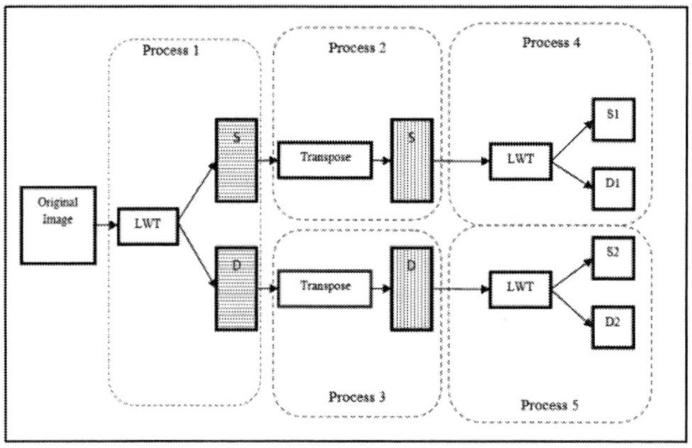

Figure 7. Parallel architecture of LWT decomposition.

4. Implementations Results

A 16×16 image was used for applying LWT. Image pixels was read row by row as shown in figure 8. After S and D band had been produced, the next LWT operation is applied column by column, hence, a transpose on S and D was needed. VHDL simulation results for the four bands are shown in figure 9.

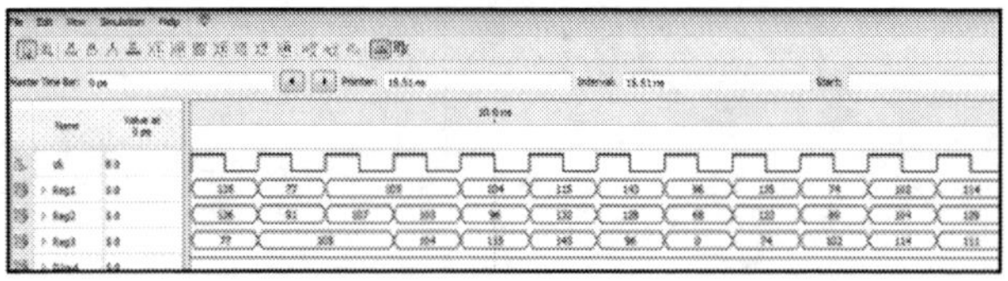

Figure 8. Reading image pixels

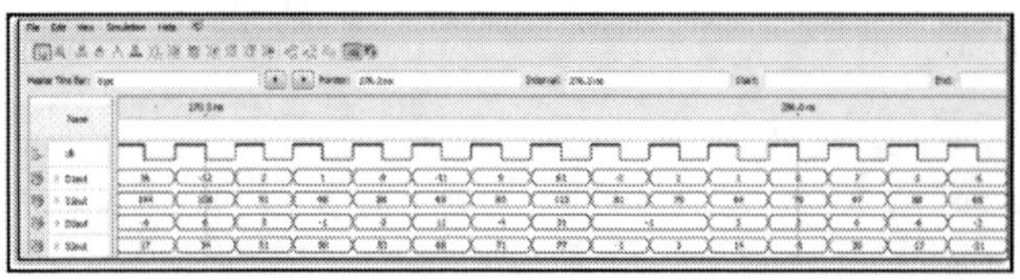

Figure 9. Produced LWT coefficients for different bands using VHDL

The LWT was implemented on Altera DE2 Cyclone II (EP2C35F672C6) FPGA device. The LWT implementation utilized 538 logic elements at maximum frequency of 151.91MHz. A total of 421 logic register was used for the 16×16 image, it had better logic utilization than the model proposed by Bao-feng & Yong [12] where 466 logic registers were utilized for 2×2 image, in addition, the implementation is done using VHDL without relying on off-shelf components; hence, the implementation results can be considered in the practical implementations and in different platforms.

5. Conclusions

In this paper, an implementation of LWT is achieved on FPGA using VHDL. The algorithm is based on parallel processing for higher execution speed. The implementation consumed 421 logic register from DE2 Cyclone II (EP2C35F672C6) FPGA device with maximum frequency 151.91MHz for 16 × 16 sample image. The utilization and the execution speed denoted to the ability of perform the transform using the proposed implementation efficiently and in real time. Different LWT-based operations can be implemented based on the proposed methods, as compression, watermarking and perceptual mapping.

References

[1] Sweldens, W. Lifting scheme: a new philosophy in biorthogonal wavelet constructions. In Wavelet applications in signal and image processing III *International Society for Optics and Photonics* **2569** 68-80.

[2] Gholipour, M. Design and implementation of lifting based integer wavelet transform for image compression applications *International Conference on Digital Information and Communication Technology and Its Applications* pp. 161-172.

[3] Wang, X., Qin, Q., & Cheng, Y. Design and implementation of digital image watermark based on FPGA. *Recent Advances in Computer Science and Information Engineering* pp. 223-229.

[4] Rodriguez-Andina, J. J., Moure, M. J., & Valdes, M. D. Features, design tools, and application domains of FPGAs *IEEE Transactions on Industrial Electronics* **54** 1810-1823.

[5] M. Thompson, "FPGAs Accelerate Time to Market for Industrial Designs," EE Times (July 2,2004). Accessed on 9-May-2019 from http://www.us.design-reuse.com/articles/8190/fpgas-accelerate-time-to-market-for-industrial-designs.html.

[6] National Instruments. (2012). Introduction to FPGA Technology: Top 5 Benefits. Accessed on 9- May, 2019, from http://www.ni.com/white-paper/6984/en/#toc2.

[7] Sulaiman, N., Obaid, Z. A., Marhaban, M. H., & Hamidon, M. N. Design and implementation of FPGA-based systems-a review *Australian Journal of Basic and Applied Sciences* **3** 3575-3596.

[8] Karmani, S., Djemal, R., &Tourki, R. Computer Standards & Interfaces Efficient hardware architecture of 2D-scan-based wavelet watermarking for image and video. Computer Standards & Interfaces, **31** 801–811.

[9] Darji, A. D., Lad, T. C., Merchant, S. N., &Chandorkar A. N. Watermarking hardware based on wavelet coefficients quantization method. Circuits *Systems, and Signal Processing* **32** 2559–2579.

[10] Joshi, A. M., Darji, A., & Mishra, V. Design and implementation of real-time image watermarking *2011 IEEE International Conference onIn Signal Processing, Communications and Computing (ICSPCC)* pp. 1-5.

[11] Karthigaikumar, P., Anumol, & Baskaran, K. FPGA implementation of high speed low area DWT based invisible image watermarking algorithm. *Procedia Engineering* **30** 266–273.

[12] Li, B. F., & Dou, Y. Fidp: A novel architecture for lifting-based 2D DWT in JPEG2000 *International Conference on Multimedia Modeling* pp. 373-382.

IOP Conf. Series: Materials Science and Engineering **767** (2020) 012042 doi:10.1088/1757-899X/767/1/012042

Performance improvement of multi access OCDMA system based on a new zero cross correlation code

M Alayedi[1], A Cherifi[2], A F Hamida[3] C B M Rashidi [4] and B S Bouazza[1]

[1]Laboratory of Scientific Instrumentation (LSI) Ferhat Abbas University Setif, Algeria.
[2]Laboratory of Technologies of communications (LTC) Dr.Tahar Moulay University Saida, Algeria
[3]Laboratory of optoelectronics and components (LOC) Ferhat Abbas University Setif, Algeria.

[4]Advanced Communication Engineering, Centre of Excellence, School of Computer and Communication Engineering, Universiti Malaysia Perlis, (UniMAP), Pauh Putra, 02600, Arau, Perlis

cherifi.abdelhamid@gmail.com

Abstract. In this paper, a novel zero cross correlation (ZCC) code is proposed for spectral amplitude coding-optical code division multiple access (SAC-OCDMA) systems. The code construction method starts from the identity matrix and has several benefits summarized in both the code simplicity and flexibility. The proposed code presents also an adapted (or accepted) code length for SAC-OCDMA systems and a high SNR value that reaches respectively 3.18 and 1.84 times of the system capacity comparing to other existed codes such as modified quadratic congruence (MQC) and modified double weight (MDW) codes.

1. Introduction

Optical code division multiple access (OCDMA) technique is characterized by multiple advantages that make researchers more attracted to investigate its features comparing to time division multiple access (TDMA) or frequency division multiple access (FDMA). It employs orthogonal codes for securing the range of spectrum without overlapping it with the others [1]. It presents a spread spectrum technique that permits sharing the channel bandwidth between users by given each of them a specific code namely a sequence code [2]. However, OCDMA systems suffer always from different noises like: thermal noise, shot noise, dark current and multiple access interference (MAI) which leads to create a phase induced intensity noise (PIIN) [1,3]. Phase induced intensity noise (PIIN) is considered as one of the dominant factors of decreasing the system performance [4]. Accordingly, several codes such as: modified double weight [5], modified Quadratic Congruence (MQC) [6] and zero cross correlation [7] are proposed for OCDMA systems to outperform this issue. However, these codes suffer from many drawbacks such as: long code length, in-phase cross correlation which is always higher to one, non-null cross correlation value and complex construction steps [4].

Accordingly, this paper presents a novel ZCC code for OCDMA systems which compared other reported codes families of MDW, MQC codes.

This paper is structured as: Section II describes the steps of constructing the novel ZCC code, sections III and IV analyzes the system performance results and section V ended by a conclusion.

2. ZCC Code Construction

Let take two different codes sequences:
$A_i = \{A_0, A_1, A_2, \dots\dots, A_N\}$ and $B_i = \{B_0, B_1, B_2, \dots\dots, B_N\}$, it can be define the auto and cross correlation as [8]:

$$\begin{cases} \lambda_a = \sum_{i=1}^{N} A_i \, . \, A_{i+t} & (1) \\ \lambda_c = \sum_{i=1}^{N} A_i \, . \, B_{i+t} & (2) \end{cases}$$

The novel ZCC code can be defined by these parameters: (L, w, C) where each of them presents code length, code weight and system capacity. It should to construct a novel code which high auto correlation and cross correlation null.

It can describe the method of ZCC code construction by the following steps:

A. Step 1

Create an identity matrix$\boldsymbol{A_N}$. For example ($\boldsymbol{N} = \mathbf{3}$), it gives the following matrix:

$$A_3 = \begin{bmatrix} 1 & 0 & 0 \\ 0 & 1 & 0 \\ 0 & 0 & 1 \end{bmatrix} \tag{3}$$

B. Step 2

Starting from shifting of rows property and for an identity matrix with order (N), the rule of required rotation number is (*N-1*). After applying this rule, the rotations number is (2). We will obtain a new matrix which also rotated by (2). Consequently, we will obtain these matrices:

$$\boldsymbol{A_3} = \begin{bmatrix} \mathbf{1} & 0 & 0 \\ 0 & \mathbf{1} & 0 \\ 0 & 0 & \mathbf{1} \end{bmatrix}, \boldsymbol{A_3'} = \begin{bmatrix} 0 & \mathbf{1} & 0 \\ 0 & 0 & \mathbf{1} \\ \mathbf{1} & 0 & 0 \end{bmatrix}, \boldsymbol{A_3''} = \begin{bmatrix} 0 & 0 & \mathbf{1} \\ \mathbf{1} & 0 & 0 \\ 0 & \mathbf{1} & 0 \end{bmatrix} \tag{4}$$

C. Step 3

Write A_3 matrix with the following form:

$$\boldsymbol{A_3} = \begin{bmatrix} r_1(A_3) \\ r_2(A_3) \\ r_3(A_3) \end{bmatrix} \tag{5}$$

Where $r_1(A_3), r_2(A_3)$ and $r_3(A_3)$: represent 1st row, 2^{nd}row and 3rd row of matrix A_3 respectively. By using the method, both A_3' and A_3'' can be written as:

$$\boldsymbol{A_3'} = \begin{bmatrix} r_1(\boldsymbol{A_3'}) \\ r_2(\boldsymbol{A_3'}) \\ r_3(\boldsymbol{A_3'}) \end{bmatrix} ; \boldsymbol{A_3''} = \begin{bmatrix} r_1(\boldsymbol{A_3''}) \\ r_2(\boldsymbol{A_3''}) \\ r_3(\boldsymbol{A_3''}) \end{bmatrix} \tag{6}$$

D. Step 4

Reshape each of matrices $\boldsymbol{A_3}$, $\boldsymbol{A_3'}$ and $\boldsymbol{A_3''}$ to row vector as:

$$\begin{cases} \boldsymbol{A_3} = [r_1(\boldsymbol{A_3}) \ , \ r_2(\boldsymbol{A_3}) \ , \ r_3(\boldsymbol{A_3})] \\ \boldsymbol{A_3'} = [r_1(\boldsymbol{A_3'}) \ , \ r_2(\boldsymbol{A_3'}) \ , \ r_2(\boldsymbol{A_3'})] \\ \boldsymbol{A_3''} = [r_1(\boldsymbol{A_3''}) \ , \ r_2(\boldsymbol{A_3''}) \ , \ r_3(\boldsymbol{A_3''})] \end{cases} \xrightarrow{yields} \begin{cases} \boldsymbol{A_3} = [\mathbf{1} \ 0 \ 0 \ 0 \ \mathbf{1} \ 0 \ 0 \ 0 \ \mathbf{1}] \\ \boldsymbol{A_3'} = [0 \ \mathbf{1} \ 0 \ 0 \ 0 \ \mathbf{1} \ \mathbf{1} \ 0 \ 0] \\ \boldsymbol{A_3''} = [0 \ 0 \ \mathbf{1} \ \mathbf{1} \ 0 \ 0 \ 0 \ \mathbf{1} \ 0] \end{cases} \tag{7}$$

E. Step 5

Finally, by combining between these reshaped matrices, we will obtain the novel ZCC code as:

$$ZCC_1 = \begin{bmatrix} \boldsymbol{A_3} \\ \boldsymbol{A_3'} \\ \boldsymbol{A_3''} \end{bmatrix} \xrightarrow{yields} ZCC = \begin{bmatrix} \mathbf{1} & 0 & 0 & 0 & \mathbf{1} & 0 & 0 & 0 & \mathbf{1} \\ 0 & \mathbf{1} & 0 & 0 & 0 & \mathbf{1} & \mathbf{1} & 0 & 0 \\ 0 & 0 & \mathbf{1} & \mathbf{1} & 0 & 0 & 0 & \mathbf{1} & 0 \end{bmatrix} \tag{8}$$

The relationship between code length (L) and number of user (K) of the code is given by:

$$\begin{cases} K = N \\ L = N^2 \end{cases} \xrightarrow{yields} L = K^2 \tag{9}$$

According to the above ZCC matrix, the codeword for each user would be:

$$codewords = \begin{cases} user1 \rightarrow \lambda_1, \lambda_5, \lambda_9 \\ user2 \rightarrow \lambda_2, \lambda_6, \lambda_7 \\ user3 \rightarrow \lambda_3, \lambda_4, \lambda_8 \end{cases} \tag{10}$$

According to Eqs. (1) and (2), and features of our code, we can write:

$$\begin{cases} \lambda_a = w \\ \lambda_c = 0 \end{cases} \tag{11}$$

Table 1. Comparison between MQC and MDW Codes.

Code	Code weight	Number of users	Code length	Cross correlation
MQC code	4	9	12	≥ 1
MDW code	4	4	18	≥ 1
Proposed ZCC	4	4	16	0

3. SYSTEM PERFORMANCE ANALYSIS

To make the system analysis easier, it take a consideration these following assumptions [9-11]:
1-/ Unpolarized light source ideally with flat spectrum over $[f_0 - \Delta f, f_0 + \Delta f]$ of bandwidth, where f_0 and Δf are the central optical frequency and the optical source bandwidth respectively and it estimated in Hertz.
2-/ The same power for each transmitter.
3-/ An identical spectral for each component of power spectral.
4-/ Each bit stream from each user is synchronized.

To evaluate the SAC-OCDMA system with ZCC code, it takes a consideration the effect of shot and thermal noises and the PIIN is neglected due zero cross correlation property. The dark current noise can be written as [9, 11]:

$$I_{noise}^2 = I_{shot}^2 + I_{thermal}^2 = 2eBI_{dd} + \frac{4K_bT_nB}{R_l} \tag{12}$$

Where e, B, I_{dd}, K_b, T_n and R_l present electron charge, electrical bandwidth, output of direct current, Boltzmann's constant, absolute temperature and load resistor respectively. The output of direct current can be expressed [9-11]:

$$I_{dd} = R\int_0^{\infty} r(v)dv = R\int_0^{\infty}\frac{P_{sr}}{\Delta f}\sum_{k=1}^{C} d(k)\sum_{j=1}^{L} C_E(j).C_F(j)\prod(v,i)dv \tag{13}$$

Where R and P_{sr} present the photo- diode responsivity and received power respectively. $d(k)$ is the data bit which can be ''0'' or ''1''. The $\prod(v,i)$ function can be defined as:

$$\prod(v,i) = u\left\{f - f_0 - \frac{\Delta f}{2L}(-L+2i-2)\right\} - u\left\{f - f_0 - \frac{\Delta f}{2L}(-L+2i)\right\} = u\left\{\frac{\Delta f}{L}\right\} \tag{14}$$

Where $u(f)$ present the unit step function defined as:

$$u(f) = \begin{cases} 1 & f \geq 0 \\ 0 & f < 0 \end{cases} \tag{15}$$

Let $C_F(j)$ presents the jth element of Cth user and according to our proposed ZCC code, the cross correlation of direct detection can be written as:

$$\sum_{j=1}^{L} C_E(j).C_F(j) = \begin{cases} w & for \quad E = F \\ 0 & else \end{cases} \tag{16}$$

Substituting Eqs. (14) and (16), Eq. (13) of output photo-current will be:

$$I_{dd} = R\frac{P_{sr}}{C} \tag{17}$$

Based on Eq. (17) and with take account the probability of transmitting ''0'' or ''1'' is the same and equal to (0.5) Eq. will write as:

$$I_{noise}^2 = eBR\frac{P_{sr}}{C} + \frac{4K_bT_nB}{R_l} \tag{18}$$

The average signal to noise ratio can be given as:

$$SNR = \frac{I_{dd}^2}{I_{noise}^2} = \frac{\left[\frac{R.P_{sr}}{C}\right]^2}{\frac{e.B.R.P_{sr}}{C} + \frac{4.K_b.T_n.B}{R_l}} \tag{19}$$

Using the Gaussian approximation, the bit error rate (BER) is given by [2, 9, 10, 11]:

$$BER = \frac{1}{2}erfc\sqrt{SNR/8} \tag{20}$$

4. Result Discussion:

This section explain the s SAC-OCDMA system performance with a novel ZCC code in two terms: signal to noise ratio (SNR) and bit error rate (BER) against number of simultaneous users. Used parameters for numerical analysis are listed in table 2.

Table 2. Used parameters in numerical analysis.

Parameter	Value
Photo-diode responsivity (R)	0.6
Effective source power (P_{sr})	-10 dBm
Data rate (R_b	622 Mb/s
Electron charge (e)	1.6×10^{-19} c
Receiver noise temperature (T_n)	300 K
Boltzmann's constant (K_b)	$1.38 \times 10^{-23} J.s^{-1}$
Receiver load resistor (R_l)	1030 ohm
Spectral width (Δv)	3.75 Thz

Figure 1 shows the SNR variation as function of number of active users for effective power of $-10\ dBm$ and 622 Mb/s of data rate. It seems that our code has higher power and that's due to zero cross correlation property which has been canceled the PIIN effect and increased from desired signal power comparing to noise power.

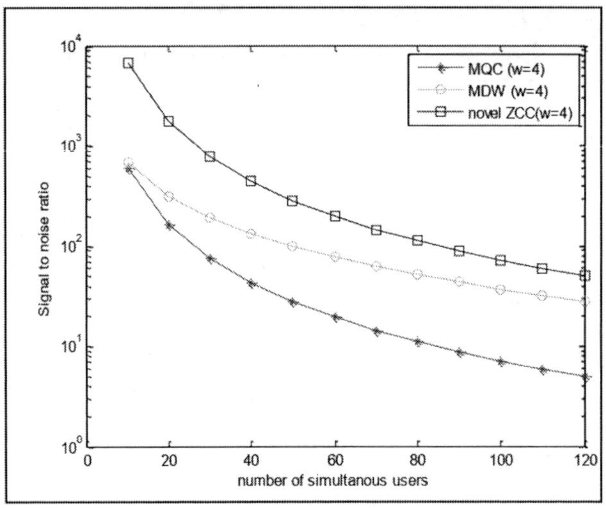

Figure 1. SNR versus number of simultaneous users for (w = 4).

In figure 2, it shows the BER variation as function of number of active users for the same code weight which equal to (4). It seems that our code has minor BER and can support higher number of users. For the same acceptable BER, the system can support 22, 38 and 70 of number users for MQC, MDW and ZCC codes. Thus system capacity has been enhanced up to 3.18 and 1.84 times comparing with MQC and MDW codes respectively.

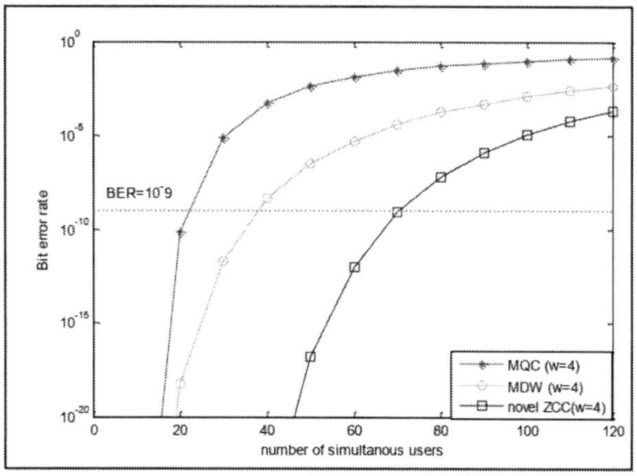

Figure 2. BER versus number of simultaneous users for $(\mathrm{w} = 4)$.

Finally, figure 3 shows the BER variation as function of power source at reciver level for 50 of number of active user. It seems that our proposed system needs minor power at reciver reaches around $-12.6\ dBm$ comparing to other systems using MQC and MDW codes where SAC-OCDMA using MQC code does not reach an acceptable BER and using MDW code, the received power reaches around $-6{,}4\ dBm$ which is about half the energy of our system.

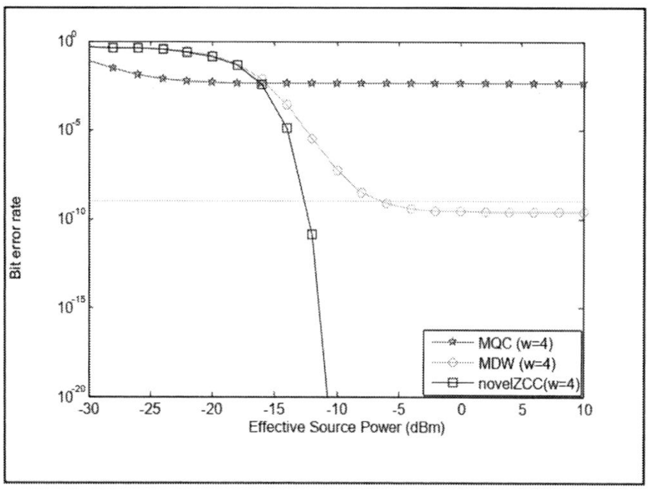

Figure 3. BER versus effective source power for $(K = 50)$.

5. Conclusion

In this paper, a novel ZCC code is developed for SAC-OCDMA systems based on shifting rule property. The codes provides several advantages presented in its weight simplicity and flexibility, an efficient zero cross correlation property that permits a large MAI and PIIN minimization, a high auto-correlation and a short code length. In addition to OCDMA system employing the new ZCC code can support a high cardinality and signal power in terms of acceptable BER comparing to MQC and MDW codes.

REFERENCES

[1] Nisar KS 2016 Construction Zero Cross Correlation Code using Permutation Matrix for SAC-OCDMA Systems *Int. J. Open Info. Technol.* **4** 51-54.

[2] Malleswari M and Murugesan K 2013 Construction and bit error analysis of zero cross-correlation codes for SAC-Optical CDMA systems *J. Opt* **42** 307–310.

[3] Ahmed N, Aljunid SA, Fadil A, Ahmad RB and Rashid MA 2013 Performance enhancement of OCDMA system using NAND detection with modified double weight (MDW) code for optical access network *Optik* **124** 1402–1407.

[4] Rashidi CBM 2011 New design of Zero Cross Correlation Codes for Spectral Amplitude Coding in OCDMA systems *2nd International Conference on Photonics*

[5] Aljunid SA, Ismail M, Ramil AR, Ali BM and M.K. Abdullah 2004 A new family of optical code sequences for spectral amplitude-coding optical CDMA systems *IEEE Photon. Technol. Lett.* **16** 2383–2385.

[6] Wei Z, Shalaby HM and Ghafouri-Shiraz H 2001 New code families for fiber-bragg-grating-based spectral-amplitude-coding optical CDMA systems *IEEE Photon. Technol. Lett.* **13**, 890–892.

[7] Wei Z and Ghafouri-shiraz H 2002 Codes for spectral-amplitude coding optical CDMA systems *J. Lightwave Technol.* **8** 1284-1291.

[8] Rashidi CBM, Aljunid SA, Ghani F, Fadhil HF and Anuar MS 2012 Cardinality Enhancement Using Flexible Cross Correlation (FCC) Code for Spectral Amplitude Coding Optical Code Division Multiple Access Systems *J. Appl. Sciences Research* **8** 5614-5626.

[9] Abd TH, Aljunid SA, Fadhil HA, Ahmad RA and Saad NM 2013 Development of new code family based on SAC-OCDMA system with large cardinality for OCDMA network *Opt. Fiber Technol.* **17** 273-280.

[10] Ahmed HY and Nisar KS 2013 Diagonal Eigenvalue Unity (DEU) code for spectral amplitude coding-optical code division multiple access *Opt. Fiber Technol.* **19** 335-347.

[11] Gupta S and Goel A 2017 Advance Method for Security Enhancement in Optical Code Division Multiple Access System *IETE J. Research* **64** 17-26.

Explosive Pressing Experiment and Its Challenges to Conduct it on University Level

A.H. Hilmi[1] and M.A. Yusuf[2]

[1]School of Manufacturing Engineering, Universiti Malaysia Perlis

[2]Faculty of Civil Engineering, National Defense University Malaysia

humaizi@unimap.edu.my

Abstract. Typical press force for hydraulic press machine for most universities are around 10 to 30 ton. For higher load, student uses Ultimate Testing Machine (UTM). Typical press force for UTM for higher learning institution is 300 ton. To obtain more press force, the specimen must be small. This is according to force equation where the smaller the surface area, the higher the force generated on it. However, the problem with small specimen is that it is difficult for researchers to do further investigations for hardness test, tensile test and microstructural analysis. Besides, there is a limit to maximum force a UTM machine can give. This paper presents steps for doing experiments using explosive pressing at university level. The explosive pressing apparatus was an improved designed by the same author. This experiment uses an amount of explosive that is place at top of the apparatus. The experiment is done underground for safety purposes. Explosive handling was done with help from authorized explosive expert. Major benefits of explosive pressing are larger samples can be pressed and the press force is very high. This paper will discuss the steps to do experiment and the challenges while doing explosive pressing using this apparatus. This paper will also show that higher pressing force can lead to finer microstructure of material being pressed.

1. Introduction

For pressing experiment, students have two options, using hydraulic press or using UTM. Typical hydraulic press machine can press around 30 ton to 50 ton while typical UTM can reach 300-ton maximum load. UTM is a testing machine that is widely used in all institutions of higher learning. UTM can do various tests such as tensile tests, three-point bending tests, compression tests and pressing. Examples of specimens tested are composites, alloys, steels and green samples. Among notable brands are Instron, Shimadzu and Santam. On average, the maximum load a UTM machine can achieve is 300kN or 300 ton, testing width 600 mm and displacement rate, 500 mm/min. [1]–[3]. Explosive research at university level is rarely done because of its danger and limited access to explosive material. Fortunately, the authors can do the experiment with help by experts from National Defence University Malaysia (NDUM). When properly channelled, energy from explosive produces higher press force greater than UTM. The explosive force produces shock wave that improves particle bonding and increases density of product. [4], [5]. Notable researches on explosive pressing was mostly done in China and Russia. Their experiment setup uses tubular cores that is enclosed with explosives. There is other research that encloses a cylindrical tube with explosives. Most of their study focuses on pressing sandwich material to see the effect at higher press force. [6]–[8]. This research places the explosive at the top of the plunger. The plunger thickness determines the press force. The thinner the plunger, more force can be press on the specimen. Previous design apparatus by the same author shows a big and bulky apparatus that is difficult for handling [4]. This paper will show improvements for the design, discussion on experimental steps for doing explosive pressing experiment and the challenges in doing experiment.

2. Experimental

Experiments involving live explosion must be done with help from experts at NDUM. The project leader will do all the arrangements with NDUM. At the arranged date, all team members travel to blast site. Explosive pressing experiments is a team effort. To ease of experiment, at least 6 persons must be involved. 2 persons handles the explosives matter along with explosive expert. 4 persons will do the digging and all physical works. First team digs the hole at demolition site. The hole depth is about 2 feet. The hole should fit a plywood that acts as the base. Second team handles the explosive preparation. They measure the thickness as well as moulding the explosive according to dimension. Explosive expert will assist them. Once the hole is prepared, the apparatus is placed on the plywood. Once ready, the apparatus is buried. One person must hold the apparatus while being buried to avoid it topple down. Then the explosive and the specimen is carefully placed on its position. Handling of explosives is done by explosive expert. To add more strength, several sandbags is placed vertically on the blast apparatus. The explosive expert will ask everyone to be at safe distance while they do the final checking. Then they trigger the explosion. After blast, researchers went to blast area. The blast area is examined thoroughly. The specimen is retrieved. If the specimen is missing or parts of the blast apparatus got thrown away due to explosive effect, perimeter check must be done to find them. The blast test is repeated until all specimen is completed. After all experiments are completed, the blast test site is cleared. Figure 1 shows the flowchart for steps in doing explosive pressing experiment. Table 1 shows list of equipment for explosive pressing test.

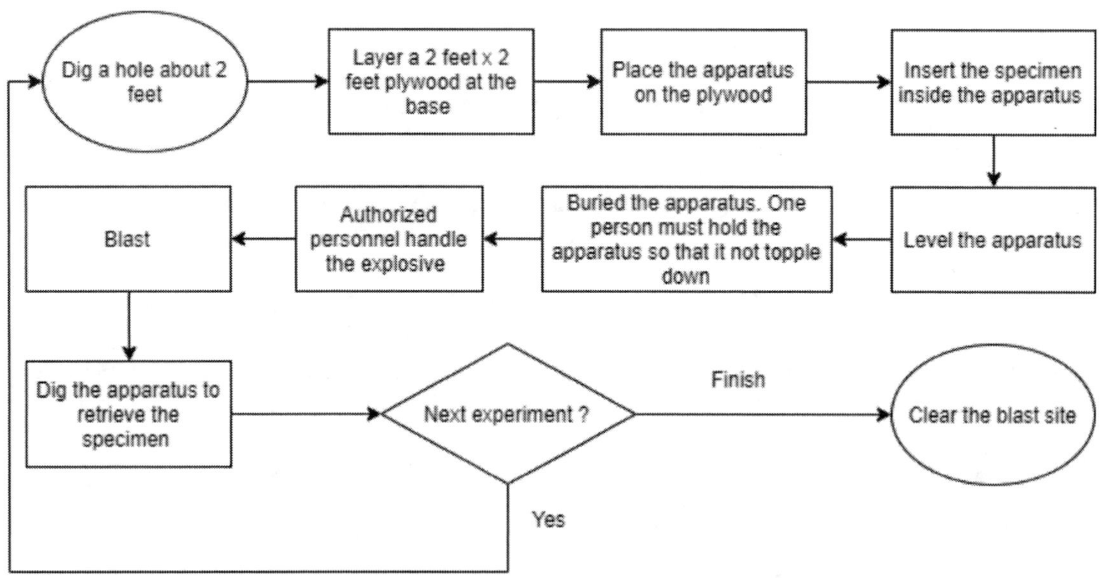

Figure 1. Flow Chart for the Explosive Pressing Experiment

Table 1. The list of equipment are as follows

No	Equipment	Purpose
1	Emulex	A type of Explosive
2	Detonator	Explosive connector
3	Sensors	For any measurements
4	Plastic container	Store the specimen
5	Plywood 2 feet x 2 feet	As a foundation to put on the ground
6	Shell cover	To cover the apparatus for underground explosion
7	Shovel	To dig hole
8	Explosive pressing apparatus	

The explosive pressing apparatus is an improved design from the same author in [4]. Figure 2 shows the apparatus. The apparatus consists of a plunger, mold, shell cover and lower base. The specimen is place inside the mold. Then plunger is put on to it. Plywood is layout on the ground to assist levelling. While buried, one person must hold the apparatus to avoid it topple down. Emulex which is a type of explosive is place on top of the plunger. The specimen in this research is iron powder.

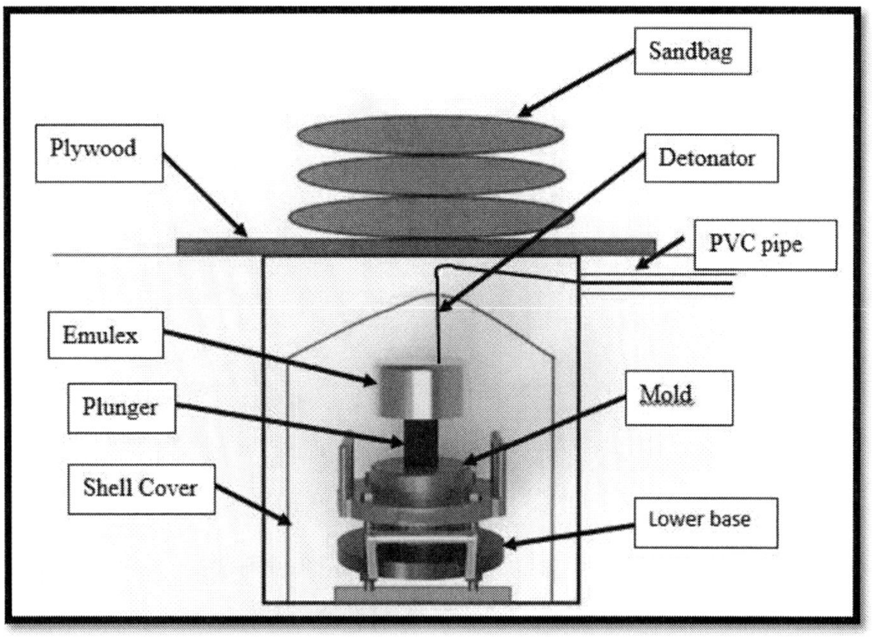

Figure 2. Explosive pressing apparatus

3. Analytical analysis

Analytical analysis shows all the equations used to calculate force impacted on the specimen. The formula stars with simple stress equation followed by energy equation, jet velocity and followed by stress wave propagation. Finally, the force is calculated by deriving the impact calculation.

3.1 Force produced by Emulex

The stress produced can be calculated based on the following equation:

From Eq. (1)

$$\sigma = \frac{m\left(\frac{v}{t}\right)}{A} \tag{1}$$

Where,

m= mass (kg)

v= velocity (m/s)

t = time(s)

A= area (m^2)

The energy produced can be calculated based on the following equation:

From Eq. (2) (Henry, 1967)

$$\mathrm{V_o} = \sqrt{2\mathrm{e}}\left[\frac{m}{c} + \frac{1}{2}\right]^{-\frac{1}{2}} \tag{2}$$

Where;

m= Mass of Casing

c= Mass of explosive

The jet velocity produced by Emulex can be calculate based on the following equation:

From Eq. (3)

$$\mathrm{v} = \sqrt{\frac{2E}{m}} \tag{3}$$

Where;

E= Internal energy produced by Emulex

m= Mass of explosive

3.2 *Stress Propagation in Plunger*

The stress propagation in plunger can be calculated using the equation:

From Eq.(9) [9]

$$C_p = \sqrt{\frac{\frac{d\sigma}{d_\varepsilon}}{\rho}} \quad (4)$$

Where,

$$d_\sigma = \frac{F}{A} \quad (5)$$

F= force produced by Emulex

A= area of plunger (m^2)

ρ = density of tool steel $\left(\frac{kg}{m^3}\right)$

Where,

$$d_\varepsilon = \frac{\varepsilon \text{ lateral}}{\varepsilon \text{ horizontal}} \quad (6)$$

Then,

$$d_\varepsilon = \frac{\Delta l_f}{l_f} - \frac{\Delta l_o}{lo} \quad (7)$$

Therefore;

$$C_p = \sqrt{\frac{\frac{\frac{F}{l \times h}}{\frac{\Delta l_f}{l_f} - \frac{\Delta l_o}{lo}}}{\rho}} \quad (8)$$

3.3 *Force produced impacted on powder metal*

From the Eq. (14) [10]

$$\frac{V_1}{V_0} = \left[\frac{1 - m(v_0)^2(1 - \gamma_{pw}}{2PV_0}\right]^{\frac{1}{1-\gamma_{pw}}} \quad (9)$$

V_0 = Volume initial $(m\wedge3)$

V_1 = Volume final (m^3)

m = Mass of plunger (N)

v= Velocity produced (m/s)

$\rho = \text{Density}$

Where $\gamma = 10^{-9}$

Therefore

$$\mathbf{P} = \frac{(\gamma_{pw-1})m(v_o\sqrt{\frac{1}{\rho}}\left(\frac{V_1}{V_0}\right)^{\gamma pw}}{2\left[(V_1 - V_0)(\frac{V_1}{V_0})\right]} \tag{10}$$

4. Results and discussion

From the research, it can be stated that the higher amount of Emulex used to give a higher value of pressing force beyond 50,000 ton pressed. The results give a good achievement for powder metallurgy method. From the analysis, the Iron powder after the explosive pressing has a high density and hardness. The higher the pressing force the higher the density of the Iron formed. The hardness of Iron also increases with the increment of pressing force. Moreover, the SEM analysis also was done to the specimen. The microstructure of Iron powder was compared by using different amount of Emulex increases from 1000g to 2000g of Emulex. The porosity shown by the SEM was decrease because the higher force can produce effective binding of the Iron particle. The increment the mass of Emulex leads to a reduction in porosity. The density of Iron increases with the reduction of pore and increment of cold- welded region. Thus, it can be stated that the finer microstructure can be obtained using a higher pressing force and the explosive pressing can produce a smooth microstructure of metal with low porosity.

4.1 Pressing results

The explosive used is Emulex (Austin Powder, 2013). The Emulex is form into shape charge due to more pressing force. The pressing material is sintered to fortify the holding between particles. Investigation of minimized powder microstructure is performed using checking Scanning Electron Microscope (SEM). The hardness of powder checked by Rockwell Digital Hardness and the outcome demonstrates the hardness of minimal powder for explosive was 50.6 HRA, while the reduced powder using Universal Testing Machine is 24.0 HRA. Microstructure comes about showed lesser porosity utilizing touchy pressing. To compare with UTM, explosive pressing gives better microstructural results. UTM can produce press force limited to 300 ton while explosive pressing can reach 50,000 ton. The differences are very high. Some basic press theory for both method but for explosive pressing, the results show better and finer microstructural results. However explosive pressing have challenges as discussed below.

4.2 Challenges

Upon doing blast test, there are points for improvements. This paper discusses the problems related to doing the experiment and improvements to the blast method for future research.

4.2.1 Interval time

The average time taken for each blast test is 25 minutes. This is because the authors have to dig the samples, asses the results, doing minor adjustments for the blast apparatus, flatten the ground, levelling the apparatus and keeping a safe distance from blast site. Therefore, for a period of 9 hours beginning at 9 a.m to 7 p.m, there can be 16 tests can be done per day. No blast is done during lunch time. Therefore, the blast session must be plan carefully. To save time, it is suggested that two set of explosive pressing apparatus is prepared. Two blast can be triggered at one time.

4.2.2 Lost parts and specimens
Due to tremendous force generated from explosive, some parts of explosive apparatus may get thrown away. Once, the specimen got thrown away also. Therefore, after each blast, the first thing to check is the perimeter to see if there are some parts of the apparatus got thrown away. Then when digging back the apparatus, make sure all parts are there. Most importantly the specimen should be intact. As a precaution, sandbags are put on the perimeter of the hole. During blast event, the sand will mitigate the blast effect while reducing possibilities of parts got thrown away.

4.2.3 Damage parts
Since explosive gives tremendous loads, some parts are expected to damage. Parts that will damage are plunger, sandbags and plywood. However, the lower base, the shell cover and the mold must not be damage. Since blast load is unpredictable, always start the experiment with lower amount of explosive then increase it incrementally. From experience, explosive weight more than 1.6 kg may damage the mold. Lower base can survive any load if it is indirect blast.

4.2.4 Cost to do explosive experiment
It is expensive to do an explosive pressing session. Usually the test is set to be two days. Beginning Friday evening and finishes on Saturday evening. It may be extended to Sunday if necessary. The cost covers lodging, meals, consultancy, fees for NDUM and also fees for explosive team. Although it is expensive, the cost is reasonable. To save cost, usually two or more separate blast session is done concurrently with other researchers. To cost is shared by them all.

4.2.5 Sensors
Due to tremendous loads by explosive, any sensors directly in contact with explosive will damage. Sensors with lower specification cannot display the peak pressure readings. High specification pressure sensors are expensive. Therefore, to get reading to calculate force on the specimen, copper crusher method is used. The method was used to examine pressure in the rifle bore during World War II. Pressure from rifle bore will crush the copper. Then the displacement measurement is taken and compare with a table that shows pressure inside the bore. The principle behind copper crusher is stress-strain theory. For explosive pressing purpose, to get the force directly impact on the specimen, mild steel is put as the target. Once impacted by plunger, the mild steel will deform. The height of mild steel is taken before and after blast. From the displacement, strain can be obtained. Modulus of elasticity of mild steel can be obtained from various sources. By calculating the area of specimen, the force directly impacted the specimen can be obtained. It is recommended that two sets of experiments are

done. One with using mild steel as a method to know the force on the sample. One with using the real specimen. Other configurations should be the same.

4.2.6 Accuracy of data
Since blast event is unpredictable, many factors will influence the blast results. For example, different arrangements of sandbags will give different results. Most importantly, the Emulex explosive should be properly compacted. There should be no pores inside the explosive as it will give different results. Therefore, configurations should be the same for all experiments. For explosive, it is suggested to replace Emulex with booster. Booster is Standard booster is 400 grams High Explosive. It has the same geometry and weight. There is no need for a person to mold the explosive to become any shape.

5. Conclusion

From the research, it can be stated that the higher amount of Emulex used to give a higher value of pressing force. The results give a good achievement for powder metallurgy method. From the analysis, the Iron powder after the explosive pressing has a high density and hardness. The higher the pressing force the higher the density of the Iron formed. The hardness of Iron also increases with the increment of pressing force. Moreover, the SEM analysis also was done to the specimen. The microstructure of Iron powder was compared by using different amount of Emulex increases from 1000g to 2000g of Emulex. The porosity shown by the SEM was decrease because the higher force can produce effective binding of the Iron particle. The increment the mass of Emulex leads to a reduction in porosity. The density of Iron increases with the reduction of pore and increment of cold- welded region. Thus, it can be stated that the finer microstructure can be obtained using a higher pressing force and the explosive pressing can produce a smooth microstructure of metal with low porosity. Explosive pressing experiment is expensive and challenging. It involves team effort. However, it is an exciting experiment to do. When properly done, explosive pressing can give better finding for further research.

Acknowledgements
The author hugely thanks to Kementerian Pelajaran Malaysia (KPT) for funded by FRGS grant "Derivation of theoretical stress and strain for a plunger exposed to blast loading used for pressing iron powder beyond 100 000 ton" (FRGS/1/2018/TK03/UNIMAP/02/21). This project was done with the help demo Blast Research Team from National Defense University Malaysia. The authors also acknowledge School of Manufacturing Engineering, Universiti Malaysia Perlis for the lab facilities.

6. References

[1] A. L. Araújo, J. R. Correia, and C. M. M. Soares, "Characterization of Al7075-B4C Composite Fabricated By Powder Compaction Techniques Under Different

Densification Rates," pp. 1–11, 2015.

[2] R. A. Gonçalves and M. B. Da Silva, "Influence of Copper Content on 6351 Aluminum Alloy Machinability," Procedia Manuf., vol. 1, pp. 683–695, 2015.

[3] H. Babaei, T. Mirzababaie Mostofi, M. Namdari-Khalilabad, M. Alitavoli, and K. Mohammadi, "Gas mixture detonation method, a novel processing technique for metal powder compaction: Experimental investigation and empirical modeling," Powder Technol., vol. 315, no. April, pp. 171–181, 2017.

[4] A. H. Hilmi, N. A. Azmi, and A. Ismail, "A method to press powder at 6000 ton using small amount of explosive," in AIP Conference Proceedings, 2017, vol. 1901.

[5] R. N. Rosdi, N. A. Azmi, N. A. Latheef, A. H. Hilmi, and A. Ismail, "Characteristics of iron (Fe) powder using underground explosion pressing," in AIP Conference Proceedings, 2018, vol. 2030.

[6] N. A. Adamenko and G. V. Agafonova, "Explosive Pressing of Heat Resistant Polymers and Their Composites with PTFE-4," Chem. Pet. Eng., vol. 52, no. 7–8, pp. 567–572, 2016.

[7] A. V. Krokhalev, V. O. Kharlamov, S. V. Kuz'min, and V. I. Lysak, "Tribotechnical properties of powder hard alloys of chromium carbide with titanium fabricated by explosive pressing," Russ. J. Non-Ferrous Met., vol. 55, no. 2, pp. 212–217, 2014.

[8] B. Wang, F. Xie, Z. Li, and H. Zhang, "Explosive compaction of Al2O3 nanopowders," Ceram. Int., vol. 42, no. 7, pp. 8460–8466, 2016.

[9] W. Walters, "Introduction to Shaped Charges," no. March, 2007.

[10] N. Yavuz, "Mechanics of Dynamic Powder Compaction," J. Eng. Sci., vol. 2, no. 2, pp. 129–136, 1996.

Potential of developing an energy self-sufficient village through mini hydro powerplant technology

B Aminullah[1] and R Wirawan[2]

[1]Labcshool Senior High, Universitas Negeri Jakarta
Jalan Pemuda Kompleks Universitas Negeri Jakarta, 13220, INDONESIA
[2]Faculty of Engineering, Universitas Negeri Jakarta
Jalan Rawamangun Muka, Jakarta, 13220, INDONESIA

rwirawan@unj.ac.id

Abstract. Nowadays, electricity is one of the most important need for the human life. However, in developing countries, there are some areas that are not fully covered by the electrical power. The high cost of distribution is one of the reasons behind the problem. More efforts are needed to distribute the electricity to a remote area, such as a far mountainous region. On the other hand, there is a potential to provide electrical power from a river through the development of mini-hydro powerplant technology. In this study, potential of water power in a mountainous region of West Java Province, Indonesia, is measured. The measurements were taken in three different places of river, divided by the distance from a waterfall. As a result, it is found that the potential power obtained from the rivers are ranged between 17-137 kW, which is enough for lighting up 37-304 houses, indicating there is a potential of developing an energy-sufficient village through mini-hydro powerplant technology.

1. Introduction

West Java Province, Indonesia, is rich in tourism potential. One of those tourist attractions are waterfalls, which are called *Curug* in the native language of Sundanese. So far, these waterfalls have solely been used as a tourist attraction through their beauty. The beauty of these waterfalls is captured through eye-catching scenery, surrounding nature, clean air or the pleasant atmosphere brought by the waterfalls. Whereas, if investigated further, within these waterfalls, there are waters that fall from a certain height. These waterworks store energy that could be converted into electricity using hydroelectric power plants. Therefore, the *Curug* in fact has two direct potentials, the first being its tourism potential with its beauty and the second one being the potential energy contained within its water stream. If the two potentials are utilized correctly and optimally, it will be beneficial.

Unfortunately, the idea of utilizing the potential energy clashes with the concerns of the *Curug*'s caretakers. The caretakers are worried that the usage of these Hydroelectric powerplants would disturb or destroy the natural beauty of the waterfalls. To solve these concerns, a new model of powerplant is needed: a powerplant that would both suffice and support both the potential tourism and potential energy. Based on those reasons, an alternative scheme for Mini Hydro Powerplants in the waterfall areas is proposed. It is expected the scheme would not only harvest the potential energy contained within the waterfalls, but it would also improve tourism potential. The waterfall that has been chosen is *Curug*Cipurut, located in the base of Mount Burangrang and is a part of the Burangrang Mountain

Nature Reserve (Figure 1). Because of that, the waterfall is still managed by the Jabar I Natural Resources Conservation Center, with operations in the field being assisted by Sumurugul Village and Wanayasa Village.

Generally, the waterfall is divided into three sectors, which are the main waterfall located in the edge of the cliff with a height of approximately 25 meters, and is located inSumurgul Village, 10 kilometres away from Wanayasa District, Purwakarta Regency, West Java Province. The area is selected because the waterfall is able to flow throughout the year, and the waterfall is located at a relatively high (700 meters) above sea levels.

Figure 1. *Curug*Cipurut.

2. Method

The measurement of water discharge is done by conventional methods, due to the current meter equipment desired to measure the water speed being absent. This method is done by taking samples of water in three different places. The determination of the specific locations is based on the height of the fallingwater and the volume of the water flow within the river channel.

The amount of electric power produced from the water current could be counted with the equation (1) [3],

$$P_{net} = gQH_{gross}\varepsilon_o \quad \text{.............................} (1)$$

where
P_{net} = net output power system (watt).
g = gravity of earth (m/s2).
Q = water discharge (m3/s).
Hgross = gross head (m).
ε_o = system efficiency.

IOP Conf. Series: Materials Science and Engineering **767** (2020) 012044 doi:10.1088/1757-899X/767/1/012044

The design of this research is designed so that the steps of research data collection consists of 2 groups of data collection. The first being the collection of data regarding potential water resources by reviewing their water discharge rate, demographics, area as well as socioeconomic factors, and the second being the collection of data on the amount of electricity

The next step is to prepare a structured and systematic study design as a framework for data and information collection on site. Doing a field survey, which consists of potential water resource data, demographics and area data, socioeconomic data and electrification with the model.

3. Results and Discussion

3.1. *Cipurut River Channels, a runoff from Cipurut Waterfall*

The Condition of Cipurut river's flow that has an 8560 m channel length gets water flow from waterfall runoff which wholly falls to the river. The water discharge sampling is done in three different places, namely the falling point, the rice field area—about 1,5 km from the waterfall, lastly the village—3,3 km from the waterfall. Given the lack of equipment—which is caused by the absence of the *current meter*—the measurement of the water discharge is done manually by collecting water in a 1-litre container. Data retrieval was done 5 times within the month of August 2019, then, the data is averaged. For each sample retrieval, the collection time is recorded (tables 1-3).

Table 1. Point 1: Cipurut Waterfall, falling point, H = 25 m.

n-th Experiment	Volume (Litre)	Time (x 10^{-6} second)
1	1	1,43
2	1	1,54
3	1	1,56
4	1	1,45
5	1	1,48

Table 2. Point 2: 1500 m from the waterfall, Ricefields, H = 10 m.

n-th Experiment	Volume (Litre)	Time (x 10^{-6} second)
1	1	2,87
2	1	3,64
3	1	3,46
4	1	3,17
5	1	3,39

Table 3. Point 3: 3300 m from the waterfall (Village). H = 15 m.

n-th Experiment	Volume (Litre)	Time (x 10^{-6}
1	1	6,37
2	1	6,62
3	1	7,41
4	1	8,33
5	1	7,04

An example of the calculation is taken in the potential point of River Cipurut's channel, located along with the river's current, with the length being +/- 3000 m from the *Curug*Cipurut Waterfall. Using the water density of 1 kg/cm^3 or equivalent to 1000 kg/cm^3, a calculation of the theoretical power in those points are done. Using the equation (1) above, with the assumed efficiency is 0.8,an example of the calculation is given below.

For point 1:

$$P_{net} = gQH_{gross}\varepsilon_o$$

= 9,8(m/s2) x 699,301 (m^3/det) x 25 (m) x 0,8
= 137.062 watt
= 137 kW

The aforementioned power is water power potential. Using the same method, the data for points 2 and 3 are calculated. The results are shown in tables 4-9.

Table 4. Discharge rate calculation results, with fall height, H = 25 meters.

n-th Experiment	Volume (lt)	Time (s)	coefficient	Discharge Rate (m3/s)
1	1	1,43	1000	699,3007
2	1	1,54	1000	649,3506
3	1	1,56	1000	641,0256
4	1	1,45	1000	689,6552
5	1	1,48	1000	675,6757

Table 5. Theoretical power calculation results with fall height, H = 25 meters.

ρ (kg/m^3)	g (m/s^2)	H (m)	Q (m^3/s)	η Efficiency	Pnet (watt)	Pnet (kW)
1000	9,8	25	699,30	0,8	137.063	137
1000	9,8	25	649,35	0,8	127.273	127
1000	9,8	25	641,03	0,8	125.641	126
1000	9,8	25	689,66	0,8	135.172	135
1000	9,8	25	675,68	0,8	132.432	132
Average Pnet					131.516.280	131.516

IOP Conf. Series: Materials Science and Engineering **767** (2020) 012044 doi:10.1088/1757-899X/767/1/012044

Table 6. Discharge rate calculation results, with fall height, H = 10 meters.

n-th Experiment	Volume (lt)	Time (s)	coefficient	Discharge rate (m3/s)
1	1	2,87	1000	348,4321
2	1	3,64	1000	274,7253
3	1	3,46	1000	289,0173
4	1	3,17	1000	315,4574
5	1	3,39	1000	294,9853

Table 7. Theoretical power calculation results with fall height, H = 10 meters.

ρ (kg/m^3)	g (m/s^2)	H (m)	Q (m^3/s)	η Efficiency	Pnet (watt)	Pnet (kW)
1000	9,8	10	348,43	0,8	27.317	27
1000	9,8	10	274,73	0,8	21.538	22
1000	9,8	10	289,02	0,8	22.659	23
1000	9,8	10	315,46	0,8	24.732	25
1000	9,8	10	294,99	0,8	23.127	23
Average Pnet					23.874.640	23.875

Table 8. Discharge rate calculation results, with fall height, H = 15 meters.

n-th Experiment	Volume (lt)	Time (s)	Coefficient	Discharge rate (m3/s)
1	1	6,37	1000	156,9859
2	1	6,62	1000	151,0574
3	1	7,41	1000	134,9528
4	1	8,33	1000	120,048
5	1	7,04	1000	142,0455

Table 9. Theoretical power calculation results with fall height, H = 15 meters.

ρ (kg/m^3)	g (m/s^2)	H (m)	Q (m^3/s)	η Efficiency	Pnet (watt)	Pnet (kW)
1000	9,8	15	156,99	0,8	18.462	18
1000	9,8	15	151,06	0,8	17.764	18
1000	9,8	15	134,95	0,8	15.870	16
1000	9,8	15	120,05	0,8	14.118	14
1000	9,8	15	142,05	0,8	16.705	17
Average Pnet					16.583.705	16.584

3.2. *Public Socioeconomic and Electricity condition*

The condition of the people surrounding the Cipurut River are mostly farmers, with a scattered and dispersed housing spread pattern, with an average of 10 households, thus with the addition of the Micro Hydro Powerplant, the distribution of loading can be in the form of loading of lighting and household appliances. The requirement of power for lighting from the calculation above from an average of 10 households where the usage in each household of 100 W is predicted to be 1000 W, or equivalent to 1 KW. Household appliances as an addition to the lighting, or for instance agricultural electrification equipment could be used with loading regulation that could be agreed upon, examples being using the power for agriculture electrification equipment in the morning and appliances at night. The table of Electricity needs parameter per household is shown in table 10.

Table 10. Electricity Needs Parameter per Household.

Electricity Parameter	
Electrical System Voltage	: 220 volt
Current limiter	: 2 ampere
Connected electric capacity	: 450 watt
Usage	
3 lighting spots @20 watt	
4 lighting spots @10 watt	: 3 x 20 watt = 60 watt
Total for lighting	: 4 x 10 watt = 40 watt
	100 watt
Backup power needs	
Home appliances	
Electronic appliances	: 200 watt
	: 150 watt
	Total 450 watt

3.3. *Data Analysis*

From table 1, it is observed that the discharge produced is relatively high compared to the data on Tables 2 and 3. This is due to the direct discharge from the source, which is Cipurut Waterfall. However, this large discharge has disadvantages, which is the distance to the villager's settlements is quite far away, around 3-4 km with a hilly land contour, so to install pipes or create canals with conditions like that would be very costly. Also, the maintenance of the installation pipes would be very costly. The advantage of the data from Table 1 is that the discharge would produce a large output power, considering the fall height is also significant—around 25 meters. The potential power generated is 132kW.

For point 2 form table 2, it is observed that the discharge is almost reduced by half. This may be the consequence of the water being partly absorbed by the soil, and additionally also going into wormholes, which are small channels located along the river channel. The height of the falling point was still around 10 meters, which produces a theoretical power potential of about 23,875 kW. Even so, the excess waters from the turbines could be utilized by the public for household uses or irrigating their fields. Point 2 was located in the middle between the waterfalls' path and the village

For point 3, the location is in the villager's settlements, which is around 3300 meters from Cipurut Waterfall. Measurement is done on the highest point, and there is a 15-meter height difference with the lowest point in the village. With an average discharge of 141,02 m3/sec, a power potential of around 17kW is generated. The advantage of a powerplant located in the villager's settlements is that the costs for power cable transmission can be cheaper. But, due to the location of the settlement—which is located on the slope of the mountain—then the excess water can't be used for the villager's showering and washing needs, because the water is directly discharged to the valley.

4. Conclusion

Based on the results of the experiment and the study of discharge obtained, the availability of sufficient water sources with the length of the main channel being less than 8560 m, has 3 potential points.

The potential water power produced at raining seasons is around the range of 17 kW to 137 kW. Using the potential power of 17 kW, it is concluded that the Mini Hydro Powerplant is categorized as a micro powerplant, not a mini powerplant.

From the power extracted from the potential energy in the water,theability of the powerplant to powerhouses could be calculated using a simple equation which is as follows:

$$P_{net} = 137.000 \text{ watt} / 450 \text{ watt} = 304 \text{ Houses}$$
$$P_{net} = 24.000 \text{ watt} / 450 \text{ watt} = 53 \text{ Houses}$$
$$P_{net} = 17.000 \text{ watt} / 450 \text{ watt} = 37 \text{ Houses}$$

Whilst, the number of houses in the Cielungsing Village +/- 150 houses. This deficiency of power has to be supplemented using other resources, for instance, adding or replacing the water supply for the Cipurut River channel. If not, then it needs to be assessed from solar, wind, or biogas energy sources.

The research on electricity self-sustainability is important because the national energy demand has not been able to cover all of its citizens. If the people can produce their own energy, then the electrification rate in Cileungsing Village could be 100% and provide positive effects to the villager's activities.

References

[1] Novi Kurniasih, dkk. 2015. Analisis mode sistem pembangkit listrik tenaga hibrid microhydro-photovoltaic array menggunakan H. Jurnal Nasional Teknik Elektro [Analysis of the hybrid microhydro-photovoltaic array power generation system mode using Homer]: Universitas Andalas. Vol: 4, No. 1.

[2] Teguh Utomo. 2018. Pengembangan Jaringan Listrik Mikrohidro Dan Solar Home System/SHS Sebagai Penopang Desa Mandiri Energi di Bawah Program PPDM Desa Mandiri Energi Kabupaten Probolinggo [Development of Micro Hydro and Solar Home System / SHS as a Support for Energy Independent Villages under the Energy Independent Villages Development Program of Probolinggo Regency]. Seminar Nasional Teknik Elektro Universitas Brawijaya. ISBN 978-602-8692-34-2.

[3] Dedi Nugroho, dkk. 2017. Desain Pembangkit Listrik Tenaga Mikrohidro di Air Terjun Kedung Kayang [Design of Micro Hydro Power Plants in Kedung Kayang Waterfall]. Jurnal Rekayasa Elektrika, Universitas Islam Sultan Agung. Vol. 13, No. 3, Desember 2017, hal. 161-171.

[4] Eduardo Heyko, dkk. 2016. Strategi pemanfaatan energi terbarukan dalam rangka kemandirian energi daerah Provinsi Kalimantan Timur [Renewable energy utilization strategy in the context of energy independence in the area of East Kalimantan Province]. Jurnal Ekonomi Keuangan, dan Manajemen, Universitas Mulawarman Indonesia. Vol : 12, No. 1.

[5] Eneng Dayu Saidah, dkk. 2017. Pengembangan desa berkelanjutan melalui konsep mandiri energi di Desa Parakanmuncang Kabupaten Bogor [Sustainable village development through an independent energy conceptin Parakanmuncang Village, Bogor Regency]. ISBN : 978-602-73463-1-4.

[6] Badan Informasi Geospasial.. 2017. Laporan Penelitian Potensi Energi Lokal Mikro Hidro Tahun 2017 [Research Report 2017 on the potential of local micro hydro energy]. Bidang Penelitian Pusat Penelitian, Promosi dan Kerja Sama.

[7] Arif Febriansyah Juwito, dkk. 2012. Optimalisasi Energi Terbarukan pada Pembangkit Tenaga Listrik dalam Menghadapi Desa Mandiri Energi di Margajaya [Optimization of Renewable Energy in Power Plants in Facing an Energy Independent Village in Margajaya]. Jurnal Ilmiah Semesta Teknika. Vol. 15, No. 1, hal: 22-34.

[8] Bono Pranoto, dkk. 2018. Potensi Energi Mikrohidro Di Daerah Irigasi (Studi Kasus Di Wilayah Sungai Serayu Opak) [The potential of microhydro energy in irrigation areas (case study in the Opak Serayu River region]. Jurnal Irigasi. Vol. 12, No. 2, hal. 77-86.

[9] Tiara Mentari. 2014. Partisipasi Masyarakat Dalam Program Desa Mandiri Energi Di Desa Kujau Kecamatan Betayau Kabupaten Tana Tidung (Studi Kasus Kelompok Tani Ternak Mandiri di Desa Kujau Kecamatan Betayau Kabupaten Tana Tidung) [Community Participation in the Energy Independent Village Program at the Village of Kujau, District of

Betayau, Regency of Tana Tidung (Case Study of the Independent Farmer group at the Village of Kujau, District of Betayau, Regency of Tana Tidung)]. eJournal Ilmu Pemerintahan : Universitas Mulawarman. ISSN 2338-3651.

[10] Achmad Syahid, dkk. 2013. Studi Teknologi Tepat Guna Pembangkit Listrik Berskala Piko Di Kabupaten Lany Jaya Kampung Malagai [Study on appropriate technology for pico scale power plant at the regency of Lany Jaya, village of Malagai]. Jurnal Teknik Mesin : ITS. Tahun 21, No. 2, Oktober 2013.

[11] Gaina Sapoetra, dkk. 2014. Studi Potensi PLTMH Kampung Nyomplong-Bogor [Study on the potential of micro hydro power plant at Kampung Nyoblong Bogor]. Journal Mandiri Energi : FTUI.

IOP Conf. Series: Materials Science and Engineering **767** (2020) 012045 doi:10.1088/1757-899X/767/1/012045

Pilot spacing controller in orthogonal frequency division multiplexing systems

M S Abu Hassan, R B Ahmad and N Shoib

School of Computer and Communication Engineering, Universiti Malaysia Perlis, Perlis Malaysia.

suhaizal@unimap.edu.my

Abstract. As the backbone of the fifth generation, 5G communications, Orthogonal Frequency Division Multiplexing, OFDM is seen to be reliable in addressing Inter-symbol Interference, ISI issues as well as providing high-speed access to broadband. Channel estimation and tracking are generally carried out by transmitting known pilot symbols in given positions of the frequency-time grid. However, the use of a fixed amount of pilot for all channel circumstances is hugely detrimental. Several methods, especially on block-type and comb-type pilot arrangement, have presented. In this paper, the method of determining the pilot space depends on the current channel quality. The proposed technique focused on improving estimation and bandwidth in time-varying and frequency-selective fading channels without increasing the complexity and sacrifice the bandwidth efficiency of the OFDM system.

1. Introduction

For various types of multiplexing techniques, OFDM has recognized as an efficient spectrum through selective frequency channels. One of the advantages of OFDM is that it can transfer data at high speeds and has high bandwidth efficiency. Besides, it can reduce the effects of Inter Symbol Interference (ISI). As a backbone for 5G systems, OFDM has been extensively applied in mobile communication systems [1]. The example of application is not limited to Extended Reality, XR, Esports, immersive gaming and toys, Ultra high definition TV, UHD TV, multi-person video call, and many more. The key to OFDM advantage is to divide the existing spectrum into several overlapping sub-channels but maintain the orthogonality each sub-channel and then convert it back to frequency selective channels to non-frequency selective channels [2]. High-speed transmission obtained by splitting the frequency-selective-fade signal band into several narrow-fade subchannels. Then level up by applying a compact constellation to each subcarrier. High spectral efficiency obtained by choosing a specific carrier frequency, which is orthogonal, as the spectra of SCs overlap with each other. Inter Symbol Interference, ISI was the main problem for multicarrier methods. However, it can evade the implementation of cyclic prefix, CP by spreading the OFDM symbol to several parts. The extended symbol is known as the head or tail for the symbol. Principally, the OFDM systems performance relied on the channel estimation process. It depends on the capability to estimate the channel behavioral precisely. In OFDM systems, both frequency and time domain represented by wireless channel transfer function existed irregularly. This irregularity happens because of the radio channels in wideband wireless communication systems typically frequency selective and time-variant.

Mostly OFDM-based communication standards offer some arrangements of reference signals individually as pilot signals or preamble. The purpose of the reference signal is to attain accurate channel estimation focus to equalize each subcarrier. Techniques used for channel estimation divided into two. First is decision direct channel estimation and secondly known as pilot-assisted channel estimation. Comparing two techniques, some drawbacks of decision direct channel estimation methods have identified. One of the weaknesses of the decision direct channel estimation method is because of its behavioral to error detection leads to the propagation of error. It is also required a massive amount of data, which can lead the convergence rate to slow down [3].

The ability of reference signals to provide good estimation for the whole frequency and time grid depends on the number of reference signals injected in the pair of the frequency domains and time domains. It comprised of the case of wireless channels, which subjected to time selectivity and high frequency. The receiver estimates the channel in frequency-domain by consuming information about the reference symbols around the location of the reference symbol. Inserting a pilot signal with a specific duration or on each OFDM symbol is known as pilot-assisted channel estimation [4]. The initial channel information then estimated with a certain number of pilots. Between OFDM symbols, interval pilot signals inserted. Using one-dimensional linear interpolation channel estimation results can be obtained [2,6]. The determining factor between the channel estimation accuracy and bandwidth efficiency refers to the number of pilot signals used. Therefore, the optimum quantity of the pilot signal needs to be determined. The challenge in designing the OFDM system is to maintain a low number of pilot signals while maintaining Bit Error Rate, BER under the desired value of Signal to Noise Ratio, SNR. The problem is the same as determining the maximum pilot distance to meet the BER requirement on a given SNR.

2. OFDM Systems Description

A conventional block diagram of the OFDM system with a pilot symbol shown in figure 1. According to the modulation for signal mapping, binary data is first grouped and then mapped. Pilot signals then inserted to each subcarrier regularly or with an exact period between the data sequence. By implementing the Fourier transform, earlier signal converted to multi-carrier modulation. Identically, Inverse Fast Fourier Transform, IFFT, and Fast Fourier Transform, FFT was chosen primarily to reduce the mathematical operations.

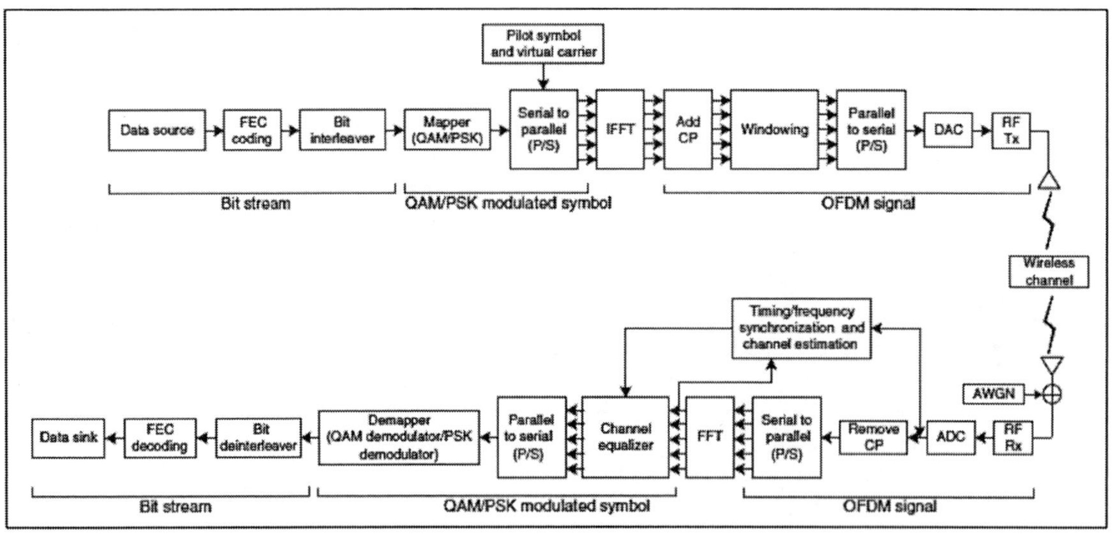

Figure 1. Block diagram of an OFDM system with a pilot symbol.

The function of IFFT block in figure 1 is to transform the frequency domain signal into time-domain signal as follow:

$$x(n) = IFFT\{X(k)\}, n = 0,1,2, \dots, N-1 \quad (1)$$

$$x(n) = \sum_{k=0}^{N-1} X(k) e^{j(\frac{2\pi kn}{N})} \quad (2)$$

The guard time mainly to avoid ISI. The parameter for guard time is preferred to be greater than the predictable delay spread. Equation (3) presented the resultant OFDM symbol mathematically:

$$x_f(n) = \begin{cases} x(N+n), n = -N_g, -N_g+1, \dots, -1 \\ x(n), n = 0,1, \dots, N-1 \end{cases} \quad (3)$$

N_g represents the length of the guard interval. Noise added to the transmitted signal after passing via the fading channel. The acknowledged signal at the receiver sensed as:

$$y_f(n) = x_f(n) * h(n) + w(n) \quad (4)$$

Where $h(n)$ is the channel impulsive response and $w(n)$ is Additive White Gaussian Noise, AWGN. FFT block then converts back the time domain, $y_f(n)$ signal toward the frequency domain signal after guard time has removed. Extraction of pilot signal executed at estimation block. This process mainly to obtain the estimation channel $H_e(k)$, aimed at the respective sub-channels data. Then the estimation of transmitted data presented as:

$$X_e(k) = \frac{Y(k)}{H_e(k)} \qquad k = 0,1,2,3 \dots N-1 \quad (5)$$

At the signal de mapper block, binary information data achieved.

3. Pilot Planning

In this section, the introduction of the famous pilot arrangement preceded. Then followed by channel estimation algorithms. Classification of channel estimation depends on different types of pilot patterns. Dedicated pilot subcarriers interlaced with data subcarriers in OFDM systems. The pilot arrangement relies on the channel state, either flat fading channels or slow fading channels. Therefore, several methods exist and characterized into two types. [7] Describe a pilot pattern base on a 2 x 2 sampling matrix $\boldsymbol{Y} = [y_1 \quad y_2]$. As $y_1 = [y_{11} \quad y_{21}]^T$ and $y_2 = [y_{12} \quad y_{22}]^T$ are the two spanning vectors that generated for all pilots in the time-frequency domain. It is also known as a doubly periodic sub-lattice.

The first pilot arrangement is exclusively for slow fading channels. The procedure is by inserting pilot signals with a specific period into all the subcarriers. Known as the block-type pilot, it is insensitive to frequency selectivity with the assumption that the channel is constant. As the pilots set at all carriers, there is also no interpolation error. The second approach for flat fading channels is a comb-type pilot arrangement. Each OFDM symbol inserted with pilot signals and the N_p pilot signals are constantly presented in $X(k)$ and shown by the followed equation:

$$X(k) = X(mL + l)\, X(k) = \begin{cases} x_p(m), & l = 0 \\ data, & l = 1,2,3 \dots L-1 \end{cases} \quad (6)$$

Where $L = number\ of\ carrier/N_p$ and $x_p(m)$ is the mth is the value of the pilot carrier. For the comb-type pilot, the interpolation technique is required to estimate channel efficiently at data subcarriers. The comb-type pilot and block-type arrangement are the two essential 1D pilot insertion, as shown in Figure 2.

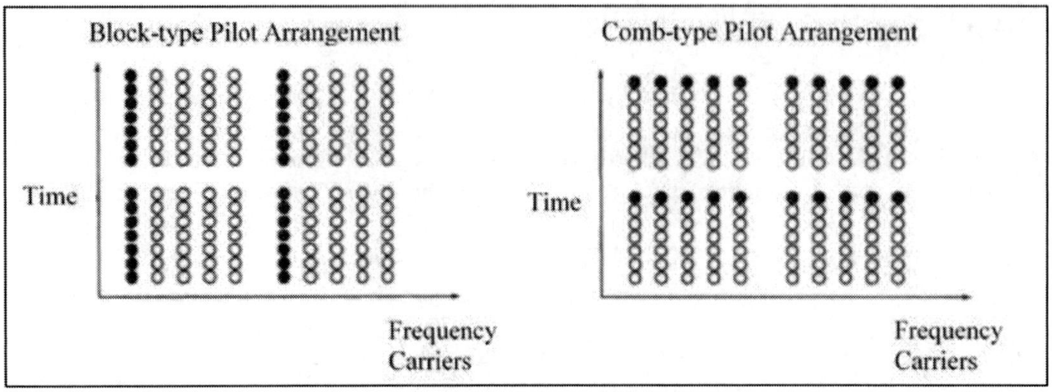

Figure 2: Pilot signal arrangement [4].

4. Block-type pilot arrangement channel estimation.

Figure 2 shows the pilot symbols are transmitted periodically in a block-type pilot arrangement. Entire subcarriers for dedicated symbols used as pilots, and it is required to estimate channel conditions. Information from the estimated channel condition process by the receiver used to interpret the received data for every block until the next block arrived together with the new pilot symbol. Commonly in the block-type pilot arrangement, MMSE, LMMSE, or LS algorithm is the recognized estimation techniques.

4.1 Minimum Mean Square Error, MMSE.

The channel impulse response, h estimated using the MMSE is given by [2,8–10]:

$$H_{MMSE} = FR_{hY}R_{YY}^{-1}Y \tag{7}$$

R_{hY} is the cross-covariance matrix between h and Y, while R_{YY} is the auto-covariance matrix of Y. The R_{hh} represents the auto-covariance matrix of h. The time-domain channel vector, h has a Gaussian distribution, and it is uncorrelated with the channel noise, W. σ^2 represents the noise variance $of\ E\{|W(k)|^2\}$.

4.2 Linear Minimum Mean Square, LMMSE.

AWGN channel added the noise variance σ^2 to receive signals. The LMMSE estimation of the channel vector g is given by [11–13]:

$$g = \Gamma_{gy}\Gamma_{yy}^{-1}y \tag{8}$$

Γ_{gy} is the cross-covariance matrix between g and y, while Γ_{yy} is the auto-covariance matrix of y. Covariance matrices should be positive definite to archive a unique minimum MSE.

4.3 Least Square, LS.

[4,14,15] discussed the LS estimation by:

$$H_{LS} = X^{-1}Y \tag{9}$$

(9) minimizes the $(Y - XFh)^H(Y - XFh)$. The channel estimation for each subcarrier inside the block can be updated using the decision feedback equalizer because the effect of fading is slow.

5. Comb-type pilot arrangement channel estimation.

Estimation of pilot subcarriers based on LS is given by [4]:

$$H_e = \frac{Y_p}{X_p}, \quad k = 0,1,2,3 \ldots N_p - 1 \tag{10}$$

$Y_p(k)$ and $X_p(k)$ are output and input at the kth pilot subcarrier, respectively. Either MMSE and LS estimation techniques have their advantages. For example, MMSE is better to compromise complexity; meanwhile, LS estimation is susceptible to ICI, and noise.

5.1 Interpolation techniques

Subsequently, the pilot signal estimation of the channel transfer functions, the channel response of data signals interpolated concerning the adjacent pilot. The data carrier, k, use linear interpolation with the value of $mL < k < m + 1)L$. Higher-order polynomial interpolation performed better in obtaining channel response compared to standard linear interpolation hypothetically.

6. Pilot spacing controller

Better BER performance achieved by applying a dense pilot mode in OFDM systems. Smaller pilot spacing leads to a more substantial complexity burden. However, utilized a sparse pilot-symbol in the OFDM system, reverse the phenomena with the dense pilot. Considered an unknown Doppler environment with the vary channel taps randomly. Standard deviation, σ_p of the received pilot signal for every channel tap obtained by equation (11).

$$\sigma_p = \sqrt{\frac{\sum_{i=1}^{k}\left(Y_{p_i} - \overline{Y_{p_i}}\right)^2}{k-1}} \tag{11}$$

$\overline{Y_{p_i}}$ is, the arithmetic mean of the Y_{p_i} and directly corresponding to the variability of the channel taps. Different channels represented by different standard deviations consequently.

In this paper, the locations of the pilots assumed known at the receiver. Information for pilot locations obtains via the control messages. It is replicated the same pilot structure as that at the transmitter. [16] proposed the number of pilot carriers varied in the OFDM symbol. The aim is to avoid the estimation of the channel on the same pilots' number. It could be unnecessary for high SNR environment. Therefore, the designed pilot controller added to the conventional OFDM systems to control the pilot interval. It is the focus of a higher bandwidth efficiency with tolerable quality channel estimation.

7. Simulation specification

A per objective is to determine channel estimation performance, a perfect synchronization assumed between transmitter and receiver. The guard interval is chosen to be large enough compared to the maximum delay spread towards eliminating the ISI. The Rayleigh fading channel is selected as the channel model to assess the fading effect.

Parameters such as channel characteristics, modulation techniques, interpolation methods, and estimation approaches affected the BER performance on channel estimation. Table 1 shows the simulation OFDM system parameters for block-type.

Table 1. Simulation parameter of block-type.

Parameter	Specification
Bandwidth	7.5kHz
Channel model	Rayleigh fading
FFT size	128, 256, 1024, 2048
Guard interval	256
Guard type	Cyclic extension
Number of active carriers	256
Pilot Ratio	1/8

While table 2 shows simulation parameters detail for comb-type pilot technique. At pilot frequencies, the LS channel algorithm applied. Performance observed for corresponding SNR values.

Table 2. Simulation parameter of comb-type.

Parameter	Specification
Channel model	Rician K=11dB
Doppler frequency	70Hz
Guard Interval	1/32 from symbol period
IFFT, FFT size	512
Modulation type	Binary PSK
Number of subcarriers	512
Pilot ratio	1/8

8. Results

Figure 3 shows the simulation performance of basic OFDM with AWGN and Rayleigh fading.

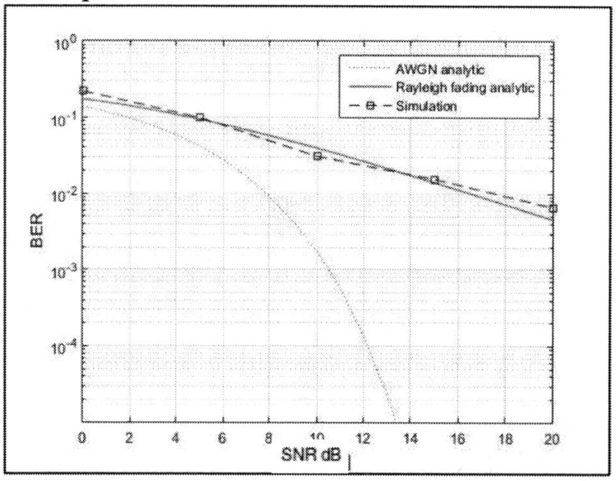

Figure 3. BER vs. SNR for AWGN and Rayleigh fading channel.

In the block-type channel estimation algorithm, all blocks contain a fixed number of symbols, and for the first symbol of each block, pilots sent in all subcarriers. Channel estimation is executed by using LS, MMSE, and LMMSE estimation algorithm. Performance analysis of block-type channel estimation shown in figure 4.

IOP Conf. Series: Materials Science and Engineering **767** (2020) 012045 doi:10.1088/1757-899X/767/1/012045

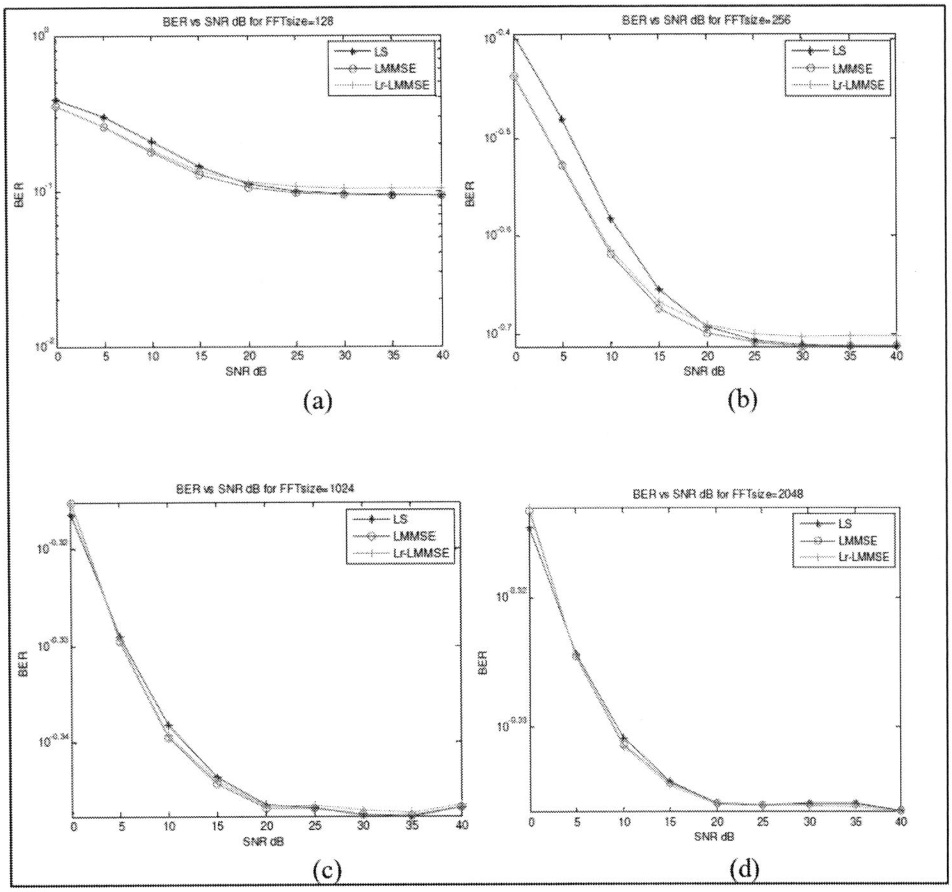

Figure 4. BER vs. SNR.

It has noted from figure 4 (a) and (b) that for FFT size, less than 1024 LS algorithm performed less as compared to LMMSE for SNR less than $25dB$. However, for higher FFT size (1024 and 2048), figure 4 (c) and (d) shown there is no significant difference in terms of their performance up to $40dB$ SNR value. Less pilot spacing is used purposely to improve channel estimation. However, it may cause the data rate reduction or bandwidth efficiency. Increasing the SNR will not improving the performance if the number of the pilot signal is comparatively undersized.

Figure 5 (a) shows the performance of the fixed pilot estimation with the number of pilots fixed at 8 and variable pilot estimation ranging from 2 to 8. The variable pilot controlled at the proposed pilot spacing controller block based on performance BER standard deviation. It shows that the improvement of $0.02dB$ compared to fixed pilot estimation across 0: $30dB$ SNR range. The total average number of pilots for every SNR range for fixed and variable is 80 and 54.59, respectively. It is shown the variable pilot spacing process at the proposed pilot spacing controller able to reduce the number of pilot signals which is known do not bring any data. Although the number of pilots varies according to BER standard deviation, the proposed technique increases the estimation performance.

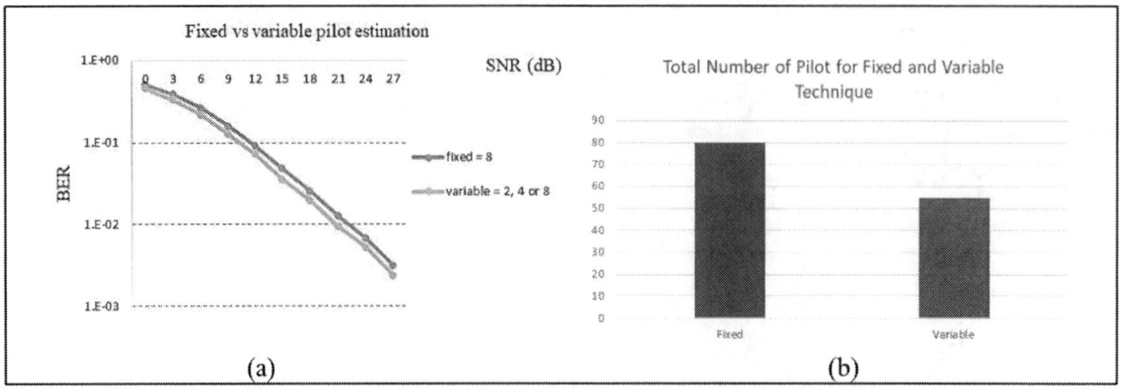

Figure 5. Performance of estimation and the total number of pilot allocation for fixed pilot and variable pilot.

9. Conclusions

A variable pilot spacing using the proposed pilot spacing controller approach has presented in this paper. In the case of block pilot insertion arrangement, LS, MMSE, and LMMSE used. It shows that the number of pilots decreased with some improvement in BER and bandwidth. This finding can stimulate the environment with high SNR and has a good line of sight between transmitter and receiver. Proposed method tracks channel variations better than the typical case with fixed pilots in terms of channel estimation accuracy also in terms of channel efficiency.

References

[1] Zaidi AA, Baldemair R, Tullberg H, Bjorkegren H, Sundstrom L, Medbo J, et al. Waveform and Numerology to Support 5G Services and Requirements. IEEE Commun Mag. 2016;54(11):90–8.

[2] Li YG, Member S. Pilot-Symbol-Aided Channel Estimation for OFDM in Wireless Systems. 2000;49(4):1207–15.

[3] Chen S, Sugiura S, Hanzo L. Semi-Blind Joint Channel Estimation and Data Detection for Space-Time Shift Keying Systems. IEEE Signal Process Lett [Internet]. 2010 Dec;17(12):993–6. Available from: http://ieeexplore.ieee.org/lpdocs/epic03/wrapper.htm?arnumber=5594618

[4] Coleri S, Ergen M, Puri A, Bahai A. Channel estimation techniques based on pilot arrangement in OFDM systems. IEEE Trans Broadcast. 2002;48(3):223–9.

[5] Zhou X, Wang C, Tang R, Zhang M. Channel estimation based on statistical frames and confidence level in OFDM systems. Appl Sci. 2018;8(9).

[6] Nee R Van, Awater G, Technologies L, Labs B, Webster M, Halford KW. New High-Rate Wireless LAN Standards. Time. 1999;

[7] Fernandez-Getino Garcia MJ, Zazo S, Paez-Borrallo JM. Pilot patterns for channel estimation in OFDM. Electron Lett. 2000;36(12):1049–50.

[8] Zhang L, He C, Mao J, Ijaz A, Xiao P. Channel estimation and optimal pilot signals for universal filtered multi-carrier (UFMC) Systems. IEEE Int Symp Pers Indoor Mob Radio Commun PIMRC. 2018;2017-Octob:1–6.

[9] Lee H, Jeon Y-S, Do H, Lee N. Concatenated MMSE Estimation for Quantized OFDM Systems. ICC 2019 - 2019 IEEE Int Conf Commun. 2019;1–6.

[10] He J, Li Z, Shi J, Zhang R, Xiang C, Chen L. A novel pilot pattern channel estimation scheme with MMSE-DFT for MIMO-OFDM VLC system. Opt InfoBase Conf Pap. 2017;Part F83-ACPC 2017(2):1–3.

[11] Xin R, Ni Z, Kuang L, Jiang C, Wu S. Low-complexity joint channel estimation and symbol detection for OFDMA systems. China Commun. 2019;16(7):49–60.

[12] Ehsanfar S, Matthé M, Chafii M, Fettweis GP. Pilot- and CP-Aided Channel Estimation in MIMO Non-Orthogonal Multi-Carriers. IEEE Trans Wirel Commun. 2019;18(1):650–64.
[13] Hsieh MH, Wei CH. Channel estimation for OFDM systems based on comb-type pilot arrangement in frequency selective fading channels. IEEE Trans Consum Electron. 1998;44(1):217–25.
[14] Wu D, Fan W. A New Channel Estimation Method Based on Pilot-aided and Local Adaptive Least Squares Support Vector Regression in Software Radio OFDM System. IJCAI Int Jt Conf Artif Intell. 2009;(1):349–52.
[15] Akbarpour-Kasgari A, Ardebilipour M. Pilot allocation approaches for channel estimation in MIMO relay networks. IET Commun. 2018;12(16):2030–7.
[16] Hassan MSA, Ahmad RB. Variable pilot channels estimation based on blocktype and comb-type pilot arrangement in OFDM system. 2016 3rd Int Conf Electron Des ICED 2016. 2017;99–102.

Development of a synchronverter for a grid connected photovoltaic system

M J Y Liaw [1] and C R Sarimuthu[1]

[1] Electrical and Computer Systems Engineering Department, Monash University Malaysia, Jalan Lagoon Selatan, Bandar Sunway, 47500 Subang Jaya, Selangor, Malaysia.

mjlia6@student.monash.edu

Abstract. As global population has grown exponentially since the past decade, the demand for a sustainable framework for cleaner, greener and stable energy production is becoming more important for a holistic society. Among multiple renewable energy sources, solar power is highly favoured to dominate the market due to improvements in solar energy conversion efficiency and reduction in production cost. Thus, many transmission grids will observe a high penetration of inverters as an interface between solar panels and the AC transmission grid. Consequently, inverter-dominated transmission grid will face issues with maintaining and achieving overall grid stability due to the characteristics of inverters which are composed of mainly solid state devices (SCR, MOSFETS or IGBTs). This is due mainly to the lack of rotational inertia, which is typically contributed and maintained by traditional synchronous generators when the stability of the transmission grid is disturbed. For this reason, conventional inverters can potentially be exploited to mimic the characteristics of synchronous generators to stabilize the transmission grid. Thus, this project is motivated to design a 'smart' inverter or synchronverter which can operate on islanded mode while integrated with solar panels. This project focuses on developing and prototyping a working three-phase inverter model using a Raspberry Pi microcontroller with Simulink support. Bipolar Sinusoidal Pulse Width Modulation (SPWM) technique is used to create the sinusoidal output waveform with a fixed frequency of 50Hz. The output waveform will also have a Total Harmonic Distortion (THD) of less than 5%. A Proportional-Integral (PI) controller is also implemented to ensure that amplitude and frequency of output sinusoidal remains constant. In this work, simulation of a three-phase inverter is done using Simulink (MATLAB) to validate the performance of the inverter (hardware prototype). The developed hardware prototype is experimentally tested by integrating with 1000W solar panels.

1. Introduction

The traditional transmission grid is commonly powered by synchronous generators that control the grid parameters and damps any unwanted oscillations. Recently, the penetration of renewable power sources, such as solar power, are given a higher priority due to cleaner energy production and cheaper photovoltaic (PV) solar panel manufacturing cost. But, solar panels generate Direct Current (DC) Power which incompatible with an Alternating Current (AC) grid, thus an inverter is required to integrate solar panels with transmission grids. Inverters differ from traditional synchronous generators because inverters are composed of solid state components, which means inverters do not possess mechanical inertial characteristics as exhibited by a rotational synchronous generators. This leads to higher frequency instability as the transmission grids transition from a rotational-dominated system to an inverter dominated system. Thus, there is a strong need to develop a 'smart' inverter that is able to provide 'virtual inertia' for the grid during critical situations, synchronize with the grid without a PLL and predict the optimal entry points to inject real and reactive power.

Synchronverters are inverters that mimic synchronous generators. The basic groundwork that models the mechanical aspect of a synchronous generator has been developed [1]. Their work has proven that synchronization of a synchronverter is successful through simulation and experimentation with a low-voltage prototype. Their work shows that grid synchronization occurs smoothly, without any transients for grid frequencies that are lower and higher than the grid frequency (i.e. 50Hz in many countries). Synchronverters also exhibits pre-set real and reactive power tracking, with a response time of less than 1s. Their results show a sudden spike in frequency but stabilizes immediately after. P drooping and Q dropping methods also show spikes, which stabilizes immediately after. Furthermore, Weiss' and Zhong's work was adapted to develop PV Synchronverter that integrates with the grid [2]. Their work includes a MPPT to generate highest power output which is conditioned to be suitable for grid connection. But, their work was proven with MATLAB simulation, but lack experimental results. [3]

An improvement on the traditional Synchronverter technology is to improve the Synchronverter controller to include virtual impedance. Virtual filter inductance was increased by modifying the control algorithm to improve stability of the system. A virtual capacitor in series with the inverter outputs is also introduced to shunt DC components from entering the grid. Furthermore, an improved field current controller, which uses an anti-windup mechanism and the new formula for reactive power will improve stability of grid-connected synchronverter, but this improvement operates under a few assumptions. A comparison of performance of inductor and resistance values of 5 times greater than original and 25 times greater shows that a higher value will remove high-frequency fluctuations by 3 times when there is a dip in grid voltage and twice when there is a voltage drop across a phase. [4]

A traditional inverter or Synchronverter relies on a Phase Locked Loop (PLL) synchronizing unit to connect to the main gird. Improvement have been made to remove PLL units which allows Synchronverters to self-synchronise with the gird. This improves performance of the synchronverter and reduces computational burden on the control system. Experimental results shows that the synchronizing unit improves performance of real and reactive power control by 83% and 70% respectively. Frequency tracking of grids before and after connections also improved by 65%. Lastly, code size of control program reduced by 71.1% and average execution time is reduced by 19.2%.[5]

Further improvements have been introduced to develop a Synchronverter that restricts its frequency and voltage to a given range. This improvement is parameterized and ensures that the Synchronverter is always stable and converges to a unique equilibrium, provided power exchange at the terminal is kept within this area. The proposed work assumes a stiff grid to derive the existence condition of an equilibrium point. Simulations have shown that frequency and voltage remains in tight bounds even with the effect of grid voltage variation. [6]

Improvements on Synchronverters has been largely focused on control algorithm of the original strategy. The current system also heavily depends on the current status of the main grid for connection conditions. With the improvements on data collection units, a database of grid conditions can be generated and stored over long periods of time. Thus, there is a potential to study and develop prediction models to improve Synchronverter operations. Artificial Intelligence (AI) is a useful tool to study and develop models which is able to evolve the current Synchronverter system to better predict faults conditions and failure. But, there has been minimal research on applying AI control methods to develop and evolve the current Synchronverter control. With a proper AI control method, the Synchronverter will be better able to predict changes in the gird and respond appropriately.

Thus, we are motivated to develop a low-cost, low-voltage, 3-phase inverter that is able to integrate with the grid and control real and reactive power flow from DC sources. The inverter will be controlled by a Raspberry Pi with MATLAB Simulink, and be integrated with solar panels. The output of the inverter will be a 50Hz sine wave with a total harmonic distortion (THD) of less than 5%, voltage ripple of less than 10% and current ripple of less than 5%. Response time of closed loop system will be less than 1 second for any sudden changes in reference wave.

The rest of the paper will be organized as follows. Section 2 will review the working principles of Synchronverter control Section 3 will discuss the methodology of this report. The results will be presented and discussed in Section 4 and the report is concluded in Section 5.

2. Working Principle of a Synchronverter

Synchronverter is an inverter that mimics the inherit characteristics of a synchronous generator. The Synchronverter will also inherit the advantages and disadvantages of a synchronous generator, but a Synchronverter can dynamically select physical parameters, such as inertia, friction coefficient, field inductance and mutual inductances. This means that the synchronverter can mimic and enhance any mechanical losses experienced by a normal synchronous generator. Most synchronverter technology have similar designs and operations, based on the work of Zhong and Weiss [1]. But, there have been improvements on the original designs over the years. For example, self-synchronisation without PLLs [7], improved stability with virtual capacitors, virtual inductors and anti-windup [4] and introducing bounded voltage and frequency ranges to improve stability [6]

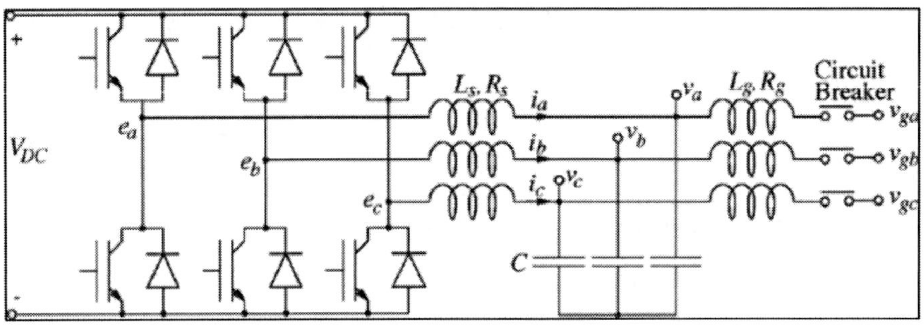

Figure 1. Three-Phase Inverter Layout.

The power part or the electronic components of a synchronverter is shown in figure 1. The schematic diagram shows a three-phase inverter, with a switching circuit that is controlled and composed of power transistors (e.g. MOSFETs) which is filtered with a LCL filter and connected to the gird via a circuit breaker. PWM signals which are generated by control unit of the synchronverter controls the operational state of the transistors.

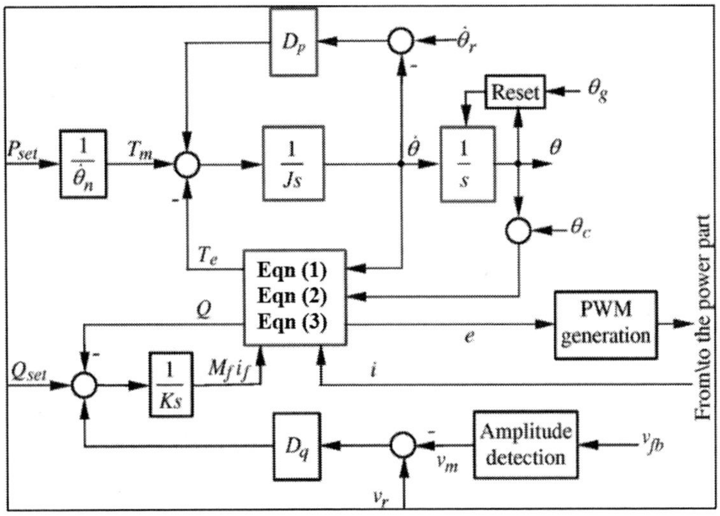

Figure 2. Synchronverter Control Scheme Layout.

$$T_e = -M_f i_f \langle i, \frac{\delta}{\delta\theta}\widetilde{\cos}\theta\rangle = M_f i_f \langle i, \frac{\delta}{\delta\theta}\widetilde{\sin}\theta\rangle \quad (1)$$

$$e = \dot{\theta} M_f i_f \widetilde{\sin}\theta \quad (2)$$

$$P = \dot{\theta} M_f i_f \langle i, \widetilde{\sin}\theta\rangle, Q = -\dot{\theta} M_f i_f \langle i, \widetilde{\cos}\theta\rangle \quad (3)$$

Constants in equation (1), (2) and (3) are defined below:

$$i = Stator\ Phase\ Current = \begin{bmatrix} i_a \\ i_b \\ i_c \end{bmatrix}$$

$$M_f = Mutual\ Inductance, T_e = Elelectromagnetic\ Torque$$

$$\widetilde{\cos}\,\theta = \begin{bmatrix} \cos\theta \\ \cos(\theta - \frac{2\pi}{3}) \\ \cos(\theta - \frac{2\pi}{3} \end{bmatrix}, \widetilde{\sin}\,\theta = \begin{bmatrix} \sin\theta \\ \sin(\theta - \frac{2\pi}{3}) \\ \sin(\theta - \frac{2\pi}{3} \end{bmatrix}$$

$$e = Electromotive\ Force = \begin{bmatrix} e_a \\ e_b \\ e_c \end{bmatrix}$$

The Synchronverter controller scheme is shown in figure 2. The control scheme incorporates both frequency drooping and voltage drooping techniques to regulate real and reactive power respectively. The proposed control scheme also relies on a PLL for the amplitude detector as shown in **Figure 2.** A PLL is still required to measure grid parameters which is transmitted to the Amplitude Detection block as shown in figure 2.

Real power reference are defined by P_{set}, which varies in proportion to the load. Frequency droop method attempts to reduce error between virtual angular speed, $\dot{\theta}$ with nominal angular frequency of grid, θ_r (which would normally be equal to nominal angular frequency of the grid $\dot{\theta}_r$). The error is multiplied with the damping factor, D_p, and summed with the difference between mechanical torque, T_m, and electrical torque, T_e. $\dot{\theta}$ is calculated and used in equation (1-3) to generate PWM signals.

Reactive power reference, defined by Q_{set}, also has similar regulation methods with voltage drooping techniques. The difference between reference voltage, V_r, and amplitude, V_m, of the feedback voltage, V_{fb} (which would normally be grid voltage), is amplified with the voltage drooping coefficient, D_q. This value is subtracted from difference between Q_{set} and reactive power, Q

IOP Conf. Series: Materials Science and Engineering **767** (2020) 012046 doi:10.1088/1757-899X/767/1/012046

(calculated from equation (3), is passed to an integrator to generate armature reaction, $M_f i_f$, which is then passed to equations (1-3) to generate PWM signals.

Virtual inertia is introduced by altering D_q, which is designed to cater to grid requirements [8]. Real and reactive power can be changed over time to meet the changes in load requirements. Output voltage and frequency of the synchronverter can also be adjusted to meet load requirements or for protection setting purposes.

3. Methodology

The closed-loop setup for a three-phase inverter is shown in figure 3. The solver used was ode23tb with a max step size of 1μs. The DC input voltage is fixed at 30V.

PI controller gain is tuned using the Zigler-Nichols method [9]. The initial proportional gain was selected and adjusted until the output amplitude matches the reference. This allows us to adjust the integral gain to ensure that the inverter output matches the reference with different load types.

The switching frequency was selected given that a higher switching frequency will push harmonic frequencies to a higher value. This allow design of the low-pass filter circuit to utilise capacitor and inductors with smaller capacity. A sinusoidal signal with a frequency of 50Hz is used as the reference voltage and the triangular wave frequency was set to be 200 times the reference signal frequency which is 10kHz. The cut-off frequency which is the product of the switching frequency and signal frequency approximates to 707Hz is used to design the low pass filter. The cut-off frequency is much higher than the reference signal frequency, which makes it easier to filter higher frequency components.

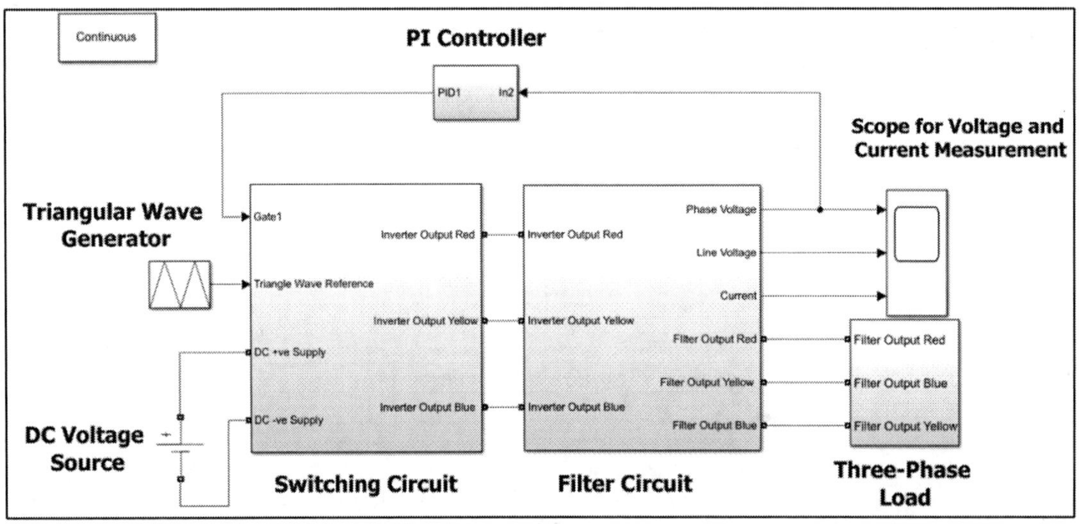

Figure 3. Simulink Layout for Closed-Loop Three-Phase Inverter Simulation.

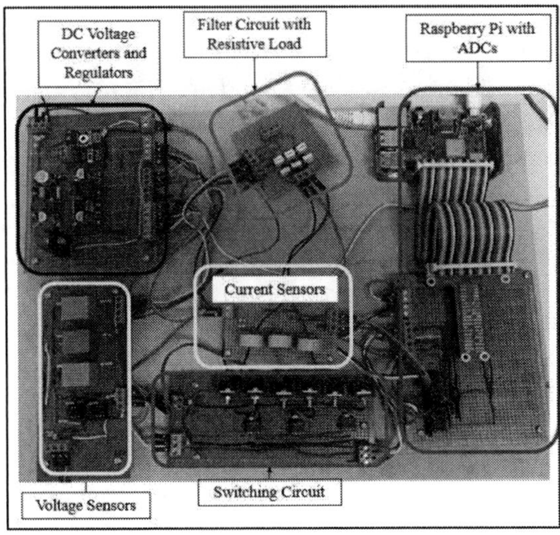

Figure 4. Three-Phase Inverter Hardware Prototype.

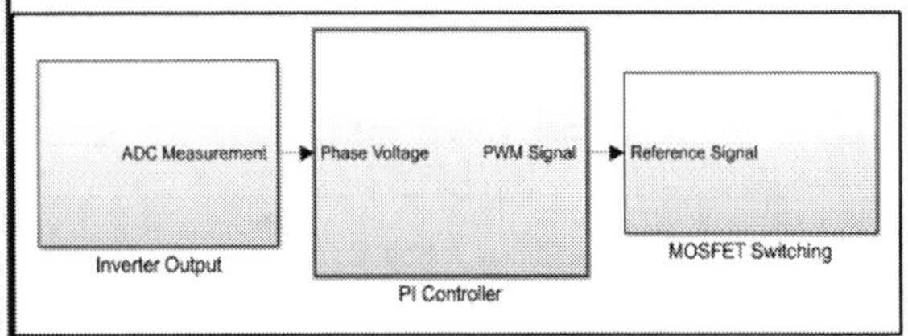

Figure 5. Simulink Control Layout for Inverter Prototype.

The inverter prototype, shown in figure 4, which is powered by a DC power supply supplying 31.3V. The Raspberry Pi microcontroller measure voltage and current from the sensors and generate a PWM signal which is transmistted to the gate drivers. The gate drivers than determine the on-off state of the Power MOSFETs. The Simulink layout which generates the PWM signal is shown in figure 5. Since Raspberry Pi does not have on-board Analogue-to-Digital Converter (ADC), an external device with Serial Peripheral Interface (SPI) is utilised.

4. Results

This section presents measurements from both simulation results and Synchronverter hardware prototype. Results shown presents the Synchronverter output for different load types. **Figure 6** shows simulation results of the output from a three-phase inverter based on the Simulink layout shown in **Figure 3**.

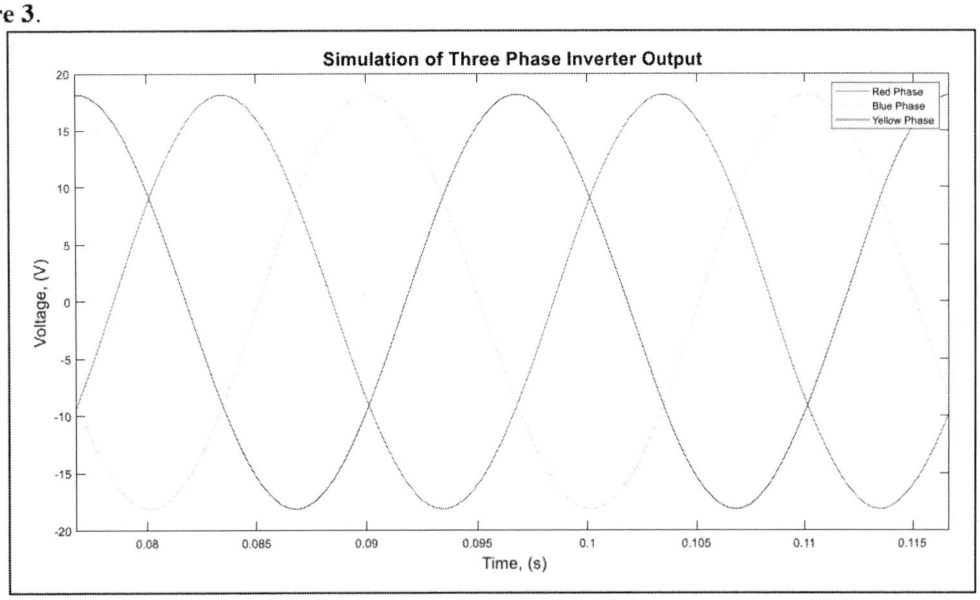

Figure 6. Simulation Results of Three-Phase Inverter Output.

It is observed that phase voltage waveforms are 120° shifted from each other in figure 6. Peak voltages are maintained for 3 cycles, with minimal distortion to their signals. Voltage ripple is kept to a minimum and THD values are approximately 2%, which meets our targeted requirements.

As shown in **Figure 7**, the output voltage is similar to the reference voltage under different load conditions. There is minimal distortion shown for resistive loads, capacitive loads and resistive and reactive loads. But, purely inductive loads shows slight distortion due to the filter circuit. The proportional gain is slightly altered to maintain output stability as shown in **Table 1**.

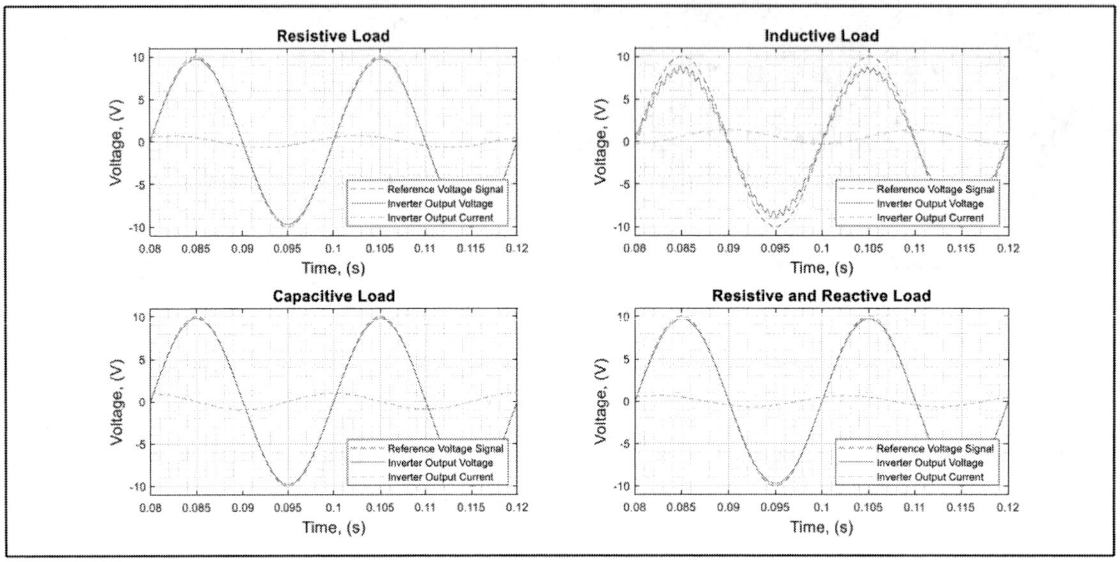

Figure 7. Simulation Results for Various Load Conditions.

Table 1. Power Load Requirement for all Simulation Cases.

Scenario	Real Power Load (W)	Reactive Power Load (VAR)	Proportional Gain	Integral Gain
Resistive Load	5	0	2	0.01
Inductive Load	0	-5	0.5	0.01
Capacitive Load	0	5	2	0.01
Resistive and Reactive Loads	2.5	2.5	2	0.01

Next, the inverter prototype is developed as shown in **Figure 4** and the output is probed and recorded with an oscilloscope. The results shown in **Figure 8** is replotted via MATLAB and analysed.

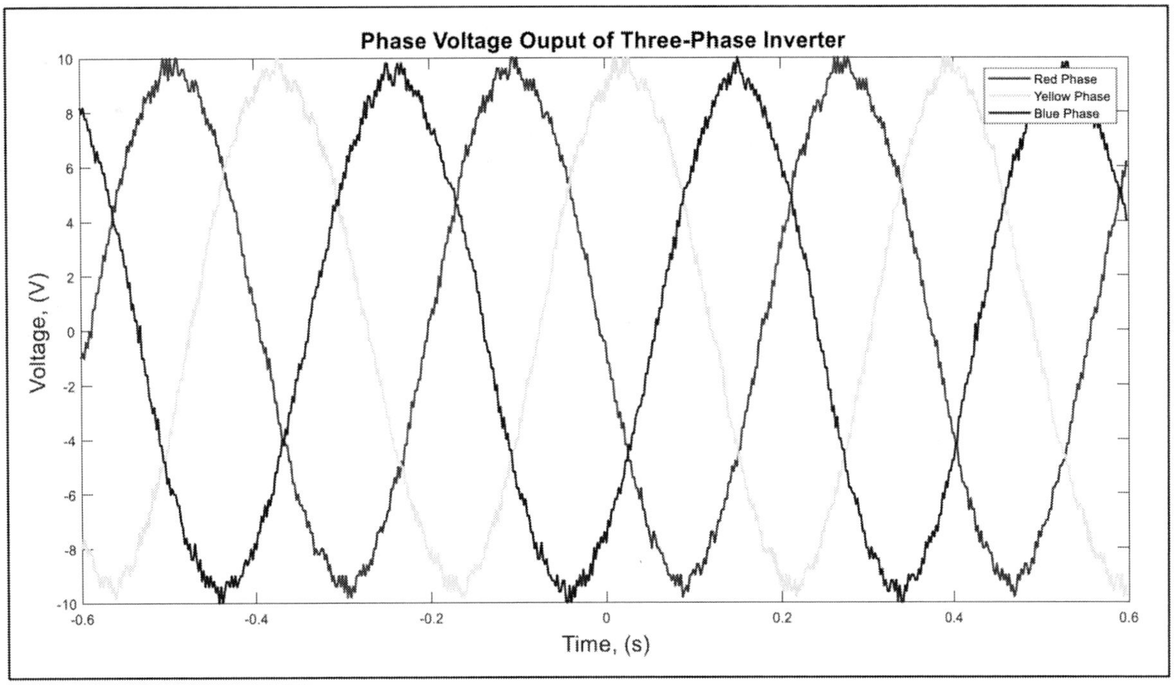

Figure 8. Inverter Operation in Real-time.

The results shown in figure 8 shows the three-phase voltage output from the inverter hardware. The results have slight distortion, while the signal waveform is maintained continuously. The amplitude of the signal reaches its target with the PI controller, but the signal frequency has dropped to 2.5Hz, which is a huge drop from the expected 50Hz.

The THD for the signal waveform shown in **Figure 8** is -3.0977dB or 62.3897%, which is much higher than the targeted THD tolerance. The high THD value is due to the low frequency of the output signal. Thus, THD calculations includes higher order harmonics multiples of the fundamental frequency (i.e. harmonics at 5Hz, 7.5Hz, 10Hz, etc.) . These higher order harmonics may also contain low-frequency noise which deterioates signal quality, thus resulting in a higher THD value.

A limitation on the current setup with Simulink and Raspberry Pi is the long computational time required to generate the PWM. This is mostly due to Simulink attempting to improve calcualtion accuracy and thus results in slower operational output. Future works will include attempting to increase computatinal time which allows for higher switching frequencies and better inverter signal output. Reducing computational time will also better integrate the Synchronverter algorithm through Simulink.

5. Conclusion

This paper has presented a simulation of a three-phase inverter and shown that a hardware prototype is feasible to be developed with a microcontroller and Simulink. Therefore, Synchronverter control can be easily established with Simulink to control the output of the inverter which can be interfaced with an islanded grid.

References

[1] Q Zhong and G Weiss, "Synchronverters: Inverters That Mimic Synchronous Generators," *IEEE Transactions on Industrial Electronics,* vol. 58, no. 4, pp. 1259-1267, 2011. [Accessed March 10th, 2019]

IOP Conf. Series: Materials Science and Engineering **767** (2020) 012046 doi:10.1088/1757-899X/767/1/012046

[2] P Chandrakar, S Saha, P Das, A Singh, and S Debbarma, "Grid Integration of PV System Using Synchronverter," in *2018 International Conference on Computation of Power, Energy, Information and Communication (ICCPEIC)*, 2018, pp. 237-242. [Accessed May 11th, 2019]

[3] E Rajan and S Amrutha, "Synchronverter based HVDC transmission," in *2017 Innovations in Power and Advanced Computing Technologies (i-PACT)*, 2017, pp. 1-6. [Accessed May 12th, 2019]

[4] V Natarajan and G Weiss, "Synchronverters With Better Stability Due to Virtual Inductors, Virtual Capacitors, and Anti-Windup," *IEEE Transactions on Industrial Electronics,* vol. 64, no. 7, pp. 5994-6004, 2017. [Accessed May 13th, 2019]

[5] Q Zhong, P Nguyen, Z Ma, and W Sheng, "Self-Synchronized Synchronverters: Inverters Without a Dedicated Synchronization Unit" *IEEE Transactions on Power Electronics,* vol. 29, no. 2, pp. 617-630, 2014. [Accessed May 14th, 2019]

[6] Q Zhong, G C Konstantopoulos, B Ren, and M Krstic, "Improved Synchronverters with Bounded Frequency and Voltage for Smart Grid Integration," *IEEE Transactions on Smart Grid,* vol. 9, no. 2, pp. 786-796, 2018. [Accessed May 14th, 2019]

[7] F L Luo, Knovel and H Ye, Eds. *Advanced DC/AC inverters : applications in renewable energy*. Boca Raton, FL. Boca Raton, Fla.: Boca Raton, FL : Taylor & Francis, 2013. [Accessed May 28th, 2019]

[8] S Wang, R Qi, and Y Li, "Fuzzy Control Scheme of Virtual Inertia for Synchronverter in Micro-Grid," in *2018 21st International Conference on Electrical Machines and Systems (ICEMS)*, 2018, pp. 2028-2032. [Accessed April 1st , 2019]

[9] M. a. Chidambaram, N. a. Saxena and SpringerLink, Eds. *Relay tuning of PID controllers : for unstable MIMO processes* (Relay tuning of proportional integral derivative controllers : for unstable multiple input multiple output processes). Singapore : Springer, 2018. [Accessed October 12th, 2019]

Modelling the statistical dependence of rainfall variables using copula: A case study of Terengganu state

M F Fauzi [1], U F Abdul Rauf [2*], R Khan [3] and N A A Jamaludin [4]

[1] Department of Defence Science, Faculty of Technology and Defence Science, National Defence University of Malaysia, 57000 Sg. Besi, Kuala Lumpur, Malaysia
[2] Department of Mathematics, Centre for Defence Foundation Studies, National Defence University of Malaysia, 57000 Sg. Besi, Kuala Lumpur, Malaysia
[3] Sypaq System Pty Ltd, 441 St Kilda Rd, Melbourne, Victoria 3004 Australia
[4] Department of Computer Science, Faculty of Technology and Defence Science, National Defence University of Malaysia, 57000 Sg. Besi, Kuala Lumpur, Malaysia

ummul@upnm.edu.my

Abstract. Extreme rainfall can have severe negative impacts possibly resulting in loss of life and property, hence the need for means of predicting the occurrence of extreme rainfall. One method that addresses extreme phenomena is the extreme value theory. In bivariate data, extreme value analysis can be approximated by the distribution of maximum likelihood estimation. This paper discusses the role of the copula in modeling the structure of dependencies between two rainfall variables, namely duration, and severity. Goodness-of-fit approaches are used to resolve whether or not to reject a parametric copula which then determines the best copula to utilise for the data set. In this paper the Akaike information criterion (AIC) is investigated for its ability to choose an appropriate copula model from Archimedean copula models. The analysis shows that the Frank copula is the best model to explain the dependency structure of the two variables discussed.

1. Introduction

Analysis of dependency between two or more variables is an important study in many disciplines such as marketing, clinical trials, economics, and business to name a few. The standard of analysis is usually displayed in the form of a correlation value, which then provides an indicator to researchers about the analyses to be conducted.

This paper presents a method that has received a lot of attention in the last few years in hydrology, namely copula modeling, which is a statistical method intended to model the dependence structure between two or more variables. It is capable of providing more information about dependencies than the Pearson, Kendall or Spearman correlation, as highlighted by several authors, e.g., [1], [2], and [3]. In practice, the precursors often ignore or do not even know the marginal distribution of the variable being analyzed. Pearson correlation is often the easiest and simplest choice for measuring dependencies however it assumes that the variable has a normal distribution.

The role of copula becomes important when one or both variables have an abnormal marginal distribution or have tail dependencies. Many bivariate or multivariate distributions have been developed as alternatives to overcome cases of abnormalities in marginal distributions such as bivariate gamma [4], bivariate exponential distribution [5], bivariate exponential distribution [6], etc. However, these distributions are limited to the same marginal and have a complex structure in its probability density function (including its dependency structure) which is certainly not preferred in practice.

Copula comes with flexibility, where the marginal distribution of variables can be different or even unknown. Copula has been well applied in many fields such as hydrology [7], [8], [9], [10], finance and insurance [11], [12] and many more.

The research herein is a continuation of work by [13] where the authors presented forecasting of shortest and longest-term timescale of the wet event using monthly precipitation index data. In this paper we discuss the application of copula in modeling the dependence structure between two climate variables, namely rainfall duration and rainfall severity. The paper is outlined as follows. Section 2 introduces the methodology used throughout this paper. Sections 3 presents the results of a case study where copula is successfully applied to determine dependencies of these two variables. Finally, Section 4 presents the conclusions.

2. Methodology

It is important to understand how marginal distributions and Akaike information criterion work to fit copula models before conducting any analysis. This is because each distribution function can influence the result differently. This section will describe this methodology in detail.

2.1 Marginal Distributions and Akaike Information Criterion

To fit a copula model to the data the characteristics of precipitation exploited in this study to determine the marginal distribution are rainfall severity and rainfall duration. Four distribution functions have been tested the Gamma, Log-normal, Weibull and Exponential distributions. Maximum likelihood estimation (MLE) method is the standard method used to estimate the parameters of these marginal distributions. The equations of the probability density functions (pdfs) of the four distributions are listed below:

Table 1. Probability density functions of the four distributions.

Distribution	Probability Density Function	
Gamma	$f(x) = \frac{1}{\Gamma(\alpha)\beta^{\alpha}} e^{-\frac{\alpha}{\beta}} x^{\alpha-1}$ for $x > 0$	(1)
Log-normal	$f(x) = \frac{1}{x\sigma\sqrt{2x}} e^{-\frac{(\ln x-\mu)^2}{2\sigma^2}}$ for $x > 0$	(2)
Weibull	$f(x) = \left(\frac{a}{b}\right)\left(\frac{x}{b}\right)^{(a-1)} e^{-\left(\frac{x^a}{b}\right)}$	(3)
Exponential	$f(x) = \lambda e^{-\lambda x}$	(4)

The Akaike information criterion (AIC) was introduced by [14] based on information theory. The goodness of fit of the copulas was evaluated by AIC. AIC provides a relative estimate of the information lost when a given model is used to represent the process that generates the data. The Akaike information criterion is defined in (5):

$$AIC = 2k - 2\ln(L_{max}) \quad (5)$$

Here L_{max} is the maximum value of the likelihood function for the model, k is the number of estimated parameters in the model dependent on the type of univariate marginal distribution and copula parameters.

2.2 Copula

Copula is a joint distribution function of some marginal distribution function hence is used to analyse the dependence of random variables in structures described by the combined function [15]. Generation of the multivariate distribution function from the join univariate distribution function are generally the function of the copula. Copulas are great tools for modelling and simulating correlated random variables that were created by [17]. In this paper, severity (S) and duration (D) are the two rainfall variables with marginal distribution functions represented as $F_S(s)$ and $F_D(d)$. For this work the copula function is modelled as a joint bivariate cumulative distribution function with these two characteristics as is given in Equation (6) below;

$$F_{S,D}(s,d) = C(F_S(s), F_D(d)) \tag{6}$$

and become unique if both $F_S(s)$ and $F_D(d)$ are continuous.

2.3 Archimedean Copula

Archimedean copulas are among the most used in practice and have been used in a wide range of applications because they are easily generated and can capture a wide range of dependence. The Archimedean copula are of the form:

$$C(u,v) = \phi^{-1}(\phi(u) + \phi(v)) \tag{7}$$

where ϕ is generator function over the interval [0,1], continuous, convex, decreasing, with ϕ (1) = 0. There are three Archimedean copulas commonly used in hydrology research namely, the Clayton, Frank, and Gumbel-Hougard copulas. These copulas and their generators are displayed below.

Table 2. Families of bivariate Archimedean copulas.

Archimedean Copula	$C(u,v)$	
Clayton	$C_\theta(u,v) = (u^{-\theta} + v^{-\theta} - 1)^{-\frac{1}{\theta}}$	(8)
Frank	$C_\theta(u,v) = -\theta^{-1}\log\left(1 + \frac{(e^{-\theta_u} - 1)(e^{-\theta_v} - 1)}{e^{-\theta} - 1}\right)$	(9)
Gumbel-Hougaard	$C_\theta(u,v) = exp\left[-(\log u)^\theta + (-(\log v)^\theta)^{\frac{1}{\theta}}\right]$	(10)

The following section will apply the method described above to a real hydrology data set to determine the dependence between rainfall severity and duration to help predict rainfall behavior.

3. Case Study Area: Terengganu State

Terengganu is a state in the East Coast which has the effects of Northeast Monsoon, from November to March every year, and the level of floods that occur is dependent on the intensity and rainfall value that falls at a time during the season. The precipitation time series used in this research consists of three rain-gauge stations across the states of Pahang consists of 45 years of rainfall observations from 1970 to 2014, collected from the Department of Irrigation and Drainage (DID) Ampang, Malaysia, retrieved to monthly time series data for tail dependence analysis. Table 3 presents the geographic locations of the selected rain-gauge stations.

Table 3. Summary of selected station in Terengganu States.

Station	Station Number	Station Name	Latitude	Longitude
ET1	5230041	SK Kuala Telemong	05°12'N	103°01'E
ET2	4334094	SK Kijal	04°19'N	103°29'E
ET3	4434093	SK Kemasek	04°25'N	103°27'E

3.1 Marginal Distributions and Akaike Information Criterion

As previously mentioned in this study rainfall severity and rainfall duration are used to find the marginal distribution before copula parametric estimation methods can be applied to real hydrological data to avoid the misspecification of the marginal distributions. The four types of distributions described in section 2.1 were fitted to monthly rain-gauge station data and the aggregates were examined to identify the most suitable distribution for rainfall duration and severity. The parameters found for each of these marginal distributions are given in table 4.

Table 4. Marginal distributions of parameter estimate for severity and duration.

Characteristics	Station	Estimates						
		Gamma		Log Normal		Weibull		Exponential
		α	$1/\beta$	μ	σ	α	β	Λ
Severity	ET1	1.85	0.46	1.10	0.75	1.33	4.44	0.25
	ET2	2.15	0.57	1.08	0.72	1.50	4.22	0.26
	ET3	1.23	0.25	1.13	0.92	1.06	5.02	0.20
Duration	ET1	2.08	0.75	0.76	0.71	1.43	3.07	0.36
	ET2	2.83	1.05	0.80	0.62	1.76	3.04	0.37
	ET3	1.60	0.48	0.86	0.81	1.24	3.59	0.30

The most suitable marginal distribution is selected based on the AIC goodness of fit test where the lower the AIC value the better the fit. The AIC results for each distribution model are presented in table 5 for all 3 stations (ET1, ET2 and ET3) for both variables (severity ad duration). The results show the Log-normal distribution as having the lowest AIC value hence is the best marginal distribution for the given data set.

Table 5. Goodness of fits test for marginal distributions-AIC.

Characteristics	Station	Marginal Distributions			
		Gamma	Lognormal	Weibull	Exponential
Severity	ET1	142.18	**138.46**	143.96	145.84
	ET2	118.10	**117.16**	119.22	123.24
	ET3	117.20	**112.44**	117.63	115.77
Duration	ET1	117.42	**114.19**	119.29	123.06
	ET2	95.57	**94.96**	96.72	105.50
	ET3	98.21	**94.90**	99.14	99.77

3.2 Copula

The following method is used to estimate the dependence parameter for copula. The calculation of parameters, θ (section 2.2), and the value of the maximum log-likelihood is calculated to determine the best dependency structure model on copula. Copula functions and the parameter space of the copulas selected in the study are presented in table 6. The entries outlined in bold in table 6 show the best model for the given station where best is defined as the largest log-likelihood value. The

Frank copula model was identified as the best copula fit for two out of the three stations. Therefore, it can be concluded that the most appropriate model for describing the dependency structure between the two variables, duration and the severity of the rainfall is from the copula Archimedean family model, Frank.

Table 6. Parameter estimate for fitted copula.

Station	COPULA	MLE	Estimate
ET1	**Clayton**	38.14	4.49
	Frank	31.95	19.41
	Gumbel	35.36	5.61
ET2	Clayton	19.05	3.96
	Frank	26.94	17.96
	Gumbel	22.99	4.01
ET3	Clayton	24.78	7.74
	Frank	27.76	24.37
	Gumbel	25.94	5.70

For future work the behavior of the upper part of the distribution should will be analysed by comparing the upper tail dependence of the observed data with the upper tail dependence provided by each copula model to select the copula that best fits the data.

4. Conclusion

To estimate the best fitted copula dependence parameter, the empirical study was analyzed to provide statistical evidence in choosing which parameter estimation methods that are more accurate and efficient. Log-normal distribution was fitted to each characteristic using the AIC goodness-of-fit method and the Frank copula was found to be the best copula estimate by maximum log-likelihood.

Acknowledgement

The funding for this research is fully supported by Ministry of Higher Education Malaysia under Fundamental Research Grant Schemes (FRGS /1/2015/SG04/UPNM/02/2).

References

[1] Pearson K 1895 *Proc. of the Royal Society of London* **59** 69–71
[2] Kendall M G 1970 *Rank Correlation Methods* (London: Griffin) p 202
[3] Zhang A 2006 *Advanced Analysis of Gene Expression Microarray Data* World Scientific Publishing Company
[4] Moran P A P 1969 *Statistical Inference with Bivariate Gamma Distributions* (Oxford University Press) p 627
[5] Conway D A 1979 Farlie-Gumbel-Morgenstern *Distributions Encyclopedia of Statistical Sciences* **3** 28-31
[6] Gumbel E J 1960 *J. of the American Statistical Association* **55** 698-707
[7] Abdul Rauf U F and Zeephongsekul P 2014 *Theor Appl Climatol* **115** 153-166
[8] Masina M, Lamberti A and Archetti R 2015 *Coastal Engineering* **97** 37-52
[9] Balistrocchi M & Orlandini S, Ranzi R and Bacchi B 2017 *Water Resources Research* **53** 9883-9900
[10] Stamatatou N, Vasiliades L and Loukas A 2018 *Proc. 3rd EWaS Int. Conf.* vol 2 (Greece: Lefkada Island) p 635
[11] Mejdoub H and Ben Arab M 2017 *Asian Development Policy Review* **5** 100-119
[12] Tamraz M 2018 *Mixture copulas and insurance applications* (Cambridge University Press) p 391

[13] Fauzi M F et al. 2017 *Proc. 1st ICoAIMS Int. Conf.* **890** 012159
[14] Akaike H 1974 IEEE **19** 716-723
[15] Nelsen R 1998 *An Introduction to Copulas* 139 (New York:Springer Verlag)
[16] Sklar A 1973 *J. of Information Theory and Automation of the American (Kybernetika)* **9** 449–460

A review of detecting outlier in a circular regression model

I M Ramlee[1], S Ibrahim[1], Leow W Z[2] and M I Yusoff[3]

[1] Institute of Engineering Mathematics, University Malaysia Perlis, Pauh Putra Main Campus, 02600 Arau, Perlis, Malaysia.
[2] School of Electrical System Engineering, University Malaysia Perlis, Pauh Putra Main Campus, 02600 Arau, Perlis, Malaysia.
[3] S2-L1-26, Center of Diploma Studies, University Malaysia Perlis, Unicity Campus Sungai Chuchuh, 02100 Padang Besar (U), Perlis, Malaysia.

intanmastura@studentmail.unimap.edu.my

Abstract. Circular data is very relevant and important application technique in many fields such as physical science, medical science and others. During the last few years, writers have shown a deep interest of outlier detection in the circular regression model. In this case, authors have a tendency to study and explore in detail about the outlier detection in circular regression model. This paper aims to review the outlier detection methods in circular regression model. Here, we concentrate the attention on the methods of identifying outlier in this model. These survey of circular regression models in which many interesting properties and is good enough to detect the occurrence of outlier. Through the survey may highlight the significant of methods to detect outliers in circular regressions model and provide guideline for future work to look into the research gap.

1. Introduction

Statistical analysis is a study the relationship between the independent variable and dependent variable in regression analysis. Research on circular variable in regression model has strong interest since four decade ago. The circular regression is referred to the relationship when both explanatory and response variable are circular. Circular data arise in many different fields such as biology, physics, geology, medical science and others. The real-life problem such the wind direction and the direction of movement of clouds, the arrival of patient (24 hours) in the emergency room in a hospital and others are the example of circular data [1].

Data circular is types of data direction whereas the data measurement take range in degree direction or unit time [2]. The analysis of circular data is the measurement value of data are repeated periodically. As a result, the statistical software have been provided including Axis, DDSTP, Oriana, MATLAB and R/S language for circular data analysis. According to [3], the presentation of circular data in the graphics is important in the analysis of circular data. The graphical form used for circular data such as transit diagram, cycle diagram and rose diagram.

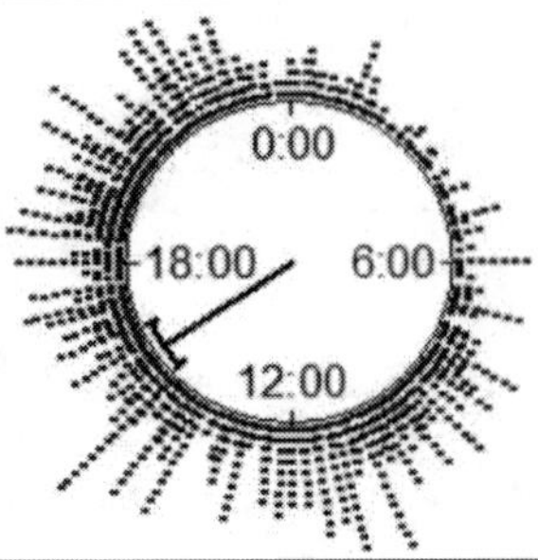

Figure 1. Transmit diagram.

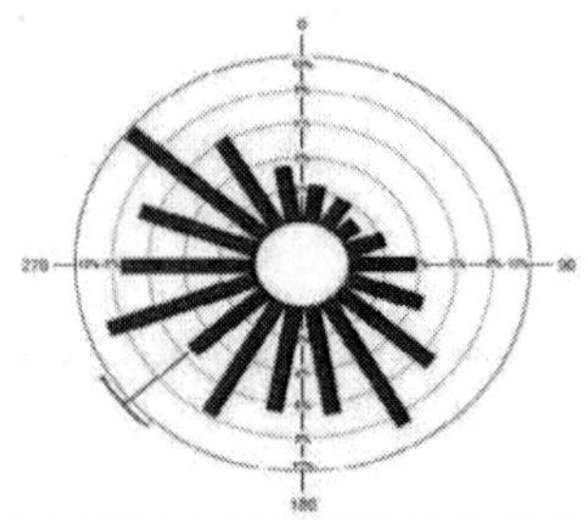

Figure 2. Cycle diagram.

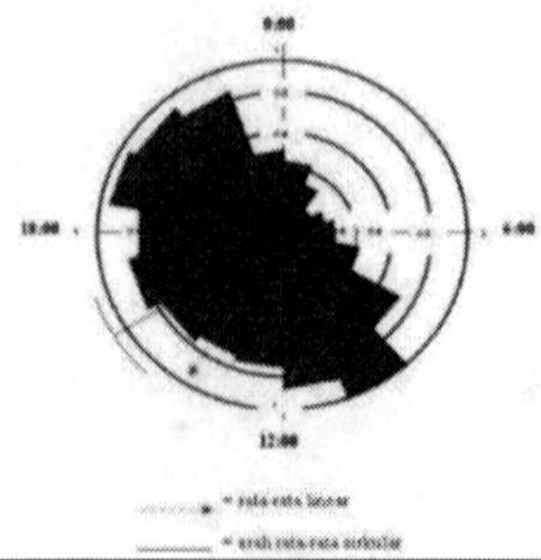

Figure 3: Rose diagram.

In the analysis of circular data, these focused on the descriptive and investigation to develop of special characteristics and descriptive measure of circular data [4][5][6][7]. The descriptive measures of circular namely; the mean direction, the median direction and the sample circular standard deviation.

One of the common problem in circular regression modelling is an outlier. Outlier is defined as extreme values that deviate from other observation on set of data [8]. In other words, an outlier is an observation that diverge from overall pattern on a sample. The discussion of outlier detection has been widely studied by many researchers. It is because their research analysis will give a significant difference in case study respectively. Thus, it is important to detect and access the observation and estimate its impact on the proposed model [9].

This paper is organized into four sections. Firstly, the following section is explained about the types of circular regression model. In section 3, some methods on outlier detection in circular regression model is reviewed accordingly by listing the research goal in their publications. Section 4 discussed the potential of outlier detection methods in their studies. The researchers are highlighted their finding and also identified how much the outlier detection in circular regression models using different methods. Finally, section concludes of the whole research.

2. Circular Regression Model

Circular regression concerns the analysis of relationship when at least one or both explanatory and response variables are circulars. Due to the special characteristics and limited property of circular observation, the special statistical methods are required to analyze this data since the conventional methods of the linear data is not suitable for it. In recent year, many researchers have been working on circular regression models. A circular regression models divided into three types, namely circular-linear regression, linear-circular regression and circular-circular regression.

2.1. Circular – Linear Regression

For circular-linear regression, the circular variable is an independent variable while the linear variable is a dependent variable. When regression model is circular-linear, [10] extended the model which was proposed earlier by [11] in estimating the mean direction of a circular response variable from a vector of linear covariates $\boldsymbol{X} = (x_1,...,x_2)$ expressed as

$$\mu = \mu_0 + \sum_{j=1}^{p} \beta_j x_j, \tag{1}$$

where μ and β_j are anonymous parameters, and x_j is a linear covariate with j=1, 2,..., p.

2.2. Linear – Circular Regression

The linear variable is an independent variable while the circular variable is a dependent variable. An earlier study on circular–linear regression was started from the 70s in the last century. [12] developed parametric models for circular–linear dependency based on a maximum entropy conditional distribution which have been generalized by [13] to encompass possibly asymmetric directional data on manifolds. In 2006, [14] proposed asymmetric circular–linear multivariate regression models which were motivated by the need to predict some environment characteristics based on some circular and linear predictors. On the basis of one of the parametric models proposed in [14], [15] in 2009 proposed a semi-parametric Bayesian hierarchical framework for handling all forms of uncertainty arising in a linear–circular data set. The linear – circular regression model is given by [3]

$$f(\theta;\mu,\kappa)=\frac{1}{2\pi I_0(\kappa)}e^{\kappa\cos(\theta-\mu)} \tag{2}$$

in which $I_0(\kappa)$ is the modified Bessel function of order zero. This distribution is parameterized by the parameter μ and κ which are called mean direction and concentration parameter.

2.3. Circular – Circular Regression

The circular variable is an independent variable while the circular variable is a dependent variable. The following models have been proposed by the researchers for different types of circular-circular regression models.

2.3.1. [5] indicated that the regression curve as a form of Mobius circle transformation. For circular random covariate, $\boldsymbol{X}$, and circular response variable, $\boldsymbol{Y}$, the researchers proposed the regression curve to be define by

$$y=\beta_0\frac{x+\beta_1}{1+\overline{\beta_1}x}\in \tag{3}$$

where β_0 and β_1 are complex parameters with $\beta_0\in\Omega$ and $\beta_1\in\mathbb{C}$.

2.3.2. Later on, [2][4] proposed a circular-circular regression model given by

$$E\left(e^{iv}\mid u\right)=\rho(u)e^{i\mu(u)}=g_1(u)+ig_2(u) \tag{4}$$

where $\mu(u)$ is the conditional mean direction of v and $u,\rho(u)$ is the conditional concentration parameter for a certain periodic functions $g_l(u)$ and $g_2(u)$.

2.3.3 Then, [16] proposed a model of

$$y_i=\alpha+\beta x_i+\varepsilon_i\ \ (\mathrm{mod}2\pi) \tag{5}$$

where ε_i is a circular random error having a von Mises distribution with mean circular 0 and concentration parameter κ. This model is extended the [17] model to analyses the wind and wave direction data.

3. Outlier Detection in Circular Regression Model

Outliers are defined as observations that appear inconsistent with other observation in data sets. Detection and investigation of outliers in circular regression is important since outliers have influence on the inferential results of the model. Currently, many researchers and statisticians from multiple scientific fields draw attention to the circular data analysis and detection of outliers in circular regression model. Therefore, to study the implications concerning the circular data analysis containing the existence of outliers in circular sample and circular regression models, more methods are expected to be available in the future.

Recently, several published works associated with the existence of outliers in the regression for circular variables or circular regression models. A number of ways to detect outliers have been proposed by various researchers. The row deletion method is one of the methods to detect the outliers. The method was suggested to investigate how the deletion of any row affects the residuals, the estimated coefficient, and the estimate covariance structure of the coefficient as well as the predicted value such as *DFBETAc*, *DFFITc* and *COVRATIO*.

3.1. DFFITc Statistic. The *DFFITc* Statistic is a diagnostic meant to show how influential a point is in a statistical circular regression. [18] proposed *DFFITc* define as

$$DFFITc_{j,i} = \frac{\hat{V}_j - \hat{V}_{j(-i)}}{\sqrt{S_{(-i)}{}^2 h_{jj}}} \tag{6}$$

where $\hat{V}_j$ are prediction for the *i*th observation from the full regression model, $S_{(-i)}$ is the standard error without the point *i* while h_{jj} is leverage for the point of $(U'U)^{-1}$.

3.2. DFBETAc Statistic. The *DFBETAc* is a measure of how much an observation has affected the estimate of a regression coefficient, see [9]. The *DFBETAc* statistic is given

$$DFBETAc_{j,i} = \frac{\hat{\lambda}_j - \hat{\lambda}_{j(i)}}{\sqrt{S_{(i)}^2 C_{jj}}} \tag{7}$$

where $\hat{\lambda}_j$ represent the estimated parameter of the full data, $\hat{\lambda}_{j(i)}$ represents the corresponding estimated parameter after the deletion of the *i*th observation, $S_{(i)}$ represents the standard error which is estimated without point *i*, C_{jj} is the *j*th diagonal element of $(U'U)^{-1}$.

3.3. COVRATIOc Statistic. The *COVRATIOc* statistic was introduced by [19], where it is a measure ratio of covariance matrix for set of data that excludes the *j*th row.

$$COVRATIO_{(-j)} = \frac{\left|\hat{\Sigma}_{(-j)}\right|}{\left|\hat{\Sigma}\right|}, \tag{8}$$

where $\hat{\Sigma}$ is the covariance of the residuals of the full data and $\hat{\Sigma}_{(-j)}$ is the corresponding covariance after the *j*th observation is deleted. Any observation with $\left|COVRATIO_{(-j)}\right|$ exceeds the cut-off points will be identified as an outlier. Later on, [20][21] used the same method to detect the outliers in others circular-circular regression model.

3.4. Bootstrapping Methods. The bootstrapping method was introduced by [15], where it is can be used to obtain the approximated variances of parameters estimation. It's also generally used to generate the estimation errors. In analysis, the bootstrapping method is a resampling method process with replacement form the set of data. Recently, the bootstrapping method is a technique to develop the kinds of statistical inference.

3.5. MCEc Statistic. The *MCEc* statistics can be considered as an arithmetic type that is not robust to the existence of outliers while, in the circular regression, it can be used to detect the possible outliers [7] [23].

$$MCEc = 1 - \frac{1}{n}\sum_{j=1}^{n}\cos\left(v_j - \hat{v}_j\right), \tag{9}$$

where n is the size, $\hat{v}_j$ is the fitted values of v_j under equation given by [7].

3.6. DMCEc Statistic. The maximum absolute difference between full-statistic values and reduction of data sets is given by *DMCEc*, where the *j*th observation is identified as an outlier if the *DMCEc* exceeds a pre-specified cut-off point [7] [14].

$$DMCEc = \max_j \left\{ \left| MCEc - MCEc_{(-j)} \right| \right\} \tag{10}$$

3.7. G_a Statistic. The G_a statistic was used to identify outliers in the data sets produced from Wrapped Normal sample of a unit circle [25]. According to [25], he defied the G_a statistic with used the one-step spacing for i^{th} sequenced sample as

$$G_{1i} = \theta_{(i+1)} - \theta_{(i)},\ i = 1,2,\ldots,n$$

and

$$G_{1n} = 2\pi - \theta_{(n)} + \theta_{(1)}\,. \tag{11}$$

3.8. Single-linkage Method. Single-linkage builds of clusters tree where the two clusters are determined to distance between a single pair element with closest to each other. This method is one of several methods to detect the presence outliers. The circular distance based on City-block distance denoted by $d_{C-Satari}$ was proposed by [26] as presented below

$$d_{C-Satari} = \sum_{k=1}^{p} \left(\pi - \left| \pi - \left| \theta_{ik} - \theta_{jk} \right| \right| \right). \tag{12}$$

It is our aim to review and exploit some method for detecting the outliers in circular regression models. The methods are listed in Table 1 together with the name of researchers and their proposed methods are presented.

IOP Conf. Series: Materials Science and Engineering **767** (2020) 012048 doi:10.1088/1757-899X/767/1/012048

Table 1. Summary of Outlier Detection Methods in Circular Regression Models

Author/s	Proposed Method	Model Used	Outlier Detection
Alkasadi et al., 2019 [18]	*DFFITc* statistic	Multiple Circular Regression Model (MCRM)	Eye data 2 Observation: No. 1 and 23
Alkasadi et al., 2018 [9]	*DFBETAc* statistic		Eye data 3 Observations: No. 1, 15 and 23
Alkasadi et al., 2016 [21]	*COVRATIOc* statistic		Eye data 2 Observations: No. 1 and 23
Abuzaid and Allahham, 2015 [27]	Bootstrapping methods	Jammalamadaka and Sarma (JS) Circular Regression Model	Wind direction data 3 Observations: No. 28, 77 and 85
Ibrahim et al., 2013 [19]	*COVRATIO* statistic		Wind direction data 2 Observations: No. 38 and 111
Ibrahim. S., 2013 [7]	*DMCE* statistic		Eye data 1 Observation: No. 2 and 15
Sidik et al., 2019 [25]	G_a statistic	Down and Mardia (DM) Circular Regression Model	Wind direction data February (2 Observations: No. 3 and 9) April (1 Observation: No. 28)
Di et al., 2017 [28]	Single – linkage method - *u*-space - *v*-space		Wind direction data Outlier in v-space: n = 30, n = 120; Outlier in u-space: n = 30, n = 120 κ = 5: SL-Satari, SL-Euc, SL-Satari, SL-Satari /SL-Euc κ = 20: SL-Satari, SL-Euc, SL-Satari, SL-Euc
Rambli et al., 2016 [29]	*DMCEs* statistic		Wind direction data 1 Observation: No. 8
Rambli et al., 2015 [20]	*COVRATIO* statistic		Wind direction data 2 Observations: No. 38 and 111
Abuzaid et al., 2013 [24]	Mean circular error (*MCE*) statistic		Wind direction data 2 Observations: No. 38 and 111

So far the detection of single outlier in the circular regression model has been investigated. However, no related study has been found for the multiple outlier detection in circular regression models. Moreover, the robust methods also have not received much attention. Therefore, the new methods regarding the detection of multiple outliers in the circular regression model will be proposed later. Furthermore, the robust method for circular cases also will be proposed.

4. Discussion

This section summarized the finding of reviewed articles regarding the research in circular regression models due to detection of outlier. By referring to table 1, the researchers' has been proposed many methods to detect the single outlier in circular regression models.

Besides, most of the researchers presented a different real data sets due to the complexity problem in circular regression models. Form the table 1, [9][18][21] proposed three different methods to detect outlier in MCRM using multivariate eye data as real data. In order to assess the detection of outliers, two from three methods have been identified three observations with numbers 1, 15 and 23. The bootstrapping method and *COVRATIOc* statistic on JS circular regression model is applied to detect the possible outliers in wind direction data. The different observation number are indicated as outliers. By applying *DMCE* statistic on JS circular regression model of multivariate eye data in order to detect any possible outliers, we obtain only one observation.

According to this reviewed [20][24][25][29], its show that the suspected outlier observations will be identified using different methods at the same real data sets. However, in DM circular regression model nowadays, it seems that more researchers interested in studies along with proposed methods. All suspected outlier observations have been removed and the models have been refitted to get the best parameter estimations.

5. Conclusion

The studies of circular regression model have become a familiar area among the researchers' to be explored. This study investigate the different methods of detecting outliers in various circular regression models. There are many outlier detection methods have been used in the literature and in practice. Also, most of them are attracted to solve these problems and intentions to achieve some objectives in their studies. The finding is contributed to the knowledge regarding the modelling of circular regression and detection of outliers' methods. The finding in this study contributed to the survey methodology development of statistics to detect the multiple outliers in circular regression model. Then, the robust method also will be proposed for circular cases.

References

[1] Mardia K V and Jupp P E 2000 Directional Statistics. New York: John Wiley and Sons
[2] Jammalamadaka S R and A SenGupta 2001 Topic in Circular Statistics. London World Scientifics Publishing
[3] Fisher N I 1993 Statistical Analysis of Circular Data. Cambridge University Press
[4] Jammalamadaka S R and Sarma Y R 1993 Circular Regression. Statistical Science and Data Analysis Utrecht p 109 – 128
[5] Kato S, Shimizu K and Shieh G S 2008 A Circular – Circular Regression Model. Statistica Sinica **18** 2 p 633 – 643
[6] Abuzaid A H and Mohamed I B 2012 Boxplot for Circular Variable. Computational Statistics **27** 3 p 381 – 392
[7] Ibrahim S 2013 Some Outlier Problem in a Circular Regression Model. PhD Thesis University of Malaya
[8] Freeman P R 1980 On the number of outliers in data from linear model. Trabajos de estadistica y de investigacion operative **31** 1 p 349 – 365
[9] Alkasadi A N, Abuzaid A H, Ibrahim S and Yusoff M I 2018 Outlier detection in multiple circulr regression model via *DFBETAc* statistic. International Journal of Applied Engineering **13** 11 p 9083 – 9090
[10] Fisher N I and Lee A J 1992 Regression models for an angular response. Biometrics **48** p 665 – 677
[11] Gould A L 1969 A regression technique for angular reponse. Biometrics **25** p 683 – 700
[12] Johnson R A and Wehrly T E 1978. Some angular–linear distributions and related regression models. Journal of the American Statistical Association **73** 363 p 602 – 606

[13] SenGupta A 2004 On the constructions of probability distributions for directional data. Bulletin of the Indian Mathematical Society **96** 2 p 139 – 154
[14] SenGupta A and Ugwuowo F I 2006 Asymmetric circular – linear multivariate regression models with applications to environmental data. Environ Ecol Stat **13** 2 p 299 – 309 DOI 10.1007/s10651-005-0013-1
[15] Bhattacharya S and SenGupta A 2009 Bayesian analysis of semiparametric linear – circular models. Journal of Agricultural Biological and Environmental Statistics **14** 1 p 33 – 65
[16] Hussin A G, Fieller N R J and Stillman E C 2004 Linear regression model for circular variables with application to directional data. Journal of Applied Science and Technology **9** 1 p 1 – 6
[17] Mardia K V 1972 Statistics of Directional Data. Academic New York
[18] Alkasadi N A, Ibrahim S, Abuzaid A H, Yusoff M I, Hamid H, Zhe L W and Abd Razak A 2019 Outlier Detection in Multiple Circular Regression Model using *DFFITc* Statistic. Sains Malaysiana **48** 7 p 1557 – 1563
[19] Ibrahim S, Rambli A, Hussin A G & Mohamed I 2013 Outlier Detection in a Circular Regression Model Using *COVRATIO* Statistic. Communications in Statistics – Simulation and Computation **42** 10 p 2272 – 2280
[20] Rambli A, Yunus R M, Mohamed, I B, and Hussin A G 2015 Outlier Detection in a Circular Regression Model. Sains Malaysiana **44** 7 p 1027 - 1032
[21] Alkasadi N A, Ibrahim S, Ramli M F and Yusoff M I 2016 A comparative study of outlier detection procedures in multiple circular regression. AIP Conference Proceedings **48** 7 p 1557 – 1563
[22] Efron B 1979 Bootrap Methods Another Look at the Jackknife. The Annals of Statistics **7** 1 p 1 – 25
[23] Abuzaid A H, Hussin A G and Mohamed I B 2008 Identifying Single Outlier in Linear Circular Regression Model Based on Circular Distance. **3** 1 p 107 – 117
[24] Abuzaid A H, Hussin A G and Mohamed I B 2013 Detection of outlier in simple circular regression models using the mean circular error statistic. Journal of Statistical Computation and Simulation **83** 2 p 269 – 277
[25] Sidik M I, Rambli A, Mahmud Z, Redzuan R S and Shahri N H N M 2019 The Identification of Outliers in Wrapped Normal Data by using Ga Statistics. International Journal of Innovative Technology and Exploring Engineering (IJITEE) **8** 4 p 181 – 188
[26] Rambli A 2015 A Half – Circular Distribution and Outlier Detection Procedure in Directional Data. PhD thesis Faculty of Science University of Malaya Kuala Lumpur
[27] Abuzaid A H and Allahham N R 2015 Simple circular regression model assuming wrapped cauchy error. Pakistan Journal Statistics **31** 4 p 385 – 398.
[28] Di N F M, Satari S Z and Zakaria R 2017 Detection of different outlier scenarios in circular regression model using single – linkage method. IOP Conference Series Journal of Physics Conference Series **890** 1 p 012127
[29] Rambli A, Abuzaid A H M, Mohamed I B and Hussin A G 2016 Procedure for detecting outlier in a circular regression model. PLoS ONE **11** 4 e0153074. doi:10.1371/journal.pone.0153074.

IOP Conf. Series: Materials Science and Engineering **767** (2020) 012049 doi:10.1088/1757-899X/767/1/012049

A cloud-based automated parking system for smart campus

O C Yee[1], N Yaakob[1], M Elshaikh[1] and F Azahar[1]

[1] School of Computer and Communication Engineering, Universiti Malaysia Perlis (UniMAP), Pauh Putra Main Campus, 01000 Arau Perlis.

Abstract. Finding a vacant parking space is becoming a real problem especially in areas with limited parking spots such as airports, shopping centres, offices, as well as universities. Searching for available parking slots is normally time consuming and always results in frustration especially when time is the major constraints. Moreover, vacant parking is hard to be noticed due to unsystematic parking system. This will result in longer searching time which can also lead to traffic congestion. In addition, lack of security enforcement on cars entering universities campus is also one of the main issues contributing to insufficient parking spaces. This might cause some unauthorized cars to take opportunity to get inside the campus without any approval and consent from security department. Therefore, A Cloud-based Automated Parking System for Smart Campus is developed in this project. It consists of a sub-system that is developed to display availability number of parking slots so that it will assist authorize users to easily find their parking spots. The proposed system can also recognize car plate number through Automated Car Plate Recognition (ACPR) mechanism which is located at the campus main entrance gate to avoid unauthorized cars from getting in. This has strengthened security level inside campus and ensure the safety of students and staffs. All the information collected are sent into the cloud and stored inside a database system. The information regarding vacant parking can also be displayed using the developed mobile apps.

1. Introduction

Rapid development in smart cities leads to the realization of many innovations that spans across wide varieties of sectors and societies. One of the dominant applications include smart parking system which taken into consideration the mechanism of finding available parking spots. The development of smart parking involves the integration of various sensors and microprocessors to assist real-time parking information updates and systematic management system. The idea of smart parking is triggered by the common difficulties in finding vacant spot especially in crowded area in cities, shopping centres and universities. Finding a vacant parking spot can be troublesome especially in a hurry situation.

In this project, we narrow down our focus to develop a smart and automatic parking system for Smart Campus which is part of Smart City development. In this project, a smart parking system is developed to provide fast and accurate real-time parking information to drivers (students, staff, visitors) even before they arrive at their destination. The aim of this project is to provide convenient parking system, as well as making parking more easy, faster to be found and thus, less stressful.

This project is developed as part of Smart Campus realization with the purpose to lessen the use of manpower with the help of electronic devices to provide an efficient parking management system. Smart Campus is aimed to achieve a remarkable reduction in energy consumption. It has been introduced in different areas inside campus, including the smart watering system, smart electric usage control, air quality control, smart parking system and other as shown in figure 1. Smart Campus can also enable the realization of Green Campus, which is based on the concept of energy saving. As an example, if the sensor detects no human presence, electricity will be automatically turn off to save energy and cost.

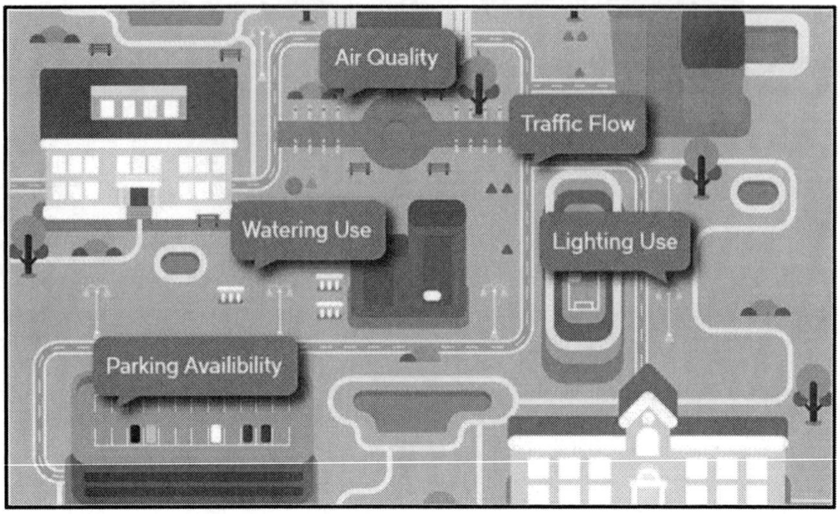

Figure 1. Smart Campus facilities [1].

Considering the significance of having an efficient parking system, this project emphasises on implementation of Smart Parking system in Smart Campus. Various types of electronic sensors are used in this project to monitor and guide users to identify the number of available parking slots in the campus. In addition to provide convenience and accurate parking information to users, the development of this smart parking also facilitates to solve security issue by enabling recognition of authorize and unauthorized vehicles.

Smart parking system in this project is designed to solve the ever increasing difficulties in finding available parking slots in campus. Due to limited spaces, available parking slots are hard to be noticed or found in campus. This is especially true throughout semesters where students and staff are rushing for class and available parking are hard to be found near facilities. Moreover, when parking area is almost full of cars, free parking spots are becoming harder to be noticed, making it more time consuming for drivers. In this case, drivers need longer time to search for the available parking slots.

Another common problem in campus is difficulties in recognizing unauthorized visitors. Until recently, most security inspection on cars entering any campus are done manually by security guards. Security guards may inspect passengers individually or just checking the car stickers whether they are belong to the university or not. Upon inspection, if they neither staff nor students, visitors' identification will be manually recorded on papers. The obvious problem with this manual inspection and recording is visitors' ID may be impersonated. This might be difficult to manually identify the authorized and unauthorized vehicles and caused them to take opportunity to get inside without the approval and the consent from security department. In addition, jotting down the car plate number and visitor ID will not solve the unauthorized visitors since it is almost impossible for the security guards to manually and compare with previous records (if there is any). Moreover, authorize car stickers can be simply stick onto cars by unauthorized visitors, which could be worsening the situation. For instance, many theft of equipment inside campus is a major problem that might be done by unauthorized visitors. These problems are hence becoming the matter of utmost importance in developing smart campus and thus require efficient solution. Therefore, in order to prevent any undesirable circumstances, a smart parking system need to be developed.

2. Objective

The main objective of this project are as follows:

- To develop a smart parking system that can detect vehicle occupancy and management of parking slots.

In order to solve the aforementioned problems, the Smart Parking system proposed in this project can assist users in two aspects. Upon arriving at university's main gate, drivers (staff, students or visitors) will be provided with information on numbers of available parking slots. This information will be displayed at the LCD screen near the main gate. For convenient purposes, this information can also be accessed through their mobile apps (for students and staff only). In addition, once drivers reach the parking compound, there is also an on-spot display at the parking area. The mobile phone application enables immediate information to disseminated about the available parking slots before they arrive at the parking area. User can choose not to install the mobile application as there is also an on-spot display that is installed at the main gate entrance on campus. Not only that this will reduce fuel consumption, but also air pollution, in accordance with providing smart campus.

- To increase security and safety level in smart campus.
  To increase the campus security level and cater the problem of unauthorized visitors, an Automatic Car Plate Recognition system is built. This part of the system is designed to prevent unauthorized cars entering campus without approval from security department.

The Automated Parking System for Smart Campus in this project is integrated with Internet of Things (IoT) that allows physical devices which are embedded with electronic components like sensors or software to collect and exchange data through internet. Various sensors are used to record the information of environment and sent to cloud for storage purpose. Information that are stored in cloud is then shared to the user through Graphical User Interface (GUI) which is either Android-based GUI or website-based GUI. Users can also send the data from GUI to cloud for certain purpose such as reservation of free parking slots. Thus, information exchange between devices and cloud through internet will take place.

3. Related Work

Research on developing a smart parking system have attracted significant attention from researchers as part of the efforts in developing smart cities. Most of the smart parking systems found in literature are based on RFID, WSN, Bluetooth, Wi-Fi and Zigbee.

3.1 Available parking information

Aydin et al., 2017 [1] had developed a system that can navigate users to the nearest free parking slots. This system also allows users to do reservation of free parking slot before they arrived at the parking area. This method consists of two parts. First part is related to the method of sending real time parking slots updates to internet. Second part concerns on the detection and determination of the nearest empty parking slots. A parking sensor is installed to send the parking status to a gateway, and this parking sensor send data by using 802.15.4 transceiver. GPS module is also used to find nearest empty parking slot with the help of the genetic algorithm.

Another smart parking system is proposed by Wang et al., 2017 [2] which adopts framework of cloud computing and Internet of Things. This proposed system is divided into four levels which are parking application, internet, devices and Internet of things platform levels. All data collected from sensors and other electronic devices are sent into cloud using Arm-based microcontroller. This system is able to point out the shortest and optimal path for user to the parking spots. LED indicators, zone controllers, signs and vehicle detectors are used to help the vehicle drives to quickly find the empty parking slots.

Moreover, few detection sensors are also deployed such as camera based detector, magnetic sensor, infrared based detector and ultrasonic sensor.

An E-parking system has been developed by P. Sadhukhan, 2017 [3]. This system is built up using various types of electronic components like ultrasonic sensor, alarm IC module and different type of software algorithm like Central Parking Management System (CPMS). Same like the other smart parking system, this system allows drivers to obtain information regarding the free parking slots. Not only that, it is also equipped with an integrated component called parking meter that can estimate the duration of the parking lots is occupied. R. Grodi et al., 2016, [4] also introduced a smart parking system that is similar with previous system. However, this new system will update users with real-time information through its website or mobile application. *Node.js* is used to set up a web server so that customer can obtain the parking information from any of their smart devices. Fang and et al., 2017, [5] also introduced a system that is using the similar concept with Grodi et al., 2016 [4].

3.2 Automatic Car Plate Recognition System

In order to allow only authorized users to enter campus and control/manage the access of unauthorized ones, Automatic Car Plate Recognition (ACPR) could be one of the necessary methods that can be implemented. Several related papers have been found and can be divided into two parts which are traditional method and the android platform method. S. H. M. Kasaei and S. M. M. Kasaei, 2011 [6] proposed an ACPR system that not only can recognize the car plate, but can also recognize motorcycle plate number. In this system, one new improvement is made as compare to other ACPR system is that the dilation operator is implemented. Dilation operator is able to separate characters even when they are too close to each other. After dilation operation, template matching will be carried on with all the characters that are stored inside a database.

An ACPR system is also introduced in the Android-based platform. According to Mutholib et al., 2012 [7] a Graphical User Interface (GUI) is possible to build up to capture image of car plate by using the built-in camera of a smart phone. After the image is captured, the pre-processing of enhancing, filtering and straightening the raw image is done. Optical Character Recognition (OCR) is used to extract texts and numbers from the car plate through neutral network. This system is tested on Malaysia's car plate. Its segmentation of characters able to get the accuracy of 83.5% and recognizing unit able to achieve 92% accuracy and this provide an 88% of great performance for the overall system on the recognition rates.

Based on literature, there are 4 steps in implementing ACPR; (i) capture the car number plate in colour, (ii) process the image into grey scale, (iii) process the grey-scale image into binary image (iv) compare with characters that are stored inside database.

4. Proposed System

A smart parking system in this project can be divided into two parts: (i) available parking detection and notification (ii) secure authorization through Automatic Car Plate Recognition (ACPR). This section presents the details of system overview and illustrates design. Brief introduction on this project will also being carried out in this section.

The overall system overview is as shown in the figure 2 while the prototype of the system is as shown in the figure 3(a) and figure 3(b). The system is consisting of two main parts: (i) the main entrance gate and (ii) the parking compound. At the main gate entrance, an ultrasonic sensor is installed to sense the presence of cars, whereby upon detection of any vehicle, the camera which are placed at an appropriate position will capture the image of the incoming cars for recognition. When the ultrasonic sensor sense any car existence at the main gate, microcontroller will give instruction to turn on the camera for recognition and authorization purposes.

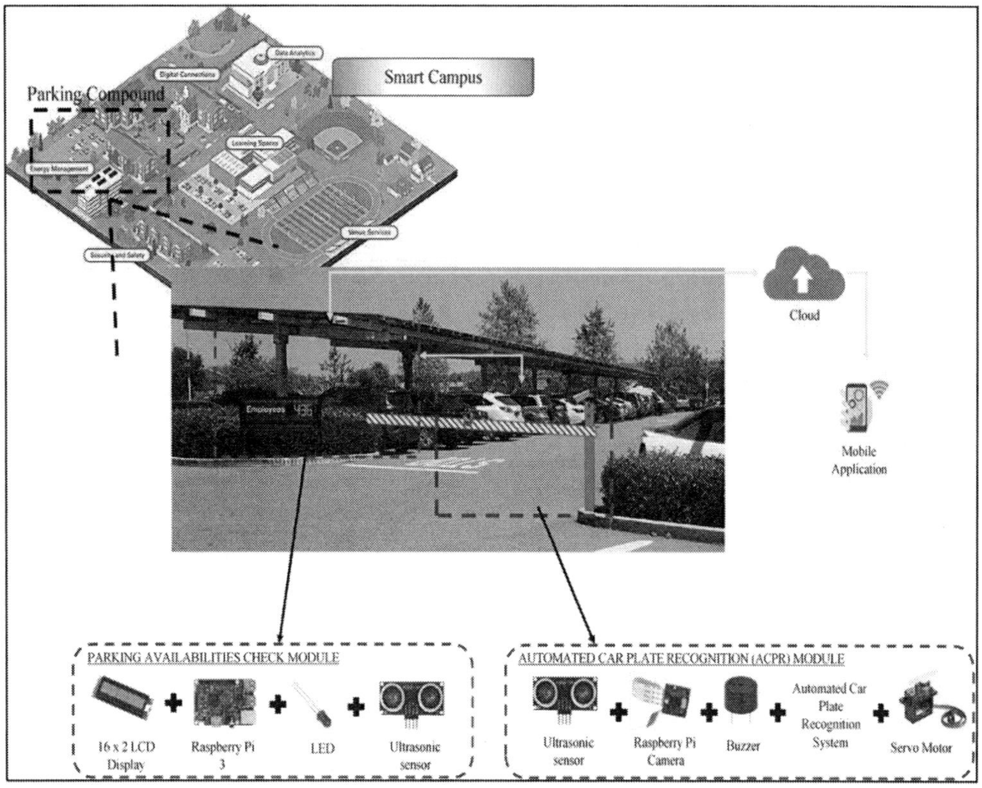

Figure 2. Smart Parking System Design.

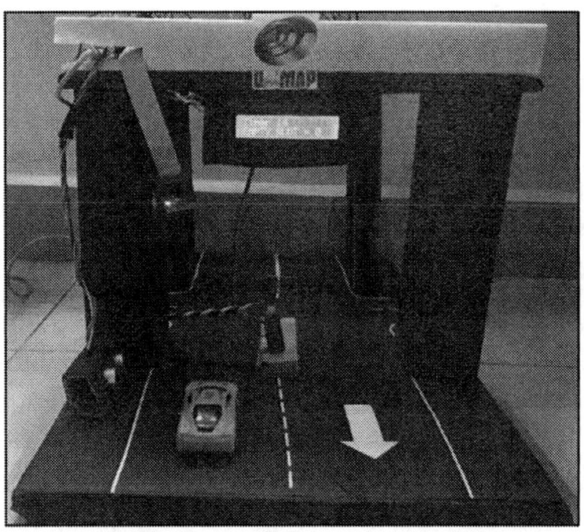

Figure 3(a). Front view of prototype.

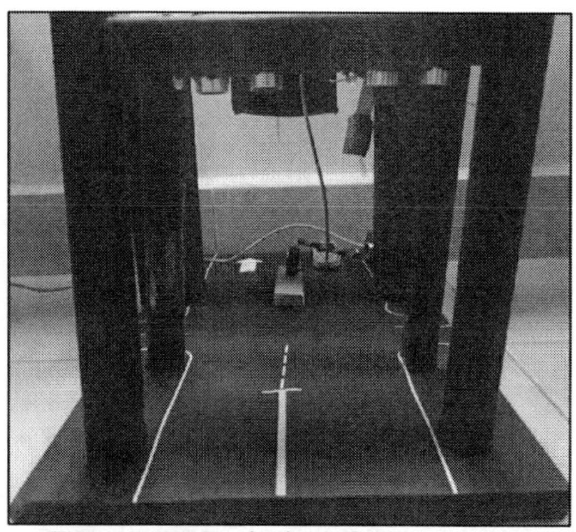

Figure 3(b). Back view of prototype.

After the image is captured, the Raspberry pi which is the microcontroller will collect the information and send it into Firebase which acts as a Cloud and database in this system. The Cloud will store the information sent by the microcontroller which will then perform data comparison with the existing car plate numbers that are already stored beforehand by the authority. After the comparison is completed, the microcontroller will instruct gate to be opened if car plate number is matching the one in the database.

On the other hand, the gate will remain close if the system fails to identify the matching car plate number. Security units will then act to check and record the visitor's ID and the purposes of visiting for security purposes.

Next, to collect information about the available parking spots, ultrasonic sensors are placed above each of the parking slot. The sensors send the information about the empty slots to the microcontroller (Raspberry pi) which will in turn calculate the total empty spaces. The LCD screen that is placed at the main gate entrance will then be updated with the latest number of empty parking slot. Information of empty parking slot is also being updated into the Firebase cloud. Users can also access the updated information regarding the remaining parking spots when they are not in the parking area through mobile application that has been developed. The overall structure of Cloud-based Automated Parking System is as illustrated in Figure 4. This project involves 4 stages.

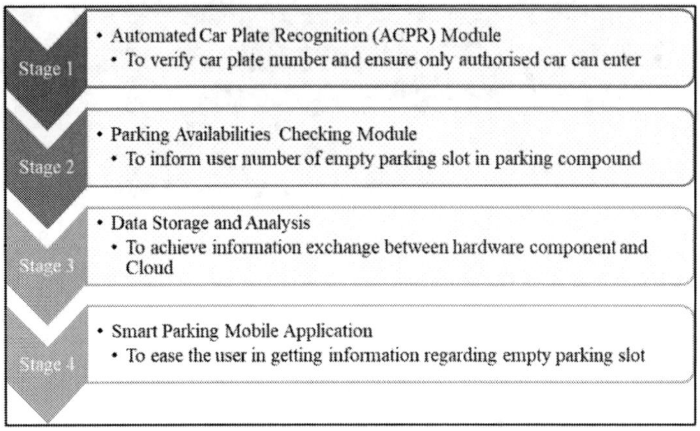

Figure 4. Overall Structure of system.

4.1 Stage 1: Automated Car Plate Recognition (ACPR) Module

This stage consists of Automated Car Plate Recognition (ACPR) System which is made up of Raspberry pi as microcontroller, camera, ultrasonic sensor and servo motor. The flow of ACPR process can be found in figure 5. An ACPR system is implemented by connecting the system to camera to fulfil the requirement of capturing and recognizing the car plate number. This system captures any car plate number from the car that is in the range of camera capturing area. The system will then process the image captured to obtain the car plate number. This is done by comparing the captured car plate number with the information stored in pre-defined database which is developed to record all the car plate numbers of all staff and students in campus.

In this module, a servo motor helps to open and close the main gate entrance. The servo motor is connected to the ACPR system to control the in and out of visiting cars. Few conditions have been set for that purpose. For instance, the system will only allow a car to enter the campus when the captured image of car plate number matches with the registered car plate number inside the database. If the car plate number matches the one inside the database, the entrance gate will be opened, and the car will be allowed to enter the campus compound. If the car plate number is not available inside the database, the entrance gate will not lift up to prevent the user from entering the campus and the buzzer will be turned on to warn and inform security for unauthorized incoming car. This system will then go back to initial state to stand-by for next car arrival. In this case, Raspberry pi is used to send information and control the pre-defined condition. This module increases the security level of campus by avoiding any unauthorised cars from entering the campus compound. Indirectly, it also ensures the safety of students and staffs.

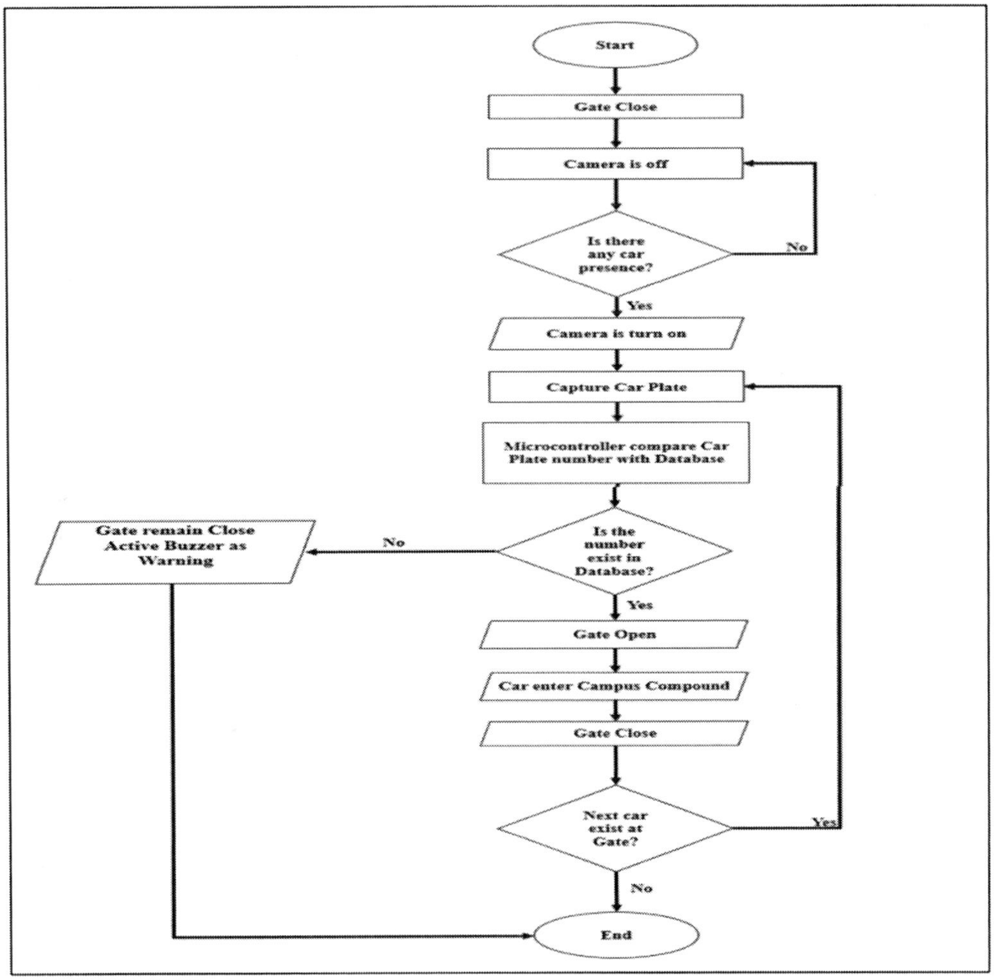

Figure 5. Flow chart for Automate Car Plate Recognition (ACPR) Module.

4.2 Stage 2: Parking Availabilities Checking Module

This stage consists of the process of checking out the available spaces for the parking compound. Several ultrasonic sensors are installed at the parking area to sense the presence of cars and empty spaces. This system provides real time information updates about the parking slot availability. The information is then sent to the Raspberry pi for processing and determination of vacant parking slots. The parking slot information is conveyed to users in two ways, which is through the LCD screen at the main gate and through mobile application (need to be installed by users). The information displayed can be used by users to plan where to park before arriving at the parking compound. Using this method, users can save time searching for available parking slots.

If there is any empty slot available, the microcontroller give instruction to LCD screen to update the number of availabilities. Green LED at specific empty parking slot will light up and numbers of availabilities in Smart Parking System application will also be updated. However, if there is no available parking slot inside the parking compound, the microcontroller will give instruction to LCD screen to display "Empty Slot: 0", and red LEDs at parking slot will light up and numbers of availabilities in Smart Parking System application will be updated to "0". The system will continue to check for any update in the parking availabilities. If there is any car leaving or incoming, the ultrasonic will immediately update the information to microcontroller which will then send updates to the cloud. If any car is leaving, the microcontroller will give instruction to LCD screen to update to the latest number of available parking slots, red LED is turned off and green LED of a particular empty slot will be lighted up. The number of availabilities parking slots is also being updated inside the Smart Parking System application. Despite that, the number of availabilities of empty parking slots will remain the same in both LCD screen and Smart Parking System application if no car is leaving the parking compound.

4.3 Stage 3: Data Storage and Analysis
As mentioned in Stage 1 and Stage 2, all the collected information are sent to cloud interface (Firebase) for information storage and analysis purposes. Few information about staffs' and students' cars are also be registered or stored before-hand inside the cloud system for the data comparing purposes. This information can be easily extracted and accessed regardless of location and time.

4.4 Stage 4: Smart Parking Mobile Application
A Smart Parking mobile application is developed to give an immediate and updated information to users. All the information regarding the parking compound are conveyed to users through this mobile application. It will update the users with the number of the available parking slots and inform users when the parking compound is full. The mobile apps is convenient to users as they do not have to be presented there to check the parking availabilities. It is designed to direct the users to the empty parking slots inside the campus parking compound. This method helps in saving the users' time to search for an empty parking slot. The mobile apps also can help the authority to monitor the list of authorized cars, and allow them to register new users.

Figure 6 shows the top-level entity for the proposed system design. All the components used in this project are connected directly or indirectly to the Raspberry Pi 3. The microcontroller is connected to the Automated Car Plate Recognition (ACPR) Module, Parking Availabilities Checking Module and Cloud, while the mobile application is connected directly to the cloud.

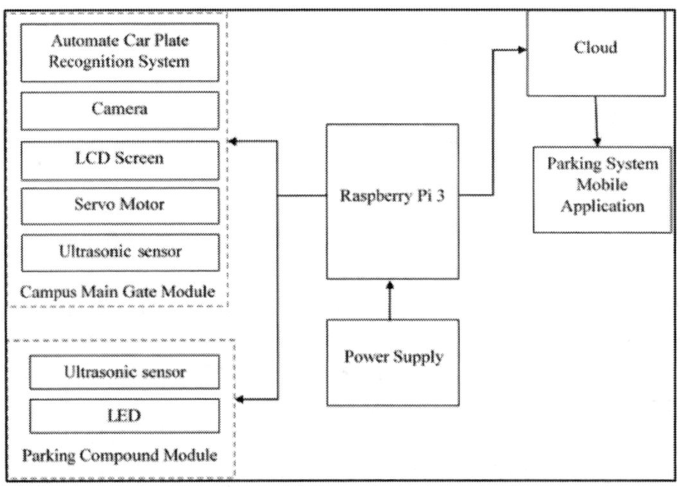

Figure 6. Block diagram for the proposed smart parking system.

5. Results and Discussion
This section discusses about results obtained from the proposed Smart Parking system. For testing purposes, the total number of parking is initially set to 2. At the initial point when the parking area is empty, all the green LED are lighting up while the red LED lights are turned off. At initial stage, the LCD that is set up at the entrance shows "Empty slot: 2". This initial condition is shown in figure 7.

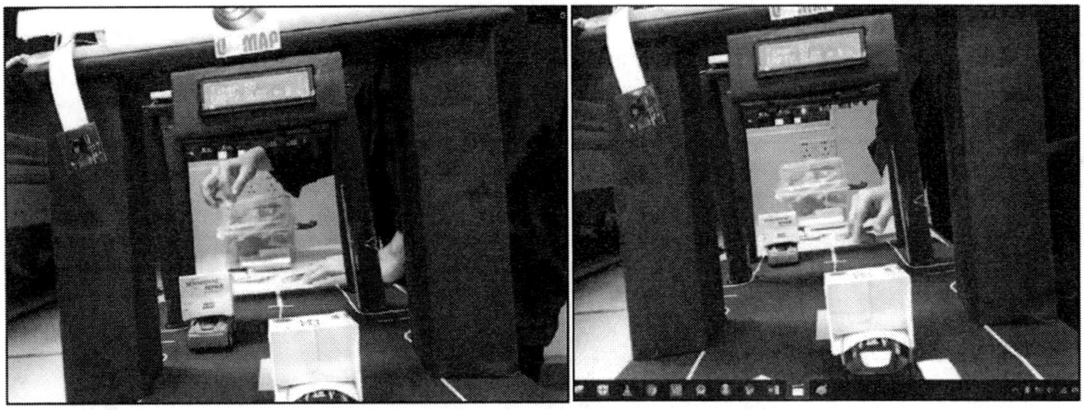

Figure 7. Initial condition: (a) Empty Parking (b) First Car entering - Red LED light up.

When the system detects the presence of any car at the gate, the ultrasonic sensor then senses the changing in distance and it will recalculate the number of available spaces and send the information to Raspberry pi. Then, the red LED at the empty slot will light up and the green LED will go dim. This latest information is sent into the database for storage and update purposes. The same process is repeated when another car is entering.

The results can also be visualized using user's mobile application as shown in the following figure. Only the registered users will be able to access this application and view the car park's information through the apps. This is to ensure that users that are not registered inside the system by the Authority cannot get any information about the parking compound via this apps. This user login page also consists a button that allow authority to enter security login.

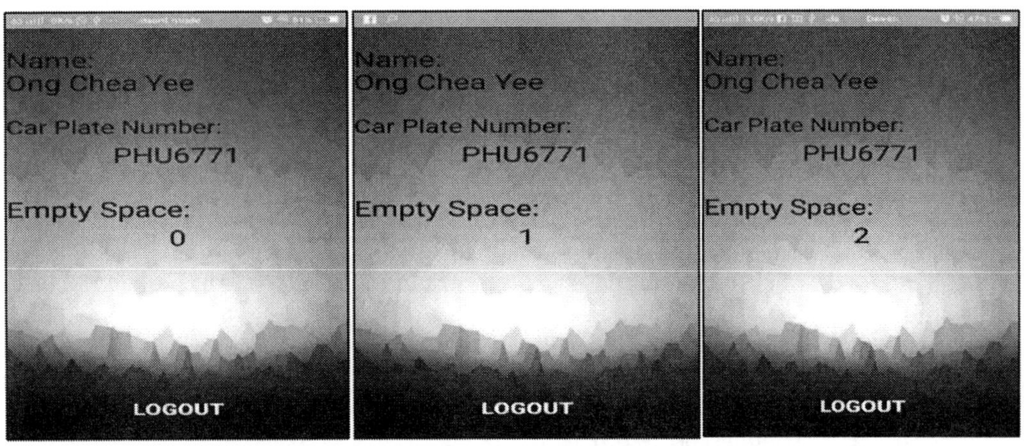

Figure 8. Updated parking information for mobile application.

After users successfully login to the apps, a User Main Page will be appeared as shown in figure 8. User Main Page shows all the details regarding the registered user which is extracted from the database. Those details included the name of user and his/her registered car plate number. The empty parking slot number that is available at current moment at the specific parking area also will be displayed in this page for users' information. The availability of parking slot will be updated real-time as illustrated in figure 8. This function fulfilled the first objective of this project which can provide convenience to users in which they can plan which parking area to park and estimate the time to arrive at the campus parking area. After finish viewing all the information needed, users can click on the logout button to quit the user main page. For ACPR, figure 9 shows the image captured by the camera. After the car plate number photo is captured by camera and processed by the Raspberry pi, it will compare the car plate number from the extracted image with the car plate number stored inside the database.

Figure 9. Image of car plate number captured by camera.

6. Conclusion

This Smart Parking system is built to provide a convenience and safety environment to users. It is able to provide users with real-time parking information using mobile application and also displayed the information to users at the main entrance of the parking area. With this service provided, users can plan their journey in advance before leaving for campus and save their time on searching around the parking area just to look for an empty parking slot. Another features of this Smart Parking system is Automatic Car Plate Recognition which enhance the security level of the campus. While allowing only the authorised cars to enter, it helps to lower down the rate of unauthorized cars or people want to sneak in campus. This feature secures the safety of students and staffs inside campus area.

References

[1] Aydin I, Karakose M and Karakose E 2017 A navigation and reservation based smart parking platform using genetic optimization for smart cities *Proc. 5th International Istanbul Smart Grid and Cities Congress and Fair (ICSG)* (Istanbul) pp 120-124

[2] Wang M, Dong H, Li X, Song L and Pang D 2017 A novel parking system designed for smart cities *Proc Chinese Automation Congress (CAC)* (Jinan) pp 3429-3434

[3] Sadhukhan P 2017 An IoT-based E-parking system for smart cities *Proc. Int. Conf. on Advances in Computing* (Udupi: Communications and Informatics (ICACCI)) pp 1062-1066

[4] Grodi R, Rawat D B and Rios-Gutierrez F 2016 Smart parking: Parking occupancy monitoring and visualization system for smart cities *Proc. SoutheastCon* (Norfolk: VA) pp 1-5

[5] Fang J, Ma A, Fan H, Cai M and Song S 2017 Research on smart parking guidance and parking recommendation algorithm *Proc. 8th IEEE Int. Conf. on Software Engineering and Service Science (ICSESS)* (Beijing) pp 209-212

[6] Kasaei S H M and Kasaei S M M 2011 Extraction and Recognition of the Vehicle License Plate for Passing under Outside Environment *Proc. European Intelligence and Security Informatics Conf.* (Athens) pp 234-237

[7] Mutholib A, Gunawan T S and Kartiwi M Design and implementation of automatic number plate recognition on android platform *Prof. Int. Conf. on Computer and Communication Engineering (ICCCE)* (Kuala Lumpur) pp 540-543

[8] Ruckus Wireless 2017 Smart Campus Learning (Sunnyvale, CA, US) Ruckus Wireless, Inc.

Development of system identification from traditional concepts to real-time soft computing based

A M Kwad, D Hanafi, R Omar and H Abdul Rahman
Faculty of Electrical and Electronic Engineering (FKEE), Universiti Tun Hussein Onn Malaysia (UTHM), 86400 Parit Raja, Batu Pahat, Johor, Malaysia

ayad_m_k@yahoo.com

Abstract. Advancement in computer and data acquisition technology gave the system identification a push forward, especially in the speed of the process and parallel calculation techniques. This article reviews the progress of system identification methods and techniques from traditional to modern methods and from offline to fast on-line analysis through previous years till these days. It takes into account part of artificial intelligence techniques (soft computing techniques), primarily artificial neural networks methods, for off-line system data analysis and real-time system identification, including the linear and non-linear issues. This study will provide a very concise survey for researchers about the development of system identification, its implementations, and techniques that have been used.

1. Introduction

System Identification is steps of logical operation which are followed to build a mathematical model that describes features of a dynamic system (or a process) depending on existing relations among its parameters and variables or between the input and corresponding measured output data [1]. Each system has parameters and variables that interconnected together to form its behavior. Sometimes, these parameters and variables could be seen clearly [2], hence it could build a mathematical model to describe a system with small error margin (although the mathematical model is involved, it cannot adequately describe a real system), and that what is named in some references as a white box [3]. In contrast, a black box [4] is a system that has unknown variables, parameter values, and their relations. All that can be seen are the input (to the system) and output (from the system); so, the system is evaluated by observing the output-input relationship. Finally, there is a gray box system [3], which is mixed from white box and black box systems; there are some features known and others unknown. It is crucial to select appropriate variables from the principle equations and tune the parameters by using regression methods to minimize the variation between the derived mathematical model and the measured data [1]. Most modeling operations are done after the data have been collected (off-line), but there is another type of modeling is done within real-time to estimate the parameters of a specific system, which is named on-line system identification [5,6]. The main equipment in the system identification are: measured data, derived model, selected cost function, and the numerical method used to minimize the cost function (it must be fast to convoy on-line system identification [7,8]).

Nowadays, researchers trend to use the Artificial Intelligence technique to solve many system problems like non-linearity, time-varying, and ambiguity or inaccessibility [9,10]. This technique based on the learning concept, where weights that belong to a specific perception start with random values, then these weights are tuned up to be suitable amounts. Whenever the learning increased (till a specific limit), it makes weights close from appropriate value. Artificial intelligence techniques can be divided into two types relative to teaching manner; supervised learning and unsupervised learning. Where the first type compares the output of the network with the desired output (that means the desired output is known), while the second type is learning without teacher, also it is can be divided into two kinds relative to the size of network; single layer and multi-layer networks [11,12], and there are other classifications according to training methods and used function.

The following sections will explain the off-line and online system identification, what is the difference between them, and how the soft-computing algorithms were used in these two types instead of the traditional methods.

2. Off-line System Identification

Off-line system identification is the analysis of the pairs of input-output data (with older information for current pairs in the dynamic system) to find the internal mathematical structure of the system (model of the system) and assess the parameters of that structure one step ahead, after collecting the data via sensors. The steps of off-line identification can be summarized by:

1- Finding the mathematical model for the system from the principle physical rules, if the system is a white or gray box. The black box has no known structure, so this step ignored.
2- Stimulating the system with a controlled input to get on system response.
3- Sensing the input and output signals and accumulating them in pairs form.
4- Implementing the system identification operation on the input/output data (in some cases, the data needs to be treated before the identification operation) to estimate the value of the parameter in the defined model.
5- After completing the identification operation, the model now ready to be compared to part of the data through operation named model validation. If the output of the model is convergent from the original output, it can consider the model represent the system relatively, if not the step (4) must be returned.

Soft computing methods can substitute the traditional methods in creating Identifying a mathematical model for either known (white box) or partially known (gray box) system's properties, and unknown system's properties (black box).

De la Sen et al. [13] designed an expert network that had been created to estimate a discrete transfer function for a linear time-invariant system with impulses sampling. They ran different methods for system identification. One of these methods relied on recursive least squares (RLS);

$$\mathbf{A}(\mathbf{q}^{-1})y_l = \mathbf{B}(\mathbf{q}^{-1})x_l + v_l$$
$$\mathbf{V}_M = \sum_{l=1}^{M} v_l \qquad (1)$$
$$\hat{\theta}_l = \hat{\theta}_{l-1} + \mathbf{k}_l(y_l - \hat{y}_l)$$

Where v_l represent the calculated error, M is the amount of the samples, $\hat{\theta}_l$ are the expected parameters at time $t = l$, y_l is the measured output at the same time t, and $\hat{y}_l$ is the predicted output. Another method relies on Leverrir's algorithm;

$$\begin{aligned}\mathbf{G(q^{-1})} &= \frac{\mathbf{Y(q^{-1})}}{\mathbf{X(q^{-1})}} = y_t + y_{t-1}q^{-1} + \cdots + y_{t-n}q^{-n} \\ &+ \sum_{r=n+1}^{\infty} y_{t-r}q^{-r} \\ &= \frac{\mathbf{B(q^{-1})}}{\mathbf{A(q^{-1})}} \\ &= \frac{b_1 + b_2q^{-1} + b_3q^{-2} + \ldots}{1 + a_1q^{-1} + a_2q^{-2} + a_3q^{-3} + \ldots}\end{aligned} \tag{2}$$

Where, q represents a shift time in the dynamic system $q\,y_t = y_{t+1}$ and $q^{-1}\,y_t = y_{t-1}$.
Artificial Intelligence is one of the best choices to solve the non-linearity problem; it has some attributes that make it more adaptable. The soft robot was used for the motion of some parts of a body during radiotherapy was used as a case study, using supervised deep learning for multilayers recurrent neural networks [14]. de Canete et al. [15] indicated how to benefit from the Labview, which is a graphical language for the data acquisition and the automation control, to Identify the distillation column process. The nonlinearity of the system was solved by neural networks tactic.

Shamsudin and Chen [16] developed an off-line dynamics system identification approach for unmanned copter depend on neural networks (NN). The Levenberg-Marquardt model form was applied to teach the NNs with autoregressive exogenous (AXR). A combination of correlation analysis, cross-validation, weights regulation, and Lipschitz criterion was used to enhance the form of the network.

Hanafi et al. [17] developed a mathematical model that relied on the physical principle laws, that ruled the system, and the attributes of the parameters. The case study was a suspension system of a car with nonlinearity in the damper and spring elements. The SI was based on the multilayers NNs to solve the complex relations between the inlet and the outlet of the system. The car was examined using the artificial bump on the road, the data were handled off-lined, and the NNs were taught by the weighted least squares (WLS) method.

Mills and Zomaya [18] examined an off-line dynamic system identification with the assist of neural networks. They utilized a backpropagation procedure for training the networks based on the linear estimator of the algorithmic least-squares for the nonlinear system. Besides NNs, the Fuzzy Logic with Gaussian member shape was used to be a six-layer fuzzy neural for improving the weights with pre-optimization step using the genetic algorithm (GA) optimization method [19].

Cong et al. [20] followed a parametric system identification, which is one of the identification approaches, to construct a nonlinear dynamic mathematical model representing a DC motor with the nonlinear effect of the Coulomb friction. The suggested approach utilized GA, which is one of heuristic and evolutionary methods, with the simplex method to solve the global optimization problems for estimating the model. The validation, which was utilized to compare between the model and the DC motor, was comparing the output of the motor with the output of the model (from the simulation).

Conforming to Herreros et al. [21], a straightforward model was employed to identify the suspension system of a car applying non-interfering input signals. The amplitude difference between the real and the estimated power magnitude of the system output for spectral density was taken as the main objective to be reduced and to reduce a discrepancy of estimation due to distorted input (because of the noise and ignoring the presence of dynamics), the GA was implemented as an optimizer tactic.

Jizhen et al. [22] suggested a model based on the gray-box principle for identifying an online parametric model of wind turbine generation system (WTGS) in a closed-loop system. The system as an aerodynamic system and a generation core of the WTGS entirely were modeled all together to obtain a total identification model relied on a nonlinear state-space model. GA was implemented to assist in identifying the system and assess the parameters of the suggested model.

As stated by Cen et al. [23], a gray-box multilayers perceptron neural networks and integrator was established for identifying a nonlinear dynamic system and estimating fault scheme. The model was formed from simple rules of the neural networks, the essential criterion is the deviation between the actual system output and the derived model output, which represents the error offset value for various fault values.

The article [24] proposed a process to find the cross-section of neutron-induced reaction in a specific condition, used weighted least squares method to regress the error and to update the neural network weights and backpropagation technique for training the network off-line.

Kurita [25] examined the teaching procedures of multilayers NNs from the maximum likelihood estimation (MLE) viewpoint. It was shown that Fisher information (FI) of the network, with subtraction expectancy values of Hessian matrix (H), was gained from a weighted covariant tensor of input variables. The training approach utilized Fisher's scoring concept, which applied FI instead of H, that can be symbolized by the weighted least squares technique for multilayers (with single hidden layer) neural networks.

Puscasu and Codres [26] proposed a non-linear system identification and system control tactic supported by structured NNs. To decrease the degree of computation complexity for whole the system, in the research, splitting the system into subsystems by separating algorithm, was offered. Thus, each nonlinear subsystem module has a nonlinear controller built from neural networks, and it was switched among the networks of controllers by a dynamic switch which was done also by neural networks.

Narendra and Parthasarathy [26] presented the principles of model representation to identify and control a non-linear dynamic system by utilizing NNs ability. The research tried to solve some problems with applying the neural networks for higher-order multi-variable nonlinear systems, and the effects of applying simpler models for identification and control. Quiroga et al. [27] explained utilizing neural networks for identifying a permanent magnet synchronous motor (PMSM). The recurrent neural network pattern used for maintenance to perform fault detection or integrity assessment. It was used Matlab/Simulink with other techniques to implement and validate the model.

Sierra and Santos [28] examined implementing alternative tactics for identifying an unmanned quad-copter system; artificial neural networks approach, parametric model estimation methods, neuro-fuzzy structure, and a hybrid of some previous methods. The comparison among these methods revealed that the hybrid method was retaining an appropriate balance between precision and computation time.

Ismeal et al. [29] investigated a permanent magnet DC motor identification using artificial neural networks. Input-output data (armature voltage-current and angular speed, respectively) were taken to use in building a non-parametric black-box model for identifying a dynamic system of motor. The neural networks with hidden three-layers depended on a non-linear autoregressive network with exogenous (NARX) input was applied. NARX network was taught by parallel constructions and scaled gradient algorithm using the error derivative (first and second order derivatives) to decrease the error. Whereas, Rahim et al. [30] examined utilizing a non-linear autoregressive moving average for exogenous input (NARMAX) method with multilayers neural networks to identify a non-linear system for the DC motor. The off-line learning of neural networks was improved by implementing the regularization method.

Lu et al. [31] built a semi on-line, locally linear model for an unmanned drone system. Angles of attack and sideslip are measured as output for the system. The observer/Kalman filter identification algorithm is used for identifying a fixed-wing unmanned aerodynamic system, built on the state-space method, which developed from both lengthways and side force-components.

Pislaru and Shebani [32] utilized the feed-forward neural network established on radial basis function, which is a function with real values that relies on the distance from the origin only, what is known as (RBFNN) technique for identifying a nonlinear system. That technique was validated by applying it on a DC motor, tank system, double tank system, and two lab's models. The clustering algorithm of K–means used to choose the RBFNN centers.

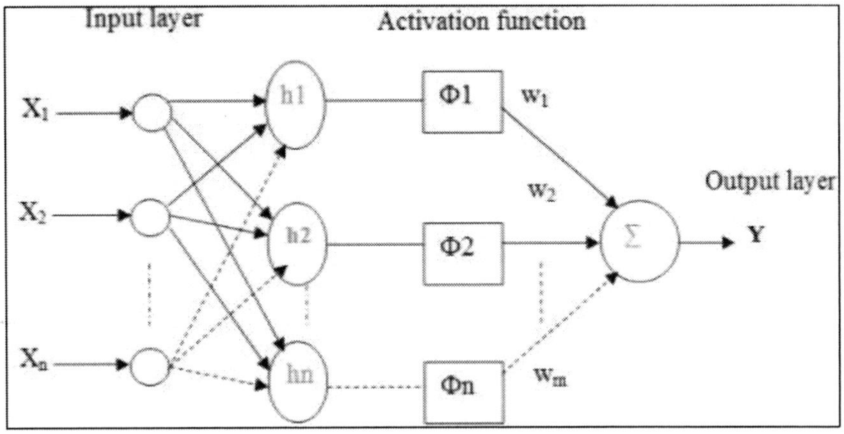

Figure 1. RBFNN [32].

3. Real-time System Identification

Real-time (online) system identification is estimating the model parameters at the current moment depending on the pairs of input/output data (with older information for the current pairs in the dynamic system) to find the internal mathematical structure of the system (model of the system) one step ahead. The steps of real-time identification can be summarized by:

1- Finding the mathematical model for the system from the principle physical rules, if the system is a white or gray box. The black box has no known structure, so this step ignored.
2- Stimulating the system with a controlled input to get on system response.
3- Sensing the input and output signals and accumulating them in pairs form. The input signal for the system is the same input for the model, and the subtraction of the model from the real system output represents the current error.
4- Implementing the system identification operation recursively on the input/output data (in some cases, the data needs to be treated before the identification operation), to estimate the parameter's value in the defined model.
5- Using the calculating error to update the parameters' value instantaneously.
6- After completing the identification operation, the model now ready to evaluate the model. If the output of the model was convergent from the original output, it could consider the model represent the system relatively; if not, it may return the whole process or start from the step (2).

The recursive methods are dominant in real-time system identification.

An approach inspired by Takagi-Sugeno model for real-time training was proposed by Angelov and Filev [33], it was constructed on supervised and unsupervised teaching algorithm that updated the model parameters and structure repetitively. It was implemented to the data obtained from airconditioning equipment operating in a certain construction.

Qian et al. [34] proposed a system identification scheme adopted a parametric online identification method to get parameters of an actuator for a range of loads using hybrid simulation. A GA was built on experimental data, which was loaded the input part to the actuator and the output used to assess the convergence of estimated values for the model's parameters from the original system behavior.

Al Seyab and Cao [35] developed an identification process based on recurrent neural networks (RNN) for nonlinear predictive and control models. The process based on neural networks was derived from a nonlinear state-space form to be a real-time process using a developed automatic differentiation algorithm. The technique was used for online optimization and predictive controller, as well. Wang et al. [36] offered an unconventional motion-mode energy method to identify moving shapes of a vehicle body (roll, bounce, and pitch motion modes) with a minimum error ratio by applying neural networks.

Pairan and Shamsudin [37] introduced experimental results based on the RBFNNs trained by minimal resource allocating network approach (MRAN) for identifying the quad-chopper system in real-time. The activity of the MRAN tactic was compared to the RBFNN tactic with a constant trace (CT) process for 2500 pairs of input-output data samples. It was found that using MRAN appending or eliminating hidden layer neuron tactic to optimize RBFNN improve weights estimation precision and minimize the time of teaching.

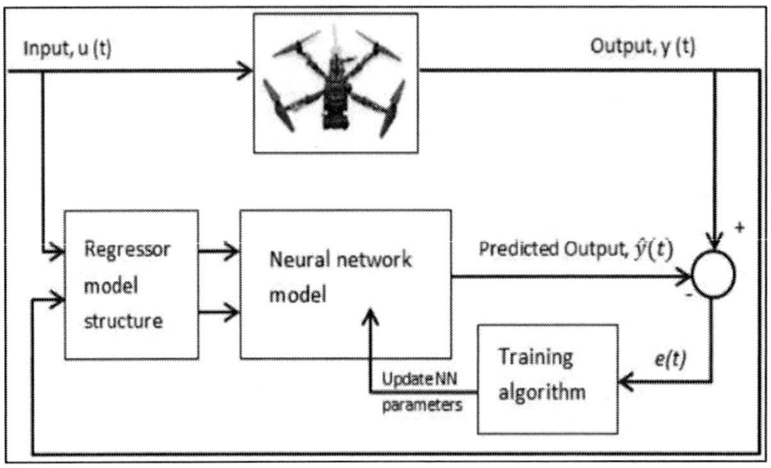

Figure 2. System identifier based on NN [37].

Kirkpatrick et al. [38] proposed a linear system identifier by using neural networks for airplane system, the identifier function was derived from state-space equations and the data used for training the neural networks, to get matrices of the state space, were obtained from previous flight operation, the same data had already been utilized for identifying the dynamic airplane system by Observer/Kalman filtering approach. The suggested identifier compared with (O/K) filter method [39], which was locally linear, for airplane dynamic nonlinear system identification in real-time.

Hussein et al. [40] introduced a model for identifying and controlling a DC motor speed; the input voltage was provided from the solar power system through a DC-DC converter utilized two universal learning networks (ULNs). The feed-forward NNs were applied to control, evaluate and update the parameters of the nonlinear model in real-time. The essence of Rubaai and Kotaru [41] research was how to deal with controlling a DC motor speed through utilized from feed-forward NNs to represent the nonlinear system of the DC motor in an on-line system identification with on-line learning for neural networks were done.

Griñó et al. [42] proposed a tactic depended on NNs structure for identifying a dynamic system which has an unknown dynamical behavior; an adjustable NNs were added to take the place of unknown dynamic part of the system. The model of NNs has linear parameters with updating weights synchronously. Two techniques were used; the first technique adopted the gradient method and analysis of the model's output attitudes against calculated parameters, the second one used variant calculation, which was leading to an offline statement and the invariant embedded way that converts it to an online solution.

Guo et al. [43] proposed a method based on (RNN) with shared and specialized memories (ShM and SpM) or (SSMs) tactic for chaotic adaptable filtering to improve a simultaneous estimation. The SpM involved weights of the last layer with linear specifications, whereas ShM designed for nonlinear weights, which was pre-learned in off-line to a particular value, then implemented as fixed weights in on-line, while the SpM is trained on-line. The method based on recursive least-squares teaching. Ibrahim [44] suggested an approach for identifying the system depended on neural networks as the auto-adjusting regulator. A real-time identification is implemented by RNNs to estimate parameters of discrete PID control, which inspired by the Zeigler-Nichole method. It was tested in simulation software and compared to the traditional PID controller.

Sadeghi and Farrokhi [45] presented a model for Identifying a nonlinear dynamic system in real-time based on Wiener oriented blocks. The Levenberg-Marquardt algorithm with multilayers NNs were used for the nonlinear static part in the Wiener model, the identifier of subspace with the multi-variable approach was used for the linear dynamic part. The proposed model was simulated on the supposed system and Lyapunov method utilized to improve the steadiness of model convergence.

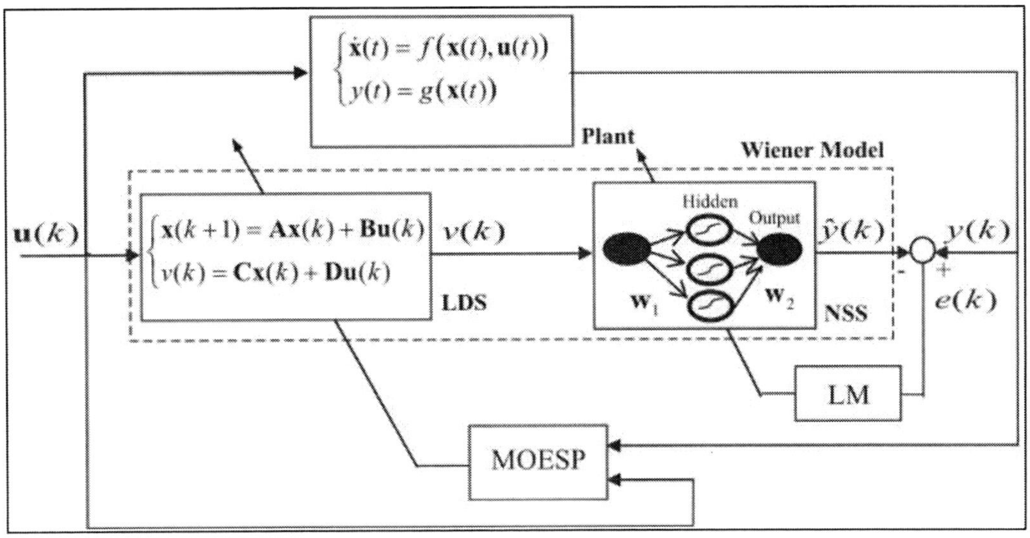

Figure 3. Suggested online identification [45].

Axial Idea of PIVOŇKA et al. [46] studied the impacts on the identification procedure in high-speed sampling and compared among three techniques of real-time identification: recursive least-squares, gradient, and system identification approach based on neural networks with damped least-squares minimizing method, then showed that the neural networks getting the better of the others. The structure of neural networks was upgraded from [47] and [48], which utilized to identify a black-box system in real-time updating parameters according to the stability of Lyapunov.

Chen et al. and Purwar et al. [51,52] utilized a model of neural networks for identifying ambiguous, dynamic, nonlinear systems in discrete time. Purwar et al. [49] applied only one layer of neural networks instead of multilayers and substituted that by distributing the input by Chebyshev-polynomials, and it used the iterative least-squares technique with forgetting factor as real-time teaching algorithm to update parameter. While, Chen et al. [50] adopted a multi-layer NNs (with a single hidden layer), and used a developed algorithm derived from prediction error formulation.

Jafari et al., and Salahshoor and Jafari [51,52] produced an approach using the consecutive growth and trimming radial basis function (RBF) to identify a nonlinear system online. It was a newly proposed method for the growing and pruning neurons criteria to be proper for identifying online with modified teaching tactic based on unscented Kalman filter (UKF) for neural networks, the method implied using a changeable forgetting factor technique embedded in UKF form to stimulate parameter assessment in nonlinear dynamic systems.

Rocha Filho and de Oliveira Serra [53] presented an identification algorithm for a black-box stochastic system in real-time built on neuro-fuzzy approach with impressive variables to model a dynamic system. A continuous online neuro-fuzzy technique with Takagi-Sugeno developed structure, which applies the maximum likelihood with iterative parameter estimation, was adopted as a procedure. Copter with two degrees of freedom was used as a case study.

Aeronautics is a rich area for applying online system identification since it needs a quick reaction and moving. By utilizing three layers unmanned airborne vehicle with fixed wings as a plant; Puttige and Anavatti [54] found the differences and similarities between the real-time identification and off-line identification using three layers of neural networks with a multiple input\output system that has a six-output, where ARX was applied to minimize the error. The same previous strategy was implemented on Eagle helicopter six degrees of freedom [55].

Feng et al. [56] determined a non-straight scientific model for the electro-water driven stage, including the impact of Stribeck friction, where the friction dissipated power was determined from a deliberate weight of pressure-driven chambers at different speeds extend. There are a few upgrades were made on GA method, like the fitness, crossover, and mutation assumption. In real-time, a noticed speed utilized to assess the actual velocity, and dynamic feed-forward remuneration for friction impact was configured.

4. Conclusion

Identifying a dynamic system allows scientists and engineers to understand the nature of the system and, as a result, gives them the ability to control that system or to gain full knowledge of the system and improve it. Conventional linear analyses for identifying systems are developed to include some systems with nonlinear behavior (using linearity assumption). The majority of the dynamic systems are not linear, and implementing linearity presumptions lead to a false mathematical model. Determining a precise nonlinear model is not a simple mission (to reach the genuine physical model), and evaluating the parameters of that condition will be somewhat hard, so utilizing artificial intelligence can deal with non-linearity and uncertainty issues. Estimating the value of parameters in the off-line analysis is a little bit different from the on-line analysis, where the first leads to get fixed values, which have a considered error in specific points, while the on-line analysis has flexible and adaptable parameters.

5. Future Work

Based on the previous, the soft-computing features could be employed in the system identification, either off-line and online mode. The noted effectiveness in dealing with nonlinear systems makes the soft-computing aspects be nominated as a perfect solver for the complexity of the system so that future work will be using this approach from developing the system algorithm for gray-box, not only representing the black-box system depending on the input-output data.

Reference

[1] Schoukens J, Pintelon R and Rolain Y 2012 *Mastering system identification in 100 exercises* (John Wiley & Sons)

[2] Noël J-P, Esfahani A F, Kerschen G and Schoukens J 2017 A nonlinear state-space approach to hysteresis identification *Mech. Syst. Signal Process.* **84** 171–184

[3] Nelles O 2013 *Nonlinear system identification: from classical approaches to neural networks and fuzzy models* (Springer Science & Business Media)

[4] Hunt K J, Irwin G R and Warwick K 2012 *Neural network engineering in dynamic control systems* (Springer Science & Business Media)

[5] Hong E Y, Meng T K and Chitre M 2013 Online system identification of the dynamics of an autonomous underwater vehicle *Underwater Technology Symposium (UT), 2013 IEEE International* (IEEE) pp 1–10
[6] Yang Q, Ren X, Zhao W, Lv Y and Wang S 2017 Optimal adaptive parameter estimation-based tracking control of motor turntable servo system *Modelling, Identification and Control (ICMIC), 2017 9th International Conference on* (IEEE) pp 139–144
[7] Gensler A, Gruber T and Sick B 2015 Fast feature extraction for time series analysis using least-squares approximations with orthogonal basis functions *Temporal Representation and Reasoning (TIME), 2015 22nd International Symposium on* (IEEE) pp 29–37
[8] Amrani N, Serra-Sagristà J and Marcellin M 2017 Low complexity prediction model for coding remote-sensing data with regression wavelet analysis *Data Compression Conference (DCC), 2017* (IEEE) pp 112–121
[9] Worden K and Green P 2017 A machine learning approach to nonlinear modal analysis *Mech. Syst. Signal Process.* **84** 34–53
[10] Nechyba M C and Xu Y 1994 Neural network approach to control system identification with variable activation functions *Intelligent Control, 1994., Proceedings of the 1994 IEEE International Symposium on* (IEEE) pp 358–363
[11] Mehrotra K, Mohan C K and Ranka S 1997 *Elements of artificial neural networks* (MIT press)
[12] Daniel G 2013 *Principles of artificial neural networks* vol 7 (World Scientific)
[13] De la Sen M, Garrido A J, Barambones O and Maseda F J 2003 An expert network for obtaining approximate discrete-time models for LTI systems under real sampling using parameter identification *Emerging Technologies and Factory Automation, 2003. Proceedings. ETFA '03. IEEE Conference* vol 1 (IEEE) pp 462–469
[14] Ogunmolu O, Gu X, Jiang S and Gans N 2016 Nonlinear systems identification using deep dynamic neural networks *ArXiv Prepr. ArXiv161001439*
[15] de Canete J F, Gonzalez-Perez S and del Saz-Orozco P 2008 Artificial neural networks for identification and control of a lab-scale distillation column using LABVIEW *World Acad. Sci. Eng. Technol.* **47** 64–69
[16] Shamsudin S S and Chen X 2012 Identification of an unmanned helicopter system using optimised neural network structure *Int. J. Model. Identif. Control* **17** 223–241
[17] Hanafi D, Tee K S, Johana E, Rahman H A, Rahmat M F, Wahid H and Ghazali R 2018 Neuro Model for Passive Suspension of a Light Car *Int. J. Integr. Eng.* **10**
[18] Mills P M and Zomaya A Y 1993 A neural network approach to on-line identification of nonlinear systems *Cybern. Syst.* **24** 171–195
[19] Attia A and Horáček P 2001 Modeling Nonlinear Systems by a Fuzzy Logic Neural Network Using Genetic Algorithms **41** 69–76
[20] Cong S, Li G and Feng X 2010 Parameters identification of nonlinear DC motor model using compound evolution algorithms *Proceedings of the World Congress on Engineering* vol 1
[21] Herreros A, Baeyens E, Perán J R and Melgar A 2002 Parameter identification of a car suspension system using non-intrusive signals *IFAC Proc. Vol.* **35** 427–432
[22] Jizhen L, Junlin G, Yang H, Juan W and Hong L 2017 Dynamic modeling of wind turbine generation system based on grey-box identification with genetic algorithm *Control Conference (CCC), 2017 36th Chinese* (IEEE) pp 2038–2042
[23] Cen Z, Wei J and Jiang R 2013 A gray-box neural network-based model identification and fault estimation scheme for nonlinear dynamic systems *Int. J. Neural Syst.* **23** 1350025
[24] Dubey B, Kataria S and Mohanty A 1997 Neural network fits to neutron induced reactions using weighted least-mean-squares *Nucl. Instrum. Methods Phys. Res. Sect. Accel. Spectrometers Detect. Assoc. Equip.* **397** 426–439
[25] Kurita T 1994 Iterative weighted least squares algorithms for neural networks classifiers *New Gener. Comput.* **12** 375–394
[26] Narendra K S and Parthasarathy K 1992 Neural networks and dynamical systems *Int. J. Approx. Reason.* **6** 109–131

[27] Quiroga J, Cartes D and Edrington C 2009 Neural network based system identification of a PMSM under load fluctuation *Dyna* **76** 273–282
[28] Sierra J E and Santos M 2018 Modelling engineering systems using analytical and neural techniques: Hybridization *Neurocomputing* **271** 70–83
[29] Ismeal G A, Kyslan K and Fedák V 2014 DC motor identification based on Recurrent Neural Networks *Mechatronics-Mechatronika (ME), 2014 16th International Conference on* (IEEE) pp 701–705
[30] Rahim N, Taib M, Adom A and Halim M 2007 Nonlinear System Identification for a DC Motor using NARMAX Model with Regularization Approach *International Conference on Control, Instrument and Mechatronics Engineering, CIM* vol 7
[31] Lu H H, Rogers C, Goecks V G and Valasek J 2018 Online Near Real Time System Identification on a Fixed-Wing Small Unmanned Air Vehicle *2018 AIAA Atmospheric Flight Mechanics Conference* p 0295
[32] Pislaru C and Shebani A 2014 Identification of Nonlinear Systems Using Radial Basis Function Neural Network *Int. J. Comput. Inf. Syst. Control Eng.* **8** 1528–1533
[33] Angelov P P and Filev D P 2004 An approach to online identification of Takagi-Sugeno fuzzy models *IEEE Trans. Syst. Man Cybern. Part B Cybern.* **34** 484–498
[34] Qian Y, Ou G, Maghareh A and Dyke S J 2014 Parametric identification of a servo-hydraulic actuator for real-time hybrid simulation *Mech. Syst. Signal Process.* **48** 260–273
[35] Al Seyab R and Cao Y 2008 Nonlinear system identification for predictive control using continuous time recurrent neural networks and automatic differentiation *J. Process Control* **18** 568–581
[36] Wang L, Zhang N and Du H 2015 Real-time identification of vehicle motion-modes using neural networks *Mech. Syst. Signal Process.* **50** 632–645
[37] Pairan M and Shamsudin S 2017 System identification of an unmanned quadcopter system using MRAN neural *IOP Conference Series: Materials Science and Engineering* vol 270 (IOP Publishing) p 012019
[38] Kirkpatrick K, May J and Valasek J 2013 Aircraft system identification using artificial neural networks *51st AIAA Aerospace Sciences Meeting including the New Horizons Forum and Aerospace Exposition* p 878
[39] Valasek J and Chen W 2003 Observer/Kalman filter identification for online system identification of aircraft *J. Guid. Control Dyn.* **26** 347–353
[40] Hussein A, Hirasawa K and Hu J 2003 Online Identification and Control of a PV-Supplied DC Motor Using Universal Learning Networks. *ESANN* pp 173–178
[41] Rubaai A and Kotaru R 2000 Online identification and control of a DC motor using learning adaptation of neural networks *IEEE Trans. Ind. Appl.* **36** 935–942
[42] Griñó R, Cembrano G and Torras C 2000 Nonlinear system identification using additive dynamic neural networks-two on-line approaches *IEEE Trans. Circuits Syst. Fundam. Theory Appl.* **47** 150–165
[43] Guo Y, Wang F and Lo J T-H 2017 Nonlinear system identification based on recurrent neural networks with shared and specialized memories *Control Conference (ASCC), 2017 11th Asian* (IEEE) pp 2054–2059
[44] Ibrahim H E 2012 Adaptive control based on neural network system identification *Proceedings of the 11th WSEAS international conference on Electronics, Hardware, Wireless and Optical Communications, and proceedings of the 11th WSEAS international conference on Signal Processing, Robotics and Automation, and proceedings of the 4th WSEAS international conference on Nanotechnology* (World Scientific and Engineering Academy and Society (WSEAS)) pp 84–91
[45] Sadeghi M and Farrokhi M 2016 Online identification of non-linear dynamic systems by Wiener model using subspace method and neural networks *Trans. Inst. Meas. Control* 0142331216663618
[46] PIVOŇKA P, VELEBA V and OŠMERA P 2007 Using of Neural Network Based Identification for Short Sampling Period in Adaptive Control *11th WSEAS Int. Conf. Syst. Crete Isl. Greece* 12–17
[47] Daachi B and Benallegue A 2003 A stable neural adaptive force controller for a hydraulic actuator *Proc. Inst. Mech. Eng. Part J. Syst. Control Eng.* **217** 303–310

[48] Goedtel A, SILVA I N D and SERNI P J A 2005 Neural network based estimation of torque in induction motors for real-time applications *Electr. Power Compon. Syst.* **33** 363–387

[49] Purwar S, Kar I and Jha A 2007 Nonlinear system identification using neural networks *IETE J. Res.* **53** 35–42

[50] Chen S, Billings S and Grant P 1990 Non-linear system identification using neural networks *Int. J. Control* **51** 1191–1214

[51] Jafari M R, Alizadeh T, Gholami M, Alizadeh A and Salahshoor K 2007 On-line identification of non-linear systems using an adaptive RBF-based neural network *Proceedings of the World Congress on Engineering and Computer Science WCECS*

[52] Salahshoor K and Jafari M R 2012 On-line identification of non-linear systems using adaptive RBF-based neural networks *Int. J. Inf. Sci. Manag. IJISM* **5** 99–121

[53] Rocha Filho O D and de Oliveira Serra G L 2016 Online identification based on instrumental variable evolving neuro-fuzzy model for stochastic dynamic systems *Fuzzy Systems (FUZZ-IEEE), 2016 IEEE International Conference on* (IEEE) pp 9–16

[54] Puttige V R and Anavatti S G 2007 Comparison of Real-time Online and Offline Neural Network Models for a UAV *IEEE Int. Conf. Neural Netw.* 412–417

[55] Samal M K, Anavatti S and Garratt M 2009 Real-time neural network based identification of a rotary-wing UAV dynamics for autonomous flight *Industrial Technology, 2009. ICIT 2009. IEEE International Conference on* (IEEE) pp 1–6

[56] Feng H, Qiao W, Yin C, Yu H and Cao D 2019 Identification and compensation of non-linear friction for a electro-hydraulic system *Mech. Mach. Theory* **141** 1–13

IOP Conf. Series: Materials Science and Engineering **767** (2020) 012051 doi:10.1088/1757-899X/767/1/012051

Implementation of Li-Fi based home automation system

S Ismail, N Ahmad and I D Arshad

School of Computer and Communication Engineering, Universiti Malaysia Perlis, 02600 Arau, Perlis, Malaysia.

sabarina.unimap@gmail.com

Abstract. This paper presents an Implementation of Li-Fi based home automation system prototype. This system was develop to verify the effect of LED types on VLC transmission. LED is used to transmit the input data information to the receiver LDR through visible light communication (VLC). Arduino microcontroller was used to retrieve information from the temperature sensor, light sensor, PIR sensor and MQ2 gas sensor. The output data of sensors were sent to LDR by LED. The result demonstrates that the colour types of LED has an effect on transmission distance and the voltage of LED is proportional to the distance. Red LED has better performance compared to others.

1. Introduction

Communication through light has been exist since ancient time. As time passed, the research on technology of communication has discovered new method to improve human communication. In 2011 during TED Global talk, Harald Haas has introduce a new optical wireless communication (OWC) to emerging wireless communication. The OWC is called Light Fidelity (Li-Fi), a coined term created by Harald Haas [1]. Li-Fi is the technology which will replaces the Wi-Fi technology.

Li-Fi is a new wireless communication technology which enables data transmission through LED light. The disadvantages of Wi-Fi will be overcome by Li-Fi such as the speed of data transfer of Li-Fi. The main differences between Li-Fi and Wi-Fi was Li-Fi using light spectrum and Wi-Fi using radio spectrum. Li-Fi visible light spectrum is 10,000 times broad compared to radio frequency (RF) spectrums. This light spectrum is illustrated in figure 1.

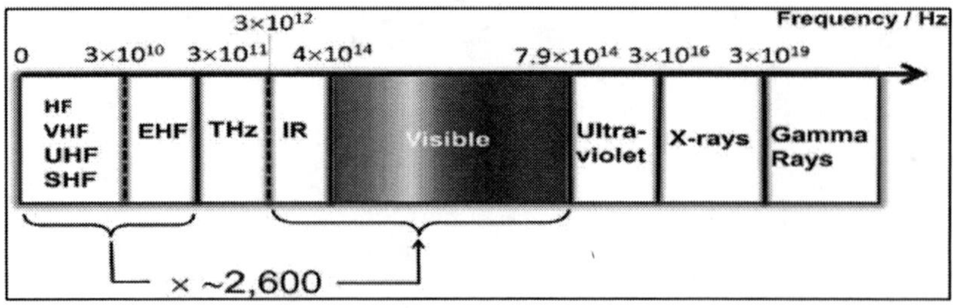

Figure 1. Light spectrum [1].

Li-Fi almost similar to Wi-Fi in term of bidirectional and fully networked wireless communication technology. It could be replacement of RF in data broadcasting. It is measured that some Wi-Fi is about 100 times slower than Li-Fi implementations [2]. Due to continuation of the trend to move to higher frequencies in the electromagnetic spectrum, Li-Fi could be classified as nm wave communication. Li-Fi technology uses white light emitting diodes (LED) for high speed wireless communication, and can achieve speeds of over 3 Gigahertz from a single micro LED [3]. This prove that Li-Fi technology is much faster compared to Wi-Fi.

Li-Fi based home automation system has the ability to control home using light intensity from one place or one device as it save time and save energy. Besides of increasing energy efficiency, the users able to monitor their home. Multiple of smart home devices has been created by integrated with different wireless medium such as Zigbee, Bluetooth and Wi-Fi. The advance of technology nowadays enhances the development of this smart devices [4]. However, this wireless medium has lacked in term of communication range, speed and security. The prominent problem of this wireless medium is the security. The security is not satisfying due to a lot of cyber hacking case on home security system. By using Li-Fi technology, the system is much secured because the light frequency cannot be hack.

Li-Fi technology ideas as the optical version of Wi-Fi based on VLC which use the visible light as the optical carrier for illumination and data transmission [5]. In this technology, the LED plays an important role. LEDs have evolved as lighting source and have potential for data transmission that justifies the high eligibility of LED lights. In Li-Fi based home system, LED is used at transmitter of VLC and photo diode at the receiver capture the light. Li-Fi relies on transmission of light that can be deployed by LED to cover entire home. In advanced LED technology, IEEE 802.15.7 can support high data rate of VLC [6]. VLC had wavelength in the range of 380 -7 80 nm and for unregulated bandwidth more than 350 Terahertz.

2. Li-Fi transmitter and receiver

Most of the Li-Fi technology comprises the LED [7]. LED device is used to emit visible light when an electric current passes through it. The data send in term of the optical spectrum wireless medium for signal propagation through the light spectrum. Data are easily encoded by adjusting the flickering on and off of the LED lighting. This action leading to the generation of different string of binary codes in the form ones and zeros. In addition, LED must in the line of sight (LOS) with the receiver for data transmission between LED and detector to ensure the receiver LOS can detect the light signals.

XLamp MX-6 LED is used as the LED transmitter in this Li-Fi communication. It has the standard lighting class for indoor LED lighting design that have reliability and color consistency. Besides of high flux and high voltage option characteristic, it can deliver high lumen output maximum to 339 lumens depends on the color of LED [8].

Phototransistor is used as detector devices that have light sensitive base region and this devices can be used for switching application or implementation. It has luminous sensitivity and spectral response that decide the longest wavelength as the sensitivity of phototransistor. However the speed of response the gain of phototransistor are inversely proportional. The output of phototransistor depends on wavelength of incident light, area of light exposed and the direct current (DC) gain of transistor.

Illumination is the spread of the light that making the LEDs can be as a media in data communication. The challenge is how the illumination of LEDs enable the data sending while the illumination is low. Therefore, it is related to distance between sender and receiver, which proportional to LEDs brightness while the Li-Fi can fulfil the user satisfaction in order the illumination of LED can stay safe for the user.

3. Implementation of Li-Fi home automation system

The implementation of Li-Fi home automation system prototype is shown in figure 2. It consist of combination Li-Fi transmitter and Li-Fi receiver. The temperature sensor, light sensor, smoke sensor and motion sensor was the input for microcontroller on the transmitter. The receiver is consist of relay circuit, fan, bulb, buzzer and LED.

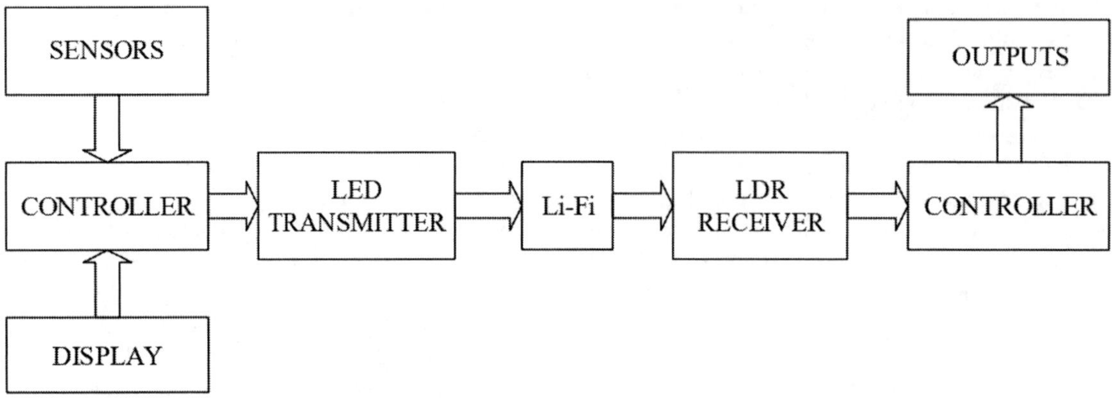

Figure 2. Block diagram of Li Fi home automation system.

The signal from sensors will sent to controller. The controller will process the signal and sent to Li-Fi transmitter (LED). This LED will emits the light signal through free space and detect by photo detector (LDR) at the Li-Fi receiver. The light signal will be process by controller and sent to relay circuit to switch on the electrical devices respective to the sensors.

Arduino Uno board microcontroller with embedded universal asynchronous receiver-transmitter (UART) as controller is use connect to Li-Fi LED transmitter to process the output in terms of flickering LED which represent data '0' and '1'. Li-Fi receiver, LDR that will capture the light and sent it to receiver microcontroller. PWM will be used to replace Pulse Position Modulation (PPM) in this microcontroller to reduce the delay and increase the data rates. PPM requires that both transmitter and receiver devices to be synchronized. PPM also have high sensitivity to multi-pathway interference, which can disrupt a transmission because it will altering the difference in arrival times of each signal.

The prototype of Li-Fi based home automation system is operated by using 9V power supply. Figure 3 shows the flow of functionality sensors of the system. It will by initiates input signal of sensors with two Arduino microcontroller. The input are LM35 temperature sensor, LDR light sensor, PIR motion sensor and MQ-2 gas sensor. If LM35 is detected high temperature, the voltage will increase. If the value of sensor higher than voltage of room temperature, that sensor will activate microcontroller. The Li-Fi transmitter will receive the output signal from microcontroller and transmit the light signal to Li-Fi receiver and will on the fan. Same goes to MQ-2 gas sensor, if it detect high smoke, the voltage will be increase, so it will sent information to microcontroller to turn on the buzzer on the receiver side. Meanwhile, for LDR light sensor, when the sensor in the dark surrounding, the voltage will be drop, so it will sent information to microcontroller and turn on the bulb on the receiver. PIR motion sensor work based on temperature. If the changing of temperature is detected, the LED will turn on at the receiver. LED transmitter will transmit signal in term of flash LED and that light signal will be received by the LDR photo detector. The signal will be process by controller. The output will sent to relay circuit to switch on the electrical devices respective to the sensors.

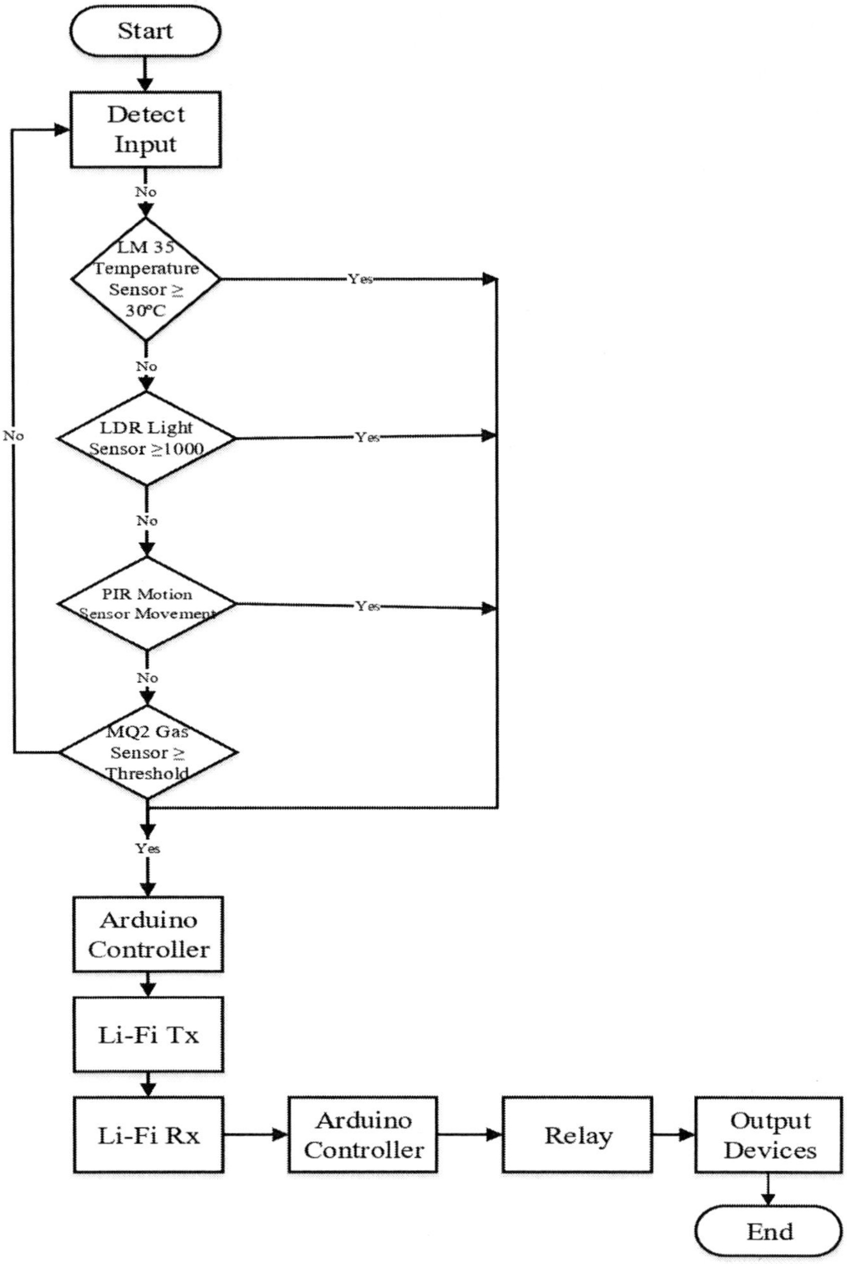

Figure 3. Signal flow of Li Fi home automation system.

Meanwhile, for LDR light sensor, when the sensor in the dark surrounding, the voltage will be drop, so it will sent information to microcontroller and turn on the bulb on the receiver. PIR motion sensor work based on temperature. If the changing of temperature is detected, the LED will turn ON at the receiver.

LED transmitter will transmit signal in term of flash LED and that light signal will be received by the LDR photo detector. The signal will be process by controller. The output will sent to relay circuit to switch on the electrical devices respective to the sensors.

In this prototype, Li-Fi LED transmission is tested by using three different colour of LED which were super bright red LED, Cree X-TE warm white LED, and super bright blue LED. The wavelength of super bright red LED was 660 nm, 580 nm for Cree X-TE warm white LED and 525 nm for super bright blue LED. Cree X-TE warm white LED is a mixed of yellow and red phosphors that allow it have more wavelength compare to cool white LED.

4. Result and discussion

The Li-Fi LED receiver received the light intensity in term of analogue input that detected by LDR as Li-Fi receiver. The result was obtained by tested the Li-Fi based home system prototype in the room with normal light intensity. Table 1 show the maximum distance between Li-Fi transmitter and Li-Fi receiver can achieved by exchanging the colour of LED with the different voltage input data to the LED.

Table 1. Transmission distance between transmitter and receiver.

Voltage level (volt)	Maximum Distance between Li-Fi TX and Li-Fi RX (cm)		
	Super Bright Red LED	Cree X-TE Warm Light LED	Super Bright Blue LED
3.5	86	42	19
4	105	48	30
4.5	121	90	80
5	165	97	86

Figure 4 show the comparison result between difference colours of LED for Li-Fi transceiver transmission. The Red LED can transmit long distance compare to Blue LED and Warm white LED for the same level of voltage. This result has proven the effect of light wavelength. The light colour of LED can give high light intensity to be received by the receiver. Its show that the wavelength of LED colour effect the transmission distance between transmitter and receiver.

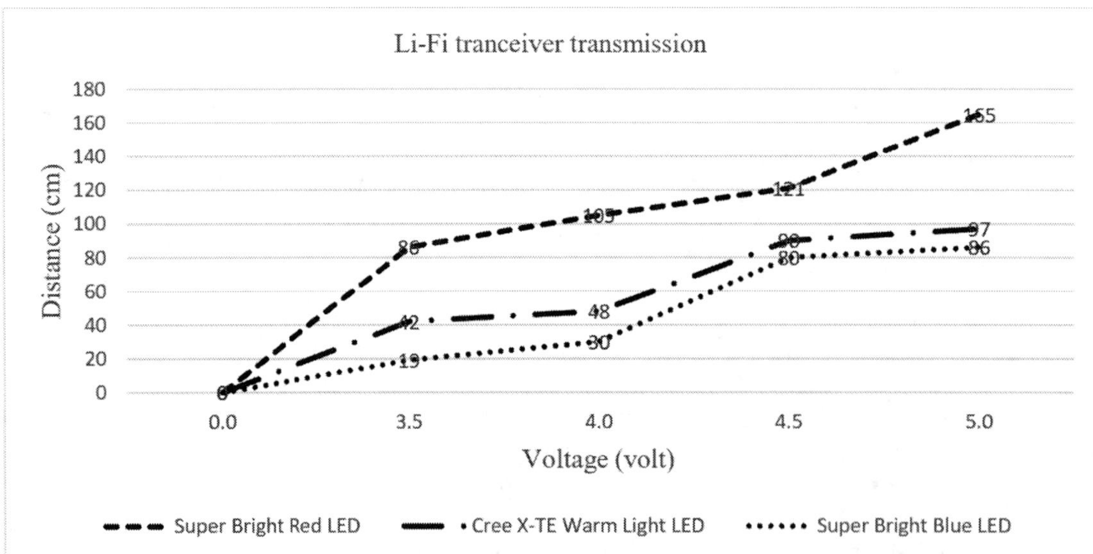

Figure 4. Distance versus different voltage level of input data and LEDs.

5. Conclusion

Through implementation of Li-Fi based home automation system, the trend and the relationships effect of types of LED with the distance between Li-Fi transmitter and Li-Fi receiver is shown. Therefore, the wavelength of LED colour on transmitter and the photo detector sensitivity has significant contribution for VLC communication. VLC through Li-Fi able to transmit the efficient information even though low power supply for the LED beam generation that could transmitted to larger transmission distance. By using Li-Fi based home automation system, the user able to control home automatically without wasting the energy. The Li-Fi based home automation system can provide higher data security as light spectrum cannot be tapped.

References

[1] S Dimitrov and H Haas, 2015, *Principles of LED Light Communications Towards Networked Li-Fi*. Cambridge CB2 8BS, United Kingdom: Cambridge University Press, 2015, 207 pp.

[2] A Chinchawade and P Sujatha, 2016, "Li-Fi Based Audio TransmissionWith Home/Office Automation System", *International Journal of Innovative Research in Science, Engineering and Technology*, vol. 5, no. 7.

[3] H Haas, L Yin, Y Wang and C Chen, 2016, "What is LiFi?" *Journal of Lightwave Technology*, vol. 34, no. 6, pp. 1533-1544.

[4] E Griffith, "How to Build Your Smart Home: A Beginner's Guide", *PC Magazine Asia*, 2016. [Online].Available:http://sea.pcmag.com/digitalhome/9916/feature/how-to-build-your-smart-home-a-beginners-guide. [Accessed: 23- Oct- 2016].

[5] P Koppperundavi, S Mageshwari and P M Priya, 2015, "Smart area-light fidelity communication", *SSRG International Journal of Electronics and Communication Engineering (SSRG-IJECE)*, vol. 2, no. 3.

[6] S Rajagopal, R D Roberts and S K Lim, 2012, "IEEE 802.15.7 visible light communication: modulation schemes and dimming support," in *IEEE Communications Magazine*, vol. 50, no. 3, pp. 72-82.

[7] K Kalidhas, J Ninan, JM Chacko, S Saseendran, Kalidhas, and U Chandran, 2016, "Implementation of Li-Fi Technology for Home Automation and Vehicle Communication." *International Journal of Science Technology & Engineering*, vol. 2, no. 10.

[8] "Cree XLamp MX-6 LEDs", *Cree.com*, 2016. [Online]. Available: http://www.cree.com/LED-Components-and Modules/Products/XLamp/Discrete-Directional/XLamp-MX6. [Accessed: 29- Oct- 2016].

IOP Conf. Series: Materials Science and Engineering **767** (2020) 012052 doi:10.1088/1757-899X/767/1/012052

Technologies of solar tracking systems: A review

A.R. Amelia[1], Y.M. Irwan[2], I. Safwati[3], W.Z. Leow[1], M.H. Mat[2] and Mohd Shukor Abdul Rahim[2], S. Md. Esa[2]

[1]School of Electrical System Engineering, Universiti Malaysia Perlis, Malaysia
[2]Centre for Diploma Studies, Universiti Malaysia Perlis, Malaysia
[3]Institute of Engineering Mathematics (IMK), Universiti Malaysia Perlis, Malaysia

Email: amelia_razak87@yahoo.com

Abstract. Solar energy is abundantly in nature and sustainable energy resources around the world. The main challenge with the solar field is less amount of sun energy captured by photovoltaic (PV) systems. The great performance of the PV systems can be achieved if the panel is kept perpendicular to the direction of the radiations of sun. Hence, solar tracker system is the method to keep the optimum position of the PV panel for always perpendicular to the solar radiation. This paper aims to review on various technologies of solar tracking to determine the best PV panel orientation. The various types of technologies of solar tracking system have been discussed which includes passive solar tracker, active solar tracker and chronological tracker system. The movement degrees of solar tracking system also have been addressed which consisting single-axis solar tracking system and dual-axis solar tracking system. This paper is also overviews the tracking techniques performance, construction, performance, advantages, and disadvantages of existing solar tracking system. The limitations of solar tracking systems are also highlighted for future action improvement. Through this research studies, the most favorable solar tracking system was identified as active solar tracker with the dual axis rotation.

1. Introduction

Nowadays, the concentration in using renewable energy instead fossil fuel has expanded throughout the world. Producing electric energy from renewable energy sources provides environmental protection and sustainability [1]. The solar energy is currently holds incredible promises for renewable energy to meet the world energy demand. Solar energy can be transformed into electricity in photovoltaic (PV) system using PV panels. However, energy generated from PV panels is related with atmospheric factors such solar radiation intensity, ambient temperature, wind speed and humidity [2]- [5]. Higher efficiency produced from PV panel will be generated when sunlight is perpendicular to the surface of the PV panel during the operation. This can be achieved by means of using a PV panel mount which tracks the movement of the sun throughout the day. Therefore, many investigations have been attempted to increase the efficient usability of PV panel using the solar tracking system [6]- [8].

In fixed PV panel operation, the tilt angle is location dependent. Instead of, using the solar tracking system, the PV panel can be keep in optimum position perpendicular to the solar radiation during daylight hours in order to increase the electrical energy generated. An ideal solar tracker would allow PV panel to align with the sun, compensating for both changes in the altitude angle of the sun and changes in the azimuth angle [9]. These both tracking angles depends on latitude and atmospheric conditions of the site location. The motion of tracking system could either be about one-degree freedom (single axis tracker) and two-degree freedom (dual axis tracker). By using either single or dual axis, the tracking system can be classified into the three main type of their technologies, namely as passive solar tracker active solar tracker and chronological tracker system [10].

Over the years, the researchers had developed of solar tracking system with various designs to maximizes the energy production from PV panel. Many investigations have been attempted to improve the solar tracking performance using effective mechanical drives and efficient control systems. This study seeks to review different types of solar tracking systems considering their technologies to implement and drive the solar tracking systems. The rest of this review paper is organized as follows. The preliminaries study on solar tracking system in Section 2. The technologies of solar tracking system will be presented in Section 3 which divided into three parts. There is passive solar tracker, active solar tracker and chronological solar tracker. Then, the limitation of the solar tracker system will be discussed in Section 4 and the paper is concluded in Section 5.

2. Solar tracking systems

Solar tracking systems are mainly designed to keep the surface of the PV panel perpendicular to direction of the solar radiation beam. The solar tracking should be positioned that can achieve the optimum angle of incidence, thereby electrical energy produced by the PV panel can be maximized [11]. Solar irradiance, solar azimuth angle, elevation angle, inclination angle, declination angle and zenith angle are the important parameters that identify the best position of solar tracking system [12]. The most important angles in determining the position of the sun is altitude (elevation) and azimuth angle.

Solar irradiance can be determined by measuring either the power of the light sources or the luminous flux. Elevation (α_s) and zenith (θ_z) angles have close definition to the declination angle [13]. Declination angle (δ) is the angle between equator and a line drawn from the centre of the sun to the centre of the earth. Meanwhile, elevation angle is the angle between solar radiation and the horizon. According to definition, an elevation can be calculated using the Eq. 1 [12].

$$\alpha_s = 90 - \theta_z \tag{1}$$

Solar azimuth angle (γ_s) is the angle showing the tilt of the sun's ray to the north in the direction of the clockwise. Meanwhile, surface azimuth angle (γ) is the angle showing the deviation of the vertical surface to the local longitude. Inclination angle (β) represents as the angle between the surface and horizon, positive value for the surface facing to the equator. Fig. 1 shows the diagram illustrating the solar angle. Moving solar tracking systems can be manually or automatically motion. In most of the cases, solar tracking system includes several considerations such one or two motors, different type of optical sensor and independent or auxiliary power supply. The classification of these aspects depends on various parameters including the force driving their moving fixtures and mode operation.

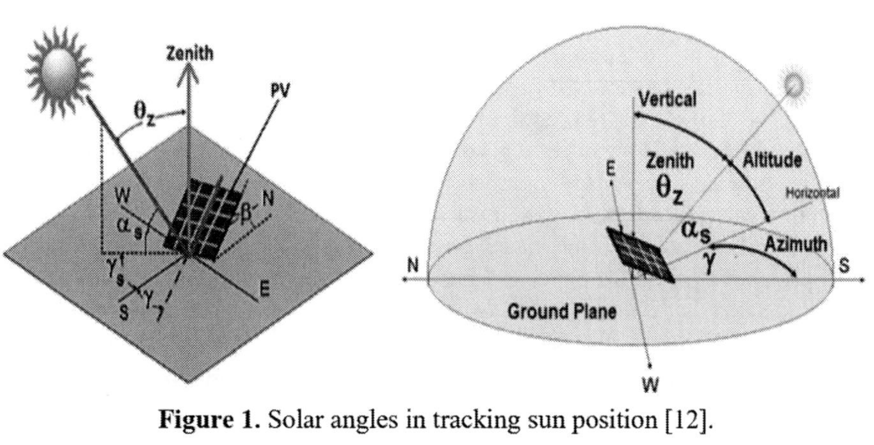

Figure 1. Solar angles in tracking sun position [12].

3. Categorization of solar trackers based on technologies

Solar tracker systems can be classified into three main categories based on the techniques that control the PV panel movement. It can be either passive, active or chronological tracker system [14].

3.1. Passive solar tracker

Passive solar trackers have capability to orient their sensing units towards the direction of the solar radiation beam without using any mechanical drives [15]. Most of these trackers consists a pair of actuators and filled with expansible gas or an on-shape memory alloy. This system utilizes the concept thermal expansion, or an imbalance in pressure between two points at the both ends of the tracker. When the PV panel is perpendicular with the sun, the two sides are at equilibrium. Once the sun moves, one side is heated and causes one side to expand and the other to contact, causing PV panel to rotate [16].

The first commercial passive solar tracking system was introduced by Zomeworks Corporationn in 1969. In Zomeworks Track Racks, the PV panel with tracking system can increase their electrical production output by 25% when compares with the fixed PV panel. Track Racks systems are highly cost-effective components for water pumping systems, industrial power systems, utility applications and cathodic protection systems [17]. In other work, Clifford and Eastwood introduced an economic passive solar tracker model mounted on a wooden frame, symmetrically on either side of a central horizontal axis. The movements of the system were activated by both aluminium/steel bimetallic strips and controlled by a viscous damper. Simulation results and real time measurements stated that more 23% energy gain by the proposed solar tracker when compares to a fixed PV panel [18].

Few years later, Narendrasinh et al. proposed a passive solar tracker system that has been implemented and tested using three types of gasses: thinner, acetone and methanol. The proposed solar tracker which working in the Zomework principle able to increase the power up to 23% when compared to the fixed PV system [19]. In the same year, Elmaged et al. evaluated the performance of solar pumping systems by installing the PV panels in two different modes. The first mode is fixed PV panel with its installation facing due to the South. Meanwhile, the other mode is the PV panel with passive solar tracker, which installed facing due to the South and inclined with the horizon. This passive solar tracker system was designed to be rotated from east to west throughout the day (daily movement). The results illustrate that when using the passive solar tracker system, the incidence of solar radiation and accumulated power can be increased by 14% and 16%, respectively [20].

Using the passive solar tracker system, the operation is less complex, most of them have a pair of actuators, filled with expansible gas or an on-shape memory alloy, working against each other and balanced by equal illumination [21]. However, this method of sun tracking is low in accuracy and relies mainly on weather conditions of the site location. The selected location for solar tracker installation is very crucial because the location must attain adequate continuous sunlight for an efficient heating process. The type of active solar tracker system can solve the problem of using passive solar trackers.

3.2. Active solar tracker

Active solar tracking system determines the position of the sun during the day with existing of the sensors, continuously [22]. Sensor will trigger the motor or actuator to move their fixture following the solar radiation. If the sunlight not perpendicular to the tracker, then there will be a different illumination on one light sensor compared to another. This difference can be used to determine the direction in which tracker has to be directed in order to be perpendicular to the sun [23]. Active solar trackers are further categorized in four categories based on their tracking strategies. There are microprocessors and electro-optical sensor, based on date and time, based on auxiliary bifacial PV cells as well as the combination of the three types above [24]. This paper is focused on the movement technologies of the solar tracker. The number of the axis that is commonly used to move the PV panels is the most crucial factor for all solar trackers system. Currently, the solar trackers are categorized into two main groups based on their movement technologies which are single-axis trackers system and dual-axis trackers system.

3.2.1. Active solar tracker with single-axis system
A single-axis tracker system provides only one degree of freedom which acts as the axis of rotation [25]. The single-axis tracking system is the simplest tracker, they usually use less energy and have minor complexity than a multi-axis system [26]. In 2012, Deb et al. proposed a solar tracker system with two sensors to measure the temperature in eastern and western locations. By using the LABVIEW, the authors calculate the light intensity with the data of the temperature collected. The movement direction of the solar tracker system through a stepper motor directly was determined when there has a difference between one direction of the light intensity into the other direction [27]. Kamala et al. designed a single-axis solar tracking driven by a PIC microcontroller to maximize energy harvesting. The two PV panels are arranged in the form of a triangle located opposite of each other. Two levels of the threshold are implemented to drive the PV panel movement. The first threshold is designed to activate tracking and indicate the availability of solar power. Meanwhile, the second threshold is utilized to switch off the peripherals during the cloudy or rainy weather condition for a long period. The authors have designed the economic system but facing with the problem when the system stops operate after sunset. It is due to the PV panel needs to be positioned back to east side for the next tracking day [28].

Recently, Dian et al. proposed and designed a GPS-based solar tracker system, that can move the vertical solar still follow the azimuth angle of the sun. They compare the power of 2 solar cells placed in different positions. The first solar cell was located upright and rotated with the solar tracker system. Meanwhile, the second solar cell was located fixed horizontally. The results show that the first solar cell generates about 22% larger than second solar cell in term of average output power. Besides that, the authors also state that solar tracking systems become faster with the use of GPS and astronomical calculations [29]. After a year, Sebastián and Pedro carried out two very simple tracking strategies, North-South horizontal single-axis tracker for latitudes up to 50°. In this study, they change the tracker positions considering 2 positions in morning/afternoon or 3 positions in morning/noon/afternoon at pre-defined daily schedules. Results indicate that the 3 position tracker can provide significant energy gains with respect to fixed photovoltaic system with the high percentage for annual irradiation gains [30].

3.2.2. Active solar tracker with dual-axis system
The earth follows a complex motion that consists of two motions which are the daily motion and the annual motion. The daily motion causes the sun to appear in the east to the west direction over the earth. Meanwhile, for the annual motion, the sun to tilt at particular angle while moving along east to west direction [31]. Therefore, extensive research has been conducted to improve the efficiency and promote the use of dual axis solar tracker. In a dual axis solar tracker, there are two degrees of freedom which act as axes of rotation. There are two variants of dual axis tracking systems namely as a polar-altitude and azimuth-altitude [10]. These solar tracking systems usually consisting of four Light Dependent Resistors (LDRs), two motors and a controller [32].

Chabuk et al. proposed a dual-axis solar tracking system based on microcontroller principle working. By using an algorithm, the microcontroller used to determine the positions and directions of the PV panel. Meanwhile, the best angle was obtained using a real time clock microcontroller then has been feed directly to the stepper motor for PV panel movement. The proposed system does not rely on the clearness of the sky or the position of the sun, thereby able to work even in bad weather. The results generated by this system are more efficient when compared to the single axis and fixed panels [33]. Fathabadi develop a novel offline sensorless dual-axis tracker with high accuracy for tracking the sun position. The simple structure with low cost tracker consists of a controller, stepper motors for adjusting altitude angle and azimuth angle and gear box that function to rotate the PV panel. This proposed solar tracker also used offline data without any feedback signal. Therefore, the solar tracker is robust to external disturbances. The experimental results verified that 24.59% more solar energy is captured during one year with the existing solar tracker [6].

Sebastijan et al. proposed a dual-axis tracking system which able to move either horizontally or vertically based the selected zenith and azimuth angles. The LDRs were used to determine the position and direction of the PV panel to trigger the DC motor. The performance of the proposed system was compared with the fixed PV panel under a few selected sunny days. The results show that the output energy of the dual-axis tracker can be improved by 27% when compared to the fixed PV panel [34]. In the similar year published, Sidek et al. introduced a dual-axis tracking system that implement the sun trajectory path algorithm to identify the solar tracker's position. The proposed system uses the GPS sensor to identify the best position of the PV panel referring to the longitude and latitude lines. Meanwhile, the azimuth and altitude angle are utilized to directly feed the positioning controller which leads the motors to move either clockwise or counter clockwise. The results indicated that in clear and cloudy sky conditions, more 26.9% and 12.8% energy output can be generated by the proposed system when compared to the fixed PV panels [35].

Rashid et al. design built and tested a hybrid dual-solar tracking system. This working operation of the solar tracker is based on the solar map combining with the light sensor based continuous tracking mechanism. The light sensors used to compare the darkness, cloudy and sunny conditions for daily tracking. Meanwhile, a real time clock device was required by the electrical controlling device to track the sun's apparent position at different months and seasons. The capability of this proposed hybrid solar tracker was compared with a static and continuous solar tracking systems. Results revealed the power saved in system operation by the hybrid solar tracking is 44.44% more when compared to the continuous tracking system even though their power gain is almost similar [31]. Sathis designed a dual-axis solar tracking system that can maximize the incident of the sun through the PV panel. This system used the LDRs reading to determine the direction of the PV panel through microcontroller. The Op-Amp comparator is used to collect the readings of the LDRs for operating microcontroller. In this proposed system, the drive and stepper motors are used to move the PV panel from one side to another side. The results were evaluated by comparing the performance between the proposed system and the fixed PV system. Results indicate that the output power can be increased by 40% and 45% for fixed and dual-axis tracking system, respectively [36].

Masoud et al. combined a dual-axis tracking system with a wind-tracking system. The wind-tracking system acts as an auxiliary system, which complements the dual-axis tracking in case of windy conditions. By using this proposed system, the power generation has been increased throughout the experimental study. The conventional dual-axis tracking system presents an increase of 39.43% in total daily collection over a fixed PV system meanwhile, the hybrid tracking system yielded an increase of 49.83% [37]. In order to improve the operation of the solar tracking system, Arlikar et al. proposed a three-dimensional solar tracker system with a special arrangement of LDRs. A PLC is used to control the position of the solar tracker system with the existing of LDRs. Meanwhile, the PV panel is fixed and reflectors are used in order to maximize the incident light intensity. The two stepper motors are used for three-dimensional tracking, one to move the plane of the PV panel and the other will move this plane in a third dimension. The results identified that the three-dimensional PV panel produces more energy than a traditional PV panel [38].

Bentaher et al. develop a simple tracking system based on LDRs working principle. A mechanism of PV and parabolic concentrator with dual-axis movement was taken into consideration to identify the difference of the LDRs against its direction towards the sun. The DC motors are used to actuate and move the two axis as well as control the movement of the two axis by the two angular sensors [39]. Juang et al. designed and built a dual-axis solar tracking system to maximize irradiance and limit the battery voltage consumption. This proposed system consisting microcontrollers, LDRs, actuators, an Arduino Uno R3, and a Pololu Dual Motor Shield. The linear actuators are adjusted when the absolute value of the calculated difference is greater than the dead band value provided by LDRs. The proposed system implemented for real-time responsiveness, and the increase efficiency of the output power at least between 15% and 20% [40].

Yao et al. used two automatic tracking strategies for their dual-axis tracking system. The first type was the normal tracking strategy and the other was a daily adjustment strategy. A hybrid strategy involving time-based control as well as sensor-based control was used in this proposed system. The solar tracker was designed to reach a calculated position based on the local time while the sun position sensor would provide a feedback to correct the tracking error. For flat PV systems, a simple daily adjustment strategy was employed since a high degree of accuracy is not necessary [41]. Assaf develop a dual axis solar tracking system to measure accumulate solar radiation. The proposed system consists of four LDRs, a programmable logic controller (PLC) device and two servo motors. The function of LDRs as usual to identify the sun position, thereby provide a voltage signal to the PLC. The Soft Comfort V6.1 software was used to develop the control software in order to identify the optimum PV panel's position during daylight. The proposed solar tracking system was tested and implemented for the duration of three days. The results show that the accumulated solar radiation hen using the two-axis tracking system was higher than the solar radiation collected using the fixed PV system [42].

H.S. Akbar et al. proposed the dual-axis solar tracker to track the to track the position of the sun in both azimuth and altitude axis. A tracker system consists mainly of the ATmega 328 controller, DC motors, LDRs sensor and relays. The authors also studied on the effect of temperature and panel covering by colored cellophane on the performance of the tracker system. The results indicated that the production output power for single axis solar tracker and dual axis solar tracker are 24.05% and 26.22%, respectively higher energy than fixed PV panel [43]. El Hammoumi et al. designed and implemented the low cost of an active dual- axis solar tracker. Two servo motors facing the highest intensity of sunlight captured by four units of LDRs sensors used to track the sun's movement. In order to minimize the cost, ATMega328 microcontroller which is integrated into a low- cost board, Arduino Uno was implemented for this proposed system. The results were evaluated by comparing the performance between the energy produced by the proposed tracker and a fixed PV panel. By using an active dual- axis solar tracker, more 36.36% energy was produced when compared to the fixed PV panel [44].

3.3. Chronological solar tracker

A chronological solar tracking system is a time-based tracking system where the structure is moved at a fixed rate and angle throughout the day as well for different months. Therefore, the motor or actuator is controlled to rotate at a slow average rate of one revolution per day (15° per hour) [31]. This method of sun tracking is more energy efficient because no energy losses at this tracking calibration due to low tracking error. Roong and Chong using the chronological method with angle of rotation 15° per hour for single axis solar tracking system. The solar tracking system plant with the total surface area is around 70m^2 was constructed in University Teknikal Malaysia Melaka (UTeM). The experiments started from 8:00 a.m. to 3:00 p.m. throughout a day for duration of 5 days with the fixed angle rotation of the solar tracker with 15° per hour [45].

Sidek et al. presents the dual axis solar tracking system using electronic control system. The working principle of the algorithm for the proposed method is based on the GPS and astronomical equation. Furthermore, the microcontroller based tracking system embedded with a PID controller. By using SolidWorks, the biaxial solar tracker structure was designed. The power generated between two systems fixed-tilted PV panel and the solar tracker system were compared with each other in order to evaluate the performance of the solar tracer system [46]. A single-axis solar tracking system was designed by Samantha et al. to maximize the efficiency of PV cells. By following the working principle of chronological solar tracking system, allow the PV panel to track the position of the sun with the assistance of a motor controlled by a microcontroller. In order to identify the position of the sun, a PIC18252 microcontroller is connected to an L293D motor driver thereby the PV panel can be moved to the desired directions. The maximum-intensity of solar energy can be produced when the rays of the sun is perpendicular to the surface of the PV panel [47].

4. Limitations of solar tracking system

As for passive and active tracker differ with way of deploying the design trade off. Passive tracking system require virtually no extra power supply because this system works alone with the solar radiation. This tracking technology is less complex and economic, the construction developed without any electronic controls or motors. However, it is low in accuracy and relies mainly on the weather conditions. Lower energy has been produced by this tracking system when compared to the active tracker system. Moreover, this technology of the solar tracker cannot be favourite for the regions hit by the extreme cold

IOP Conf. Series: Materials Science and Engineering **767** (2020) 012052 doi:10.1088/1757-899X/767/1/012052

weathers because they easily stop working at very low temperatures. Therefore, they have not yet been widely accepted on their application and a few bit literature available about this type of technologies of solar tracking system. Meanwhile, the active tracking is based on accuracy while there is consumption of generated energy. This technology of sun tracking is reasonably accurate except on very cloudy days. It is because the sensors facing the problem in determining the position of the sun in the sky. In addition, the active solar tracking system are more accurate and frequently used, but they require the extra power used for sensors, microcontroller or motors. For the chronological tracking system, the continuous rotation of the motor or actuators leads the system used more power consumption. Besides, this system also unnecessary when operate under the very cloudy day.

By considering the movement degree of freedom in solar tracking system, the single-axis solar tracker is the simplicity when compared to the dual axis rotation. In addition, the single-axis solar tracker used lower cost and energy consumption since this system only use one motor for tracking the sun horizontally or vertically. However, this system can only track the sun during the daily movement and not the yearly movement. In addition, the efficiency of the tracking system is reduced by a large amount due to the rotation around only single-axis caused by the cloudy days. Although single-axis solar tracker is less complexity construction, but their efficiency is lower than that of dual-axis solar tracker. The main advantage of using dual-axis tracking in comparison with single one is tracking the sun movement not only during the day, but also for the yearly movement. Although dual-axis solar tracker is more expensive construction, they deliver efficient performance when compared to the single axis solar tracker. Table 1 conclude the operation, advantages and disadvantages for each type of technologies for solar tracking system.

Table 1. Passive, active and chronological solar tracker

Technology	Descriptions	Advantages	Disadvantages
Passive	-Thermal expansion in material or imbalance in pressure between two points at the both end of tracker.	- work without using motors or actuators - easy installation - low maintenance cost	- strong dependence on weather conditions - low in accuracy
Active	- use sensors and motors	- more accurate - efficient in tracking the position of the sun	- require the extra power consumption - not very accurate under cloudy day
Chronological	- timed based tracking system - rotate at 15° per hour	- no energy losses - low tracking error	- continuous rotation requires more energy - unnecessary work under cloudy day

5. Conclusion

In conclude, the power generated and efficiency of the PV panel could be improved with the existing of the solar tracker systems. This work present and categorize different types of solar tracking systems based on their technologies and degree movement of rotation. Active solar tracking systems use sensors and motors to track the position of the sun and control PV panel's position. Meanwhile, the passive tracking systems use a low-boiling-point compressed gas fluid that originates from solar heat. Available previous literature introduced the various performance components, use of controller, system efficiency and tracking control strategies. The review reveals that the active trackers were more commonly used in PV panel application as compared to passive trackers. The conclusion from the literature review that active solar tracker with dual-axis system are vital for implementation to PV system with maximum efficiency gained.

References

[1] Yasser M. Safan, S. Shaaban and Mohamed I. Abu El-Sebah 2018 Performance evaluation of a multi-degree of freedom hybrid controlled axis solar tracking system, *Solar Energy* vol. 170 pp 576-585

[2] Swapnil D., Jatin N S and Bharath S. 2013 Temperature dependent photovoltaic (PV) efficiency and its effect on pv production in the world- A review, *Energy Procedia* vol. 33 pp 311-321

[3] W.Z. Leow, Y.M. Irwan, M. Asri, M. Irwanto1, A.R. Amelia, Z. Syafiqah and I. Safwati 2016 Investigation of solar panel performance based on different wind velocity using ANSYS, *Indonesian Journal of Electrical Engineering and Computer Science*, vol. 1 pp 456-463

[4] Chegaar, M., Hamzaoui, A., Namoda, A., Petit, P., Aillerie, M., and Herguth, A. 2013 Effect of illumination intensity on solar cells parameters, *Energy Procedia* vol. 36 pp 722-729

[5] W.Z. Leow, Y.M. Irwan, M. Irwanto, M. Isa, A.R. Amelia and I. Safwati 2016 Temperature Distribution of Three-Dimensional Photovoltaic Panel by Using Finite Element Simulation, *International Journal on Advanced Science, Engineering and Information Technology* vol. 6 pp 607-612

[6] Hassan Fathabadi 2016 Novel high efficient offline sensorless dual-axis solar tracker or using in photovoltaic systems and solar concentrators, *Renewable Energy* vol.95 pp 485-494

[7] W. Batayneh, A. Owais and M. Nairoukh 2013 An intelligent fuzzy based tracking controller for a dual-axis solar PV system, *Automation in Construction* vol. 29 pp 100-106

[8] Nadir Barsoum 2011 Fabrication of Dual-Axis Solar Tracking Controller Project, *Intelligent of dual-axis solar tracking controller project* vol. 2 pp 57-68

[9] C.S. Chin, A. Babu and W. McBride 2011 Design, modelling and testing of a single axis active tracker using MATLAB/Simulink, Renewable Energy vol.36 pp 3075-3090

[10] R. Singh, S. Kumar, A. Gehlot and R. Pachauri 2018 An imperative role of sun tracker in photovoltaic technology: A review, *Renewable and Sustainable Energy Reviews* vol. 82 pp. 3263-3278

[11] Adarsh S., Abhishek Anand and Jayant Singla 2015 Increasing the efficiency of a PV system using dual axis solar tracking, *Proceedings of 11th IRF International Conference* pp. 37-41

[12] S. Yilmaz, H.R. Ozcalik, O. Dogmus, F. Dincer, O. Akgol and M. Karaaslan 2015 Design of two axes sun tracking controller with analytically solar radiation calculations, Renewable *and Sustainable Energy Reviews* vol 43 pp. 997-1005

[13] Nadia Al-Roslan, Nor Ashidi Mat Isa and Mohd Khairunaz Mat Desa 2018 Advances in solar photovoltaic tracking system: A review, *Renewable and Sustainable Energy Reviews* vol. 82 pp. 2548-2569

[14] Dhanabal. R, Bharathi.V, Ranjitha. R, Ponni. A, Deepthi. S, Mageshkannan. P 2013 Comparison of Efficiencies of Solar Tracker systems with static panel SingleAxis Tracking System and Dual-Axis Tracking System with Fixed Mount *International Journal of Engineering and Technology (IJET)*, vol.5 no. 2 pp. 1925-1933

[15] J. Ya'u Muhammad, M. Tajudeen Jimoh, I. Baba Kyari, M. Abdullahi Gele, and Ibrahim Musa 2019 A review on solar tracking system: a technique of solar power output enhancement, *Engineering Science*, vol. 41, pp 1-11

[16] R.S. Zulkafli, A.S. Bawazir, N.A. M.Amin, M.S.M. Hashim, M.S.A. Majid and N.F.M. Nasir 2018 Dual axis solar tracking system in Perlis, Malaysia, *Journal of Telecommunication, Electronic and Computer Engineering* vol. 10 pp 91-94

[17] M.S. Sabry and B.W. Raichle, 2014 Characteristics of residential tracker accuracy in quantified direct beam irradiance and global horizontal irradiance, *Journal of Technology Innovations in Renewable Energy* vol. 3 pp 44-57

[18] Clifford M.J and Eastwood D. 2004 Design a novel passive solar tracker, *Solar Energy* vol.77 pp 269-280

[19] Narendrasinh J, Parmar ANP and Gautam Vinod S. Passive solar tracking system. *International Journal Emerging Technology and Advanced Engineering* vol.5

[20] Elmaged A, and Kamal H 2015, Passive solar tracking system.
[21] S. Degeratu, S. Rizescu, L. Alboteanu, C. Caramida, P. Rotaru, I. Boncea and C. Iancu 2014 Using a shape memory alloy spring actuator to increase the performance of solar tracking system, Annals of the University of Craiova, Electrical Engineering series no. 38 pp 116-121
[22] A. Razif Hamid, A. Khusairy Azim and M. Hafizuddin Bakar 2017 A review on solar tracking system, *e Proceeding National Innovation and Invention Competition Through Exhibition (iCompEx'17)*, pp 1-9
[23] W. Indrasari, R. Fahdiran, E. Budi, L. Jannah, L. Kadarwati, Ramli 2018 Active solar tracker based on the horizon coordinate system, *The 8th International Conference on Theoretical and Applied Physics*, pp 1-6
[24] Ching Chuan Wei, Yu-Chang Song, Chia-Chi Chang and Chua-Bi Lin 2016 Design of a solar tracking system using the brightest region in the sky image sensor, Sensors, vol.16
[25] V. Sumathi, R. Jayapragash, Abhinav Bakshi and Praveen Kumar Akella 2017 Solar tracking methods to maximize PV system output – A review of the methods adopted in recent decade, Renewable *and Sustainable Energy Reviews,* vol 74 pp. 130-138
[26] M. Salaheldin Elsherbiny, Dr. Wagdy R. Anis, Dr. Ismail M. Hafez and Dr. AdelR. Mikhail 2017 Design of single-axis and dual-axis solar tracking systems protected against high wind speeds, *International Journal of Scientific & Technology Research*, vol.6 issue 9 pp 85-89
[27] Deb G and Roy AB 2012 Use of solar tracking system for extracting solar energy, *International Journal Computer Electrical Engineering*, vol.4 pp 42-46
[28] Kamala J and Joseph A. 2014 Solar tracking for maximum and economic energy harvesting, *International Journal Engineering Technology*, vol.5 pp 5030-5037
[29] Dian Artanto, A. Prasetyadi, D. Purwadianta and Rusdi Sambada 2016 Design of a GPS-based solar tracker system for a vertical solar still, *International Conference on Smart Green Technology in Electrical and Information Systems (ICGTEIS)*, pp 140-143
[30] Sebastián Gutiérrez and Pedro M. Rodrigo 2017 Energetic analysis of simplified 2-position and 3-position North-South horizontal single-axis sun tracking concepts, *Solar Energy*, vol. 157, pp 244-250
[31] R.A. Ferdaus, M. Asif Mohammed, S. Rahman, S. Salehin and M.A. Manan 2014 Energy efficient hybrid dual axis solar tracking system, *Journal of Renewable Energy*, vol. 2014 pp 1-12
[32] Jing-Min Wang and Chia-Liang Lu 2013 Design and implementation of a sun tracker with a dual-axis single motor for an optical sensor-based photovoltaic system, *Sensors*, vol. 13, pp 3157-3168
[33] A. Chabuk, A. Shinde, M. Narale, P. Gonjari and Prof. Mr. P. S.Magdum 2017 Dual axis solar tracking system using microcontroller, *International Research Journal of Engineering and Technology (IRJET)*, vol. 4 pp 796-800
[34] Sebastijan Seme, Gregor Srpčič, Domen Kavšek, Stane Božičnik, Tomislav Letnik, Zdravko Praunseis, Bojan Štumberger and Miralem Hadžiselimović 2017 Dual-axis photovoltaic tracking system- Design and experimental investigation, *Energy*, vol. 139 pp 1267-1274
[35] M.H.M. Sidek, N. Azis, W.Z.W. Hasan, M.Z.A.Ab Kadir, S. Shafie and M.A.M. Radzi 2017 Automated positioning dual-axis solar tracking system with precisiom elevation and azimuth angle control, *Energy*, vol. 124 pp 160-170
[36] Sathis and Vatumalai 2016 Arduino based solar tracking system, Faculty of Manufacturing Engineering, University Malaysia Pahang
[37] R Masoud, B Meisam, T Yaghoub and Valeh-e-Sheyda Peyvand 2015 An insight on advantage of hybrid sun–wind-tracking over sun-tracking PV system. *Energy Conversion Management*, vol.105 pp 294–302.
[38] A Pratik, B Abhijit, P Manoj and D Amruta 2015 Three dimensional solar tracker with unique sensor arrangement, *In: Proceedings of the international conference on smart technologies and management for computing, communication, controls, energy and materials (ICSTM)*. pp 509–513

[39] Bentaher H, Kaich H, Ayadi N, Ben Hmouda M, Maalej A and Lemmer U. 2014 A simple tracking system to monitor solar PV panels, *Energy Conversion Management*, vol. 78 pp 872-875

[40] Juang J-N, and Radharamanan R. 2014 Design of a solar tracking system for renewable energy, *In: Proceedings of Zone 1 Conference of the American Society for Engineering Education (ASEE Zone 1)*

[41] Yao Yingxue, Hu Yeguang, Gao Shengdong, Yang Gang and Jinguang Du 2014 A multipurpose dual-axis solar tracker with two tracking strategies. *Renewable Energy*, vol. 72 pp 88–98.

[42] Assaf EM 2014 Design and implementation of a two axis solar tracking system using plc techniques by an inexpensive method. *International Journal Academic Science Research*, vol. 2, pp 54–65

[43] Akbar HS, Siddiq AI and Aziz MW 2017 Microcontroller based dual axis sun tracking system for maximum solar energy generation. *American Journal Energy*, vol. 5 pp 23–27.

[44] A.El Hammoumi, S. Motahhir, A.El Ghzizal, A. Chalh and A. Derouich 2018 A simple and low-cost active dual- axis solar tracker, *Energy Science & Engineering*, vol. 6 pp 607–620

[45] Roong ASC and Chong S-H. 2016 Laboratory-scale single axis solar tracking system: Design and implementation, *International Journal Power Electronic Drive System* vol. 7 pp 254–264

[46] M.H.M Sidek, W.Z.W Hasan, M.Z.A.Ab. Kadir, S. Shafie, M.A.M. Radzi, S.A. Ahmad and M.H. Marhaban 2014 GPS based portable dual-axis solar tracking system using astronomical equation. *IEEE International Conference Power Energy*, pp 245–249

[47] Arnab Samanta, Rachit Varma and Shrikant Bhatt 2015 Chronological single axis solar tracker, *International Journal of Engineering Trends and Technology (IJETT)*, vol. 21 pp 204-207

Voice-Based Malay Commands Recognition by Using Audio Fingerprint Method for Smart House Applications

M Abu[1], A Amir [1], N A H Zahri [1], R Ngadiran[1]

School of Computer and Communication Engineering
Universiti Malaysia Perlis
Perlis, Malaysia

amizaamir@unimap.edu.my

Abstract. Voice-based command recognition is commonly used in security systems, phones, household appliances and hardware designed for handicapped people. Most of the current research study the voice command recognition for the smart home in English. Lack of study for voice command recognition in Malay makes it challenging to apply the voice command services for the smart home in Malaysia. Also, voice recognition is a non-trivial task in natural language processing. This project is to identify the command used for smart home appliances using Malay and design the algorithm for this system. Then, the proposed algorithm will be deployed on a Raspberry Pi to see the performance of Malay command in accuracy and the suitability of the algorithm to be deployed on low cost embedded devices. Light, fan, and television had been chosen as electrical appliances to build the command. An algorithm that previously used to recognize songs, the robust quad algorithm, is used in this project for voice command recognition. The proposed method has two main processes, known as feature extraction and voice recognition. In the feature extraction process, the audio fingerprint will extract data from the command spectral peak. For voice recognition, audio fingerprint matching will be used to analyze the audio commands. The outcome of this project is when the voice command is given by the user by activate or deactivate the target home appliance. The second outcome is the background noise that affects the system is reduced by using robust quad algorithm and increase the accuracy of the system. The results of this project have shown that the proposed algorithm is suitable to be implemented on a Raspberry Pi and achieve a high recognition rate with 87%. In the presence of noise with 15 dB, the proposed algorithm can maintain the high recognition rate with 82%.

Keywords—peak detection, audio fingerprint method, noise resistance

1. Introduction

Human voice had been used in many technologies for example voice dialling, data entry, ID verification, detect identity in the crowd, smart house, and security [1], [2]. Voice recognition for controlling house appliances has been widely studied, but most of the available applications are developed in English. This limits the applicability of these applications for non-English speakers. Hence, the voice recognition system based on other languages is vital to ensure the scalability of the application is widely used in different communities. Recently, research in voice recognition using different languages such as Japanese and Thai have also attracted attention among research communities[3].

XXX-X-XXXX-XXXX-X/XX/$XX.00 ©20XX IEEE

This paper proposes the Malay language as a command since it can help those critically ill, aged and disabled people in Malaysia who cannot speak English. The Malay-based voice recognition is significant to ensure more users from Malaysia can use the voice-based control smart home systems. Lately, most of the research and project for voice recognition are based on English speech recognition, and in comparison, other language speech recognition is still limited [4]–[6]

Based on Aripin and Othman (2014) research, voice-based command recognition is susceptible to detect unnecessary noise like screaming, dog barking and crying sound[2]. Leechor *et al.* (2010) stated that voice recognition accuracy and response speed could be affected by environment, because when the commands are given in a noisy environment, low recognition rate will occur, and when the command is given in a quiet environment, a high recognition rate will occur. Therefore, using a suitable microphone and a noise filter or noise handling algorithm can improve the accuracy rate for the proposed voice commands recognition [7], [8].

Oulkar *et al.* (2017) [9], Gyulyustan and Enkov (2017) [10], and Upadhye and Khan (2017) [11] choose Raspberry Pi to implement their system on board. The authors built their system in English and tested the performance of their systems in Raspberry Pi. Oulkar et al. stated that by using Raspberry Pi and connecting it with a suitable microphone, the accuracy could achieve 99% based on the ideal environment. Abd Manan (2006) developed his Malay-based voice commands in MATLAB but not implement it on board. The implementation of a Malay-based voice command system on a Raspberry Pi is desirable since to apply such system in a smart home system requires it to be implemented on an embedded device [9]–[11].

There are voice command recognition for smart house that have been proposed by other researchers. For example, Aripin and Othman (2015) chose fan and light for their "on" and "off" command [2]. Samah and Osman (2015), stated that light, fan, air conditioner, television, and radio are suitable for "on" and "off" command [12]. These commands built by the researchers is in English. Therefore, in this project, three devices —fan, light, and television— are chosen to build Malay commands for smart houses application.

2. Literature Review

Hidden Markov models (HMM) used a statistical approach to characterize the spectral properties frames of speech in the context of statistical methods. The advantage of HMM is provides a natural and highly reliable way of recognizing speech for a wide variety of applications as a stochastic modelling tool. HMM can integrates well into systems incorporating information for both acoustics and syntax [3], [13].

Artificial Neural Networks (ANN) often used as a classifier for tasks in speech recognition, and have several advantages over parametric classifiers. However, there are disadvantages in terms of the amount of training data required and length of training time. Second disadvantage, the length of time required to train the networks can present problems, particularly when investigating a variety of feature sets to represent speech data. ANN has three architecture, Feedforward Perceptrons Trained with BackPropagation, Radial Basis Function Networks and Learning Vector Quantization. Based on three architectures, Feedforward Perceptrons Trained with BackPropagation is the most popular model [3].

Support Vector Machine (SVM) have the ability to transform data to high dimensional space implicitly and to construct a linear binary classifier in this high dimensional space without perform any computations in the high dimensional space. The disadvantages of using SVM is the number of training samples it can deal with is limited. Second disadvantages, the formulation of SVM can only work with vectors of fixed dimension input and it does not give a reliable measure of the probability for correctness during classification. This can affect the recognition performance because, without a concrete value of probability, some algorithms cannot carry out to search for the most probable sequence of recognition units [5].

Differently, this research investigates the potential of an audio fingerprint approach, the Robust Quad-Based Audio Fingerprint Method [14] for voice command recognition. Audio fingerprint algorithm is usually used for music recognition. The advantages of using this algorithm are noise and distortion resistant. This algorithm is also able to recognize a short audio sample of music that had been broadcast that means timing differences between train samples and test samples will not affect. The time to recognize a large database is faster compared to another recognition method [14]. The robustness to noise may not be the main concern for this algorithm to recognize audio material that was modified in tempo or pitch. The other advantages of using audio fingerprint method is it can efficiently identify audio in a large database. It is also robust to noise and audio quality degradation, as well as to severe distortions of speed, tempo, and frequency [15].

3. Methodology

This project studies the performance of voice-based Malay command recognition in the normal environment by using Robust Quad-Based Audio Fingerprint Method. Firstly, All the voice command will be recorded using a mobile phone. Each speaker will record their voice three times for each command. The command will be divided into 80:20 ratio for the training and testing datasets. The recorded voice command will undergo feature extraction to extract the audio fingerprint. The feature extraction will be done for both training and testing datasets. Then, the extracted audio fingerprint for the testing dataset will be matched together with training dataset to get the result. The performances of the system will be tested on PC; then it will be deployed in Raspberry Pi. The time-efficiency of Raspberry Pi and computer will be compared.

3.1. Survey to Identify Common Malay Voice Commands for Smart Home

Firstly, we selected three electrical appliances: fan, light, and television. In previous studies[5], the usual commands are 'Buka Kipas', 'Tutup Kipas', 'Buka Lampu', and 'Tutup Lampu'. In addition to these commands, this paper investigates the common Malay commands to turn on and turn off a television and control television channel. To decide the usual commands for these tasks, a survey has been conducted by asking the respondent what are the common Malay commands used to control televisions. The survey questions were generated by using Google form and shared with 55 respondents through WhatsApp messaging system and Facebook.

For turn on and turn off the television, from 55 respondents 45 (85%) respondents chose 'Buka TV' and 'Tutup TV', respectively to turn on and to turn off the television. Only 10 (18%) of the respondents selected 'Buka Televisyen' and 'Tutup Televisyen', which shows that these commands are not commonly used for the task to turn on and turn off the television. For control television channel, from 55 respondents 48 (87%) chose 'Siaran 1', 'Siaran 2', and 'Siaran 3' to control the television channel. Only 4 (7%) of the respondents chose 'Siaran Naik' and 'Siaran Turun' and the other 3 (5%) of the respondents chose others as voice command that not commonly used for controlling television channel. Others are the voice command suggested by the respondent. Hence, 'Buka TV', 'Tutup TV', 'Siaran 1', 'Siaran 2', and 'Siaran 3' is used for data collection. Hence, nine commands are selected to control for the electrical appliances.

3.2. Audio Data Collection

For audio data collection, 24 speakers from different backgrounds and genders were asked to record their voices in the normal environment, for example, a quiet room with fan and light are activated. From the 24 speakers, 12 males and 12 females with age ranging between 20 to 30 years old recorded their voice based on the command given in Table 1. Each speaker recorded their voice three times for each command, so there are 648 samples overall. The samples were divided into 80:20 ratio for the training and testing datasets. The distance between the speaker and the phone is 0.5 meters, based on a previous study [16]. The time range for the speaker to record their voice depends on their timing. The voice command sample rate is 8 kHz because this sample rate is adequate for human speech. Recording channel for the device is mono. Figure 1 shows an example of the voice command signal.

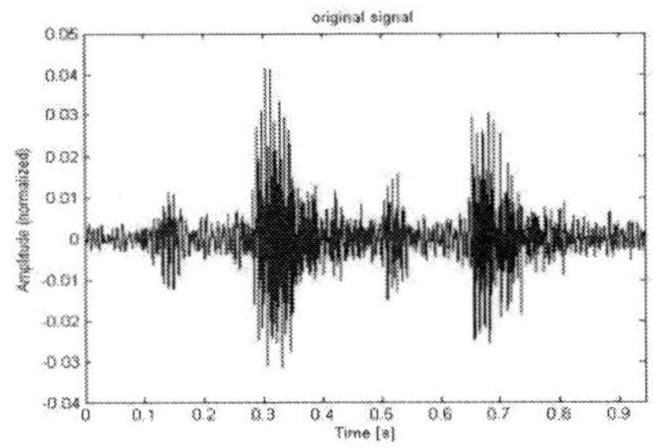

Figure 1. An example of a voice command signal

3.3. Software Design

Based on the flowchart as shown in Figure 2, in feature extraction, voice command is converted into a spectrogram, and the local maxima (spectral peak) will be searched to build quads. Two types of filters are applied in robust quad-based audio to obtain the actual spectral peak in the spectrogram. Quad is a continuous geometric hash representation of quadruples of a point. The quad will be converted into the hash and stored as an audio fingerprint. Then, the audio fingerprint for training and the testing dataset is used during classification model for decision-making.

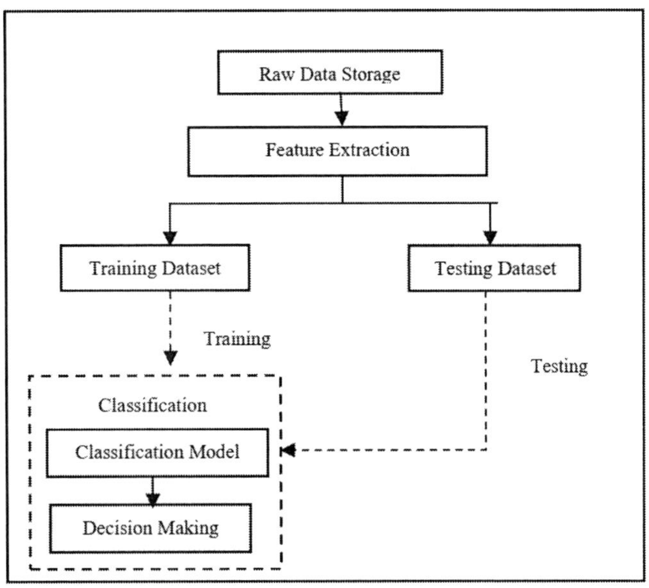

Figure 2. Flowchart of voice-based Malay command recognition for smart house applications

The detail to obtain the hashes for the matching process is explained in Robust Quad Based Audio. Figure 3 shows the training process and Figure 4 shows the classification process.

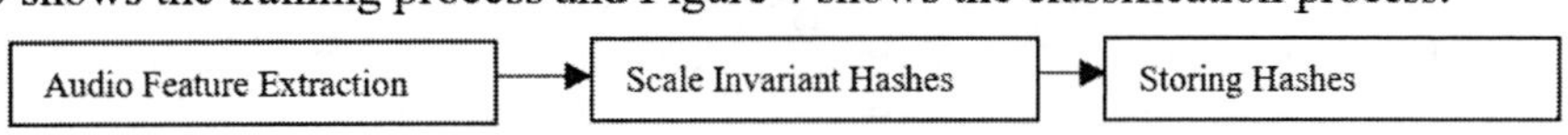

Figure 3. The training process

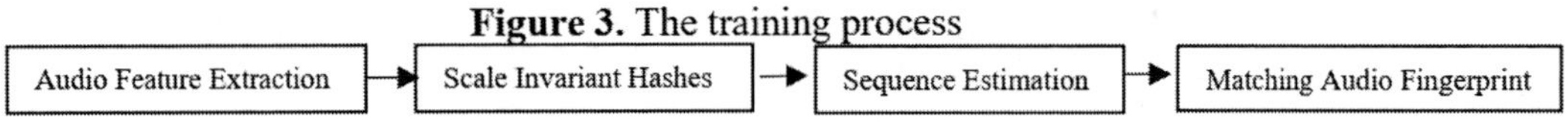

Figure 4. The classification process

Audio Feature Extraction

To extract audio fingerprint, voice command quads need to be construct as shown in Figure 5. This algorithm works on translation and scale invariant geometric hashes by combinations of spectral peaks. To produce quad the command signal will be converted into spectrogram in two-dimension time-frequency. Then, the local maxima of voice command need to be found and will be shown in the scatter plot. The local maxima are represented as time for the x-axis and the peak of frequency for the y-axis in spectrogram space. The spectrogram is using a Hann window of size 1024 samples and hops size of 32 samples. The advantage of using windows is that the frequency can be seen as a narrow peak in the spectrogram [18]. To find the quad, max filter and min filter will be used. The local maxima that are detected by both filter will be removed from the spectrogram.

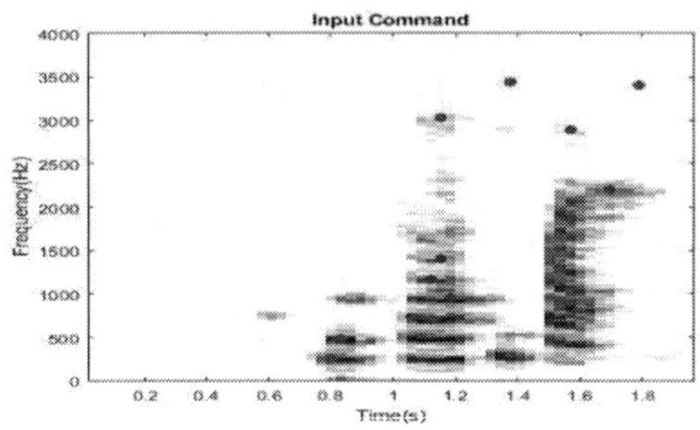

Figure 5. Examples of the quad for voice command.

Each command shows a different quad pattern because voice command is extract based on frequency and time. Pre-processing is removed in this algorithm because the spectral peak of the speaker is higher than the background noise and it is neutral for the speaker to speak louder than the noise. In this paper, the noise is stationary, but the timing of the command is nonstationary so the system will start to detect the spectral peak as the signal is fluctuating. The offset time for this system to detect the peak spectral is within 0.3 seconds because the command time is too short, and the system will give output even though only part of the command is given.

Scale-Invariant Hashes

Each command will produce different hash because of frequency and the tempo of the speakers' speak. The quads will be converted into Translation and Scales Invariant after it has been extract from each command. To make sure the test samples quad match the train samples quad, the quad needs to be summarized by converting it into hashes. To convert quads into a hash, the property of rotational invariant and the resulting mode is released to compute translation and scale-invariant to the hash.

Storing Hashes

The hashes will be stored in four data structures *peakfile*, *refrecords*, *fidindex* and *searchtree*. *Peakfile* contains the continuous of two-dimensional coordinates of spectral peaks that obtained from train samples. *Refrecords* stores quads and quad hashes along with the file-ID of all train samples in the reference collection. A quad record consist of spectral peak and where denotes the height and width of the quad in spectrogram space. *Fidindex* will map each voice command file for train samples to a unique file-ID. The number of extracted peaks and quads also is stored by *fidindex* along with other metadata. Given a specific voice command file-ID, the *fidindex* is used to find the corresponding record range in the *peakfile*. *Searchtree* is used to perform an efficient fixed-radius near neighbour searches of quad hashes. Memory-bounded tree construction is simple to establish by using tree variant.

Audio Fingerprint Classification

IOP Conf. Series: Materials Science and Engineering **767** (2020) 012053 doi:10.1088/1757-899X/767/1/012053

There are three stages to matching test sample with train samples: the first stage performs the selection of match candidates, followed by a filtering stage to remove the false positive candidate. Second is the match sequence estimation stage, using the result from the first stage, the sequences are searched within matched candidates. The last stage is the verification step. This step is to maintain high identification precision on extensive database collections, especially if the voice command is repeated many times.

Match Candidate Selection and Filtering

A fixed-radius near neighbour search in the *searchtree* is performed for each test sample quad hash. This lookup returns a set of raw match candidates that consists of those quad records with hashes that are similar to the test sample quad-hashes. It is called the set of raw candidates because it will most likely be a mixture of true positives and a large number of false-positive matches. For matching audio fingerprint, all the process will be done on spectral quads.

Sequence Estimation and Match Verification

This stage is on a per-file-ID basis, so the match candidates are group by file-ID, and the groups will be sorted by the number from the most match candidates to less match candidate. Per file-ID, by following the step in Shazam's algorithm, the matching process is done by using the histogram method [17]. The method is adapted such that the test sample time is scaled according to the uncovered time scale factor. The file-ID for the largest histogram bin is returned together with the match position. Finally, if match sequences are found for a given file-ID, and their number of matched candidates is larger than a threshold value, these sequences will be verified match-by-match. If the testing samples match, from the beginning of the training dataset matching feature, it will occur at the similar relative offset. In one of the training samples, sequences of hashes should occur in matching training samples with the same relative time sequences. Figure 6 shows a diagonal line is performed within the quad.

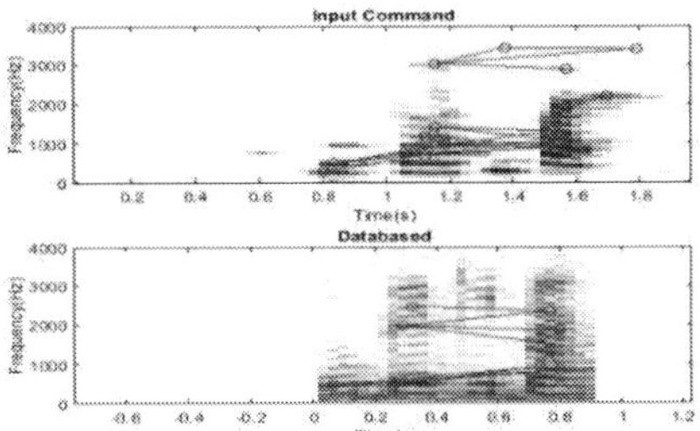

Figure 6. The diagonal line is performed within the quad.

3.4. Hardware Design

The algorithm will be deployed in Raspberry Pi (as shown in Figure 7) to test it accuracy and efficiency of the algorithm when it deploy on board. Raspberry Pi is chosen because its size is small and easy to carry compared to the computer. For real-time test, the Raspberry Pi will be connected to the microphone to record the Malay command given by the user. The LED that connected with the Raspberry Pi will show the output.

Besides using the recorded dataset, Raspberry Pi also can connect online and offline with Google Speech Recognition API. In this case, not all Malay command can be obtained in the Google Speech API. If the command given by the user matches with any of the recorded datasets or match with the API, the LED will be switched on for 'Buka' command and switched off for 'Tutup' command. For controlling the television channel, RGB LED is used so the colour of the LED will be changed according to the number of channels chosen by the user. Command 'Siaran 1' blue colour, 'Siaran 2' green colour and 'Siaran 3' red colour.

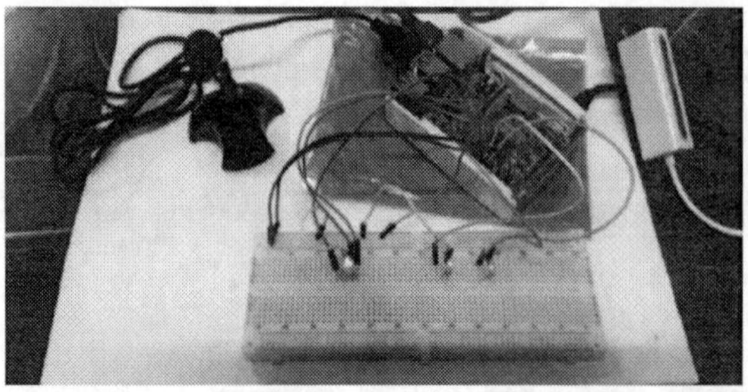

Figure 7. The connection of Raspberry Pi

4. Result and Discussion

The voice-based Malay command recognition will be tested with the different signal of noise ratio (SNR) in range of -10 dB to 30 dB (noise that been implement in MATLAB). Then, the system will be deployed in Raspberry Pi, and the performance of the system to will be compared between computer and Raspberry Pi. The accuracy and performance of each command will be recorded.

4.1 Accuracy Results

The accuracy of the system will be measured based on its recognition rate. To get the recognition rate, the number of commands had been recognized by the system will be divided by the total number of testing samples. The recognition rate of voice-based command recognition is defined as follows Equation 1:

$$Recognition\ Rate = \frac{\#correctly\ recognized}{total\ number\ of\ testing\ samples} \times 100\% \tag{1}$$

The precision of this system also will be recorded. Precision is the proportion of cases out of all cases where the system claim to have identified the reference, where the prediction is correct. Given *tp* is number of cases where the commands are correctly recognized and *fp* is the number of cases when the command are incorrectly recognized, the precision for the system is defined in Equation 2:

$$Precision = \left({}^{tp}/_{tp + fp}\right) \times 100\% \tag{2}$$

The recognition rate and precision of this system are obtained by calculating the mean of the 5-fold cross-validation process. The recognition rate of this system is 87%. The recognition rate of each command are 86% for 'Buka Kipas', 80.56% for 'Tutup Kipas', 83.33% for 'Buka Lampu', 83.33% for 'Tutup Lampu', 87.50% for 'Buka TV', 87.50% for 'Tutup TV', 91.67% for 'Siaran 1', 90.28% for 'Siaran 2', 90.28% and 'Siaran 3'. The recognition rate for 'Tutup Kipas' and 'Tutup Lampu' is low because their command signal has a similar pattern but different amplitude. The command signal for 'Lampu' is slightly smaller than 'Kipas'.

The precision for voice command is 88.22%. The precision for each command are 87.32% for 'Buka Kipas', 82.86% for 'Tutup Kipas', 84.51% for 'Buka Lampu', 86.96% for 'Tutup Lampu', 88.93% for 'Buka TV', 91.30% for 'Tutup TV', 91.67% for 'Siaran 1', 90.28% for 'Siaran 2', 90.28% and 'Siaran 3'. Precision was calculated because there is some command that cannot be predicted by the system. To see the performance of the algorithm in the noisy background, each command is tested with different levels of noise. Table 1 shows the recognition rate and the precision for each command before adding SNR.

Table 1. The recognition rate and the precision for each command before add SNR.

Malay Command	Recognition Rate	Precision
Buka Kipas	86.57	87.32
Tutup Kipas	80.56	82.86
Buka Lampu	83.33	84.51
Tutup Lampu	83.33	86.96
Buka TV	87.5	88.73
Tutup TV	87.5	91.30
Siaran 1	91.67	91.67
Siaran 2	90.28	90.28
Siaran 3	90.28	90.28

Then, the background noise with signal to noise ratio (SNR) from range -10 dB to 30 dB is added with test samples. The mixed test sample will be compared again with the training dataset to see the recognition rate and precision of voice-based Malay command recognition. The recognition rate of voice-based Malay command begins to drop 50% at approximately -5 dB SNR because the noise is too loud. When the mixed sound is played, only some part of the command is heard. Figure 8 shows the recognition rate and the precision for all command.

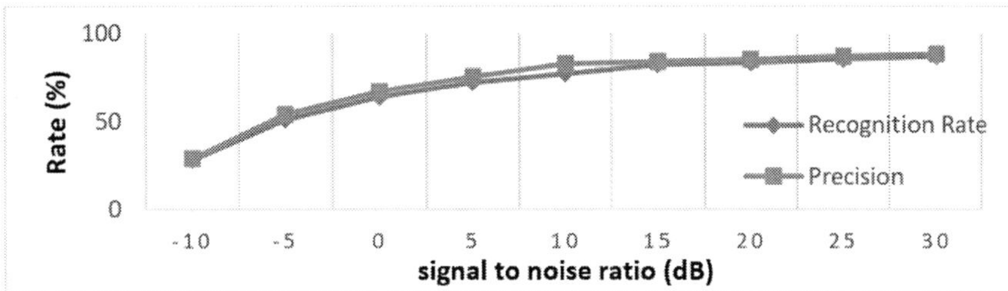

Figure 8. The recognition rate and the precision for all command

After that, the algorithm will be deployed in Raspberry Pi for real-time test. To test in real-time, the speaker needs to train their voice three times in the system first before test on the system five times. The accuracy of real-time recognition is 68.5%, and the accuracy starts to decrease when the SNR reaches 15 dB approximately. The accuracy of real-time is less than the already recorded voice command because the tempo for the speaker to record their voice is different. There also some disturbance while recording the command in real-time, for example, other speakers voice and their movement. The device used to record the voice command also can be the reason why the accuracy of real-time in low. The time taken to record voice command in real-time is 2 seconds compare to recorded voice most of the command is recorded in 1 second because 1 second is not enough time for real-time. Some speaker starts late, so some of the commands are cropped.

4.2 Recognition Time

The efficiency of this system will be tested on the computer and Raspberry Pi based on the performance of the system to recognize the voice command in a different platform by changing the number of train and test the sample. The specification used for the PC is Intel core i5, RAM 6 and the operating system is Windows 8.1. The training dataset and the testing dataset is divided into 80:20 ratio by choosing the dataset randomly. The samples will be trained from 100 samples until 500 samples. It takes approximately 1 second to train 100 samples, 2 seconds to train 200 samples, 3.8 seconds to train 300 samples, 5 seconds to train 400 samples and 6 seconds to train 500 samples. Figures 9 and 10 show the performance for computer and Raspberry Pi.

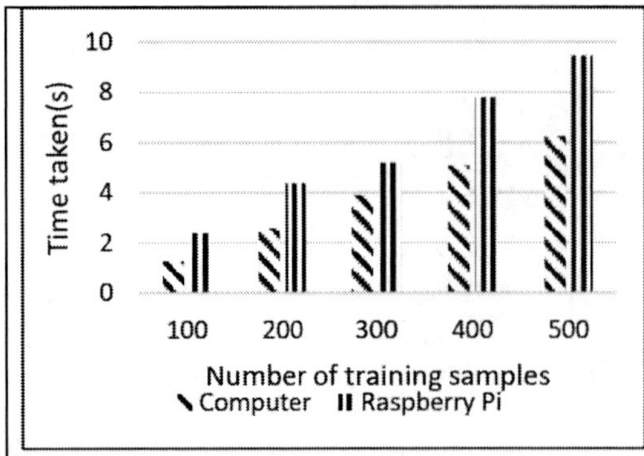

Figure 9. The training performance for computer and Raspberry Pi

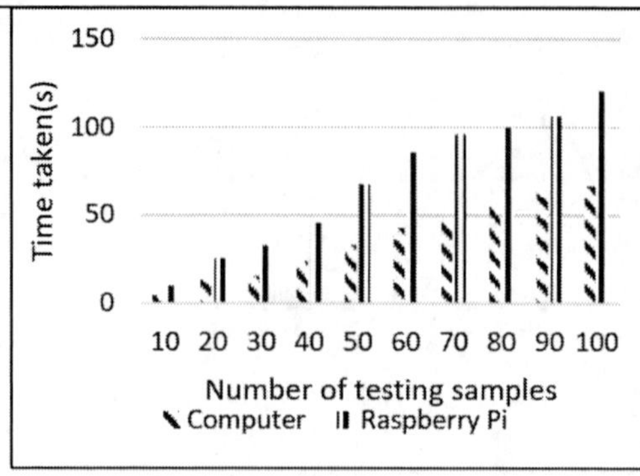

Figure 10. The testing performance for computer and Raspberry Pi

The training performance for Raspberry Pi is approximately two times the performance of the computer. For testing, the samples will be test from 10 samples to 100 samples. The performance of the system for testing dataset also increases as the samples of testing dataset increase. It only take approximately 4 seconds to train 10 samples, 13 seconds to train 20 samples, 15 seconds to train 30 samples, 24 second to train 40 samples, 33 second to train 50 samples, 43 second to train 60 samples, 47 seconds to train 70 samples, 56 second to train 80 samples, 63 second to train 90 samples and 67 second to train 100 samples. As shown in Figure 9 and Figure 10, the training time and testing time of the proposed algorithm on the Raspberry Pi is higher than the computer due to the higher computational capability of the computer. However, the training and testing time of the algorithm on Raspberry Pi is linear with the number of samples. These results show the efficiency of the algorithm to be deployed on a Raspberry Pi.

4.1. Comparison MFCC+NN and Audio Fingerprint Method
In this paper, comparison had been made between robust quad based audio fingerprint method and the neural network (NN) method with Mel-frequency Cepstral Coefficient (MFCC). The accuracy for audio fingerprint method is 86% higher than the neural network method only 40%. The accuracy of the NN method low because the time taken for the speaker to record their voice is difference, so the timing for the training data and testing data are different. For the NN method, the timing for all training and testing datasets must be the same. The audio fingerprint method focus on the spectral peak from the frequency and time domains so time differences and background noise can be ignore. Even in real-life application, the louder the noise, the louder voice command given. Besides that, the time taken for the NN method to train the data is longer than the audio fingerprint method. The time taken for the neural network to train 500 samples is approximately 17 seconds while audio fingerprint method only takes approximately 5 seconds.

5. Summary
Voice-based Malay command recognition for smart house applications has very limited Malay commands compared to English. Therefore, a preliminary work on voice-based Malay commands recognition was designed by using an audio fingerprint method. Three electrical appliances (light, fan, and television) had been chosen as target appliances for smart house applications. In this paper, an audio fingerprint method that detects the highest peak of frequency was used. By detecting the highest peak, the background noise can be ignored because human voice will become louder if the noise is louder. Then, the algorithm was deployed on a Raspberry Pi board and we evaluate its performances based on accuracy and efficiency. The performance was evaluated in terms of recognition time by running the

system on the computer and by deploying it on Raspberry Pi. The experiment result shows that besides using low pass filtering and high pass filtering, peak detection also can be used to increase the accuracy of voice command recognition. This paper has shown that the proposed algorithm is lightweight, and it can be implemented on a Raspberry Pi. From all experiment that has been done, some further works can be added to see the performances of this method. In this paper, the test sample is added with SNR and the highest accuracy is 87%. Besides adding background noise to the voice command, by using the same algorithm the voice command can be tested in different distances. This algorithm is originally used for music recognition. Nonetheless, it has the potential to be applied for detecting a long speech recognition.

Acknowledgements
Firstly, I would like to dedicate my acknowledgement of gratitude toward my supervisor, Dr Amiza Amir, for her guidance and support as I make my research paper. I also want to thank my parents, family and friends, who provide advice and financial support during my research.

References

[1] A. Kumar, "Voice Based Control Command Signal Generation for Intelligent System," *2015 IEEE UP Sect. Conf. Electr. Comput. Electron.*, pp. 1–6, 2015.
[2] Norhafizah and M. B. Othman, "Voice control of home appliances using Android," *Proc. - 2014 Electr. Power, Electron. Commun. Control Informatics Semin. EECCIS 2014. conjunction with 1st Jt. Conf. UB-UTHM*, pp. 142–146, 2014.
[3] U. Shrawankar and V. Thakare, "Techniques for Feature Extraction in Speech Recognition System : A Comparative Study," 2013.
[4] H. F. Ong and A. M. Ahmad, "Malay Language Speech Recogniser with Hybrid Hidden Markov Model and Artificial Neural Network (HMM / ANN)," *Int. J. Inf. Educ. Technol.*, vol. 1, no. 2, pp. 114–119, 2011.
[5] Abd Manan, "The Design and Development of a System for Controlling Automotive Functions Using Speech Recognition," 2006.
[6] P. Barua, K. Ahmad, A. Anam, S. Khan, and M. Sanaullah, "Neural Network Based Recognition of Speech Using MFCC Features," *3rd Int. Conf. Informatics, Electron. Vis. 2014*, pp. 1–6, 2014.
[7] P. Leechor, C. Pompanomchai, and P. Sukklay, "Operation of a radio-controlled car by voice commands," *ICMEE 2010 - 2010 2nd Int. Conf. Mech. Electron. Eng. Proc.*, vol. 1, no. Icmee, pp. 2–5, 2010.
[8] H. L. Zhe Gong, "Voice Controlled Home Automation System," no. 13, 2014.
[9] S. Oulkar, R. Bamane, S. Gulave, and P. Kothawale, "Voice Controlled Home Automation Using Raspberry Pi 3," *Int. J. Recent Innov. Eng. Res.*, vol. 2, no. 1, pp. 28–32, 2017.
[10] H. Gyulyustan and S. Enkov, "Experimental speech recognition system based on Raspberry Pi 3," *IOSR J. Comput. Eng.*, vol. 19, no. 3, pp. 107–112, 2017.
[11] Upadhye and R. A. Khan, "Raspberry Pi Home Automation with Voice Control," *Int. J. Adv. Res. Ideas Innov. Technol.*, vol. 3, no. 6, pp. 1297–1301, 2017.
[12] A. Samah and A. A. Osman, "Controlling Home Devices for Handicapped People via Voice Command Techniques," *Int. Conf. Comput. Control. Networking, Electron. Embed. Syst. Eng.*, pp. 374–378, 2015.
[13] L. Pan, "Research and simulation on speech recognition by Matlab," 2013.
[14] A. L.-C. Wang, "An Industrial Strength Audio Search Algorithm.," *Proc. 4th Int. Soc. Music Inf. Retr. Conf. (ISMIR 203), Balt. Maryl. (USA), 26-30 Oct. 2003*, pp. 7–13, 2003.
[15] R. Sonnleitner and G. Widmer, "Robust quad-based audio fingerprinting," *IEEE/ACM Trans. Audio Speech Lang. Process.*, vol. 24, no. 3, pp. 409–421, 2015.
[16] N. Mulhern, N. Mccaffrey, N. Beretta, E. Chabot, and Y. Sun, "Designing Android Applications using Voice Controlled Commands For Hands free interaction with Common Household Devices," *2013 39th Annu. Northeast Bioeng. Conf. Des.*, pp. 265–266, 2013.

IOP Conf. Series: Materials Science and Engineering **767** (2020) 012054 doi:10.1088/1757-899X/767/1/012054

Comparative Study of Parallelism and Pipelining of RGB to HSL Colour Space Conversion Architecture on FPGA

Phaklen Ehkan, Soon Voon Siew, Fazrul Faiz Zakaria, Mohd Nazri Mohd Warip, Mohd Zaizu Ilyas

Embedded, Networks and Advanced Computing (ENAC) Research Cluster
School of Computer and Communication Engineering, Universiti Malaysia Perlis, Pauh Putra Campus, 02600 Arau, Perlis, Malaysia

Corresponding author's : phaklen@unimap.edu.my

Abstract. RGB colour model is a basic colour model and complements together to produce full colour range but it is unable to produce sufficient information for digital image analysis. However, HSL is capable to provide other useful information such as colour in degree, saturation of the colour and brightness of colour. In this work, RGB to HSL mathematical conversion algorithm is implemented on FPGA chip. Parallelism and pipelining capabilities of FPGA helps to speed up conversion performance. The RGB to HSL equation is implemented by using two architectures which are parallel and 7-stages pipeline architectures. The designed parallel and pipeline converters have one clock and seven clock cycle of data latency respectively. The parallel and pipeline architectures for RGB to HSL converter have been achieved rate of accuracy by hardware verification up to 99% and 98% and possessed maximum operating frequency merit of 50 MHz and 120 MHz respectively.

Key Words: FPGA, RGB (Red, Green, Blue), HSL (Hue, Saturation, and Luminance), Pipelining, Parallelism.

1. Introduction

Colour space conversion is an important process when performing digital image processing. This conversion process can be performed in either software or hardware domains depending on the processing system. For non-time crucial system, this conversion process commonly implemented in software basis, however for real time system hardware implementation presents a good alternative. The trend of the digital image or digital video resolution is getting higher and higher from Standard Definition (SD) to Ultra High Definition (UHD). This booming trend has directly increased the digital image data volume for processing. While handling such huge data volume, a specific processing core is to be considered during system design in order to reduce the processing time and the output quality.

To date, many existing RGB to HSL converter methods used in image processing but most of them are software-based. The problem with the software is that its implementation is not suitable applied in the real time digital video stream. Besides, the RGB colour model is an additive colour model in which red, green, and blue lights are added up together in numerous ways to reproduce a broad array of colours but there is limitation for RGB colour model to provide sufficient information for image analysis. On the other hand, HSL provides the information such as colour in degree, saturation of the colour and brightness of the colour. These information are much more usable during image analysis. Parallelism and pipelining capabilities of FPGA has simplified the processing steps and only consumes of one clock cycle to produce the HSL value for each pixel without any latency. Hence, the embedded FPGA-based

IOP Conf. Series: Materials Science and Engineering **767** (2020) 012054 doi:10.1088/1757-899X/767/1/012054

RGB to HSL converter with less computational time and capable to work on images analysis is developed.

2. Hue, Saturation, and Luminance (HSL) Colour Space

RGB is one of the well-known colour spaces due to its implementation in colour liquid crystal display (LCD) screen, meanwhile HSL usually employed in image processing for specific purpose. Each of the colour space has their own attributes as RGB is categorized as hardware-oriented space while HSL consider human oriented space [1]. HSL is similar with RGB which represented by three components. Hue component can be defined as all the possible colour mixture generated by three primary colour red, green and blue which also defined as rainbow colour [2]. Saturation component is defined as the purity of the particular hue colour and luminance is defined as the amount of light reflected with the particular hue colour [3]. Hue is measured in degree of angle value from 0 to 360 degree. Both saturation and luminance components are measured in percentage from 0% to 100% [2][4-5]. The initial colour of hue is beginning from primary colour pure red to green and green to blue [6]. The generation of hue from primary colour red, green and blue which each has a phase shift of 120 degree to each other in the circular disk with a series of variation each of them from 0 to 360 degrees to produce a rainbow spectrum. Hue and saturation conversion equation is widely available but most of them are for hue, saturation, intensity (HSI) and hue, saturation, value (HSV) colour spaces, but not for HSL colour space. Although the hue conversion equation for all three colour spaces is the same but some works use the equation with trigonometry operation [6-7]. Therefore, those equations are not suitable to implement in this work. Trigonometry operation is difficult and tedious to be implemented in digital architecture, therefore it is not considered. As observed from works by [8-11], they used hue conversion with trigonometry and all worked fine. They implemented this conversion process in software based programming which did not pose any obstacles. The work done by [7] has suggested to adopt the hue conversion equation from [12] which without any trigonometry operation and digital implementation friendly.

This work chooses the RGB to HSL conversion pseudo code from [13]. The conversion steps described by Agoston's pseudo code did not contain any trigonometry operation. It only involve basic arithmetic add, subtract, multiply and division operation and selection operation. Hence, it is feasible to implement the operation that described by the pseudo code using digital architecture. Equations (1), (2), (3) and (4) are the formularized mathematical RGB to HSL conversion from Agoston's algorithm where r,g,b is the 8-bits unsigned binary integer input and R,G,B is the range of 0 to 1. These equations shown direct attributes helps to reduce the digital architecture design complexity. Before these equations are implemented in digital domain, a simplification process is performed to eliminate some mathematical operations and to normalize the equation to be functional in direct 8-bits unsigned binary input data to avoid floating point operation in the architecture. Floating point architecture is more complex to be implemented and also resources consuming. Fixed point multiplication and division operation is much easy to be implemented by using Look-Up-Table (LUT).

$$(R,G,B)=\left(\frac{r}{255}\ ,\ \frac{g}{255}\ ,\ \frac{b}{255}\right) \tag{1}$$

$$L=\frac{\max(R,G,B)+\min(R,G,B)}{2} \tag{2}$$

$$S=\begin{cases}\dfrac{\max(R,G,B)-\min(R,G,B)}{\max(R,G,B)+\min(R,G,B)}, L<0.5\\[2ex] \dfrac{\max(R,G,B)-\min(R,G,B)}{2-\max(R,G,B)-\min(R,G,B)}, L\geq 0.5\end{cases} \tag{3}$$

$$H = \begin{cases} 60\left(\dfrac{G-B}{\max(R,G,B)-\min(R,G,B)}\right), R_{max} \\ 60\left(2.0+\dfrac{B-R}{\max(R,G,B)-\min(R,G,B)}\right), G_{max} \\ 60\left(4.0+\dfrac{R-G}{\max(R,G,B)-\min(R,G,B)}\right), B_{max} \end{cases} \quad (4)$$

Hue-saturation colour space is a popular choice when comes to the application. Its application covers from object detection and tracking, image segmentation, and also industrial applications. Most of the applications have make use of hue-saturation colour space because of its advantage in representing colour in one dimension (1-D) which is called Hue. Object detection is an important operation in machine vision and real time image processing. This operation is to detect a desired object whether exist in the image or not. Several researches have used hue-saturation colour space for object detection operation [4][10-11][14-16]. Besides, [17-19] applied this colour space to perform image segmentation in their work. [20] had applying hue and saturation colour in agriculture industry to judge the ripeness of the oil palm fruit while [21] has utilizing hue-saturation colour space to perform segmentation in the real time video from abdominal surgery instrument.

3. Implementation of RGB to HSL Conversion Digital Architecture

There are two RGB to HSL conversion architecture (parallel and pipelined architectures) were designed in this work. This design process was performed by using Intel-Altera Quartus II v9.2 and a programming language called Very High Speed Hardware Description Language (VHDL).

3.1 Parallel RGB to HSL Conversion Architecture

As can be observed from the conversion equations (2)-(4), all of them required a maximum, minimum, summation of maximum and minimum, difference of maximum and minimum, and value from the input RGB. These values are then passed to the HSL module to perform HSL conversion operation. Figure 1 shows a hue conversion module which is the most complex module in the parallel converter architecture. Figure 2 displays a saturation conversion module while Figure 3 is the luminance conversion module. The input of RGB value is fixed to 8 bits wide for each component. The hue output value is designed into 10 bits wide while saturation and luminance outputs are 8 bits wide for each of them. In between of this architecture, there is none of register required to store temporary data. All signals are directly fed to the arithmetic operator.

There are several constant values exist in the RGB to HSL conversion operation but do not appeared in the conversion Equations (1), (2), (3) and (4). These constant values are inserted to normalize the whole conversion process and the output value is maintained in integer value. The arithmetic operation before the MUX1 is to prepare the dominator and numerator of the division operation for all three conditions of hue equation in Equation (4). The multiplexer select a correct denominator to be used base on the maximum RGB component. This denominator value is then normalized and fed into a division operator. The second multiplexer MUX2 is used to ensure the final output Hue value is lie within 0 to 360 positive integer ranges. Parallel saturation architecture is less complex if compared to the parallel hue architecture. This saturation conversion architecture is directly implementing according to Equation (3). Both saturation condition are prepared and calculated simultaneously using two division operators. Luminance value then determines which calculated saturation value is being employed via MUX1. The saturation value becomes zero if the input RGB is a grey scale value and achromatic. This is because gray value does not contain any colour information, so the saturation of the colour does not exist. Luminance architecture is the simplest architecture. It constructed by a multiplier and a divider. The summation of maximum and minimum value has been scale up by 10 times and divided with value 51. The value of 51 is produced by the constant 255 from Equation (1) multiplied with the divisor of 255 from Equation (2) and scale down by 10. This parallel architecture serve the purpose to generate the result as fast as possible once the RGB value is supplied to it regardless of the clocking trigger.

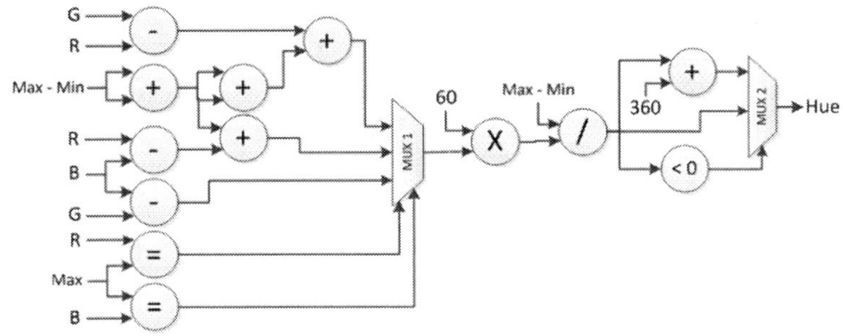

Figure 1. Parallel hue architecture.

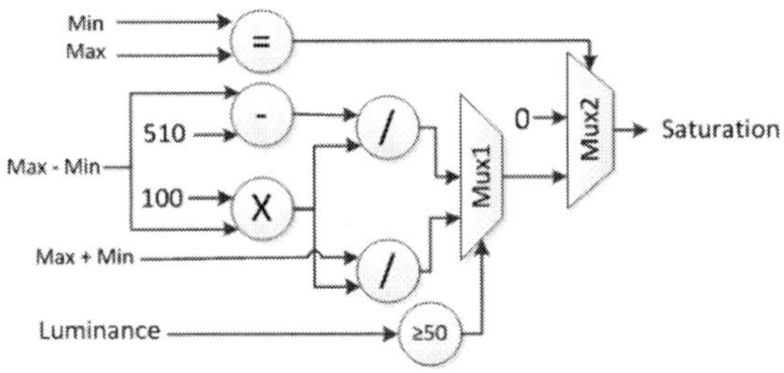

Figure 2. Parallel saturation architecture.

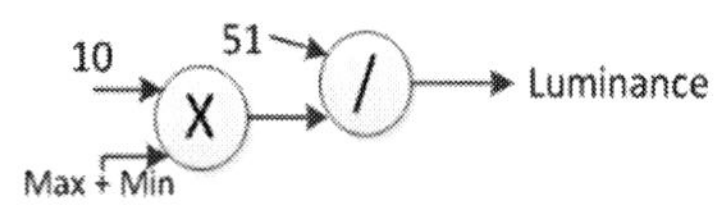

Figure 3. Parallel luminance architecture.

3.2 Pipelined RGB to HSL Conversion Architecture

This RGB to HSL conversion process is divided into 7 stages pipeline process. During development of this pipeline converter, the most complicated Hue equation is broken down into its individual sequential operation and arranges them into 7 stages. It then followed by saturation equation which is also break into sequential steps. Any step that existed in the hue process and shared variable is removed from the saturation equation to avoid redundancy operation and also reduce hardware logic resources. As the result, the saturation also requires a total of seven steps in order to produce the final saturation value. Lastly, the luminance is structured into only five sequential steps. Each step from these three equations is arranged according to the variable dependency in each stage so that all operations are able to process accordingly without any delay. Finally, this pipeline converter contains total of seven stages which are the same as the number of steps in hue equation.

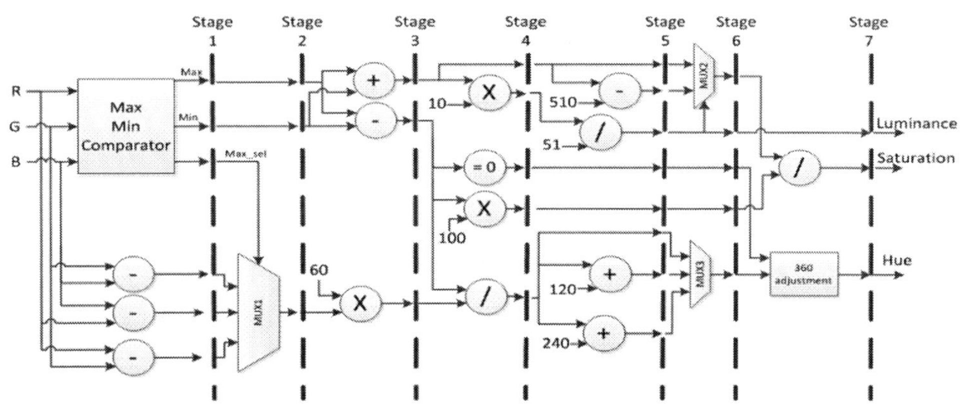

Figure 4. Pipelined of RGB to HSL conversion architecture.

During implementation of hue, saturation and luminance equation in the pipeline, the sign bit is handling using external register. This sign bit is carry through the beginning of the pipeline until the end of the pipeline. This sign bit is responsible for switching the arithmetic operation for calculating the variable to avoid overflow or negative value occurs. If negative value occurs during the calculation, it will severely disturb the calculation accuracy.

As differ from the parallel architecture, this pipeline architecture requires a clock input pin. This clocking signal used to drive the register in every stage of the pipeline. As depicted in Figure 4, the vertical dashed line represents the registers for each stage. Every intermediate signal and variable produced in each stage of the pipeline is then stored in the register and passing to the next stage. This helps to stabilize the signal before further process.

4. Test Result and Performance Evaluation

A set of standardize testing RGB value is required in order to test and evaluate the designed RGB to HSL conversion architecture. By using random() function in the C code compiler, a total of 1024 sets of random RGB integer values are generated and listed in a text file. This set of random RGB values is then used for C code conversion to generate a reference HSL value and also to test the hardware parallel and converter. During testing process, conversion output accuracy and maximum operating conversion frequency is the main criteria in this work.

4.1 C Code Conversion Algorithm and Data Set Generation

An C code based RGB to HSL conversion based on the Agoston's algorithm is developed by using Dev-C C code compiler. The purpose is to verify RGB to HSL conversion result and accuracy. After this software based converter is developed, it is used to convert the 1024 sets of predefine random generated RGB values to obtain the 1024 sets of HSL values as a reference for later process of conversion accuracy evaluation. All the C code converted 1024 sets of HSL values are defined into 2 decimal places for accuracy.

4.2 Simulation Results of Parallel and Pipeline Converter

This section is taking both of the converters for functional simulation verification. The main purpose for this simulation test is to verify the theoretical functional performance for both converters. Both simulations use the same set of RGB input value as the C code reference RGB input value. However, only the first 50 sets of RGB value are used in both simulations. Figure 5 represents the simulation outcome for parallel architecture RGB to HSL converter. It clearly shows that the HSL value is produced instantly without any delay when the RGB value is given. This simulation phenomenon has proved its main parallelism attribute.

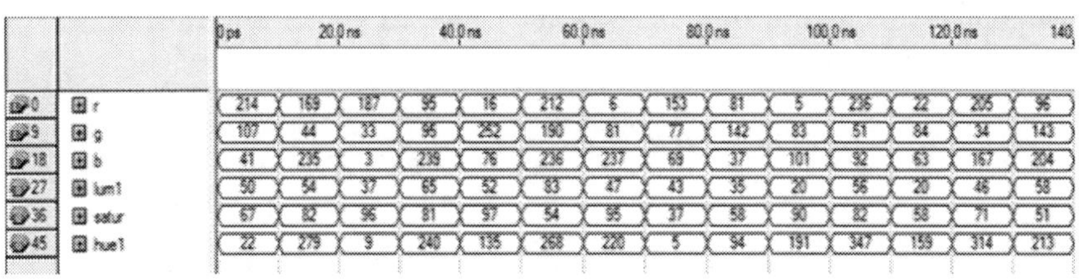

Figure 5. Simulation results for parallel architecture.

For the pipeline simulation, there are two additional signals are needed. The first signal is the clock signal labeled as *clock_in*, and a second signal is the *reset signal*. The *clock_in* signal is used to drive the registers in each stage. Since the developed pipeline converter has seven stages, this means that the pipeline has seven clock cycles of data latency. The data latency has proved in the simulation result as shown in Figure 6. The first HSL result is produced after seven clock cycles since the input of RGB value. The HSL output data before the seventh clock cycle is considered invalid. As comparison for both simulation results, it is noticeable that both parallel and pipeline architecture converters are capable to produce the similar HSL output value. Therefore, both of the converters architecture is working properly.

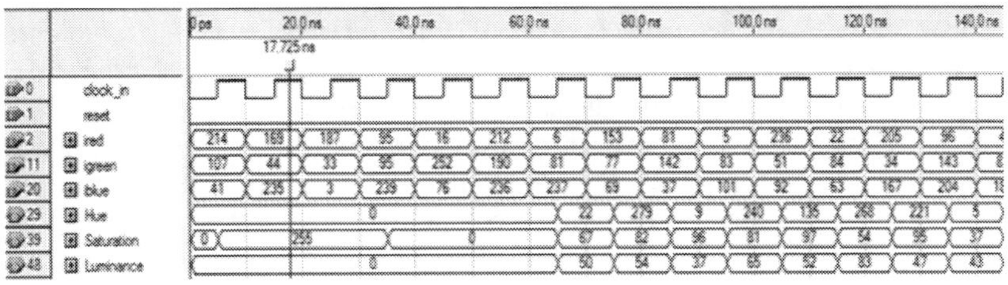

Figure 6. Simulation results for pipeline architecture.

4.3 Hardware Conversion Testing Architecture

To perform this pure hardware conversion testing, a specific testing architecture is developed specifically for RGB to HSL conversion process. Figure 7 is the hardware testing architecture developed in this work. It consists of five 8bit x 1024 word embedded Cyclone II M4K RAM and one 9bit x 1024word M4K RAM. The first three RAM is used to store the C code predefine 1024 sets of RGB values. The remaining two 8-bit and 9-bit RAM are used to store the converted HSL output values. All 1024 sets of predefine RGB values are written into RAM during initialization of the test architecture in FPGA.

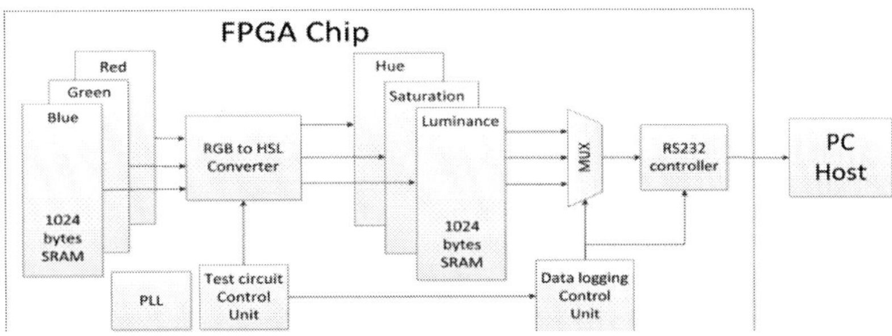

Figure 7. Hardware testing architecture.

After all 1024 sets of data are done with conversion; all the HSL output values stored in the RAM are then read and transmitted to the host computer by using RS232 serial communication protocol. The RS232 controller firstly convert the HSL value in binary into ASCII format before transmit to host computer. The test circuit control unit is responsible to handle and signal the conversion process step and other functional module. This testing architecture is also implemented with a Phase-Lock-Loop (PLL). This PLL is important and used to generate difference clock speed to drive this testing architecture and also the control unit. It used to investigate both parallel and pipeline converter output accuracy in respond to difference clock speed and to investigate their maximum operating frequency.

4.4 Slow Speed Conversion Accuracy

Before both converters are undergoing variable operating frequency testing, both of them had gone thru a slow speed test which is running at 10Hz clock. This test is to ensure that both converters are capable to produce the correct HSL output value when implementing in hardware. This output result also used as reference result when both converters are tested at high clock speed. Table 1 represents the accuracy test for 1024 RGB samples at the data rate of 10Hz clock.

Table 1. Accuracy of 10Hz data rate for 1024 RGB samples.

Converter	Accuracy
Parallel	99.00%
Pipeline	98.96 %

4.5 High Speed HSL Conversion Accuracy between of Parallel and Pipeline Architecture

The clock frequency used begins from 10MHz up to 150MHz. It is generated by the PLL. The maximum test frequency ceiling at 150MHz because it has fulfil the High Definition (HD) display clock speed. The accuracy of both converters is compared and analysed by individual output component of hue, saturation and luminance. This is because each of hue, saturation and luminance output values are produced by individual module; therefore, each of component accuracy is different.

Figure 8 shows comparisons of hue conversion accuracy. Both converters are capable to sustain at 99.20% accuracy at the beginning of the test until 50MHz. At 55MHz, the hue accuracy of parallel converter begins to fall. It shows that hue component from parallel converter has a maximum operating frequency at 50MHz. When both converters are tested with 70MHz clock, the parallel converter hue output accuracy start to fall to 19.38% which can considered as useless while the pipeline still able to maintain 99.07% accuracy. The pipeline hue conversion is capable to sustain above 90% up to 150MHz.

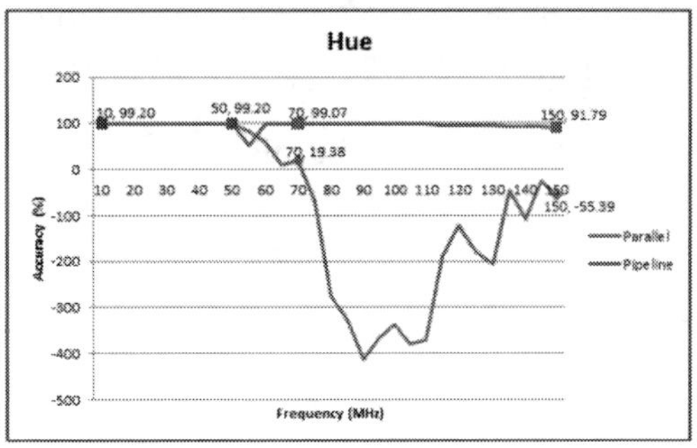

Figure 8. Hue conversion accuracy comparisons.

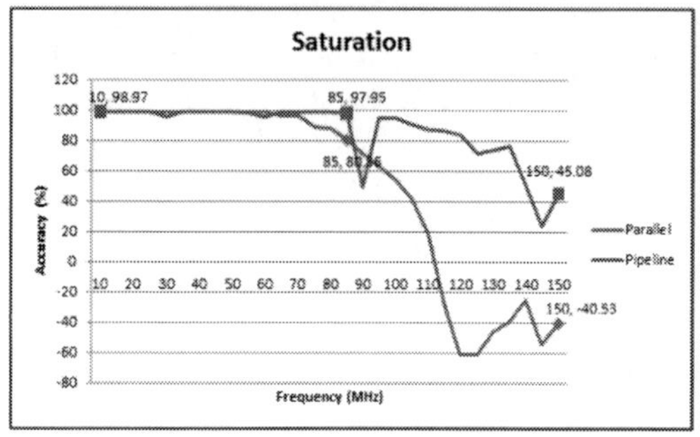

Figure 9. Saturation conversion accuracy comparisons.

From Figure 9, the parallel saturation conversion accuracy sustains at 98.97% started from 10MHz until 70MHz. When frequency is pushed up to 85MHz, the accuracy for parallel architecture dropped below level 80%. The parallel saturation conversion accuracy started to fall at the operating frequency of 75MHz. The pipeline saturation conversion accuracy is sustaining until 85MHz. When the frequency reaches 90MHz, the accuracy for pipeline architecture begin to drop. Significantly, when the frequency is pushed up to 150MHz, the pipeline converter still maintains the accuracy level of 45% and -40% for the parallel architecture.

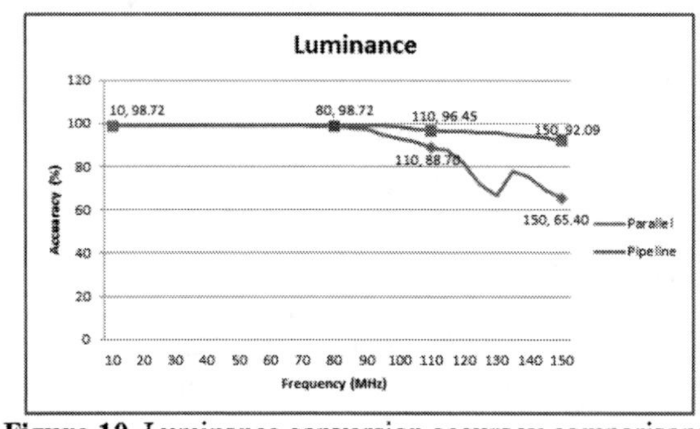

Figure 10. Luminance conversion accuracy comparisons.

Luminance conversion posed the most stable conversion output value. This can be seen in Figure 10. Its accuracy is able to sustain at level of 98.72 % started from frequency 10MHz until 80MHz. But the pipeline luminance is able to keep the accuracy until 100MHz, which is 20MHz higher than parallel converter. Parallel luminance conversion accuracy starts falling when the frequency reaches 85MHz. When frequency goes beyond 120MHz, the accuracy drops below 80% at which the output of the converter is no more reliable and useless. With the comparison to the pipelined architecture, the accuracy is capable to sustain at 96.45% up to 110MHz which 30MHz higher than the parallel architecture. The accuracy of the pipelined converter begin to fall when frequency pushed up beyond 110MHz. Significantly, when the frequency tested until 150MHz, the pipeline converter still maintain the accuracy level of 92% compared to 65.4% for the parallel architecture.

It is obviously that the pipeline hue conversion able to withstand high operating frequency up to 120MHz but the parallel converter only can properly function up to 50MHz, which is only half of the pipeline converter.

4.6 Real time Hardware Implementation

Both converters are implemented in the DE2_TV sample project provided by Altera to perform real time conversion operation. DE2_TV sample project is interfaced with a Close Circuit Television (CCTV) imaging sensor to capture real time digital video stream. The video stream is converted into RGB value to be displayed in VGA screen. By making use of this RGB stream, both tested RGB to HSL converters are implemented into the DE2_TV to produce hue, saturation and luminance image of the real time video stream. Figure 11 represents the comparisons of original RGB colour frame with a real time converted video frame of both converter and also with C code converted video frame.

The hardware converted HSL image in Figure 11 did not pose any major difference as compared to the C code converted HSL image. The saturation and luminance image can be considered identical for all hardware and C code conversion image. There is only a minor difference in the hue value located at the center of Figure 11(a) the word 'K'. It is darker as compared to the Figures 11(e) and (h). This difference is caused by the cyclic properties of the hue. The word 'K' is in red colour as depicted in Figure 11(a). Red colour in hue degree is range from 0° to 30° and also 330° to 360° degrees. This large difference of red colour causes the hue value to fluctuate between dark and bright region.

(a) Original RGB colour video frame

(b) Parallel converter of hue image

(c) Parallel converter of saturation image

(d) Parallel converter of luminance image

(e) Pipeline converter of hue image

(f) Pipeline converter of saturation image

(g) Pipeline converter of luminance image.

(h) C code converted of hue image.

(i) C code converted saturation image

(j) C code converted luminance image.

Figure 11. Comparisons of original RGB colour frame with a real time converted video frame of both converters and C code.

4.7 Hardware Resources Consumption
The parallel and pipeline converter hardware resources consumption on the FPGA has been summarized as shown in Table 2. Both of pipeline and parallel architecture converters are only utilize of 3% and 4% from total logic element of the Cyclone II EP2C35F672 FPGA chip respectively.

Table 2. Accuracy of 10 Hz data rate for 1024 RGB samples

Resources	Available	Parallel	Pipeline
Total Logic Element	33216	1247 (4%)	950 (3%)
Total Combination Functions	-	1247 (4%)	895 (3%)
Dedicated Logic Elements (Register)	-	0 (0%)	262 (<1%)
Total Memory Bits	483840	0(0%)	48 (<1%)
PLLs	4	0	0
Total I/O Pins	475	50 (11%)	51 (11%)
Embedded Multiplier 9-bits element	70	0	3 (4%)

5. Conclusion
This work has been successfully developed on both pipeline and parallel hardware architectures for RGB to HSL converter on FPGA. Both of the pipeline and parallel converter architecture functionalities have been verified using Intel-Altera Quartus II functional simulation. Both are further tested in hardware to obtain its accuracy and it has gone through 10Hz slow speed conversion testing to obtain the conversion accuracy. The achievement of accuracy by the parallel and pipeline converters approach are 99.00% and 98.96%, respectively. The HSL output values for these converters are accurate enough to be applied in machine vision system and image processing algorithm. Besides that, HSL output of pipeline architecture is more established than parallel architecture when tested with variable clock frequency from 10MHz to 150MHz. It can be concluded that the parallel converter only capable to handle the data rate up to 50MHz while the pipeline converter is able to handle up to 120MHz. The pipeline architecture is able to speed up twice of parallel architecture. Another performance merit is the parallel converter has consumed 1247 logic elements, 4% of the FPGA resources available while the pipeline converter is only consuming 950 logic elements (3%), 291 logic elements less compared to parallel architecture in the FPGA chip. Therefore pipeline RGB to HSL converter is a better choice with less resources consumption and twice the speed up compared to the parallel converter.

References
[1] Wu, D., and Sun, D.-W. Colour measurements by computer vision for food quality control – A review. *Trends in Food Science & Technology* 2013; 29(1), 5-20.
[2] Wang, M.-J., Lee, G., and Lin, H.-Y. Extraction of Perceptual Hue Feature Set for Colour Image/Video Segmentation. *Advances in Multimedia Information Processing: Springer Berlin Heidelberg* 2008; 5353, 875-878.

[3] Aptoula, E., and Lefèvre, S. On the morphological processing of hue. *Image and Vision Computing* 2009; 27(9), 1394-1401.

[4] Tsai, S.H. and Tseng, Y.H. A novel colour detection method based on HSL colour space for robotic soccer competition. *Computers and Mathematics with Applications* 2012; **64(5)**, 1291-1300.

[5] Hanumantharaju, M. C., Vishalakshi, G. R., Halvi, S., and Satish, S. B. A Novel FPGA Based Reconfigurable Architecture for Image Colour Space Conversion. *Global Trends in Information Systems and Software Applications: Springer Berlin Heidelberg* 2012; 270, 292-301.

[6] Gonzalez, R. C., and Woods, R. E. Digital image processing. *Upper Saddle River, N.J.: Prentice Hall*, 2018

[7] Nnolim, U. A. Design and implementation of novel, fast, pipelined HSI2RGB and log-hybrid RGB2HSI colour converter architectures for image enhancement. *Microprocessors and Microsystems* 2015; 39(4–5), 223-236.

[8] Nashat, S., Abdullah, A., Aramvith, S., and Abdullah, M. Z. Support vector machine approach to real-time inspection of biscuits on moving conveyor belt. *Computers and Electronics in Agriculture*, 2011; 75(1), 147-158.

[9] Khamar Basha Shaik, P. Ganesan, V. Kalist, B.S. Sathish, and J. Merlin Mary Jenitha, Comparative Study of Skin Colour Detection and Segmentation in HSV and YCbCr Colour Space, *Procedia Computer Science* 2015; **57**, 41-48 .

[10] YanJie, Z., and Li, X. A new object detection method in colour image processing. *4th International Congress on Image and Signal Processing (CISP),* 2011.

[11] Zhong, L., Weihai, C., Yuhua, Z., and Cun, H. Regions of interest extraction based on HSV colour space. *10th. IEEE International Conference on Industrial Informatics (INDIN)*, 2012.

[12] Swenson, R. L. (2000). A real-time high performance universal colour transformation hardware system (BL). (U132261 Ph.D*.), University of Kent at Canterbury (United Kingdom)*, Ann Arbor.

[13] Agoston, M. K. Computer graphics and geometric modelling. Retrieved from http://dx.doi.org/10.1007/b138805, 2015

[14] Pérez, P., Hue, C., Vermaak, J., and Gangnet, M. Colour-Based Probabilistic Tracking. In A. Heyden, G. Sparr, M. Nielsen, & P. Johansen (Eds.), Computer Vision: Springer Berlin Heidelberg. 2002; **2350**, 661-675

[15] Jung Uk, C., Seung Hun, J., Xuan Dai, P., Dongkyun, K., and Jae Wook, J. IEEE International Conference on FPGA-based real-time visual tracking system using adaptive colour histograms. Paper presented at the *Robotics and Biomimetics,* 2007.

[16] Tsung-Ying, S., Shang-Jeng, T., and Chan, V. HSI colour model based lane-marking detection. *The Intelligent Transportation Systems Conference*, 2006.

[17] Vadivel, A., Mohan, M., Sural, S., and Majumdar, A. K. Segmentation Using Saturation Thresholding and Its Application in Content-Based Retrieval of Images. *Image Analysis and Recognition* 2004; 3211, 33-40.

[18] Na, L., Jiajun, B., and Chun, C. Real-time video object segmentation using HSV space. *International Conference on Image Processing* 2002

[19] Jin-liang, Y., Yi-bo, G., Liang-jun, M., and Yi-Ping, Y. Scene Text Extraction Based on HSL. *International Symposium on.Computer Science and Computational Technology* 2008

[20] Abdullah, M. Z., Mohamad-Saleh, J., Fathinul-Syahir, A. S., and Mohd-Azemi, B. M. N. Discrimination and classification of fresh-cut starfruits (Averrhoa carambola L.) using automated machine vision system. *Journal of Food Engineering* 2006; 76(4), 506-523.

[21] Doignon, C., Graebling, P., and de Mathelin, M. Real-time segmentation of surgical instruments inside the abdominal cavity using joint hue saturation colour feature. *Real-Time Imaging* 2005, 11(5–6); 429-442.

Plastic optical fibre power splitter for surface profiling of 3-D object

N Syafiqah Mohamed-Kassim and M Kamil Abd-Rahman

School of Physics and Material Studies, Faculty of Applied Sciences, Universiti Teknologi MARA, 40450 Shah Alam, Selangor, Malaysia

drkamil@uitm.edu.my

Abstract. Single-mode fibre has established the intensity-based surface profiling application with the notable results such as surface topography of ancient stones and three-dimensional rigid-body shapes. However, cost and experimental handling has been the most common issue for single mode fibre related-applications. Thus, plastic optical fibre (POF) has become an alternative for surface structure monitoring due to its highly-multimode and larger cross-sectional area. Millions of light rays can be gathered and enhanced the light detection, aside from lower installation cost and offer easier handling of the fibre system. In this work, the operating principle and experimentation of the surface profiler on a reflective 3-D object is presented. The measurement of the profiler was investigated via a 0.7 dB low excess-loss, non-branching POF Y-shape power splitting coupler. The variation in light intensity reflected from the target surface provides details on the change in depth and thickness of the target. The 1-mm diameter POF end-facet provides excellent probing tip in measuring the diameter of 3-mm reflective spherical surface with good accuracy.

1. Introduction

POF power splitter device has been studied since 1990s, where it focused on plastic optical fibre and cable device. Thus far, POF is still an exploratory fibre due to its low-cost and reliable for short-distance network. However, the advantages among passive devices in the market nowadays not only focuses on performance of the device but also the cost of the product. POF power splitter device is a passive component that found applications in optical communication systems, photonics integrated circuits and optical signal channelling. POF power splitter is very popular among multimode fibre in the development of optical short-distance networks, such as local area network [1], home-network application [2], automation system [3] and optical sensors.

Various types of POF sensor have been reported [4][5][6]. However, due to the large diameter of POF, the intensity distribution corresponds to the modal structure of the POF [7]. The excitation light propagates at varying degrees as it exits the fibre end-facet. Hence, reshaping of fibre tip is realised to rectify the condition, especially for better light guiding and uniform power distribution [8]. POF sensor based on probe tip deformation has been widely used in physical and chemical measurements. It can be used to improve light focusing, light coupling, elimination of Fresnel back reflection, improvise focus distance and adjust light collimation [9].

In this paper, we demonstrate a non-contact optical surface profiler with the ability for thickness measurement of reflective surface object. The 3-dB 1x2 splitter separates the input fibre from the output fibre. The splitter has zero-branching angle at the junction, which is to minimise the bending and propagation loss between the two fibres in the splitter. One end of each 1-m long three connecting POFs

IOP Conf. Series: Materials Science and Engineering **767** (2020) 012055 doi:10.1088/1757-899X/767/1/012055

placed at three ports of the splitter converts the circular cross-sectional profile of POF to rectangular via thermal-compressed moulding technique to minimise optical loss in the splitter [10]. As input light incidents on the target via the input fibre and splitter, some of the light will be reflected back towards the fibre tip and propagates to the receiver via splitter and output fibre. The intensity of the reflected light was analysed to determine the surface profile and thickness of the target. The optical system exhibits low optical loss due to the reshaping of the POF cross-section along the mold-insert waveguide slots and by employing the zero-branching angle of the splitter.

2. Experimental Setup

Surface profiling of an object was performed using a home-made 3-dB 1x2 optical power-splitting coupler, where the shape of circular cross-section profile of the POF was molded into rectangular cross-section along the mold-insert waveguide splitter device. The conversion from circular to rectangular profile of the POF fibre is to reduce the propagation and coupling loss between POF and waveguide power-splitter. The experimental configuration of surface profiling is shown in Figure 1. It consists of a He-Ne laser as the light source, a 3-dB POF power-splitter, a reflective target, an XYZ translational stage, a universal motion controller (UMC), a photodetector, a digital multimeter (DMM), and a computer for data acquisition and storage. The 3-dB splitter is to separate the input (laser light) from the output (photodetector) with minimal splitting loss at the junction of the splitter. The 3-dB power-splitter can produce better output resolution for surface profiling as compared to bundle fibres (6). Fibre at port 1, which functioned as a detection probe was set perpendicular to the target surface under study. The other two fibres were connected to the laser light source and photodetector accordingly. The reflectivity measurements were accomplished by placing the reflective target on the XYZ translational stage and moved horizontally under the probe. The micrometer stage movement was controlled by a UMC (model ESP300). The probe was fixed and secured into a position while the target was made to move under the POF probe (from port 1). This is to ensure that any vibration generated from UMC due to the movement does not affect the POF probe and to provide consistency and repeatability of measurement.

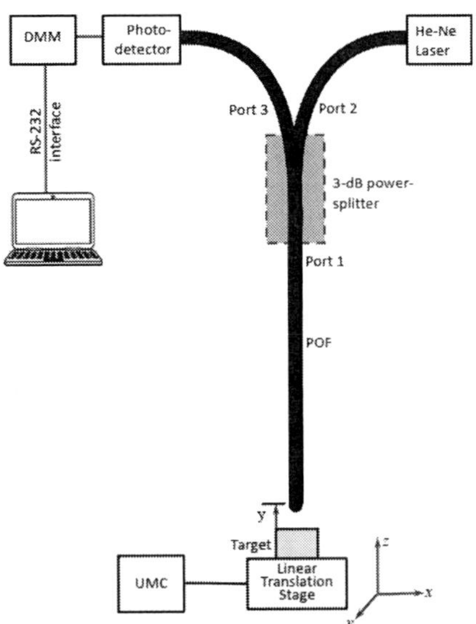

Figure 1. Experimental setup of the intensity-based POF surface profiler system.

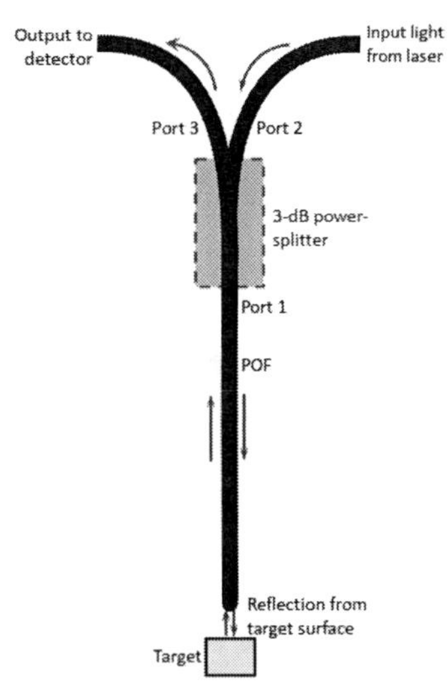

Figure 2. Principle of surface profiling via reflectivity from reflective surface target.

The embedded POFs in the Y-junction splitter have similar core cross-section of 0.785 mm^2. The POF numerical aperture is 0.5, while the refractive index of the core and the claddings are 1.490 and 1.402, respectively. The coherent light source used in the setup is a He-Ne laser operating at 632.8 nm wavelength with laser power of about 3.9 mW. The transmitted power to the surface of target object under study is about 0.39 mW, measured from an optical power meter. The reduced in optical power was due to coupling loss between POF and the light source, and 3-dB splitting-loss at the junction of the splitter. The position of the light coupling was maintained throughout the experiments by locking the position of the splitter using a customized holder.

The scanning of samples for surface profiling was performed in the x and y direction (length and width), while the z-direction detects the thickness of the target via relative intensity of reflected light from samples. The universal motion controller movement has about 13-mm span, thus the scanning path for all targets was limited to less than 13 cm in length. The POF photodiode (IFD91) detects the intensity of reflected signals. The POF photodiode converts the optical signals to electrical (voltage) signal and was measured using a digital multimeter (DMM - Pro's Kit MT-1860). Acquisition of data was recorded and saved in a computer via RS232 interface card. The DMM was set to provide high accuracy and resolution of ±(0.5% reading +4 digits) display and 1.0 mV/0.1 mV AC/DC voltage resolution. The speciality of the DMM is that it can acquire continuous data in real time. The performance of the power-splitting coupler was tested as part of profiler component as demonstrated in Figure 2. The He-Ne laser provides light to POF probe from port 2 and transmitted to port 1. Light reflected from the target enters the POF probe and the signal is then transmitted to the receiver from POF in port 3.

The surface profiler was calibrated to ensure accuracy and reliability of measurement. The calibration was done by analysing the reflected signal from the target object within a 3-mm range. Figure 3 shows the calibration configuration where the probe and sphere were near contact. The stage was gradually moved away from the probe to a distance of 3.0 mm. The reflected light outputs to photodiode and DMM, where the measured data was recorded in millivolts.

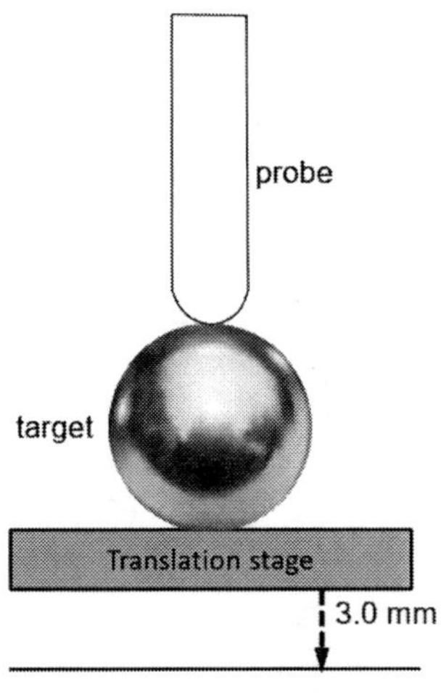

Figure 3. Setup for the calibration of the profiler.

Figure 4 shows the variation of output voltage with a displacement of 3-mm of the target object from the probe (POF tip). The amount of reflected light is at the maximum as the POF tip is in the close proximity to the target's surface. As the probe moves away from the target, the amount of light decreasing polynomially. The slope follows closely to inverse square law relationship and calibration curve will be used as a reference to all measurements. With numerical aperture of 0.5 (30° diffraction angle from POF end-facet), the power density of light reduced due to increase in the size of the cone of light where it spreads to much larger area from POF end-facet as it hits the target. The light diffused away and only small amount of light couple back into the probe. From the calibration curve, the voltage recorded 42.7 mV for the distant of 3.00 mm between POF probe tip and target.

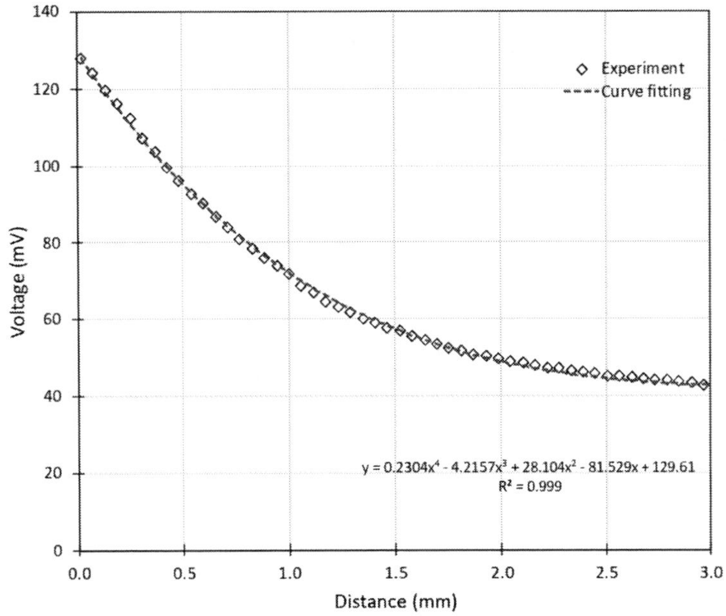

Figure 4. Calibration curve of 3-dimensional sphere surface profiling.

3. Results and Discussion

A 3-dimensional surface profile of a sphere was obtained by moving the target's stage consecutively along x- and y-directions. The initial scanning of the sphere profile was along the x-direction and gradually moved along the y-direction to cover the whole target's surface as shown in Figure 5. The probe was set at $z = 3.0$ mm from the stage's surface.

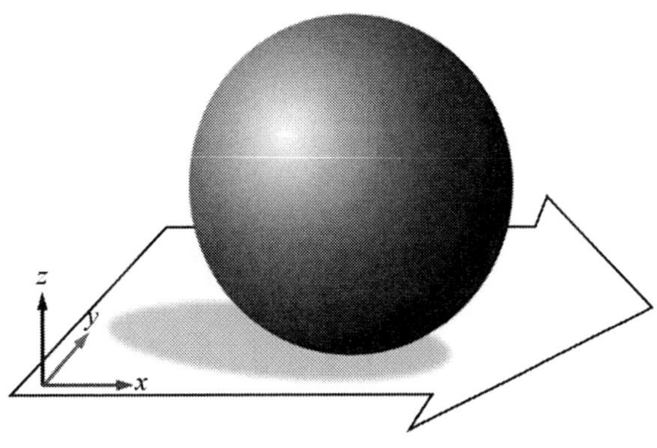

Figure 5. Scanned area covering the whole surface across the x- and y-axes.

From Figure 6, at $x = 0$ mm and $x = 3$ mm, the light intensity recorded the background value of about 41.9 mV, where the light was reflected back from surface of the stage that support the sphere and none of the light incident the sphere. The reflected light intensity has a maximum value at $x = 1.5$ mm where the probe and the target are close to contact. The signal decreases by scanning x-values in both directions and reaches to a signal near to the background signal. The reason for such decreases is due to the distance from the sphere's surface to the probe has increased and the spherical curvature changed along the movement, which result a lower reflected signal. The measurement on the profile of the sphere demonstrates the significance of the thickness of the target. Thus, using this method we can determine the diameter of the sphere, where the voltage at 3.0 mm distance from the probe showed almost similar values of 41.9 mV from Figure 6 as compared to 42.7 mV from the calibration curve. The main factor that can affect the thickness measurement is related to the contact distance of the target to the probe, whose imperfections and alignment errors can reduce the sensing accuracy.

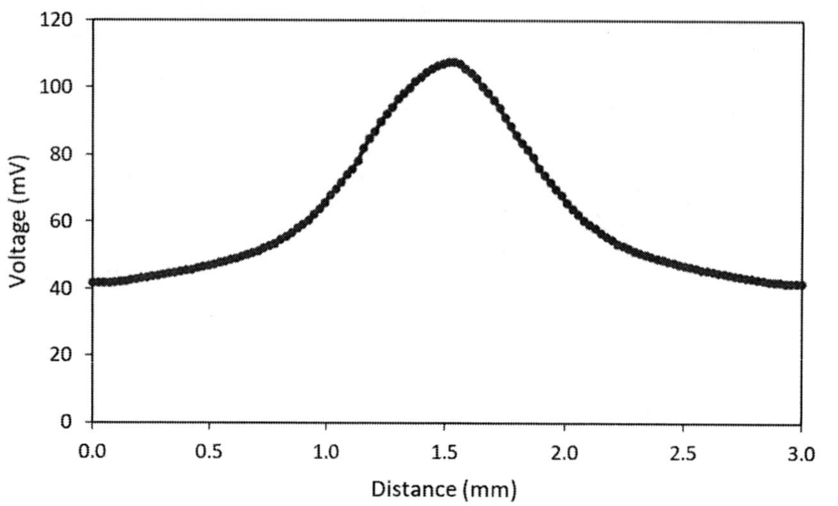

Figure 6. Surface profile of x-scan of the sphere target.

Figure 7 (a) and (b) show the 2D and 3D views of the sphere surface profile. The sphere profile was formed by the deviations of the reflected light. The intensity of the reflected light from the sphere surface depends upon the target's surface shape and standoff distance between the surface and fibre tip. The signal shows a maximum at scan length of 1.75 mm and reaches to a signal near to the background signal at the target's sides. However, the surface profile diameter obtained was 3-mm with accuracy of ±1.74%. The reason is because at some point, POF is no longer able to receive any signal due to the too far distance between the target and the probe. Some of the light rays are scattered out of the probe, and some that is able to reach the target is deflected out and did not return back to the probe as shown in Figure 8.

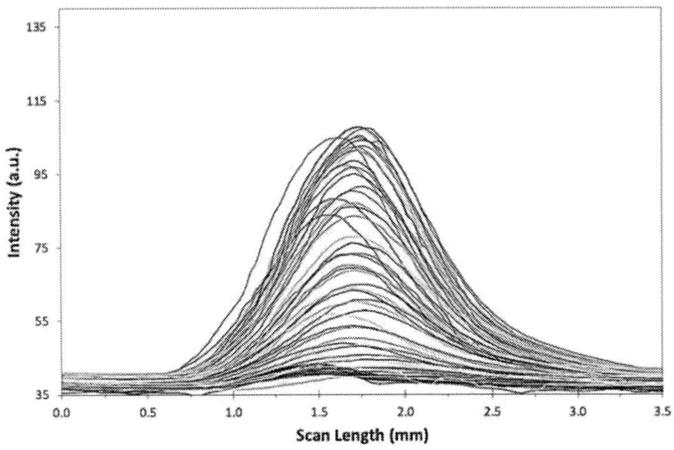

Figure 7. (a) 2-Dimensional view of a 3-mm diameter sphere.

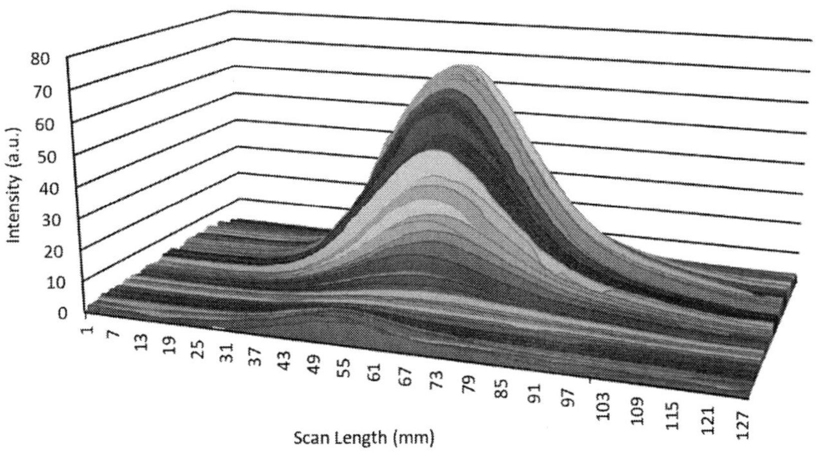

Figure 7. (b) 3-Dimensional view shows intensities obtained corresponding to target's shape with scan length scale of 19:0.5 mm.

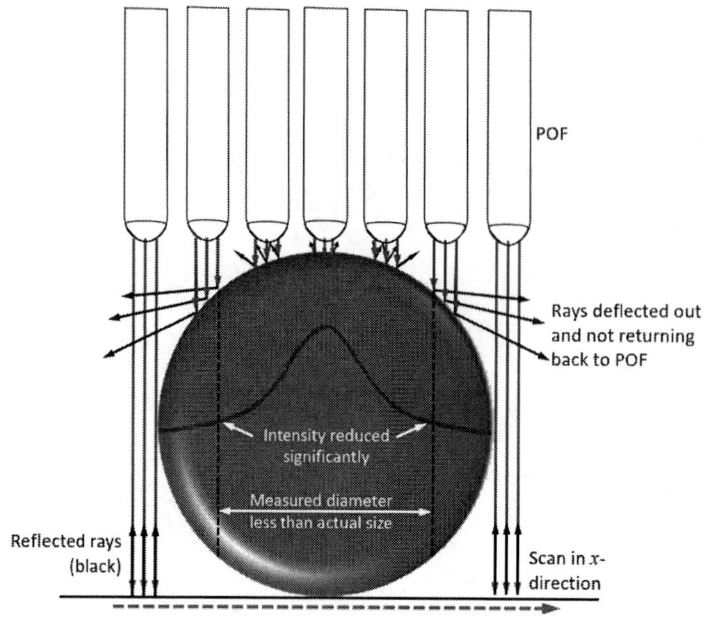

Figure 8. The measured diameter was less by about 33% from the actual diameter of the sphere due to deflection of light out from the sides of the sphere and not reflecting back to POF.

The stability of the device was investigated using similar configuration as in Figure 9. A reflective steel block was used as the target. Marker ♦ denotes the reflected signal of the laser reflected of the steel block over 10 minutes. The average reflectance of the steel block was measured about 437.90 mV with standard deviation of 0.51 mV. For a better comparison, the reflection in photodetector output with no laser light source also shown with marker ■. The average background signal was 3.84 mV with standard deviation of 0.13 mV.

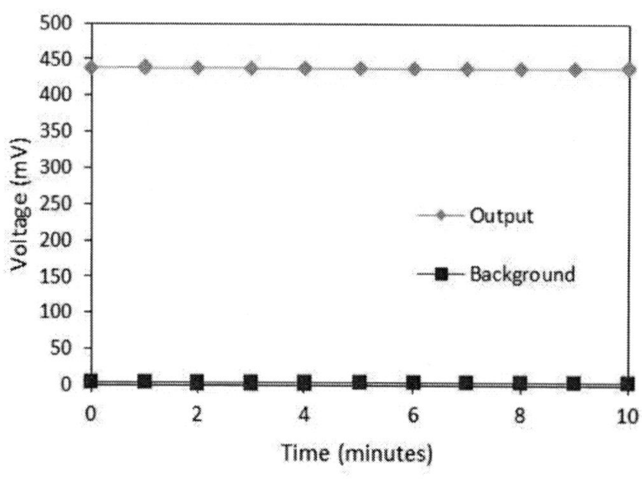

Figure 9. Stability of scanning device when the laser turned on and off as a function of time. The output and the background denote when the laser turned on and off, respectively.

In verifying the x-y plane to be of uniform state, a steel block was placed on the translational stage. The laser emits light to the target block, while the stage moved in x- and y-directions with velocity of 0.2 mm/s. In order to verify the stage is on the horizontal position, the output signal for every point scanned across x and y must be similar. Hence, uniform beam distribution and precision of surface profiling can be achieved. Figure 10 presents the reflected signal as a function of x- and y-scan distance. For each scan, the stage movement was set to 10 mm. The plot shows the signal slightly deviated along the distance. Average output signals for x- and y- directions were 443 mV and 445 mV with standard deviation of 0.6 mV and 0.4 mV, respectively. The deviated signals could be as a result of surface non-uniformity which introduce the light fluctuation and affect the reflection characteristics of the target's surface. The small output signal differences demonstrate that the surface profiling setup is right and proper for surface profiling measurement.

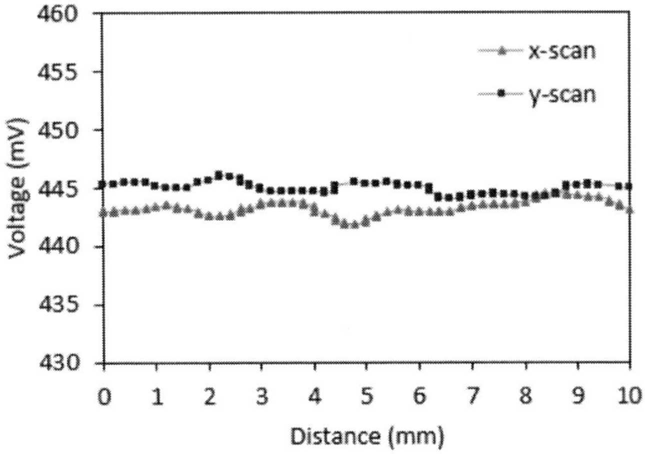

Figure 10. The x-y scan of target's stage for 10 mm scanning.

4. Conclusion

An optical surface profiler device was developed to scan the surface of a reflective target in a non-contact mode. The device relies on the intensity of the reflected light from the target object, where it comprised of a polymer optical fibre tip as the probe attached to a 3-dB 1x2 optical coupler and a coherent light source from a laser. The He-Ne laser can be replaced with a semiconductor laser diode for a more compact and robust device setup. The diameter of the target sphere was found to be 3.0 mm with ±1.74% accuracy, based on the voltage measured from calibration curve (42.7 mV) and scanned profile of the sphere (41.9 mV). The POF-based optical profiler can be used to measure the width, depth and roughness of reflective surfaces. Even though the device could detect width changes throughout the scanning without desired surface profiling, this method could still be exploited for related applications. It would be a valuable device in the future with some development for improvement, as the intensity-based device offers simplicity, portability and compact in size, besides faster scanning time as compared to single mode fibre over the same scanning area, since light emitted from POF has larger spot size.

References

[1] Mohamed-Kassim N S, Ehsan A A, Yusoff M H M and Rahman M K A, "Fabrication and simulation of multimode symmetrical and asymmetrical Y-junction optical power splitters", *Int. Conf. Photonics,* 2010, p 1–5.

[2] Shi Y, Okonkwo C, Boom, Den H V, Tangdiongga E, Koonen T, "480Mb/s UWB distribution over large-core POF based bus and tree P2MP systems for in-home networks", *Opt. Fiber Commun. Conf. Expo.* 2012.

[3] Ab-Rahman M, Guna H, Harun M, Zan M, Jumari K, "Fabrication and characterization of

customer-made 1x3 POF based optical coupler for home networking", *Int. J. Comput. Sci. Netw. Secur.*, 2008, **8,** p 43–48.
[4] Golnabi H, "Surface profiling using a double-fiber optical design", *Opt. Lasers Eng.* 2010, **48** p 421–426.
[5] Rahman H A, Rahim H R A, Ahmad H, Yasin M, Apsari R, Harun S.W., "Fiber optic displacement sensor for imaging of tooth surface roughness", *Measurement,* 2013, **46,** p 546–551.
[6] Rahman H A, Harun S W, Batumalay M, Muttalib F A, Ahmad, "Fiber optic displacement sensor using multimode plastic fiber probe and tooth surface", *IEEE Sens. J.*, 2013, **13**, p 294–298.
[7] Chandrappan J, Mohan R V, Gomez P O, Aung T A., Ramana P V, Lau J H, "Optical coupling methods for cost-effective polymer optical fiber communication", *IEEE Trans. Components Packag. Technol.*, 2009, **32**, p 593–599.
[8] Chandrappan J, Zhang J, Mohan R V., Gomez P O, Aung T A, Xiao Y, "Cost effective optical coupling for polymer optical fiber communication", *Proc. SPIE*, 2008, **6899**, p 68991A
[9] Sanghera J, Florea C, Busse L, Shaw B, "Reduced Fresnel losses in chalcogenide fibers by using anti-reflective surface structures on fiber end faces", *Optics Express*, 2010, **18**, p 26760-26768.
[10] N. S. Mohamed-Kassim, A. M. A. Ibrahim, and M. K. Abd-Rahman, "Low Loss and Dynamically-Tunable Power Splitter Based-on Circular-to-Rectangular POF Converter," *Journal of Lightwave Technology*, **34**(4), p 1–1, 2016.

IOP Conf. Series: Materials Science and Engineering **767** (2020) 012056 doi:10.1088/1757-899X/767/1/012056

Automatic vehicle identification system using number plate recognition in POLIMAS

Nui Din Keraf[1], Phaklen Ehkan[2], Virakwan Hai Kelian[3], Norulhuda Mohd Noor[1], Hafifah Darus[1] and Eunice Loke[2]

[1]Department of Electrical Engineering, Politeknik Sultan Abdul Halim Mu'adzam Shah (POLIMAS), 06000 Jitra, Kedah, Malaysia
[2]School of Computer and Communication Engineering, Universiti Malaysia Perlis (UniMAP), Pauh Putra Campus, 02600 Arau, Perlis, Malaysia
[3]Department of Information Technology and Communications, Politeknik Seberang Perai (PSP), 14000 Permatang Pauh, Penang, Malaysia

phaklen@unimap.edu.my

Abstract. The security and management of transportation system becomes an important key in controlled place such as campus area. With an increased number of vehicles, there is a need for vehicle identification mechanism that is effective, affordable and efficient. This paper presents the development of automatic vehicle identification system using NP recognition in POLIMAS. Number Plate (NP) Recognition is an image processing technology in computer vision which captures the image of the vehicle and recognizes its NP. The system is installed at the main entrance to ensure that only the authorized vehicles can enter the campus area. Once the vehicle is detected by the input sensors, a system will capture the image of vehicle plate number. An image is then extracted and investigated character segmentation by using optical character recognition. The method used for detection of a plate number is by pre-processing of the image and using a combination of Sobel Edge Detection and Laplacian Edge Detection Techniques. Bounding Box technique is used to find the NP and character recognition. The accuracy of NP recognition has an average of 87%. The system is sustainable as the camera will only be switched on when a vehicle is present.

Keywords: Vehicle Identification, Number Plate Recognition, Image Processing, Optical Character Recognition.

1. Introduction

Number Plate (NP) Recognition is a highly accurate system that performs optical character recognition (OCR) on images to read the NP on vehicles [1][2]. The system allows for real time recognition of a vehicle's NP and commonly uses infrared lighting to allow the camera to take the picture at any time of day or night. Security at a highly restricted area such as academic institution is highly important. However, with just security guards guarding the perimeters, human error may likely to occur. In academic campus such as Politeknik Sultan Abdul Halim Mu'adzam Shah (POLIMAS), there are strict laws required to prevent intruders or robbers which may affect the safety of the students, lecturers and assets. Many challenges must be considered in order to develop an efficient automatic authorized vehicle

IOP Conf. Series: Materials Science and Engineering **767** (2020) 012056 doi:10.1088/1757-899X/767/1/012056

identification system such as the light intensity is caused by the weather. When it is sunny it may be too bright and when it rained it may be dark or distortion may occur. Accuracy may change during the night especially when raining. It may cause an error while reading the NP. Besides that, car number plates (NPs) have different formats such as the single line NPs and double line NPs. It is an issue in Malaysia that the NPs are not uniform and some are written in cursive words [3]. The position of the NP could sometimes be at the side but mostly centre.

The objective of this work is to implement an automatic vehicle identification system that will detect and capture the image of vehicle NPs. With the increase of security risk, capturing and extracting NP can help to reduce human error.

2. System Model

The vehicle identification system is separated into two features; hardware and software configurations. Hardware configuration is the combination of a prototype and single board computer (SBC). The hardware is integrated with the software. Two different approaches are carried out in this project. The first approach is connecting the software directly to the camera to ensure that simulation could work. Once it is successful, the second approach is integrating SBC with the prototype and software. Software configuration is the method used to capture the image of the NP and process it through mathematical formulae using MATLAB. Two stages of the method used in the software; (i) extraction of a NP and segmentation, and (ii) recognition of characters.

2.1. Hardware Configuration

A Raspberry Pi 2 B+ board is a main controller of the system. The system is using a camera to capture image when an infrared sensor (IR) senses the presence of a car. Detection of the car's presence is done using an IR sensor and it will then switch on the camera. The camera will capture image of the car. Figure 1 shows the input and output through the Raspberry Pi 2 B+ board. The IR sensor and camera are the inputs to the board while the output is the extraction and recognition of NP characters. A camera installation is a crucial step as it changes due to the surrounding and location that is being installed. Many challenges such as the position of the sun shine or when it is too dark is considered.

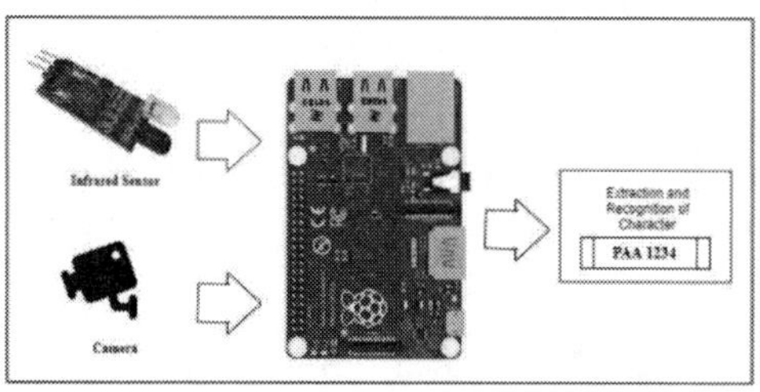

Figure 1. Diagram of vehicle identification system.

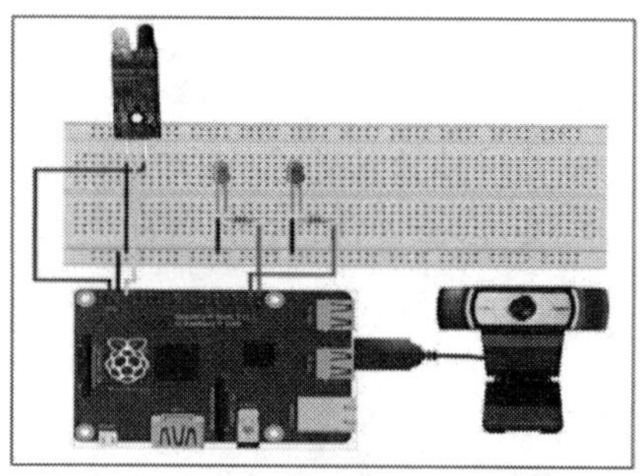

Figure 2. Layout of the circuit diagram.

Figure 2 shows the layout of the circuit diagram. The IR sensor connected to GPIO 14 for transmitter input is used to sense the present of a car. Two LED lights are connected to GPIO 20 and 21, Green and Red LEDs are used to show if it is authorized vehicle for entry or access is being denied respectively. Web Camera Logitech c930e is an input at Universal Serial Bus (USB) port.

2.2. Software Configuration

Software configuration is used for NP detection and vehicle identification. A simulation on NPs will be carried out to identify each step of a configuration where a certain result is expected to be shown for different techniques.

2.2.1 Pre-processing of Image. Once an image has been captured, the pre-processing can begin as it is critical in order to continue to the edge detection phase. The pre-processing consists of converting from RGB to grayscale followed by contrast and intensity adjustments. Removal of RGB is done by eliminating the saturation and hue data of the image. The RGB image comprises 30% of red, 60% of green and 11% of blue. Equation (1) is used for converting RGB to grayscale [4].

$$\text{Grayscale} = 0.2090{*}R + 0.5870 * G + 0.1140 * B \quad (1)$$

The intensity and contrast are adjusted to reduce noise of image. Histogram Equalization technique is used to adjust the intensity and helps to improve the image contrast when the value of intensity image is transformed. Huge amount of data detected in grayscale image. The general way of edge detection of the image is through filtering, differentiation and detection which have higher chances of detecting the image.

2.2.2 Extraction of Number Plate. Detection of NP consists of two methods; detecting rectangular images according to the size of NP in the image and by comparison if a size matches image of NP. An edge of image may have a drastic change in the shade of colour and texture which is used to identify the dimensions and objects in image.

a) Edge Detection Technique
Sobel Edge Operator is a first order differential operator in the edge extraction techniques. Sobel mask in the Sobel operator is usually used to distinguish the boundary of objects because the processing is quicker in terms of speed and the extraction gives out thick edge operation. The extraction includes straight, horizontal and diagonal directions. It has an advantage for brightness of the boundary of the character and background. This operator consists of a pair of 3*3 convolution kernels as shown in Figure 3. Gy shows the image of Gx rotated by 90°.

-1	0	1
-2	0	2
-1	0	1

Gx

1	2	1
0	0	0
-1	-2	-1

Gy

Figure 3. Convolution masks for Sobel operator.

Both Gx and Gy can be combined to find the magnitude of the gradient at each point (Equation (2)),

$$|G| = \sqrt{Gx2 + Gy2} \quad (2)$$

The approximation magnitude is given as Equation (3),

$$|G| = |Gx| + |Gy| \quad (3)$$

The angle of orientation of the edge giving rise to spatial gradient as shown in Equation (4),

$$\theta = \arctan\,(Gy \;/\; Gx) \quad (4)$$

Canny edge operator performs in multi-stage process. The image is first smoothed by Gaussian convolution. Next, 2-D derivative operator used to smoothen and highlight regions of the image. The edge gradient and direction can be determined using the same equation in Sobel operator. Thirdly, non-maximum suppression is applied to remove false response. Then, double threshold to decide for potential edges is performed. Lastly, track edge by hysteresis.

IOP Conf. Series: Materials Science and Engineering **767** (2020) 012056 doi:10.1088/1757-899X/767/1/012056

The Laplacian of Gaussian (LoG), the second derivative is very responsive to noise and it plays an important role. The rapid intensity change in the Laplacian is usually used for edge detection. Gaussian smoothing filter is still applied to reduce its sensitivity to noise. The operator of 2-D function produces the scalar function (Equation (5)):

$$\Delta f(x,y) = \frac{\partial^2 f}{\partial x^2} + \frac{\partial^2 f}{\partial y^2} \quad (5)$$

Zero crossing is found as edges are examined on those pixels on the image which have locally maximum gradient. The zero crossing of the second derivative selected image edges has some threshold. The 2-D LoG function centred on zero crossing with Gaussian standard deviation is using Equation (6).

$$LoG(x, y) = -1/\pi\sigma^4 \,[1 - (x^2 + y^2)/2\sigma^2]\, e^{(x2+y2)/\,2\sigma 2} \quad (6)$$

The Laplacian edge detector computes the second order derivative mask using only on mask unlike the Sobel operator. The mask used is 2-D convolution kernel.

b) Combination of Edge Detection Techniques
After convert the image to grayscale, the technique which is introduced will be used as a combination. An operation is processed where both are combined to create a final detection image. There are two combinations taken in this process to identify which one is more suitable for edge detector. A combination of Canny edge and Laplacian edge detector, and another would be Sobel edge and Laplacian.

2.2.3 Segmentation and Recognition of Character. Segmentation of character is the isolation of each character on the NP. This begins with the image being divided into different subparts. The technique used for character segmentation is bounding box (Figure 4). The bounding box method is the label region enclosed with rectangular boxes followed by the determination of corner coordinates, height and width. Each alphanumerical character is covered in another bounding box. After identifying each character, it is then cropped and sent for segmentation.

Figure 4. Bounding box technique.

Figure 5. OCR using eigenvector.

After segmentation of character, comes the character recognition step. The goal is by converting an image text to characters. To develop this part, OCR algorithm is used to recognize the NP. An image that was cropped is now inverted, this is done by all white pixels are inverted to black and all black pixels are inverted to white as shown in Figure 5. The NP is 25 separated, and OCR is used to compare each character individually according to the alphanumerical database. Correlation method is used to identify each character and the result is stored in a string format.

3. Analysis and Discussion

The results on each process for vehicle identification system have been recorded. Analysis of practicality has been taken into consideration while building and carrying out this project. This section is basically about the simulation carry out to process a captured photo by a camera. The captured image is then checked in the database system to prove that the car has clearance to enter the controlled area.

3.1. Detection and Capturing of Image

As far as extraction of the plate region is concern, detection and capturing are a crucial part of this process. This part is divided into the detection of car and capturing of the NP and the image undergo pre-processing which consist of converting RGB to grayscale and enhancement of the image using histogram equalization. However, the position of the camera do co-relate with capturing of the ideal image.

3.1.1 Comparison of Position of the Camera. The comparison of the position of the camera is to identify which position is best to install the NP camera. The two options are the front and the rear of the car. If it is place in this position, the camera would be position in a gantry position. Table 1 shows the comparison and accuracy of each position.

Table 1. Comparison of front and rear position.

Vehicle		
Position	Front	Rear
Accuracy	43%	86%

Table 2. Comparison of front and rear skewed position

Vehicle		
Position	Front and Skewed	Rear and Skewed
Accuracy	70%	66%

Few images have been processed and analyzed. It shown that placing a camera in front will be less accurate compared to placing it behind. However, it would be a challenge to place the camera directly behind a car. It would be easier to put at the front because it can be fixed at the boom barrier. The accuracy is lower because of surrounding of the NP. Another scenario is that placing the camera by the side. The position of the camera would be either left or right and front or rear. In short, the image taken would be slightly skewed due to the placing of the camera which is at the end of the boom barrier. The image taken would be as shown in Table 2. According to a few images processed and analyzed, it is shown that both accuracies are close together. However, the reasons are quite similar to the previous scenarios which many boxy parts on the camera cause this issue. Besides that, some accuracy is caused by the environmental factor. Therefore, the angle of the camera may contribute to the accuracy of reading the NP.

3.1.2 Comparison of Environment Factor. The surrounding environment contributes to the challenges of reading the NP. Environment issues such as lighting issues and rain may cause the brightness to change. Table 3 shows the comparison of a vehicle during day and night. The accuracy of day and night comparison is similar as long as a reasonable amount of light is available. For the night situation, a bright light is present during the simulation of this experiment. The bright light is placed shone on the vehicle to ensure the sufficient amount of light is given before capturing the image. However, the light source should not be shone too direct as it will cause reflection on the NP which contributes to an accuracy of the reading NP.

Table 3. Comparison of day and night

Vehicle		
Day/Night	Day	Night
Accuracy	80%	80%

The angle of a camera is very crucial as the placing of the camera is based tailored according to every environment of the surrounding. If the camera is pointed where sunlight is shone directly or the light source used at night is shone directly on the car plate, the system would misread or unable to detect the character on the NP. Figures 6 and 7 show the cause and effect of reflection cause by light source and natural sunlight respectively. The effect of the character is either blurred out or it could not be recognized. For instance, Figure 6 could not be read even character beside character '4' while Figure 7 the reflection was on the character '6' and 'U' so both of the character could not be processed.

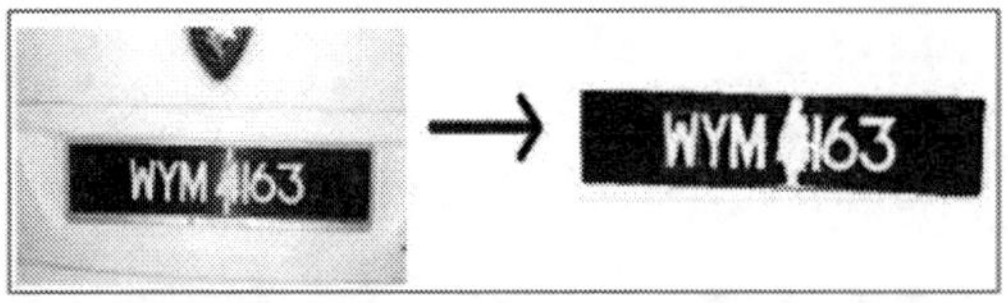

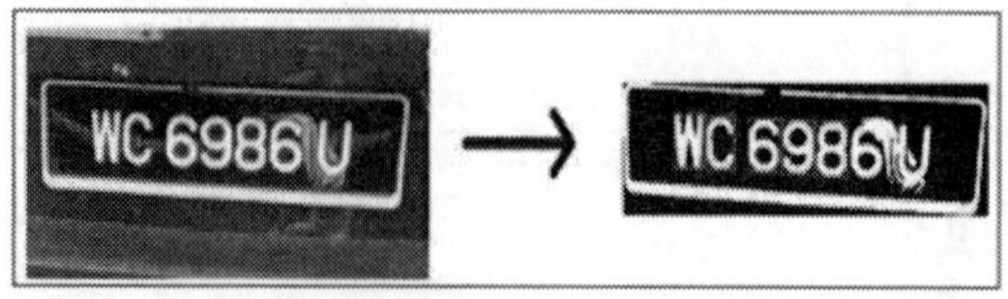

Figure 6. Reflection caused by direct light source. **Figure 7.** Reflection caused by natural sunlight.

3.1.3 Grayscale Technique. Figure 8 shows the image captured as a sample for this system. Figure 9 shows the image of Figure 8 being converted into grayscale. The reasons behind these conversions are signal to noise, the difficulty of visualization and speed. Signal to noise in a colour image will not help much in detecting the edge while grayscale images can provide higher stand out on the edges. An RGB image has difficulty in visualizing because of the pixel and various types of colour which consist of multiple dimensions while a grayscale image only has two special dimensions and one brightness dimension as a 3D image. Speed is something that is considered when using it in an application such as OCR and facial recognition as shifting or resizing the image will change the segmentation. After converting the image to grayscale, histogram equalization is performed on it to enhance the contrast of the image. Figure 10 shows the image of histogram equalization on a grayscale.

Figure 8. RGB image. **Figure 9.** Grayscale image. **Figure 10.** Histogram equalization.

3.1.4 Comparison of Grayscale and Histogram Equalization. Figure 11 shows the comparison of the grayscale image and histogram equalization image. Histogram equalization image shows a clearer contrast. Figure 12 shows the histogram image before and after histogram equalization. This method is used to increase the contrast of the overall image.

(a) (b)

Figure 11. Comparison of (a) grayscale and b) histogram equalization.

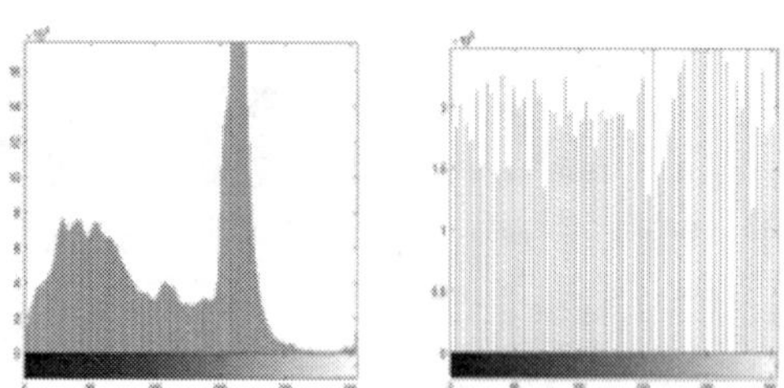

(a) (b)

Figure 12. Comparison of histogram (a) grayscale and (b) histogram equalization.

The advantage of this method is to improve the background and foreground by adjusting the brightness and the darkness in an image. Based on both histograms, the original grayscale image shows the intensity is uneven while for histogram equalization shows how the contrast value is spread out and distributed. The darker spot will be brighter while the brighter part is darkened.

3.2. Extraction of Number Plate

After the pre-processing stage comes the stage to extract the NP. To extract the NP, few methods were proposed form combination of both edge detection techniques as it would give higher accuracy [5]. Simulation for comparison between Canny edge and Sobel edge detection was carrying out as well as comparison between Sobel edge and Laplacian edge detection. The results will show which technique combination is more suitable for this system.

3.2.1 Comparison of Canny Edge and Sobel Edge Detection Technique. Figure 13 represents the comparison between Canny edge and Sobel edge [6][7]. The result of Canny edge has given many challenges to providing an accurate reading as the edge detection detected more noises compared to the Sobel edge detection. As a solution, the Sobel edge detection technique is used to provide a more reliable reading. Combination of edge detection techniques is the Sobel edge detection technique and Laplacian edge technique. Both convert from histogram equalization image respectively. A combination technique is better and the edge will be more precise [8].

(a) (b)

Figure 13. Comparison of (a) Canny edge detection and (b) Sobel edge detection.

3.2.2 Comparison of Sobel Edge and Laplacian Edge Technique. Figure 14 shows comparison between Sobel edge detection and Laplacian edge detection after histogram equalization. These techniques are used to detect the edge extraction technique [9][10]. Sobel edge uses a gradient-based method which consists of two kernels (horizontal and vertical) while the Laplacian edge uses only one kernel. Sobel calculates the first derivative of the X and Y axes separately.

The image then undergoes convolution shows strong vertical and horizontal line in the image. On the other hand, Laplacian calculates the second order derivatives in a single pass. Laplacian will be very sensitive to noise because it undergoes second order derivatives. Laplacian usually focuses on the contour of image. The Laplacian is generally used to find whether a pixel is on the dark or light side of the edge.

(a) (b)

Figure 14. Comparison of (a) Sobel edge detection and (b) Laplacian edge detection.

IOP Conf. Series: Materials Science and Engineering **767** (2020) 012056 doi:10.1088/1757-899X/767/1/012056

3.2.3 Combination Edge Detection Technique. Figure 15 shows the combination of Sobel edge detection and Laplacian edge technique. Sobel edge provides a standout on vertical and horizontal lines while Laplacian provide a standout on curves and contours. When both are combined the result reduces the horizontal and vertical lines as well as curves which then only the bold lines will remain. Thus, this will provide a sharper image to detect the edge [11].

Figure 15. Combination of edge detection technique.

Figure 16. Binarization.

3.2.4 Binarization. Figure 16 shows binarization of an image after the process of combination edge detection technique. Binarization is the technique used to convert a pixel image to a binary image. Figure 17(b) shows each pixel is converted into either 0 or 1. It is turning the image into a black-and-white image. Black stands for 0 while white is 1. Figure 17(a) shows the image before binarization where it is harder to determine the edge than Figure 17(b) which only shows the lines as '1'. The usage of this technique is to make it easier determining the lines of the image by just using binary numbers.

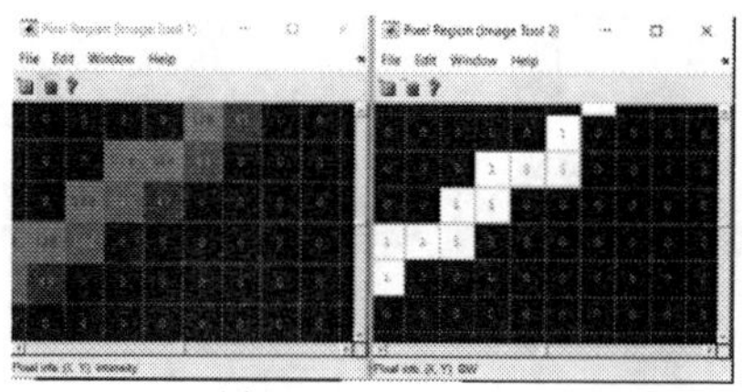

(a) (b)

Figure 17. Image (a) before and (b) after binarization.

3.2.5 Dilation, Fill Holes and Erode. A figure 18 show the image undergo dilation process which helps to improve the line by dilating it to make it looks thicker. It dilates binary images by enlarging the boundaries of a region. This process will grow the pixels and the region becomes smaller. Next, is filling of the holes on the binary image as shown in Figure 19. The filling of holes is filling up background pixel that is not reached as well as edges. As compared to image in Figure 19, the NP can no longer be seen as that pixel has been filled up.

Figure 18. Dilation

Figure 19. Fill holes.

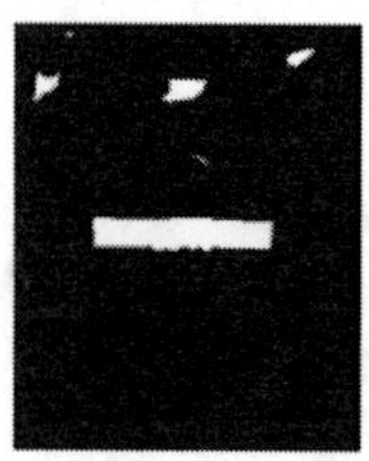

Figure 20. Erode.

Figure 21. Bounding box on detected NP.

Figure 20 shows the image after eroding process which is the second mathematical morphology other than dilation. This process is applied to a binary image to erode white pixels which then leaves the image with shrink pixels leaving larger holes in those areas. Figure 21 shows the image of the bounding box on the detected NP. There are two bounding boxes which require the process to choose which the suitable one is. The process would pick the one which is usually smaller. Therefore, the NP is selected in this image. The NP with the bounding box is cropped on and the following process is character segmentation and character recognition.

3.3. Segmentation and Recognition of Character

Before Segmentation, compliment of the image is necessary in order to use bounding box technique on the image (Figure 22). Compliment of the image is like placing XOR logic where '0's' turn to '1's' and '1's' turn to '0's'. When this method is done, the bounding box of each character is performed.

Figure 23 shows the final step before using the bounding box method to segment out the character while Figure 24 shows the bounding box on segmented character. The segmented character is then being compared to the created template. Once the comparing of each character is done, each character is read and the NP data is stored. Figure 25 shows the image after the bounding box technique is used. The bounding box of character uses the smallest rectangle to surround the image as tight and close as possible. The image is then cropped and compared to the template which is provided. Last but not least, the systems read and recognize the NP as LA2715 and it is checked in the database for accessibility.

Figure 22. Compliment of number plate.

Figure 23. Number plate before bounding box.

Figure 24. Bounding box on segmented character.

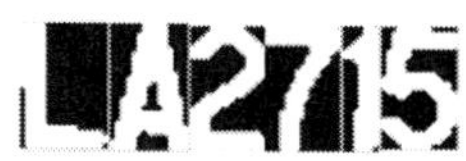

Figure 25. Bounding box cropped.

3.4. Database for Access

After reading the NP it is compared with the database in Excel. Excel is link to MATLAB to check if the NP is in the system. The following Figure 26 shows the database where it stores the NP to access into POLIMAS area. After checking if the NP is in the system, the LED light will turn green which mean the car is allowed to enter and if the LED is red it means the access is denied. It has been put to trial in MATLAB to check if for access as shown in Figure 27.

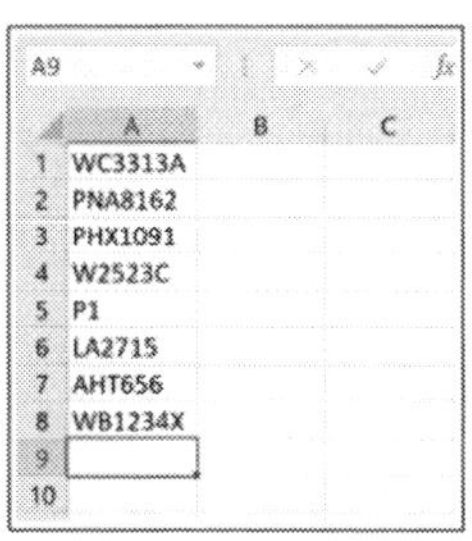

Figure 26. Database created for access.

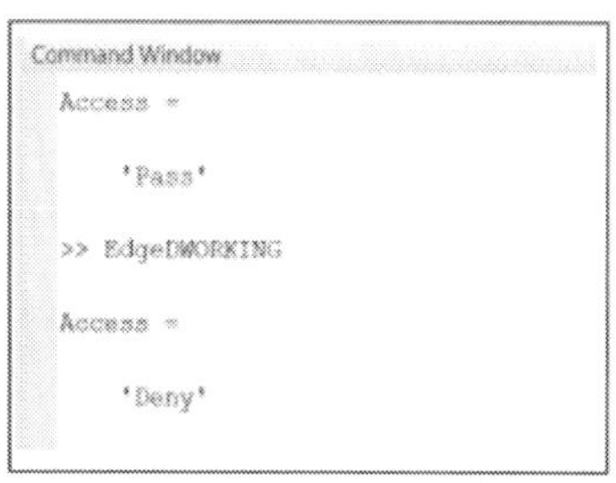

Figure 27. System in MATLAB.

4. Result

Table 4 shows the analysis of the testing results. There are 23 samples used and the success rate of detecting the NP is 95.83%. Character Segmentation is the segment of each NP on the car. There are 151 numbers of characters in total from 23 numbers of samples and 135 characters are detected from the images. The percentage of character segmented is 89.40%.

Table 4. Testing results.

Process	No. of Samples	No. of Correct Result	Success Rate (%)
Detection	23	22	95.83
Character Segmentation	151	135	89.40
Character Recognition	135	118	87.41

Lastly for character recognition, it is based on the total of character segmented as for character that is not segmented is not considered. About 135 characters undergo character recognition process and 118 characters are identified correctly. The percentage of correctly recognized character is 87.41%.

5. Conclusion

The operation of vehicle identification system was analyzed in this paper. The objectives which had been proposed in this project have been achieved. The system is able to capture the image using a web camera when a car is detected. The extraction of NP has quite an accuracy using the proposed algorithm. The techniques preferred for edge detection are the combination of Sobel edge and Canny edge detection technique. Before extraction of NP, a few factors must be considered such as, position of the camera, height and distance and surrounding environment. However, they vary from one-to-one situation and require survey before installation. Lastly, segmentation and character recognition by using bounding box and optical recognition using template matching provide satisfactory and the accuracy of the system is up to the mark. The system is able to segment and recognize the character on the NP.

References

[1] Anagnostopoulos, C. N. E. (2014). License Plate Recognition: A Brief Tutorial. *IEEE Intelligent Transportation Systems Magazine*, **6**(1), p 59–67

[2] Jain, A., & Sharma, J. (2014). Classification and Interpretation of Characters in Multi Applications OCR System. *2014 IEEE International Conference on Data Mining and Intelligent Computing*. p 1–6.

[3] Chai, H. Y., Woon, H. H., Meng, L. K., & Li, Y. S. (2014). Non-Standard Malaysian Car License Plate Recognition. *2014 IEEE Symposium on Computer Applications and Industrial Electronics*. p 152–157.

[4] Layolin, J., Angelin, M. P., & Chennai, T. (2014). Automatic Multi Style Licensed Plate Detection by using Fuzzy Logic Classifier. *International Journal of Advanced Research in Computer Engineering & Technology*. **3**(4), p 1194-1202

[5] Saleem, N., Muazzam, H., Tahir, H. M., & Farooq, U. (2016). Automatic License Plate Recognition using Extracted Features. *IEEE 2016 4th International Symposium on Computational and Business Intelligence*, p 221–225.

[6] Mokayed, H., Meng, L. K., Woon, H. H., & Sin, N. H. (2014). Car Plate Detection Engine Based on Conventional Edge Detection Technique. *Proceedings of the International Conference on Computer Graphics, Multimedia and Image Processing*, p 101–106.

[7] Vikram Mutneja, D. (2015). Methods of Image Edge Detection: A Review. *Journal of Electrical & Electronic Systems*. **04**(02)

[8] Ahn, C.-S., Lee, B.-G., Yang, S.-S., & Park, S.-C. (2017). Design of Car License Plate Area Detection Algorithm for Enhanced Recognition Plate. *4th IEEE International Conference on Computer Applications and Information Processing Technology*, p 1–4.

[9] Kaur, S., & Kaur, S. (2014). An Efficient Approach for NP Extraction from Vehicles Image under Image Processing. *International Journal of Computer Science and Information Technology*, **5**(3), p 2954–2959.

[10] Pechiammal, B., & Renjith, J. A. (2017). An efficient approach for automatic license plate recognition system. *3rd. IEEE International Conference on Science Technology Engineering & Management (ICONSTEM),* p 121–129.
[11] Beibut, A., Magzhan, K., & Chingiz, K. (2014). Effective algorithms and methods for automatic NP recognition. In *2014 IEEE 8th International Conference on Application of Information and Communication Technologies (AICT),* p 1–4.

Indoor Navigation System for Vision Impairment People through Visible Light Communications

Muhamad Nurnajib Rithuan, 'Aini Syuhada Md Zain, Norizan Mohamed Nawawi

School of Computer and Communication Engineering, Universiti Malaysia Perlis, Pauh Putra Campus, 02600 Arau, Perlis, Malaysia

ainisyuhada@unimap.edu.my

Abstract. Indoor navigation is opportune for everyone, and it is particularly requisite for the visually impaired. The paper proposes an indoor navigation system for the Visually Impaired to attain audio output through visible light communications. The transmission of data occurs after identifying the person who has entered the room and the technology light fidelity is used for presenting information to the user. Light fidelity technology is used by the proposed system for the wider range of transmission of data. The usage of visible light for transmitting the data encompasses a wide variety of advantages and eliminates most problems of communication caused by the electromagnetic waves outside the visible spectrum. Light fidelity technology is an evolving branch of Optical Wireless Communication and can be beneficial in coming years for indoor communication because it can afford higher data rate transmission along with the capability to utilize additional users since it uses a broader spectrum bandwidth. The main idea of our project is to create internal navigation system by using Arduino software and Arduino hardware. The transmission of data can act as transceiver, where LED and solar cell are used as the medium to transmit and receive the data.

1. Introduction

Vision is one of the God's gift to the human. The sense of vision gives us to enjoy the beautiful scenery around us. It is one of the most important sense to the human in fulfilling our daily tasks. Human would have difficulties to complete their task properly without it. In this case, people without the ability of seeing will find it as a trouble to do anything, including in decision making.

Visually impaired people also known as the blind people are those who are not able to identify the obstacle and smallest details with healthy eyes. According to the website of World Health (WHO), there are estimated 253 million people who live with vision impairment, 36 million are blind and also 217 million others have moderate to severe vision impaired [1]. From the total amount, about 81% of people who are blind or have moderate or severe vision impaired are aged 50 years and above [1].

The main problem for visual impaired people is they face difficulty to move from one place to another because they do not know what obstacles are they going to face and it definitely will limit their movement. Besides that, they have to be more dependent to people surrounding them or people who are close to them like their family members, relatives, friends or other people who are close to them. These people will guide their movement. Apart from that, it is difficult for them to know their current location especially when they are indoor places like in the house or shopping mall. Due to the problem, this project is carried out with its purpose is to solve the problem with the integration of the selections of hardware component and software.

There are several research works that have been done to help the visually impaired people. There is a research work proposed by Shin & Lim previously in [2], which aim of designing a cane that detects obstacle in all directional sides from head level to floor level using ultrasonic sensors. The positional direction in an indoor environment is given by light fidelity (Li-Fi) technology. From this paper, they are using three ultrasonic sensors that have been fitted around the cane so it can detect any obstacle with three different directions. Furthermore, by using water detection sensor that gives the user any information like whether it is wet or dry floor. It also provides battery to supply the power needed to glow the LED and transmit data that makes the project is user friendly.

By referring to the work in [3], the authors proposed to create a navigation system that uses radio frequency identification (RFID) as the key technology to guide people with visual impairment in an unfamiliar environment, both inside and outside. These project will help the blind people in travelling with ease and confidence as the people with sense of vision. This project is using RFID technology (reader/tags), Bluetooth and android devices. The Bluetooth channel will receive the data from RFID tags ID and sent it to the headsets which finally produce the voice that will guide the visually impaired people. The advantages of this system is it can be used indoor and outdoor. Besides, the where about of the blind person can also be tracked to ensure their safety. Then, the disadvantage of this project is on the voice playback module which produce a delay in the feedback audio.

Another work in [4], proposed to develop the navigation assistance for the visually impaired people using RGB-D sensor. As we know by using the ultrasonic sensor is cheaper and easier. But, for this project they are using RGB-D sensor which is good at miniaturization perspective and cheaper. The RGB-D device provides a variety of information from active sensors using infrared sensors and intensity images of passive sensors such as standard cameras. RGB-D sensor detect the obstacle ahead the user through colour information. Then, from this information it will proceed by using image information. The system is assessed in real scenario and with a set of public data gives good results from floor segments with 99% accuracy and 95% of returns. From the result shows, the algorithm supports the lighting changes, glows, and reflections.

Next related work studied in [5] that proposed a device that is fully automated, easy to maintain, cheap and comfortable to use walking stick. The walking stick is also low on power consumption and easily operated. The aim is to make the visually impaired person self-dependent by informing to the user about the environment condition surrounding them like humidity, obstacle or dark area through the buzzer. Besides that, this project is using water sensor, ultrasonic sensor, buzzer and RF transmitter/receiver to read any information about the presence of obstacle ahead the user. This project is suitable to use for indoor and outdoor navigation because it provides the information environment surrounding the user not only that it also has battery to power up the circuit.

Another recent work in [6] is similar to previous work in [2] which is also aimed to help the blind people to move in indoor place easily and safe with the use of the sensor. This project is using visible light communication because it is more accurate to calculate and get the information of the user position. Three ultrasonic sensors are used to detect three different path like front, right and left that contain possible obstacles. Apart from that, this project is divided into two major part which is transmitter and receiver part. The transmitter part contains LED and PIC microcontroller PIC16f877A, it will generate the binary data and sent it through the LED. Then, ultrasonic sensor, speaker, DF mini player and photodiode are part of the receiver module. The photodiode will receive the data from the LED and convert it into the electrical energy, then DF mini player will get the signal and the output will send to the speaker. Lastly, by using this method it will reduce the cost and the data transfer is more secure.

Hence, creating the visible light communication (VLC) based indoor navigation system will help visually impaired people to solve their major problem in navigating around without being help by the third person. As we believe that nothing is impossible in this era of technology, the system will help the user to detect the obstacles and give the current location of the user independently. The proposed VLC based indoor navigation system is using the Arduino microcontroller. A program is coded using the Arduino IDE into the microcontroller.

This project will use the ultrasonic sensor that can detect any obstacles ahead the user. Upon sensing an obstacle, the sensor will pass this data to the microcontroller. The microcontroller will calculate and determine the obstacle's distance. Later, the microcontroller will give feedback and send signal in the form of sound via buzzer. In addition, this project has one more feature that necessary implemented along with the other features. The feature is Li-fi module, which allows the user to know their current location through LED. In addition, voice playback circuit functions as to record voice that will give the instructions to the users. Concisely, the system implemented in this project allows for obstacle detector and the voice command to tell the actual location of the user.

2. Project Development

This project involves both software and hardware development. Some of the major hardware components and software used are highlighted and discussed in this section.

2.1. Hardware requirements

In this project, an Arduino Mega and Uno is used. Arduino Mega is connected to all of the components that required in implementing VLC Based Indoor Navigation System which is as the obstacles detector while Arduino Uno using as the Li-fi transmitter module. In addition, Arduino has been used in this project as a backbone.

2.1.1. Arduino Mega Board. The microcontroller of this project is from the Arduino families, which is the Arduino Mega. Arduino Mega is a microcontroller board based on the ATmega1280. It has 54 digital input/output pins (of which 14 can be used as PWM outputs), 16 analogue inputs, 4 UARTs (hardware serial ports), a 16 MHz crystal oscillator, a USB connection, a power jack, an ICSP header, and a reset button. It contains everything needed to support the microcontroller; simply connect it to a computer with a USB cable or power it with a AC-to-DC adapter or battery to get started.

2.1.2. Ultrasonic sensor. In this project, ultrasonic sensor is used to detected the obstacle by listening sound wave to bounce back and sending out a sound wave at specific frequency through sound wave. The model of the ultrasonic sensor used in this project is HC-SR04. The ultrasonic sensor can detect an object in the range of 2cm until 400cm. In this project, a threshold limit is set within that particular range. The transmitter of an ultrasonic sensor can emit up to 40kHz, while the receiver of an ultrasonic sensor can accept up only 28 to 40kHz sound waves. This frequency is almost cannot be heard by human ears. Upon detecting an obstacle, the transmitted sound waves hit the obstacle and is reflected back to the detector or receiver. Then, the output will be given out through the buzzer. The general equation that is used to calculate the distance of the obstacle is:

$$distance = \frac{speed\ of\ sound \times time\ taken}{2} \tag{1}$$

2.1.3. Voice Recorder Module. The voice recorder module uses in this project is ISD1820. It will give the current location of the user through the speaker. For this project, three ISD1820 have been used to create three different data that will be presented by the LED. The ISD1820 or voice recorder module can offer true single-chip sound recording, volatile storage, and playback capability for 8 to 20 seconds. With the push button on the board, it makes more easily to be used and controlled it. However, the button function is not use in this project because it will be controlled by a microcontroller.

2.1.4. Buzzer. Buzzer is used as one of the output component for this project. The buzzer will produce beeping sound that will serve as a warning to the user based on the threshold set in the Arduino coding. It has two terminals which are the voltage supply and ground terminal.

2.2. Software requirements

The software requires for this project are Arduino Software (IDE). An Arduino has own programming language which Arduino Integrated Development Environment (IDE) is known as Arduino Software It is used to write programs and upload them to board via USB cable. This software works on Windows, Mac OS X, and Linux. An Arduino has own programming language which Arduino Integrated Development Environment (IDE) is too known as Arduino Software. The environment is written in Java and it is based on processing and other open-source software. This Arduino IDE supports all types of existing Arduino Board such as Arduino Uno, Arduino Nano, Arduino Micro, and Arduino Mini. The software itself is available for download on the Arduino website for free.

2.3. System architecture

The system architectures of VLC Based Indoor Navigation System is illustrated in Figure 1 and Figure 2 which consist of two Arduino Board, three Li-Fi transmitter (LED), a Li-Fi receiver (solar cell), an ultrasonic sensor and three voice recording module (ISD1820). This prototype will function when the ultrasonic sensors are detected any obstacle ahead the user, then it will send the signal to the Arduino Board and it will sound the buzzer to warn the user. Then, by using three LED as the transmitter to send the data, each light contains different location data that will be sent through the light which is by using the Li-fi. In addition, when the user passes through under the lights (LED), they will get the command through the speaker that help the user to know their current location.

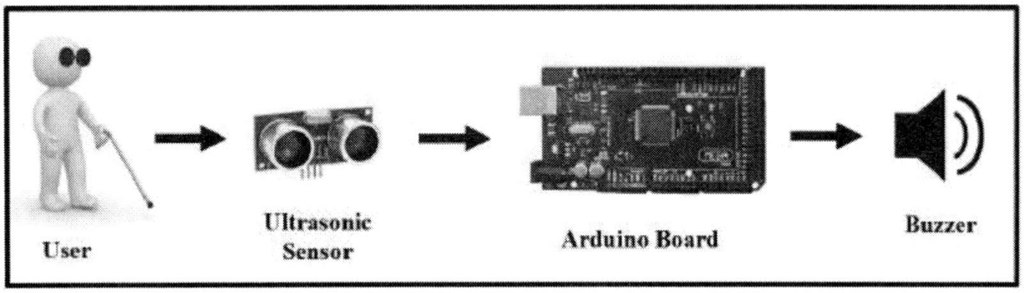

Figure 1. System architecture of obstacles detector.

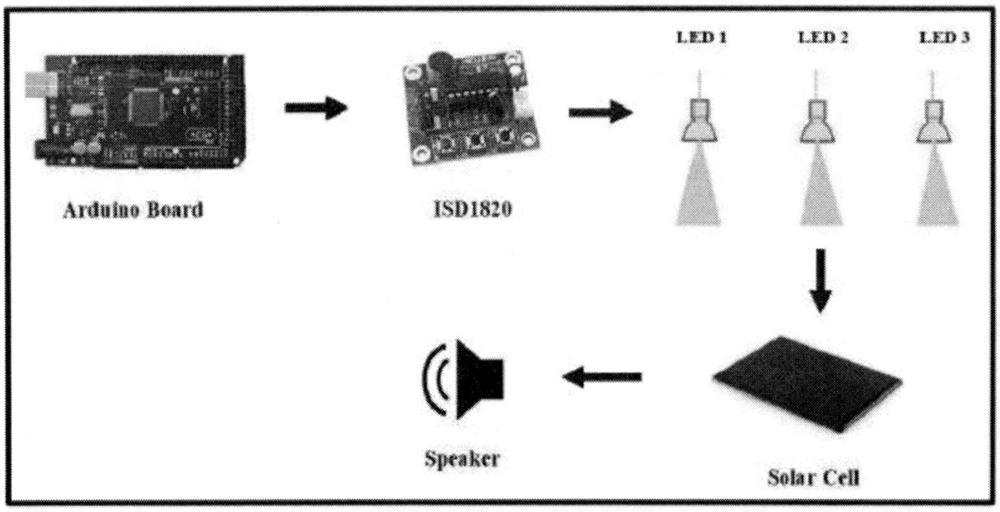

Figure 2. System architecture of Li-Fi transmitter and receiver.

2.3.1. Li-Fi transmitter. The transmitters used in VLC are used inside areas to provide lighting. The transmitter part comprises of the data input which is provided to a switching circuit. Based on the data, the switching circuit creates a string of 1s and 0s thereby encoding the data in binary. The arrangement of LEDs turns OFF and ON at extremely high speeds because of the output obtained from the switching circuit. This ON OFF modulation of the LED light transfers the data. LED is the optimal for light source since it consumes very low power when compared to the fluorescent lamp or a light bulb. Each LED has different location data; it will send the data through the visible light by sending the data in binary form. With different ISD1820, it will record and send the different location data through different LED.

2.3.2. Li-Fi receiver. The receiver involves of an optical section to detect and receive the radiation onto the solar cell receiver. The solar cell converts visible light into an electrical signal biased the solar cell operates in the photoconductive mode generating a current proportional to the light that is received. The current obtained is of a smaller value and a preamplifier converts it into a voltage. The final voltage signal should resemble to the received light pulses which are then demodulated, to obtain the digital data. The data will be processed by the solar cell and convert it into the electrical energy then the speaker will convert it into the sound energy. This method will help the visually impaired people in indoor navigation by providing an audio outputs.

2.3.3. Obstacle detector. The microcontroller that is being used in this project is Arduino which consisted of a software and a hardware. For this project, the Arduino reads, calculates and processes the distance of the obstacle that is detected by the sensors. The programming code that will be generated by the Arduino IDE software contained the conditions fixed for the Arduino to be able to give out beep and audio command as an alert. The conditions will include a beep for detection of an object ahead of the user, and the audio command that gives the current location of the user. In this project, the ultrasonic sensors are used to detect the presence of the obstacle. A few conditions are set as a limit whereby if any obstacle is detected within a fixed range it will send out alert to tell the blind person that there is an obstacle ahead. Once the distance of the obstacle is processed then the conditions are checked. After the conditions are met, the signal is then sent to the microcontroller to be operated by the buzzer.

3. Results and Discussion

Figure 3 shows the implementation of the VLC based navigation system. Figure 3(a) shows the prototype of the VLC based navigation system, (b) and (c) shows the Li-Fi transmitter and receiver circuits respectively.

This project needs functionality testing. Two main components that need practical testing are Li-Fi transmitter and receiver circuits, as well as the ultrasonic sensor. A few tests have been done in order to test the efficiency of the Li-Fi module. The efficiency of the Li-Fi transmitter and Li-Fi receiver is also investigated by testing it at different range between the transmitter and the receiver. The distance varied for the testing are 90 cm, 60 cm, 30 cm and 0 cm from the transmitter to the receiver. The Li-Fi module data obtained by using the application LUX Meter to measure the light intensity and the Sound Meter to measure the volume of the command. Figure 4 shows the output result of LUX meter application measuring light intensity with the variation of transmitter and receiver range. Figure 5 shows the output of Sound Meter application. The ultrasonic sensor tested with random distance to check the functional of distance calculation via serial monitoring. From the testing, ultrasonic sensor is working and functioning as expected. To work as obstacles detector, few obstacles such as chair, wall and human are visualized in the testing scenario. The buzzer will beep long buzzer sound if the obstacles detected below 10cm distance and the buzzer will keep beeping sound for the obstacles below than 30cm distance.

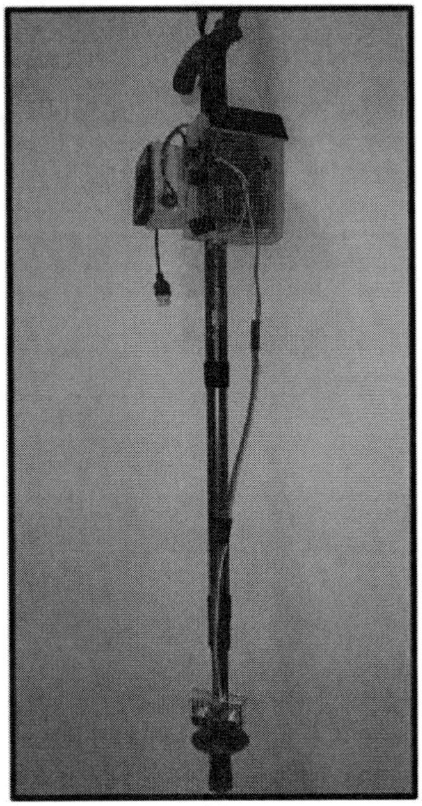

(a)

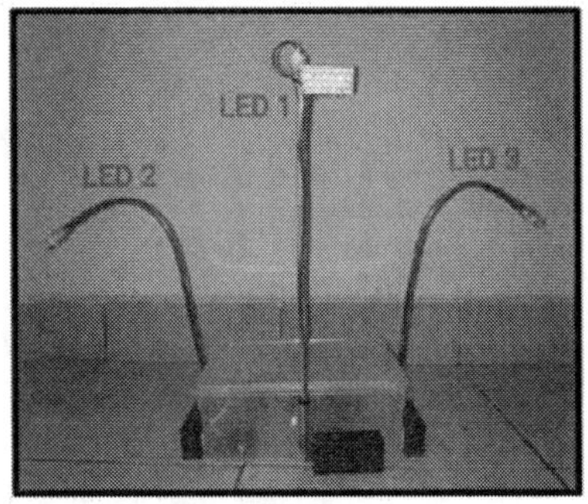

(b)

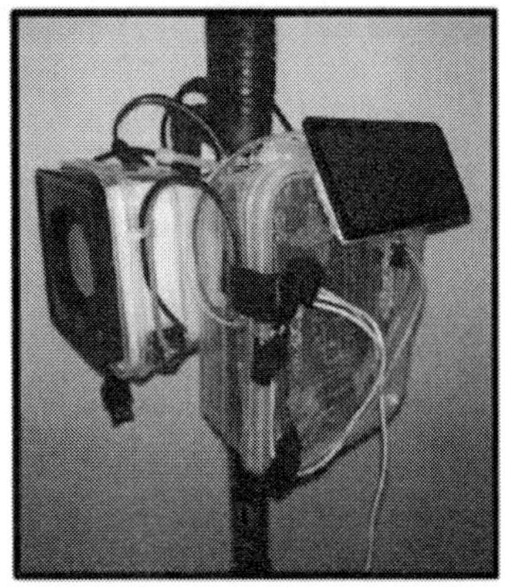

(c)

Figure 3. (a) Prototype of VLC based Indoor Navigation System. (b) Li-Fi transmitter circuit. (c) Li-Fi receiver circuit.

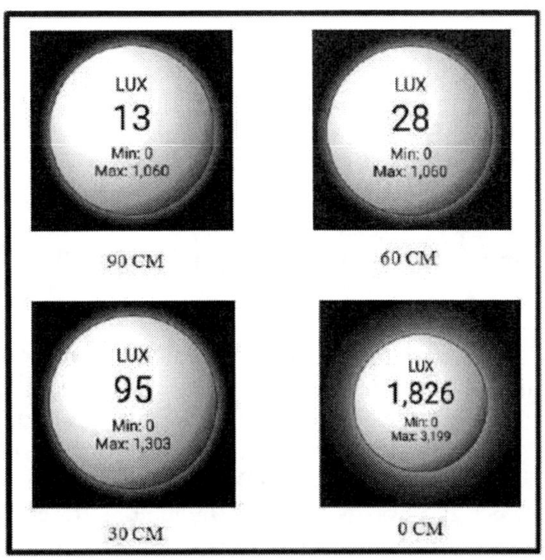

Figure 4. Light intensity reading with LUX meter application.

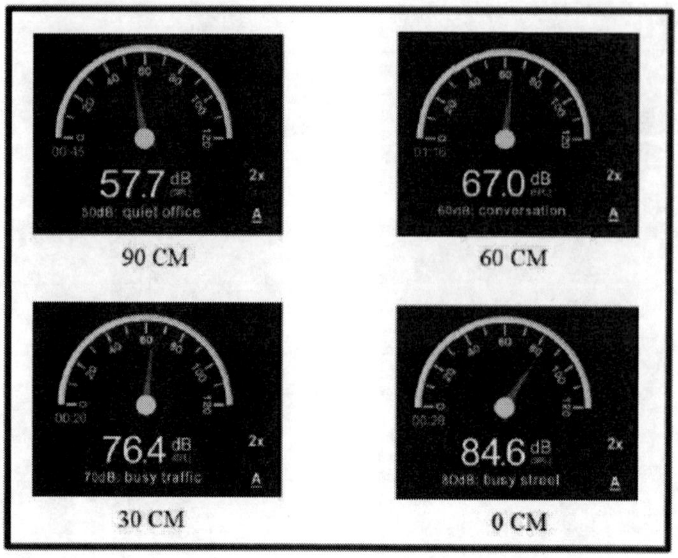

Figure 5. Volume command testing with Sound Meter application.

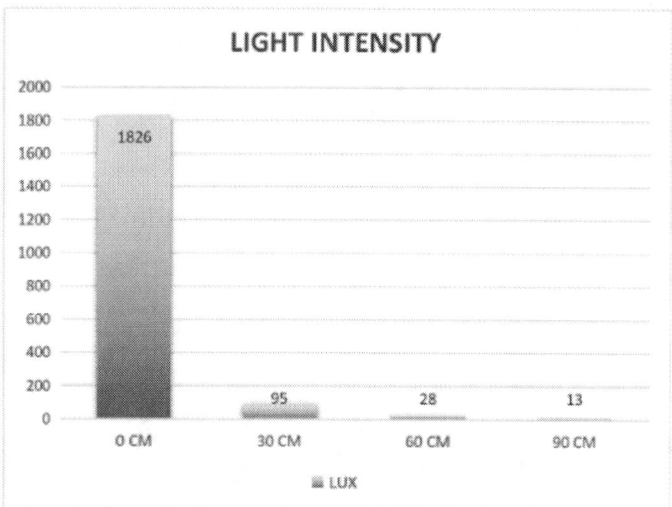

Figure 6. Light Intensity reading varies with distance.

Based on Figure 6 and 7, it can be concluded that light intensity and volume of the sound are depend by the range and the distance between the light source. If the distance is further, the light intensity and the volume will decrease. This situation happens because if the light intensity is low the volume also will be low. For example, when the distance is 0 cm, its mean the receiver is close to the light source so the light intensity is really high which is 1826 LUX then the volume also will be increase when the receiver gets more light which is 84.6 decibel (dB). In addition, to create a good Li-Fi module, the transmitter need to produce high light intensity to make sure the receiver will read the data clearly.

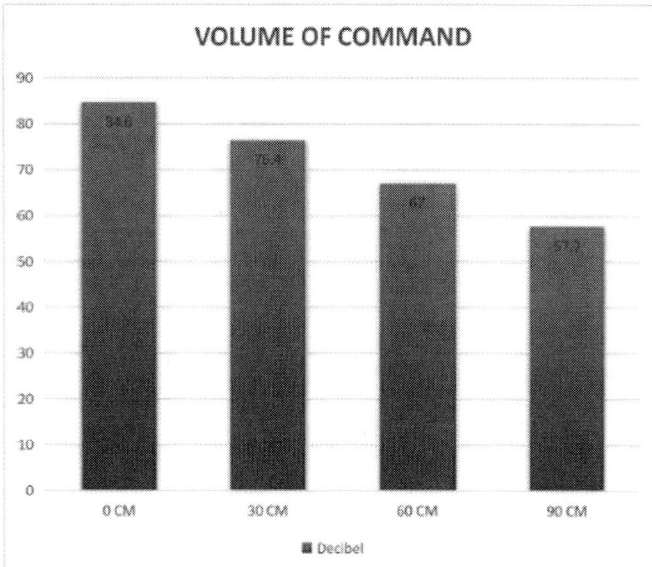

Figure 7. Voice command variation with distance.

Table 1 shows the output result of the ultrasonic sensors. The sensors that are equipped at the Li-Fi Based Indoor Navigation System is able to detect objects or obstacles ahead of the user. Upon detecting objects or obstacles, the buzzer will give the alert to the user.

Table 1. Ultrasonic sensors output result.

Subject of detection	Detection range	Output
Wall Chair	Below than 10 cm range	Buzzer will keep long beep buzzer sound
Human Vehicle	Below than 30 cm range	Buzzer will keep beeping sound

Table 2 shows the result of output command for Li-Fi module testing. Each light source is placed at different location. Once the user passes through the specified location, Li-Fi receiver will process the location data and then provide output in the form of voice command from the speaker.

Table 2. Output command for Li-Fi module testing.

Location data	Light source	Output command
Toilet	LED 1	The speaker will give the command **"Now you are in Toilet"**
Kitchen	LED 2	The speaker will give the command **"Now you are in Kitchen"**
Bedroom	LED 3	The speaker will give the command **"Now you are in Bedroom"**

4. Conclusion

In conclusion, the objectives of the project achieved successfully. The navigation system for visually impaired people basic functionality has been elevated by implementing modern technology sensors such as the ultrasonic sensor, voice recorder module ISD1820 and buzzer. Apart from that, the main goal that make the Li-Fi as the medium transmit and receive the data are successfully achieve.

Despite meeting the objectives stated beforehand, there are still rooms for the improvement of the Li-Fi Based Indoor Navigation System. The research done on this particular project has discovered a lot more of future development that can be done for this prototype. One of the major improvements that can be done for this project is used a GSM Module and GPS Module to make the system is more user friendly. The GPS module will be used as a tracker to monitor the whereabouts of the user and the GSM module is used to send message containing the user's location to the close relatives or guardian in case of emergency.

References

[1] Teplin, S. W., Greeley, J., & Anthony, T. L. (2009). Blindness and Visual Impairment. In *Developmental-Behavioral Pediatrics*. https://doi.org/10.1016/B978-1-4160-3370-7.00071-7

[2] Shin, B., & Lim, C. (2007). Obstacle detection and avoidance system for visually impaired people. *Haptic and Audio Interaction Design*, p 78–85. https://doi.org/10.1007/978-3-540-76702-2_9

[3] Chumkamon, S., Tuvaphanthaphiphat, P., & Keeratiwintakorn, P. (2008). A blind navigation system using RFID for indoor environments. *2008 5th International Conference on Electrical Engineering/Electronics, Computer, Telecommunications and Information Technology*, **13**, p 765–768. https://doi.org/10.1109/ECTICON.2008.4600543

[4] Aladren, A., Lopez-Nicolas, G., Puig, L., & Guerrero, J. J. (2016). Navigation Assistance for the Visually Impaired Using RGB-D Sensor with Range Expansion. *IEEE Systems Journal*, **10**(3), p 922–932. https://doi.org/10.1109/JSYST.2014.2320639

[5] Gbenga, D.E., Shani, A.I., & Adekunle, A.L. (2017). Smart Walking Stick for Visually Impaired People Using Ultrasonic Sensors and Arduino. *International Journal of Engineering and Technology (IJET)*, **9**(5) p 3435 – 3447.

[6] Reddy, S. S., & Rao, Y. B. (2018). Indoor Navigation System for Blind People Using VLC, *International Journal of Engineering & Technology*, **7**(3.27), p 77–79.

IOP Conf. Series: Materials Science and Engineering **767** (2020) 012058 doi:10.1088/1757-899X/767/1/012058

Design of Compact UHF-RFID Tag Antenna with Meander Line Technique

A F M Fazilah[1], M Jusoh[1], A Zakaria[2], T Sabapathy[1], M F Ibrahim[3], M N Osman[1], M N Yaasin[1], S Malhotra[4] and H A Rahim[1]

[1]Bioelectromagnetics Research Group (BioEM), School of Computer and Communication Engineering, Universiti Malaysia Perlis (UniMAP), Pauh Putra, 02600 Arau, Perlis, Malaysia
[2]Centre of Excellence of Advance Sensors Technology (CoE-CEASTech), School of Mechatronics Engineering, Universiti Malaysia Perlis.
[3]School of Bioprocess Engineering, Universiti Malaysia Perlis.
[4]Chitkara University Institute of Engineering and Technology, Chitkara University, , Rajpura, Punjab, India.

muzammil@unimap.edu.my

Abstract. In this paper, a compact UHF RFID tag antenna that can operate at Malaysia UHF RFID frequency (860 MHz – 960MHz) is proposed. The antenna with a common geometry structure without any shorting pin consists of 90degree angled line (meander line) and double T-match structure. The proposed antenna has been designed and simulated using CST Simulation software. The proposed tag antenna design shows a good performance in terms of size, gain and impedance with a dimension of 36 mm x 25 mm x 1.6 mm. The simulated gain of antenna obtained is -0.135 dB at 910 MHz with 19 MHz bandwidth.

1. Introduction

RFID is a technology which is working on radio frequency or radio wave. This technology is used to automatically identifying or tracking the objects. The detectable object can be anything such as book in library, an item which is purchased from shopping mall or a car such as implemented with RFID in Malaysia which is known as MyRFID. In RFID technology, the RFID tag will be stickered on the object's surface and tracked by the reader. Therefore, to detection and tracking, the RFID reader will continuously send the radio waves. When the object is in the range of the reader, RFID tag will send their feedback to the reader and the object can be tracked easily.

RFID tags have a potential to replace barcode due to its long reading range, non-line-of-sight reading and automated identification and tracking. However, due to its highly price compared to the barcodes technologies, the current RFID technology are still not fully accepted, especially in the field of item-level tagging of cheap item [1]. Thus, the reduces the cost, a modern RFID tag with a duo-triangle using paper as a substrate was developed [2]. To achieved a good impedance, the author in [2] had chosen the short dipole dimension, the length of duo-triangles and T-matching network for designing the tag.

T-matching has been a favourite method that used by the researcher while modelling the tag antenna. This is because to easily match the chip impedance could be easily matched with the impedance of the antenna without much affecting the whole dimension of the tag antenna [3]. T-matching structure have also enhanced the robustness of the tag detuning effects from material loading [4].

For the past two decades, increasing the substrate permittivity and meandering the radiation arms were among the popular techniques used to achieve miniaturization. Those techniques have been explored to achieved high compactness in designing tag antenna [5]. In that work, a small slot antenna with a single-layer structure for passive metal mountable was designed using high permittivity ceramic $BaTiO_3$ as a substrate. $BaTiO_3$ was used for size-reduction purpose. In another related work, it has shown that by increasing the electrical separation, lower antenna resonance frequency could be achieved [6]. Meandering line with arc structure also found to be useful for reducing the size of the tag antenna [7].

This paper presents the design of the passive RFID tag antenna using T-matching technique in order to achieve conjugate matching condition between the impedance of chip tag antenna. A simple meander line structure is also implemented when designing this tag antenna to ensure the compactness of the antenna structure. The antenna is designed to operate at 910MHz UHF band. Since this work considers low-cost antenna design, FR-4 is used as a substrate. The tag was designed and simulated using CST Simulation software.

2. Antenna Design

Figure 1 shows the 2D view structure of the proposed antenna. The dimension of this tag antenna is 36 x 25 x 1.6 mm. It consists of two layers which are copper trace and FR-4. The substrate used for this project is FR-4 with a relative permittivity 4.3, loss tangent of 0.03 and a thickness 1.6 mm. In the copper layer, several structures are formed, including double T-match (red) and meander line (blue) as shown in Figure 1. SOT1040-1 chip with impedance value of 14.8-j127.1Ω from NXP Semiconductor is deployed at the center of the tag antenna. In the CST Simulation software, the chip had been represented as discrete port.

Table *1* shows the parameters location with the dimension of tag antenna.

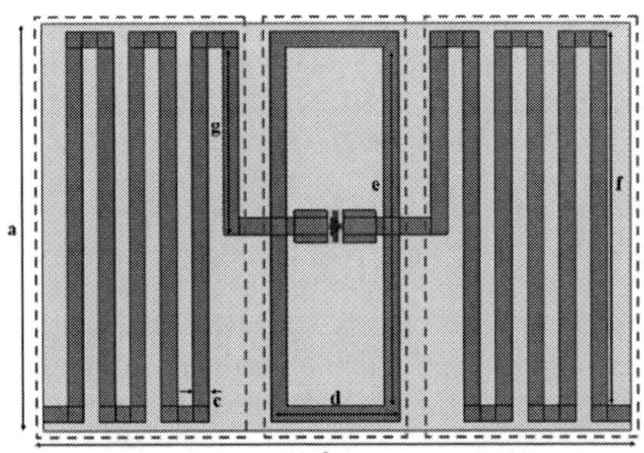

Figure 1. 2D view structure of antenna created.

Table 1. Parameters and dimensions value of tag antenna.

Parameters	Dimensions (mm)
a	25
b	36
c	1
d	8

e	24
f	24
g	12

To scale down the tag antenna size, 90-degree angled line is embodied in the physical structure of the tag antenna. By folding the antenna in 90 degree angled line structure, the antenna will resonate at much lowest frequency than a single element antenna of equal size. This is due to disruption of the current flow that leads to the inductance and capacitance of antenna increase. Thus, this lead with an increasing additional charges at the corner of the 90 degree angled line [3]. Frequency shifting will be affected to the resonant frequency from the highest to the lowest frequency. Therefore, to optimize back the resonant frequency near to the desired frequency, scaling down the antenna size is a must. Hence, a compact RFID tag antenna can be attained.

Impedance matching technique become compulsory while modelling the tag antenna to ensure that the waves are radiated in the most efficient. In this regard, the antenna can be matched and highly inductive with the chip impedance. In this work, a double T-match structure is implemented when designing the antenna structure. The double T-match structure easily match the impedance of the antenna without affecting the whole dimension properties of the tag antenna. The double T-match structure can be modelled by joining a slotted rectangular shape to the meandered feed line which is indicated with parameter *e* and *d* as shown in Figure 1. T-matching method technique with the proposed antenna circuit arrangement is illustrated as in Figure 2.

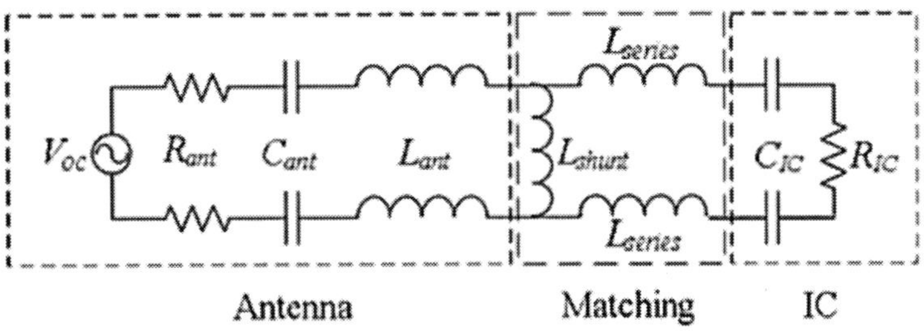

Figure 2. T-matching method technique with the proposed antenna circuit arrangement [8].

The impedance from the input of the purposed antenna need to be highly inductive and equal to the conjugate impedance of the tag chip, in order to achieve the maximum threshold of power from the RFID chip to the tag antenna [9]. The input impedance of the antenna can be calculated as in (1), where characteristic impedance of both conductor's transmission line is a Z_o, the current splitting factor between the conductors and Z_a is the impedance of the meandered feed line antenna without the T-match connection [4].

$$Z_{in} = \frac{2j\, Z_0 tan\frac{ka}{2}(1+a)^2 Z_a}{2j\, Z_0 tan\frac{ka}{2} + (1+a)^2 Z_a} \tag{1}$$

The power reflection coefficient is adapted to deal with the complex impedance of the chip and proposed antenna. The power reflection coefficient formula is given as follows [10]:

$$\Gamma^* = \frac{Z_a - Z^*_{chip}}{Z_a + Z_{chip}} \tag{2}$$

3. Results and Discussion

For RFID systems, the performance of tag antenna directly affects the range and the quality of use of these systems. Based on the results of these characteristics such as reflection coefficient (S_{11}), gain, directivity and radiation pattern of the tag antenna, the quality of adaptation between antenna and the chip and the read range can be estimated.

To make sure the tag antenna has a better performance, the reflection coefficient must be less than -10 dB. The resonant frequency can be controlled by the width of the meander line. In the simulation, a parameter sweep is carried out for the value of c need to optimise the S_{11} until the desired frequency range is obtained. The parameter sweep result for the meander line width, *c* is shown in Figure 3. Figure 4 shows the final optimized S_{11} result of the RFID tag antenna.

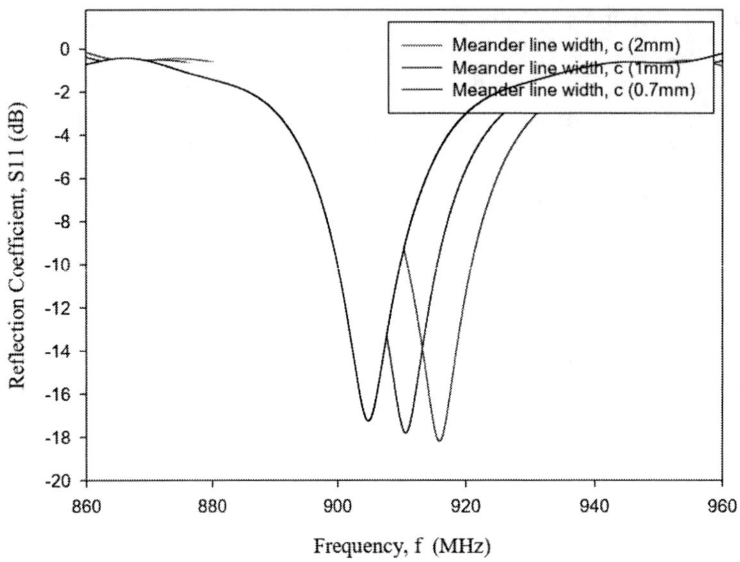

Figure 3. Simulation results of the parameter sweep for the width of the meander line.

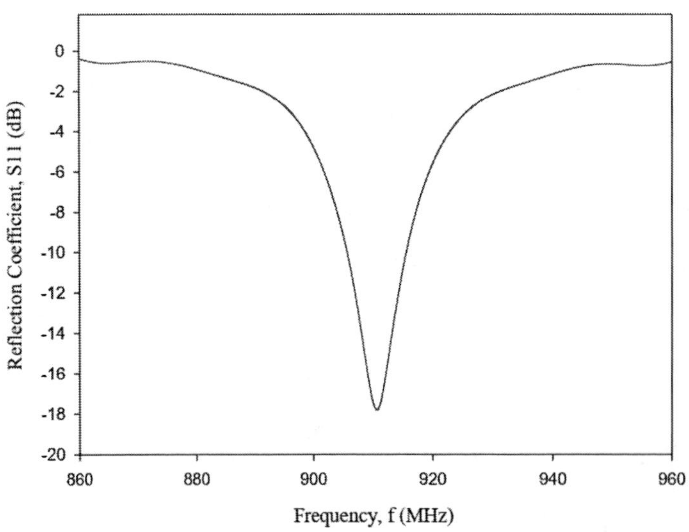

Figure 4. Simulated reflection coefficient, S_{11} in CST.

The simulated bandwidth obtained by CST microwave software is 19 MHz (901 MHz to 920 MHz) which covers the operating frequency of UHF RFID in Malaysia with S_{11}=-17.80 dB at 910MHz.

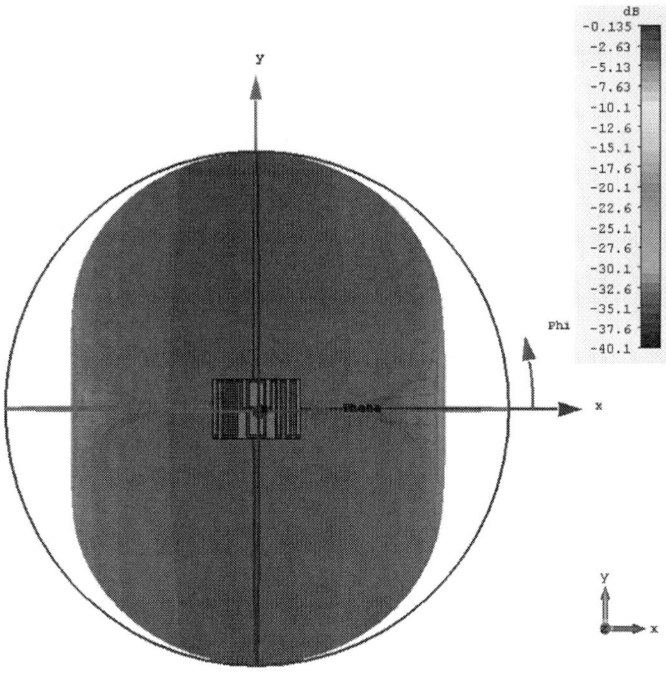

Figure 5. 3D radiation pattern for tag antenna at 910 MHz.

Figure 5 shows the 3D radiation pattern of the RFID tag antenna with almost omni-directional radiation pattern. The negative gain is obtained due to the small size of the tag antenna, which is definitely smaller than a wavelength at the operating frequency 910 MHz. The realized gain for tag antenna at 910 MHz is -0.135 dB as shown in Figure 5. The simulation read range can be calculated theoretically using Friss free space equation as listed in (3)and (4) [11]:

$$r = \frac{\lambda}{4\pi}\sqrt{\frac{P_t G_t G_r \tau}{P_{th}}} \qquad (3)$$

$$\tau = \frac{4R_c R_a}{|Z_c + Z_a|^2} \qquad (4)$$

where P_t is the power transmitted from RFID reader. While for G_r and G_t is the gain obtained from the RFID reader and tag, P_{th} is the threshold power and τ is the power transfer coefficient. The power transfer coefficient measures how much the antenna impedance is matched to the chip impedance. The calculated simulated read range obtained is estimated at 3m using equation in (3).

4. Conclusion

A compact UHF RFID tag antenna has been successfully designed and optimized using CST Microwave Studio software. The simulated result obtained shows that the proposed antenna operates in the frequency band (860 MHz to 960 MHz), which covers the UHF RFID operating frequency in Malaysia. The proposed antenna is very compact, easy to fabricate and low-cost. In future work, the tag antenna with flexible substrate will be proposed for bending purposes and the for the use on non-planar objects.

References

[1] Islam M A and Karmakar N, "On a compact printable dual-polarized chipless RFID tag using slot length variation encoding technique for barcode replacement," *2015 IEEE MTT-S Int. Microw. Symp. IMS 2015*, p 1–4, 2015.

[2] El Hamraoui A, Abdelmounim E H, Zbitou J, Bennis H, Latrach M, and Errkik A, "A low cost miniature UHF RFID tag antenna using paper substrate," *2017 Int. Conf. Wirel. Technol. Embed. Intell. Syst. WITS 2017*, p 1–6, 2017.

[3] Faudzi N M, Rashid A N A, Ibrahim A, Khyasudeen M F, and Ali M T, "Microstrip Dipole UHF-RFID Tag Antenna for Metal Object Tagging," in *Proceedings - 6th International Conference on Computer and Communication Engineering: Innovative Technologies to Serve Humanity, ICCCE,* 2016, p 36–41.

[4] Choo J, Ryoo J, Hong J, Jeon H, Choi C, and Tentzeris M M, "T-matching networks for the efficient matching of practical RFID tags," *Eur. Microw. Week 2009, EuMW 2009 Sci. Prog. Qual. Radiofreq. Conf. Proc. - 39th Eur. Microw. Conf. EuMC 2009*, October, p 5–8, 2009.

[5] Bong F L, Lim E H, and Lo F L, "Compact Folded Dipole with Embedded Matching Loop for Universal Tag Applications," *IEEE Trans. Antennas Propag.*, **65**(5) p 2173–2181, 2017.

[6] Björninen T and Ukkonen L, "Small Slot Antenna for Metal Mountable UHF RFID Tags," *Proc. 2012 IEEE Int. Symp. Antennas Propag.*, p 1–2, 2012.

[7] Wongsiritorn P, Pumpoung T, Phongcharoenpanich C, Nuangwongsa K, and Kosulvit S, "Meander-Line Antenna with Arc Structure for UHF- RFID Tag," *2011 Int. Symp. Intell. Signal Process. Commun. Syst.*, p 7–10, 2011.

[8] Dobkin D M, *The RF in RFID : passive UHF RFID in practice LK - https://unimap.on.worldcat.org/oclc/762090082*. Burlington, MA SE - Elsevier/Newnes, 2008.

[9] Ennajih A, Zbitou J, Latrach M, Errkik A, Tajmouati A, and El Abdellaoui L, "A New Design of UHF RFID Tag Antenna Based on Negative Index Metamaterial," in *Proceedings of the 2nd International Conference on Computing and Wireless Communication Systems* p 1–6, 2018.

[10] Shi Y, Fang C, Qi K, and Liang C H, "A broadband design of UHF Fractal RFID Tag antenna," *Prog. Electromagn. Res. Lett.*, **58** p 45–51, 2016.

[11] Rao K V S, Nikitin P V, and Lam S F, "Antenna design for UHF RFID tags: A review and a practical application," *IEEE Trans. Antennas Propag.*, **53**(12), p 3870–3876, 2005.

Acknowledgement

This work is supported by Universiti Malaysia Perlis, under Internal Social Project via Research grant scheme (INSPIRE 2019- 9001 00582).

Shared Cache Partitioning Based on Performance Gain Estimations

N Mahrom[1] and M J Liebelt[2]

[1] School of Computer and Communication Engineering, Universiti Malaysia Perlis, Arau, 02600 Perlis, Malaysia
[2] School of Electrical and Electronic Engineering, The University of Adelaide, South Australia, 5005 Australia

norfadila@unimap.edu.my

Abstract. In multiprocessor systems, dynamic cache distribution has been used to increase system performance by effectively partitioning the cache resources. However, different performance metrics used at runtime used to dynamically decide the partition sizes can give different impacts on performance, as well as varying impacts on the hardware cost of the system. In this paper, we propose an Adaptive CPI-based Cache Partitioning (ACCP) scheme to provide better utilisation of the shared cache resources among the competing applications in the system. ACCP uses performance gain estimations of the cache, without incurring significant hardware overhead. It aims to allow all applications in the system to run at approximately the same speed by accelerating the slowest application without significantly decelerating the others. We evaluated the ACCP on a quad-core system on which it achieved on average 23% reduction in miss rate, compared to an unpartitioned shared cache. ACCP also yields a similar IPC throughput improvement to a well-known UCP scheme, and better performance compared to the CPI by Muralidhara et al. Overall, the throughput of the system is improved at minimal complexity without yielding significant additional hardware cost. Hence, ACCP shows better overall performance in managing the hardware overhead compared to the UCP scheme.

1. Introduction

Different applications competing for shared memory resources in multiprocessor systems have different behaviours and requirements throughout the execution process. The performance of the applications can be characterised by many factors, such as the speed of the slowest application, or the maximum number of misses incurred by an application in the system. There are several trade-offs between these factors in optimising system performance. Consequently, various techniques have been proposed to enhance the performance of multiprocessor systems and to improve the overall shared cache throughput [1][2][3][4]. An Adaptive CPI-based Cache Partitioning (ACCP) scheme is proposed to modify the cache space allocation of applications in the system by up to four ways at the end of every execution interval. At the end of each interval, the shared cache space is partitioned based on the cycles per instruction (CPI) values in such a way that the application with the highest CPI (the slowest application) will receive extra ways from the application with the lowest CPI (the fastest application). However, to ensure that the slowest application will get benefit from the additional cache ways, its miss counter associated with the most recently used (MRU) position is compared with the MRU miss counter of the fastest application. If the value of the counter belonging to the slowest application is less than the value in the miss counter of the fastest application, another application in the system will be selected as the slowest application (i.e., the application with the highest MRU miss count).

In a case having four applications running on different cores, the application with the next highest CPI value will not be selected. This is because there is a possibility that the MRU miss counts of this application is less than the MRU miss counts of the fastest application or the remaining applications in the system. The application with the highest MRU miss count is used to avoid an iterative search through applications with successively lower CPI values. The process of determining the number of cache ways to be given to the slowest application is described in Algorithm 1. The miss counters of the fastest and the slowest applications are read to estimate the performance gain. Note that the miss counts in the counters are correlated with the number of hits, had the respective application been given additional cache ways.

Algorithm 1. Adaptive CPI-based Cache Partitioning scheme.

- Start with equal partition for each competing application *i*, at the first interval.
  $allocation\ [i] = \frac{total\_cache\_ways}{number\_of\_cores}$
- At the end of each execution interval, do
  - ✓ Read instruction counts, $C_{INS}[i]$ and cycle counts, $C_{CYC}[i]$ for each application and calculate the CPI values.
  - ✓ Determine the slowest application, *slowest_app* and the fastest application, *fastest_app* based on the calculated CPI values.
    - Check the first miss counter, $C_{MISS}[1]$ of the *slowest_app* and the *fastest_app*.
    - If $C_{MISS\_slowest\_app}[1] < C_{MISS\_fastest\_app}[1]$, then another application in the system that has the highest value of $C_{MISS}[1]$ will be assigned as the new *slowest_app*.
  - ✓ Next, evaluate *j* miss counters, $C_{MISS}[j]$ that belongs to the *slowest_app* and the *fastest_app* to determine the number of ways to modify.

```
function PARTITION THE SHARED CACHE
  max_gain = 0 // To check if adding one more way can provide further benefits
  extra_ways = 0 // Number of ways to modify.
for each miss counter j = 1 to 4, do
      new_max_gain = max_gain + CMISS_slowest_app[j]
      if (new_max_gain > max_gain) and
        (CMISS_slowest_app[j] > CMISS_fastest_app[j]), then
                                        max_gain = new_max_gain
                                        extra_ways += 1
      else
            break;
      end if
  end for

  if (allocation[fastest_app] – extra_ways > 0)
    allocation[fastest_app] -= extra_ways
    allocation[slowest_app] += extra_ways
end if

return allocation
end function
```

In this paper, we do not estimate the performance loss that will cost the fastest application because several more counters would be required to record the number of hits incurred at each priority position of the lines in a set. The number recorded in these counters will represent the number of misses that will be received by an application if some ways are taken away from it. That is, the hit counts at the least recently used (LRU) position represents the number of additional misses that will be incurred by an application if a cache way is removed from it. If two cache ways are taken away from the application, the number of hits incurred at the second LRU position will represent the additional number of misses to be received by the application. Although the use of hit counters to estimate the performance loss of each application seems beneficial, it will cost additional hardware and processes at every execution interval. We chose to implement an algorithm that does not require this additional overhead.

To estimate performance gain from additional cache ways, the miss counters of the slowest application are compared with the counters of the fastest application, one at a time. The first counter to be checked is the one associated with the MRU position that represents the performance gain if the application were to be given an extra one way. In sequence, the counters are compared from the one associated with the MRU position to the one associated with the LRU position. Each time a counter is checked, if the miss count of the slowest application is greater than the miss count of the fastest application, an extra way is given to the slowest application. This process is repeated for all four counters, or until the value of the miss count belonging to the slowest application is less than the miss count of the fastest application. Note that this scheme guarantees that each application owns at least one cache way by making sure that the total number of ways given to the slowest application by the fastest application will leave at least one cache way to the fastest application. Otherwise, the cache will not repartition. All the miss counters of all applications are reset to zero after each partitioning interval.

2. Experimental Methodology

2.1. Configuration
Our baseline architecture is based on the Tensilica Xtensa toolset [5] and the Xtensa Xplorer is used to configure and customise the Xtensa LX4.0 processor cores. The Xtensa design tool provides a cycle accurate system simulator to perform software emulation of multiprocessor systems. We configured the Xtensa cores by using Xtensa Processor Generator (XPG) without local data and instruction caches. The Xtensa cores used in our work are configured with the Processor Interface (PIF) in which the read and write busses are of equal width, 32-bit. This is because PIF supports external communication for inter-core communications in an Xtensa multiprocessor system. Additionally, PIF does not limit the size of external peripherals that can be connected to the core. We created all memories (e.g. caches) in our system as custom memory-mapped devices using XTMP_device objects. In Xtensa Modeling Protocol (XTMP), a memory-mapped device refers to any component in the system that is accessed by load or store transactions from a simulated Xtensa core [6]. We used 4KB level 1 private cache and 512KB level 2 shared cache. The level 2 cache requires 15 cycles hit latency, while the main memory requires 300 cycles access latency.

2.2. Metrics
For all of the ACCP evaluations, the performance metrics used to quantify the performance of the system are the *MPKI* (misses per 1000 instructions), *IPC throughput*, *weighted speedup,* which indicates reduction in execution time, and the *harmonic mean* of weighted speedup, which accounts for both fairness and performance of the system [7][8].

Let IPC_i denote the IPC of the ith application, $SingleIPC_i$ the stand alone IPC of the same application if it is executed in isolation, and N the number of processes the system executes concurrently. The formulae for the aforementioned metrics are:

$$IPC\,Throughput = \sum_{i=0}^{N} IPC_i \tag{1}$$

$$Weighted\,Speedup = \sum_{i=0}^{N} IPC_i / SingleIPC_i \tag{2}$$

$$Harmonic\ Mean = N/\sum_{i=0}^{N}(SingleIPC_i/IPC_i) \quad (3)$$

2.3. Benchmarks

The performance evaluation of different cache replacement schemes has been undertaken using sets of applications that were selected from the SPEC CPU2000 benchmark suite. As the Tensilica Xtensa development tool includes the Xtensa C/C++ compiler (XCC), which supports compiling C and C++ applications, only applications that are written in C were selected to be used for evaluation purposes. The applications were compiled with Xtensa's optimising compiler xt-xcc at the -O3 optimisation level. The chosen applications are the combinations of three different applications categories, which are classified based on the demand for L2 cache space. They are named *high utility*, *low utility*, and *saturating utility* applications [9].

Low utility applications are benchmarks that do not benefit from the allocation of any additional cache space. Applications named mcf, bzip2, and equake are under this category. The argument could be that these applications fit in the L1 cache and suffer frequent compulsory misses in the L2 cache. *High utility* applications improve their performance whenever more L2 cache space is allocated. They are twolf, vpr, parser, art, and ammp. However, if increasing the available cache space can improve the performance of applications up to a point at which their performance saturates, these applications were classified as *saturating utility*. The applications are gzip, mgrid, and mesa. All the applications from different categories were combined to represent eleven workload mixes as shown in Table 1.

Table 1. Workload summary.

Workload	Applications	Utility
Mix-1	art, bzip2, parser, vpr	HLHH
Mix-2	vpr, parser, mcf, bzip2	HHLL
Mix-3	ammp, twolf, vpr, art	HHHH
Mix-4	equake, bzip2, twolf, vpr	LLHH
Mix-5	bzip2, ammp, vpr, art	LHHH
Mix-6	mcf, vpr, twolf, gzip	LHHS
Mix-7	parser, mesa, vpr, art	HSHH
Mix-8	twolf, art, bzip2, mgrid	HHLS
Mix-9	art, vpr, twolf, equake	HHHL
Mix-10	equake, bzip2, art, vpr	LLHH
Mix-11	ammp, vpr, twolf, bzip2	HHHL

3. Simulation Results

We evaluated the ACCP scheme on a quad-core system sharing a 512KB, 32-way associative L2 cache. The shared cache was warmed up for 10 million instructions and the partitioning algorithm was invoked after every execution interval of 10 million instructions. Note that the size of the execution intervals of utility cache partitioning (UCP) scheme [10] and CPI cache partitioning scheme [11] are 5 million and 15 million, respectively. Therefore, we decided to use 10 million instructions for comparability with the previously published schemes. We analysed and compared the results of ACCP against two existing cache partitioning schemes, namely the UCP scheme [10] and CPI [11]. We forced the entire system including all the cores to shut down when any one of the cores reached the end of its execution. For a system employing UCP, the traditional LRU policy is modified so that it works in a thread-wise manner, in which the victim selection must be among the cache lines belonging to the miss-causing application

[10]. However, in the case where the number of allocated cache lines is less than the number of cache lines currently belonging to the miss-causing application, a cache line belonging to another application is replaced.

For all of the performance comparisons, the conventional unmanaged/unpartitioned shared cache using the LRU policy is used as the baseline. Figure 1 shows the miss rate (in misses per 1000 instructions, MPKI) of ACCP, alongside the performance of UCP and CPI. While ACCP was proposed to simplify the complex partitioning decision logic in UCP and to improve the partitioning conditions in CPI, ACCP on average has yielded a performance rate similar to both schemes. However, ACCP in a few cases is able to provide miss rate improvements over UCP and CPI by up to 5% and 9%, respectively. Even though for the majority of the workloads ACCP incurred approximately similar miss counts to UCP and CPI-based, the results show that our approach is able to achieve cache miss reductions as much as UCP, with less hardware overhead. Furthermore, comparison with CPI shows the ability of ACCP to improve the miss rates by considering both CPI values and miss rate.

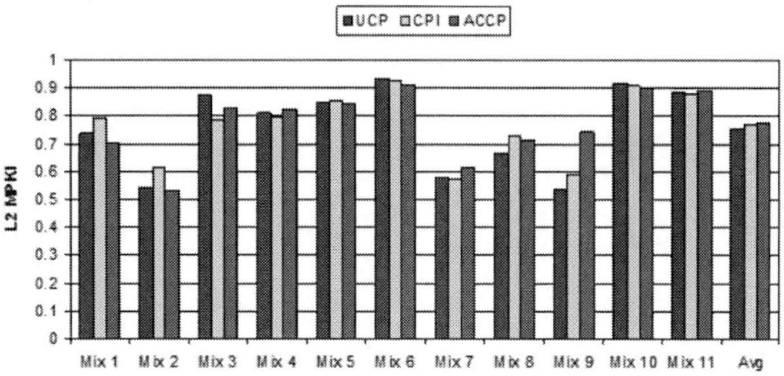

Figure 1. Miss rate improvements relative to an unmanaged traditional LRU cache.

Our findings in Figure 1 are supported by the IPC throughput observed across the simulated workloads, shown in Figure 2. The IPC throughput of Mix3 when using ACCP was 4.3% better than the IPC throughput of CPI, although it was observed that ACCP has incurred a 5% additional miss rate. A similar pattern is demonstrated by Mix11, for which ACCP yields a 2% improvement in IPC over CPI, whereas the MPKI figures recorded by both schemes show a negligible difference. ACCP performs better than CPI because while the use of IPC values can give good predictions in partitioning the shared cache, the estimation of performance gain due to the increasing of cache resources can result in better partitioning decisions of the slowest application and further improve the system performance. Hence, ACCP has demonstrated that for the majority of the workloads, the combination of CPI values and performance gain estimation could be used effectively to partition the shared resources.

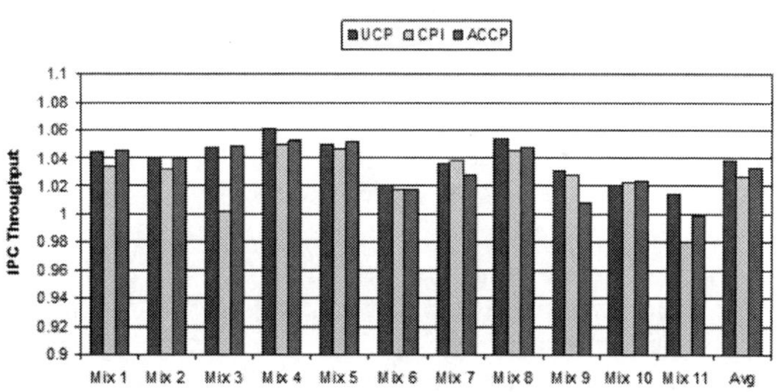

Figure. 2. Performance comparison of UCP, CPI and ACCP relative to an unmanaged LRU L2 shared cache.

Additionally, we found that a simpler partitioning algorithm and a lower implementation overhead do not limit the ability of ACCP to achieve similar performance to UCP. ACCP also has an advantage over UCP, in that it can be employed with any type of cache replacement policy. To investigate the effect of ACCP on each application within a workload, Figure 3 shows speedups achieved by ACCP compared to UCP and CPI.

We observed that although the patterns for both weighted speedup and harmonic means of the IPC are similar to the achieved IPC throughput reported in Figure 2, there are further improvements exhibited by ACCP. In addition, ACCP performs similarly to or slightly better than UCP for the fairness performance of several mixes as illustrated in Figure 3 (a). ACCP compares more favourably when performance is measured with the harmonic mean, as in Figure 3 (b) which shows that ACCP in some cases outperforms UCP, and has a positive impact on the applications' throughput of the workload mixes. Comparison with CPI shows that the fairness speedup of ACCP is better than its weighted speedup by up to 2%. Thus, we show that ACCP is able to assist the fairness of the workload mixes, and hence meets the objective of the scheme which is to reduce the speed difference of all running applications.

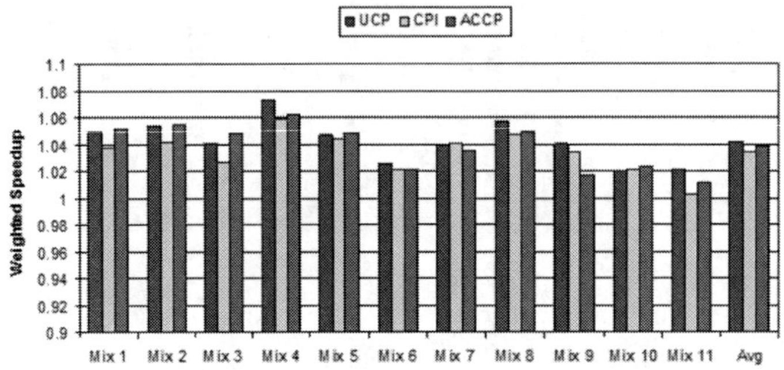

(a)

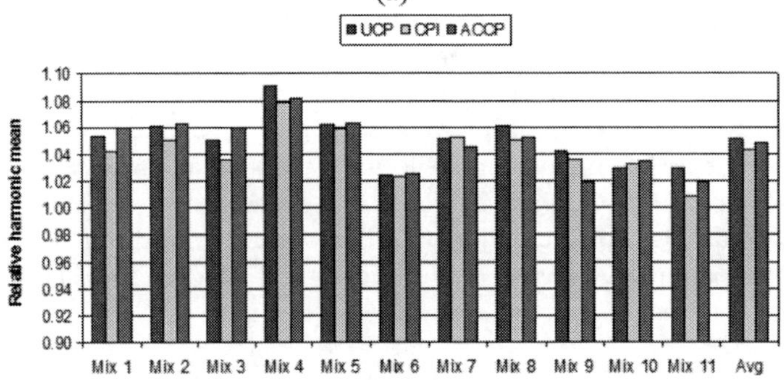

(b)

Figure 3. Performance comparison using the (a) weighted speedup, and (b) harmonic mean of weighted IPC metrics.

4. Performance Analysis on Cache Scaling and Sensitivity

We further consider the performance of ACCP by analysing its effect on 1MB and 4MB L2 shared caches. Figure 4 shows the performance of ACCP is largely unaffected by changes in cache sizes. This means that ACCP retains similar performance trends, and consistently achieved similar performance to UCP. Also, ACCP consistently outperforms the CPI for the majority of the simulated workloads, with the highest improvement recorded in a 1MB cache at 1.5% over UCP and 3.7% over the CPI, for Mix11. Meanwhile, with a cache size of 4MB ACCP for Mix10 outperforms the CPI by the maximum value of 5.6% and UCP by a maximum of 6.9%. This better performance could be due to the total benefits of the larger cache size received by all high utility applications in the mix that contribute to the significant overall performance compared to other workload mixes. Hence, ACCP has provided sufficient cache space to the high utility applications in Mix10 to achieve higher overall performance improvement compared to other schemes.

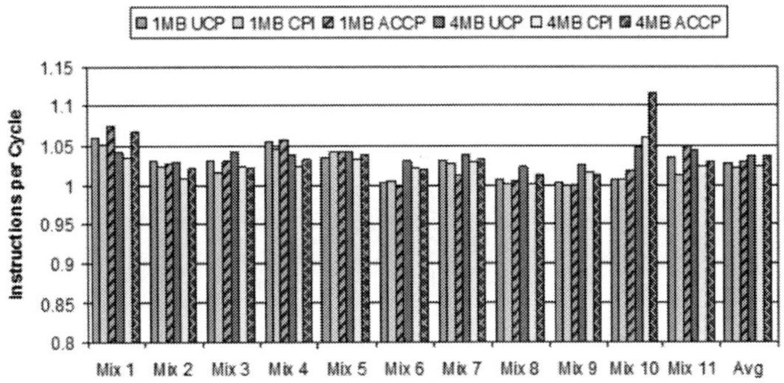

Figure 4. Performance comparison of ACCP, UCP and CPI on different sizes of L2 shared cache, relative to baseline LRU.

Next, the effect of different cache sizes is investigated in terms of the MPKI metric. Figure 5 depicts the MPKI of ACCP normalised to the MPKI of the UCP and CPI. An interesting point to observe in the figure is that in the majority of cases, ACCP performs better over CPI than it does over UCP. For the 1MB cache, the reduction over CPI is between 5% and 10% greater than the reduction relative to UCP. However, the reduction of ACCP recorded over CPI for the 4MB cache is more than 10% better than the reduction over UCP. The most significant reduction exhibited by ACCP in the 4MB cache size occurs for Mix10, in which the recorded MPKI is less than 20% that of UCP. This is correlated with the performance value recorded in Figure 4, in which ACCP achieved the highest improvement over UCP in the case of Mix10. ACCP shows a reduction of less than 5% on average in the 1MB cache and about 5% to 10% in the 4MB cache over the UCP and CPI, respectively. However, we observed in Figure 5 that ACCP performed poorly in Mix7 (for the 1MB cache), in which it incurred around 20% more misses compared to the UCP and CPI. For the 4MB cache, ACCP achieved the highest improvement in the miss rate in Mix10 compared to UCP and CPI.

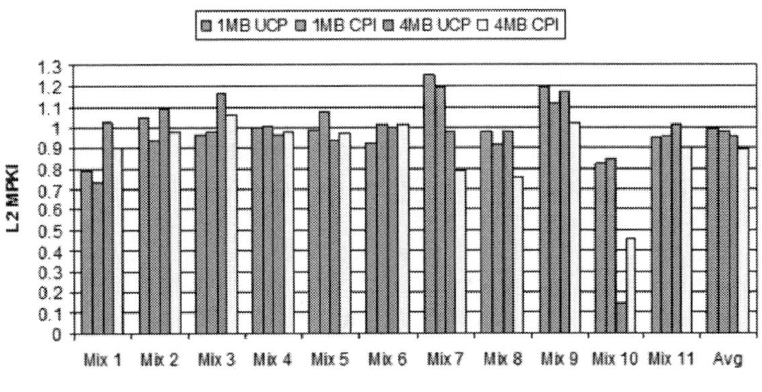

Figure 5. Miss rate reduction of ACCP over UCP and CPI.

To identify the contributing factor to the significant performance variation of ACCP in the two mixes, the distribution of the miss counts among the cores in Mix7 (for 1MB cache) and Mix10 (for 4MB cache) is shown in Figure 6. For Mix10 in the 4MB cache, the ACCP has significantly reduced the number of misses in core C2, which executed a high utility application. This application could have received most of the 4MB cache space from ACCP, causing the significant reduction of the miss counts in this core to achieve the highest overall miss reduction of Mix10, compared to the UCP and CPI.

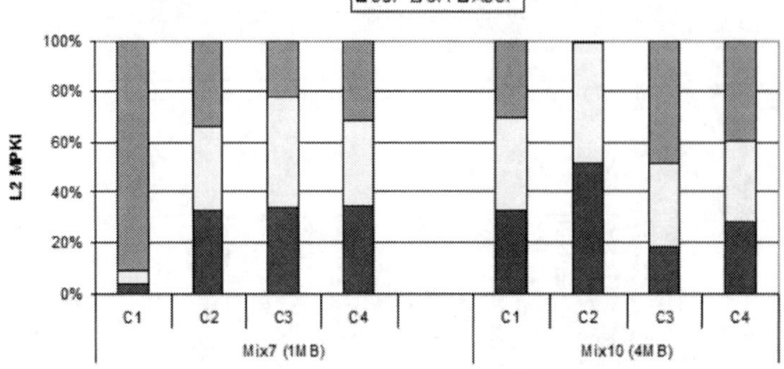

Figure 6. Distribution of L2 miss rate among UCP, CPI and ACCP.

For the 1MB cache, we found that core C1 of Mix7 incurred the highest number of misses compared to other cores in the system. Even though core C1 also executed a high utility application, the fact that the core did not get much benefit from the 1MB cache could be due to competition with other high utility applications executed by cores C3 and C4. Hence, the significant number of misses incurred by core C1 has contributed to the worse performance of Mix7 compared to the UCP and CPI. Note that an application could perform better or worse due to competition with other cores in the system, since the application could be treated differently from its type. This means that a high utility application could be treated as a streaming process in one mix, but could get many advantages from the available cache space in another workload mix. Therefore, this could account for the different results produced by an application in different mixes, leading to variation of performance improvements among mixes.

5. Complexity Management and Hardware Implementation

In Algorithm 1, the fastest and the slowest applications are chosen based on the CPI values, then the four miss counters of the two applications are read and compared. The values of the counters indicate how much the applications could benefit from the allocation of another 1, 2, 3 or 4 ways. Primarily, the partitioning algorithm of ACCP requires the implementation of addition, subtraction, and comparisons. However, ACCP does not employ any complex control logic due to the use of few *if* statements, in which the *if* statements can be directly implemented in hardware by using a compare-and-multiplex operation. Furthermore, the number of iterations that the miss counters need in order to decide the number of ways to reallocate can be determined in advance. Hence, comparisons between each miss counter can be done in parallel. Additionally, there is no control-flow dependence on the input values of the iteration counts, except the dependence on the slowest application identifier. This scenario therefore permits the implementation and parallelisation of simple control logic for the algorithm to make the partitioning decisions.

The strategy of examining each application's performance gain by increasing up to four cache ways has reduced the hardware cost of monitoring and gathering information on all applications in the system. Comparison with a well-established scheme, namely UCP, shows that the ACCP scheme uses considerably less storage and fewer counters. In order to monitor the behaviour of each competing application, had the applications received four additional cache ways, ACCP needs only four tag registers per set and four miss counters assigned to each application. The counters used in this work were assumed to be 32 bits long, but from our observation, the size can be reduced in practice for the following reasons. Firstly, the monitoring period was reset every ten million instructions. Hence, 24-bit counters are sufficient to be used for the instruction counters. Secondly, the miss counts of the miss counters recorded in the simulation results showed that the numbers of misses could be contained well within the 24-bit counters.

Now consider a 4MB, 32-way associative cache assuming a 32-bit physical address space. The storage overhead that will be incurred for a quad-core system if all sets are monitored and evaluated by ACCP is presented in Table 2. Note that, as ACCP requires each cache line to have a core ID, the length for the identification is log_2N bits. Thus, the overhead due to the identification fields is $8192 \times 16 \times 2bits = 32KB$, an increase of 0.66%. In total, the storage overhead for a quad-core system is $4cores \times 68.02KB + 32KB = 304.08KB$, an increment of 6.3% over the L2 cache area. Moreover, the aforementioned adders and comparators are needed to update the counters and to perform partitioning decisions in the ACCP. This will increase the total hardware overhead slightly over the indicated value even though the additional overhead is relatively small. However, by using DSS to reduce the hardware overhead, the actual hardware cost for monitoring 256 sets by ACCP is reduced. To achieve such a low hardware overhead, ACCP will cost each core 2.14KB of additional hardware due to the shadow tag arrays and counters to record the CPI values, which is an increment of 0.04% in the L2 cache area. As a result, implementation of ACCP will add an extra 0.2% area overhead towards the total shared cache area of a quad-core system.

Table 2. Hardware overhead per processor.

Type of Overhead	Value
Shadow tag array entry ((15 bits tag + 2 bits LRU) * 4 positions)	8.5B
Total shadow tag array overhead (8192 sets * 8.5B)	69,632B
Miss counters overhead (4 counters * 3B)	12B
CPI counters overhead (2 counters * 3B)	6B
Total overhead	68.02KB
Area of baseline L2 cache (736KB tags + 4MB)	4832KB
% increase in L2 cache area due to a shadow tag array	1.4%

6. Conclusion

ACCP attempts to improve the utilisation of a shared cache among multiple cores and to manage the progress of running applications in the system at approximately the same speed, by accelerating the slowest application without significantly slowing any others, while maintaining low hardware overhead. From the simulations that have been conducted, ACCP has achieved similar or better performance than existing cache partitioning schemes. Comparison with the complex UCP scheme, which requires higher hardware overhead than ACCP, has shown that the ACCP is capable of achieving similar performance, even though ACCP adjusts the cache allocation of only two cores in the system. The performance gain estimation used in ACCP also has outperformed the CPI in implementing effective partitioning decisions. While it was proven by the CPI that CPI values could be used as the sole metric to determine better partition sizes, it is found from our investigations that the performance gain estimation can produce significantly more effective partitioning decisions. This is because the performance gain estimation reflects how much an application would benefit from additional cache ways.

Overall, ACCP shows that referring solely to CPI values to determine the cache partition sizes is less effective than also using information provided by historical memory accesses made by each application in the system. Evaluation on the lines that have been evicted from the cache in order to make better partitioning decisions can give more flexibility to the partitioning scheme and the system to gain further benefits from the employment of both an optimal partitioning scheme and an optimal cache replacement policy. This is because the applications' behaviours and their cache activities will not have any impact on the estimation of performance gain. Finally, the analysis made on the hardware cost has demonstrated that the new ACCP scheme incurs a very low hardware overhead. Hence, the proposed scheme can be effective in improving CMP system performance without incurring the significant hardware overhead of UCP.

References

[1] Sato M, Nishimura S, Egawa R, Takizawa H and Kobayashi H, (2016). "A cache partitioning mechanism to protect shared data for CMPs" in *IEEE Symposium in Low-Power and High-Speed Chips (COOL CHIPS XIX)*, 2016, p 1-2.

[2] Shen F, He Y, Zhang J, Li Q, Li J and Xu C (2019). Reuse locality aware cache partitioning for last-level cache. *Computers & Electrical Engineering*, 74, p 319-330.

[3] Gupta S and Zhou H (2015) "Spatial locality-aware cache partitioning for effective cache sharing" in *44th International Conference on Parallel Processing* p 150-159.

[4] Mittal S (2017). A survey of techniques for cache partitioning in multicore processors. *ACM Computing Surveys (CSUR)*, **50**(2), 27.

[5] Cadence Design Systems, Inc. (2014). *Tensilica Software Development Toolkit (SDK)*, 2014 Release. Cadence Design Systems, Inc., Santa Clara California, USA.

[6] Tensilica Inc. (2010). *Xtensa Modeling Protocol (XTMP) User's Guide*. Tensilica Inc., Santa Clara California, USA.

[7] Mahrom N, Liebelt M, Raof R A A, Daud S and Ghazali N H (2018). "An Adaptive Insertion and Promotion Policy for Partitioned Shared Caches" in *IOP Conference Series: Materials Science and Engineering* **318**(1), p 012048.

[8] Selfa V, Sahuquillo J, Gómez C and Gómez M. E (2015). "Methodologies and performance metrics to evaluate multiprogram workloads" in *23rd Euromicro International Conference on Parallel, Distributed, and Network-Based Processing*, p 150-154.

[9] Jain P and Surve S K (2019) "Resource Centric Characterization and Classification of Applications Using KMeans for Multicores" in *2019 International Conference on Information Networking (ICOIN)* p 25-30.

[10] Qureshi M K and Patt Y N (2006) "Utility-based cache partitioning: A low-overhead, high-performance, runtime mechanism to partition shared caches," in *Proc. of the 39th Annual IEEE/ACM International Symposium on Microarchitecture*, p 423-432.

[11] Muralidhara S P, Kandemir M and Raghavan P (2010) "Intra-application cache partitioning," *in IEEE International Symposium on Parallel & Distributed Processing (IPDPS),* p 1-12.

IOP Conf. Series: Materials Science and Engineering **767** (2020) 012060 doi:10.1088/1757-899X/767/1/012060

Miniaturized and high gain RFID reader antenna at 13.56MHz

Y.M.Saleh [1], F.H.Wee [1], B.S.Yew[2], S. Z. Ibrahim [3], Y.S. Lee[4]

[1]*Bioelectromagnetic Research Group, School of Computer and Communication Engineering, Universiti Malaysia Perlis, Pauh Putra Main Campus, 02600 Arau Perlis,*
[2]*Faculty of Innovative Design and Technology, Universiti Sultan Zainal Abidin (UniSZA), Kampus Gong Badak, 21300 Kuala Terengganu, Terengganu,*
[3]*Advanced Communication Engineering (ACE) Centre of Excellence, School of Computer and Communication, Universiti Malaysia Perlis, 02600 Pauh, Perlis, Malaysia.*
[4]*Department of Electronic Engineering Technology, Faculty Engineering Technology, Universiti Malaysia Perlis, Malaysia*

fhwee@unimap.edu.my

Abstract. Technology for radio frequency identification (RFID) is commonly used worldwide. RFID is a developing wireless technology which make new vital challenge. The ability to precisely read a set of RFID tags become major concern in RFID networks coverage Detailed analysis of the design based on the 13.56 MHz radio frequency was suggested to design a kind of RFID reader antenna. Previously, researchers did facing problem with the size of the antenna which was to maintain the compact size at low frequency is challenging. The proposed antenna will be designed in the shape of rectangular spiral which contribute into miniaturize of the antenna size. The antenna will be fabricated after the optimum length and width had been achieved. The goal of this project is to design a small size antenna with 55mm x 55 mm at frequency resonance of 13.56 MHz. to reduce the antenna gain on the future. We are going to design and add the metamaterial to the RFID reader antenna and combine them together and that will help to reduce the antenna gain.

1. Introduction

Radio Frequency Identification (RFID) collects identity information from RFID tags to RFID readers and in the current world this a promising wireless identification technology [1]. RFID has the features of long distance, no-sight-of-see, fast identification and, large volume, and that is suitable for the surveillance of mobile objects as well [2]. Much attention has grown-up into the 13.56 MHz frequency in the last era of technology more than the VLF, LF and UHF bands. It is proven that the use of 13.56 MHz is more useful compare to others band that discussed in [3].

The RFID is a tag (or transponder) system that includes an electronic microchip and a reader (or interrogator) to read and write information.[4].The magnetic and electromagnetic fields are used to exchange data with the reader.The antenna will help move the production data that the microchip has stored to several meters or more than 10 meters. The host computer receives the data from reader after the tag data is decoded. RFID has comparatively multi-tag recognition and extensive transmission range, conventional bar-code system can be replaced as it is a core technology that is used in the field requires security, safety, and logistics [5].

Numerous issues affect coverage in RFID networks; thus, some of these issues are problem area width, tag density, overlapping into readers and number of deployed readers [6]. There are several matters make the antenna designer need to concern about and those are size, shape, and weight, [7]. The author in [8] discussed the HF antenna that operates at 13.56 MHz and to reduce the size of the antenna for HF antenna there is several approaches has introduced in [7]. Hence the objective of this project is to design a small size antenna at frequency resonance at 13.56 MHz and to verify the simulated design by fabricating and testing the RFID reader hardware antenna.

Now a passive system is 13.56MHz RFID network, its operating mode is combined with inductance. It is possible to exchange energy and pass information between the reader and the tag via electromagnetic coupling between the antenna coil[7]. The antenna of the RFID device is the antenna of the user and the antenna of the mark. To order to improve the performance of the RFID system, two things can be considered, including improving the reader antenna quality and improving the tag antenna efficiency. Nevertheless, the performance of the tag's antenna is hardly improved to preserve the tag's advantages, such as low cost, small size, long life and so on[8]. So, the keyway to improve RFID system performance is to enhance reader's antenna quality.

2. RFID Reader Antenna Design Configuration

The most effective antenna design on any antenna with small size is the loop type antenna that will be printed in circuit board (PCB). PCB is made from the dielectric material named FR-4 with thickness of 1.6mm. The designed antenna is known as rectangular shape spiral that has multiples rectangular loop structure consisting of stripped copper lines with curved corners [9]. In the Figure 1, it shows the loops of the antenna and how is it follow with the same gap between each loop with distance 0.3mm.

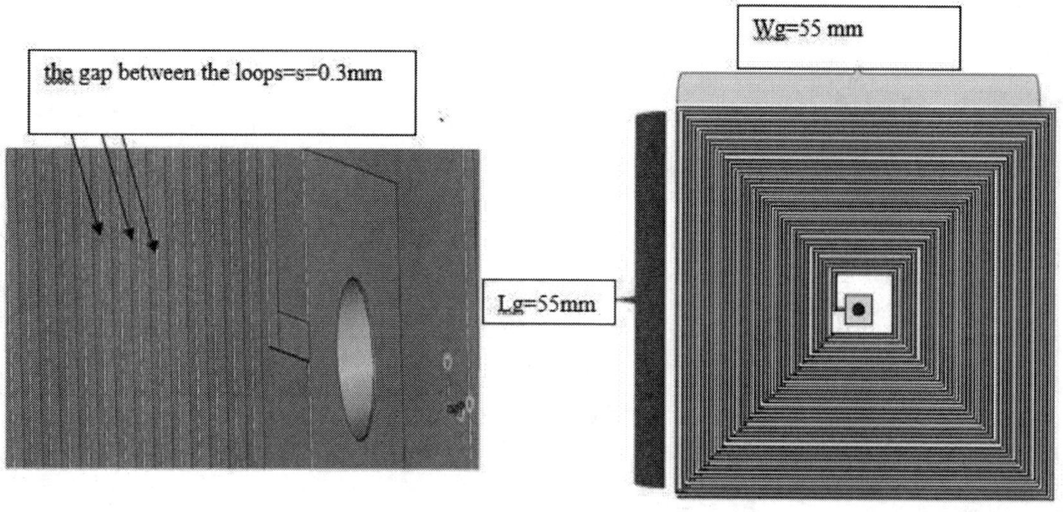

Figure 1. The design of RFID patch antenna

The determination of the antenna parameter's design has been found using some equation which is mention bellow: -

To Find Width (W):-

$$W=\frac{C}{2f_{\circ}\sqrt{\varepsilon r \cdot 1/2}}$$

f = operating frequency
εr = dielectric constant.
C = speed of light ($3*10^8$) m/s

To find the effective dielectric constant: -

$$\varepsilon reff = \varepsilon r + \frac{1}{2} + \varepsilon r - \frac{1}{2}\left(1 + 12\frac{h}{w}\right)^{\wedge} - ½$$

To find Effective length: - $Leff = c/2f \circ \sqrt{\varepsilon reff}$

To find the fringing length (ΔL):-

$$\Delta L = 0.412h\ (\varepsilon reff0.3)\frac{\frac{w}{h} + 0.264}{(\varepsilon reff - 0.258)(\frac{w}{h} - 0.8)}$$

Where; h= substrate thickness

To find the actual length: - L= $Leff - 2\Delta L$

For Finding The Length and Width Of The Ground: - Lg = 2* and Wg = 2 * W
By using those equation, we got the value that is showing below:

Table 1. Antenna Parameters

Parameters	Mm
W	55
L	55
Er	4.3
H	1.6
Ht	0.035
S	0.3

3. Antenna Fabrication

The design methods of 13.56MHz reader's antenna is according to the basic formula of the antenna. The design of the RFID reader antenna started with one loop and extended to two and multiple loops until 13.56 MHz is achieved which we put high frequency during the design and then start with one loop at 500 MHz and the others basically will help the s11 graph to shift from high frequency to the low frequency by adding loop. The simulated design has been fabricated according to the actual size in the simulation design and process of fabrication include developer, etching and photoresist have been done.

(a) (b)

Figure 2. (a) antenna fabricated (b) fabricated with SMA connector

IOP Conf. Series: Materials Science and Engineering **767** (2020) 012060 doi:10.1088/1757-899X/767/1/012060

After the measure prosses we have got the data from the measurement and then also from the CST and it will be showing in the figure 3 .we can see that the data from the CST is operating at the required frequency which about 13.56MHz and then the data from the measurement have operate at 17 MHz and that shows a bite different of the frequencies range and that because the much of looping that have been use on the design.

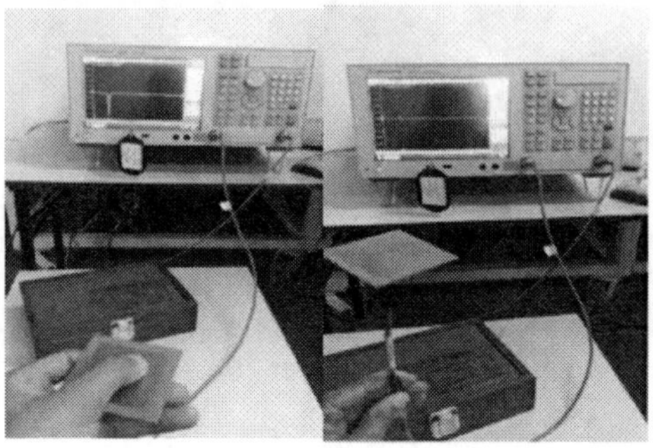

Figure 3. RFID reader antenna measurement

4. RFID Reader Antenna Design Results Analysis

In order to determine the characteristics of the designed antenna either it achieved the objective of this project which is to design and minimize the size of the antenna at the low frequency w 13.56 MHz .The characteristics that will cover are return loss response (S11), and the size of the antennas to reveal the ability of the antenna performance. Return loss is a factor used to calculate the antenna's reflected power due to the antenna's discrepancy.

Lower value of the return loss provides the high performance of the antenna. The return loss of the antenna is -20.8 dB at 13.56 MHz as performed in the Figure 4.

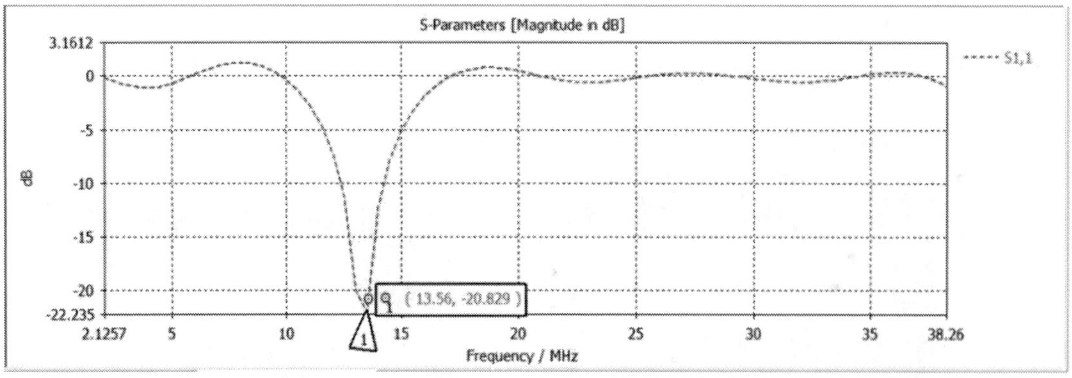

Figure 4. Return loss of proposed RFID Reader Antenna

5. Conclusions

The RFID reader antenna has become an expeditiously developing area of research. RFID reader antenna is design with small size and ease of fabrication. RFID reader antenna must fulfil the requirements of antenna with suitable antenna specifications. In this research project, the RFID reader antenna has been designed and got the simulation result by using CST software tools to obtain good performance. The return loss (S11) obtained at resonant frequency of 13.56 MHz with small size of 55mm x 55 mm.

6. Acknowledgement

Authors wishing to acknowledge assistance or encouragement from colleagues, special work by technical staff or financial support from organizations should do so in an unnumbered Acknowledgments section immediately following the last numbered section of the paper.

7. References

[1] Xiao G, Aflaki P, Lang S, Zhang Z, Tao Y, Py C, and Change S. 2018 Printed UHF RFID reader antennas for potential retail applications IEEE Journal of Radio Frequency Identification, **2** 1 p 31-37
[2] Zhu W, and L, M 2018 RFID Reader Planning for the Surveillance of Predictable Mobile Objects Procedia Computer Science **1** 29 p 475–481
[3] Basat S S, Lim K, Laskar J and Tentzeris M M 2015 Design and modeling of embedded 1356 MHz RFID antennas In IEEE Antennas and Propagation Society, AP-S International Symposium (Digest) p 64–67
[4] Chen H, Wang Z, Xia F, Li Y and Shi L 2018 Efficiently and completely identifying missing key tags for anonymous RFID systems IEEE Internet of Things Journal p2915-2926
[5] Nak-Gwon C, Hyuek-Jae L, Sang-Hoon L, and Seong-jeen K 2006 Design of a 1356 MHz RFID system Advanced Communication Technology, The 8th International Conference, **1** 4 p-843
[6] Ali K, Alsalih W, and Hassanein H 2011 Set-cover approximation algorithms for load-aware readers placement in RFID networks IEEE International Conference on Communications p 1–6
[7] Wang J J H 2007 Spiral antennas in RFID and their size reduction and performance enhancement IEEE International Workshop on Anti-Counterfeiting, Security, Identification, ASID: p 62
[8] Mayer L W, and Scholtz A L A 2018 dual-band HF/UHF antenna for RFID tags IEEE Vehicular Technology Conference p 1–5
[9] Eken F 2004 RFID Receiver Antenna Project for 1356 Mhz Band Simulation, TE 401 Microwave Course Term Project, Fall p 1–5

IOP Conf. Series: Materials Science and Engineering **767** (2020) 012061 doi:10.1088/1757-899X/767/1/012061

LaundryMama: Humanising Laundry Tasks using Laundry Management System and Laundry-On-Demand Mobile Applications

Leong Yi Mei[1], Ku Nurul Fazira Ku Azir[1], Siti Zuraidah Ibrahim[1] and Saidatul Norlyana Azemi[1]

[1]School of Computer and Communication Engineering,
Universiti Malaysia Perlis,
02600 Arau, Perlis, Malaysia.

sitizuraidah@unimap.edu.my

Abstract. Laundry Management System and Laundry-On-Demand Mobile Applications are presented in this paper. Using conventional laundry service method, customer is not informed about the laundry process stage, does not have option to arrange the preferred laundry pick up time for the deliveryman to pick up the unwashed laundry from the address provided by customer and the laundry ordering paper forms are often lost in transit between customer and admin. Therefore, a laundry management system software and laundry on demand mobile application is demanded to solve the problems. The software development is performed using an open source developing platform Android Studio IDE and Firebase Real-time Database, Authentication, Cloud Messaging and Cloud Storage. The method used to develop the software is waterfall modelling and two characters are involved, which is admin and customer. The two characters functions are separated in two different applications. A Laundry Management System Software is developed for admin to manage, make order and monitor the business. A Laundry-On-Demand Mobile Application is developed for customer to make order and monitor the order. These both applications can receive notification from each other. The data can be correctly written and read from Firebase Real-time database, Firebase Authentication, Firebase Cloud Storage. The developed software and mobile application are evaluated in term of its functionality.

1. Introduction

Nowadays, most of the laundry firms are using traditional way to manage business which is manual system for management. Laundry management system and laundry on demand mobile application is to ease the customer and launderette business owner to manage and monitor their orders in a smart way. Laundry management system software is an application to help launderette business owner to manage and monitor the administrative activities effectively, such as checking the status of orders, receiving orders from customers, billing and inventory cost control measures, providing customer service and managing the shipment. In the other hand, laundry on demand mobile application is to help customer to manage and monitor their order such as schedule a pick up time, making order, tracking shipment and get informed when the clothes have been cleaned or ready for shipping [1, 2, 3, 4, 5].

2. System Description

Laundry Management System Software and Laundry-On-Demand Mobile Application are introduced in this paper as shown in Figure 1. As seen in Figure 1, Laundry Management System Software is dedicated for admin while Laundry-on-demand Mobile Application is dedicated for customer. These applications have the different functions and features which are developed with Firebase as back-end database. Android Studio is used as framework to build the both Laundry Management System Software and Laundry-On-Demand Mobile Application. Android Studio is the recommended IDE to develop Android apps and is available for free for anyone who develops professional Android apps [1].

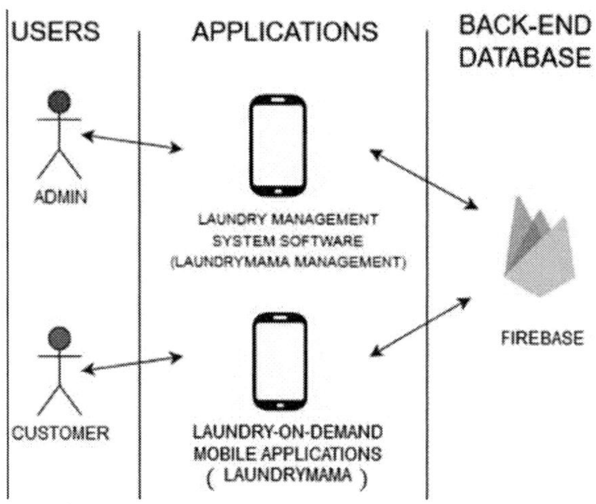

Figure 1. Overall Case Diagram

Figure 2 shows the four functions of firebase (Cloud Storage, Cloud Messaging, Real-time Database, Email Authentication) are used to develop the two applications, Laundry Management System and Laundry-on-demand Mobile Application.

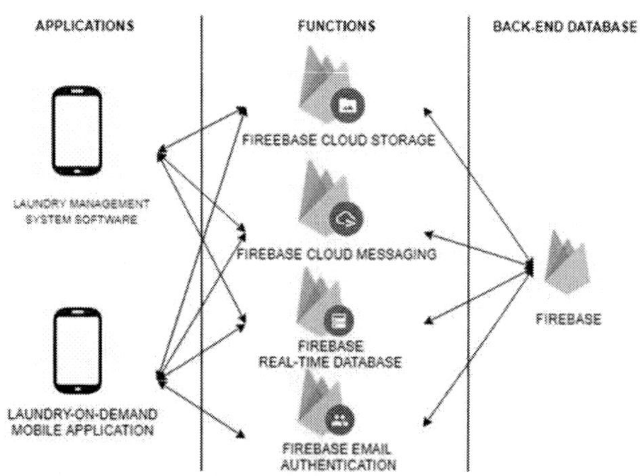

Figure 2. Functions Used in Firebase

IOP Conf. Series: Materials Science and Engineering **767** (2020) 012061 doi:10.1088/1757-899X/767/1/012061

SQLite is an open source embedded relational database. Rather than running independently as a standalone process, it symbiotically coexists inside the application it serves within its process space [2]. Figure 3 shows the SQLite that used in the development to create the database file, and then imported into the Android Studio project to store the laundry item data which added into the cart by user and then show in the data in the cart. These applications have the different functions and features which are developed with Firebase as back-end database as shown in Figure 4.

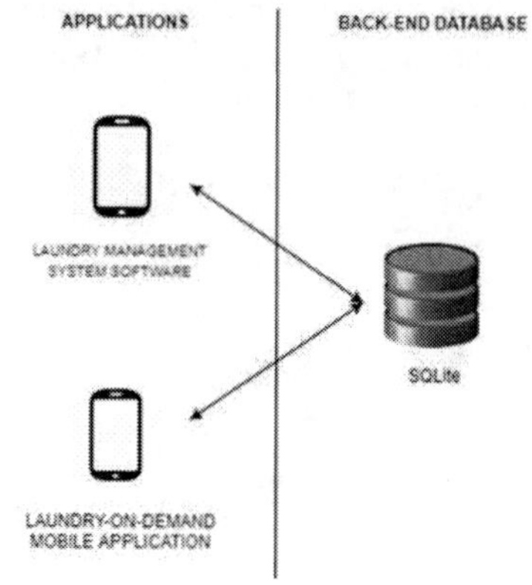

Figure 3. SQLite in Applications

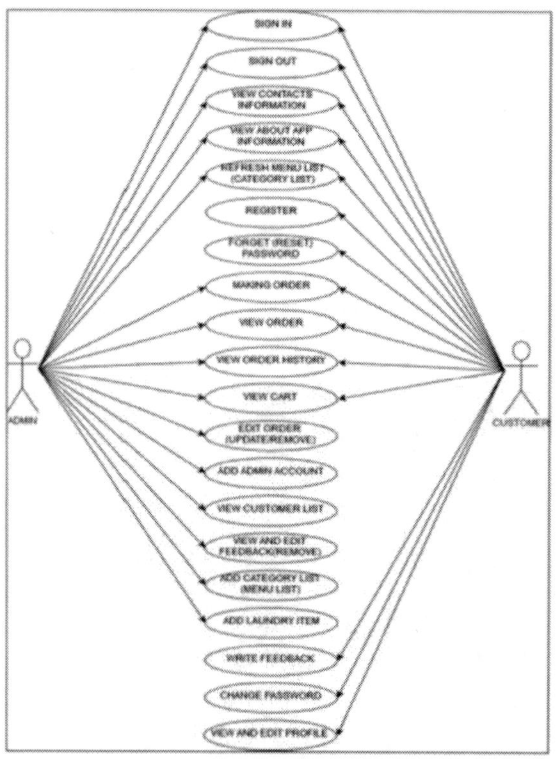

Figure 4. Unified Modeling Language (UML)

2.1. Main Flowchart of Laundry Management System App

By referring to Figure 5 and Figure 6, all the functions and features in each flowchart can only be used after successfully signing in with the valid admin account in Laundry Management System Software application and signing in with the registered account with email in Laundry-on-demand Mobile Application. Laundry-on-demand Mobile Application also has the features of register account and reset password (forget password), in order for customer to register a new account and reset the password by receiving a unique link in the email.

The Android build system compiles application resources and source code, and packages them into Android Package (APKs) that you can test, deploy, sign, and distribute. There are two ways to configure the build which are package the project into APK or activate developer mode in the testing android phone and install directly by Android Debug Bridge (ADB). The way to configure the build by packaging the project into APK, through "Build APK(s)" in Android Studio. Then, copy the APK to the android phone and install in the android phone.

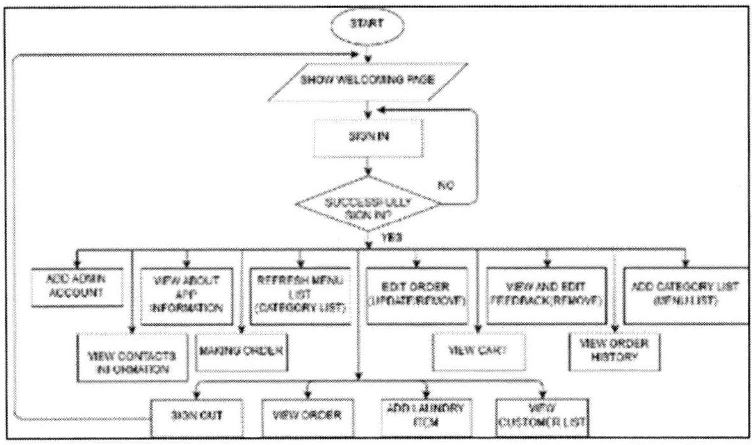

Figure 5. Flowchart of Laundry Management System Software without Signed in

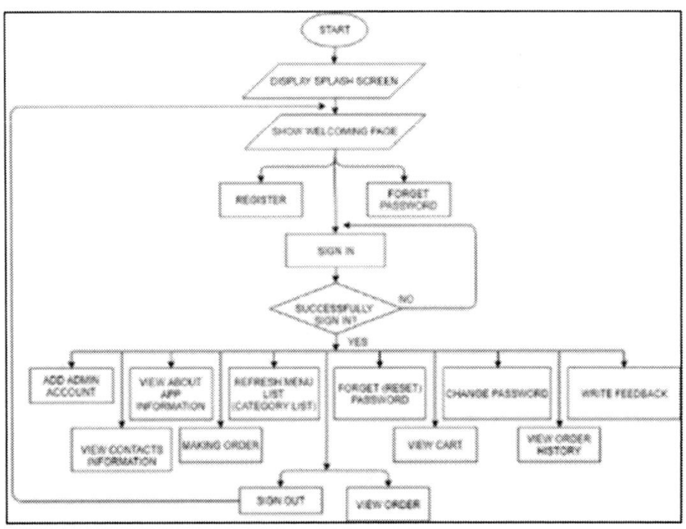

Figure 6. Flowchart of Laundry Management System Software with Signed in

3. Using LaundryMama

3.1. Testing

For testing stage, User Acceptance Testing (UAT) will be conducted. In the UAT, the actual software users test the software application to make sure it can handle required tasks in real-world situations, according to the specifications. The one of the purposes doing UAT is to find out the error of the software application by setting several different scenarios. Several test cases are created in order.

Table 1: Test Case for Main Page in Laundry Management System Software

Test Case ID	Test Scenario	Expected Result
TC1	User clicks category list	Navigate to laundry item list page
TC2	User clicks float button	Display dialog of adding category
TC3	User clicks YES in the dialog of adding category without entering name	Display error message
TC4	User clicks YES in the dialog of adding category without selecting an image	Display error message
TC5	User clicks YES in the dialog of adding category without uploading an image	Display error message
TC6	User clicks YES in the dialog of adding category with entering name, select and upload an image	Display message successfully adding new category
TC7	User clicks action bar	Display options Refresh and Contacts
TC8	User clicks Refresh in action bar	Reload and display latest category list
TC9	User clicks Contacts	Navigate to contact page
TC10	User clicks navigation bar button	Display navigation bar and the user's name
TC11	User clicks add admin account button in navigation bar	Navigate to Check Secured Word page
TC12	User clicks Making order for client button in navigation bar	Navigate to make order for client main page
TC13	User clicks order button in navigation bar	Navigate to order status page
TC14	User clicks order history button in navigation bar	Navigate to order history page
TC15	User clicks customer list button in navigation bar	Navigate to customer list page
TC16	User clicks feedback button in navigation bar	Navigate to feedback page
TC17	User clicks about LM. Management button in navigation bar	Navigate to about LaundryMama page
TC18	User clicks sign out button in navigation bar	User sign out and display welcoming page

IOP Conf. Series: Materials Science and Engineering **767** (2020) 012061 doi:10.1088/1757-899X/767/1/012061

3.2. Result from Laundry Management System Software

As shown in Figure 7, after starting the Laundry Management System Software, the application should have shown the welcoming page with title, business slogan and a button to proceed to the next page which is sign in page as shown in Figure 8.

Figure 7. Welcoming Page of Laundry Management System Software

Figure 8. Login Page of Laundry Management System Software

After admin successfully sign in with the correct ID number and password, the main page of the application is shown in Figure 9. In the Main page, a Recycle View category list is shown with title, action bar and float button. The action bar in the main page has two functions which are Refresh and Contacts. Refresh function is to load again the category list and the Contacts function is shown the contact information of the developer as shown in Figure 10.

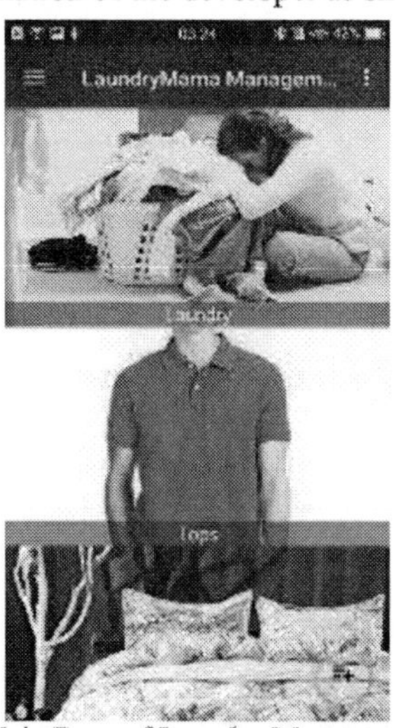

Figure 9. Main Page of Laundry Management System Software

Figure 10. Contact Page of Laundry Management System Software

The float button in the main page has the function to let admin add new category as shown in Figure 11. In the dialog as shown in Figure 11, the new category name and an image are needed to add new category into the list. The image can be chosen from the gallery and upload to Firebase Cloud Storage as shown in Figure 12. After successfully uploaded, the new Uniform Resource Locator (URL) is provided to link the location of the image and turn the button of "SELECT" to "IMAGE SELECTED" then the button "YES" in the dialog is to proceed the new category adding step while "NO" is to cancel adding new category.

Figure 11. Add Category Dialog of Laundry Management System Software

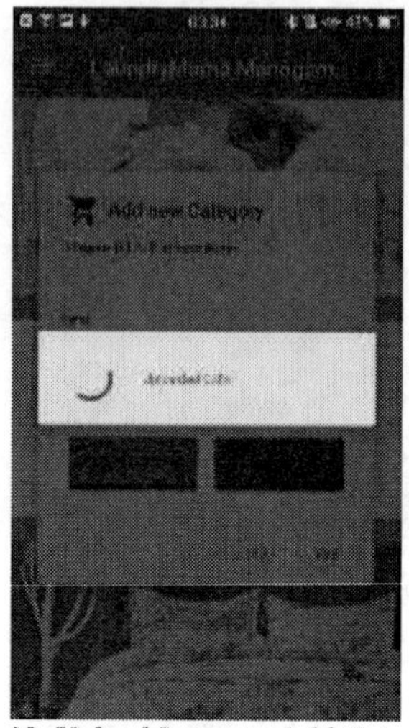

Figure 12. Upload Image to Add Category of Laundry Management System Software

The features in the navigation bar as shown in Figure 13, including Making Order for Client. In this feature, the admin can create order without customer account, this is meaning that the customer needs to approach to the laundry shop without registering any account in the Laundry-on-demand Mobile Application. In the category list as shown in Figure 14, the admin can choose the category and laundry item.

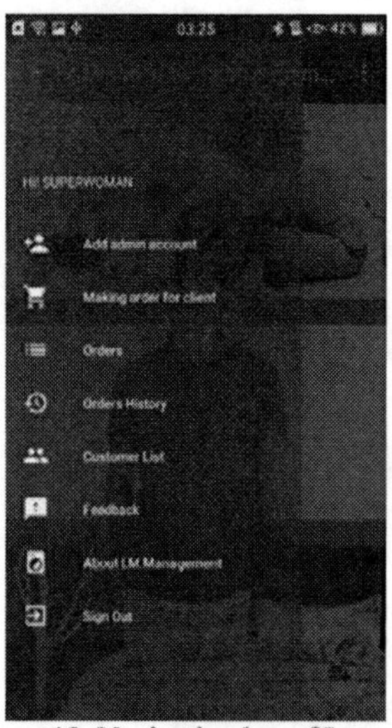

Figure 13. Navigation bar of Laundry Management System Software

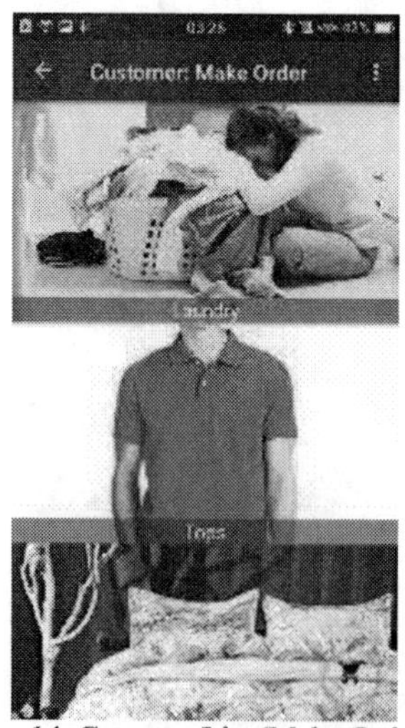

Figure 14. Category List (Make Order for Client) of Laundry Management System

To add the laundry item to the cart, the admin is required to choose the laundry item and the laundry item detail is shown in Figure 15. The admin may increase or decrease the item number and the float button is to add the chosen item and the item number to the cart. The admin can now view the added item in the cart page as Figure 16 shown by pressing the float button in the category list.

Figure 15. Laundry Item Detail (Make Order for Client) of Laundry Management System Software

Figure 16. Cart Page (Make Order for Client) of Laundry Management System Software

After successfully place the order for customer, the admin can view the order list and order history. The both of the order list and order history can proceed to view the order detail and order history detail. The admin can edit the orders by updating the order status and remove the order from the list. Only the order list can be edited, the order history is not allowed to be edited. The orders made by customer using Laundry-on-demand mobile application can be viewed in the feature of Orders and Order History. As the previous feature, the admin can edit the orders by updating the order status and remove the order from the list. Only the order list can be edited, the order history is not allowed to be edited. Once the order status has been updated by the admin, the customer should have received a notification to remind the customer that the order status has been updated.

The rest of the features includes Add Admin Account, Customer List, Feedback, About LM Management and Sign out. The feature of Add Admin Account is to add a new admin to the system. To add a new admin to the system, a secure word is needed to proceed to the next step which is to add new admin by ID number, name and password. After successfully adding the new admin account, the current admin can now sign in to the new account after signing out from the current account. For the feature of Customer List is to view the list of current registered customer. The Feedback feature in Laundry Management System Software is to view or remove the feedback given by the customer. The feature of About LM Management is for admin to view the detail information of the application. Last but not least, the sign out feature is to sign out from the current account then it will be redirected to the main page of the application as Fig.6 shown.

3.3. Result from Laundry-on-demand Mobile Application

After starting Laundry-on-demand Mobile Application, the application should have shown the welcoming page with title, business slogan and a button to proceed to the next page which is page for sign in or page for register. In the Register page, after valid information filled in and hit the button of "Register", an email will be received and it needs customer to verify the account through the given link in the email. After succeed verified, the customer can sign in to the account using the Laundry-on-demand Mobile Application as shown in Figure 17. The forget password button in the sign in page is for customer to reset password by entering the registered email. An email is sent to customer's mailbox after successfully entering the registered email in the page of forget password. An email is also sent to customer's mailbox after successfully registered the Laundry-on-demand Mobile Application account to verify the account.

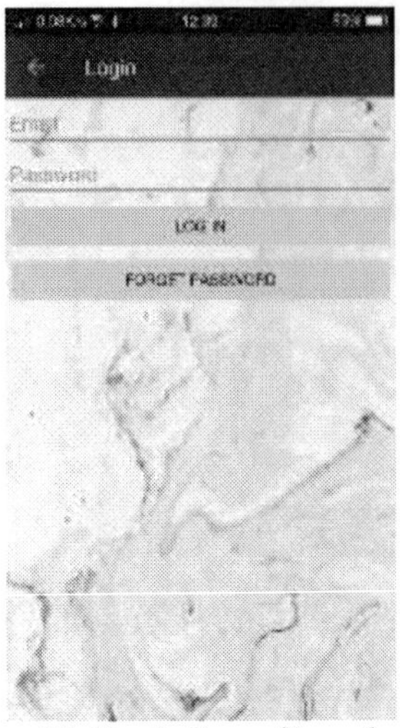

Figure 17. Sign In Page of Laundry-on-demand Mobile Application

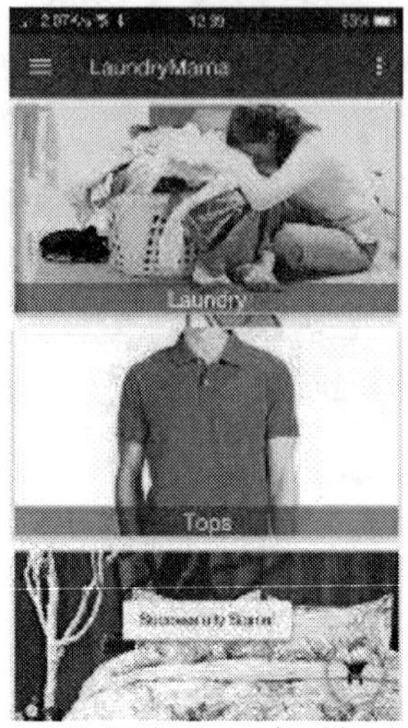

Figure 18. Main Page after Sign in of Laundry-on-demand Mobile Application

Customer can make order by using Laundry-on-demand Mobile Application with choosing the category list in the main page as shown in Figure 18, then choose the laundry item in the next page (laundry list). Choose the wanted laundry item and add the item to the cart by pressing the float button in the laundry detail page. Before adding the laundry item to the cart, the customer can increase or decrease the wanted item in the laundry item detail page. The items which are added to the cart can be viewed in the cart page. The unwanted items in the cart can be deleted by long press the specific item in the list then the action dialog is shown for customer to delete the item. To place order of the chosen item in the cart, some information is needed to enter in the dialog before place the order. After entering the information into the dialog, the "YES" button in the dialog is bringing to another dialog with asking whether customer agree to pay surcharge. The order is successfully made after customer pressing "YES" button to agree with the surcharge in the dialog.

When the order is successfully made by the customer in Laundry-on-demand Mobile Application, the notification is received by the admin in Laundry Management System Software. The navigation bar in the main page of Laundry-on-demand Mobile Application. The first feature in the navigation bar is "Order" which is navigated the customer to the order list page. In the order list, the customer can view the order detail by pressing the "Detail" button in the order list page then navigated to the order detail page. The customer can also view order history page which navigated from the navigation bar. From the view order history page, the customer can also proceed to view order history detail page.

The new orders which made by customer in Laundry-on-demand Mobile Application are shown in the order list page which is navigated from the navigation bar in the main page of Laundry Management System Software. The order list is shown in the order status page. The order editing function is only dedicated for Laundry Management System Software. The admin can view the specified order detail by pressing the "Detail" button in each order. The admin can also update and remove the order. After admin successfully updated the order status, the notification is received by the customer in Laundry-on-demand Mobile Application. Besides, the admin can also view the order history and order history detail.

The rest of the features in Laundry-on-demand Mobile Application including View and edit profile, change account password, About LaundryMama, Feedback and Sign out. The third feature shown in the navigation bar of main page is view and edits profile. From the navigation bar in the main page the customer can view the profile page. In each of the edit button beside the customer personal information, it is brought to each page to edit the personal information such as name, phone number, IC number, address and postcode.

In change password page, the customer can change password by entering the new password. After successfully changing password, the customer can sign in with the new password after signing out from the current account. The customer can view about application information in the page of About LaundryMama which is including the detail information of the application and the application release date. The customer can also give feedbacks. The feedbacks can be viewed by the admin and the action to delete feedbacks can only be taken by the admin in the Laundry Management System Software.
before.

4. CONCLUSION

A Laundry Management System Software is developed for admin to manage, make order and monitor the business. A Laundry-On-Demand Mobile Application is developed for customer to make order and monitor the order. These both applications can receive notification from each other. The data can be correctly written and read from Firebase Real-time database, Firebase Authentication, Firebase Cloud Storage. The developed software and mobile application are evaluated in term of its functionality.

References

[1] Akanksha Gupta, Debendra Kumar Panda, Mayank Pande (2018). Development of Mobile Application for Laundry Service Using Android Studio, *International Journal of Applied Engineering Research,* Volume 13, Number 12 (2018) pp. 10623-10626

[2] Lu, Y., & Yu, H. (2010). A Flexible Architecture for RFID Based Laundry Management Systems. *2010 International Conference on Computational Intelligence and Software Engineering.*

[3] O. Shoewu, N.T. Makanjuola, D.A. Phillips, A. Emmanuel (November 2016). Design and Implementation of a Laundry Management System, *The Pacific Journal of Science and Technology*, Volume 17, Number 2

[4] Ufuk CELIKKAN, Kaan KURTEL. (June 2017). Application of Service-Oriented Context-Aware Architecture to Laundry Management System, *Studies in Informatics and Control,* pp 1220-1766

[5] Van, N. T., Lee, S. J., Lee, C. W., Eom, K. H., & Jung, K. K. (2012). An implementation of Laundry Management System based on RFID hanger and wireless sensor network. *2012 Fourth International Conference on Ubiquitous and Future Networks (ICUFN).*

Mobile Augmented Reality (AR) Marker-based for Indoor Library Navigation

Rusnida Romli[1], Amir Firdhaus Razali[1], Nur Hafizah Ghazali[1], Nik Adilah Hanin[1] and Siti Zuraidah Ibrahim[1]

[1] Centre of Excellence Advanced Communication Engineering Centre *(ACE),* School of Computer and Communication Engineering, Universiti Malaysia Perlis (UniMAP), Pauh Putra Campus, 02600 Arau, Perlis, Malaysia

rusnida@unimap.edu.my

Abstract—This paper presents the development of Augmented Reality (AR) for smart campus urbanization using library as the environment for the demonstration of the AR prototype. The main goal of the AR development is to help users to get information and direction easily using AR based mobile application when walking inside the library. In normal circumstances, users typically walk around and explore the library area before reaching their targeted destinations. Depending on the library size and number of reading corners, exploration and walking in the library can be time consuming. Therefore, an AR technology is introduced in this paper to improve the user experience inside the library in the right direction and information instantly. This application is developed using Vuforia software to set the image marker-based and process the output into Unity3D software, Android Studio for the Main Menu interface and IBM Watson for voice recognition. The final form of the application is successfully generated from the development of Augmented Reality (AR) application for the smart campus by using a library for the demonstration of the AR prototype. A series of application tests are conducted in each corner of the library to evaluate the effectiveness of developed AR.

1. Introduction

A smart campus is a campus that has innovative frameworks in relatively all aspects of lives. A smart campus presents industrial answers for understudies in a considerable measure of fields that are the sustainable power source, digital resistance, innovative strength, wi-fi focuses, e-government and e-region applications, portable applications and different fields. These facilities have been made for students to make their daily activities easier. Smartphones and mobile applications are increasing the time by time, and better applications emerge evolving with mobile technology. New mobile application growing while becoming the primary usage on campus every day. These applications are very important for the campus to become a smart campus. One of the most prominent mobile technology for the smart campus is Augmented Reality [1].

Augmented Reality (AR) is a new technology that implemented the view of the physical world and computer graphics overlay, which is one of the significant advancements combining the real world and computer at a time. That technology can generate almost every content such as video, 2D or 3D image. Content activated when an image or location information is collected. Meanwhile, the virtual content and real environment will exist on the same screen or in the same space, and this technology is making impossible things happened in the real world [2]. Most of any devices such as smartphones nowadays contain a camera and screen to display, which is one of the elements for visualizing augmented reality. Smartphones technology has been refined with higher level making it production parallel with the technology of computer functions. Therefore, making it prime for the AR application.

Augmented Reality integrates the surrounding environment of a user with a digitized live image or image in real-time, generating a spectacular virtual outcome. Recognition using images is an indispensable component of augmented reality systems. By recognizing visual markers that are already integrated within the scheme, real-world objects are identified to superimpose the virtual objects. IN order tom implement the AR, the camera's orientation and position need to be estimated through the detection of actual-world image. Generally, the optical square marker is a black square within a predefined size white box. It is the black square encoded with the marker's identity. There is a range of methods that decode the marker by matching it.

By applying adequate and high-quality technology to both ordinary and online facilities in the library, and by developing 3D library and AR navigation would help users find books accurately and rapidly. Users can see where the books are located, and the application scheme will help user to navigate in the unfamiliar area inside the library. [3]. In addiditon, AR application can be a tool to increase the usability of the library, where interactive applications can enhance the user experience and encourage users to visit the library more often. For instance, the usage of virtual banner in library area and interactive book searching system could attract more users such as students to go to the library as the main destination of attractive and exciting learning environment for the students to learn and gain knowledge.

1.1 Marker-Based AR
Once the developer starts using a device with the marker-based augmented reality scheme, the picture of the physical environment captured by a camera is transformed into a grey-scale picture to accelerate the algorithm for image processing [4]. Marker-based AR is a type of augmented reality that uses the recognition of images, eyes, or other real-world objects in real-time to provide the user with additional digital data. A handheld device, such as a smartphone with the right software, could use a marker to read product bar codes and provide relevant information, such as reviews and prices, or to read faces and then connect to an item[5].

1.2 Markerless Based AR
The Markerless Augmented Reality technique enables any and all sections of the physical setting to be used as the goal or base for the positioning of virtual objects superimposed. Markerless AR depends not on the identifying markers, but on the natural characteristics of an environment. Some markerless systems have the capacity to collect and store data and features about the areas on which they are used for subsequent use. Typically, when used in smartphones and other digital devices, the markerless AR system uses the GPS function built-in to find and communicate with the accessible augmented reality resources [3]. Markerless Based Augmented Reality is used to indicate an AR application that does not require an understanding of a user's surroundings in order to overlay 3D material into a scene and keep it at a fixed spatial location [4].

1.3 Location-Based AR
Location-based AR is also known as Augmented Reality without marks, position-based and geo-based. It depends mainly on GPS, accelerometer, digital compass and other techniques to define the place and position of the telephone with a high precision level. Augmented truth is accessible for every mobile device owner because all moderndevices are fitted with these sensors.

2. Related works
Table 1 shows the comparison of the previously published works related to Augmented Reality

application. As can be deduced in table 1, the AR applications are developed for various applications with different software and hardware tools. For example, in [7], AR is developed for construction project using Unifeye SDK and 4D CAD. In [8], AR for campus navigation is developed using a few software tools such as AutoNavi MAP API, Vision-based AR API, and OpenGL. Then, Vuforia and Unity3D with C language are used as software tools to develop the design of VR (Virtual Reality) headset using AR. In turn in [10], the AR is implemented to See-Through Cars application, and the program is controlled by the Robot Operating System (ROS) software, and its hardware consists of D-GPS/LIDAR, PointGrey Flea, and Picostation M. Finally, for the AR of 3D Ping-Pong Game System, the application is developed on Android Mobile Platform using Eclipse with the assistance of Wi-Fi Direct to connect to other devices. In summary, software tools that are suitable for creating AR Application are Vuforia and Unity 3D, as it can improve the views of virtual and real objects that displayed onthe device. Furthermore, Android Studio also suitable to be used for developing the applicationas it supports the C language. Thus, using this software to create the application would create a realistic graphic of virtual objects.

Table 1: Comparison of Previously Published Works on AR

No	Titles	Software/Hardware Implemented
1	Application of Augmented Reality Object in Construction Project[7]	-Unifeye SDK -4D CAD
2	Campus Navigation System Based on Mobile Augmented Reality[8]	-AutoNavi MAP API -Vision-based AR API -OpenGL
3	Design of VR Headset using Augmented Reality[9]	-Vuforia -Unity3D -C language
4	A Real-time Augmented Reality System to See-Through Cars[10]	Hardware: -D-GPS/LIDAR -PointGrey Flea -PicoStation M Software: -ROS (Robot Operating System)
5	An Augmented Reality 3D PingPong Game System on Android Mobile Platform[11]	-Eclipse with Wi-Fi Direct

3. Proposed Framework

The tools used to develop and register the marker and how the AR application is designed are described in this section. figure 1 shows the flowchart of the AR application. Since the camera is the main component of this AR technology; hence, it needs permission to access and activate the camera located on the Smartphone. After that, it gains access to the data from the camera and gains control to change the display of the camera. The virtual objects will be triggered when the tag with markers are detected by written codes. Once the marker is identified by the camera, the 3D objects will directly display on the device's screen to indicate the marker is now recognized. If the marker is not detected, the camera will continue capturing images, and the application will try to search for the edges or marker to trigger the AR objects. The block diagram of the AR application is shown in figure 2.

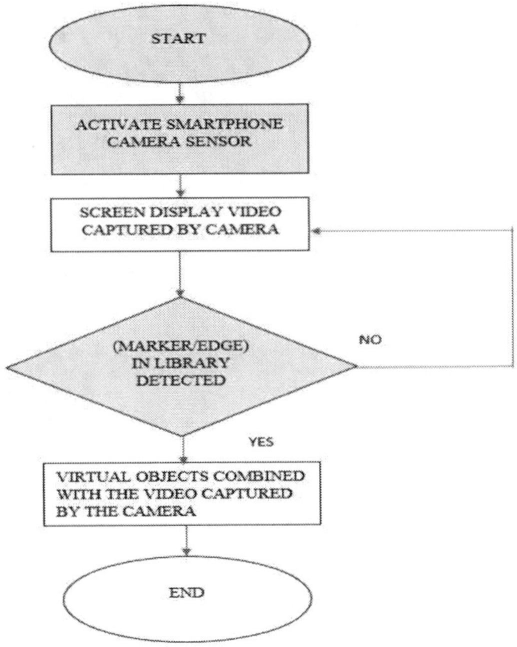

Figure 1: Augmented Reality Application Flowchart

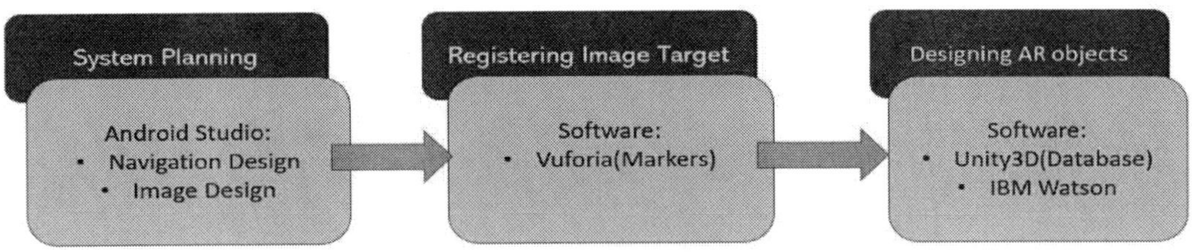

Figure 2: Block Diagram of Project

3.1 System Planning

The application includes a menu interface to ease the user for their direct guidance. Figure 3 shows the main menu interface of the AR application. It includes navigation to Lobby, Books, First or Second Floor, Direction and Open Camera. Icon is used for navigation so that the interface looks more straightforward and easy to use. Then, the image is snapped to be registered as markers.

Figure 3: Main Menu Layout of the Application

3.2 Registering Image Target

AR-markers are visual indicators that trigger virtual information display. Markers are standard images or small objects that are trained in advance to be recognizable later in the stream of the camera. Upon recognition of a marker, its position, scale and rotation are derived from visual indications and transferred to practical information. Different images at different library areas are chosen as the image marker-based for the AR objects to be triggered. The collected images are the input images and set as a tag. Figure 4 shows one of the examples of the image target registration to trigger AR objects.

Figure 4: Image Target Registration to Trigger AR Objects

3.3 Designing AR Objects

By registering a target image to the Vuforia, the pinpoint for the virtual object to trigger was made and what left is to design the virtual object that will appear after scanning the target image. For examples, figure 5 shows the virtual objects that appeared above the image target. Vuforia database needs to be downloaded first and imported into the Unity3D libraries. Assets also need to be imported outsource since Unity3D provides limited 3D objects, and some come with a price. Animation needs to be scripted into the movement of the 3D model, making the presentation of the AR more advances.

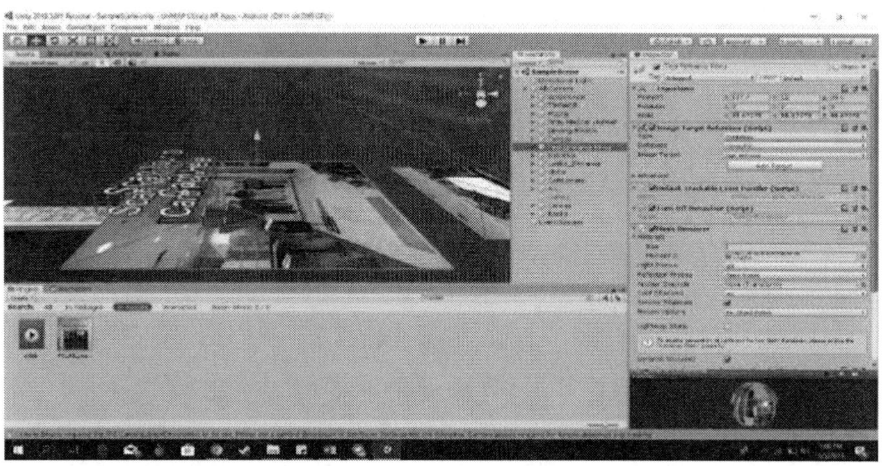

Figure 5: Virtual objects designated to appeared above image markers

Figure 6 is included with the audio source, and if the user tilts their device's camera to the markers towards the image target, a voice starts to sound as such "Please scan your ID here" and prompted them the correct machine to scan their ID card. This scene is scripted in the Unity3D with an audio source to enhance the presentation of the augmented reality.

Figure 6: Directional Guide to User with Audio Source

4. AR Application Result

This project is focused on the development of the AR application for urbanization area by using Unity3D, and the library is chosen as a demonstration environment of the AR prototype. Vuforia has been used to set the image markers for the scenes, and each of the markers is being rated on how it can easily be triggered.

Several tests have been conducted to make the AR application more appealing and the virtual objects to stay robust from the original positions. As such, a test of multi-image to trigger the virtual objects have been conducted, and the results are shown in figure 7. The test is to validate the effectiveness of actual virtual objects in real surrounding with different scenes.

Figure 7: Multi-Image Target

16 students and 4 librarians have been chosen to experience the AR applications at different scenes using their own smartphone device. This application only supports the Android Operating System Oreo 8.0 and above as it has the new features from the Unity3D SDK and some of the participants does not meet the specific requirement thus need to use developer's device to experience the AR application. There are 14 questions for them to answer directly using provided google form after the test are conducted. The questionnaires include the user profile background, the usefulness of the application, the ease of using the application, satisfaction of the application, representation of the user interface, and comment on how to improve it. The result of the questionnaires is shown in table 2 where it indicates a positive response.

Table 2: Questionnaire Result

Features	Percentage of strongly agree
Usefulness	75%
Easy to use	80%
Satisfaction	60%
Representation	70%
Average	**71.2%**

Three from four evaluated features achieved more than 70% of agreement from respondents, while Satisfaction features only achieved 60%. This is due to lack of creativity of AR design in a certain area, where the user satisfaction is lower compared to other features. Most of the respondents suggested to put extra 3D models so that the scene looks more entertaining to users. They also suggested to increase the usage of voice recognition features to all area, as it will be more helpful and appear more attractive for the user to control the animation using their voice. More scenes using voice are demanded to make a virtual object appeared in various area. In addition, visual motion is also suggested as additional feature that can be implement in future to enhance the navigation system. Overall average percentange achieved from the conducted survey is 71.2% for all evaluated features, which indicates positive impact of implementing marker-based AR in indoor navigation system in mobile application.

5. Conclusion

This project main contribution is to design and produce mobile application consist of interfaces of Augmented Reality on Android open-source system to testify the proficiency of virtual information shown on the screen and the viability of augmented object plotted at every section of the library. After applying the user testing such as multi-image target and choosing random participants to try the application with several questionnaires provided, the outcome is generated accurately as the virtual image are displayed with a sharp movement and stay at the right location. Hence, implementing this system into the library could reduce the usage of plastic things such as banner and stickers leading toward a go green campaign. Users can easily find sections that they wanted to go, as every area already stated in the AR apps. Hence, a more entertain way of presenting information could lead more students to try new features in the library. A variety of background and colours of the information is presented with more visibility to users includes with 3D effect for user to move freely. Furthermore, the resolution of the application is adjusted to perfectly fit into the user device display so that users able to view the virtual object working fully functional.

The access of the camera also required as it is the leading hardware that been used to detect the scene and every device nowadays includes with a decent quality camera which is making the scene easier to be detected. User will experience a different surrounding and more entertain when using this application to explore the library. It includes an interface where users can react to it and gain information from it. The virtual objects are triggered by detecting the pinpoint of every scene and its edges of 2D images from the camera. The markers that been register at the Vuforia developer web are the primary image processing that generated the virtual objects to appear. The image for the markers needs to choose wisely as it will create a tag for the base on edge inside the image. Then the image will be rated, and perfect markers will be rated as 5 stars. Based on evaluation done through users' written survey, overall result shows positive response from the respondents where average response achieved is 71.2 % for four different features, which indicates promising potential for AR based indoor navigation system.

Acknowledgement

This research was supported in part by Internal Social Project Via Reseach (INSPIRE) 2019, Universiti Malaysia Perlis.

References

[1] Musa S (2016), *Smart City Roadmap,* [Online]. Available: http: //www.academia.edu/21181336/SmartCityRoadmap (accessed on January 14, 2017).
[2] Mehdi M and Lemieux A (2014) "Augmented Reality : Applications , Challenges and Future Trends." *Applied Computational Science*, p 205-214.
[3] Jomsri P (2018) "Implementing Virtual 3D Model and Augmented Reality Navigation for Library in University," **8**(6), p 315–317.
[4] Levski Y (2019) "[Online] Available: https://appreal-vr.com/blog/markerless-vs-marker-based-augmented-reality/". [Accessed: 09- Jul- 2019].
[5] Fleck S, Hachet M, and Bastien J M C, "Marker-based augmented reality," **2**(5), p 21–28, 2015.
[6] Schechter S (2019), "What is markerless Augmented Reality?", Marxent, [Online]. Available: https://www.marxentlabs.com/what-is-markerless-augmented-reality-dead-reckoning/. [Accessed: 09- Jul- 2019].
[7] Kim H, Kim C, and Kim Y (2013) "Application of Augmented Reality Object in Construction Project", *Third World Congress on Information and Communication Technologies (WICT)* p 117–120.
[8] Qin Y-X, Liu J-M, Qiu H, Yan B, and Jin C-X (2013) "Campus Navigation System Based on Mobile Augmented Reality," *2013 6th International Conference on Intelligent Networks and Intelligent Systems (ICINIS),* p 1–4.
[9] Aloor J J and Thomas S (2016) "Design of VR Headset Using Augmented Reality," *2016 International Conference on Electrical, Electronics, and Optimization Techniques (ICEEOT)*

p 3540-3544.

[10] Rameau F, Ha H, Joo K, Choi J, Park K and Kweon I S, "A Real-Time Augmented Reality System to See-Through Cars," *IEEE Transactions on Visualization and Computer Graphics*, **22**(11) p 2395-2404.

[11] Gao X, Tian J, Liang X, and Wang G (2014) "ARPP: An augmented reality 3D ping-pong game system on Android mobile platform," *23rd Wirel. Opt. Commun. Conf. (WOCC 2014)*, p 1–6, 2014.

Effect of acids formation on the breakdown voltage properties of transformer insulating oil

Imran Sutan Chairul[1], Sharin Ab Ghani[1], Norazhar Abu Bakar[1], Nor Hidayah Rahim[1] and Nur Hakimah Ab Aziz[1]

[1]High Voltage Engineering Research Laboratory (HVERL), Faculty of Electrical Engineering, Universiti Teknikal Malaysia Melaka (UTeM), Hang Tuah Jaya, 76100 Durian Tunggal, Melaka, Malaysia.

sharinag@utem.edu.my

Abstract. Studies show that acids influence the electrical insulation strength of transformer insulating oil. Thus, this study was conducted to verify the influence of Low Molecular Acid (LMA) and High Molecular Acid (HMA) on the breakdown voltage of transformer insulating oil. Different weight of LMA and HMA were added into transformer insulating oil samples and then mixed using hotplate magnetic stirrer at 50°C and 750 rpm for 3 hours. Prior to the mixing process, mineral insulating oil samples were filtered, dried and its moisture content were determined according to ASTM D1533 standard test condition. After the mixing process, total acid number (TAN) and AC breakdown voltage (BdV) of the oil mixtures were measured according to standard test conditions of ASTM D974 and ASTM D1816 respectively.. The obtained results verified that LMA significantly reduced the breakdown voltage of mineral insulating oil compared to HMA for acidity lower than 1 mg KOH/g.

1. Introduction

Generally, there are three (3) main functions of insulating fluid used in oil-immersed power transformer (Saha & Purkait, 2017); (1) providing electrical insulation among active parts (*i.e.* between high voltage (HV) coils and low voltage (LV) coils as well as between core and transformer's tank); (2) enhancing heat dissipation process because as transformer's core and coils heated up during transformer operation, insulating fluid will carries the generated heat by conduction and transfer it to surrounding areas, thus ensuring low and controlled temperature in a transformer; and (3) providing indicator to monitor the condition of a transformer (*i.e.* furanic compounds dissolved in oil are indicators specifically used to monitor the cellulosed-based material (Heywood, Xiao, Ali, & Emsley, 2000)). Generally, transformer's condition monitoring can be divided into three (3) tiers (Ghazali, Talib, & Rosli, 2009). Tier 1: Oil analysis (*i.e.* dissolved gas analysis, breakdown voltage, moisture content, acidity, power factor, furanic compounds), Tier 2: Basic electrical tests (*i.e.* turns-ratio, winding resistance, dielectric dissipation factor/tan delta, excitation current, insulation resistance and polarization index), and Tier 3: Advanced electrical tests (*i.e.* frequency response analysis and partial discharge). Noted that the oil acidity is a part of Tier 1: Oil analysis. Acidity is being monitored because few studies (Lars E. Lundgaard, Hansen, Linhjell, & Painter, 2004), (Yunguang et al., 2013) show that acids in transformer oil will affect cellulosed-based material thus at the same time the performance of a transformer as well. Therefore, this paper is describing the formation of acids in mineral insulating oil as well as reporting the effect of acids on breakdown voltage of mineral insulating oil. The rest of this paper is organized as follows. The formation of acids in mineral insulating oil is described in the following section, followed by a brief description on the methodology adopted in this study. The findings of this study are then discussed with emphasis on two (2) different types of acids, and lastly, this paper provides verification on the effect of acids to breakdown voltage of mineral insulating oil.

2. Formation of acids in insulating oil

Figure 1 shows the molecule structure of acids that can be found in aged mineral insulating oil. The acids are categorized into two (2) types: (1) Low Molecular Acid (LMA) (*i.e.* formic, acetic and laevulinic acids) and (2) High Molecular Acid (HMA) (*i.e.* naphtenic and stearic acids). These acids are by-products formed due to ageing of transformer insulating materials (mineral insulating oil and cellulose) during a transformer's services. The ageing mechanisms for these insulating materials are described in the following section.

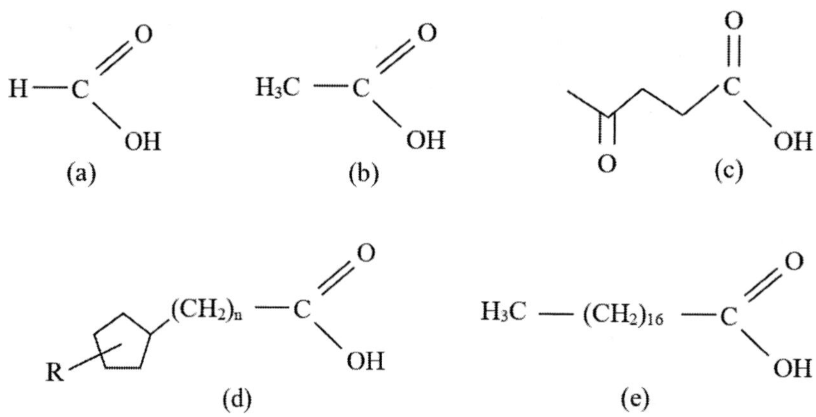

Figure 1: Molecular structure of carboxylic acids: (a) formic, (b) acetic,(c) laevulinic,(d) naphtenic and (e) stearic acids (L. E. Lundgaard, Hansen, & Ingebrigtsen, 2008)

3. Ageing mechanism of cellulose-based materials

Ageing mechanism of cellulose-based materials are due to (1) heat, (2) presence of oxygen, (3) moisture and acids. As cellulose being heated without any influences either from oxygen or moisture, to a maximum of 200°C, the cellulose's glycosidic bonds tend to break and in due course open the glucose rings. Thus, free glucose molecules, moisture, carbon oxides and organic acids are produced out of this reaction. In the presence of oxygen, cellulose's hydroxyl groups are oxidized to carbonyl (aldehydic) and carboxyl (acidic) groups. While existence of water and acids will cleave the cellulose glycosidic bonds yielding free glucose.(Unsworth & Mitchell, 1990).

4. Ageing mechanism of mineral-based insulating oil

Ageing mechanism of mineral insulating oil is normally associated with oxidation process; with high temperature and presence of moisture act as catalysts. Oxidation of oil resulting in formation of by-products such as acids, sludge and other polar compounds. In general, oxygen attack hydrocarbon molecule of mineral insulating oil, which generates hydro-peroxides. These hydro-peroxides can be further decompose to form ketones and water. Ketones can either be oxidized to form carboxylic acids or cleaved to make aldehydes. The carboxylic acids that are produced will either dissolve in the oil or volatilize into the headspace (Sanghi, 2003),(Meshkatoddini, 2008).

5. Methodology

The methodology adopted in this study is summarized in a flow chart, as shown in Figure 2.

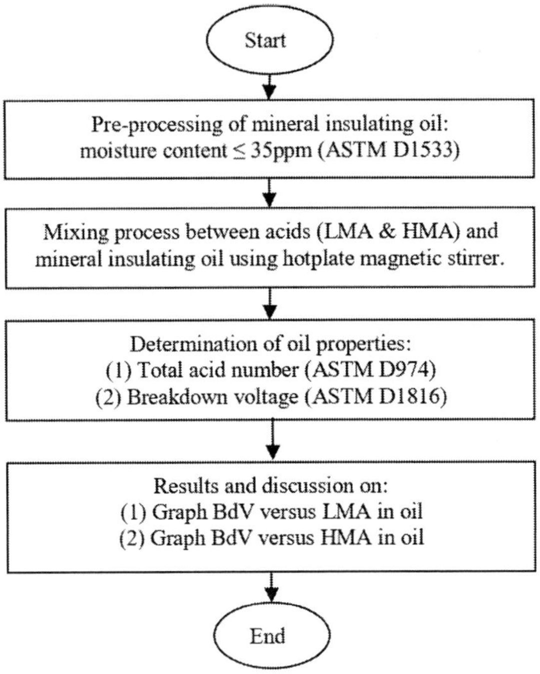

Figure 2: Flow chart of the methodology adopted in this study

5.1. Mixing process of acids and mineral insulating oil

Initially, mineral insulating oil was filtered using filter paper with a pore size of 0.2μm. Then, the filtered oil was dried in oven for 24 hours at 105°C so that the moisture content will be less than 35 ppm. After that, different weight of LMA (formic) and HMA (stearic) were added into separate samples. Each sample consist of 500 ml mineral insulating oil that previously being filtered and dried. The acids and oil samples were mixed using hotplate magnetic stirrer at 50°C and 750 rpm for 3 hours. Following the mixing process, each oil samples was prepared for breakdown voltage measurements as well as titrated for level of acidity.

5.2. Water content, acidity and AC brekakdown voltage

Water content, total acid number and breakdown voltage for all oil samples was measured in accordance with (ASTM D1533, 2012), (ASTM D974, 2014) and (ASTM D1816, 2012) standard test method, respectively. Water may be present in insulating oils either as free water or in dissolved form. Measurement of water content in oil is done in terms of milligrams/kilogram or parts per million (ppm). The water content in oil was determined using a coulometer based on the Karl Fischer titration method.The Karl Fischer titration method is based on the oxidation of sulphur dioxide by iodine in methanolic hydroxide solution. Acidity or total acid number (TAN) is a measure of acidic components present in the oil. New and unused mineral insulating oil is expected to be neutral and free from any acidic compound. However, during service, acidic by-products are formed due to oil's oxidation and cellulose's degradation. TAN of oil sample was measured using a TAN analyser, an equipment that measured basically based on the amount of potassium hydroxide (in mg) required to neutralize hydrogen ions (H+) in 1 g of oil. Ability of insulating liquid to withstand electric stress without failure is called dielectric breakdown voltage (BdV). A low value of BdV indicates the existence of contaminants (*i.e.* air bubbles, moisture, and suspended solid particles). The BdV measurements were conducted using a portable oil tester. The minimum volume of oil used to perform this measurement is 350 ml. The portable oil tester used two (2) cleaned electrodes (semi-sphere shape) with the gap distance between these two (2) electrodes was kept fixed at 1 mm while the rate of voltage rise was 0.5kV/s.

6. Results and discussion

Figure 3 shows a graph of BdV (kV) versus LMA in oil (mg KOH/g). Notice that, breakdown voltage decreased 37.5% from its initial value once 0.07 mg KOH/g of LMA in oil was detected. Next, as 0.50 and 0.70 mg KOH/g of LMA in oil were detected, the breakdown voltage drop 62.5% and 70.8% from its initial value respectively. Thus in general, the increment of LMA in oil resulting on decrement of breakdown voltage. Next, Figure 4 shows a relationship between BdV (kV) and HMA in oil (mg KOH/g). Notice that, there is fall and rise effect on breakdown voltage within the range of 0.10 to 1.5 mg KOH/g of HMA in oil. The breakdown voltage is steadily decrease as HMA in oil is bigger than 1.5 mg KOH/g. Noted that the breakdown voltage of mineral insulating oil drop 12.5% and 60% from its initial value as 1mg KOH/g and 9 mg KOH/g of HMA detected in oil respectively.

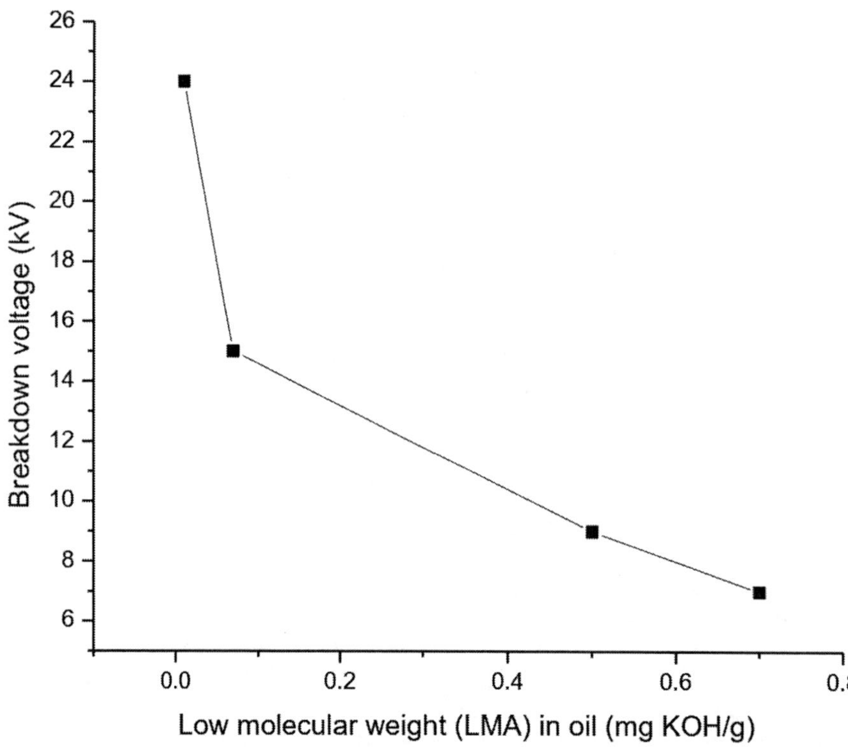

Figure 3: Graph Breakdown voltage (kV) versus Low molecular acid in oil (mg KOH/g)

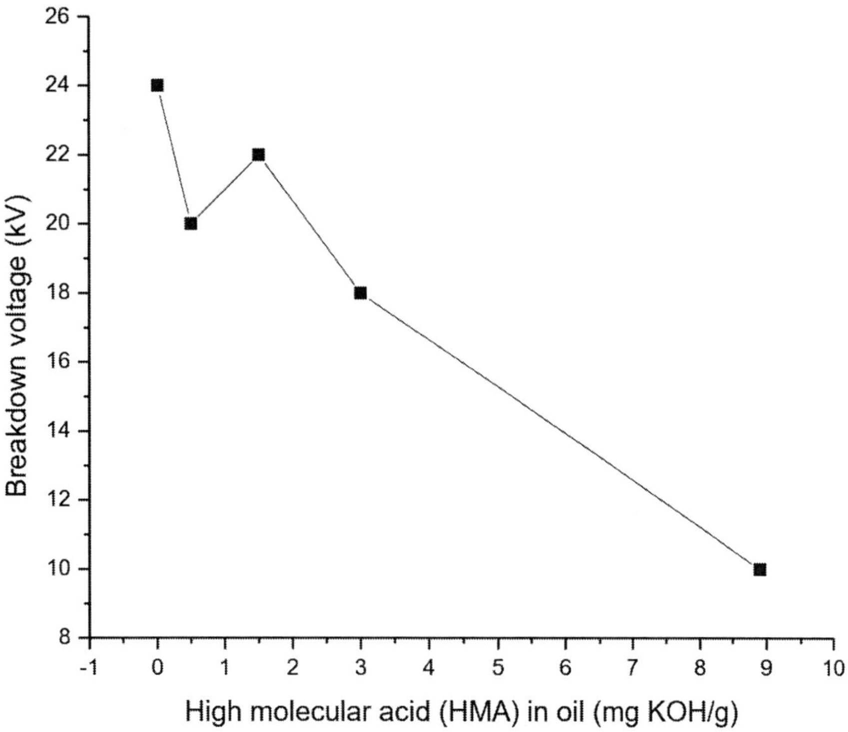

Figure 4: Graph Breakdown voltage (kV) versus High molecular acid in oil (mg KOH/g)

7. Conclusions

As acidity of mineral insulating oil for in-services transformer rarely reach 1 mg KOH/g (Azis, 2012), it can be concluded that the presence of LMA in oil gives more significant reduction on breakdown voltage of mineral insulating oil compared to HMA.

Acknowledgement

The authors gratefully acknowledge the financial support provided by the Ministry of Education Malaysia (MoE) and Universiti Teknikal Malaysia Melaka (UTeM) under the grant: (FRGS/1/2017/TK04/ FKE-CERIA/F00332). The authors amiably thank Ms. Norasmah Ismail from Faculty of Electrical Engineering, UTeM, Malaysia for providing assistance on the preparation and measurements of the materials used in this study.

References

[1] ASTM D1533 (2012). Standard Test Method for Water in Insulating Liquids by Coulometric Karl Fischer Titration. https://doi.org/10.1520/D1533-12.2

[2] ASTM D1816 (2012). Standard Test Method for Dielectric Breakdown Voltage of Insulating Liquids Using VDE Electrodes. https://doi.org/10.1520/D1816-12.2

[3] ASTM D974 (2014). Standard Test Method for Acid and Base Number by Color-Indicator Titration. https://doi.org/10.1520/D0974-08.2

[4] Azis N (2012). *Ageing Assessment of Insulation Paper with Consideration of In-Service Ageing and Natural Ester Application*. https://doi.org/10.1017/CBO9781107415324.004

[5] Ghazali Y Z Y, Talib M A, and Rosli H A (2009). TNB experience in condition assessment and life management of distribution power transformers. In *20th International Conference on Electricity Distribution* (p. 0686(1-4)). https://doi.org/10.1049/cp.2009.0919

[6] Heywood R J, Xiao X, Ali M, and Emsley A M (2000). Degradation of cellulosic insulation in power transformers. Part 2: Formation of furan products in insulating oil. *IEE Proc. - Science, Measurement and Technology*, **147**(3), p 115–119. https://doi.org/10.1049/ip-smt:20000021

[7] Lundgaard L E, Hansen W, and Ingebrigtsen S (2008). Ageing of mineral oil impregnated cellulose by acid catalysis. *IEEE Transactions on Dielectrics and Electrical Insulation*, **15**(2), p 540–546. https://doi.org/10.1109/TDEI.2008.4483475

[8] Lundgaard L E, Hansen W, Linhjell D, and Painter T J (2004). Aging of Oil-Impregnated Paper in Power Transformers. *IEEE Transactions on Power Delivery*, **19**(1), p 230–239. https://doi.org/10.1109/TPWRD.2003.820175

[9] Meshkatoddini M R (2008). Aging Study and Lifetime Estimation of Transformer Mineral Oil. *American J. of Engineering and Applied Sciences*, **1**(4), p 384–388. https://doi.org/10.3844/ajeassp.2008.384.388

[10] Saha T K, and Purkait P. (2017). Transformer Insulation Materials and Ageing. In *Transformer Ageing: Monitoring and Estimation Techniques* (p 1–33).

[11] Sanghi R (2003). Chemistry behind the life of a transformer. *Resonance*, **8**(6), p 17–23. https://doi.org/10.1007/BF02837865

[12] Unsworth J, and Mitchell F (1990). Degradation of Electrical Insulating Paper Monitored with High Performance Liquid Chromatography. *IEEE Transactions on Electrical Insulation*, **25**(4), p 737–746. https://doi.org/10.1109/14.57098

[13] Yunguang C, Guangning W, Xiaojian W, Qian P, Ling C, and Bo G (2013). The influence of organic acid on thermal aging of power transformer oil-paper insulation. *2013 Annual Report Conference on Electrical Insulation and Dielectric Phenomena*, p 116–119. https://doi.org/10.1109/CEIDP.2013.6748119

Automated Detection of Printed Circuit Boards (PCB) Defects by Using Machine Learning in Electronic Manufacturing: Current Approaches

S S Zakaria[1], A Amir[1], N Yaakob[1], S NAzemi[1]

[1] School of Computer and Communication Engineering, Universiti Malaysia Perlis (UniMAP), Pauh Putra Main Campus, 01000 Arau Perlis.

amizaamir@unimap.edu.my

Abstract. The manufacturing of a printed circuit board in the SMT assembly line goes through multiple phases of automatic handling. To ensure the quality of the board and reduce the number of defects, inspection tasks such as solder paste inspection and automatic optical inspection are conducted. The inspection tasks are carried out at various phases of the assembly line. The paper aims to answer the questions of how machine learning technology can contribute for better PCB fault detection in the assembly line and at which parts of the assembly line this technology has been applied. The paper discusses the PCB defect detection by using machine learning and other approaches. The current research shows that PCB defect detection using machine learning are miniscule. Early detection is still unexplored and experimented in the industry.

1. Introduction

Surface-mount technology (SMT) is a method for producing electronic circuits in which the components are mounted or placed directly onto the surface of Printed Circuit Boards (PCB)s. PCB are boards with electronically connected lines, tracks and paths to connect electronic components. These paths are connected by a substrate laminated by copper material which allow electric to conduct from one component to another. To stick the component on the board, solder paste solution is printed on the board via solder paste printing process. Assembling those fine pitch components poses significant challenges to solder paste printing. The semiconductor industry reports that the 60% to 70% of defects are associated with solder paste printing process depends on various generalized factors like process, squeegee, substrate, printer machine, stencil and solder paste [1].

Each and every factor affects the solder volume deposition either directly or indirectly. The solder paste printing process is a crucial part of the assembly line as 50%- 70% of production defect is contributed from volume related defect in this process as stated in the journal authored by Burr et al (1997)[2]. It is a widely believed fact that 52-71% of SMT defects can be attributed to the solder paste printing process [3] as shown on the chart in figure 1.

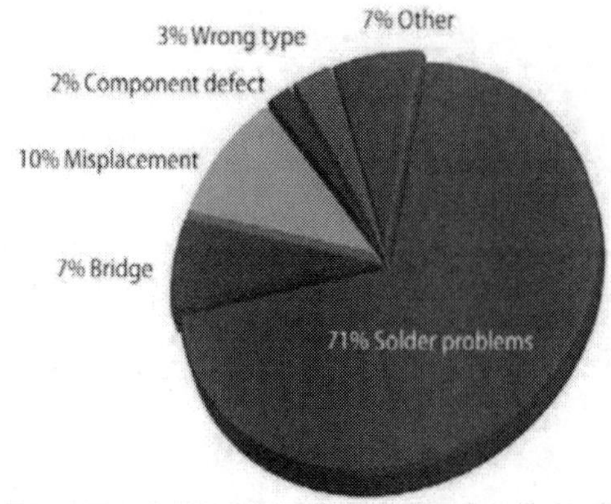

Figure 1. The Frequency of SMT Defects [3]

As the SMT pads corresponding to the part leads have become smaller and denser, the corresponding aperture in the stencil used to apply the solder paste have also become smaller and denser [4]. With reduced width-to-thickness aspect ratio, the stencil aperture is prone to clogging with dried solder paste, resulting in poor application of solder paste to the circuit board pad. Human visual inspection is the traditional method of choice to reveal defects in the physical quality of solder paste bricks for most manufacturers of SMT assemblies. Unfortunately, this method is inherently inefficient and subject to operator variability. Human visual inspection only provides an approximation of actual quality levels due to the limitations imposed by normal fatigue an eye strain. Other variability is associated with differences between operators.

The research production defects is important in today's world where consumerism is at its peak. To help the industry meet its supply and demands, catching a defect at an early stage such as on the assembly line, where the PCB is assembled, can increase yields by prevention of defects and ensure the company does not have a loss by having to rework defective products. Quality control has advanced from determining defective boards during final inspection to using process control to reduce the number of defects occurring. Thus, making in-process inspection become more important than ever. Manufacturing engineers are turning to in-line, post-print inspection equipment for implementing process control systems for their board assembly line.

This paper aims to investigate the available machine learning approaches that have been applied for PCB defect detection. This helps us to understand the available gaps we aim to identify the potential points in the SMT assembly line that can be improved by using machine learning.

The paper is organized as follows, section 2 introduces the backgrounds of SMT assembly line and two main stream methods on defects detection. Section 2 details the available works that utilized machine learning for PCB defect detection. Other PCB defect detection works that do not apply machine learning approaches are given in section 3. Finally, conclusions is shown in section 4.

2. SMT Assembly Line

This section discusses the main processes in the SMT Assembly line. Then, two inspection concepts that common in the SMT processes: Solder Paste Inspection and Automatic Optical Inspection, are briefly explained.

2.1. Main Processes

Generally, the SMT assembly line in electronic industry consists of six main processes (as described in figure 2): the PCB board registration, solder paste printing, component placement, mounting, repaneling and testing.

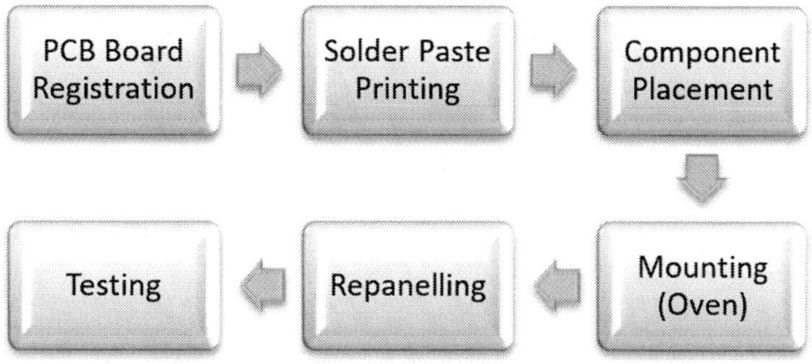

Figure 2. Processes in the SMT Assembly Line

Start with the PCB Board Registration, the board's serial number is recorded to ensure the board is traceable in the assembly line and can be monitored. Then, the PCB is printed with the solder paste (in Solder Paste Printing) and then the solder paste inspection will inspect the quality of the printing. Any rejected PCB with an error in the solder paste printing will produce a warning to alert the operators in charge. The error free PCB will go through the Component Placement where the components will be arranged. The PCB is then scanned again with the line scanner and before it undergoes the Mounting (Oven). The oven ensures the component sticks to the PCB board by mounting the component to the PCB board. Automatic Optical Inspection (AOI) machine is used to check if the components are error free. After all this, the PCB's goes to the Repanelling, where the PCB is cut into separate pieces. Finally, the repanelled PCB's undergoes Testing to confirm the components are working properly.

2.2. Solder Paste Inspection

It is now generally recognized that solder paste printing is a crucial process in controlling the solder joint quality on finished boards. The solder paste depositing inspection is very important in the process of surface mounting for printed circuit board (PCB). It is essential to develop the process that assists in depositing the correct volume of solder paste on the solder pads. Nowadays, most of the electronic manufacturing using solder paste inspection system to inspect and study the sider paste deposition on Printed Circuit Board after printing. The defects related to solder paste printing are captured using inspection system. It helps to find an optimum process setting that makes our printed circuit board as defect free. Solder paste inspection is conducted right after the solder paste in printed on the PCB. During the inspection, electronic assemblies which does not meet the criteria are sorted after solder paste printing. This is to avoid unnecessary cost due to placement of components.

2.3. Automatic Optical Inspection

The automatic optical inspection is an important tool which aids in detecting faulty component on a printed circuit board. This technique utilizes high performance scalable modular camera technology with four colour illuminations from all spatial directions to produce optimum contrast value for many defects during feature value extractions. An example of the colour illuminations technique in AOI inspection is shown in figure 3.

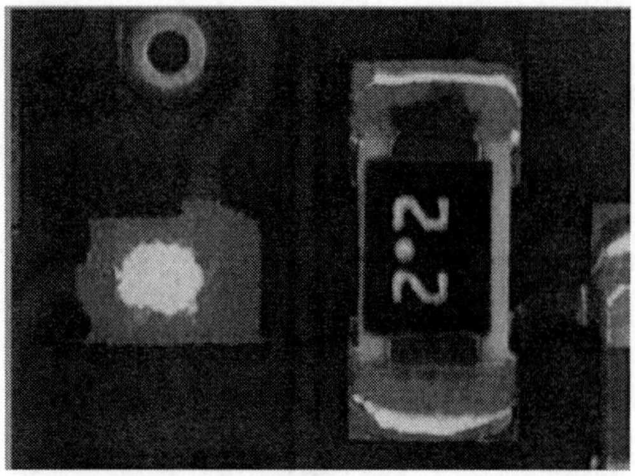

Figure 3. Color Illumination Technique

3. PCB Defect Detection by Using Machine Learning

3.1. Defect Detection By Using Artificial Neural Networks

Wu, Zhang, Kuang, & Lu (2008) argued that the currently inspection for solder paste which are mainly performed by laser-based systems is not practical due to high cost and low inspection speed[6]. They proposed a real-time inspection approach based on the machine vision. First, a new fast image matching method was applied to align the PCB and determination of region of interest. Second, the images of the PCB are analysed, and the two-dimensional feature and a pseudo-3D feature are obtained. Finally, the artificial neuron network was established for classification the solder pastes.

Acciani, Brunetti, and Fornarelli (2006) [12] introduced a diagnostic process to detect solder joint defect by using Neural Network System which processes the image of the solder joints of the integrated circuits mounted on the board. The PCB tested images are pre-processed by means of several methods to reduce the amount of data to feed to the neural networks. The board images are acquired and then pre-processed to extract the regions of interest for the diagnosis which are the solder joints of the integrated circuits. First, image segmentation was performed for board extraction, followed by I.C extraction. Finally, image segmentation was performed for pins extraction. During feature extraction, two feature vectors have been extracted from each region of interest, the "geometric" feature vector and "wavelet" feature vector. Both vectors feed the neural network system constituted by two Multilayer Perceptron neural networks and a Linear Vector Quantization network for the classification. The experimental results are devoted to comparing the performances of a Multi-Layer Perceptron network of a Linear Vector Quantization network, and of the overall neural network system, considering both geometric and wavelet features.

3.2. SPI Inspection Using Probabilistic Approaches

A probabilistic approach called, the Hierarchical Marked Point Process (HMPP) framework was proposed by Benedek (2011)[9], It is able to handle paste and scooping extraction problems simultaneously, so that the solder pastes and scoops have a parent-child relationship. A global optimization process attempts to find the optimal configuration of entities, considering the observed data, prior knowledge, and interactions between the neighbouring circuit elements. The proposed method is evaluated on a real PCB image set containing more than 3000 solders paste data with 600 scooping artefacts [9]. A morphology-based baseline method is also introduced for the problem and used as reference for qualitative and quantitative comparison against the proposed model.

3.3. PCB Defect Detection by Using Ensemble Machine Learning

With the continuous development of the electronics industry, the number of PCB has grown at a rapid rate, and the requirements for the detection, the main reference is the comparison method. However, in

real scene, here are a series of problems such as non-uniform illumination, tilting of the camera angle, and the like, resulting in a less satisfactory effect of the reference comparison method.
Lu, He, Xiang, & Liu (2018) proposed a cross-reference comparison framework of PCB defect detection [11]. They intended to solve the problem in the reference comparison method such as non-uniform illumination and tilting of the camera angle that result in a less satisfactory. This framework achieves good result in speed and accuracy. The authors extract the histogram oriented gradient and local binary pattern features for each PCB image respectively, put into the support vector machine to get two independent models. Then, according to Bayes fusion theory, the authors fuse two models for defect classification. The authors have established a PCB data set that includes both defective and defect free. It has been verified that the accuracy of the verification set is improved compared to the individual features using the fused feature. The authors also illustrate the effectiveness of Bayes feature fusion terms of speed

3.4. PCB Defect Detection by Using Deep Learning

Robust and precise defect detection is of great significance in the production of the high-quality PCB. Due to the complexity of PCB production environments, most previous works still utilize traditional image processing and matching algorithms to detect PCB defects. In Zhang, Shi, Li, Zhang, & Liu (2018) [13], an improved bare PCB defect detection approach is proposed by learning deep discriminative features, which also greatly reduced the high requirement of a defect dataset with some artificial defect and affine transformation to increase the quantity and diversity of defect data. Then, a deep pre-trained convolutional neural network is employed to learn high-level discriminative features of defects. They fine-tune the base model on the extended dataset by freezing all the convolutional layers and training the top layers. Finally, the sliding windows approach is adopted to further localized the defects. Extensive comparison with three additional shallow features-based methods demonstrate that the proposed approach is more feasible and effective in PCB defect detection area. Srimani and Pratibha (2016)[13] also presented a deep learning neural network and hybrid genetic algorithm for PCB defect detection. Deep learning neural network was used for classification. For feature selection, genetic algorithm was applied to optimize the feature reduction process. This work focused on feature generation by applying filtering scheme which provides feature ranking procedure by considering gain ratio, GINI index, and correlation. In addition, two stage optimal solution selection was adopted.

3.5. PCB Defect Detection by Using Speed-Up Robust Feature and Random Forest

Yuk, et al (2018)[15] proposed defect detection method by using speeded-up robust features (SURF) and Random Forest. SURF was used to extract the features from the PCB images and then Random Forest learned the fault pattern and obtained the probabilities. By using the probabilities to set the weights, the Weighted kernel density estimation (WKDE) map is generated and this provides the density of the features. Based on the density information, the area of the defects can be identified.

4. Other PCB Defect Detection Approaches

4.1. SPI Using Multi-Frequency Moiré technique

Stansfield (2014) [5] discussed the SPI machine that measures the values of the solder paste deposition by using a dual projection technology, where by two projection light is shone on the solder paste and a camera sits directly on top. A technique known as Multi-Frequency Moiré technique is used in providing a 3D image acquisition with a true and easy to use production tool. This technique is a method where a single or multiple projector are used to project a shifting pattern of lines on a given region of interest to which a digital camera captures the image of deformed lines as it shifted across the solder paste deposits' surface [5]. Then, after applying phase shift analysis and unwrapping techniques, the 3D profile of the solder paste surface is reconstructed and measured.

4.2. SPI Using Machine Vision

An integrated inspection approach based on the machine vision was presented in Shenglin, Xianmin, & Yongcong (2008)[7]. The method developed can identify the major defects of the solder paste depositing, such as displacement, deficiency, excess, bridge and reflow. Firstly, position compensation is applied to improve the section accuracy. Secondly, an image enhancement algorithm based on the texture is developed, which can enlarge the difference of the grayscale values between the PCB images and the solder paste images. This makes the multi-threshold method more stable and precise. Thirdly, the images of the printed PCB are analyzed by using the particle analysis method and the two-dimensional 2D inspection results are obtained. A pseudo three-dimensional 3D inspection approach is further proposed to identify the defects in the 3D state. The 2D and pseudo 3D inspection approach can be used to inspect the solder paste depositing defects effectively.

Jiang, Cheng, & Tao (2012) [8] proposed a new defect detection scheme for solder paste based on learning the colour biological feature sub-manifold. The biologically inspired colour feature (BICF) is applied to represent the solder paste images and introduced a sub manifold learning method to extract the intrinsic low-dimensional BICF manifold embedded in an extrinsic high-dimensional ambient space. This scheme mimics the function of human visual cortex in recognition tasks and can separate poor quality solder pastes from good quality ones.

Wu and Chen (2011) introduced image-array-based automatic optical inspection system captures multiple images of the inspected printed circuit board through synchronous exposure [10]. Therefore, a correct mosaic of images is very import to subsequent solder paste inspection. However, design error of the mechanism always leads to geometric errors in the image such as translation, rotation and scaling distortion. These errors make severe displacements in the image. To solve such problems, a concise and rapid error correction algorithm is proposed base on multiple point interpolation. Experiments show that the algorithm can assure rapid and correct mosaic of images of the whole printed circuit board [10].

4.3. Solder Joint Detection by Using X-Ray Imaging

(Kong, Su, & Scale, 2005) deals with the detection of defects at BGA solder joints in PC boards by using the X-ray imaging. It is possible to detect defects of solder joints by visual inspection, but only by utilizing X-ray method could inspect all BGA joints within the area array because they are hidden under the IC package. In traditional production, the inspection of BGA depends on the fiction test of electric circuits or boundary lighted inspection. Now with the employment of X-ray in solder joints inspection, it becomes very important about the detecting algorithm based on X-ray imaging (Kong, Su, & Scale, 2005). This paper discusses the flow of detecting defects based on the X-ray imaging and describes an approach to automatic inspection of BGA solder joint defects by using seed filling and contour extracting. Experimental result reveal that the proposed method shows practical usefulness in BGA solder joints inspection.

5. Conclusion

In this paper, we have presented, in detail, a current approaches of PCB defect detection. Focussing on the machine learning approaches, we have discussed several techniques of PCB defect detection by using machine learning techniques. Our review shows that deep learning, random forests, neural networks, and probabilistic approaches have been applied for PCB defect detection with automatic optical inspection. Although the methods have shown good result for PCB defect detection, it is desirable to apply machine learning in forecasting the potential of future PCB defect before the solder printing takes place.

References

[1] Manickam N (2017). Investigation of pre and post reflow solder paste depositions using theoretical, simulated and experimental techniques.

[2] Burr D (1997), Solder paste inspection: process control for defect reduction, *Proceedings International Test Conference*, Washington, DC, USA, p1036-. doi:

10.1109/TEST.1997.639726
[3] Chintamaneni K C (2015). *Determination Of Solder Paste Inspection Tolerance Limits For Fine Pitch Packages (Dissertation).* State University of New York at Binghamton, Thomas J. Watson School of Engineering and Applied Science, Department of Systems Science and Industrial Engineering
[4] Bryant S M and Loewenthal K H (1995). *Neural Network Solder Paste Inspection System*. US Patent US5751910A.
[5] Davies C (2014). SPI inspection techniques in the SMT line: The new Paradigm. *Electronic Production and Test. (4, April), Konradin Publishing Group.*
[6] Wu H, Zhang X, Kuang Y, and Lu S (2008). A real-time machine vision system for solder paste inspection. *IEEE/ASME International Conference on Advanced Intelligent Mechatronics (AIM)* (50375051), p 205–210. https://doi.org/10.1109/AIM.2008.4601660
[7] Lu S, Zhang X, and Kuang Y (2008). An integrated inspection method based on machine vision for solder paste depositing. *IEEE International Conference on Control and Automation (ICCA)* p 137–141.
[8] Jiaang J, Cheng J, and Tao D (2012). Color biological features-based solder paste defects detection and classification on printed circuit boards. *IEEE Transactions on Components, Packaging and Manufacturing Technology*, **2**(9), p1536–1544. https://doi.org/10.1109/TCPMT.2012.2205149
[9] Benedek C (2011). Detection of soldering defects in Printed Circuit Boards with Hierarchical Marked Point Processes. *Pattern Recognition Letters*, **32**(13), p1535–1543. https://doi.org/10.1016/j.patrec.2011.06.006
[10] Wu J and Chen Z (2011). A concise and rapid error correction algorithm in automatic optical inspection of solder paste. *Proceedings - 2011 International Conference on Internet Computing and Information Services (ICICIS)* p325–329. https://doi.org/10.1109/ICICIS.2011.85
[11] Lu Z, He Q, Xiang X, and Liu H (2018). Defect detection of PCB based on Bayes feature fusion. *The Journal of Engineering*, **2018**(16), p 1741–1745. https://doi.org/10.1049/joe.2018.8270
[12] Acciani G, Brunetti G and Fornarelli G (2006). Application of neural networks in optical inspection and classification of solder joints in surface mount technology. *IEEE Transactions on Industrial Informatics*, **2**(3), p 200–209. https://doi.org/10.1109/TII.2006.877265
[13] Zhang C, Shi W, Li X, Zhang H, and Liu H (2018). Improved bare PCB defect detection approach based on deep feature learning. *The Journal of Engineering*, **2018**(16), p1415–1420. https://doi.org/10.1049/joe.2018.8275
[14] Srimani P K and Prathiba V, . (2016) *Adaptive Data Mining Approach for Pcb Defect Detection and Classification.* Indian Journal of Science and Technology **9** (44). p 1-9. ISSN 0974-5645
[15] Yuk E H, Park S H, Park C-S, and Baek J-G (2018), "Feature-Learning-Based Printed Circuit Board Inspection via Speeded-Up Robust Features and Random Forest", *Appl. Sci* , **8**(6), p 932.

Hybrid Animation: Implementation of Motion Capture

Syed Muhammad Hazry Asraf[1, a], Nazreen Abdullasim[1, b] and Rusnida Romli[2, c,]

[1] Faculty of Information & Communication Technology, Universiti Teknikal Malaysia Melaka, Hang Tuah Jaya, 76100 Durian Tunggal, Melaka, Malaysia

[2] Centre of Excellence Advanced Communication Engineering Centre (ACE), School of Computer and Communication Engineering, Universiti Malaysia Perlis (UniMAP), Pauh Putra Campus, 02600 Arau, Perlis., Malaysia

[a] hazrysyed@gmail.com; [b] nazreen.abdullasim@utem.edu.my; [c]rusnida@unimap.edu.my

Abstract. In the 21st century, much progress has been made following the industrial revolution 4.0. Therefore, the field of animation has also been injected into technological advances. The transmission of information is difficult to convey accurately and interactively to users. In providing users with more exposure and convenience in a more attractive and convenient way. Therefore, animation and motion capture are the best options for use today. As a result, the results of this project can have a huge impact on the animation industry in making video or animation more interesting.

1. Introduction

Animation of two dimensional (2D) and three dimensional (3D) known as hybrid animation is one of the most popular techniques used by many designers or animators to make cartoons, hybrid animation short story and many more. The creation and presentation of hybrid animation is not a new thing in multimedia industry. Hybrid animation is a 2D and 3D measurement of X, Y and Z scale on a flat plane. It is also the creation of moving pictures in a two-dimensional environment, such as through "traditional" cell animation or in computerized animation software[1]. Although nowadays, 3D animation is the new type of animation in industry, yet there are more advantages in using hybrid animation. One of the advantages is the process to create and manipulate the objects are easier. Besides that, hybrid animation cartoons are more simple and funny look.

The two most exhausting processes in 2D animation production are, the generation of key-frames and in between frames. The enormous amounts of still animation are produced by most cartoon studios manually, which is time-consuming and heavy. Compared to 2D animation, three dimensional animation take advantages such as, ease of camera motion, complex lighting and shading, realism and high reusability of assets from scene to scene[2]. The mixing of 2D and 3D assets is a kind of special approach for hybrid animation. The blending should be consistent throughout the film when combinating the two types of elements.

Capturing movement (sometimes referred to as Mo-Cap or MoCap, for short) is the process of recording the movement of an object or person. It is used in military, entertainment, sports, medical applications, and for computer and robotics vision verification. In filmmaking and video game development, it refers to the recording of human actors, and uses that information to turn digital character models into 2D or 3D computer animations. When it involves face and fingers or captures subtle expressions, it is often referred to as performance capture. In many areas, motion capture is sometimes referred to as motion tracking, but in film and game making, motion tracking usually refers more to matching motion[3].

Figure 1. Innovation of texture from 2D animation to 3D animation (Animasia Studio, 2014)

2. Related works

Motion capture has been widely used in the movie and game industry to achieve maximum realism and fidelity and is seen as the Holy Grail in modern character animation, even though it usually requires intensive man power for clean up and postprocessing. Recent research and development has targeted automating and reducing the humans factors from the loop[4].

To address the problem of developing a better method for collecting motion capture data, the focus on video motion capture. A new factoring method is presented that allows one to solve a subject frame model from a series of video images. No special marking or summons is required. The remainder of this thesis discusses new techniques for using flexible motion capture data after they are collected. For cases of cyclic motion such as walking[5].

The experience working on the animation pipeline with a small art team at ICT makes believe that motion capture is certainly not going away anytime soon. However, this is not to say that high quality animations must rely on motion capture and animation artists. In fact, over the past years many advances based on motion capture have been made, concerning slight but precise modifications of an original motion or the parameterization of large motion databases[6].

Keyframe animation has been used by traditional animators (animators who draw the frames of the animation by hand) long before the advent of computers. In traditional animation, one normally draws the extremes, or important landmarks in the motion, called " keyframes", and then draws the intermediate frames using the keyframes as a guide. With the advent of computers and 3D graphics, people began using the computer as a tool to assist in creating an animation. A 3D model of a character is created in the computer, and the animator again specifies keyframes, this time not by drawing, but by posing the model in the computer[7].

Creating realistic character animations is still a great challenge in computer graphics. Currently, there are three methods by which this animation can be produced. The first is the main frame, where the animator provides a key key to the character in a particular frame. The second one uses physical simulation to drive the movement of the character as a result, because of the lack of control that is difficult to use and expensive and to the character it is not very successful. The last one is motion capture, which has been widely used to revive characters. It uses sensors placed on people and collects data that describes their movements as they perform the desired movements[8].

The main challenge facing animators is to produce character animations with realistic looks. Some seemingly simple behaviors like human walking are difficult to model because of their inherent instability. Motion capture is useful in many areas such as biomechanics, gait analysis, computer animation, signal recognition, sign language, music and fine arts performance. In sports science, motion capture data is used to analyze and refine the mechanics of prime athlete sequence, as well as monitor the progress of physical therapy recovery. Computer animation of human locomotives has become trending in recent years due to the need to use humans as artificial actors in a three-dimensional simulation environment. It also helps in the pursuit of action to learn the true patterns of robots. In the field of animation and the gaming industry, it is common that motion information is captured in different formats and is limited to specific tasks[9].

TABLE I. COMPARISON OF THE PREVIOUS PROJECTS

Index	Title	Implementation techniques in MoCap	Implementation techniques in Animation
[5][7]	Motion Capture Assisted Animation: Texturing and Synthesis	- Optical System - Magnetic systems - Passive optical system - Physical Simulation	- 3D graphic - 3D modeling - Keyframe
[4][6]	Motion Capture Based Animation for Virtual Human Demonstrators: Modeling, Parameterization and Planning	- Active Marker - Optical System - Motion Parametrization - Kinematic-based Models	- Keyframing - 2D animation - 3D animation
[8][9]	A framework to create humanoid robot animations through motion capture	- Optical System - Physical Simulation	- 3D animation - Humanoid robot

3. The structural design of the hybrid animation

In the development of 2D and 3D modeling, it goes through some of the processes commonly used by some animation industries.

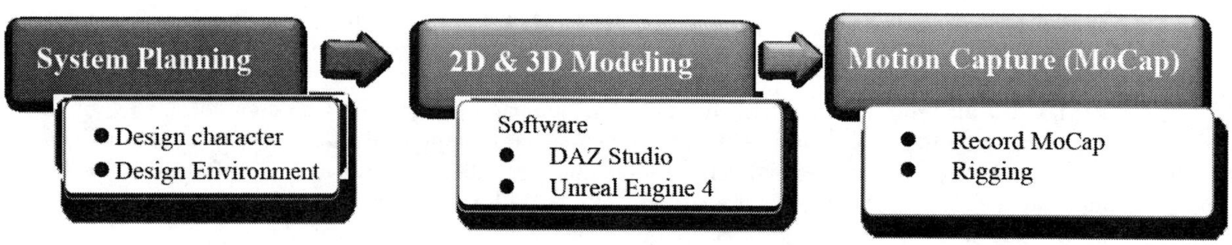

Figure 2. The structural design of the hybrid animation.

3.1. System planning

In animation design, the main factor is the design of the character and the environment. Character design plays an important role in the presentation of animations. It does not matter from the shape of the body to the facial expression of the character. Figure 3 shows the sketches and storyboards in the making of an animation. Storyboard is important as an introductory idea before doing character modeling.

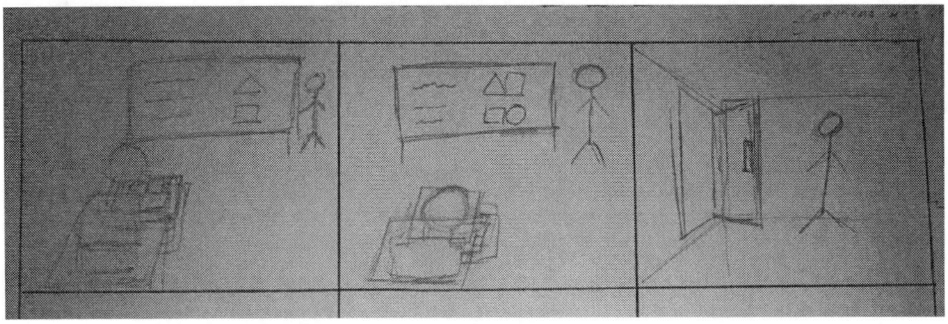

Figure 3. The storyboard in designing character and environment

3.2. 2D and 3D Modeling

The 3D Modeling is "the process of manually creating geometrical object models which meet desired design criteria" and DAZ Studio software are used. The features in the DAZ Studio software has a significant impact on the product development process, allowing improved quality, reduced cost, and aids products to get to market faster. Nowadays, there are many different 3D modeling software packages for different purposes are available such as Cinema 4D, Unity, and Unreal Engine 4. In this study Unreal Engine 4 software package from Epic Games has been chosen. It is one of the world's most advanced and user friendly game engine. The software has been used for building and visualizing the prototype modeling, Figure 4 and Figure 5 shows the modeling character using DAZ Studio and modeling environment using Unreal engine 4 software.

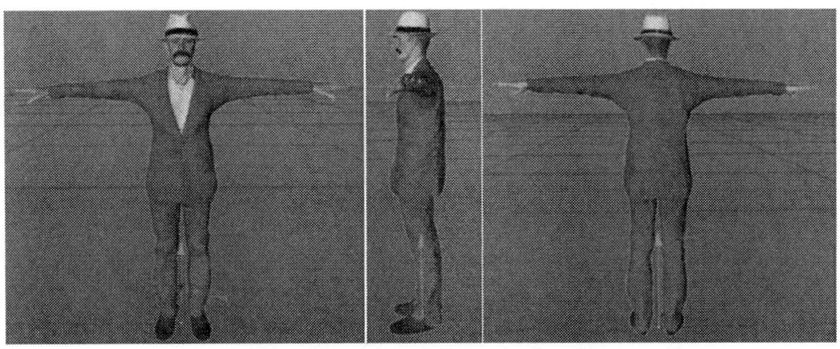

Figure 4. The character modeling.

Figure 5. The environment modeling

3.3. Motion Capture (MoCap)

There are several issues associated with the proper development of virtual simulation environments including the motion capture data transfer, rigging the motion capture skeleton with the modeling character. The Perception Neuron hardware will be used include with Axis Neuron software to capture the data about the physical movement. To create the intended Hybrid Animation using motion capture the steps shown in Figure 6 and Figure 7 will be followed:

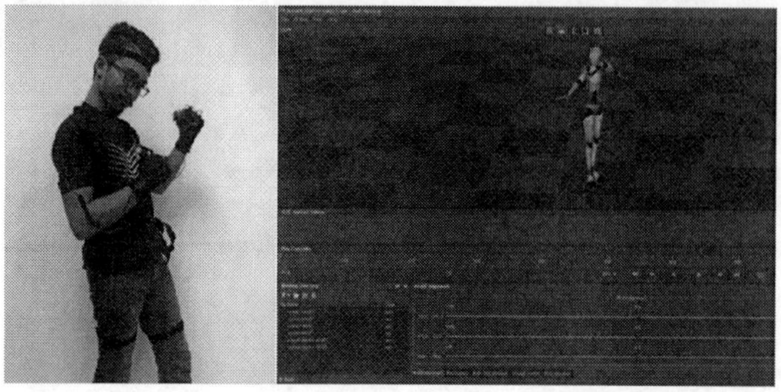

Figure 6. Perception Neuron hardware and Axis Neuron software.

- Put Perception Neurons into the body and ready to record the physical movement.
- Make move based on the storyboard.
- Save the data of the movement before transfer it to the Axis Neuron software.
- Open and play back the movement using the Axis Neuron software.
- Export the motion capture movement to .bvh format.

Figure 7 Rigging

- Import the .bvh format into Unreal Engine 4.
- Make sure the modeling character is .fbx format
- Rigging the motion capture skeleton with the modeling character.
- Export the overall project to the video format.

4. Result

Based on the final decision of the project, the techniques used in motion capture can be implemented in a hybrid of video and animation. Binding the actor with the bone and also how to animate the characters involved. This requires a lot time to complete this project. What is expected in the future is to improve the movement of characters and increase the number of characters in order for this animation to look more interesting.

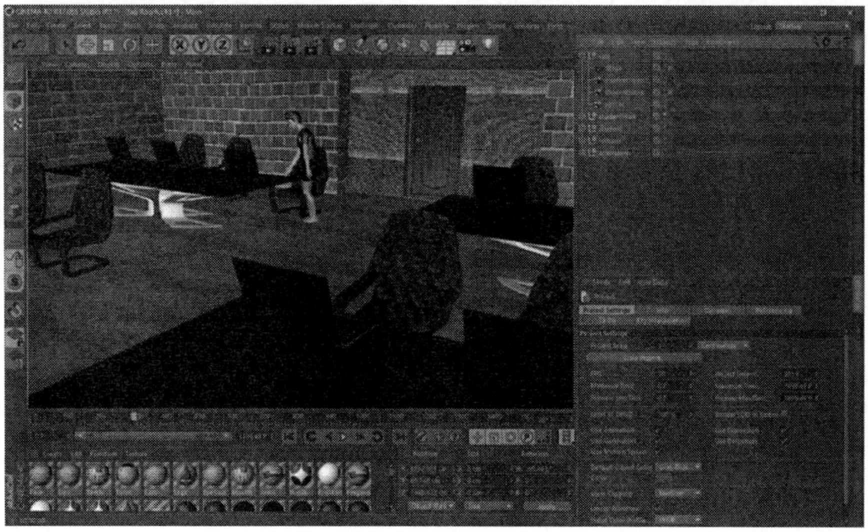

Figure 8. The hybrid animation environment

Integration is a process of combine close and smooth coordination between multiple departments, groups, organizations, systems, etc. Multimedia integration of animation, audio, graphics, text, and full motion video through computer hardware and software for education, entertainment, or training of a few group. Integration of animation, audio, graphics, text, and full motion video through computer hardware and software for education, entertainment, or training. Video a visual multimedia source that combines an array of images to form a moving image. Computer general purpose machines, usually consisting of digital circuits, which receive input, save, manipulate, and video files, or an electric signal, according to the instructions called the program. Hardware physical equipment that make up computer systems, such as circuit boards, keyboards and mice, monitors, printers, power supplies, storage devices. Unlike the software.

5. Conclusions

After being observed in this hybrid animation project. Among them, the animated shows are very short but full of the good content. This makes audiences feel satisfied when watching. In addition, the animation shown is in silent form, only motion and background sound. This animation is very interactive to the user due to the suitable color selection and lighting in this animation. Multimedia consists of orderly instructions and codes written by the programmer in any special computer language. Before using a software, the individual needs to evaluate more about a software in a subject to give a deep understanding.

References

[1] O'Hailey, T. (2010) Hybrid animation: Integrating 2D and 3D assets. Amsterdam: Elsevier Science.

[2] Lee,H-C., Chang, C-M., Chao, J-S., and Lin, W-T. (2017) 'Realistic Character Motion Response in Computer Fighting Game' Multimedia 2007. ISM 2007. Nineth IEEE International Symposium on, pp. 169-175, 2007

[3] David Noonan, Peter Mountney, Daniel Elson, Ara Darzi and Guang-Zhong Yang. A Stereoscopic Fibroscope for Camera Motion and 3D Depth Recovery During Minimally Invasive Surgery. In proc ICRA 2009, pp. 4463-4468.

[4] M. Mori, K.F. MacDorman, and N. Kageki. "The Uncanny Valley [From the Field]." Robotics Automation Magazine, IEEE, 19(2):98–100, june 2012.

[5] Pullen, Katherine Ann.Stanford University, ProQuest Dissertations Publishing, 2002. 3067919.

[6] Michael Gleicher. "More Motion Capture in Games – Can We Make Example-Based Approaches Scale?" In Arjan Egges, Arno Kamphuis, and Mark Overmars, editors, Proceeding

of the International Conference on Motion in Games, pp. 82–93. Springer-Verlag, Berlin, Heidelberg, 2008.

[7] K. Perlin and A. Goldberg. Improv: a system for scripting interactive actors in virtual reality worlds. Proc. SIGGRAPH 1996, pages 205-216, 1996.

[8] Fei, S. L. (2011). A framework to create humanoid robot animations through motion capture (Order No. 10514271). Available from ProQuest Dissertations & Theses Global. (1869988871). Retrieved from https://search-proquest-com.ezproxy.unimap.edu.my/docview/1869988871?accountid=33397

[9] Feng, Andrew W.; Xu, Yuyu; Shapiro, Ari. An example-based motion synthesis technique for locomotion and object manipulation. Proceedings - I3D 2012: ACM SIGGRAPH Symposium on Interactive 3D Graphics and Games (2012): 95-102.

IOP Conf. Series: Materials Science and Engineering **767** (2020) 012066 doi:10.1088/1757-899X/767/1/012066

Real-Time Drowsiness Detection System for Driver Monitoring

M Arunasalam[1], N Yaakob[1], A. Amir[1], M Elshaikh[1], N F Azahar[1]

[1] School of Computer and Communication Engineering, Universiti Malaysia Perlis (UniMAP), Pauh Putra Main Campus, 01000 Arau Perlis.

naimahyaakob@unimap.edu.my

Abstract. Nowadays, the rate of road accidents due to microsleep has been alarming. During microsleep, people might doze off without realizing it. For many decades, drowsiness detection system for vehicles was not among the major concerns though it turns out as one of imperative features that could have avoid microsleep and thus should be implemented in all vehicles in order to ensure safety of drivers and other vehicles on the road. To the best of our knowledge, enforcements on driving restriction during drowsiness state is yet to be implemented. The absence of such system in the current transportation systems expose drivers to great danger especially at night because accidents are highly likely to happen at night due to drowsy and fatigue drivers. Therefore, this project proposes a real-time drowsiness detection system for vehicles, featuring ignition lock to reduce accidents. An eye blink sensor is embedded in a wearable glasses and heart beat sensor is used to detect drowsiness level of drivers. The system also includes SMS notification system to relatives or friends with location details of the drowsy driver. This project is able to detect and react based on 3 levels of drowsiness by alerting the driver through buzzer. Ignition lock will be applied when high level of drowsiness is detected. Consequently, the vehicle will be slowed down and eventually stopped when dangerous level of drowsiness is detected as a safety precaution.

1. Introduction

Accidents involving vehicles on road are becoming unpreventable nowadays. World Health Organization (WHO) once had revealed that over 3400 people die on the world's road every day and billions of people are suffering non-fatal injuries and disabilities as a result of their injuries [1]. Traffic Safety Foundation study found that 37% of drivers have fallen asleep behind wheels. This might be due to drowsiness or mircosleep. Accidents could cause fatal injuries and increase mortality rates.

According to statistic of Indonesia National Police, from 2007 to 2010, 65% of accidents is due to human carelessness and human error. For instance, in 2011 during EID festival, a statistic reveals that most of the accidents in that country happened due to drowsiness of drivers (1018 cases) [2]. Studies also shows that most of the accidents around the world are due to drowsiness [3][8]. Drowsiness is commonly known as tiredness or sleepiness which leads to falling asleep at inappropriate times or situation. Drowsiness causes the driver to loss control on speed or become unaware of obstacles on road and this could end up in fatal accidents. Statistics show that fatal accidents happen every 52 seconds around the world [4]. Driving after working hours might cause drivers to be drowsy and careless due to tired of working or driving for long time. Some important problems have been outlined as follows:

- **Inefficient drowsiness detection systems**
  Many drowsiness detection systems have been developed and implemented but some of the systems do not have an effective system to alert the drivers. Those systems do not alert the driver by triggering buzzer but just display warnings on LCD using different colour of LEDs that light up according to drowsiness levels. This causes the driver to be unaware of the dangerous situation. This still putting drivers at high risk of accidents.
- **Existing systems does not provide real-time drowsiness detection**
  Moreover, only few of the existing systems provide real time drowsiness detection. Real time in this system means that once drowsiness is detected, there should be an immediate action taken without delay to save the drowsy driver from getting into an accident. Systems with real-time is an important factor for designing drowsiness detection.
- **Lack of notification in current drowsiness detection mechanisms**
  Besides that, most of the previous proposed systems do not send notification (SMS, email, etc) with location details of the driver to the driver's next of kin. This notification is critically important to ensure the safety of the associated driver. If nobody comes to pick up the driver from the location where the car has been stopped, the driver might continue to drive in drowsy state which is fatal to the driver and also other vehicles around. By sending out notification to their relative or friends, the safety of the driver can be ensured.

2. Objectives
Based on the stated problems stated, this project has outlined the following objectives.

- To develop a system that can automatically detect and perform appropriate immediate reaction based on different level of drowsiness.
- To design a system that will automatically control the speed of vehicles when drowsiness is detected and apply ignition lock when highest level of drowsiness is detected.
- To develop a drowsiness detection system with SMS notification that includes the location of the vehicle by using GSM and GPS modules.

3. Related Work
The tremendous increase in road accidents around the world has led to serious investigation among researchers. Many different solutions have been proposed to improvise safety of drivers. However, most of existing drowsiness detection systems still suffer some drawbacks. Most of the systems can only detects drowsiness and gives out warnings to alert drivers while still allow them to ignore the warnings and continue driving in drowsy condition. In this case, road accidents can still happen and the menaces imposed by these drivers to other road users are still high.

Most of the existing works are focusing on both internal and external vehicles to prevent accidents due to drowsiness. Studies carried out on the internal unit of vehicle are mainly concern on drivers. This is because the main reason of 20 % of the crashes and 30 % of fatal crashes around the world is caused by drowsiness of drivers and lack of driver's concentration [5]. The system inside will alert the driver by giving some warnings when the driver is already drowsy. Figure 1 shows various techniques on drowsiness detection methods.

3.1 Physiological Approach
Physiological approach refers to the method of detecting drowsiness by using data collected through sensors fixed on driver's body such as wearable device [6]. According to Anilkumar et al., 2017 [7], drowsiness can be detected using camera which is placed at an angle where drivers's face can be clearly captured. Then the video input is converted into frames and analysed using frame difference algorithm.

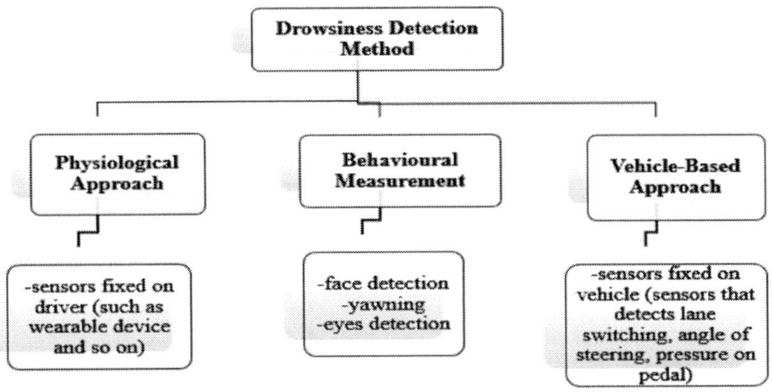

Figure 1. Various techniques to detect drowsiness (A. Lemkaddem et al., 2018)

Heart rate of driver is analyzed using ECG sensor for simulation purposes. Sample of ECG signal is taken to make comparison with normal heart rate (60 > rate < 100). Finally, the decision is made whether the driver is drowsy by matching both outputs from camera and heart beat rate. Driver is monitored continuously and warnings through audible speaker is given to prevent accidents if the driver seems to be drowsy. However, some limitation can be seen when generating audible warning is just not enough to ensure safety of driver.

Riztiane et al., 2017 [8] have proposed a system in which driver have to wear a wearable device to obtain the normal heart rate and the value is saved on the device. Once it is saved, the driver could turn on their front camera of mobile phone which will detect driver's face, mainly the eyes. Then, the state of eyes, whether it is opened or closed will be detected and the heart rate of driver is also monitored for making comparison between current heart rate while driving and the heart rate. Both the data will be combined to determine drowsiness state of driver. As the drowsiness value is indicated, three levels of alarm warnings such as normal, drowsy or sleepy will be determined. Moreover, if the eye blinking duration of driver exceeds the pre-defined condition and if drowsiness is also detected through heart beat rate, then the warning alarm will be triggered to alert the driver. However, driver needs to restrict his or her activities to a minimum movement so that the device can produce measurement accurately.

Another physiological approach is proposed by Gobhinath et al., 2017 [9] which includes a few sensors such as eye blink sensor, heart-beat sensor and body temperature sensor. In this project, eye blink rate, heart beat rate and body temperature of driver are monitored by using sensors separately and the measured values are sent to microcontroller and compared with the reference values which were recoded previously. If the values do not match the reference value, warning will be displayed on LCD. The car engine is controlled by microcontroller. If the sensors detect no output, the engine will be locked or stopped by microcontroller. Similar like [9], Leng et al., 2015 [10] also proposed a system which is mainly based on biomedical and motion sensors. This project exposes wristband consists of a few sensors such as *Photoplethysmogram* sensor and galvanic sensor. All the data collected will be sent to driver's smart phone through Bluetooth. Those data are analyzed with the motion sensor which are built-in accelerometer and gyroscope sensors. Sensors data including heart rate, pulse rate variability, respiratory rate, stress level and adjustment counter aid to detect drowsiness of driver. If drowsiness level reach 5 which indicate 80% drowsiness, a vibration of smart watch (wristband) will be triggered.

3.2 Behavioral Measurement

Behavioral measurement refers to the changes in reactions of driver's face which usually could be detected through yawning, monitoring state of eyes and mouth. Manu, 2017 [11] focused on face detection and skin segmentation method which only considers chromatic components. In this method, eyes of driver are tracked and the position of eyes is detected by Viola Jones and Sobel, a technique to detect edge in frame. Next, yawning detection is implemented as it detects the mouth region during yawning. Figure 2 shows the picture after the face and skin region of driver are detected while figure 3 shows the flow of eye detection.

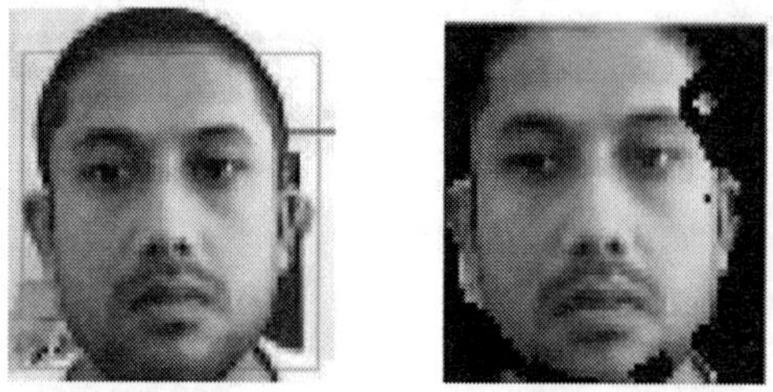

Figure 2. Face detection and skin region found on the picture (Manu, 2017)

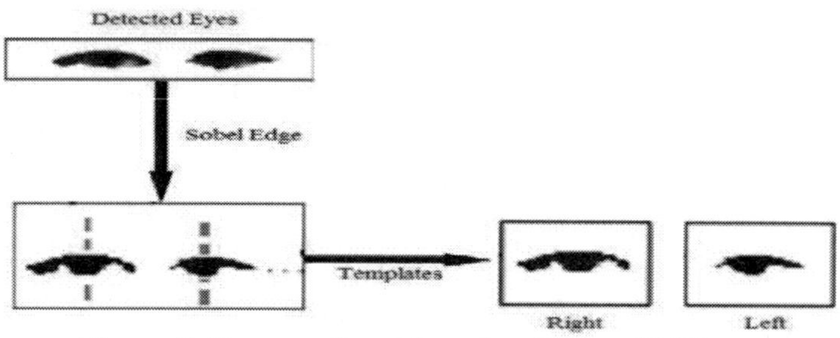

Figure 3. Process of eye detection (Manu, 2017)

This method eliminates influences of luminosity since it considers only chromatic components. It also use invariant to illumination and have a high drowsiness detection rate which reduces false alarms. However, this method is unable to predict drowsiness when head of driver was tilted right or left.

Another research by Naz et al., 2017 [12] analyses yawning, eye closure and head movement of driver to detect drowsiness. An alarm sounds badly if the driver leans his/her head when eyes are closed and a three level warnings are also blasted. In this method, warnings by level are given to irritate and alert the driver. Sequentially, the researchers in another journal used faced detection and eyes detection method to detect drowsiness of driver [13]. According to the system, it starts off with initialization phase in which face and eyes detection is conducted. Then, tracking of face and eyes in different frames carried out as to proceed to the next steps which are eyes state identification and driver state.

Besides, in another related journal by Akrout et al., 2016 [14], only yawning detection method is used for drowsiness detection. Akrout used Viola Jones technique for the stages of localization of face and mouth. Then, the stage of lips extraction is carried out followed by extraction of spatio-temporal descriptor in which calculation of the tracked lips surface and internal zone of mouth take place. However, yawn detection method fails most of the time due to bad localization of mouth. Other than that, if the drivers hide their mouth while yawning, the system will not function because yawning will not be detected.

3.3 Vehicle-based Approach

Vehicle-based approach refers to the method which detects drowsiness by using the data collected from the sensors fixed on the car itself. Those sensors monitors the changes such as lane deviation, angle of steering and acceleration on the pedal of the car [6]. An interesting research on safe driving by detecting lane discipline and driver drowsiness is proposed by Katyal et al. (2015) [15] which is a vehicle-based approach method to detect drowsiness. Firstly, the authors used two cameras in the vehicle, one is to detect driver's face and another is to detect road lane. When the car changes lane without signal, error is detected, and a warning is sent to the driver to avoid accidents. Next, another camera which monitors driver's face capture one frame of video, performs Viola-Jones method to find eye localization, performs edge detection of the eyes and performs Hough circle to detect drowsiness. This method is known as face detection. This is an outstanding method because it warns the driver by audible sound (alarm) when the driver crosses lane without signaling and it immediately applies brake to slow down the vehicle. Another advantage is that when the system detects more than 10 frames of closed eyes, alarm sounds or brake applied and the vehicle is slowed down. Through face detection method, eyes are also detected even in the low light intensity condition. However, the method also will not detect drowsiness if the eyes are detected less than 10 frames which is obvious that it took long time to decide drowsiness and this might lead to fatal accidents. This vehicle warning system had become one of the widely used method of real-time face detection. Nevertheless, this system could not detect drowsiness if the driver wear sun glasses while driving.

4. Proposed Real Time Drowsiness Detection System for Vehicles

This project includes some improvisation to overcome some limitations in the existing systems by including some extra features. This project is separated into 4 phases as illustrated in figure 4.

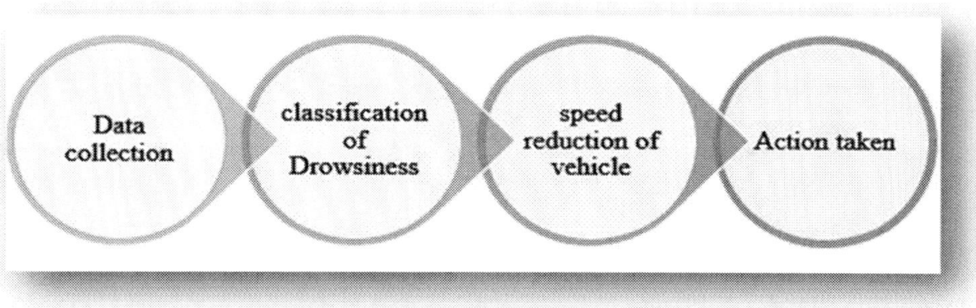

Figure 4. Phases in detecting drowsiness

The first phase involves data collection from all the sensors which are heart beat sensor and eye blink sensor. Once the driver is in the car, the eye blink sensor and heart beat sensor start to collect data by counting the eye blinking and heart pulse rate of driver. These sensors generate data for every 5 seconds to detect level of drowsiness. The data collected by eye blink sensor and heart beat sensor are used for further processing in order to detect driver's drowsiness which can be categorized into 3 types: awake, drowsy and sleepy.

Subsequently phase involves classification of drowsiness levels. As previously head movement, there are 3 levels of drowsiness which are awake, drowsy and sleepy. Normal alert for the driver is only effective when the driver is in less or drowsy state. However, if the driver is already in a sleepy state, the possibility for involving in fatal accidents is likely to be high, hence the alert alone might not be effective for drivers. In such case, extra precaution need to be implemented to ensure higher safety level for driver. Hence, the subsequent phase involves speed reduction of vehicle. When the danger level of drowsiness which is sleepy state is detected, an alarm will be triggered for 12 seconds and the speed of the vehicle will be slowly reduced. The speed will be reduced by 20km/h for every 4 seconds until the car is completely stopped. This phase implies the application of ignition lock which means the power, and engine of the vehicle will stop functioning once the ignition is locked. This is to avoid the driver from continuing driving in dangerous condition.

The last phase of the project is to implement the necessary auto-immediate action after the driver is found to be sleepy. This is to ensure the safety of driver. After the car has completely stopped, the system tracks the location of the vehicle through GPS module and sends message notification through GSM module to a mobile phone of next of kin of driver. This is to inform the driver's current condition and location. This will help them to pick up or fetch the driver from the current location. figure 5 shows an architecture of the proposed system.

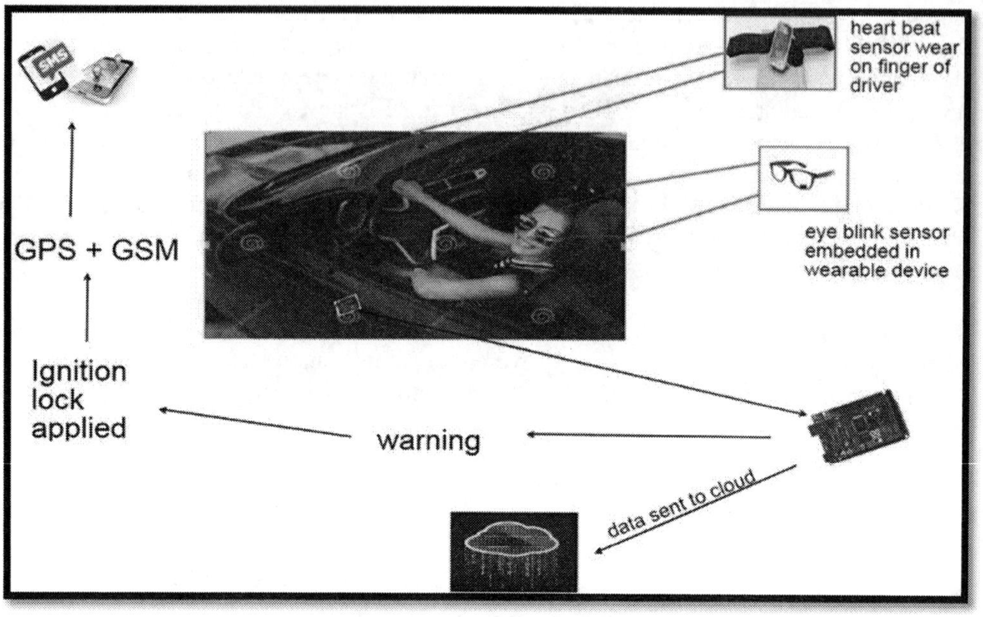

Figure 5. System Architecture

As shown, eye blink sensor is fixed in the wearable device (glasses) of driver and heart beat sensor is attached on driver's finger. The driver must wear the glasses and the heart beat sensor for safety reason. The enforcement to wear these two devices is as important as wearing seat belt in the car. Practically, in real implementation, the eye blink sensor will be embedded in the glasses so that it will not disturb the driver's view. However, for testing purposes, the eye blink sensor is visible in this project. Then, all the sensors are connected to Arduino Mega which receives and verifies all the data from the sensors and sends them to cloud data environment. The microcontroller also sends warning to driver every time when abnormalities or danger is encountered. Upon the last warning, where the driver is considered as sleepy, the GPS module generates the location of driver and a message is sent to their relatives or friends with location details to let them know the status of driver. Overall process flow and block diagram of the proposed system is shown in figure 6.

4.1 Detecting drowsiness by counting eye blinking

There are 3 conditions for eye blinking sensors which have been set to differentiate different levels of drowsiness. Setting many conditions like in this system will be helpful in monitoring each and every activities/ behaviors of driver in order to reduce the chance for accidents due to silly mistakes of drivers. Eye blink sensor starts to collect data from driver every 5 seconds. The detection of eye blinking is based on the presence of light or proximity. When the eyes are closed, the proximity is less whereas when they are opened, the proximity ranges up to 4.6.

At first, eye blink sensor checks whether the proximity is between 3.00 and 4.00 (3.00>=Proximity<=4.00). If it is not within this range, the eye blink sensor checks for further conditions such as proximity between 2.00 and 2.90 (2.00>=Proximity<=2.90) and proximity between 1.00 and 1.90 (1.00>=Proximity<=1.90). Each conditions refers to each level of drowsiness. First condition is considered as less drowsy (awake), while second and third condition is referred as drowsy and sleepy state respectively. The last two conditions need to be analyzed with another sensor as an extra confirmation of the driver's condition.

In first level of drowsiness (3.00>=Proximity <=4.00), an audible warning will be alarmed for about 4 seconds. In second level of drowsiness, if the proximity is 2.00>=Proximity<=2.90, the alarm will sound for about 8 seconds. In third level which represents high level of drowsiness, the proximity will be in the range of 1.00>=Proximity<=1.90 where the buzzer will be alarmed non-stop. When this third level of drowsiness (sleepy) is reached, it could be danger to driver since the driver would have fall asleep by that time. In such situation, the process of controlling speed of vehicle that is process C will be engaged (ignition lock control).

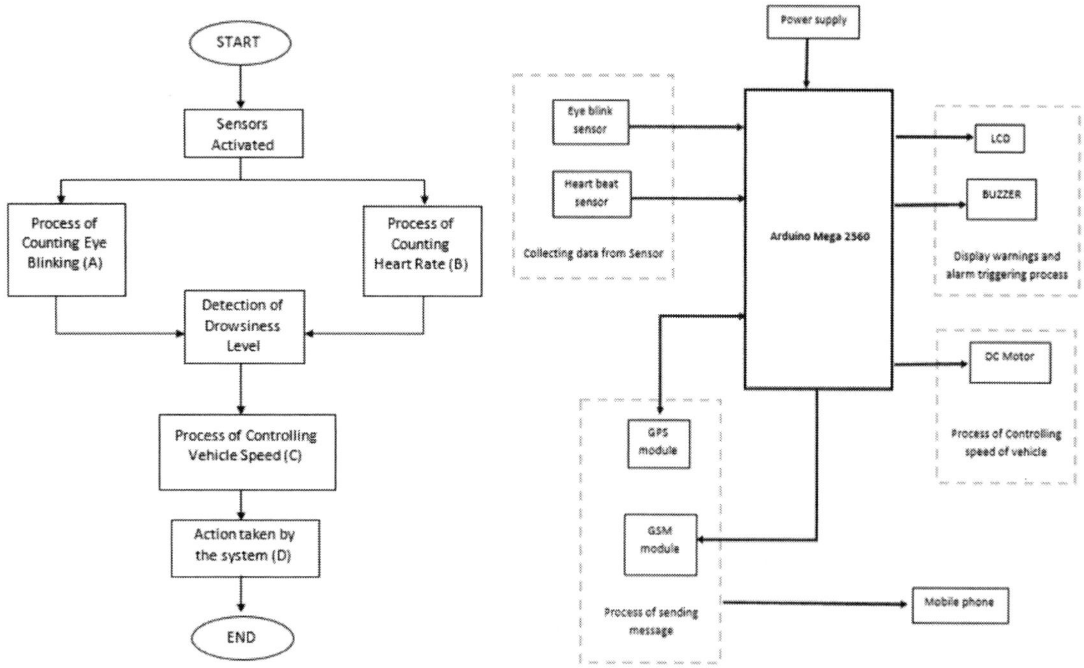

Figure 6. Overall flowchart and block diagram

4.2 Detecting drowsiness through heart beat sensor

The heart beat sensor collects pulse rate of driver every second. Normal heart beat rate of a person is between 60 to 100 per minute. The heart beat per minute varies depending on how active the person is. For example, heartbeat of athletes per minute will be faster and ranges from 40 to 60 under normal condition (PICKUT, 2017). Since this system is only covers normal people who has the range of heart beat rate between 60 to 100, the threshold level to detect heart rate is set to range of 60 to 100 (60>= r <=100). Therefore, if the heart rate of driver falls between 70 to 84, a 4 seconds alarm will be triggered, otherwise the sensor carries on collecting the heart beat rate of the driver. Next, if the pulse rate of the driver is between 65 to 69, the second alarm will be triggered for about 8 seconds. If the system does not activate alarm on time as per conditions set, the driver might drive very fast and rush in order to reach their destination as early as possible despite being tired.

However, at this stage, the first alarm only sounds when the eye blink sensor proximity is 2.00>=proximity<=2.90 and heart rate is 70>=r<=84. This indicates that the data of eye blink sensor and heart beat sensor need to be analyzed together to decide the state of the drive. Similarly, the second alarm will be activated when the eye blink sensor proximity is 1.00>=proximity<=1.90 and heart rate is 65>=r<=69. This is because the closing and opening of eyes might vary for different person. Therefore, eye blink sensor alone cannot accurately detect drowsiness as it need another sensor to be more accurate in detecting drowsiness. Heart beat sensor is selected to aid eye blink sensor as heart beat sensor detects internal changes of body (heart beat) and eye blink sensor detects external changes (eye blinking) of driver. After the second alarm which sounds for 8 seconds, sleepy stage is detected and the system proceeds with process C which is about controlling the speed of vehicle and ignition lock.

4.3 Ignition Lock Control
When the third level of drowsiness is detected (sleepy state), the speed of the vehicle is reduced gradually until the speed reaches 0km/h. This causes the car to stop slowly where ignition lock is applied. For safety reason, the speed is reduced for every 4 seconds. This is to maintain a gradual reduction in speed and to alert the other drivers on the road to be more cautious in order to avoid accidents. If the speed is immediately reduced to 0km/h without delay, there is high chance for fatal accidents to happen. The delay is also for the other vehicles to be aware about the current situation of the car. The subsequent step involves with process D which is the action taken by the system to ensure safety of driver.

4.4 Sending Notification
After the car is stopped (due to ignition lock), GPS module is activated in the car to detect the location of the vehicle and GSM module is used to send a short message containing the current situation of the driver with detailed location of the car to their relatives or friends so that they will be alerted with the current situation of driver.

5. Results and Discussion
This real-time drowsiness detection system is designed using Arduino IDE platform. The hardware used in this project involve eye blink sensor (QRD1114), heartbeat sensor or pulse sensor, Arduino Mega 2560 board, 16X2 LCD display, a piezo buzzer, LED, a GSM SIM900A module, DC motor which are attached to the mini smart car kit and a GPS (GY-NEO6MV2) module as shown in figure 7.

5.1 Data Collections
Figure 7 shows a prototype and device connection for the whole project. As shown in figure 7(c) and figure 7(d), an eye blinking sensor is attached with the wearable glass and must be worn by the driver while driving according to the scope of the project. The heart beat sensor will measure and sense the heartbeat of the driver and sent it to microcontroller for further processing. In actual implementation, the heart beat sensor will be placed on the seat belt so that when the driver fastens the seat belt, the sensor will be close to the heart and can get the heart rate easily. Other than that, this sensor also can be fixed around one of the fingers of the driver to read his/her heart beat.

5.2 Warning and Alerts
Some conditions have been set for the eye blink and heart beat sensors to detect driver's drowsy conditions. When the conditions are fulfilled, the buzzer will be triggered to alert the driver so that they can get awake and the same time warnings will be displayed on the LCD screen in the car. This will avoid the driver to sleep while driving and at the same time avoid accident from happening. At the very least, this alert system can wake them up to freshen up and remind them to be more careful while driving.

According to the pre-defined conditions, there are three conditions of eye blink sensor which detects drowsiness of driver. At lowest level of drowsiness which is when 3.00>=Proximity<=4.00, the driver will still be awake, but need to be more careful since this is initial indicator of drowsiness that could lead to worse condition. At the second level when 2.00>=Proximity<=2.90, the driver feels drowsy and the possibility of the driver to close his eyes frequently is very high. At this point, the buzzer is triggered for 8 seconds to alert the driver. Next, at the highest level of drowsiness (level 3) when 0.00>=Proximity<=1.59, the driver will feel sleepy as he might close the eyes for even longer. Since this is a very dangerous and critical condition, the buzzer will be alarmed non-stop and the speed of the car will be slowed down to finally stop the car. The level of drowsiness is determined using the amount of presence light on the eye blink sensor which is attached on the spectacles. When eyes are closed, it indicates the absence of light whereas the presence of light is indicated by the opened eyes of driver. Upon detection of each level, a buzzer sound will be triggered at different frequency to alert the driver. This is to alert and awake the driver from sleeping.

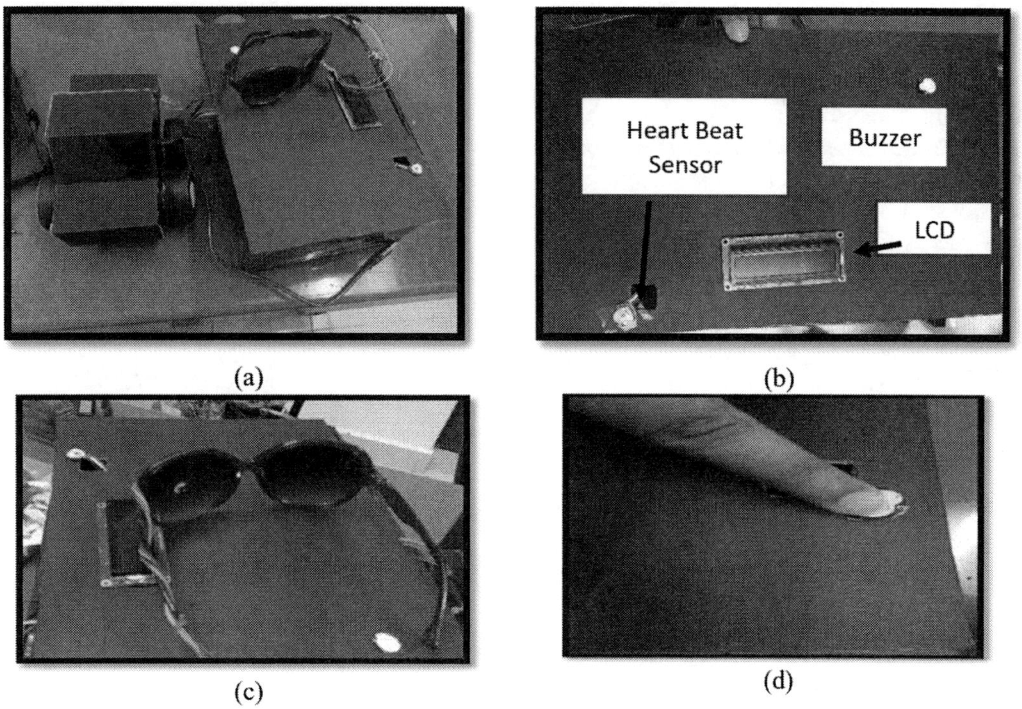

Figure 7. Prototype of the proposed work

Figure 8. GPS location and SMS

5.3 GPS tracking and SMS

When the third level of drowsiness is detected, the motor will slow down and the engine stops functioning. Then the GPS module detects the location of the vehicle and a message will be sent to the driver's next of kin through GSM module with the location details. Figure 8 shows the outcome of GPS module latitude, longitude of driver's current location (e.g. 6.459648 and 100.361457). When these latitude and longitude are checked on Google MAPS, it clearly shows the current tested location of PFi2, Universiti Malaysia Perlis (UniMAP). Figure 8(b) shows an SMS resulted at 3rd level of drowsiness which includes detail location of the vehicle. The module was tested at Kolej Kediaman Tuanku Tengku Fauziah (PFi2) and when the location is tracked, a message which includes the location of the driver was sent to the driver's relative or friends.

6. Conclusion

This real-time drowsiness detection system for vehicles has overcome the problems and limitations of various similar projects that had been proposed earlier. By developing this system, basic road security against drowsy drivers can be established with low cost and effective. This project includes drowsiness detection, alert system for drivers (triggering alarm), speed control once for high level of drowsiness and sending message with the details location of driver. The real implementation of this project can indirectly reduce the number of accidents due to drowsiness or sleepiness of drivers.

References

[1] Rizwan O, Rizwan H, and Ejaz M (2013). "Development of an efficient system for vehicle accident warning", *IEEE 9th Int. Conf. on Emerging Technologies (ICET)*. https://doi.org/10.1109/ICET.2013.6743484

[2] Gao Z, Le D, Hu H, Yu Z and Wu X (2017). "Driver Drowsiness Detection Based on Time Series Analysis of Steering Wheel Angular Velocity", *9th Int. Conf. on Measuring Technology and Mechatronics Automation (ICMTMA)* p 99–101.

[3] Colic A (2014). *Design and implementation of driver drowsiness detection system*, p 133. Retrieved from https://search-proquest com.etechconricyt.idm.oclc.org/docview/1673145106?accountid=189277

[4] Khunpisuth O, Chotchinasri T, Koschakosai V and Hnoohom N (2016). "Driver Drowsiness Detection Using Eye-Closeness Detection". *12th Int. Conf. on Signal-Image Technology & Internet-Based Systems (SITIS)*, p661–668. https://doi.org/10.1109/SITIS.2016.110

[5] Sigari M H, Fathy M and Soryani M (2013). "A driver face monitoring system for fatigue and distraction detection". *Int. Journal of Vehicular Technology*, **2013**(1).

[6] Lemkaddem A, Delgado-Gonzalo R, Turetken E, Dasen S, Moser V, Gressum C, Sola J, Ferrario D and Verjus C (2018) "Multimodal Driver Drowsiness Detection: A Feasibility Study", *IEEE EMBS Int. Conf. on Biomedical & Health Informatics (BHI)*, Las Vegas, NV p 9-12.

[7] Anilkumar C V, Mansoor A, Sahana R, Thejashwini R and Anisha P S (2017). "Design of drowsiness, heart beat detection system and alertness indicator for driver safety". *IEEE Int. Conf. on Recent Trends in Electronics, Information and Communication Technology (RTEICT 2016) - Proceedings*, p 937–941.

[8] Riztiane A, Hareva D H, Stefani D and Lukas S (2017) "Driver Drowsiness Detection Using Visual Information On Android Device", *Int. Conf. on Soft Computing, Intelligent System and Information Technology (ICSIIT)*, p 283–287.

[9] Gobhinath S, Apama V and Azhagunacchiya R (2017) "An Automatic Driver Drowsiness Alert System By Using GSM", *Proceedings of 11th Int. Conf. on Intelligent Systems and Control (ISCO)*, p 125–128.

[10] Leng L B, Giin L B and Chung W Y (2015) "Wearable driver drowsiness detection system based on biomedical and motion sensors", *2015 IEEE SENSORS - Proceedings*.

[11] Manu B N (2017) "Facial features monitoring for real time drowsiness detection", *Proceedings of the 12th Int. Conf. on Innovations in Information Technology (IIT)* p 78–81.

[12] Naz S, Ahmed A, ul ain Mubarak Q and Noshin I (2017) "Intelligent driver safety system using fatigue detection", *19th Int. Conf.on Advanced Communication Technology (ICACT)*, p 89–93.

[13] Alshaqaqi B, Baquhaizel A S, Ouis M E A, Boumehed M, Ouamri A and Keche M. (2013) "Driver drowsiness detection system", *8th Int. Workshop on Systems, Signal Processing and Their Applications (WoSSPA)*, p 151–155.

[14] Akrout B and Mahdi W (2016) "Yawning Detection by the Analysis of Variation Descriptor for Monitoring Driver Drowsiness", *Int. Image Processing Applications and Systems Conference*, p 1–5.

[15] Katyal Y, Alur S, and Dwivedi S (2015) "Safe driving by detecting lane discipline and driver drowsiness", *Proceedings of the IEEE Int. Conf. on Advanced Communication, Control and Computing Technologies (ICACCCT)*, p 1008–1012.

Machine Learning Algorithms for Optic Pathway Disease Diagnostics: A Review

M Abu[1], N A H Zahri[1], A Amir[1], M I Ismail[2]

[1]Advanced Communication Engineering Centre (ACE), School of Computer and Communication Engineering, Universiti Malaysia Perlis, Malaysia
[2]School of Mechatronic Engineering, Universiti Malaysia Perlis, Malaysia

adilahhanin@unimap.edu.my

Abstract. Most of people are unaware that some of the indicators of optic pathway diseases such as stroke or tumor can be detected from the loss part of human vision, or referred as visual field defect. Ophthalmologist will manually examine the site, size and margin of the lesion from patient's visual field points mapped by Humphrey Field Analyzer. Different site, size and margin of lesion indicates different type of defects and disease that associated with it. Therefore, an effective automated detection mechanism of multi class visual field defect is in demand to help decision making by ophthalmologist. In this paper, we review multiple techniques of supervised and unsupervised learning method for detection of optic pathway disease.

1. Introduction

Studies on optic pathway disease detection have been conducted for many years not only by medical researchers but also by researchers from the engineering field. Many people are unware that some of these optic pathway disease are one of many indicators of serious illness such as stroke and tumors. An optic pathway consists of a series of cells and synapses that carry visual information from the environment to the brain for processing [1], and the defect to the optic pathway caused by tumors or any other brain related illness can interrupt the higher order processing of visual input. Other than detection of abnormalities in retina structure, researchers are developing a diagnosis system for specific eye disease such as glaucoma [2]–[5], diabetic retinopathy [6]–[8], and age-related macular degeneration [9], [10].

Most of the previous works are working on automated detection of eye disease in order to provide more accurate and consistent mechanism to aid ophthalmologists to diagnose and prioritize based on the severity of the patient's conditions. Therefore, effective decision making mechanism for medical diagnostic can be produced to deal with currently rapid growing medical data and improve patient healthcare.

In this work, existing machine learning methods for eye disease detection (using visual field data, fundus images and OCT images as a dataset) will be reviewed, analysed from a different perspectives and finally concluded to summarize the current advancement of detection mechanism for eye disease.

2. Existing Machine learning Algorithms

Machine learning is a part of an Artificial Intelligent application that provides systems ability to learn automatically and improve from experience without being explicitly programmed [11]. Machine learning also learns the information directly from data without relying on a predetermined equation as a model. The performance of machine learning can be improved by adding more data [12], [13]. Machine learning has several main categories, for example, supervised, semi-supervised, unsupervised, deep hierarchical learning/ deep structured learning, reinforcement and active learning algorithms [14], [15].

2.1. *Supervised Learning*

There are two types of feature learning frequently used to train data set for example supervised learning and unsupervised learning. Supervised learning is the machine-learning procedure of deriving a function from supervised training data. Training data in supervised learning include a set of model that match input subjects and desired output. A supervised learning algorithm analyses the training data and produces a deriving function, which is called classifier or a regression function. The function should predict the correct output value for any valid input object. This requires the learning algorithm, to sum up from the training data to concealed circumstances in a sensible manner [15], [16]. For supervised learning, the inputs data for the training process will be labeled in different classes using classification methods. Then, the trained system generates actions that allocate hidden inputs to these classes. This process is called multi-labeling process in supervised learning [14], [15], [17].

2.2 *Unsupervised Learning*

For unsupervised learning, it is different compared to supervised learning because it will not label the data into any class. Therefore, the systems are intended to outline information to itself with the purpose of the systems to learn better features by themselves. This is called feature or representation learning [14], [18]. One advantage of this approach is there is no need to manually label the data and as the consequences, the networks can be trained with a large number of unlabelled data compare to supervised learning that only can train a small amount of data [18]. Besides that, this method also can suffer from misprediction in some regions that have various conditions [19].

2.3 *Deep Learning*

Neural network works a similar way as the human brain performs a given task [20]–[23]. Deep learning is an advanced phase of machine learning that utilized in neural networks for learning and prediction of data [14], [24]. It a procedure that avoids most predictive features to learn directly from the data in the large data set [6], [12]. For image classification, feature extractors obtained from a pre-trained deep learning model perform well on different types of tasks, including tasks that are very different from the original task for which the feature extractors were trained [25]. Deep Learning performed very well in terms of accuracy and time processing to process a mass amount of information compared to human. However, as the samples and parameter increase, the artificial selection of features will restricted the performance of classification. Therefore, deep learning is used to find the most distinguished features from thousands of parameters automatically [26]. There are several types of neural network models that had been used for deep learning [27].

- Multilayer Perceptron (MLP): It is a feed-forward neural network and it has multiple hidden layers between the input layer and output layer.
- Autoencoder (AE): An unsupervised model, it reconstruct the input data in the output layer. The middle layer is used as a salient feature that represent the input data. Types of autoencoder are denoising autoencoder, marginalized denoising autoencoder, sparse autoencoder, contractive autoencoder and variational autoencoder [28], [29].
- Convolutional Neural Network (CNN): A feed forward neural network, which build with convolution layers and pooling operations. It can capture the global and local features. It significantly enhancing the efficiency and accuracy. It performs well in processing data with grid-like topology. [14], [19], [30]–[32]

- Recurrent Neural Network (RNN): Suitable for modelling sequential data. RNN has loops and memories so it can remember former computations different from feedforward neural network. Long Short Term Memory (LSTM) and Gated Recurrent Unit (GRU) network is used to overcome the vanishing gradient problem [33]–[35].
- Restricted Boltzmann Machine (RBM): A two layer neural network consisting of a visible layer and a hidden layer. Restricted here means that there are no intra layer communications in visible layer or hidden layer [36].
- Neural Autoregressive Distribution Estimation (NADE): An unsupervised neural network built atop autoregressive model and feedforward neural networks. It is a tractable and efficient estimator for modelling data distribution and densities.
- Adversarial Networks (AN): A generative neural network which consists of a discriminator and a generator. The two neural networks are trained simultaneously by competing with each other in a minimax game framework.
- Attentional Models (AM): Differentiable neural architectures that operate based on soft content addressing over an input sequence. Attention mechanism is typically ubiquitous and was incepted in Computer Vision and Natural Language Processing domains. However, it has also been an emerging trend in deep recommender system research.
- Deep Reinforcement Learning (DRL): Reinforcement learning operates on a trial-and-error paradigm. The whole framework mainly consists of the following components: agents, environments, states, actions and rewards. The combination between deep neural networks and reinforcement learning formulate DRL, which have achieved human-level performance across multiple domains. Deep neural networks enable the agent to get knowledge from raw data and derive efficient representations without handcrafted features and domain heuristics.

3. Detection of Eye Disease by Using Different Machine Learning Algorithm

There are many existing reviews and surveys on some diseases and techniques such as CNN, SVM, GEM, VIM, PNN and Bayesian classifier [14], [37]–[39]. In this paper, we highlight the works published on supervised and unsupervised techniques. We divide all these algorithms into categories of eye disease detection, which are glaucoma, diabetic retinopathy, cataract, hard exodus, and macular-edema.

3.1 Supervised Learning

Convolution Neural Network is a popular algorithm that had been used in supervised deep learning. Li *et al.*, 2018 had build 22 layers of CNN, which contain Inception-V3 CNN model to detect GON in 48116 fundus images, 8000 from the images is used for validation. The accuracy is 92.9%[40]. Raghavendra *et al.*, 2018 used eighteen layers of CNN to detect glaucoma. The accuracy is 98.13% by using 1426 (589: normal and 837: glaucoma) fundus images [41]. Abbas, 2017 using ORIGA and SCES dataset and the accuracy obtained is 99% for both ORIGA and SCES datasets. The method used by Abbas, 2017 is Glaucoma-deep, which is the integration of CNN, DBN and Softmax deep-learning classifiers [42]. Asaoka *et al*, 2016 used deep feed-forward neural network (FNN) classifier to detect glaucoma in 279 visual field images and the accuracy obtained is 92.6%[43]. Chen *et al*, 2015a and 2015b had built six layers deep CNN for their research. The dataset used by them is ORIGA and SCES. For the ORIGA dataset, the training set contains a random selection of 99 images from the whole 650 images, and the remaining 551 images are used for testing. For the SCES dataset, 650 from the ORIGA dataset will be used for training and 1676 is used for testing. The accuracies for 2015a are 83.1% for ORIGA dataset and 88.7% for SCES dataset and for 2015b are 83.8% for ORIGA dataset and 89.8% for SCES dataset [30], [44]. The summarization of works on glaucoma diseases is shown in **Table 1**.

Table 1. Supervised Machine Learning Method for Glaucoma Detection

Machine Learning Techniques	Author	Year	Disease	Dataset	Accuracy
Deep CNN with Inception-V3 architecture	Li *et al* [40]	2018	Glaucomatous Optic Neuropathy Disease	48116 fundus images	92.9%
Deep CNN 18 layers	Raghavendra *et al* [41]	2018	Glaucoma Disease	1426 fundus images	98.13%
Glaucoma-deep(CNN,DBN and Softmax deep learning)	Abbas [42]	2017	Glaucoma Disease	ORIGA & SCES	99%
Deep Feed Forward Neural Network	Asaoka *et al*[43]	2016	Glaucoma Disease	PPGVFs	92.6%
Deep CNN with 6 layers- 5 multilayer perceptron layer,1 fully-connected layer	Chen *et al* [44]	2015b	Glaucoma Disease	ORIGA SCES	83.8% 89.8%
Deep CNN with 6 learned layer- 4 convolutional layer,2 fully-connected layer	Chen *et al* [30]	2015a	Glaucoma Disease	ORIGA SCES	83.1% 88.7%

For diabetic retinopathy, many of the supervised learning algorithms used Kaggle/ Messidor/Eyepacs images as training dataset. Liu et al had proposed WP-CNN method, where this method applies multiple weighted path into CNN. In WP-CNN, multiple path weight coefficients are optimized by back propagation, and the output features are averaged for redundancy reduction and fast convergence. The dataset they used is taken from Kaggle and the accuracy obtain are 94.23%[45]. Lam *et al.*, 2018 used 22 layers GoogleNet CNN model and Kaggle's dataset to perform their research. They also do transfer learning pre-trained for the model to improve the classification. The high accuracy achieve in this research is 96%[46]. Dutta et al, 2018 also used Kaggle's dataset. The dataset will undergoes some image processing process and Fuzzy C-Means to identify the proper class of severity of the dataset. Then, it will be trained and classified using CNN model(VGGNet) and obtain accuracy 72.5%[47]. Xu *et al.*, 2017 also using Kaggle's dataset. The method they used is CNN and Softmax layer is added at the end of their CNN layers. The accuracy they obtain is 94.5%[48].

Kanungoet *et al.*, 2017 used Inception V3 CNN model techniques to evaluate different number of batch size, epoch and dataset. The highest accuracy they obtain are 88% by using 128 batch size, 200 epoch and 40,000 DR dataset that obtain from Kaggle[49]. Zeng et al, 2016 used EyePACS dataset that obtain from Kaggle. It contains 35126 high resolution fundus photographs taken under a range of imaging conditions. Inception V3 CNN model will be used as their method to automatically detect RDR. The accuracy of their research are 82.9% [50]. Pratt *et al.*, 2016 used Kaggle's dataset, 80000 images for training and 5000 images for validation. They using eight layers CNN for detection by using Keras model (Theano) to build their CNN. The accuracy obtained is 75% [51]. Abramoff *et al*, 2016 compare performance of a deep-learning enhanced algorithm for automated detection of diabetic retinopathy (DR), to the previously published performance of that algorithm, the Iowa Detection Program (IDP) without deep learning components on Messidor taken from Kaggle. It was shown that the algorithm with CNN, a deep learning method can achieve a high performance of 96.8% [52]. Gulshan *et al*, 2016 used 9963 images from EyePACs and 683 images from Messidor as their dataset taken from Kaggle. The method they used Inception V3 CNN model to detect diabetic retinopathy. The accuracy for EyePACs is 98.1% and for Messidor is 98.5% [6]. The summarization of works on diabetic retinopathy is shown in **Table 2**.

Table 2. Supervised Machine Learning Method for Diabetic Retinopathy Detection

Machine Learning Techniques	Author	Year	Disease	Dataset	Accuracy
WP-CNN (applies multiple weighted path into CNN)	Liu *et al* [45]	2019	Diabetic Retinopathy	Kaggle	94.23%
22 layers of CNN following GoogleNet CNN model	Lam *et al*[46]	2018	Diabetic Retinopathy	Kaggle	96%

FCM+CNN(follow VGGNet CNN model)	Dutta *et al*[47]	2018	Diabetic Retinopathy	Kaggle	72.5%
Deep CNN adding Softmax at the last layers.	Xu *et al* [48]	2017	Diabetic Retinopathy	Kaggle	94.5%
Inception V3 CNN model (Test on different size of batch size, epoch and dataset)	Kanungo *et al* [49]	2017	Diabetic Retinopathy	Kaggle	88%
Inception V3 CNN model	Zeng *et al* [50]	2016	Diabetic Retinopathy	Kaggle (EyePACs)	82.9%
8 layers Deep CNN using Keras Model (Theano)	Pratt *et al* [51]	2016	Diabetic Retinopathy	Kaggle	75%
Iowa Detection Program (IDP) + CNN	Abramoff *et al* [52]	2016	Diabetic Retinopathy	Kaggle	96.8%
Deep CNN Inception V3	Gulshan *et al*[6]	2016	Diabetic Retinopathy	Kaggle (EyePACS & Messidor)	98.1% 98.5%

Dong *et al.*, 2017 had split 7851 fundus images into 70% for training and 30% for testing. They compared a deep learning method with classic machine learning. For feature learning, Dong *et al* 2017 compare Caffe and Wavelet feature. They combined this feature learning with classification method, SVM and deep learning classifier, softmax classifier to detect cataract on the eyes. In their research, the highest accuracy is 94.07% by using Caffe as feature learning and Softmax as classifier [26]. Gao *et al*, 2015 used 5378 images from the ACHICO-NC dataset. The dataset is categorised based on grade 1=94 images, 2=1874 images, 3=2476 images, 4=897 images, and 5=37 images for training and the remaining will be used as testing. The method used in their research is Convolution-recursive Neural Network. In this method, the patches will be randomly selected using the pre-training convolutional neural network (CNN), then, feeding the filters into a CNN layer by generating local representations of each image, and lastly, multiple recursive neural networks (RNNs) with random weight will learn hierarchical feature representations. The result obtained are 99.0% when the decimal grading error less than one [33].

Table 3 shows the supervised machine learning method to detect cataracts.

Table 3. Supervised Machine Learning Method to detect Cataract

Machine Learning Techniques	Author	Year	Disease	Dataset	Accuracy
Caffe+CNN Caffe+SVM Wavelet+CNN Wavelet+SVM	Dong *et al*[26]	2017	Cataract	7851 fundus image	90.82% 84.17% 82.94% 84.7%
Convolution-recursive Neural Network(CRNN)	Gao *et al*[33]	2015	Grade nuclear cataract	ACHICO-NC	99%

3.2 Unsupervised Learning

For unsupervised learning, Amil *et al.*, 2019 used a large database of anterior chamber OCT images. The algorithm used is t-SNE with the Helinger technique as feature extraction. The accuracy obtained is 81.0% [54]. Montuoro *et al.*, 2017 used 20,000 OCT B-scans from 100 scans of patients. The method used to detect macular edema in OCT images is the PCA method. The accuracy achieved is 78.0% [55]. Yousefi *et al.*, 2016 and Yousefi *et al.*, 2014 used 999 visual field images as their dataset. They had compared two methods in their research, Gaussian Mixture Model with Expectation Maximization (GEM) and Variational Bayesian Independent Components Analysis Mixture Model (VIM) in 2016 and only using GEM in 2014 to cluster the dataset. The accuracy is 93.8% for GEM and 97.0% for VIM in

2016 and the accuracy of GEM is 93.8% for 2014 [56], [57]. Bowd *et al.*, 2014 used 1190 eyes with normal FDT and 786 eyes with abnormal FDT. The algorithm used is VIM and the accuracy obtained for normal is 93.1% and for abnormal is 82.8% [2]. Priya & Arjuna, 2013 used 350 fundus images to detect diabetic retinopathy, 100 images used for training and 250 images used for testing. In their research, three methods will be used to detect glaucoma in fundus images. The methods used are robabilistic Neural Network (PNN), Bayes classifier and SVM. The accuracy obtain are 89.6 % for PNN, 94.4% for Bayes Classifier and 97.6% for SVM [58]. **Table 4** shows the unsupervised machine learning method for multiple eye disease detection.

Table 4. Unsupervised Machine Learning Method for Eye Diseases Detection

Machine Learning Techniques	Author	Year	Disease	Dataset	Accuracy
t-SNE+Helinger technique	Amil *et al*[54]	2019	Glaucoma Disease	Anterior chamber OCT images	81.0%
PCA	Montuoro *et al*[55]	2017	Macular edema	20 000 OCT images	78%
GEM & VIM	Yousefi *et al*[56]	2016	Glaucoma Disease	Visual field images	93.8% 97.0%
GEM	Yousefi *et al*[57]	2014	Glaucoma Disease	Visual field images	93.8%
VIM	Bowd *et al*[2]	2014	Glaucoma Disease	1190 fundus images	93.1%
PNN Bayesian classification SVM	Priya & Arjuna[58]	2013	Diabetic retinopathy	350 fundus images	89.6% 94.4% 97.6%

4. Discussion and Analysis of Machine Learning Algorithm

For the diagnosis of glaucoma, diabetic retinopathy, cataract, and macular edema, several machine-learning algorithms are observed to perform very well. From existing literature, it shows that deep learning is the most frequent method used in recent researches to detecting eye disease. By using this technique, the performance of the systems detected have improved significantly. By using the CNN method, the accuracy of the system can be improved by increasing the layer of CNN model. In addition, there also other researchers build their CNN by following the existing CNN model, for example, VGG, ResNet, Inception-V3 and GoogleNet and combine CNN algorithm with other deep learning algorithm for example RNN. In this review paper, there are seventeen works used deep learning as their method. 11 from those works achieved accuracy above 90% and others achieved accuracy that varies between 80% to 90%. For unsupervised learning, most method used to build this system is Bayesian classifier. It shows that the performance of unsupervised learning can also achieved high accuracy. There are six works of unsupervised learning have been reviewed. Three works using Bayesian Classifier, and achieve above 90% and another three works used PCA, Gaussian and t-SNE with Helinger techniques and achieve accuracy 78% above. Figure 1 shows the performance of previous works mentioned in the above regardless the type of disease, dataset used and experimental parameters. This figure is constructed to generally illustrate the effectives of deep learning method, supervised and unsupervised learning that have been studied and experimented in many previous works of eye disease detection.

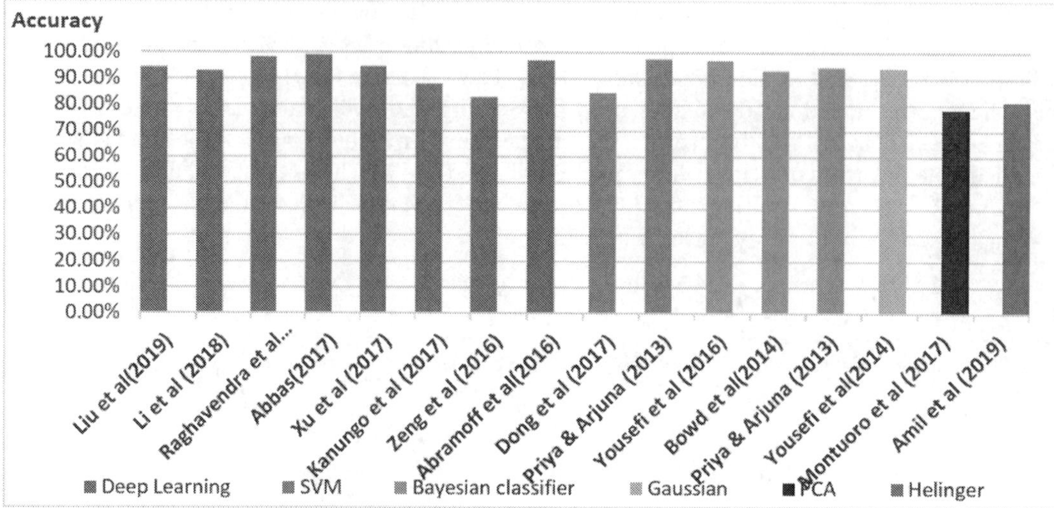

Figure 1. Accuracy of Machine Learning techniques

5. Conclusion

Machine Learning had been used in many applications, for example, image detection, data mining, natural language processing, and disease diagnostics. This paper provides a survey/review of different machine learning techniques (algorithm) for diagnosing eye diseases such as glaucoma, diabetic retinopathy, cataract, hard exodus, and macular edema. Many algorithms have shown good results because they identify the attribute accurately. From Figure 1, we can conclude that eye disease detection performs well on deep learning regardless of its dataset and method used. From our investigation, it has shown that by using deep learning, the lowest accuracy achieved are 75% and can reach maximum accuracy of 99%. Abbas, 2017 had integrates CNN with DBN and adds the Softmax layer in his algorithm to improve their accuracy. For the unsupervised method, the Bayesian classification are mostly used for detecting multiple eye diseases. From examination of the previous works, the accuracy of using Bayesian classifier method achieved above 90%. It had shown by Yousefi *et al*, 2016, by using Bayesian Classifier, they have improved previous work that used Gaussian method and increase the performance from 93.8% to 97.0%. This paper have reviewed many of detection techniques to aid the diagnosis process of many types of eye disease. From our observation, many of recent works are focused on enhancing hyper parameter of deep learning method to improve accuracy and the obtained result are significantly high and promising for many types of eye disease detection.

Acknowledgement

This research was supported by Fundamental Research Grant Scheme (FRGS 1/2018). We are thankful to our colleagues who provided expertise that greatly assisted the research, although they may or may not agree with all of the interpretations provided in this paper. We are also grateful to Advanced Communication Engineering Centre (ACE) under School of Computer and Communication Engineering, Universiti Malaysia Perlis, which has provided laboratory facilities during the course of this research.

References

[1] L. A. Remington, Clinical Anatomy and Physiology of the Visual System, Third Edit. 2012.

[2] C. Bowd et al., "Glaucomatous patterns in Frequency Doubling Technology (FDT) perimetry data identified by unsupervised machine learning classifiers," PLoS One, vol. 9, no. 1, 2014.

[3] S. Verma et al., "Visual Field Progression in Patients with Primary Angle-Closure Glaucoma Using Pointwise Linear Regression Analysis," Ophthalmology, vol. 124, no. 7, pp. 1065–1071, 2017.

[4] E. Atalay et al., "Pattern of Visual Field Loss in Primary Angle-Closure Glaucoma Across Different Severity Levels," Ophthalmology, vol. 123, no. 9, pp. 1957–1964, 2016.

[5] S. Yousefi et al., "Detecting glaucomatous change in visual fields: Analysis with an optimization framework," J. Biomed. Inform., vol. 58, pp. 96–103, 2015.

[6] V. Gulshan et al., "Development and Validation of a Deep Learning Algorithm for Detection of Diabetic Retinopathy in Retinal Fundus Photographs," Am. Med. Assoc., vol. 316, no. 22, pp. 2402–2410, 2016.

[7] S. S. Rubini and A. Kunthavai, "Diabetic retinopathy detection based on eigenvalues of the hessian matrix," Procedia Comput. Sci., vol. 47, no. C, pp. 311–318, 2015.

[8] J. Amin, M. Sharif, and M. Yasmin, "A Review on Recent Developments for Detection of Diabetic Retinopathy," Scientifica (Cairo)., vol. 2016, 2016.

[9] A. García-Floriano, Á. Ferreira-Santiago, O. Camacho-Nieto, and C. Yáñez-Márquez, "A machine learning approach to medical image classification: Detecting age-related macular degeneration in fundus images," Comput. Electr. Eng., vol. 75, pp. 218–229, 2019.

[10] A. Wood, T. Margrain, and A. M. Binns, "Detection of early age-related macular degeneration using novel functional parameters of the focal cone electroretinogram," PLoS One, vol. 9, no. 5, pp. 1–9, 2014.

[11] D. S. W. Ting et al., "Deep learning in ophthalmology: The technical and clinical considerations," Prog. Retin. Eye Res., no. December 2018, pp. 1–24, 2019.

[12] A. Fourcade and R. H. Khonsari, "Deep learning in medical image analysis : A third eye for doctors," J. Stomatol. Oral Maxillofac. Surg., 2019.

[13] D. S. W. Ting et al., "Progress in Retinal and Eye Research Deep learning in ophthalmology : The technical and clinical considerations," Prog. Retin. Eye Res., no. April, pp. 1–24, 2019.

[14] J. Latif, C. Xiao, A. Imran, and S. Tu, "Medical Imaging using Machine Learning and Deep Learning Algorithms : A Review," 2019 2nd Int. Conf. Comput. Math. Eng. Technol., pp. 1–5, 2019.

[15] M. Maliha, A. Tareque, and S. S. Roy, "Diabetic Retinopathy Detection Using Machine Learning," 2018.

[16] S. Guo, K. Wang, H. Kang, Y. Zhang, Y. Gao, and T. Li, "BTS-DSN : Deeply supervised neural network with short connections for retinal vessel segmentation," Int. J. Med. Inform., vol. 126, no. March, pp. 105–113, 2019.

[17] Y. Pu et al., "Variational autoencoder for deep learning of images, labels and captions," Adv. Neural Inf. Process. Syst., no. Nips, pp. 2360–2368, 2016.

[18] H. F. Pardede, E. Suryawati, R. Sustika, and V. Zilvan, "Unsupervised Convolutional Autoencoder-Based Feature Learning for Automatic Detection of Plant Diseases," 2018 Int. Conf. Comput. Control. Informatics its Appl., pp. 158–162, 2018.

[19] Y. M. Kassim and K.Palaniappan, "Extracting Retinal Vascular Networks Using Deep Learning Architecture," pp. 1170–1174, 2017.

[20] O. Isaac, A. Jantan, and A. Esther, "State-of-the-art in artificial neural network applications : A survey," Heliyon, no. June, p. e00938, 2018.

[21] M. S. Solanki, "Diabetic Retinopathy Detection Using Eye Images," vol. 8, no. 12419, p. 12419, 2015.

[22] T. Erić, S. Ivanović, S. Milivojša, and N. Smiljković, "Voice Control for Smart Home Automation : Evaluation of Approaches and Possible Architectures," pp. 140–141, 2017.

[23] J. Schmidhuber, "Deep Learning in Neural Networks : An Overview," vol. 61, pp. 1–88, 2015.
[24] A. S. Ahuja and L. S. Halperin, "Understanding the advent of artificial intelligence in ophthalmology," vol. 31, pp. 115–117, 2019.
[25] Y. Kim, "Convolutional Neural Networks for Sentence Classification," 2011.
[26] Y. Dong, Q. Zhang, Z. Qiao, and J. Yang, "Classification of Cataract Fundus Image Based on Deep Learning," IEEE Instrum. Meas. Soc., pp. 1–5, 2017.
[27] S. Zhang, L. Yao, A. Sun, and Y. i Tay, "Deep Learning based Recommender System : A Survey and New Perspectives," vol. 1, no. 1, pp. 1–35, 2018.
[28] Y. Bengio, A. Courville, and P. Vincent, "Unsupervised Feature Learning and Deep Learning : A Review and New Perspectives," pp. 1–30, 2012.
[29] W. Liu, Z. Wang, X. Liu, N. Zeng, Y. Liu, and F. E. Alsaadi, "Neurocomputing A survey of deep neural network architectures and their applications," Neurocomputing, vol. 234, no. December 2016, pp. 11–26, 2017.
[30] X. Chen, Y. Xu, D. Wing, and K. Wong, "Glaucoma Detection based on Deep Convolutional Neural Network," pp. 715–718, 2015.
[31] R. Yamashita, M. Nishio, R. K. G. Do, and K. Togashi, "Convolutional neural networks: an overview and application in radiology," Insights Imaging, vol. 9, no. 4, pp. 611–629, 2018.
[32] Y. Lecun, Y. Bengio, and G. Hinton, "Deep learning," 2015.
[33] X. Gao, S. Lin, and T. Y. Wong, "Automatic Feature Learning to Grade Nuclear Cataracts Based on Deep Learning," IEEE Trans. Biomed. Eng., pp. 1–9, 2015.
[34] O. Irsoy and C. Cardie, "Deep Recursive Neural Networks," pp. 1–9, 2014.
[35] M. Langkvist, L. Karlsson, and A. Loutfi, "A Review of Unsupervised Feature Learning and Deep Learning for Time-Series Modeling," vol. 42, pp. 11–24, 2014.
[36] G. Litjens et al., "A survey on deep learning in medical image analysis," vol. 42, no. December 2012, pp. 60–88, 2017.
[37] E. Pead et al., "Automated detection of age-related macular degeneration in color fundus photography : a systematic review," Surv. Ophthalmol., vol. 64, no. 4, pp. 498–511, 2019.
[38] Y. Hagiwara et al., "Computer Methods and Programs in Biomedicine Computer-aided diagnosis of glaucoma using fundus images : A review," Comput. Methods Programs Biomed., vol. 165, pp. 1–12, 2018.
[39] N. Salamat, M. M. S. Missen, and A. Rashid, "Diabetic retinopathy techniques in retinal images : A review," Artif. Intell. Med., vol. 97, no. January 2018, pp. 168–188, 2019.
[40] Z. Li, Y. He, S. Keel, W. Meng, and R. T. Chang, "Efficacy of a Deep Learning System for Detecting Glaucomatous Optic Neuropathy Based on Color Fundus Photographs," Ophthalmology, pp. 1–8, 2018.
[41] U. Raghavendra, H. Fujita, S. V Bhandary, and A. Gudigar, "Deep convolution neural network for accurate diagnosis of glaucoma using digital fundus images," Inf. Sci. (Ny)., vol. 441, pp. 41–49, 2018.
[42] Q. Abbas, "Glaucoma-Deep: Detection of Glaucoma Eye Disease on Retinal Fundus Images using Deep Learning," Int. J. Adv. Comput. Sci. Appl., vol. 8, no. 6, 2017.
[43] R. Asaoka, H. Murata, A. Iwase, and M. Araie, "Detecting Preperimetric Glaucoma with Standard Automated Perimetry Using a Deep Learning Classi fi er," Ophthalmology, pp. 1–7, 2016.
[44] X. Chen, Y. Xu, S. Yan, D. W. K. Wong, T. Y. Wong, and J. Liu, "Automatic Feature Learning for Glaucoma Detection Based on Deep Learning," in Springer International Publishing Switzerland, 2015, vol. 9351, pp. 669–677.
[45] Y. Liu, Z. Li, C. Xu, J. Li, and R. Liang, "Referable Diabetic Retinopathy Identification from Eye Fundus Images with Weighted Path for Convolutional Neural Network," Artif. Intell. Med., 2019.
[46] C. Lam, D. Yi, M. Guo, and T. Lindsey, "Automated Detection of Diabetic Retinopathy using Deep Learning," AMIA Jt. Summits Transl. Sci. proceedings. AMIA Jt. Summits Transl. Sci., pp. 147–155, 2018.

[47] S. Dutta, B. C. S. Manideep, S. M. Basha, R. D. Caytiles, and N. C. S. N. Iyengar, "Classification of diabetic retinopathy images by using deep learning models," Int. J. Grid Distrib. Comput., vol. 11, no. 1, pp. 89–106, 2018.
[48] K. Xu, D. Feng, and H. Mi, "Deep Convolutional Neural Network-Based Early Automated Detection of Diabetic Retinopathy Using Fundus Image," 2017.
[49] Y. S. Kanungo, B. Srinivasan, and S. Choudhary, "Detecting Diabetic Retinopathy using Deep Learning," Int. Conf. Recent Trends Electron. Inf. Commun. Technol., pp. 801–804, 2017.
[50] X. Zeng, H. Chen, Y. Luo, and W. Ye, "Automated Diabetic Retinopathy Detection Based on Binocular Siamese-like Convolution Neural Network," IEEE Access, vol. 4, pp. 1–10, 2016.
[51] H. Pratt, F. Coenen, D. M. Broadbent, S. P. Harding, and Y. Zheng, "Convolutional Neural Networks for Diabetic Retinopathy," Int. Conf. Med. Imaging Underst. Anal., no. July, pp. 1–6, 2016.
[52] M. D. Abramoff et al., "Improved Automated Detection of Diabetic Retinopathy on a Publicly Available Dataset Through Integration of Deep Learning," Investig. Ophthalmol. Vis. Sci., vol. 57, no. 13, 2016.
[53] N. Ayuni, M. Asyraf, W. Mimi, and D. Wan, "An automated glaucoma screening system using cup-to-disc ratio via Simple Linear Iterative Clustering superpixel approach," Biomed. Signal Process. Control, vol. 53, p. 101454, 2019.
[54] P. Amil et al., "Unsupervised feature extraction of anterior chamber OCT images for ordering and classification," Sci. Rep., vol. 9, no. 1, pp. 1–9, 2019.
[55] A. Montuoro, S. M. Waldstein, B. S. Gerendas, U. Schmidt-Erfurth, and H. Bogunović, "Joint retinal layer and fluid segmentation in OCT scans of eyes with severe macular edema using unsupervised representation and auto-context," Biomed. Opt. Express, vol. 8, no. 3, p. 1874, 2017.
[56] S. Yousefi et al., "Unsupervised Gaussian mixture-model with expectation maximization for detecting glaucomatous progression in standard automated perimetry visual fields," Transl. Vis. Sci. Technol., vol. 5, no. 3, 2016.
[57] S. Yousefi, M. H. Goldbaum, L. M. Zangwill, F. A. Medeiros, and C. Bowd, "Recognizing patterns of visual field loss using unsupervised machine learning," Med. Imaging 2014 Image Process., vol. 9034, no. March 2014, p. 90342M, 2014.
[58] R. Priya and P. Aruna, "Diagnosis of Diabetic Retinopathy Using Machine Learning Techniques," vol. 6956, no. July, pp. 563–575, 2013.

A Proposal of Low Cost Home Automation System Using IoT and Voice Recognition

Phaklen Ehkan[1], Nui Din Keraf[2], Virakwan Hai Kelian[3], Fazrul Faiz Zakaria[1], Sin Zhen Bei[1]

[1]School of Computer and Communication Engineering, Universiti Malaysia Perlis, Pauh Putra Campus, 02600 Arau, Perlis, Malaysia.
[2]Department of Electrical Engineering, Politeknik Sultan Abdul Halim Mu'adzam Shah, 06000 Jitra, Kedah, Malaysia.
[3]Department of Information Technology and Communications, Politeknik Seberang Perai, 14000 Permatang Pauh, Penang, Malaysia.

phaklen@unimap.edu.my

Abstract. Home Automation System is becoming more popular day by day due to its numerous benefits. This project proposes an idea in the design of low cost home automation system by using the Internet of Things (IoT) and voice recognition. The layout of the home divided into four areas and each area has own function and system. The Raspberry Pi 3 (RPi) Model B+ used as the main controller for the processing and transmitting the input data. IoT provided huge storage for data collection from sensors and home appliances. An Android application is developed to monitor the home environment and remotely control the home devices by using the button or voice. The speaker-independent recognition system by using Google Voice to Text on Android embedded in this project for physically challenged people to control the electrical appliances without moving. All the data will be stored in Firebase and can be retrieved at any time by the application and the RPi board. There is a side view of a prototype model with two floors and divided into four home areas. This Low-Cost Home Automation System using IoT and Voice Recognition is successfully achieved the project's objective.

1. Introduction

Home is a place that people spend most of their time in it. Thus, home automated is developed to increase the comfort and quality of people's life. The home automation is also known as domotics, and home automation system (HAS) called smart home or smart house [1-4]. The system will also include the security parts such as the fire alarm system, surveillance system, motion detection, and more. It is able automated and controlled all of the electronic appliances in home through remote control units, touch devices, electronic devices, and even the homeowner voice over the Internet. Hence, home automation is a step toward what is referred to as the Internet of Things (IoT) [7-8].

The "Internet of Things" phrase which is shortly known as IoT. It is one of the technologies which can be implemented for connecting, controlling and monitoring the smart devices over the Internet [5]. The goal of the IoT is not only focused on giving us the ability to control the things, but also allows the users to keep monitoring the status of the things. IoT based HAS typically involving climate control, light controls, smart door and so on.

IOP Conf. Series: Materials Science and Engineering **767** (2020) 012068 doi:10.1088/1757-899X/767/1/012068

About 650 million people or around 10% of the world's population live with a disability based on the Disabled-World website [6]. The physically challenged person encountered a lot of problems in moving from one place to another. Therefore, this project proposes an IoT based HAS that consists of voice recognition for physically challenged people to control the electrical appliances. The main objective of this project is to develop a low cost home automation system using IoT and voice recognition. This project is addressed at the following: (i) develop a low cost home automation system ease of installation and configuration, (ii) develop an Android application user interface for the user to monitor and control the home appliances over the internet, and (iii) embed the voice recognition system for physically challenged people to control the home appliances.

2. System Model

The overall system architecture for low cost home automation system using IoT and voice recognition is divided into different block; remote control system, voice recognition system [10-13], home automation system architecture, and cloud connection system. The first block is the remote control system - Android App. It receives two types of input, user input control and voice command. The UI includes a home control, voice recognition, and monitoring system page that allow users to remotely control and monitor home appliances. Voice commands are received via smartphone microphone and then processed by the voice recognition system which shown in the remote control system block. Android Apps use the Google Speech to Text to process the voice commands. Once the conversion is complete, the voice output is uploaded to Firebase and processed by RPi to perform the operation. The next block is the home automation system architectures, the RPi3 B+ as the main board [9]. The green colour box indicates the input to the system and the yellow colour box indicates the output of the system. The final part of the overall system is the cloud connection system. Each data read by the sensor is progressed by the board and uploaded to Firebase via the board's built-in Wi-Fi. The remote control system and HAS architecture are applied two-way wireless communication with the Firebase Real-time Database to update and retrieve the status of the home appliances. Figure 1 shows the model of home automation system block diagram.

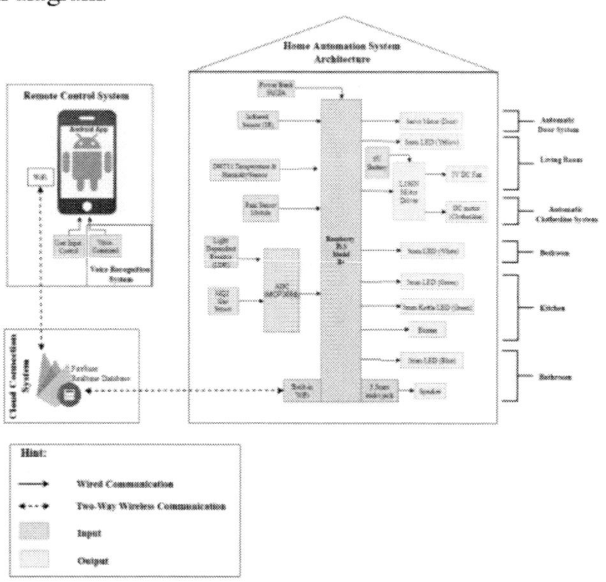

Figure 1. Home automation system block diagram

2.1 Low Cost Home Automation System Architecture

IOP Conf. Series: Materials Science and Engineering **767** (2020) 012068 doi:10.1088/1757-899X/767/1/012068

The HAS architecture divided into several parts; automatic door system, automatic clothesline system, living room, kitchen, bathroom, and bedroom. Each area has its own function and household appliances as shown in Table 1.

Table 1. Home automation system functions

Location	Sensors	Devices
Automatic door system	IR Sensor	Door (Servo Motor)
Automatic clothesline system	Rain sensor module	Clothesline (DC motor)
Living room	DHT11 Sensor	5mm Yellow LED and 5V DC Fan
Kitchen	MQ2 Gas sensor	5mm Green LED, 3mm Kettle Green LED and Alarm (Buzzer)
Bathroom		5mm Blue LED and Speaker
Bedroom	LDR	5mm White LED

2.1.1 Automatic Door System. Figure 2 shows the sub-block diagram for Automatic Door System. The user input from the remote control system sent to Firebase and RPi retrieve it. An IR sensor for detecting obstacle and a servo motor for operating the door. The IR sensor has 3 pin interfaces. The left pin (OUT) is the output pin of the sensor which provides digital output (high or low). The centre pin is the ground (GND) pin and the last pin is for input voltage (VCC) connect to either +3.3V or +5V. The sensor contains a potentiometer to adjust the obstacle detection sensitivity range (2cm - 10cm). In this project, the obstacle detection range was set to approximately 5cm. TowerPro SG90 micro servo has 3 wiring which are GND, 5V and Signal. The maximum rotation angle of the servo motor is 180 degrees. In this project, 0 degrees is used as the initial door closing angle and rotated 90 degrees as to open the door.

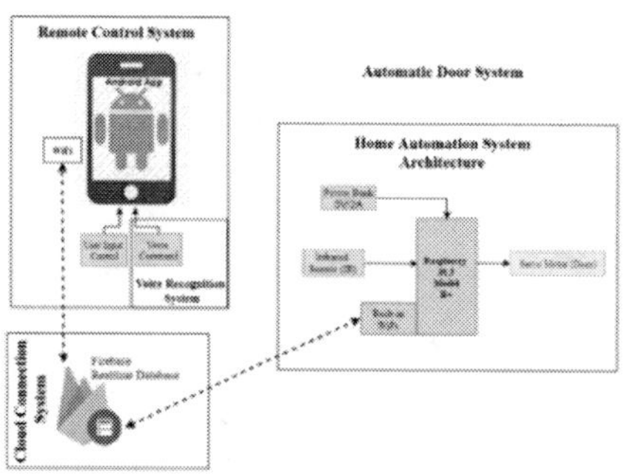

Figure 2. Sub-block diagram of automatic door system

RPi will keep to reading the door status from Firebase. When the door status is on, the servo motor will rotate 90 degrees to open the door and delay for 3 seconds to allow the people to enter the house before closing the door. If an object is detected after the completion of the time delay, the door will remain open and continue to delay more 3 seconds. Otherwise, when no object is detected, the door will close. The door status will always be uploaded to the Firebase.

2.1.2 Smart Power System (Living Room). Figure 3 shows the sub-block diagram for smart power system located in the living room area. The user input from the remote control system and the sensor value are sent to Firebase and RPi retrieve it. DHT11 sensor for detecting temperature and humidity of the room and a L298N motor driver connected with the 5V DC fan power on by 9V battery. The DHT11 sensor can sense the temperature from 0 to 50°C with +/- 2 accuracy and humidity range between 20% and 80%

with +/- 5% accuracy. The VCC pin of the sensor connected to the RPi's 5V pin and GND pin connected to RPi's GND. The OUT pin is connected to one of the GPIO pin in the RPi (GPIO 17) to get the sensor data. The sensor collects one reading every second.

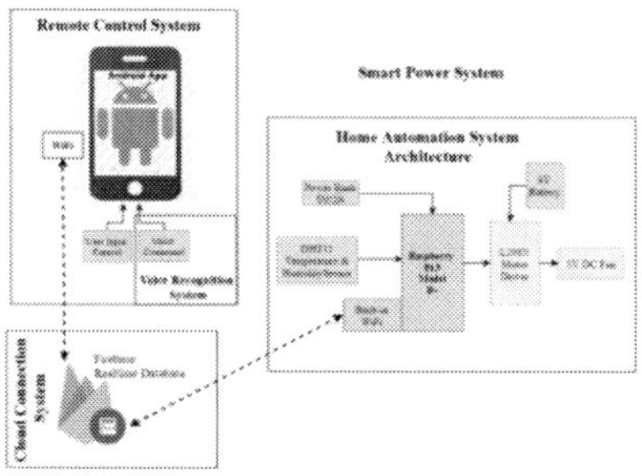

Figure 3. Sub-block diagram of smart power system

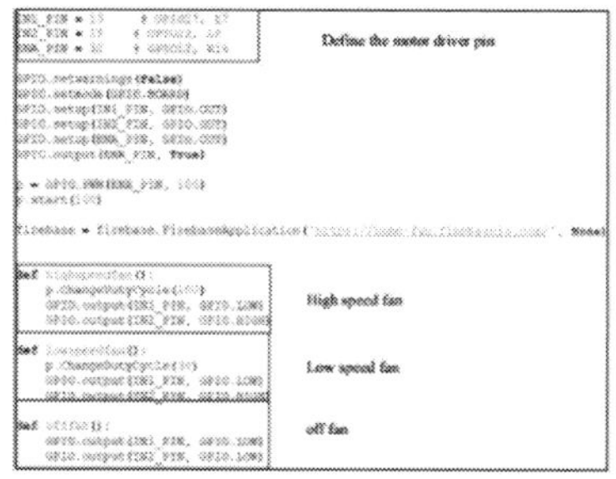

Figure 4. Adjust the fan speed using python code

There was one 5V DC fan used in this smart power system. A L298N dual motor controller module 2A is implemented with RPi to control the speed of the fan. The motor driver can receive the external power source from 7V to 30V which used to supply the power to connected fan. The input ports are to provide the PWM signal for control the fan speed while the output pins are connected to the VCC and GND pin of the fan. The motor driver +12V pin is connected with the 9V battery to power on it and the GND pin of RPi and battery are linked together to common the GND.

Figure 4 shows the method of using the Python programming language to generate PWM to control the speed of the fan. GPIO 27 and GPIO 22 of the RPi are to give the signal to the fan. The ENA pin (GPIO 12) is used to enable the generate signal send to the output port. The OUT1 pin is connected to the positive pin of the fan and OUT2 pin is connected to the negative pin of the fan. Hence, to turn on the fan is required to set both input ports with LOW and HIGH signal respectively. To stop the fan by set both inputs pin with the same signal (LOW and LOW).

2.1.3 Smoke Detector System (Kitchen). Figure 5 shows the MQ2 gas sensor and buzzer are implemented in this project for gas leakage and fire detecting in the kitchen. The gas sensor value is send to Firebase via RPi's built-in WiFi. The buzzer is act as the alarm and be activated by the gas sensor when detected gas leakage.

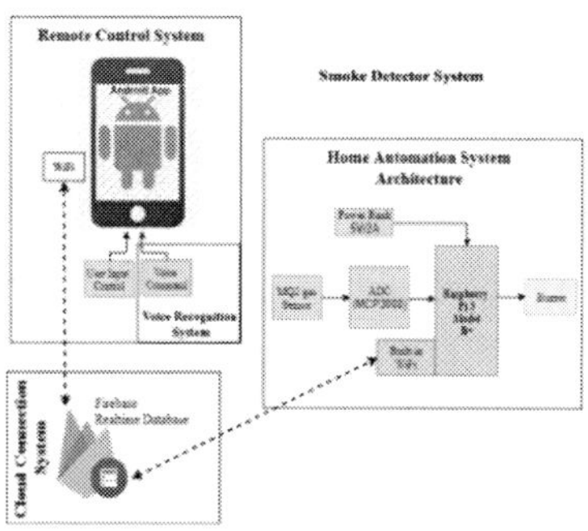

Figure 5. Sub-block diagram of smoke detector system

MQ2 gas sensor is suitable for detecting of LPG, i-butane, propane, methane, alcohol, Hydrogen, smoke and carbon monoxide. In this project, the gases used for gas leakage detection are LPG and smoke. The analog output plays the essential role in this project by giving the concentration of surrounding gas present to RPi. However, this is analog output data type, it required an MCP3008 ADC to convert the analog output voltage into digital before send into RPi. ADC module contains 8 channels to read and convert up to 8 devices analog output into digital. The pins at the right side of the ADC module are linked with SPI bus of RPi for sending the converted data into RPi.

2.1.4 Smart Lighting System (Bedroom). Figure 6 shows the sub-block diagram for Smart Lighting System located in the bedroom. The user input from the remote control system and the LDR sensor value are sent to Firebase and RPi retrieve it. The LDR is analog type output data and similar to the MQ2 gas sensor, it required an ADC to convert the analog output voltage into digital. The MCP3008 ADC module is shared for LDR with different channel. LDR is used to detect the presence or the level of the light. The 5mm white colour LED as the output for the smart lighting system. This benefit of the system is can reduce the energy and the money in electricity bill.

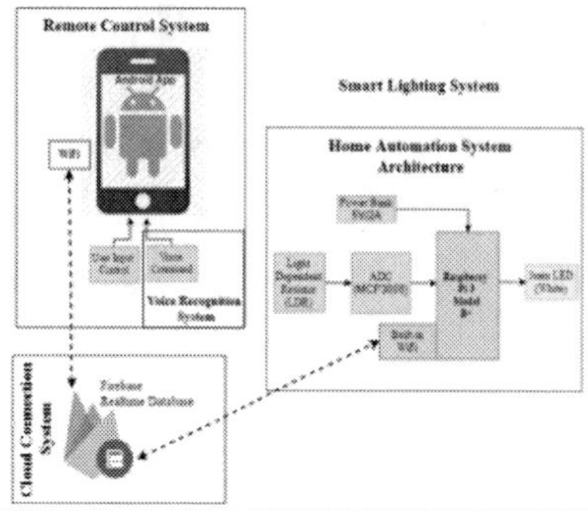

Figure 6. Sub-block diagram for smart lighting system

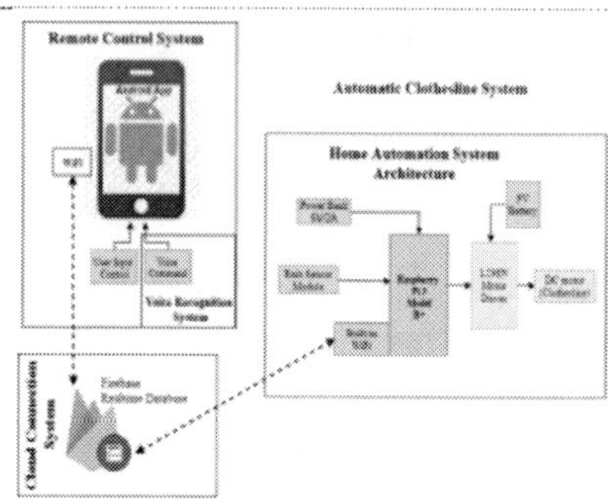

Figure 7. Sub-block diagram for Automatic Clothesline System

2.1.5 Automatic Clothesline System. Figure 7 shows the sub-block diagram for Automatic Clothesline System. The rain sensor as the input to detect the raindrop and the L298N motor driver is used to control the DC motor whether to draw back clothes to indoor or hang the clothes to outdoor. This system able to saves the manpower and the money on dryer the clothes.

The wiring connection for the Automatic Clothesline System is similar to the previous connection mentioned in the Smart Power System. It required 2 input pins, 2 output pins and 1 enable pin (IN3, IN4, OUT3, OUT4, and ENB). The input ports are to provide the signal for control the direction of the motor while the output pins are connected to the VCC and GND pin of the DC motor.

2.2 Remote Control System.

The instruction from the Android App UI is transmitted to the RPi3 B+ board for performing the action through the wireless medium such as Wi-Fi. This system allows the user to remotely control the home appliances that shown in figure 8 far away from the home via App as long as there has an Internet connection. In addition, the voice recognition system shown in the remote control system also enables the physically challenged person remotely to control the home appliances by using the voice command.

The speaker connected with RPi by using a 3.5mm audio jack cable also can be controlled by user input to play and stop the music. The connection of the door and fan which has been discussed in the previous section also can control by the user via Android App. The Home Control page in the application is shown in figure 9. The user can click the bottom bar button to switch to another UI page. The voice recognition page is for physically challenged people to control the home electrical appliances displayed in Home Control UI page.

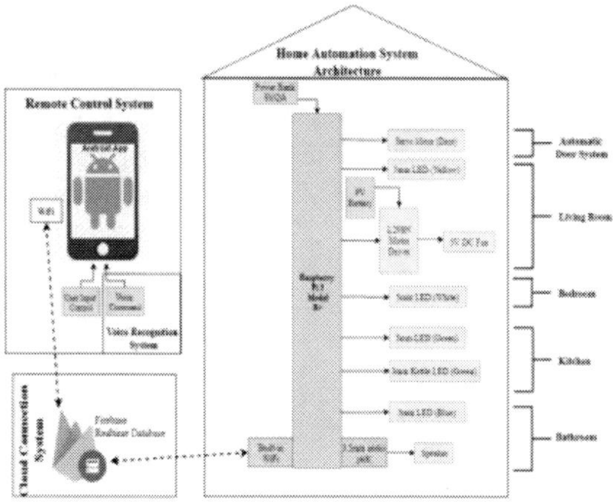

Figure 8. Sub-block diagram for remote control system

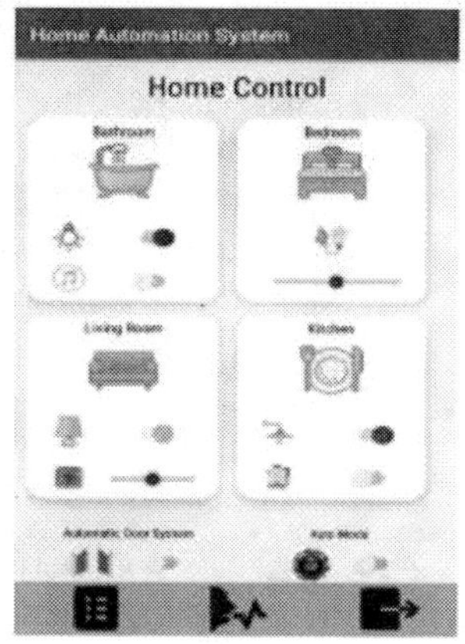

Figure 9. Home Control Page in Android App

2.3 Voice Recognition System

The users can control the home devices via the App whereas the system has provided an alternative way for the physically challenged people by using voice command to perform the same action. Total 19 voice commands that can be recognized to perform the actions. The voice commands, actions, and the voice response for HAS are shown in Table 3.2.

2.4 Cloud Connection System

The Cloud Connection System is used to store, display and monitor the collected data. The data is saved for future analysis and send to the Android App and RPi. As shown in figure 10, Firebase Real-time Database as the platform between Remote Control System, Voice Recognition System and Home Automation System to wireless communication and transmission all sensors data, home device status, and translated text output in this project. The stored data can be viewed via the Firebase console and Android App.

The Python code is used to keep upload the sensor reading and retrieve the data from Firebase. Each data is declared to be sent into the specified child so that the data do not mix up and display in different field and child in the Firebase console. The Java code is written to read the data from the Firebase by declaring the child path. After retrieving the data from the Firebase, it will send the value to the specific text view box to display it. It can obviously see that the Firebase as a good platform to interact with the mobile.

3. Results and Discussion

The prototype of the project is successfully developed. There is a side view of a prototype model with two floors and divided into four home areas which are the living room, kitchen, bathroom, and bedroom. The clothesline that implements the Automatic Clothesline System was set up on the right-hand side of the prototype. Figure 11 displays the prototype for the low cost HAS.

Furthermore, the prototype is linked with the Android App and Firebase. The user not only can observe the data at the Firebase console, an Android App also able to monitor the data anytime anywhere as long as have a stable network connection. When the system has any abnormal condition, the RPi enables an alert system by sending the email to the user. User is able to control the door by using the Android App. The servo motor is used to operate the door and IR sensor to detect the object. The open door button is in the bottom left side in the Home Control page shown in figure 12.

When the user presses the door open switch, the servo motor will rotate 90 degrees to open the door. The door status will upload to the Firebase, "1" for open and "0" for close. When the IR sensor detects an object, the output value is "0", otherwise it is "1". The RPi retrieve the door status from Firebase, and the door in the prototype is performed according to the status.

The Smart Power System is functional when the Auto Mode and Fan are turned ON. This can be viewed and controlled via the Home Control page in the Android App, as shown in the right side of figure 13. Smart Power System will adjust the fan speed based on the temperature reading that sensed by the DHT11 Temperature and Humidity sensor. The application also contains a Monitoring System page that displays the DHT11 sensor readings.

Table 2. Voice Commands of the HAS

Location	Voice Commands	Actions	Voice Responses
Living room	Ok home turns On/Off the living room light	Yellow LED On/Off	Sure living room light On/Off
	Ok home turns On/Off the living room fan	Fan On/Off	Sure living room fan On/Off
Kitchen	Ok home turns On/Off the kitchen light	Green LED On/Off	Sure kitchen light On/Off
	Ok home help me make some hot water	Kettle LED On after3s Off	Sure (before On the kettle your hot water is ready after Off the kettle
Bathroom	Ok home turns On/Off the bathroom light	Blue LED On/Off	Sure bathroom light On/Off
	Ok home plays some music	Music On	Sure music is now playing
	Ok home stop music	Music Off	Sure music is now stopping
Bedroom	Ok home turn On the bedroom light	White LED On	Sure bedroom light On
	Ok home adjusts the bedroom light to 50%	White LED 50% brightness	Sure bedroom light now 50%
	Ok home is tome to sleep	White LED Off	Sure bedroom light Off good night
Home	Ok home turns On/Off all the lights	All LEDs On/Off	Sure all lights On/Off
Door	Ok home open the door	Door open	Sure welcome home
		After 3s close (if no detected object)	
Auto Mode	Ok home auto mode On/Off	Auto mode On/Off	Sure auto mode On/Off

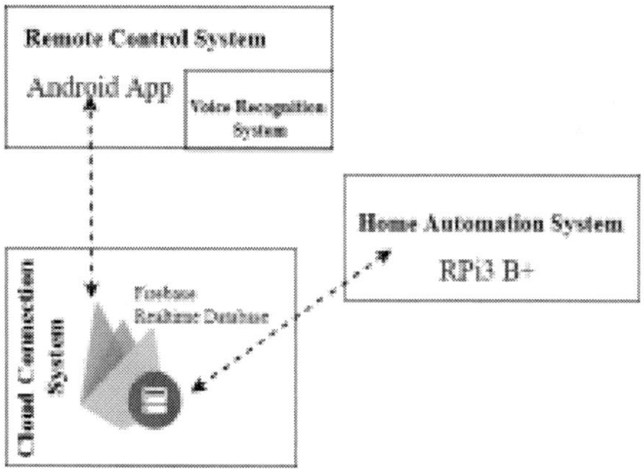

Figure 10. Sub-block Diagram of Cloud Connection System

Figure 11. Prototype for the Low Cost HAS

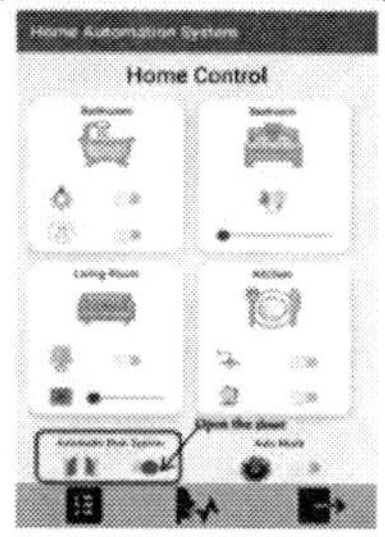

Figure 12. Automatic Door System UI

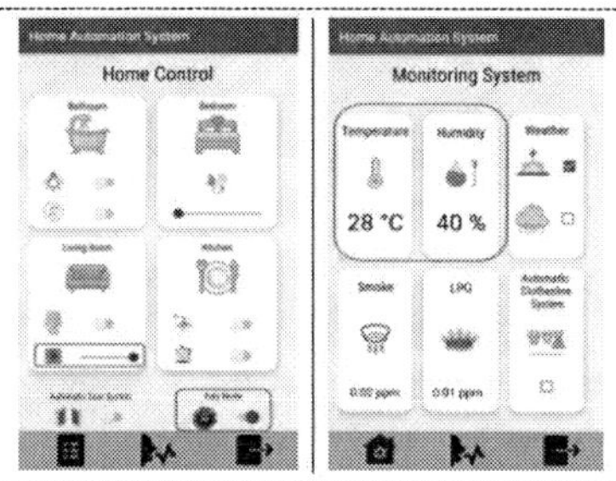

Figure 13. Smart Power System UI

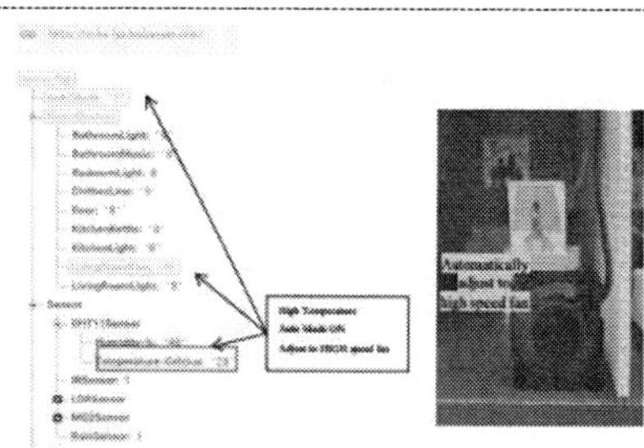

Figure 14. Smart Power System Operation

When Auto Mode and Fan are turned on and temperature is greater and equal to 25°C means it is hot day, the system will automatically adjust the fan to high speed. RPi retrieve the data from the Firebase to perform the particular action in the prototype shows in Figure 14.

Figure 15 shows the MQ2 gas sensor is tested with the lighter for LPG and smoke detection to prove the gas sensor is able to detect gas leakage in the home. The gases concentration reading of LPG and smoke from the gas sensor is increasing when provide LPG gas from the lighter. When in clean air condition, the concentration of gases is very low around 0 to 2 ppm for each gas, however, the lighter provide the LPG gas into the sensor, the gas concentration reading is raised up until 532.04 ppm and 104.07 ppm for Smoke and LPG respectively. RPi will keep transmit the LPG and Smoke value into Firebase shows in figure 16. The Alarm is activated when the gas value over the threshold value. If one of the gas values exceeds the threshold value, the alert system is triggered and the RPi will send the email to the user.

IOP Conf. Series: Materials Science and Engineering **767** (2020) 012068 doi:10.1088/1757-899X/767/1/012068

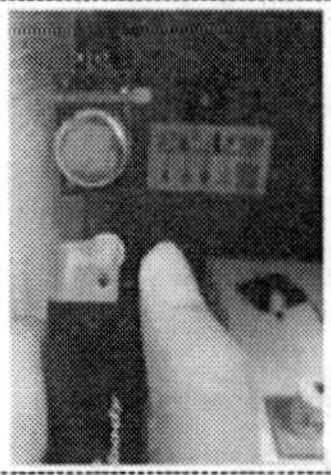

Figure 15. MQ2 Gas Sensor is tested with Lighter

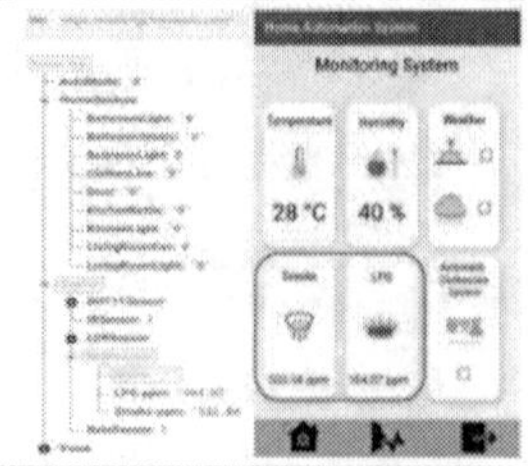

Figure 16. Result of MQ2 Gas Sensor when tested with Lighter

The seekbar that displayed in the Home Control page allows users to adjust the brightness of the bedroom's light when Auto Mode OFF via the Android App. The brightness can be adjusted to 50% and 100%. When Auto Mode ON and it is Day Mode, the bedroom's light will automatically OFF to save power. The Day Mode is determined by the LDR value. Figure 17 displays the before and after adjusting the bedroom light when Auto Mode ON.

The child paths that store the light status and LDR readings are shown in figure 18. The Day Mode equal to "1" means it is morning, the bedroom light will automatically OFF and Firebase update the latest value to the application to display. The RPi will retrieve the latest data from the Firebase and perform the automatically OFF operation in the prototype when it is Day Mode.

Android App contains one Monitoring System page to display the Automatic Clothesline System status shown in figure 19. In this system, the clothesline operates based on the rain sensor module and the Day Mode status. When the rain sensor detected raindrop or it is night mode, the DC motor will turn on to keep the hanging clothes under the roof. In the other hand, when it is sunny day and morning, the clothes are hanged at the outdoor.

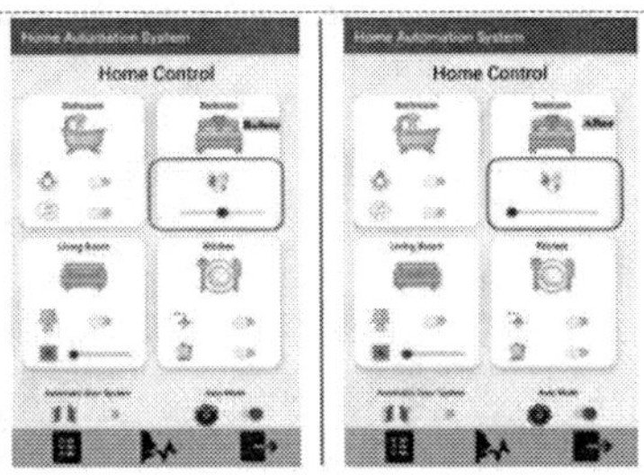

Figure 17. Smart Power System UI

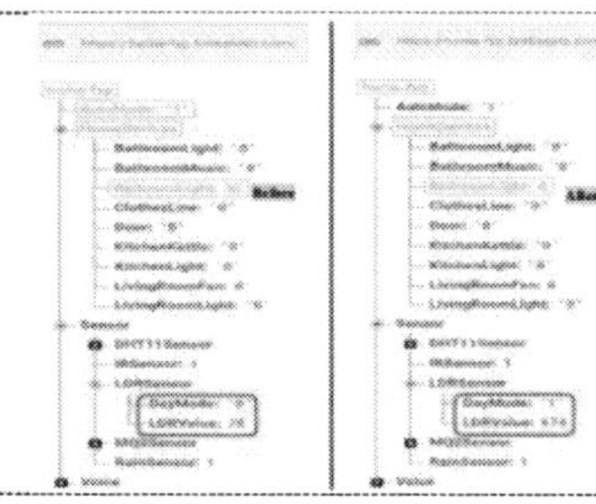

Figure 18. Smart Lighting System Data

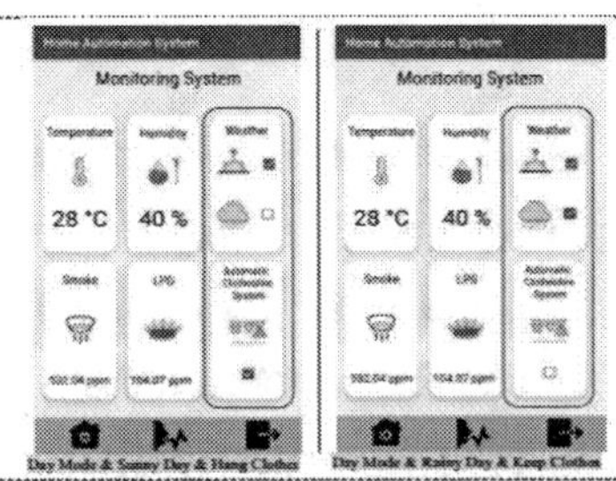

Figure 19. Automatic Clothesline System UI

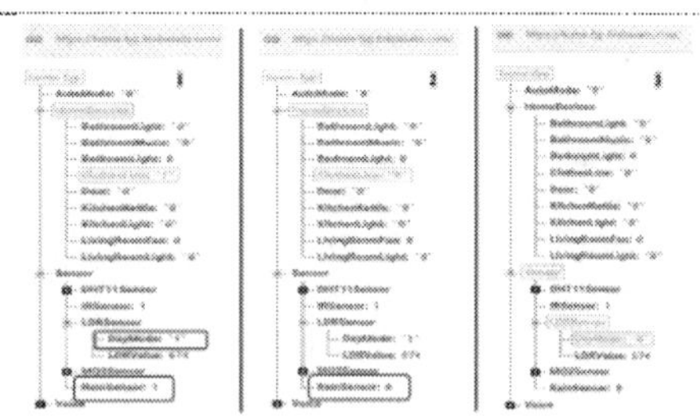

Figure 20. Data for the Automatic Clothesline System

The status of the system will always upload into Firebase to allow the RPi retrieve it to perform action. Figure 20 shows different condition that will trigger the clothesline. Rain sensor value equal to "0" means detected raindrop. The rain sensor module was placed on the side roof of the home. This system also will trigger the alert system to send the email to the user to notice about it is raining day and the clothes were kept inside under the roof.

This remote Control System total contains 3 page, namely Voice Recognition System page, Home Control page, and Monitoring System page. Each page contains its own features. The details of the button are shown in figure 21. The home devices status and voice output are uploaded into Firebase, and sensor values are retrieved from Firebase. It is obviously seen that the Remote Control System is fully relies on the network connection.

Voice Recognition System page contains a microphone button to press, as shown in figure 22. After pressing the button, it will pop up a Google Voice dialog and begin recording voice command. The physically challenged people can use this voice system to control home devices without moving. The speech is converted directly to text and displayed in the same page and then uploaded into Firebase. For instance, when the user says "ok home turn on all the lights", the voice will convert to text and uploaded to the Firebase under the Voice path.

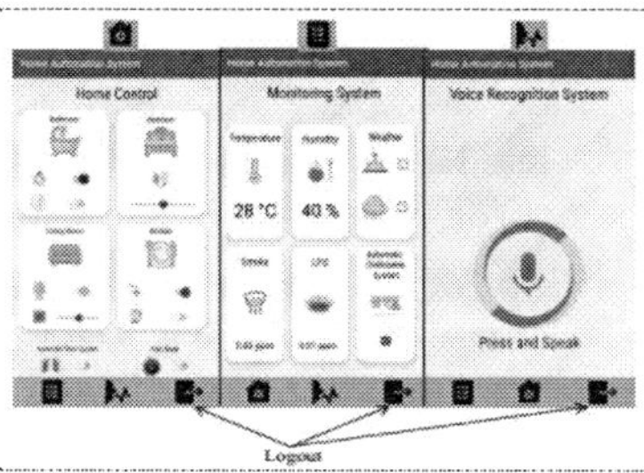

Figure 21. Different Pages of the Remote Control System

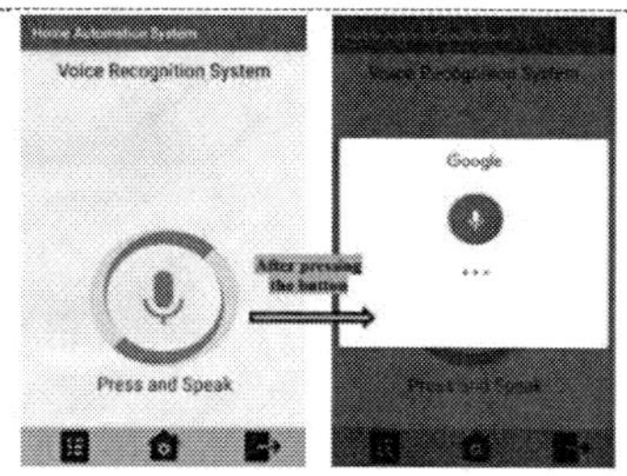

Figure 22 Voice Recognition Systems UI

The Firebase Real-time Database as the Cloud Connection System used to communicate between Remote Control System and Home Automation System Architecture. The Home Devices parent is used to store all of the home appliances status, door status, and clothesline status. All sensor values are stored under the Sensor parent node. For the Voice parent is used to save the converted text output from the application. Firebase is used in this project as it is a platform that able to link with mobile and have a real-time database. In addition, all the data collected in Firebase can be view through the mobile application or Firebase console. Thus, the limited memory space of RPi can be avoided to save all the collected data for further analysis.

In this project, the overall system is completely dependent on the network connection. If RPi is not connected to a network, it is unable to execute the Python code that contains the process of sending and retrieving data from Firebase. For the Remote Control System, all of the data unable synchronized with the Firebase cause no display value or status due to no network issues.

4. Conclusion

The integration of hardware and software utilized in this project demonstrates the reliability and effectiveness of the development system as the Remote Control System is developed. The limitations of HAS is fully dependent on the network connection to synchronizes data. In addition, there is a Monitoring System page in the Remote Control System that is developed to provide the users with a vision. This page allows the user to get updated sensor readings. The sensors used in this project have been successfully programmed and are capable of detecting, acquiring and transmitting data. All collected data is uploaded to Firebase. The developed system allows users to wirelessly control and monitor the HAS.

Nonetheless, the system is becoming easier and convenient to monitor as all the data have been stored in Firebase and displayed in the application. As a result, users can view the data by browsing the Firebase console or mobile application anywhere, as long as they have an internet connection. In addition, the system automatically sends an alert message via email when fire or rain is detected. The messages allow the user take further actions for the scenarios. The last objective of the project was achieved by embedded a Voice Recognition System in the Remote Control System for physically challenged people to control the home electrical appliances. They need to mention "ok home" at the beginning of the voice command and then specify the action and specific home area. The voice is converted to text by using Google Speech to Text on Android and the text is uploaded to Firebase. RPi retrieves the text output and based on the matched output to perform the action. The voice response of the particular action is played by the connected speaker to the physically challenged people.

References

[1] Hidayat S and Firmanda S F (2015). Scheduler and voice recognition on home automation control system. *3rd IEEE Int. Conf. on Information and Communication Technology*, p 150–155.
[2] Kodali R K and Mahesh K S (2017). Low cost implementation of smart home automation. IEEE *Int. Conf. on Advances in Computing, Communications and Informatics (ICACCI)*, p 461-466.
[3] Pavithra, D and Balakrishnan R (2015). IoT based monitoring and control system for home automation. *IEEE Global Conf. on Communication Technologies (GCCT)*, 169–173.
[4] Vikram N, Harish K S, Nihaal M S, Umesh R, Shetty A and Kumar A (2017). A Low Cost Home Automation System Using Wi-Fi Based Wireless Sensor Network Incorporating Internet of Things (IoT). *IEEE 7th Int. Advance Computing Conf. (IACC)*, p 174–178.
[5] Pawar A, Sharan R, Patil R and Chavan S (2018). Home Automation using Bluetooth and IoT. *Int. Journal of Innovative Research in Computer and Communication Engineering*, **6**(2), p 896-902.
[6] World Health Organization (2011). World Report on Disability. *WHO Library Cataloguing-in-Publication* Data, The World Bank, p 1-325.
[7] Venkatesh K, Rajkumar P, Hemaswathi S and Rajalingam B (2018). IoT Based Home Automation Using Raspberry Pi Multimodal Medical Image Fusion Using Raspberry Pi. *Journal of Advanced Research in Dynamical and Control Systems*, **10**(7), p 1721-1728
[8] Patel K K and Patel S M (2016). Internet of Things-IOT: Definition, Characteristics, Architecture, Enabling Technologies, Application & Future Challenges. *Int. Journal of Engineering Science and Computing*. **6**(5), p 6122-6131
[9] Vikram A J and Das S (2017). IoT based Simple Home Automation using Raspberry Pi. *Int. Journal of Engineering Trends and Technology*, **53**(3), p 124-125.
[10] Ajantha Devi V and Suganya V (2016). An Analysis on Types of Speech Recognition and Algorithms. *Int. Journal of Computer Science Trends and Technology*, **4**(2) p 350-355
[11] Arriany A A and Musbah M S (2016). Applying voice recognition technology for smart home networks. *IEEE Int. Conf. on Engineering & MIS*, p 1–6.
[12] Panja P A, Bagchi M, Das N, Mazumder R M and Ghosh S (2016). Voice recognition based wireless room automation system. *IEEE Int. Conf. on Intelligent Control Power and Instrumentation (ICICPI)* p 84– 88.
[13] Sen S, Chakrabarty S, Toshniwal R and Bhaumik A (2015). Design of an Intelligent Voice Controlled Home Automation System. *Int. Journal of Computer Applications*, **121**(15), p 39-42.

Textile Bowtie Patch Antenna for Potential RF Energy Harvesting Application at GSM 1800

N M Saaid, P J Soh, S A Baharuddin, H M A Said, M K Abdi, A Hassan Hirey, M F Che Mansor

School of Computer and Communication Engineering, Universiti Malaysia Perlis, Pauh Putra Campus, 02600 Arau, Perlis, Malaysia.

dianams@unimap.edu.my

Abstract. This paper presents the design and development of a textile bowtie patch antenna for potential RF energy harvesting application using the Global System for Mobile Communication (GSM 1800) band. The size of the antenna is 80 mm x 40 mm, and is designed on a 6 mm thick Felt textile as the substrate and 0.08 mm-thick ShieldIt as its conducting material. The simulated and measured reflection coefficients justified that the bowtie antenna can be used for RF energy harvesting application at 1.8 GHz. This antenna has realized gain ranging from 2.56 dB to 2.55 dB and a total antenna efficiency of -4.3 dB.

1. Introduction

Recently, there is growing popularity in the use of wearable devices such as in health monitoring, security services, emergency rescue application, sports and entertainment. These wearable devices are small, low-powered and lightweight. In order to increase the battery lifetime of these devices to ensure continuous and autonomous operation, there is a need for them to be integrated with energy harvesting mechanism. Energy harvesting has attracted much attention in research for the last decades with many types of energy harvesting technologies are available in the market. One of the promising energy harvesting technologies is the RF energy harvesting. Due to the existence of abundant ambient RF signals that have the ability to penetrate buildings, there is great opportunity for the RF signals to be harvested to generate sufficient power to power up low-powered wearable devices. One of the main challenges for the far-field RF energy harvesting is to overcome the losses due to long distance and obstacles in between the transmitter and the receiving antenna [1], which is an essential part of the RF energy harvesting building block. Thus, the antenna must possess high gain and high efficiency to receive the weak RF signals from the surrounding.

In recent years, various receiver antennas for RF energy harvesting antennas have been proposed including Yagi-Uda antenna [1-2], bow-tie antenna [3] and dipole antenna [4]. However, few researches have focused on textile antenna for RF energy harvesting. Among the papers are, in [5] Yagi textile antenna at 60 GHz was developed with the measured antenna gain about 9 dBi and radiation efficiency of 74%. A UHF wearable patch antenna was developed in [6] with measured gain of 4.6 dBi and antenna efficiency of 41.4%. While in [7], the textile Yagi antenna on jeans fabric with the gain of 6.32 dBi to 7.26 dBi and radiation efficiency from 90.52% to 91.34% has been simulated, respectively.

IOP Conf. Series: Materials Science and Engineering **767** (2020) 012069 doi:10.1088/1757-899X/767/1/012069

2. The design of a bowtie patch antenna

The bowtie dipole antenna is chosen due to its small size, lightweight, design simplicity and low cost. In this work, a bowtie patch antenna for RF harvesting application is designed at 1.8 GHz and simulated in CST Studio Suite. The antenna is then fabricated and assessed. The size of the antenna is 80 mm x 40 mm. The substrate used for this antenna is Felt textile with relative permittivity of 1.45, a thickness of 6mm and loss tangen (tanδ) of 0.044. Generally, low dielectric constant substrates are preferred for maximum radiation [9]. The ground can take any shape but the most common shape used are rectangular and circular. Meanwhile, the conducting material used is ShieldIt with a thickness of 0.08 mm and a conductivity (σ) of 1.8×10^5 S/m. For simplification in simulations, ShieldIt is considered as lossy material [8]. A waveguide feed has been designed on the antenna in order to connect to 50Ω SMA connector. Figure 1(a) and 1(b) illustrates the dimension of the bowtie patch antenna and top view of the bowtie patch antenna respectively. The antenna parameters and dimensions are as summarized in Table 1. The top view of the fabricated antenna is shown in figure 2.

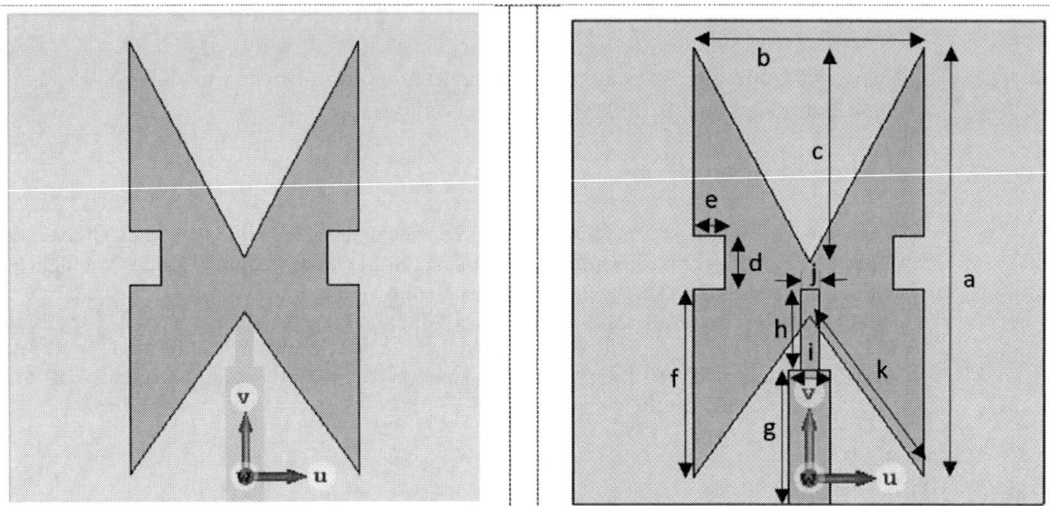

(a) (b)

Figure 1. (a) Dimensions of the bowtie patch antenna (b) top view of the bowtie patch antenna

Table 1. Antenna parameters and dimensions.

Parameter	*a*	*b*	*c*	*d*	*e*	*f*	*g*	*h*	*i*	*j*	*k*
Dimension (mm)	80	44	40	10	5.5	35	25	15	8.8	3.5	30

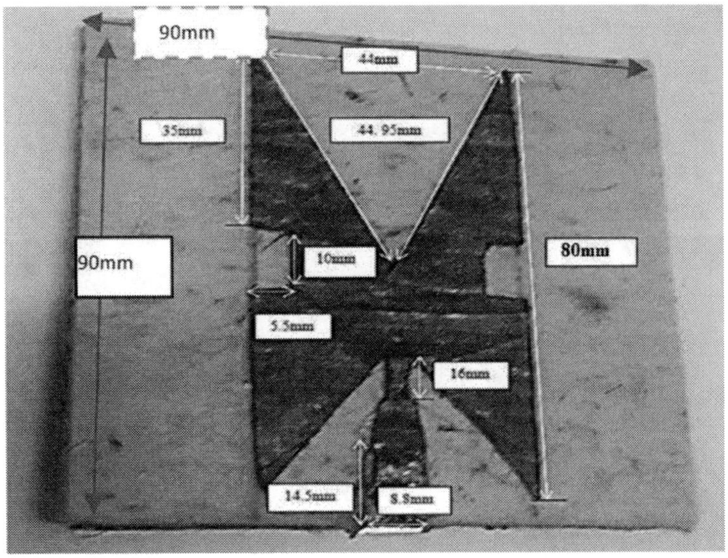

Figure 2. Top view of the fabricated bowtie patch antenna

3. Results and Discussions

All simulations have been performed using CST Studio Suite and the fabricated bowtie antenna have been tested using Vector Network Analyzer (VNA) to verify the simulation result. Figure 3(a) and 3(b) shows the simulated and measured reflection coefficients versus frequency for the textile bowtie antenna, respectively. The simulated reflection coefficient shows that the antenna is working well at 1.8 GHz, but the measurement shows slight deviation of the resonant frequency to 1.74 GHz due to the larger tolerance of fabrication when using manual cutting tools. Besides that, the ShieldIt conducting fabric is not perfectly attached to the felt substrate resulting in small air gaps. Moreover, the misalignment and soldering of the SMA conductor causes the deviation in the resonant frequency.

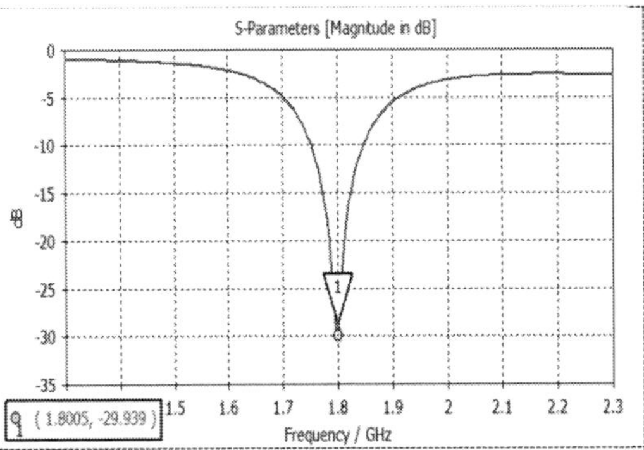

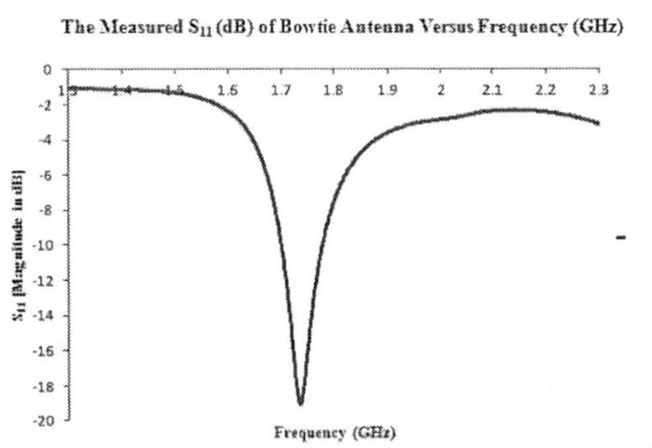

(a) (b)

Figure 3. (a) The simulated reflection coefficient versus frequency (b) measured reflection coefficient versus frequency

The simulated radiation pattern of the antenna at 1.8 GHz is as illustrated in figure 4(a) and 4(b) respectively, at $\varphi = 0°$ and $\varphi = 90°$ cuts. The antenna has half-power beamwidths of 94.6° at phi = 0° and 71.5° at $\varphi = 90°$, while the far-field realized gains of 2.55 dB at $\varphi = 0°$ and 2.56 dB at $\varphi = 90°$. The radiation efficiency and total antenna efficiency is as depicted in Figure 5(a) and 5(b), respectively.

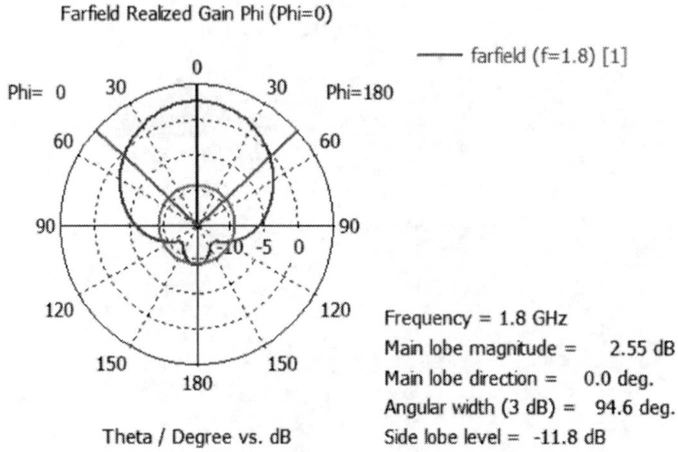

(a)

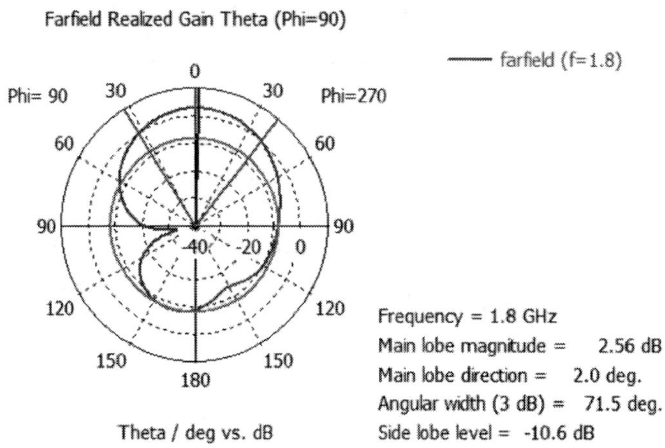

(b)

Figure 4. (a) Radiation pattern at $\varphi = 0°$ and (b) at $\varphi = 90°$.

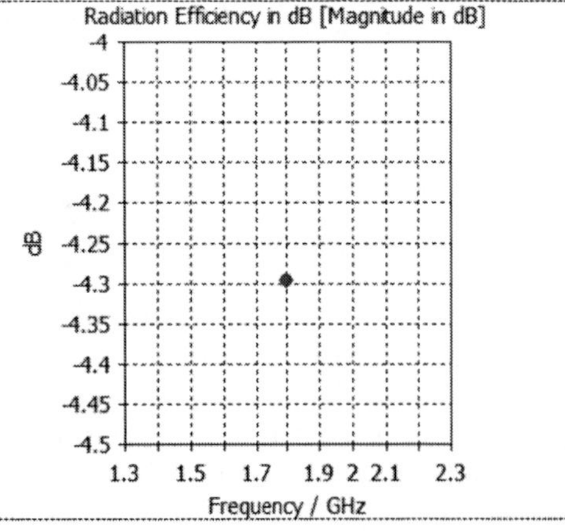

(a)

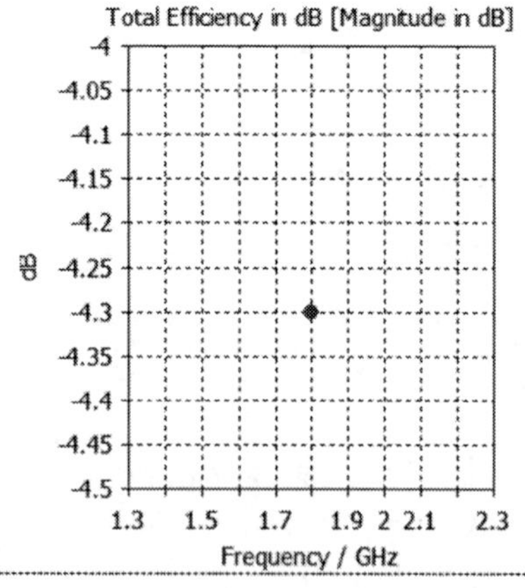

(b)

Figure 5. (a) Radiation efficiency (b) total efficiency of the bowtie patch antenna

4. Conclusions

The simulation and measurement results have validated that the textile bowtie patch antenna in this work is operating at 1.8 GHz, despite some deviation in measured performance caused by fabrication tolerances. The performance indicated that this antenna is suitable for RF energy harvesting application at GSM band since it has sufficient realized gain of 2.55 dB and 2.56 dB at $\varphi = 0°$ and $\varphi = 90°$ respectively, and a total antenna efficiency of -4.3 dB.

References

[1] Keyrouz S, Perotto G and Visser H J, "Novel broadband Yagi-Uda antenna for ambient energy harvesting," *42nd European Microwave Conf.*, Amsterdam, 2012, p 518-521.

[2] Salsabila S and Munir A, "1.8GHz printed bow-tie dipole rectenna with voltage quadrupler for RF energy harvesting," *IEEE Region 10 Conf. (TENCON)*, Penang, 2017, p 2739-2742.

[3] Nguyen N, Tuan N N, Nguyen Q C, Truong V B G, Krairiksh M and Thuy Le M, "Multiband Antenna for RF Energy Harvesting," *Int. Symp. on Antennas and Propagation (ISAP),* Busan, Korea (South), 2018, p 1-2.

[4] Wen J, Xie D, Liu X, Guo H, Liu C and Yang X, "Wideband collar-shaped antenna for RF energy harvesting," *Asia-Pacific Int. Symp. on Electromagnetic Compatibility (APEMC)*, Shenzhen, 2016, p 253-255.

[5] Chahat N, Zhadobov M and Sauleau R, "Yagi-Uda textile antenna for on-body communications at 60 GHz," *7th European Conf. on Antennas and Propagation (EuCAP)*, Gothenburg, 2013, p 411-413.

[6] Monti G, Corchia L and Tarricone L, "UHF wearable rectenna on textile materials," *IEEE Trans. Antennas Propag.* **61**(7) p 3869–3873 2013.

[7] Fernandez-Garcia R and Gil I, "Textile Yagi antenna at 1.8 GHz," *Progress in Electromagnetics Research Symposium - Spring (PIERS)*, St. Petersburg, 2017, p 652-654.

[8] Soh P J, Vandenbosch G A E, Ooi S L and Nurul H M R, "Characterization of a Plain Broadband Textile PIFA", *Radioengineering*, **20**(4) 2011 p 718 – 725.

[9] Wong K L, *Compact and Broadband Microstrip Antennas*, John Wiley and Sons Inc., New York, NY, 2002.

IOP Conf. Series: Materials Science and Engineering **767** (2020) 012070 doi:10.1088/1757-899X/767/1/012070

Estimation of Harumanis (*Mangifera indica L.)* Sweetness using Near-Infrared (NIR) Spectroscopy

M S Amirul[1,2], R Endut[1,2], C B M Rashidi[1,2], S A Aljunid[1,2], N Ali[3], M H Laili[4], A R Laili[1,4] and M N M Ismail[2]

[1]School of Computer and Communication Engineering (SCCE), Universiti Malaysia Perlis, 02600, Arau, Perlis, Malaysia.
[2]Centre of Excellence Advance Communication Engineering (ACE) Optics, SCCE, Universiti Malaysia Perlis, 01000, Kangar, Perlis, Malaysia.
[3]School of Microelectronic Engineering (SoME), Universiti Malaysia Perlis, 02600 Arau, Perlis, Malaysia.
[4]Photonics R&D, Department of Industrial Electronics Technology (IET), MIMOS Berhad, Technology Park Malaysia, 43300 Kuala Lumpur, Malaysia.

rosdisham@unimap.edu.my

Abstract. Harumanis mango quality demanded by consumers is depending on the sweetness level of the fruit. The sweetness is evaluated by brix percentage using refractometer as a representative factor correlated with near-infrared (NIR) spectroscopy spectral absorbance. NIR spectroscopy method of sampling have been tested to overcome the time consuming, complex chemical analysis more importantly invasive sampling methods in order to determine the sugar content in mangoes.Spectral absorbance data from range 941 nm to 1685 nm of mango skin is correlated with Brix reading then tested through five pre-processing techniques. Data calibration and prediction of both data is evaluated using Partial Least Square Regression (PLSR) model. In the final analysis, Unit vector normalization (UVN) technique has achieved as a best pre-processing technique for predicting results, with the coefficient of determination (R^2) values of 0.9836 and root mean square error (RMSE) values of 0.3131. Overall, the correlation of NIRS absorbance data and Brix data can be obtained using PLSR model with UVN pre-processing technique. Henceforth, we can conclude that the NIRS method of sampling can be used to identify sugar content in Harumanis mango by using time saving, non-invasive and less laborious method of sampling.

1. Introduction

Harumanis mango (Mangifera indica L.) is one of the most popular tropical fruit in Malaysian markets, which is mainly cultivated and produced in the state of Perlis. The Harumanis prefer the lime structure of the soil and climax changes in Perlis and Kedah, which make it the only state that suited the mango plant growth [1]. Therefore, the limited size of cultivation, quality produced and seasonal has made this fruit really expensive and consumer prefer the best quality they ever produced. Consumer preference is mainly driven by sweetness, which is dependent on physiological parameters described by various researchers [2–4]. Increase of total soluble solids (TSS), carotenoid pigments and decrease in acidity are some indicators of the sweetness of mango [5]. Currently, most consumers determine these by experiencing surface firmness, gloss, aroma, flavour, etc., which is often misleading [6]. Even for other quality demands fruits and vegetables, many industries have been conducting studies for maturity and quality indices, but most of them are of a chemical or destructive method, and their determination involve very laborious techniques [7–9].

Currently, near-infrared (NIR) technology is a non-destructive method for analysing the ingredients of organic products, and it is being successfully implemented throughout the food industry [7]. NIR systems are applied both in laboratories and in on-line production processes and has been used for quality evaluation [10]. These techniques are rapid and easy to conduct. Many physical characteristics of fruits and vegetables have been determined non-destructively [11, 12]. The objective of this study was to examine the potential of NIR spectrometer to predict the sweetness of Harumanis using non-invasive, rapid and non-laborious method of sampling. Besides that, this study will analyse the performances of Partial Least Square Regression (PLSR) model with different data pre-processing techniques for the prediction of mango sweetness using NIR spectroscopy. Last but not least, to find the range of wavelength in NIR spectrum, which is highly correlated with sweetness properties.

2. Materials and methods

2.1. Mango skin and juice sampling

For sampling, 30 Harumanis mangoes is taken at the peak time of the fruit season, which is between April to June [13]. The mango is selected randomly covering different size, firmness, maturity and color, but plucked from the tree at the same time and same farm. The mango then will be kept at room temperature for about 1 week to achieve the required maturity which will consequently increase the sucrose content in the fruit [14]. The mango sample was split into two types of sample which is the juice and skin. Firstly, the mango is labelled as 24 sections at both sides, including the bottom and upper part of the fruit as shown in figure 1, giving in 1440 readings for overall juice and skin data.

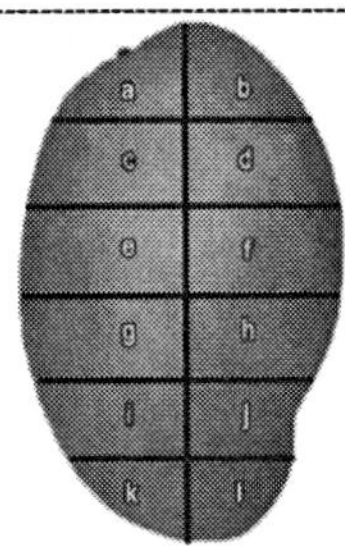

Figure 1. For every sides of mango,12 section is labelled on the skin

The juice sample will also be taken at the same spot according to the skin section labelled. Juice sample will be used by refractometer for sweetness calibration data, while skin sample will be used by NIR spectrometer for prediction in further correlation between both data.

2.2. Calibration and spectral data measurement

Hanna HI96801 (Hanna Instrument Inc., Woonsocket, USA) portable refractometer has been used to identify the sugar content of the Harumanis mango juice samples. The device converts the refractive index of food sample to % Brix and ideal for the analysis of fruits and various foods. The measurement is required to be prepared in liquid form. The conversion is based on the tables found in the International Commission for Uniform Methods of Sugar Analysis (ICUMSA) Methods Book that documents the changes in refractive index with temperature for a percent by weight sucrose solution [15]. The refractometer has a Brix range of 0% to 85% with an accuracy of ± 0.2% and this will be used as calibration data.

For spectral data, the absorbance will be measured using the DLPNIRscan Nano Evaluation Module (EVM) (Texas Instrument, Texas, USA). NIRscan data are presented in absorbance absorbance contrary to reflectance given by the light flashed to the surface of the mango skins and juice. Absorbance is detailed in spectrum and measured from 900 nm to 1700 nm for every data reading [16]. Before taking the measurement, NIRscan need to calibrate with Polytetrafluoroethylene (PTFE) compound, which giving the highest reflectance and lowest absorbance in order for the device to give best result with less calibration error [17]. For every measuring point, 10 readings are taken consequently just in about 3 second by the NIRscan. Then, the reading is presented as one reading only after averaging automatically by the NIRscan systems itself.

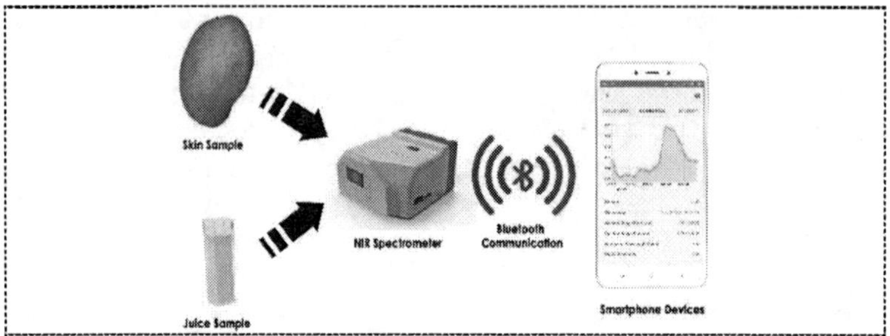

Figure 2. Technique of sampling the mango skin and juice using NIR spectrometer.

Firstly, NIRscan is used to measure the spectral data of mango skin which have been labeled. This data will be used to correlate with Brix data in order to achieve this study objective. After that, the labelled section of the mango is sliced one by one, blended and sieved to prepare the juice samples for refractometer measurement. Besides, the spectral data of the juice samples are also taken using NIRscan with a specialized setup which needed the juice samples to be prepared in a cuvettes. Figure 2 shows the technique of sampling the mango skin and juice. This data will be used to check the correlation between Brix data and mango juice spectral data. Finally, after saving both calibration and spectral data, the spectral data are plotted to analyze the pattern and identify measurement error form the absorbance response of the soil samples specifically to avoid further error in the correlation.

2.3. Pre-processing technique

Several first and last data points were removed from original absorbance spectra to avoid data interruption and noise [18], leaving remaining 941 nm to 1685 nm from overall NIR spectrum wavelength. Besides that, the data are also going through the pre-treatment process to remove negative, null and harsh data. Moreover, this process also can improve performance of the PLSR model [4]. Five out of various pre-processing techniques have been selected after being tested and analyzed consequently. Unit vector normalization (UVN), Multiplicative scatter correction using common amplification (MSCCA), multiplicative scatter correction using common offset (MSCCO), multiplicative scatter correction (MSC) and mean normalization (MN) technique is selected to treat the spectral data in this study because of their optimum performance in the PLSR model testing. The comparison of these pre-processing has been carried out to determine the best pre-processing technique and has been concluded in the results by evaluating the coefficient of determination (R^2) and root mean square error (RMSE).

2.4. Prediction model

Partial least square regression (PLSR) is a method for constructing predictive models when the factors are many and highly collinear [19]. In other words, the model emphasis on predicting the responses and not necessarily trying to understand the underlying relationship between variables. The X and Y-scores are selected with the goal that the connection between progressive sets of scores is as solid as could be expected under the circumstances. Fundamentally, this resembles a robust form of redundancy analysis, looking for direction in the factor space that is related to high variability in the reactions yet biasing them toward a direction that is accurately predicted [19].

PLSR generalizes and combines features from principle component analysis (PCA) and multiple regression (MR) to predict or analyze a group of dependent variables from a group of variables or predictors. Eventually, a prediction is retrieved by extracting it from the predictors set of orthogonal factors called latent variable (LV) which consisting the best predictive power. The orthogonal basis of LV is constructed by PLSR, aligned along the direction of maximal covariance between the spectra matrix X and the response Y. As for this analysis, the maximum number of LV is set at 10 because any number higher than this is resulting in non-actual data [20]. For this reason, PLSR will be preferred as the prediction model to validate the correlation between calibration data of Brix and NIR spectral data. Before developing the calibration model, two groups of data consisting of 75% for calibration and 25% for prediction are randomly distributed from the overall sample data. This ratio is the ideal ratio preferred for the prediction model analysis that is performed in previous studies[21]. Full cross-validation is selected to evaluate the quality and avoid over-fitting of the graph to achieve the calibration models while the prediction set validates the models. PLSR analysis and pre-processing techniques are tested and evaluated using statistical software named 'The Unscrambler X' (version 10.4, Camo Process AS, Oslo, Norway).

3. Result and discussion

Figure 3 shows a spectral plot for the overall 720 reading of Harumanis mango skins in the NIR spectral range of 941 nm to 1685 nm. The curve show respond of absorbance varies with a wavelength between all of the skin samples. From this graph, skin sample absorption is intensively high at range 1424 nm to 1560 nm. Henceforth, these absorbance were then analyzed using a prediction model to evaluate the correlation between absorbance and Brix data reading.

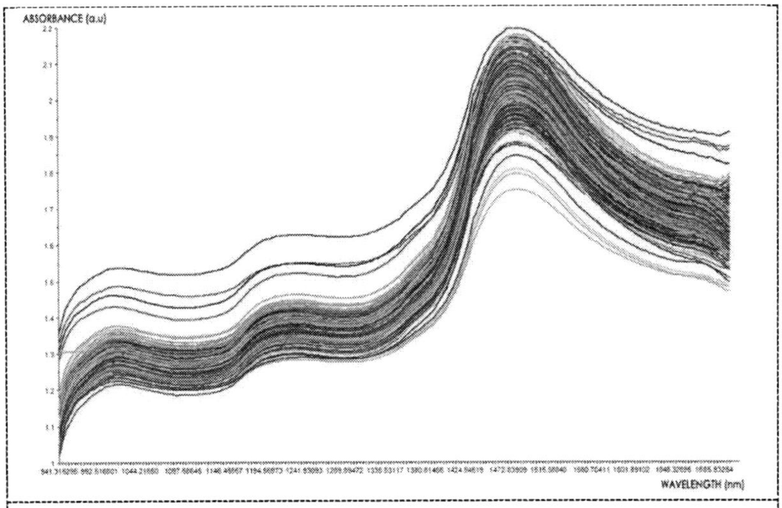

Figure 3. Spectral plot of mango skin samples absorbance.

As a result, both Brix and spectral absorbance data is then evaluate using the PLS regression prediction model and the top performer's pre-processing technique is then selected. Out of 20 pre-processing techniques, five that shown interest are UVN, MSCCA, MSCCO, MSC, and MN. In table 1, the evaluation of Brix data and spectral absorbance is presented with a different pre-processing technique that has been tested.

The selected absorbance spectral is spilled into 7 range section that is distributed evenly, and section 7th (1578 nm to 1684 nm) is selected as the observation range to be analyzed because this range shows a highly anticipated correlation for both data. For this analysis, R^2 and RMSE will be the benchmark of performance for the correlation between Brix data and spectral absorbance data. In spite of R^2, the best performance for the correlation to happen is between the ranges of 0.70 to 0.90, or even better if possible, nearer to 1.00. On the contrary, best RMSE are evaluated in ranges of 0.10 to 0.90 or even better if possible to reach nearer 0.00 [22]

Table 1. PLSR evaluation result with optimum pre-processing technique

Pre-processing technique	Calibration		Prediction	
	R^2	**RMSE**	**R^2**	**RMSE**
Unit Vector Normalization (UVN)	0.9856	0.2926	0.9836	0.3131
Multiplicative Scatter Correction using Common Amplification (MSCCA)	0.9859	0.2831	0.9832	0.3175
Multiplicative Scatter Correction using Common Offset (MSCCO)	0.9214	0.6843	0.9096	0.7366
Mean Normalization (MN)	0.9072	0.7439	0.8865	0.8265
Multiplicative Scatter Correction (MSC)	0.8830	0.8353	0.7962	1.1075

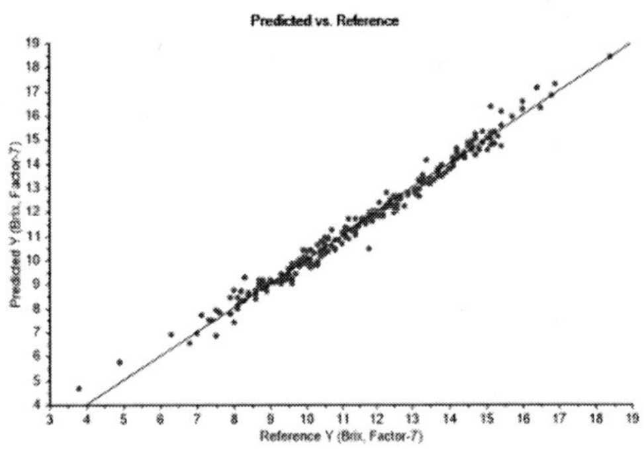

Figure 4. PLSR prediction using UVN technique

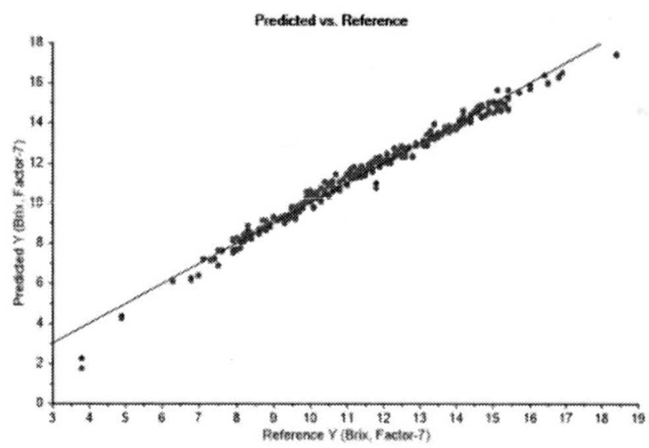

Figure 5. PLSR prediction using MSCCA technique

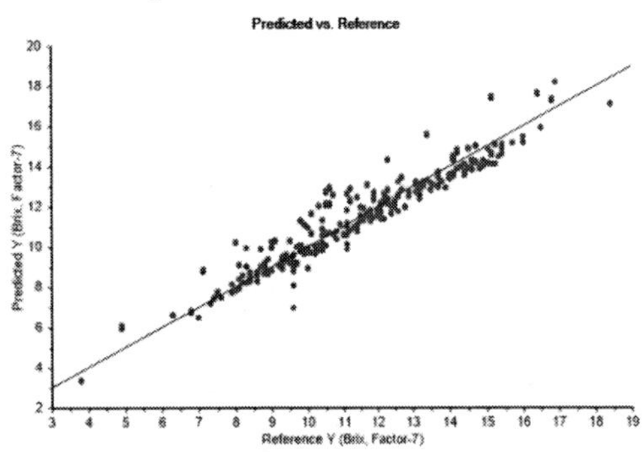

Figure 6. PLSR prediction using MSCCO technique

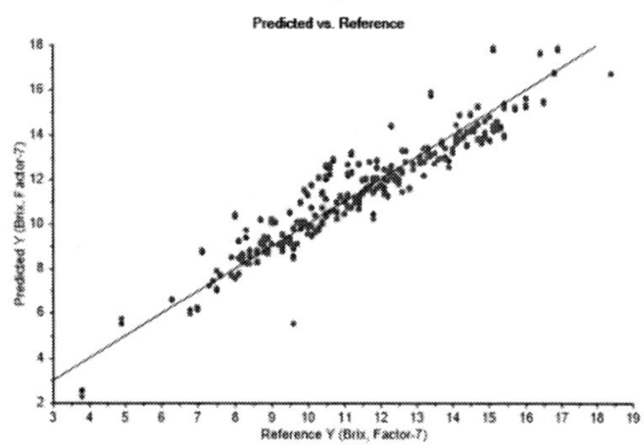

Figure 7. PLSR prediction using MN technique

IOP Conf. Series: Materials Science and Engineering **767** (2020) 012070 doi:10.1088/1757-899X/767/1/012070

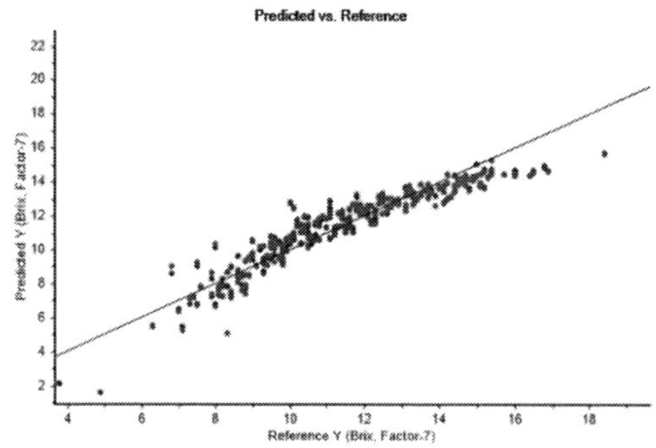

Figure 8. PLSR prediction using MSC technique

Graph plot in figure 4 to figure 8 shows the correlation of both calibration and prediction of Brix and spectral data. The red plot represent predicted data and the blue plot represents calibrated data. The line represent regression line for the correlation. From table 1, the MSCCA technique shows the most anticipated and intended calibration with R^2 values of 0.9859 and RMSE of 0.2831. However, the prediction result is at the second-best with R^2 values of 0.9832 and RMSE values of 0.3175 as shown in figure 5. More importantly, the prediction result of UVN is in the best-expected range as shown in figure 4, in which R^2 is at the values of 0.9836 and RMSE of 0.3131 which little lower compared to calibration. For calibration, the UVN technique able to achieve R^2 values of 0.9856 and RMSE values of 0.2926 This evaluation shows that prediction for Brix data is highly correlated to specified spectral absorbance data using this technique.

As we can observe from figure 7, the MN technique has achieved less linear plot if compared with MSCCA and UVN technique. With the values of 0.9072 for R^2 and 0.7439 for RMSE, it is considered 2nd lowest performance for calibration after the MSC technique. Looking at the prediction, it can only achieve R^2 values of 0.8865 with RMSE values of 0.8265. This is not even close is compared to the UVN technique, but nearly following the prediction values for the MSCCO technique.

Comparatively, the MSCCO technique as shown in figure 6 have accomplished as 3rd highest calibration performance with R^2 values of 0.9214 and RMSE of values 0.6843. Although, compared to the predicted result, the MSCCO technique also achieved 3rd highest performer with R^2 values of 0.9096 and RMSE values of 0.7366.

Meanwhile, observing the MSC technique in figure 8, the predicted result is still acceptable even though the prediction RMSE is out of the targeted range. MSC technique achieved calibration R^2 values of 0.8830 and RMSE values of 0.8353, while for prediction R^2 values of 0.7962 and RMSE values of 1.1075. Given this point, we can assure that the UVN technique has scored the best pre-processing technique so far compared to the other four techniques tested in this study for correlation of Brix and spectral absorbance data. As a suggestion for future research, a lot of other variables should be considered for observation which can also affect the performance of the prediction model such as firmness of the mango and also the color intensity of mango. Besides that, to increase correlation and the accuracy of the prediction model, a larger number of samples are expected.

4. Conclusion

In brief, after going through overall testing and observation, UVN pre-processing technique has been identified as the best performer for the analysis of Brix data correlation with spectral absorbance data of the skin samples of the Harumanis mango. Moreover, we can summarize that using NIRS, sweetness level of Harumanis mango can be identified more efficiently using time-saving, non-invasive and less laborious methods of sampling, enhanced with the UVN pre-processing technique and PLSR model. For future research, numerous fruits can be proposed and tested for sweetness estimation using the NIRS instrument as an efficient tool.

Acknowledgements
The author would like to acknowledge the support from the Fundamental Research Grant Scheme (FRGS) under a grant number of FRGS/1/2015/TK04/UNIMAP/03/2 from the Ministry of Education Malaysia.

References

[1] Dermawan A. Harumanis takes root in Kedah. *New Straits Times*, [available online]: https://www.nst.com.my/news/nation/2019/06/493403/harumanis-takes-root-kedah (2019).

[2] Jha S N, Jaiswal P, Narsaiah K, et al. Non-destructive prediction of sweetness of intact mango using near infrared spectroscopy. *Sci Hortic (Amsterdam)* 2012 **138** p 171–175.

[3] Taira E, Nakamura S, Hiyane R, et al. Development of a nondestructive measurement system for mango fruit using near infrared spectroscopy. *Eng Appl Sci Res* 2017 **44** p 189–192.

[4] dos Santos Neto J P, de Assis M W D, Casagrande I P, et al. Determination of 'Palmer' mango maturity indices using portable near infrared (VIS-NIR) spectrometer. *Postharvest Biol Technol* 2017; **130** p 75–80.

[5] Wanitchang P, Terdwongworakul A, Wanitchang J, et al. Non-destructive maturity classification of mango based on physical, mechanical and optical properties. *J Food Eng* 2011; 105: 477–484.

[6] Jha S N and Ruchi G. Non-destructive prediction of quality of intact apple using near infrared spectroscopy. *J Food Sci Technol* 2010; **47** p 207–213.

[7] Khodabakhshian R, Emadi B, Khojastehpour M, et al. Non-destructive evaluation of maturity and quality parameters of pomegranate fruit by visible/near infrared spectroscopy. *Int J Food Prop* 2017 **20** p 41–52.

[8] Magwaza L, Tesfay S. A Review of Destructive and Non-destructive Methods for Determining Avocado Fruit Maturity. *Food Bioprocess Technol*; 2015; **8**. DOI: 10.1007/s11947-015-1568-y.

[9] Kumari P, Feza M and Mir H. Non-destructive quality evaluation by sensing maturity and ripening of fruits and vegetables. 2018 p 84–89.

[10] Porep J U, Kammerer D R and Carle R. On-line application of near infrared (NIR) spectroscopy in food production. *Trends Food Sci Technol* 2015; **46** p 211–230.

[11] Beghi R, Giovenzana V, Tugnolo A, et al. Application of visible/near infrared spectroscopy to quality control of fresh fruits and vegetables in large-scale mass distribution channels: a preliminary test on carrots and tomatoes. *J Sci Food Agric* 2018; **98** p 2729–2734.

[12] Wang H, Peng J, Xie C, et al. Fruit Quality Evaluation Using Spectroscopy Technology: A Review. *Sensors* 2015; **15** p 11889–11927.

[13] Zakaria A, Shakaff A Y M, Masnan M J, et al. Improved Maturity and Ripeness Classifications of Magnifera Indica cv. Harumanis Mangoes through Sensor Fusion of an Electronic Nose and Acoustic Sensor. *Sensors* 2012, **12**. DOI: 10.3390/s120506023.

[14] Saad F S A, Shakaff A Y M, Zakaria A, et al. Bio-inspired Vision Fusion for Quality Assessment of Harumanis Mangoes. In *2012 Third International Conference on Intelligent Systems Modelling and Simulation*. 2012, p 317–324.

[15] Hanna Instruments Inc. *HI 96801 Refractometer for Sucrose Measurements*. Woonsocket, USA: Hanna Instruments Inc., 2014.

[16] Thangappan A and Pathangay V. Non-Invasive Blood Glucose Estimation from Near Infrared Spectrum using Gradient Boosted Tree Models. *IEEE-EMBC* 2016; **3**.

[17] Inácio M R C, de Lima K M G, Lopes V G, et al. Total anthocyanin content determination in intact açaí (Euterpe oleracea Mart.) and palmitero-juçara (Euterpe edulis Mart.) fruit using near infrared spectroscopy (NIR) and multivariate calibration. *Food Chem* 2013; **136** p 1160–1164.

[18] Hervé A. Partial least squares regression and projection on latent structure regression (PLS Regression). *Wiley Interdiscip Rev Comput Stat* 2010; **2** p 97–106.

[19] Tobias R D. An introduction to partial least squares regression. *SAS Conf Proc SAS Users Gr Int 20 (SUGI 20)* 1995; p 2–5.

[20] Anusia H, Jayaselan J, Nawi N M, et al. Application of Spectroscopy for Nutrient Prediction of Oil Palm. 2017; **15** p 1–9.
[21] Mat Nawi N, Chen G, Jensen T, et al. Prediction and classification of sugar content of sugarcane based on skin scanning using visible and shortwave near infrared. *Biosyst Eng* 2013; **115** p 154–161.
[22] Kyprianidis K G and Skvaril J. *Developments in Near-Infrared Spectroscopy*. Croatia: Janeza Trdine 9, 51000 Rijeka, Croatia, 2017. DOI: http://dx.doi.org/10.5772/62932.

IOP Conf. Series: Materials Science and Engineering **767** (2020) 012071 doi:10.1088/1757-899X/767/1/012071

Low Cost Autonomous Robot Cleaner using Mapping Algorithm based on Internet of Things (IoT)

R J Ong[1] and K N F Ku Azir[1]

[1] School of Computer and Communication Engineering, Universiti Malaysia Perlis, Pauh Putra Campus,02600 Arau, Perlis, Malaysia.

fazira@unimap.edu.my

Abstract. The objective of this study is to design and develop a low-cost Arduino autonomous robot cleaner using mapping algorithm to clean floor area of houses or offices. The idea is basically to detect any obstacles with the help of sensor and sent its output to microcontroller that will control the autonomous vacuum cleaner movement. A low-cost solution is proposed in this study by using HC-SR04 ultrasonic sensor for obstacle avoidance and control by Arduino UNO. By using an autonomous vacuum cleaner, user can turn ON the autonomous vacuum robot to clean without any help of a human operator. Wall mapping and random mapping is being applied in this study to find the effective mapping algorithm for autonomous robot cleaner. Additionally, instead of using the traditional button or switch to activate the robot, voice recognition through Google Assistant implemented in this project. Hence, this provides a more user friendly platform not only for normal user but also help visually impaired people to activate the robot.

1. Introduction

A robotic vacuum cleaner often referred to as a RoboVac, is an autonomous robotic vacuum cleaner with intelligent programming and an electromechanical control system with high stability and rapid functionality. In 2002, iRobot, an American advanced technology company, launched a robotic vacuum cleaner called Roomba. The Roomba was able to change direction when it encountered an obstacle, spot dirty spots on the floor, and detect steep debris to prevent the stairs from falling down. Additionally, an autonomous robot can also learn new knowledge, e.g. adapting new methods to perform its tasks or adapting to changing environments. The vacuum cleaner robot should have an artificial intelligence mechanism to solve the problem of cleaning the entire environment area, taking into account certain factors such as the number of turns and the length of the trajectory [1]. There have been new variations in robotic vacuum cleaners on the market since 2002. For example, the Canadian bob sweep robot vacuums cleaner, which use both mops and vacuum, or the Neato RoboticsXV-11 robot vacuum cleaner, which uses laser vision instead of conventional ultrasonic models. In 2014, Dyson announced the release of its new robotic vacuum cleaner, Dyson 360 Eye, equipped with a 360-degree camera mounted on the top of the robotic vacuum cleaner or better navigation than other brands. The company is researching robotics and AI to create better robotic vacuum cleaners in the future.

Four requirements should be met by a fully autonomous robot. First, the robot should be able to gain environmental information. For a robot to associate behaviour with a location, it must know where it is and be able to navigate from point to point. Current commercial robots navigate autonomously based on the recognition of natural features. Second, the robot can move wholly or partially through its operating environment without human support. The advantage of a robot vacuum cleaner compared to conventional vacuum cleaners is more convenient due to its small size and can also be vacuumed by itself. Thirdly, an autonomous cleaner should be able to avoid situations which are harmful to persons, property or itself unless this is part of its design specifications.

In a nutshell, this study aims to develop an autonomous IoT-based vacuum cleaner that uses a low-cost sensor to detect any obstacle and send its output to a microcontroller that controls the robot's movement. With the standalone vacuum cleaner, the user can turn on the robot for cleaning using the remote control or voice recognition through Google Assistant on any Android phones. Once this machine is ON mode, it moves throughout the floor and efficiently covers an entire floor area. In addition, the suction power of this robot must be sufficient to collect dust, sand, human or pet hairs and some tiny rocks that are possible find out indoors. Hence, a small high-speed DC motor with 3000rpm – 10000rpm is used to rotate a fan in order to produce strong air flow from one point to another and archive vacuum function.

2. Literature Review

As robotic technologies have evolved and become imperative, people have tried to alternate human work with new mechanical advances in autonomy, especially where people can take risks at work [1]. Robots have recently emerged as household appliances as human demands increase [10]. Robotics research for household appliances is becoming more active than ever. The market for cleaning robots is expected to grow from USD 1.83 billion in 2017 to USD 4.34 billion in 2023, with CAGR (Compound Annual Growth Rate) at 16.21% between 2018 and 2023, according to semiconductor and electronics market research. Besides, The Medium & Long Term Science and Technology Planning of China (2006 to 2020) have been clearly established that the autonomous will be a primacy enhancement of technology in the future.

It is well known that cleaning a large area is an endlessly boring task that requires a lot of time and effort [3]. Currently, the cleaning robot can be divided into the vacuum cleaner, the wiping robot, etc. according to various functions [4]. The robotic vacuum cleaner was introduced earlier among all these robots. Electrolux introduced the first autonomous vacuum cleaner in 1996. The early robotic vacuum cleaners failed to avoid obstacles and low cleanliness. As a result, this cleaner cannot be successfully brought to market. The UK technology company Dyson developed a robotic vacuum called DC06 in 2001. Due to its high price, however, it was never released on the market.

Since 2002, most of the cleaning robot's research has focused on reducing development costs and developing a highly efficient cleaning robot. There are many improvements made using various sensor and path planning algorithms to archive the goal. In [4], suggested a multifunctional floor cleaner in which can mop and also vacuum the floor. In [5], has developed an independent cleaning robot to remove dust from the solar panel to maximize the solar panel's life and efficiency. In [6], proposed an autonomous robotic vacuum cleaner with 3D vector coordinates for the planning of robot routes. Then, came the idea of Smart Dustbin is to place sensors in the container to detect the level of garbage. When the garbage reaches the threshold, the status of the bin in the cloud is updated and a notification is sent to the user. The garbage that accumulates in garbage bins is collected at the time of day, even if the garbage bins are filled at any time. Therefore, the existing technology is not fruitful as the emerging technology IoT is considered. IoT components such as sensors, detectors and actuators are specifically integrated with the Intelligent System (IS) and inspection system to ensure efficient waste disposal [7].

There are many smart waste disposal projects that are described in the literature today. In [8], Waitkus has implemented a waste bin and garbage collection monitoring system to monitor the capacity of the containers, to notify the garbage trailer, and to schedule the collection of garbage based on customer preferences. In [9], the ultrasonic sensor measures the levels of waste in the dustbin and sends the data through a wireless ad hoc network to the server. In [10] used RFID tags and load cell sensor technology for waste treatment.

Most of the works above require inconvenient operations and/or costly solutions. Therefore, there is essentially to reduce the system design complexity, cost-effective and universal availability of the infrastructure [11]. The growth of information technology and the use of pervasive computing technology have occurred in many applications, including waste disposal [12]. Therefore, in [2] propose fully automated indoor waste disposal to replace conventional disposal facilities by using a mobile waste collection robot. This method is definitely a time effective process than currently used methods.

Mapping and localisation are two closely related activities in the autonomous robot cleaner. The map created can be used in a variety of applications, such as route planning, safety, cleaning, monitoring and just to name a few. Since the autonomous robot does not require any human input for its work. Therefore, the localization and navigation of the robot will be the main concern in the design. High-priced, highly functional sensors such as cameras and laser scanners provide a wealth of information about surrounding environments and are therefore widely used. Both sensor types were successfully used for the creation of raster maps. In particular,' Kinect ' sensors have recently come to the fore, as they are cheaper but provide rich 3D information. In [2] et.al presented a robot navigation system based on the Kinect device and encoder to create a 2D map. Ronny Mardiyanto's system is tested in a laboratory setting. The effective distance is between 60 and 600 cm. Their rotary encoder sensor has an average error of 5.05% on the created 2D map and an error distance of 2.23%. This 2D map creator took 1.46 seconds to create each map in one place. They are disadvantageous, however, because they do not recognize transparent objects like windows. Additionally, high information quality is difficult to develop an algorithm to manage errors and noise. It is therefore not currently possible to promote such sensors to the market. In contrast, ultrasonic sensors are practicable for obstacles avoidance and robot mapping. The HC-SR04 ultrasonic sensor with a range of up to 4 meters is a cost-effective solution for mapping robots, rather than using an expensive laser sensor or camera. In [20] suggest a new approach to creating a raster map for a mobile robot with sonar data. Experimental results and evaluations in the home environment demonstrate the validity of the methods proposed. Complete raster maps, which were created in the way suggested by the research, probably reduce the development cost. In the meantime, Atanas Dimitroy and others presented a prototype of a mobile robot suitable for mapping [16]. It uses a HC-SR04 ultrasonic sensor that is a cost-effective alternative to sensor mapping instead of a costly laser sensor. The proposed robot was tested in the university laboratory and can be improved by visualizing real-time data. In fact, sonar data may not always be accurate because of specular reflective indoor operation. In [17] explains that ultrasonic sensors and compass sensors are used in the mapping algorithm. The sensors used in this document are limited to the environment. Ultrasonic sensors can only detect the distance between 3 cm and 400 cm. Ultrasonic sensors, however, is cheaper compared to another same functionality sensor, and the raster maps created with ultrasonic sensors can most often be reduced by the raster map making methodology proposed in the development of smart robots for low cost.

The current autonomous robotic vacuum cleaner production methods can be classified into three types, namely random walk, video camera location and laser location. In [18] present a new effective global localization approach for residential environments that adapts the notion of free space density to a ceiling - mounted camera. This vacuum cleaner has a camera which captures the ceiling image and uses the lens during transportation. The use of image processing technology creates a map that allows for more efficient cleaning. On the other hand, iRobot Roomba, one of the most popular robotic vacuum cleaners, used an algorithm called iAdapt and used a laser rangefinder to scan around and under furniture, avoid stairs, and clean up very efficiently. In contrast, the random walk algorithm only avoids obstacles with the help of some sensors, and the robot itself does not recognize the work environment. The first two methods, however, are not cost efficient due to the camera module and the cost of the laser rangefinder is relatively higher. In [19] assess a cost-efficient, lightweight, low noise and low maintenance robotic system. It is said that mapping technology is absent in its design because of the cost effective problem. Data shows that the robot covers the whole area more quickly with combined motion algorithms (random walk, s shape and spiral) based on numerous tests. The test result shows that their design is fully adapted to the planning of an effective mode for performing the cleaning task in an unknown environment [20] They also suggest a dynamic return to recharge, including the relative coordinates of the starting point and the current position. Based on the above research results, path planning is one of the keys to better cleaning robot performance. In this cyber-physical age, the autonomous cleaner can be improved by implementing Internet of Things (IoT) with a number of sensors, with some calculations, to develop a new algorithm to solve the problem in path planning or robotic navigation. For that reason, the motto of this project is to propose a highly efficient and cost-effective cleaning robot using an odometer and with the help of several sensors.

3. Method

The overall of this study is to design an autonomous vacuum cleaner that can perform well without human assistance. In the early stages of design, network connections must be established to ensure that all hardware is connected with each other. This can be done by either check the serial monitor in Arduino IDE or check the list of devices connected to the same local network. Once the system is online, the robotic vacuum cleaner can be either turn on or off by using voice command via Google assistant as shown in Figure 1. Meanwhile figure 2 shows the block diagram of the study. First and foremost, voice command has been implemented with Google Assistant to help the visually impaired. However, there is no way for Google Assistant to connect directly to NodeMCU. Therefore, Blynk and IFTTT are implemented to act as an intermediate bridge to allow the Google Assistant to control the microcontroller to perform as the ON / OFF switch of the robot. To achieve this must have some pre-setting in both IFTTT and also Blynk.

There are two algorithms can be chosen by the user which are wall follow and random walk. By selecting different algorithm, the decision to take will be different when the robotic cleaner duel with obstacle avoidance. After the device is activated, the vacuum pump will start to work. Moreover, all the sensors will begin to collect environment data and process within Arduino UNO. The robot will keep moving around the workplace until it runs out of battery or receives an OFF command from the user.

Figure 3 shows that the subroutine will be called when the random walk algorithm is chosen by the user. First and foremost, the robot will check is the obstacle right in front of the robot. If yes, then the robot will stop for a while in order for the DC motor to take action. The robot is programmed to reverse a bit and following by turn left (reverse the left wheel and forward the right wheel). If no, then the robot will check either the obstacle is at left or at right and apply the same algorithm but rotate in the opposite direction. While there is no obstacle, the robot will keep moving forward.

Figure 4 shows that the subroutine will be called when the wall follow algorithm is chosen by the user. First and foremost, the robot will check is the obstacle right in front of the robot or the robot reaching a corner. If yes, then the robot will stop for a while in order for the DC motor to take action. The robot is programmed to reverse a bit and following by turn right (reverse the right wheel and forward the left wheel). If no, then the robot will check is the robot reaching an edge or not. If yes, the robot will stop for a while and follow by slightly forward and turn left. Furthermore, the ultrasonic sensor at left will always make sure the robot moves along the wall with constant speed.

The backbone of this project is to connect the sensor to the microcontroller to control the actuator to perform certain tasks under certain conditions. Ultrasonic sensor used to avoid obstacles in this project. For instances, the robot is programmed to take action when there is an obstacle 10cm in front of the ultrasonic sensor. The red colour board at the top left in figure 5 is L298N H- bridge motor driver. There are two DC motor connected to this board and power supply by two 3.7V 18650 3000mAh rechargeable li-ion battery. There are two EN pins which are ENA and ENB have to connect to PWM pins from Arduino UNO in order to control the speed of rotation.

IOP Conf. Series: Materials Science and Engineering **767** (2020) 012071 doi:10.1088/1757-899X/767/1/012071

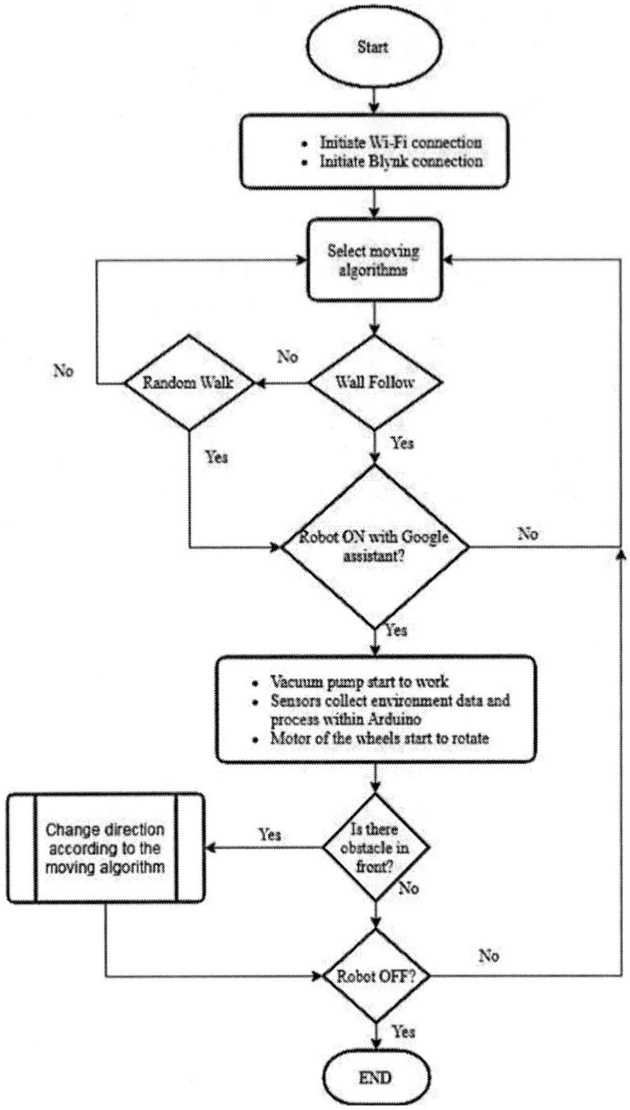

Figure 1. Flow chart of the study.

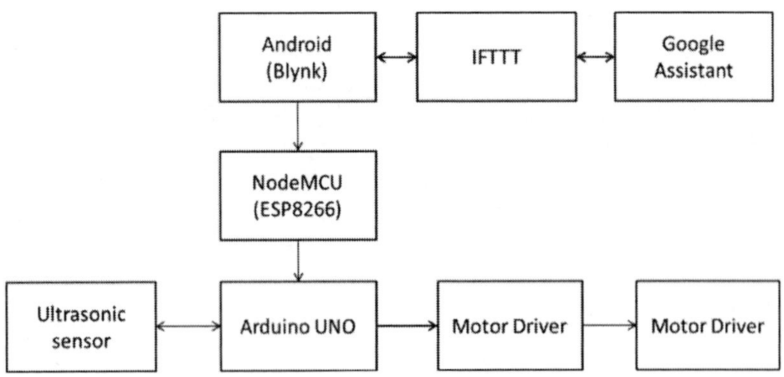

Figure 2. Block diagram of the study.

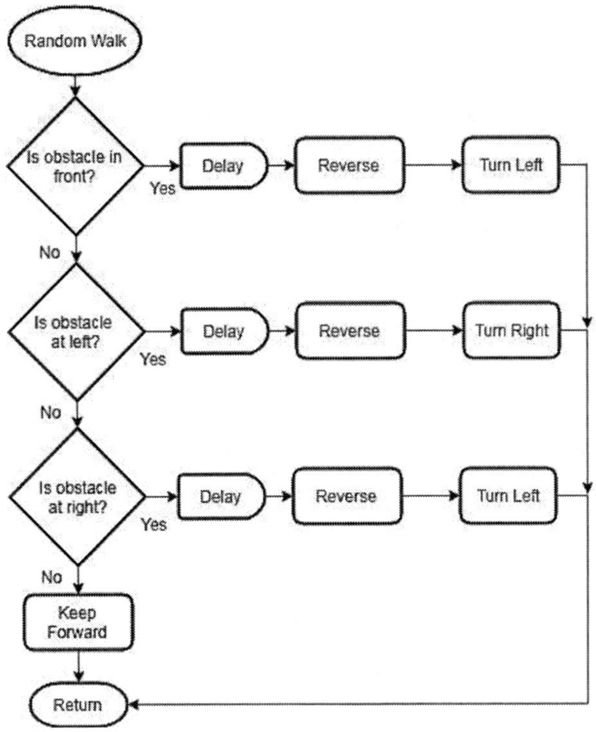

Figure 3. Flow chart of random walk.

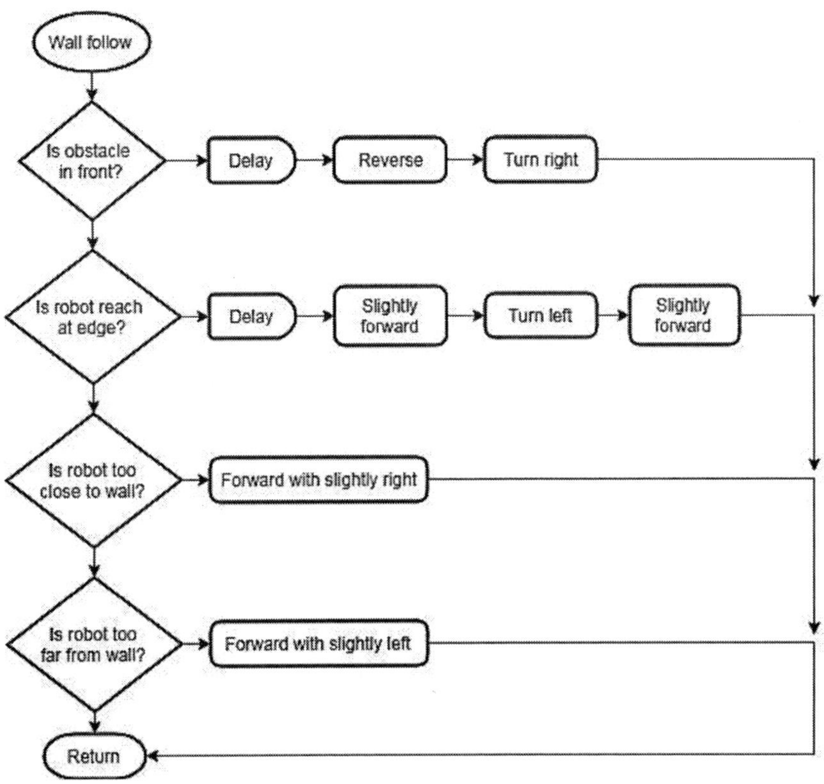

Figure 4. Flow chart of follow algorithm.

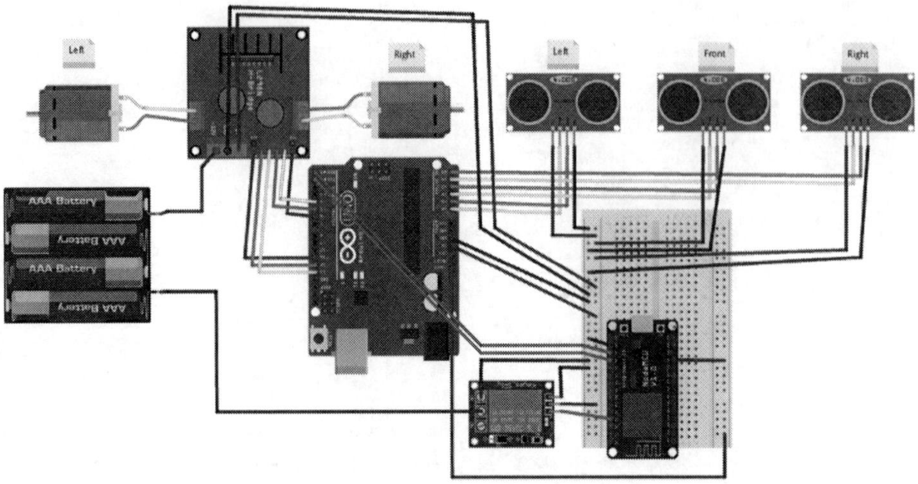

Figure 5. Circuit diagram of the study.

4. Results and Analysis

The testing of the automous robot cleaner was done by implementing both of the mapping algorithm to the automous robot cleaner. The movement of a robot can be divided into several simple movements as in Table 1. The truth table is generated based on L298N H-bridge motor driver after a few times testing programmed by Arduino UNO. The combinations of these movements are successfully achieved both wall follow and random walk algorithm.

Table 1. Truth table of motor driver

Action	Left Motor	Right Motor	ENA	ENB	IN1	IN2	IN3	IN4
Forward	On	On	Const	Const	1	0	1	0
Reverse	On	On	Const	Const	0	1	0	1
Turn Left	Off	On	0	Const	0	1	1	0
Turn Right	On	Off	Const	0	1	0	0	1
Stop	Off	Off	0	0	0	0	0	0

There are four situations considered during design. Figure 6 shows that the four situations that happened during testing and the robot effectively overcome the problem by itself without human assist. First, the robot is doing well and moving along the wall with a gap of 4cm with the wall. The robot is capable to adjust itself automatically when either going too far from the wall or too close to the wall. Second, the robot escapes from a corner and continue its path to the right intelligently. Third, the robot is doing well when come to the wall edge so that it can close back to the path and continue its task.

For a random walk algorithm, the robot proved that it is smart enough to avoid obstacle before hitting on it. In addition, there are not only static obstacles but also dynamic obstacles such as people walk through the path of the robot as shown in figure 7. Thus, the robot will stop for a while and take action before hitting on such dynamic obstacles.

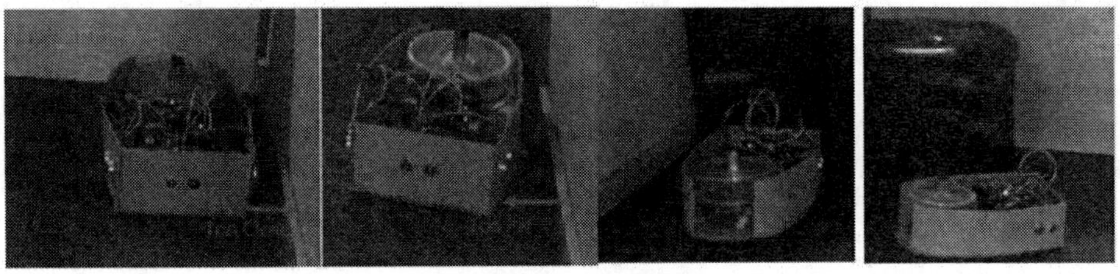

Figure 6. Four situations to overcome by wall follow mapping algorithm.

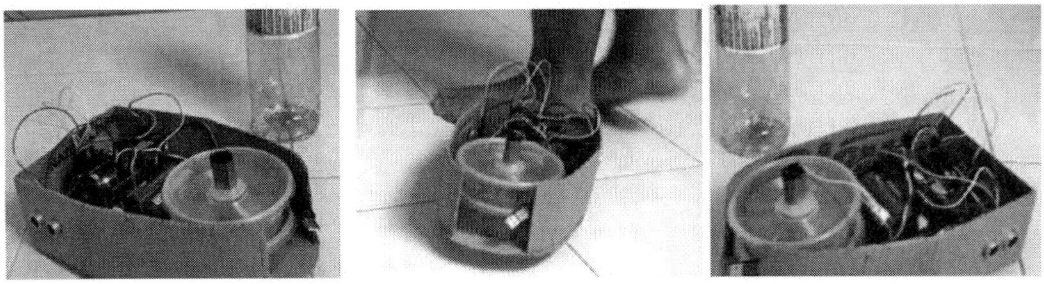

Figure 7. Random walk mapping algorithm and obstacles avoidance.

Figure 8 shows the screenshot of Google Assistant and Blynk when using voice command. As mentioned earlier in the previous chapter, there are three optional commands in order to activate the robot. When Google Assistant receives a valid command such as "turn off" in the figure 8 then it will reply "Yes sir turning off vacuum cleaner" as an acknowledgement receive command successfully.

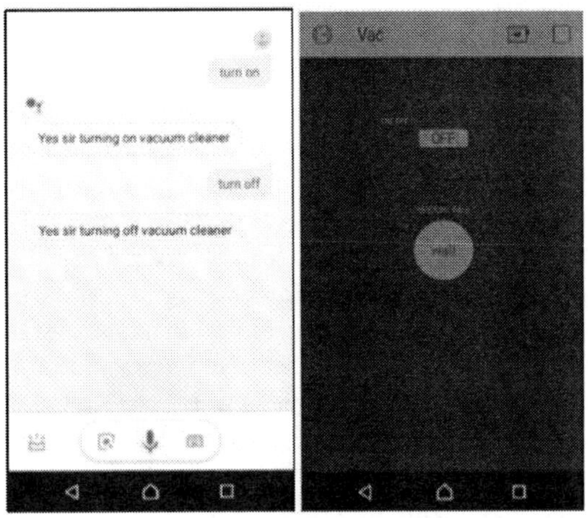

Figure 8. Google Assistant and Blynk for Voice command.

5. Conclusion

The objective of this project was to plan and deployment a cheap and smart Internet of Things (IoT) based autonomous vacuum cleaner that could work in an unknown environment. Although the robot has a minimum number of sensors, this goal has been met by deployed algorithms. Moreover, voice command fulfilled successfully and the output is same as expected. Google Assistant are available on every android smart phone, hence, this can be an ease for development and also provide a more user friendly platform not only for normal user but also visually impaired people. Walk follow and random walk algorithm implemented successfully which that all the situation considered can be overcome by the robot automatically. Even so, robot development provides endless possibilities to improve the performance and to overcome more situations for unknown environment. Future robot development therefore includes mapping technology, automatic charging algorithm and automatic dirt disposal.

Acknowledgments
Authors wishing to acknowledge assistance or encouragement from colleagues, special work by technical staff and financial support from Ministry of Higher Education Malaysia and Universiti Malaysia Perlis.

References

[1] Maddukuri S V P K, Renduchintala U K, Visvakumar A, Pang C and Mittapally S K (2017). A low cost sensor based autonomous and semi-autonomous fire-fighting squad robot. *Proceedings - 6th Int. Symp. on Embedded Computing and System Design (ISED)*, p 279–283. (doi: 10.1109/ISED.2016.7977097).

[2] Mardiyanto R, Anggoro J and Budiman F (2015). 2D map creator for robot navigation by utilizing Kinect and rotary encoder. In *Int. Seminar on Intelligent Technology and Its Applications (ISITIA)* p 81–84. (doi:10.1109/ISITIA.2015.7219957).

[3] Ravindhiran P, Gopal P, Gladwin S J and Rajavel R (2017). Automated indoor waste management system employing wavefront algorithm and received signal strength indicator values-based mobile robot. In *IEEE Region 10 Humanitarian Technology Conf. (R10-HTC)* p 284–289. (doi: 10.1109/R10-HTC.2017.8288958).

[4] Wickramaarachchi W H C, Chamikara M A P and Ratnayake R A C H (2018). Towards implementing efficient autonomous vacuum cleaning systems. *Proc. IEEE Int.l Conf. on Industrial and Information Systems (ICIIS)*, p 1–6. (doi: 10.1109/ICIINFS.2017.8300385).

[5] Hong Y, Sun R, Lin R, Yu S and Sun L. (2014). Mopping module design and experiments of a multifunction floor cleaning robot. In *Proc. of the 11th World Congress on Intelligent Control and Automation*. p 5097–5102. (doi: 10.1109/WCICA.2014.7053581).

[6] Aravind G, Vasan G, Kumar T S B G, Balaji R N and Ilango G S (2014). A Control Strategy for an Autonomous Robotic Vacuum Cleaner for Solar Panels, *Texas Instruments India Educators' Conf. (TIIEC)*, Bangalore, p 53-61. (doi: 10.1109/TIIEC.2014.18\)

[7] Bai Y W and Hsueh M F (2012). Using an adaptive iterative learning algorithm for planning of the path of an autonomous robotic vacuum cleaner. In *1st IEEE Global Conf. on Consumer Electronics (GCCE)*. p 401–405. (doi: 10.1109/GCCE.2012.6379640).

[8] Murugaanandam S, Ganapathy V and Balaji R (2018). Efficient IOT Based Smart Bin for Clean Environment. *Int. Conf.on Communication and Signal Processing (ICCSP)*, p 715–720.

[9] Folianto F, Low Y S and Yeow W L (2015). Smartbin: Smart waste management system. IEEE 10th Int. Conf.on Intelligent Sensors, Sensor Networks and Information Processing (ISSNIP) p 1–2. (doi: 10.1109/ISSNIP.2015.7106974).

[10] Chowdhury B and Chowdhury M U (2008). RFID-based real-time smart waste management system. *Australasian Telecommunication Networks and Applications Conf. (ATNAC)*. p 175–180. (doi: 10.1109/ATNAC.2007.4665232).

[11] Talha M, Upadhyay A, Shamim R and Beg M S (2018). A cloud integrated wireless garbage management system for smart cities. *Int. Conf. on Multimedia, Signal Processing and Communication Technologies (IMPACT)*. p175–179. (doi:10.1109/MSPCT.2017.8363999)

[12] Wahab M H A, Kadir A A, Tomari M R and Jabbar M H (2014). Smart recycle bin: A conceptual approach of smart waste management with integrated web-based system. *Int. Conf. on IT Convergence and Security (ICITCS)*. p1–4. (doi: 10.1109/ICITCS.2014.7021812).

[13] Yakoubi M A and Laskri M T (2016). The path planning of cleaner robot for coverage region using Genetic Algorithms. *Journal of Innovation in Digital Ecosystems*. **3**(1) p 37 – 43, (doi: 10.1016/j.jides.2016.05.004).

[14] Hasan K M, Al-Nahid A and Reza K J (2014). Path planning algorithm development for autonomous vacuum cleaner robots. In *Int. Conf. on Informatics, Electronics & Vision (ICIEV)* p 1–6. (doi: 10.1109/ICIEV.2014.6850799).

[15] Dimitrov A and Minchev D (2016). Ultrasonic sensor explorer. In *19th Int. Symp. on Electrical Apparatus and Technologies (SIELA)*. p 1–5. (doi: 10.1109/SIELA.2016.7542987)

[16] Fathan A M, Jati A N and Saputra R E (2016). *Mapping algorithm using ultrasonic and compass sensor on autonomous mobile robot*. In *Int. Conf. on Control, Electronics, Renewable Energy and Communications (ICCEREC)*. p 86–90. (doi:10.1109/ICCEREC.2016.7814986)

[17] Ribacki A, Jorge V A M, Mantelli M, Maffei R and Prestes E (2018). Vision-based Global Localization using Ceiling Space Density. *2018 IEEE International Conference on Robotics and Automation (ICRA)*, Brisbane, p 3502-3507. (doi: 10.1109/ICRA.2018.8460515)

[18] Hasan K M, Al-Nahid A and Reza K J (2014). Path planning algorithm development for autonomous vacuum cleaner robots. In *Int. Conf. on Informatics, Electronics & Vision (ICIEV)*. p 1–6. (doi:10.1109/ICIEV.2014.6850799).

[19] Bai Y W and Hsueh M F (2012). Using an adaptive iterative learning algorithm for planning of the path of an autonomous robotic vacuum cleaner. In *1st IEEE Global Conf. on Consumer Electronics (GCCE)*. p 401–405. (doi:10.1109/GCCE.2012.6379640).

[20] Lee S-J, Lee K and Song J-B (2014). Development of advanced grid map building model based on sonar geometric reliability for indoor mobile robot localization. In *11th Int. Conf. on Ubiquitous Robots and Ambient Intelligence (URAI)*. p292–297. (doi:10.1109/URAI.2014.7057437).

AUTHOR INDEX